Forest Microbiology

Forest Microbiology
Tree Microbiome: Phyllosphere, Endosphere, and Rhizosphere, Volume 1

Edited by

Fred O. Asiegbu
Faculty of Agriculture and Forestry, University of Helsinki, Helsinki, Finland

Andriy Kovalchuk
Faculty of Agriculture and Forestry, University of Helsinki, Helsinki, Finland
Industrial Biotechnology and Food Solutions, VTT Technical Research Centre of Finland, Espoo, Finland

ELSEVIER

ACADEMIC PRESS
An imprint of Elsevier

Academic Press is an imprint of Elsevier
125 London Wall, London EC2Y 5AS, United Kingdom
525 B Street, Suite 1650, San Diego, CA 92101, United States
50 Hampshire Street, 5th Floor, Cambridge, MA 02139, United States
The Boulevard, Langford Lane, Kidlington, Oxford OX5 1GB, United Kingdom

Notices
Knowledge and best practice in this field are constantly changing. As new research and experience broaden our understanding, changes in research methods, professional practices, or medical treatment may become necessary.

Practitioners and researchers must always rely on their own experience and knowledge in evaluating and using any information, methods, compounds, or experiments described herein. In using such information or methods they should be mindful of their own safety and the safety of others, including parties for whom they have a professional responsibility.

To the fullest extent of the law, neither the Publisher nor the authors, contributors, or editors, assume any liability for any injury and/or damage to persons or property as a matter of products liability, negligence or otherwise, or from any use or operation of any methods, products, instructions, or ideas contained in the material herein.

Library of Congress Cataloging-in-Publication Data
A catalog record for this book is available from the Library of Congress

British Library Cataloguing-in-Publication Data
A catalogue record for this book is available from the British Library

ISBN 978-0-12-822542-4

For information on all Academic Press publications
visit our website at https://www.elsevier.com/books-and-journals

Publisher: Charlotte Cockle
Editorial Project Manager: Michelle Fisher
Production Project Manager: Omer Mukthar
Cover Designer: Matthew Limbert

Typeset by SPi Global, India

Working together
to grow libraries in
developing countries

www.elsevier.com • www.bookaid.org

Contents

Section A
Introduction

1. An introduction to forest biome and associated microorganisms

Fred O. Asiegbu and Andriy Kovalchuk

2. Wood as an ecological niche for microorganisms: Wood formation, structure, and cell wall composition

Uwe Schmitt, Adya P. Singh, and Yoon Soo Kim

3. Methods for studying the forest tree microbiome

Kathrin Blumenstein, Eeva Terhonen, Hui Sun, and Fred O. Asiegbu

Section B
Phyllosphere microbiome

6. The phyllosphere mycobiome of woody plants

Thomas Niklaus Sieber

7. Tree leaves as a habitat for phyllobacteria

Teresa A. Coutinho and Khumbuzile N. Bophela

8. Microbiome of reproductive organs of trees

Fei Ren, Dong-Hui Yan, and Wei Dong

Section C
Endosphere microbiome

9. Bacterial biota of forest trees

Bethany J. Pettifor and James E. McDonald

10. Fungi inhabiting woody tree tissues

Gitta Jutta Langer, Johanna Bußkamp, Eeva Terhonen, and Kathrin Blumenstein

Section D
Rhizosphere microbiome

16. Microbiome of forest soil

Zhao-lei Qu and Hui Sun

Section E
Archaea and viruses in forest ecosystem and microbiota of forest nurseries and tree pests

17. Mycobiome of forest tree nurseries

Marja Poteri, Risto Kasanen, and Fred O. Asiegbu

18. Microbiome of forest tree insects

Juliana A. Ugwu, Riikka Linnakoski, and Fred O. Asiegbu

19. Archaea as components of forest microbiome

Kim Yrjälä and Eglantina Lopez-Echartea

20. Viruses as components of forest microbiome

Jarkko Hantula and Eeva J. Vainio

Section F
Challenges and potentials

21. Translational research on the endophytic microbiome of forest trees

Johanna Witzell, Carmen Romeralo, and Juan A. Martín

22. Forest microbiome: Challenges and future perspectives

Fred O. Asiegbu

Contributors

Numbers in parentheses indicate the pages on which the authors' contributions begin.

Fred O. Asiegbu (3, 35, 75, 305, 327, 395), Department of Forest Sciences, Faculty of Agriculture and Forestry, University of Helsinki, Helsinki, Finland

H. Umair Masood Awan (75), Department of Forest Sciences, Faculty of Agriculture and Forestry, University of Helsinki; Helclean Consultancy Services, Helsinki, Finland

Kathrin Blumenstein (35, 175), Forest Pathology Research Group, Department of Forest Botany and Tree Physiology, Faculty of Forest Sciences and Forest Ecology, University of Göttingen, Göttingen, Germany

Khumbuzile N. Bophela (133), Department of Biochemistry, Genetics and Microbiology, Centre for Microbial Ecology and Genomics/Forestry and Agricultural Biotechnology Institute, University of Pretoria, Pretoria, South Africa

Marc Buée (231), French National Research Institute for Agriculture, Food and the Environment (INRAE), Lorraine University, Department of "Tree-Microbe Interactions", Champenoux, France

David J. Burke (257), The Holden Arboretum, Kirtland, OH, United States

Johanna Bußkamp (175), Section Mycology and Complex Diseases, Department of Forest Protection, Northwest German Forest Research Institute (NW-FVA), Göttingen, Germany

Sarah R. Carrino-Kyker (257), The Holden Arboretum, Kirtland, OH, United States

Teresa A. Coutinho (133), Department of Biochemistry, Genetics and Microbiology, Centre for Microbial Ecology and Genomics/Forestry and Agricultural Biotechnology Institute, University of Pretoria, Pretoria, South Africa

Wei Dong (145), China Electric Power Research Institute, Beijing, China

Jarkko Hantula (371), Natural Resources Institute Finland (Luke), Helsinki, Finland

Risto Kasanen (59, 305), Department of Forest Sciences, Faculty of Agriculture and Forestry, University of Helsinki, Helsinki, Finland

Mee-Sook Kim (277), USDA Forest Service, Pacific Northwest Research Station, Corvallis, OR, United States

Yoon Soo Kim (17), Department of Wood Science and Engineering, Chonnam National University, Gwangju, South Korea

Ned B. Klopfenstein (277), USDA Forest Service, Rocky Mountain Research Station, Moscow, ID, United States

Andriy Kovalchuk (3), Department of Forest Sciences, Faculty of Agriculture and Forestry, University of Helsinki, Helsinki; Industrial Biotechnology and Food Solutions, VTT Technical Research Centre of Finland, Espoo, Finland

Bradley Lalande (277), USDA Forest Service, Forest Health Protection, Gunnison, CO, United States

Gitta Jutta Langer (175), Section Mycology and Complex Diseases, Department of Forest Protection, Northwest German Forest Research Institute (NW-FVA), Göttingen, Germany

Björn D. Lindahl (231), Department of Soil and Environment, SLU, Uppsala, Sweden

Riikka Linnakoski (327), Natural Resources Institute Finland (Luke), Helsinki, Finland

Eglantina Lopez-Echartea (357), University of Chemistry and Technology, Prague, Czech Republic

Juan A. Martín (385), School of Forest Engineering and Natural Resources, Technical University of Madrid (UPM), Madrid, Spain

James E. McDonald (161), School of Natural Sciences, Bangor University, Bangor, Gwynedd, United Kingdom

Leticia Pérez-Izquierdo (231), Department of Soil and Environment, SLU, Uppsala, Sweden

Bethany J. Pettifor (161), School of Natural Sciences, Bangor University, Bangor, Gwynedd, United Kingdom

Marja Poteri (305), Natural Resources Institute Finland (Luke), Helsinki, Finland

Zhao-lei Qu (293), Department of Forest Protection, College of Forestry, Nanjing Forestry University, Nanjing, China

Fei Ren (145), Forestry Experiment Center in North China, Chinese Academy of Forestry, Beijing, China

Ana Rincón (231), Institute of Agricultural Sciences (ICA), Spanish National Research Council (CSIC), Madrid, Spain

Carmen Romeralo (385), Swedish University of Agricultural Sciences, Southern Swedish Forest Research Centre, Alnarp, Sweden

Uwe Schmitt (17), Thünen Institute of Wood Research, Hamburg, Germany

Thomas Niklaus Sieber (111), ETH Zurich, Department of Environmental Systems Science, Forest Pathology and Dendrology, Zurich, Switzerland

Adya P. Singh (17), Scion, Rotorua, New Zealand

Mike Starr (223), Department of Forest Sciences, University of Helsinki, Helsinki, Finland

Jane E. Stewart (277), Department of Agricultural Biology, Colorado State University, Fort Collins, CO, United States

Hui Sun (35, 293), Department of Forest Protection, College of Forestry, Nanjing Forestry University, Nanjing, China

Eeva Terhonen (35, 175, 207), Forest Pathology Research Group, Department of Forest Botany and Tree Physiology, Faculty of Forest Sciences and Forest Ecology, University of Göttingen, Göttingen, Germany

Juliana A. Ugwu (327), Forestry Research Institute of Nigeria, Ibadan, Nigeria; Department of Forest Sciences, Faculty of Agriculture and Forestry, University of Helsinki, Helsinki, Finland

Eeva J. Vainio (371), Natural Resources Institute Finland (Luke), Helsinki, Finland

Harri Vasander (223), Department of Forest Sciences, University of Helsinki, Helsinki, Finland

Johanna Witzell (385), Swedish University of Agricultural Sciences, Southern Swedish Forest Research Centre, Alnarp, Sweden

Dong-Hui Yan (145), Research Institute of Forest Ecology, Environment and Protection, Key Laboratory of Biodiversity Conservation of National Forestry and Grassland Administration, Chinese Academy of Forestry, Beijing, China

Kim Yrjälä (357), Zhejiang A & F University, Hangzhou, China

Preface

Microorganisms constitute an integral component of all terrestrial and aquatic ecosystems. They are indispensable for the global nutrient cycling and for the existence of higher multicellular forms of life. In fact, all macroorganisms (e.g., plants and animals) live in close association with a diverse range of microbial symbionts. The ecological community of bacteria, archaea, fungi, and protists associated with a given organism, its organs and tissues is referred to as microbiota. Microbiome by definition refers to the entire assemblage of all microbial genomes of the microbial community (microbiota) associated with certain environment or organism (human, animal, or plant including forest trees). Research on microbiome of forest biomes has attracted much attention in recent years but still lags far behind comparable knowledge on human and agricultural crop microbiomes. Our interest in compiling the volume 1 of this Forest Microbiology book stems from the paucity and great disparity of information on forest tree microbiome, including phyllosphere, rhizosphere, and endosphere. Our understanding and perception of forest tree microbiome have recently been facilitated due to novel technological advances using metabarcoding, metagenomics, and metatranscriptomics approaches. In this book, recent advances in the study of tree microbiome were highlighted. An overview of our current understanding of taxonomic and functional diversity of microorganisms (fungi, bacteria, archaea, and viruses) associated with tissues of various broad-leaf and conifer trees was provided. Microbial communities associated with various forest insects, host trees, and different tree organs were compared, and generalists and specialists among tree-associated microbes were identified. Biotic and abiotic factors determining the composition and the structure of forest tree microbial communities were presented. However, despite significant progress so far achieved in our understanding of the factors affecting the composition of microbial communities associated with plants, very little is known about the effect of plant pathogens (pathobiomes) on their structure, particularly the least studied forest trees. In several chapters of this book, studies that unravel the potential functional roles of these microbes and their impact on forest tree health were uncovered.

The research on the integration of beneficial microbiomes into forest production is increasingly attracting attention. The rational engineering of microbial communities of forest trees is expected to be of great significance for the sustainable wood and timber production, for the improved tolerance of forest ecosystems against environmental stressors, and for the management of forest tree diseases and pests. New insights on how to harness and link the acquired knowledge on microbiomes of forest biomes for translational forest management were highlighted and discussed in this book. Translational aspects of forest microbiome study are however not without challenges. Despite the challenges, one of the emerging research directions in microbiome study is represented by metagenome-wide association studies (MWAS). In this approach, a relative abundance of a certain gene in the metagenome is used to establish an association with an occurrence of a disease of interest. The success of MWAS in human and animal models suggests that its applications can be extended to analyze associations between tree microbiome and diseases. Network modeling also represents an alternative approach to establish a link between microbiome composition and its function.

In this book, our wish was to highlight the microbiota inhabiting forest trees and their potential impact on the health and fitness of, and disease progression in, forest biomes. Additionally, to uncover the nature and structure of niches occupied by forest microbes together with their functional roles in the decomposition of wood debris and forest litter, uptake and nutrient cycling in forest ecosystems, mutualistic symbiosis and microbial parasitic interactions with trees. The book also addressed current advances in these fields of study, made possible by use of novel and modern biological techniques. We hope that this book will fill a major knowledge gap, serve as a single information source as well as prove valuable for students of biology and forest sciences, forest pathologists, other practitioners in specific areas of forestry, and everyone interested in the microbiota of forest biomes.

Fred O. Asiegbu,
Andriy Kovalchuk
University of Helsinki, Finland

Section A

Introduction

Chapter 1

An introduction to forest biome and associated microorganisms

Fred O. Asiegbu[a] and Andriy Kovalchuk[a,b]

[a]*Department of Forest Sciences, Faculty of Agriculture and Forestry, University of Helsinki, Helsinki, Finland,* [b]*Industrial Biotechnology and Food Solutions, VTT Technical Research Centre of Finland, Espoo, Finland*

Chapter Outline

1. Introduction

This book deals with various microorganisms associated with terrestrial plants, in particular trees in forest ecosystems. The assemblage of the microbial community associated with plants is referred to as plant microbiota or the plant microbiome. Plant microbiota constitute the ecological community of archaea, bacteria, fungi, and protists (Figs. 1.1 and 1.2) inhabiting various plant organs and tissues (Turner et al., 2013). The plant microbiome (phytobiome) in turn comprises all microbial genomes associated with host plant phyllosphere, rhizosphere, and endosphere (Fig. 1.1) (Guttman et al., 2014). The phyllosphere refers to the aerial or above ground parts of plants colonized by microbial communities, the rhizosphere is the microbial communities inhabiting the root surface and soil zone around the root, and endosphere is the microbial communities residing within plant tissues (Turner et al., 2013). All these various components of plant or tree microbiota have important potential supporting functions in growth and health of forest biome.

2. Forest biome

Forest simply refers to large expanse of land dominated by trees. According to Pan et al. (2013), forests are globally distributed and are pre-dominant terrestrial ecosystem on Earth. The total land area of the world covered by forests is estimated at 31% or approximately 4 billion hectares according to Food and Agricultural Organization (FAO) (FAO, 2010). Pan et al. (2013) reported that forests contain about 80% of the Earth's plant biomass with a gross primary production of 75%. The forest biomes are composed of boreal, temperate, and tropical forest with a net annual primary production estimated at 2.6 gigatonnes, 8.1 gigatonnes, and 21.9 gigatonnes, respectively (Pan et al., 2013). Forests have considerable ecological and economic value. According to Food and Agricultural Organization, the gross value added by forest sector in 2011 is estimated at 606 billion US dollars (FAO, 2015). Forests are important in climate change protection as they act as a sink for carbon (Canadell and Raupach, 2008). Every year forests absorb billions of tons of carbon dioxide stocked permanently

Forest Microbiology. https://doi.org/10.1016/B978-0-12-822542-4.00009-7

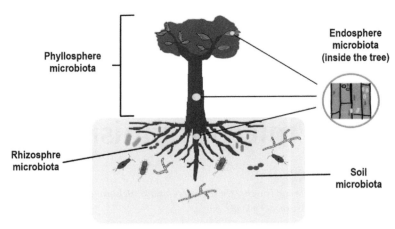

FIG. 1.1 Phyllosphere, rhizosphere, and endosphere microbiome. *(Illustration by Artin Zarsav.)*

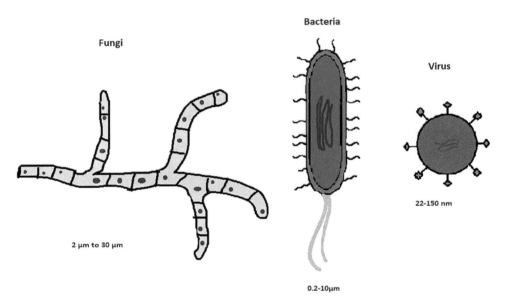

FIG. 1.2 Fungi, bacteria, and viruses as components of forest microbiota.

in their biomass (289 gigatonnes of carbon) (FAO, 2010). Forest is also beneficial not only for mitigating climate change effects, but equally for conserving biological diversity as well as potential source of bio-energy.

Forest biomes are generally divided into two main categories: the nonliving components (abiotic) and the living components (biotic). The living components of forest consist of trees, shrubs, nonwoody plants (e.g., mosses, herbs, etc.), insects, animals, fungi, and microorganisms associated with the plants or trees and present in the soil. Ecologically, trees and plants in every forest biome share overlapping habitat with a diverse range of microbial symbionts and maintain balanced dynamic relationships. These relationships could vary from pathogenic infections or mutualistic symbiosis to latent commensal or endophytic association. Many of these microbes have important impacts on the ecology and evolution of forest biome and are key components in maintaining vital ecosystem processes.

3. Forest trees and their symbionts

The word "Symbiosis" was first used in 1879 by the German mycologist Antoine de Barry to describe the "living together of unlike organisms" (Sapp, 2010). In the symbiosis association, the partners are referred to as either symbionts or host and symbionts. In the latter case, the host is the macroorganism, whereas the symbiont is the microorganism. The nature of the symbiosis can be complex and varied and could be either mutualism, commensalism, or parasitism. In the mutualistic symbiosis both partners benefit and in commensalism, one symbiont benefits and the other is unaffected. However, in parasitic association, one symbiont benefits and the other is harmed. The term "parasite" has also often been used to describe both a pathogen and a symbiont. Plant parasitic microbes are also capable of acquiring carbon through diverse nutritional modes such as saprotrophy, necrotrophy, or biotrophy (Koide et al., 2008). These categories are however not mutually exclusive as there is a continuum of possible nutritional modes from saprotrophy to biotrophy which may be determined by genetical traits and environmental conditions (Cooke and Whipps, 1993). Necrotrophs by definition are able to kill living cells for subsequent saprotrophic colonization, whereas saprotrophs subsist upon dead organic matter (Cooke and Rayner, 1984). Obligate saprotrophic wood decayers are not expected to provoke the same kind of host reactions as their necrotrophic counterparts. Evolutionarily, a necrotroph is considered to be more specialized than a saprotroph through its ability to overcome the resistance of a potential host (Isaac, 1992). Uroz et al. (2019) reported that symbionts play crucial role in structuring of plant microbiome composition and its functioning. Metabolites secreted during the microbiome interaction (e.g., bacteria-fungi) could have impact on plant disease outcome as well as contribute to shaping the taxa and assembly of phytomicrobiome.

4. Microbiome: The three domain system

According to the nearly universally accepted view, all cellular organisms are currently classified into three principal groups, known as domains of life: Bacteria, Archaea, and Eukarya (Fig. 1.3) (Woese et al., 1990).

Both Bacteria and Archaea are lacking membrane-bound nuclei. Members of both groups are predominantly unicellular microscopic organisms; however, some bacteria (e.g., streptomycetes, cyanobacteria, and myxobacteria) developed more complex structural organization. Despite their small physical sizes, Bacteria and Archaea play essential roles in the nutrient turnover and in the functioning of both terrestrial and aquatic ecosystems. Although superficially similar, Bacteria and Archaea display fundamental differences in their ribosomal RNA (rRNA) sequences, in the organization of their RNA transcription and protein translation machinery and in their biochemistry (e.g., in the composition of their membrane lipids) (Zillig, 1991). In some aspects, Archaea, in fact, are more similar to Eukarya than to Bacteria.

Certain groups of Archaea adapted to extreme habitats with hostile conditions (high temperatures, very acidic or alkaline pH, high salinity), where members of other domains of life are scarce or absent. Their unique metabolic capacities allow them to use energy sources not easily available to other living organisms. However, Archaea are not confined exclusively to extreme environments, as they recently have been found in a wide range of habitats, and some Archaea are common commensals of animals and plants (Moissl-Eichinger et al., 2018).

The domain Eukarya encompasses all cellular forms of life with membrane-enveloped nuclei. Eukaryotes exhibit enormous diversity, ranging from simple unicellular forms to complex multicellular organisms. They include animals, plants, fungi, and protists. Eukarya shape most of Earth's ecosystems, with plants constituting principal producers in terrestrial habitats. As components of terrestrial microbiomes, Fungi play an outstanding role in the decomposition of biogenic polymers (in particular, plant cell wall components). In addition, many of them are important plant pathogens. Another group of Eukarya with a prominent role in plant-associated microbial communities is Oomycota. It encompasses filamentous microorganisms, which are superficially similar to Fungi, but not closely related to them. This group includes some important plant pathogens, e.g., species of *Phytophthora*. Forests host a broad diversity of other groups of protists, but we are just starting to understand their role in the functioning of forest ecosystems. Representatives of all three domains of life are ubiquitous components of forest microbiomes. They inhabit forest soil and water bodies, occur on plant surfaces and within

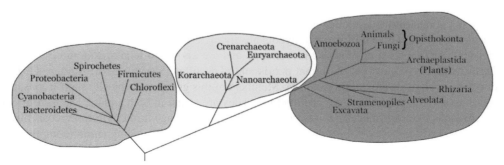

FIG. 1.3 Scheme illustrating the relationships between the three domains of life. Selected representative high-rank groups mentioned in the text are depicted. The length of the branches is not to scale and does not reflect evolutionary distances between individual groups.

plant tissues, and many of them establish close interactions with varying outcomes with their plant hosts (Baldrian, 2017). Various aspects of those interactions are discussed in this book.

5. Fungi: Morphological and structural features

Fungi are unicellular or multicellular organisms. Fungi belong to a large group of eukaryotic organisms which include yeasts, molds, as well as mushrooms. Molecular evidences suggest that fungi are more closely related to animals than they are to plants. They are one of the most important organisms in our environment both in terms of economic and ecological roles. According to Hawksworth (1991), over 70,000 species of fungi have been described and it is estimated that more than 1.5 million species remain undescribed. With current advances in next generation sequencing technology (NGS), it is likely that this number could be much higher. Multicellular filamentous fungi show polarized apical growth and are characterized by having branching thread-like filaments called hyphae. The apical growth and extension of this vegetative part of the fungus forms a mass of branching thread-like hyphae termed mycelium. The long thin thread-like hyphal filaments are divided along their length by cross-walls called septa that are not present in lower fungi. Fungi are typically nonmotile and reproduce by means of spores, through asexual or sexual process. Unicellular fungi are mostly single-celled microscopic yeasts. In yeast, vegetative growth is mostly by asexual reproduction through a process called budding, where a daughter cell develops from the parent cell. Some fungi are also dimorphic, capable of changing their growth morphology from yeast-like phase to filamentous or mycelial forms. Fungi unlike plants are nonvascular, are capable of decomposing organic matter as well as direct absorption of the nutrients through their cell membrane and wall. As heterotrophs, fungi cannot synthesize their own food from carbon dioxide, thereby depending on available sources of organic carbon for their nutrition. To access organic matter prevalent in plant or animal sources for nutrients, fungi produce a diverse range of extracellular enzymes for decomposition of the complex substrates. Fungi unlike plants do not have cellulose in their cell wall, rather the wall is composed primarily of three components (chitin, glucans, and associated proteins).

5.1 Classification of the fungal kingdom

The kingdom Fungi is composed of seven "Phyla": Basidiomycota, Ascomycota, Glomeromycota, Neocallimastigomycota, Blastocladiomycota, Chytridiomycota, and Microsporidia. This new classification was partly based on recent studies (Hibbett et al., 2007; James et al., 2006a,b). Two of these were considered as main phyla or higher fungi (Ascomycota, Basidiomycota) and both are contained within the subkingdom "Dikarya." The members of phylum Basidiomycota are filamentous except for basidiomycota yeast-like forms. They are characterized by club-shaped fruiting cells known as basidia. The basidium produces specialized sexual spores called basidiospores. This phylum includes most edible mushrooms as well as several important plant pathogens such as rusts and smuts and human pathogenic yeasts (e.g., *Cryptococcus* spp.). The phylum Ascomycota contains members commonly known as ascomycetes or sac-fungi. Typical feature of members of this phylum is the formation of ascus which contains nonmotile sexual spores called ascospores. Many members of this group are of biotechnological and commercial relevance. Examples include species that play roles in wine fermentation, baking (e.g., yeasts), cheese, and antibiotics production (e.g., *Penicillium* spp.). The edible sac fungi morels and truffles are also familiar members of this phylum. The phylum Glomeromycota also called Glomeromycetes is a newly established phylum which contains species that form mutualistic symbiotic association with roots of many plant species. Members of this group include arbuscular mycorrhizal fungi. Neocallimastigomycota are symbionts and the phylum comprises mostly of anaerobic fungi found in the digestive tract of herbivorous animals. These fungi produce enzymes that aid in the breakdown of complex polysaccharides into simpler carbohydrates, which serve as a source of nutrition in ruminant animals. The Blastocladiomycota is one of the seven newly recognized fungal phyla. Members of the phylum Blastocladiomycota together with Neocallimastigomycota and Chytridiomycota are characterized by presence of flagellate zoospores (Watson et al., 2016). Members of Blastocladiomycota are saprotrophs and are known to inhabit soil and freshwater environment where they facilitate in the decomposition of animal debris and plant litter (Money, 2016; Naranjo-Ortiz and Gabaldón, 2019). Chytridiomycota or chytrids are zoosporic fungi. Members of this group are typically saprotrophs, their spores possess whiplash flagellum and the cell walls are made of chitin. They are found in both aquatic and terrestrial environments. Some chytrids are parasites like *Batrachochytrium dendrobatidis*, causative agent of chytridiomycosis disease of amphibians (Longcore et al., 1999). Members of Chytridiomycota are also capable of degrading chitin (component of exoskeleton of arthropods) and keratin (a scleroprotein found in the hair of vertebrates). The phylum Microsporidia includes spore-forming single-celled parasites. They are known to infect a wide range of hosts, including fish, humans, and arthropods (insects, crustaceans).

6. Bacteria

Bacteria encompass prokaryotic, predominantly unicellular microorganisms lacking membrane-bound nuclei (Fig. 1.2). Bacteria are widespread and abundant in most of known habitats, including forest ecosystems. One gram of soil from a temperate forest contains about 10^7–10^9 bacterial cells (Baldrian, 2017). Most bacteria have relatively simple morphological organization and exist as single spherical, rod-shaped, or spiral cells. Many bacterial species are capable of active movement due to the possession of flagella. In some cases, bacterial cells can form diverse aggregates of varying complexity. There are examples available of more complex morphology, in particular, among Actinobacteria. For instance, species of the genus *Streptomyces* are characterized by filamentous growth resulting in the formation of pseudohyphae and, in some cases, by further differentiation of those hypha-like structures into substrate and aerial mycelium. Bacterial cells are generally much smaller than cells of eukaryotes. Most bacterial cells are surrounded by rigid peptidoglycan cell wall and, in case of Gram-negative bacteria, there is also an external lipid layer (outer membrane). Inner organization of bacterial cells is rather simple, with no intracellular organelles. However, in case of photosynthetic bacteria (e.g., cyanobacteria) the plasma membrane can form numerous invaginations, in which light-harvesting complexes are located.

Our current knowledge about bacteria is largely based on study of those species that can be propagated under laboratory conditions (in vitro). However, applications of molecular techniques and direct sequencing of environmental samples indicate that only a small fraction of bacterial species can be grown in pure cultures, and their actual diversity is largely underestimated. About 10,000 species of Bacteria have been formally described until now, but available estimates suggest existence of over 10 million bacterial species. Domain of Bacteria is further subdivided into large systematic groups called phyla. The modern classification of bacteria is primarily based on results of molecular systematics studies, taking into account the sequences of genes encoding bacterial ribosomal RNA. Currently, 30 bacterial phyla containing culturable species are recognized. Additionally, several dozens of so-called candidate phyla are proposed. Members of candidate phyla have not been isolated and characterized in pure culture, but their existence is deduced from the sequencing analysis of various environmental samples.

Bacteria play an important role in nutrient cycling as decomposers of organic matter. Their metabolic capacities allow them to degrade a broad spectrum of organic compounds. Certain bacterial taxa are capable for the fixation of molecular nitrogen and thus can contribute to the supply of nitrogen to plants and to soil fertility in general. Among nitrogen-fixing bacteria, both free-living and symbiotic species are known. The latter ones often show a high degree of host specificity (e.g., certain species of *Rhizobiales* are engaged in symbiotic relationships with legume plants). Symbiosis with nitrogen-fixing bacteria might be an important factor for plants growing in nitrogen-poor conditions (e.g., interactions between actinobacteria of the genus *Frankia* with sea-buckthorn and alder trees). Certain species of bacteria are well-known plant pathogens and can cause plant diseases. In case of trees, bacterial diseases are usually not as severe as the fungal ones (Agrios, 2005). There are, however, some examples of bacterial diseases having devastating impact, e.g., olive quick decline, which is believed to be caused by the bacterium *Xylella fastidiosa* and is currently threatening olive plantations in Italy and elsewhere in the Mediterranean region.

Even if the abundance of bacterial cells within plant tissues is lower than the one of fungi, bacteria still constitute an important component of plant microbiome. They are particularly numerous in rhizosphere, but less so in endosphere and phyllosphere. Members of most accepted bacterial phyla were reported from plant microbial communities. However, diversity of bacteria in plant microbiome is usually lower than in the surrounding soil, and there seem to be certain selection mechanisms in place, which drive the composition of plant-associated bacterial communities. The available data indicate that four bacterial phyla (Proteobacteria, Actinobacteria, Bacteroidetes, and Firmicutes) are clearly overrepresented within plant microbiome (Bulgarelli et al., 2013). Due to the availability of complete genome sequences, ease of manipulations, and good tractability, a number of bacterial species are used in the experimental studies on the reconstitution of synthetic microbial communities. This approach relies on the reconstitution of artificially designed microbial communities consisting solely of a few selected species in germ-free plant seedlings. The synthetic communities created in this way have lower complexity compared with the natural ones and are more tractable. They are widely considered as promising models to improve our understanding of interactions within plant microbial communities and mechanisms driving and maintaining their composition (Carlström et al., 2019).

7. Protists

Protists, or Protista, historically encompassed all eukaryotic organisms that were neither animals, plants, or fungi. Modern studies have clearly demonstrated that Protista in their traditional sense were an artificial group, including a number of independent evolutionary lineages. Thus the term "protists" can no longer be applied as a taxonomic category. Nevertheless, it remains in use to describe broadly defined microscopic unicellular (or unicellular-colonial) eukaryotes. Tens of thousands of species of protists have been formally described. However, similar to other microorganisms, sequencing data obtained from environmental samples indicate an existence of numerous undescribed taxa. Traditionally, protists have been classified based on their overall morphology and nutritional mode into protozoans (heterotrophic organisms adopting phagocytosis as primary food uptake mode include flagellates, ciliates, and amoebae), "lower fungi" (heterotrophic protists with absorptive nutrition mode), and algae (autotrophic protists) (Geisen et al., 2018). Modern classification schemes of protists are built on the principles of molecular systematics, and many of the traditional groups have been drastically reshuffled. The only higher-rank group that proved to be monophyletic and could be maintained in its traditional sense is Ciliates. All other groups of protists were shown to be polyphyletic, e.g., artificial assemblages of organisms representing different evolutionary lineages. Traditional circumscription of protists spans all major lineages of eukaryotes, so-called supergroups, which are currently recognized: Archaeplastida, Excavata, Amoebozoa, Opisthokonta and SAR group (consisting of Stramenopiles, Alveolata, and Rhizaria). This classification encompasses the vast majority of Eukaryotes, but relationships of some smaller groups are still a matter of debates. In this classification, plants are assigned to the supergroup Archaeplastida, both animals and fungi belong to the supergroup Opisthokonta, whereas protists are spread over all five supergroups. Overview of some of the representative and better studied groups of protists is provided in Table 1.1.

The heterogeneity of protists is reflected in their great structural diversity. Many of them are represented by microscopic unicellular forms, but some (e.g., slime molds) can form larger colonies or aggregates easily detectable by a naked eye. Protists lack tissue organization. Cells of some (but not all) protists are surrounded by rigid cell wall of varying composition. Many protists are motile due to the presence of flagella or cilia, or are capable of amoeboid movement. Others, like oomycetes, have motile forms only at certain stages of their lifecycle.

Various groups of unicellular eukaryotes inhabit forest ecosystems. Tens of thousands of protists can be found in a gram of forest soil (Geisen et al., 2018). It is recognized that protists are important components of soil food webs, contributing to nutrient cycling (in particular, nitrogen) by feeding on bacteria and other microbes. However, their ecological role in forest ecosystems is still insufficiently understood. One of the few groups of protists that received significant attention in the past is Oomycota, or oomycetes. Many members of Oomycota are filamentous saprotrophic microorganisms, superficially similar to fungi (and thus often called "water molds" in older literature), but belonging to the lineage of Stramenopila (Heterokonta), a large and heterogenic assemblage of eukaryotic organisms, including, among others, brown algae and diatoms. Some of the oomycetes (e.g., species of *Phytophthora* and *Pythium*) are notorious plant pathogens of considerable economic importance for agriculture and forestry.

Other groups of protists inhabiting forest ecosystems remain considerably underexplored. Recent sequencing efforts indicate that members of Cercozoa and Ciliophora are among the most abundant protists in soil samples (Howe et al., 2009),

TABLE 1.1 Overview of some representative groups of Protista occurring in forest ecosystems.

Unranked group	Phylum	Subphylum/class	Representatives
Amoebozoa			Naked and testate amoeboid organisms (genera *Chaos*, *Entamoeba*, *Pelomyxa*, *Amoeba*, etc.); heterotrophic; free living, parasitic or symbiotic
Excavata	Euglenozoa	Euglenoida	Flagellate free-living photosynthetic protists (genera *Euglena*, *Trachelomonas*); species-rich group
Rhizaria	Cercozoa	Filosa	Filose naked and testate amoeboid protists (genus *Euglypha*), some members are flagellate (e.g., *Cercomonas*); mostly heterotrophic; free living, very common and abundant in various types of soils
Rhizaria	Cercozoa	Phytomyxea	Plant parasites (genera *Plasmodiophora*, *Spongospora*, etc.) forming plasmodia (multinucleate cells)
Alveolata	Ciliophora		Protists characterized by the presence of numerous cilia (genera *Paramecium*, *Tetrahymena*, *Vorticella*, etc.); heterotrophic, free living; very common in wet habitats
Stramenopiles/Heterokonta		Oomycota	Hyphae-forming filamentous organisms (genera *Phytophthora*, *Pythium*, *Albugo*, *Peronospora*, etc.); heterotrophic, parasitic or free-living; widespread, include some important plant pathogens

but many of the detected sequences belong to undescribed lineages within those groups. Quite surprisingly, obligate parasites of animals from the group Apicomplexa were found as the most abundant group of protists in neotropical forests (Mahé et al., 2017). Protists contribute to nitrogen remobilization and affect the composition and dynamics of microbial communities via preying on bacteria and small eukaryotes (Flues et al., 2017); parasitic protists influence dynamics of populations of soil invertebrates; however, many important aspects of their ecological interactions remain poorly understood.

8. Viruses

Viruses are submicroscopic infectious agents that replicate exclusively inside living cells (Fig. 1.2). Viruses are able to infect virtually all living organisms, from bacteria and archaea to plants and animals. When outside of host cell, viruses are represented by individual particles (virions), consisting of viral genetic material (nucleic acids, either DNA or RNA) and protective protein coat (capsid), surrounding nucleic acids. Some types of viruses, in addition, possess an external lipid layer. Viruses are classified in several larger groups based on the type of their nucleic acids (DNA or RNA), strandedness (single-stranded or double-stranded molecule), sense (sense or antisense molecule), and mechanisms of replication. Lower-level taxonomic categories (orders, families, and genera) are recognized based on the level of their sequence similarity. Viruses are integral components of all ecosystems; however, their ecological role is imperfectly known. As viruses do not have their own metabolisms, their ecological effects are always indirect. Viral infections rarely have a devastating effect on forest trees. However, viruses may have a significant impact on other components of forest microbiome. For instance, the abundance of bacteriophages in soil system can be very high, and they may affect bacterial mortality, causing significant shifts in the composition of soil microbial communities. Viruses also might affect virulence of pathogenic microorganisms. One of the best-known examples is the reduced pathogenicity of the chestnut blight fungus *Cryphonectria parasitica*, caused by the presence of a mycovirus. Furthermore, certain insect viruses were proposed to be used as biocontrol agents against herbivorous insects. At the same time, many viruses are cryptic and have seemingly little effect on their host fitness.

9. Ecology, biochemistry, physiology, and biotechnological features of microorganisms

Microorganisms have extremely broad metabolic capacity and can use a wide range of energy and nutrient sources. The vast majority of microorganisms (bacteria, archaea, protozoans, and fungi) occurring in terrestrial ecosystems are decomposers that rely on the availability of organic material. They can be characterized as heterotrophic (or, more precisely chemoorganoheterotrophic), i.e., they use organic compounds generated by other organisms as their source of carbon, energy, and reducing equivalents. The organic compounds can originate either from organic litter (in case of saprotrophs or saprobes) or from living organisms (in case of parasites or certain mutualistic interactions).

Like many microorganisms, fungi are metabolically versatile and of huge biotechnological relevance. Several of the most important fungal groups belong to ascomycetes and basidiomycetes. Some fungi are of beneficial and industrial value and have been used in the production of pharmaceutical products (e.g., antibiotics) and industrial chemicals (e.g., bioethanol, citric acid, enzymes, etc.). Fungi have also been used as a source of food (e.g., mushroom) and in fermentation of various beverages and food products (e.g., cheese, soya sauce, wine, beer, bread, etc.) and also as biological control agents of plant and forest tree pathogens or pests (Deacon, 2006). Some fungal species produce toxic compounds (e.g., alkaloids), which are harmful to humans and animals. In recent years, our understanding of the diverse nature and regulation of fungal proteome, metabolome, and transcriptome has greatly being enhanced with the advent of "omics" technology. The use of high-throughput phenotype micro-arrays has facilitated a much broader insight on how genes important for adaptation to ambient temperature, pH, carbon and nitrogen source utilization are regulated in fungi (Wilson and Talbot, 2009). Ecologically, many fungi obtain their nutrients by breaking down dead organic matter and in the process contribute to nutrient cycling. Equally, many vascular plants also form mutualistic associations with symbiotic mycorrhizal fungi that supply them with essential mineral nutrients. At the other extreme are fungi that cause a number of animal and human diseases as well as infections on agricultural crops and forest trees. In terms of plant health, several fungi and their plant partners share overlapping life histories in the context of a dynamic standoff, each seeking after its own interests. The consequences of such interactions could range from latent endophytic relationship to fatal pathogenic infections or mutualism.

Bacteria and Archaea are equally versatile and possess some metabolic pathways that are not known from Eukaryotes. Many microbes can use different compounds as a terminal electron acceptor in electron transport chain. Aerobic organisms commonly use molecular oxygen (O_2). However, alternative acceptors can be used when oxygen availability is limited. Many bacteria and archaea can use inorganic electron acceptors, e.g., nitrate (NO_3^-), sulfate (SO_4^{2-}), carbon dioxide (CO_2), and ferric ion (Fe^{3+}). Occasionally, organic molecules can be used as electron acceptors. All those compounds, both inorganic and organic, have lower reduction potential than molecular oxygen, making their use less energy-efficient and resulting in lower growth rate. Some facultative anaerobes can switch between oxygen and alternative electron acceptors depending on oxygen availability. Another type of anoxic metabolism is fermentation. In this case, various organic compounds serve as electron acceptors, whereas electron transport chain is not used. Fermentation occurs both in obligate (e.g., Clostridia) and in facultative anaerobes (e.g., brewing yeast). This process has very low energy efficiency, and facultative anaerobes switch to aerobic respiration when oxygen becomes available.

A number of microorganisms (cyanobacteria, some protozoans (e.g., Euglenozoa), microscopic algae, and algal components of lichens) are capable of using energy of sunlight and of fixing atmospheric carbon dioxide (CO_2) in the process of photosynthesis. The ones that use water or inorganic compounds as a source of reducing equivalents can be characterized as chemolithoautotrophic. Some of them are obligate autotrophs, whereas others have a mixed nutrition mode (mixotrophic), and can switch to heterotrophy when no light is available. Phototrophic organisms commonly occupy habitats with sufficient light level, e.g., soil surface, shallow waters, rocks and large stones, tree bark or leaf surface.

Chemolithotrophic bacteria and archaea present another modification of metabolism. They gain their energy and reducing equivalents from the oxidation of inorganic compounds. Many (but not all) of them are autotrophic and can fix carbon dioxide. Compounds used as an energy source include molecular hydrogen (H_2), reduced forms of sulfur (H_2S, molecular sulfur (S), thiosulfate ($S_2O_3^{2-}$)), ammonia (NH_4^+), nitrite (NO_2^-), and ferrous ion (Fe^{2+}). Chemolithotrophic organisms are usually restricted to special habitats, but can be abundant under appropriate conditions (e.g., nitrifying bacteria in sewage treatment plants).

Two metabolic pathways of outmost ecological importance, nitrogen fixation and methanogenesis, occur exclusively in prokaryotes. Nitrogen fixation is a process of reduction of atmospheric molecular nitrogen to ammonia. Fixation of nitrogen plays an essential role in the global nitrogen cycling and greatly contributes to soil fertility. It is performed by dedicated enzymes, nitrogenases. Nitrogen-fixing microorganisms (diazotrophs) occur in several bacterial lineages (e.g., in cyanobacteria, rhizobia, species of *Frankia, Azotobacter, Beijerinckia*) and in some archaea. Both free-living and symbiotic diazotrophs are known. Symbiosis of rhizobia with legume plants (*Fabaceae/Leguminosae*) provides a textbook example of mutualistic relationships between plants and bacteria. Rhizobia in a broad sense is a collectively used term that refers to symbiotic diazotrophic members of several genera within the order *Rhizobiales* (*Bradyrhizobium, Mesorhizobium, Rhizobium, Sinorhizobium*, etc.). Rhizobia are able to interact with legumes, and this interaction results in the formation of root nodules. Bacteria establish within host cells and differentiate morphologically into bacteroids. Remarkably, expression of bacterial genes required for nitrogen fixation starts only within the host, and free-living rhizobia are unable to fix nitrogen. Rhizobia provide their host with nitrogen in the form of amino acids (glutamine and asparagine) and in turn receive from them carbon source in the form of organic acids. Host plants also produce oxygen-binding proteins (leghemoglobins), which reduce the concentration of free oxygen within nodules and, in this way, prevent the inhibition of nitrogenase by molecular oxygen.

Similar mutualism exists between nitrogen-fixing actinobacteria *Frankia* and vascular plants. Symbiosis of *Frankia* with 24 plant genera (so-called actinorhizal plants including, among others, alder (*Alnus*), bayberry (*Myrica*), sea-buckthorn (*Hippophae*), and *Casuarina*) belonging to eight different families has been documented. The interaction of *Frankia* with their hosts also results in root nodules formation, with symbionts established within host cells. Symbiosis with nitrogen-fixing bacteria provides plants with a selective advantage in nitrogen-poor habitats, and many of actinorhizal plants colonize young soils or disturbed habitats where available nitrogen is scarce.

Methanogenesis is another pathway of outstanding importance. It is a process of methane formation by certain microorganisms (methanogens). All known methanogens belong exclusively to the domain Archaea. They are anaerobic, and their growth is inhibited by free oxygen. In nature, methanogens occur in habitats where oxygen is depleted, e.g., in peatlands, bogs, lake sediments, but also in digestive tract of animals, e.g., ruminants. Biochemically, methanogenesis is a form of anoxic respiration, in which carbon dioxide or low-molecular-weight organic compounds (e.g., acetate) are used as terminal electron acceptors in respiratory chain. Methanogenesis is a final step of organic matter decay under anoxic conditions.

Viruses are nonautonomous infection agents, which do not have their own metabolism and entirely depends on metabolic machinery of their host cells.

10. Lifestyles of microbiome

10.1 Fungi as saprotrophs

As saprotrophs, fungi are able to obtain nutrients by extracellular digestion of dead organic matter. Saprotrophic fungi play an important ecological role in decomposition of wood debris and plant litter, thereby facilitating nutrient cycling. Wood decay is the breakdown of the three major chemical constituents (lignin, cellulose, hemicellulose) by enzymatic activities of microorganisms, primarily fungi. Wood-decay fungal species are classified into three main groups depending on the type of decay they cause: white rot, brown rot, or soft rot fungi. The major agents of white rot decay belong to the fungal phyla Basidiomycota (e.g., *Rigidoporus microporus*, *Phanerochaete chrysosporium*, *Heterobasidion annosum*). The nature of the decay is such that all the major wood components (lignin, cellulose, hemicellulose) are degraded leaving a bleached or white color with fibrous texture (Fig. 1.4; Daniel, 2016). The brown rot decay is also mediated by members of the group Basidiomycota (e.g., *Serpula lacrymans*, *Fomitopsis pinicola*). In brown rot decay, the polysaccharide (cellulose, hemi-cellulose) components of wood are preferentially degraded. The decayed wood has a brownish discoloration (Fig. 1.5), a cracking pattern or cubical appearance, and without its fibrous texture. Soft rot decay is caused by members of the fungal

FIG. 1.4   White rot decay.

FIG. 1.5   Brown rot decay.

phylum Ascomycota (e.g., *Ceratocystis* spp., *Chaetomium* spp.). In soft rot decay, the carbohydrates are preferentially degraded and the color of the decayed wood is brown or bleached (https://forestpathology.org/wood-decay/). Soft rot fungi usually attack wood with lower lignin content and higher moisture (Goodell et al., 2008). Unlike brown rot and white rot, not much is known about the degradative enzyme systems of soft rot decayers.

10.2 Fungi as endophytes

According to Wilson (1995), "Endophytes are fungi or bacteria which, for all or part of their life cycle, invade the tissues of living plants and cause unapparent and asymptomatic infections entirely within plant tissues but cause no symptoms of disease." Endophytes have been found in almost all plant species studied (Saikkonen et al., 2004). The nature of the relationships between endophytes and their plant hosts is still not fully understood. Endophytes play beneficial and protective roles by enhancing plant resistance against biotic agents (phytopathogens, herbivores, insects), promoting plant growth through nutrient acquisition as well as enhancing tolerance against abiotic stressors (Arnold et al., 2003; Clay, 2009; Rodriguez et al., 2009; Terhonen et al., 2019). Terhonen et al. (2014) showed that a major root endophyte *Phialocephala sphaeroides* was able to protect seedling roots of *Picea abies* from infections by the conifer pathogen *Heterobasidion parviporum*. *Phialocephala scopiformis*, a foliar endophyte significantly reduced the herbivory and detrimental effect of eastern spruce budworm *Choristoneura fumiferana* (Miller et al., 2008). The mutualistic association between the fungal endophyte *Neotyphodium coenophialum* and the forage grass, tall fescue (*Festuca arundinacea*) has been reported to facilitate the ability of the plant to adapt to growth in acidic, nutrient-poor soils as well as mediate tolerance to other stresses caused by drought and overgrazing (Siegel and Bush, 1997).

10.3 Fungi as mutualists (mycorrhiza)

Mutualism is a form of symbiotic association between two biological organisms where both partners derive benefit. Mycorrhizas have been shown to play vital roles in plant growth and nutrition (Read and Perez-Moreno, 2003). Typical example is the ecological mutualistic association between mycorrhizal fungi and vascular plants. The two major types of plant-mycorrhizal associations are endomycorrhiza and ectomycorrhiza. In endomycorrhiza association, the hyphae of the fungal partner are able to penetrate the host cell wall but not the cell membrane. There are several different types of endomycorrhiza, including orchid, arbutoid, ericoid, monotropoid, and arbuscular mycorrhizas. Arbuscular mycorrhizas are formed primarily by members of the fungal phylum Glomeromycota. Arbuscular mycorrhizas are characterized by the production of special structures, vesicules and arbuscules, that facilitate nutrient exchange with the host plant (Li et al., 2006). Ectomycorrhizas are mutualistic association formed between many woody plants (conifer trees, eucalyptus, oak, birch) and fungi belonging mostly to the fungal phyla Basidiomycota and Ascomycota. The hyphae of ectomycorrhizas unlike endomycorrhizas do not penetrate the root cells of the host. In cases where the hyphae are able to penetrate the cells, it is called ectendomycorrhiza. The ectomycorrhiza is composed of a mantle and Hartig net. A mantle is a sheath-like mass of hyphae that surrounds the root tip while Hartig net are hyphae within the intercellular spaces of the root cortex. The extensive network of filamentous hyphae outside the root originating from the ectomycorrhiza referred to as extramatrical mycelium further facilitates in the transport of nutrients and carbon cycling (Lindahl and Tunlid, 2015).

10.4 Fungi as tree pathogens (necrotrophs, biotrophs, or hemibiotrophs)

During the long coevolution with their hosts, tree pathogens have developed a wide spectrum and diversity of lifestyles ranging from necrotrophic, hemibiotrophic to biotrophic strategies in order to breach plant defensive barriers and to penetrate plant tissues. Necrotrophs are able to rapidly kill their host cells and feed saprotrophically on the contents of the dead cells (e.g., *Heterobasidion annosum*) (Asiegbu et al., 2005). Biotrophic pathogens by contrast feed on nutrients provided by living host cells over extended time and, therefore, depend on their integrity (e.g., rust fungus *Melampsora larici-populina*). The hemibiotrophs (e.g., *Phytophthora* spp.) at the initial stages of infection feed as biotrophs, but later change their behavior to a necrotrophic lifestyle (Horbach et al., 2011; Moore et al., 2011). Irrespective of their lifestyle and mode of infection, plant pathogens possess genes that are essential for causing disease or for increasing virulence on one or few hosts (Agrios, 2005).

11. Lifestyles of bacteria, archaea, and protists

The great evolutionary plasticity of bacteria is reflected in the diversity of their lifestyles. Many bacteria inhabiting forest ecosystems are free-living saprotrophs, receiving their nutrients from the decomposition of organic matter. This is particularly true for numerous soil-dwelling bacteria. A comparatively small group of bacteria (e.g., Myxobacteria, *Bdellovibrio*, *Cytophaga*) can be characterized as predators that kill other microorganisms and consume them as nutrient source. A considerable fraction of bacteria is engaged in interactions with eukaryotic organisms (first of all, plants and animals). The nature of those interactions can range from mutualism to commensalism to parasitism, and they can be either permanent (obligate) or temporary (facultative). One of the best-known examples of mutualistic interactions between bacteria and plants is the symbiosis of nitrogen-fixing rhizobia with legume plants. Remarkably, rhizobia can fix nitrogen only when colonizing their hosts, whereas free-living rhizobia cells do not express genes required for nitrogen fixation. This interaction is clearly beneficial for both partners, as bacteria supply their host with nitrogen and get plant-derived carbon sources in turn. Many bacterial plant endophytes can be considered as commensals, as they mostly remain asymptomatic and do not provide clear benefits or harm to their hosts. At the same time, there are also plant-pathogenic bacteria, which parasitize their hosts and cause various diseases. A few bacterial plant pathogens, e.g., phytoplasma, are obligate biotrophs, which require a living host for their growth and cannot be maintained on artificial nutritive media in vitro (outside of their hosts). However, vast majority of known plant-pathogenic bacteria are facultative parasites that can also grow outside of a living host (Agrios, 2005). Some plant-pathogenic bacteria (like *Agrobacterium*, *Ralstonia*, or plant-pathogenic *Streptomyces*) are typical components of soil microbial communities. Abundance of other plant pathogens in the environment is rather low, as they cannot efficiently compete with specialized soil saprotrophs and their numbers gradually decline after they are released from host tissues. A number of plant-pathogenic bacteria (e.g., *Erwinia*, *Xylella*, and phytoplasma) have developed a close relationship with herbivorous insects and use them as vectors for the host-to-host transmission, minimizing in this way their exposure to the environment (Orlovskis et al., 2015).

The lifestyles and functional roles of archaea are still imperfectly known, as many of them cannot be grown under laboratory conditions, and only a small fraction of their actual diversity has been studied in vitro. It is assumed that most of archaea are free-living saprotrophs, but many are known as symbionts of animals (residing, in particular, within their gastrointestinal tract) and a few are plant endophytes. No pathogens or parasites have been identified among archaea so far.

As a heterogenic group including several evolutionary lineages, protists display an extremely diverse range of lifestyles (Geisen et al., 2018). Many of them (e.g., representatives of Cercozoa, Ciliophora, Amoebozoa) are free-living predators feeding on bacteria and small eukaryotes. Their primary nutrient uptake mode is phagocytosis. Saprotrophs occur among protists with rigid cell walls (e.g., free-living Oomycota), which are incapable of phagocytic nutrient uptake, and instead adopted absorptive nutrition mode. At the same time, there are numerous free-living photosynthetic protists (e.g., Euglenozoa and microscopic algae). Photosynthetic protists are more abundant in water bodies, but they also occur in uppermost soil layers. Many of photosynthetic protists are mixotrophic, e.g., they fix atmospheric carbon dioxide when exposed to light, but can switch to heterotrophic nutrition mode and rely on organic carbon sources when not enough light is available.

Numerous protists are engaged in interactions with animals, plants, fungi, and other protists. Those interactions can range from mutualism to parasitism, and can be either obligate or facultative. For instance, flagellate protozoans (class *Parabasalia*) are mutualistic symbionts of insects (in particular, termites and cockroaches) playing an essential role in the cellulose breakdown in insects' gastrointestinal tract. At the same time, numerous protozoans (trypanosomes, *Plasmodium* (causative agents of malaria), *Entamoeba*, etc.) are notorious parasites of humans and animals. Analysis of environmental samples revealed unexpected abundance of parasitic Gregarines (Apicomplexa) in neotropical forest soils (Mahé et al., 2017), where they are likely to play a major role as parasites of soil invertebrates.

Important plant pathogens can be found within groups Oomycota and Phytomyxea. Numerous species of genera *Pythium*, *Phytophthora*, *Peronospora*, *Plasmodiophora*, and *Spongospora* are obligate plant pathogens, and some of them cause diseases of significant economic importance, such as seedling damping off, potato leaf blight, sudden oak death, and eucalyptus dieback, to name just a few. At the same time, some species of *Pythium* (e.g., *P. oligandrum*) are mycoparasites used as biological control agents against phytopathogenic fungi and oomycetes. In general, lifestyles of oomycetes can range from obligate biotrophic parasites to facultative pathogens to obligate saprotrophs. At the same time, many phytomyxids can exist at least for a part of their lifecycle as asymptomatic plant endophytes.

Whereas plant pathogens received a lot of attention due to their economic importance, much less is known about the role of protists as nonpathogenic plant symbionts. Recent high-throughput studies have convincingly demonstrated that protists constitute an integral part of plant microbiome. Many of the identified sequences belong to poorly characterized or even undescribed lineages of protists. At the same time, members of Cercozoa, a large and diverse group of protists with high abundance in soil, were repeatedly found in plant phyllosphere (Ploch et al., 2016). A number of novel species of Cercozoa isolated from phyllosphere show specific adaptations to their ecological niche, indicating that they might indeed constitute resident components of phyllosphere microbial communities (Dumack et al., 2017; Flues et al., 2018). Despite the recent progress in studies of protists associated with plants, many aspects of their ecology, functional role, and interactions with other components of plant microbiome remain poorly understood and await further discoveries.

All viruses are obligate intracellular parasites, which cannot replicate outside of their host cells. For their replication, they entirely depend on their host, and all building blocks of viral particles are produced in the infected cells.

12. Coevolution of plants (trees) and their microbial symbionts

The complex interactions between plants and their microbial symbionts likely involve mutual genetic exchanges, which are accompanied by reciprocal and parallel evolutionary changes between the taxa (Cairney, 2000; Naranjo-Ortiz and Gabaldón, 2019). Augspurger (1988) noted that interactions between pathogens and their host plants have impact on the assembly of plant communities. Other authors have reported that the coevolved interactions between microbiomes and plants are important for maintenance of the functioning of ecosystems and the plant communities (Keath, 2008). The successful adaptation of plants on land is equally tightly linked to the evolution of mycorrhizal interactions several million years ago (Cairney, 2000; Naranjo-Ortiz and Gabaldón, 2019; Pirozynski and Malloch, 1975; Selosse and Le Tacon, 1998). It is therefore possible that genetic and molecular programs regulating symbiosis association are present to a varying degree within all phytomicrobiome and their plant (tree) hosts.

References

Agrios, G.N., 2005. Plant Pathology. Elsevier Academic Press, Boston, Amsterdam.

Arnold, A.E., Mejia, L.C., Kyllo, D., Rojas, E.I., Maynard, Z., Robbins, N., Herre, E.A., 2003. Fungal endophytes limit pathogen damage in a tropical tree. Proc. Natl. Acad. Sci. U. S. A. 100, 15649–15654.

Asiegbu, F.O., Adomas, A., Stenlid, J., 2005. The root and butt rot fungus—*Heterobasidion annosum*. Mol. Plant Pathol. 6, 395–409.

Augspurger, C.K., 1988. Impacts of pathogens on natural plant populations. In: Davy, A.J., Hutchings, M.J., Watkinson, A.R. (Eds.), Plant Population Ecology. Blackwell Scientific, Oxford, UK, pp. 413–434.

Baldrian, P., 2017. Forest microbiome: diversity, complexity and dynamics. FEMS Microbiol. Rev. 41 (2), 109–130. https://doi.org/10.1093/femsre/fuw040.

Bulgarelli, D., Schlaeppi, K., Spaepen, S., Van Themaat, E.V.L., Schulze-Lefert, P., 2013. Structure and functions of the bacterial microbiota of plants. Annu. Rev. Plant Biol. 64, 807–838.

Cairney, J.W.G., 2000. Evolution of mycorrhiza systems. Naturwissenschaften 87, 467–475.

Canadell, J.G., Raupach, M.R., 2008. Managing forests for climate change mitigation. Science 320, 1456–1457.

Carlström, C.I., Field, C.M., Bortfeld-Miller, M., Müller, B., Sunagawa, S., Vorholt, J.A., 2019. Synthetic microbiota reveal priority effects and keystone strains in the *Arabidopsis* phyllosphere. Nat. Ecol. Evol. 3 (10), 1445–1454.

Clay, K., 2009. Defensive mutualism and grass endophytes: still valid after all these years? In: Torres, M., White Jr., J.F. (Eds.), Defensive Mutualism in Symbiotic Association. Taylor and Francis Publications, pp. 9–20.

Cooke, R.C., Rayner, A.D.M., 1984. Ecology of Saprotrophic Fungi. Longman, London.

Cooke, R.C., Whipps, J.M., 1993. Ecophysiology of Fungi. Blackwell Scientific Publications, Oxford.

Daniel, D., 2016. Fungal degradation of wood cell walls. In: Secondary Xylem Biology Origins, Functions, and Applications. Elsevier Publishers, pp. 131–167, https://doi.org/10.1016/B978-0-12-802185-9.00008-5.

Deacon, J., 2006. Fungal Biology. Blackwell Publishers, fourth ed. 371 pp.

Dumack, K., Flues, S., Hermanns, K., Bonkowski, M., 2017. Rhogostomidae (Cercozoa) from soils, roots and plant leaves (*Arabidopsis thaliana*): description of *Rhogostoma epiphylla* sp. nov. and *R. cylindrica* sp. nov. Eur. J. Protistol. 60, 76–86.

FAO, 2010. Global forest resources assessment 2010: main report. In: FAO Forestry Paper. 163. Food and Agriculture Organization of the United Nations, Rome.

FAO, 2015. Global forest resources assessment 2015: main report. Food and Agriculture Organization of the United Nations, Rome. ISBN 978-92-5-108821-0.

Flues, S., Bass, D., Bonkowski, M., 2017. Grazing of leaf-associated Cercomonads (Protists: Rhizaria: Cercozoa) structures bacterial community composition and function. Environ. Microbiol. 19 (8), 3297–3309.

Flues, S., Blokker, M., Dumack, K., Bonkowski, M., 2018. Diversity of cercomonad species in the phyllosphere and rhizosphere of different plant species with a description of *Neocercomonas epiphylla* (Cercozoa, Rhizaria) a leaf-associated Protist. J. Eukaryot. Microbiol. 65 (5), 587–599.

Geisen, S., Mitchell, A.E.D., Adl, S., Bonkowski, M., Dunthorn, M., Ekelund, F., Fernández, L.D., Jousset, A., Krashevska, V., Singer, D., Spiegel, F.W., Walochnik, J., Lara, E., 2018. Soil protists: a fertile frontier in soil biology research. FEMS Microbiol. Rev. 42 (3), 293–323. https://doi.org/10.1093/femsre/fuy006.

Goodell, B., Qian, Y., Jellison, J., 2008. Fungal Decay of Wood: Soft Rot—Brown Rot—White Rot. ACS Symposium Series. vol. 982, pp. 9–31. Chapter 2.

Guttman, D.S., McHardy, A.C., Schulze-Lefert, P., 2014. Microbial genome-enabled insights into plant-microorganism interactions. Natl. Rev. 15, 797–813.

Hawksworth, D.L., 1991. The fungal dimension of biodiversity: magnitude, significance, and conservation. Mycol. Res. 95, 641–655.

Hibbett, D.S., Binder, M., Bischoff, J.F., Blackwell, M., Cannon, P.F., Eriksson, O.E., et al., 2007. A higher-level phylogenetic classification of the Fungi. Mycol. Res. 111 (Pt. 5), 509–547. https://doi.org/10.1016/j.mycres.2007.03.004.

Horbach, R., Navarro-Quesada, A.R., Knogge, W., Deising, H.B., 2011. Infection strategies of plant pathogenic fungi. J. Plant Physiol. 168, 51–62.

Howe, A.T., Bass, D., Vickerman, K., Chao, E.E., Cavalier-Smith, T., 2009. Phylogeny, taxonomy, and astounding genetic diversity of Glissomonadida ord. nov., the dominant gliding zooflagellates in soil (Protozoa: Cercozoa). Protist 160 (2), 159–189.

Isaac, S., 1992. Fungal-Plant Interactions. Chapman & Hall, London.

James, T.Y., Kauff, F., Schoch, C., Matheny, P.B., Hofstetter, V., Cox, C., Celio, G., Gueidan, C., Fraker, E., Miadlikowska, J., Lumbsch, H.T., Rauhut, A., Reeb, V., Arnold, A.E., Amtoft, A., Stajich, J.E., Hosaka, K., Sung, G.H., Johnson, D., O'Rourke, B., Crockett, M., Binder, M., Curtis, J.M., Slot, J.C., Wang, Z., Wilson, A.W., Schüßler, A., Longcore, J.E., O'Donnell, K., Mozley-Standridge, S., Porter, D., Letcher, P.M., Powell, M.J., Taylor, J.W., White, M.M., Griffith, G.W., Davies, D.R., Humber, R.A., Morton, J.B., Sugiyama, J., Rossman, A.Y., Rogers, J.D., Pfister, D.H., Hewitt, D., Hansen, K., Hambleton, S., Shoemaker, R., Kohlmeyer, J., Volkmann-Kohlmeyer, B., Spotts, R.A., Serdani, M., Crous, P.W., Hughes, K.W., Matsuura, K., Langer, E., Langer, G., Untereiner, W.A., Lücking, R., Büdel, B., Geiser, D.M., Aptroot, A., Diederich, P., Schmitt, I., Schultz, M., Yahr, R., Hibbett, D., Lutzoni, F., McLaughlin, D., Spatafora, J., Vilgalys, R., 2006a. Reconstructing the early evolution of the fungi using a six gene phylogeny. Nature 443, 818–822.

James, T.Y., Letcher, P.M., Longcore, J.E., Mozley-Standridge, S.E., Porter, D., Powell, M.J., Griffith, G.W., Vilgalys, R., 2006b. A molecular phylogeny of the flagellated Fungi (Chytridiomycota) and a proposal for a new phylum (Blastocladiomycota). Mycologia 98, 860–871.

Keath, K.D., 2008. The coevolutionary genetics of plant–microbe interactions. New Phytol. 180, 268–270.

Koide, R.T., Sharda, J.N., Herr, J.R., Malcolm, G.M., 2008. Ectomycorrhizal fungi and the biotrophy-saprotrophy continuum. New Phytol. 178, 230–233.

Li, H., Smith, S.E., Holloway, R.E., Zhu, Y., Smith, F.A., 2006. Arbuscular mycorrhizal fungi contribute to phosphorus uptake by wheat grown in a phosphorus-fixing soil even in the absence of positive growth responses. New Phytol. 172 (3), 536–543.

Lindahl, B.D., Tunlid, A., 2015. Ectomycorrhizal fungi—potential organic matter decomposers, yet not saprotrophs. New Phytol. 205, 1443–1447.

Longcore, J.E., Pessier, A.P., Nichols, D.K., 1999. *Batrachochytirum dendrobatidis* gen. et sp. nov., a chytrid pathogenic to amphibians. Mycologia 91 (2), 219–227.

Mahé, F., de Vargas, C., Bass, D., Czech, L., Stamatakis, A., Lara, E., et al., 2017. Parasites dominate hyperdiverse soil protist communities in Neotropical rainforests. Nat. Ecol. Evol. 1 (4), 0091.

Miller, J.D., Sumarah, M.W., Adams, G.W., 2008. Effect of a rugulosin-producing endophyte in *Picea glauca* on *Choristoneura fumiferana*. J. Chem. Ecol. 2008 (34), 362–368.

Moissl-Eichinger, C., Pausan, M., Taffner, J., Berg, G., Bang, C., Schmitz, R.A., 2018. Archaea are interactive components of complex microbiomes. Trends Microbiol. 26 (1), 70–85.

Money, N.P., 2016. Fungal diversity. In: The Fungi, third ed. Elsevier, pp. 1–32.

Moore, D., Robson, G.D., Trinci, A.P.J., 2011. 21st Century Guidebook to Fungi, second ed. Cambridge University Press, pp. 367–391.

Naranjo-Ortiz, M.A., Gabaldón, T., 2019. Fungal evolution: major ecological adaptations and evolutionary transitions. Biol. Rev. Camb. Philos. Soc. 94 (4), 1443–1476.

Orlovskis, Z., Canale, M.C., Thole, V., Pecher, P., Lopes, J.R., Hogenhout, S.A., 2015. Insect-borne plant pathogenic bacteria: getting a ride goes beyond physical contact. Curr. Opin. Insect Sci. 9, 16–23.

Pan, Y., Birdsey, R.A., Phillips, O.L., Jackson, R.B., 2013. The structure, distribution, and biomass of the World's forests. Annu. Rev. Ecol. Evol. Syst. 44, 593–622. https://doi.org/10.1146/annurev-ecolsys-110512-135914.

Pirozynski, K.A., Malloch, D.W., 1975. The origin of land plants: a matter of mycotrophism. Biosystems 6, 153–164.

Ploch, S., Rose, L.E., Bass, D., Bonkowski, M., 2016. High diversity revealed in leaf-associated protists (Rhizaria: Cercozoa) of Brassicaceae. J. Eukaryot. Microbiol. 63 (5), 635–641.

Read, D.J., Perez-Moreno, J., 2003. Mycorrhizas and nutrient cycling in ecosystems—a journey towards relevance? New Phytol. 157 (3), 475–492.

Rodriguez, R.J., White Jr., J.F., Arnold, A.E., Redman, R.S., 2009. Fungal endophytes: diversity and functional roles. New Phytol. 182 (2), 314–330. https://doi.org/10.1111/j.1469-8137.2009.02773.x.

Saikkonen, K., Wäli, P.R., Helander, M., Faeth, S.H., 2004. Evolution of endophyte-plant symbioses. Trends Plant Sci. 9, 275–280. https://doi.org/10.1016/j.tplants.2004.04.005.

Sapp, J., 2010. On the origin of symbiosis. In: Seckbach, J., Grube, M. (Eds.), Symbioses and Stress, Joint Ventures in Biology, vol. 17 (Part I), Cellular Origin and Life in Extreme Habitats and Astrobiology. Springer Science, New York, pp. 3–18.

Selosse, M.A., Le Tacon, F., 1998. The land flora: a phototroph-fungus partnership? Trends Ecol. Evol. 13, 15–20.

Siegel, M.R., Bush, L.P., 1997. Toxin production in grass/endophyte associations. In: Carroll, G., Tudzynski, P. (Eds.), The Mycota V, Part A: Plant Relationships. Springer, Berlin, Heidelberg, pp. 185–207.

Terhonen, E., Kerio, S., Sun, H., Asiegbu, F.O., 2014. Endophytic fungi of Norway spruce roots in boreal pristine mire, drained peatland and mineral soil and their inhibitory effect on *Heterobasidion parviporum* in vitro. Fungal Ecol. 9, 17–26.

Terhonen, E., Blumenstein, K., Kovalchuk, A., Asiegbu, F.O., 2019. Forest tree microbiomes and associated fungal endophytes: functional roles and impact on forest health. Forests 10, 42. https://doi.org/10.3390/f10010042.

Turner, T.R., James, E.K., Poole, P.S., 2013. The plant microbiome. Genome Biol. 14, 209.

Uroz, S., Courty, P.E., Oger, P., 2019. Plant symbionts are engineers of the plant-associated microbiome. Trends Plant Sci. 24 (10), 905–916.

Watson, S.C., Boddy, L., Money, N.P., 2016. The Fungi, third ed. Elsevier. 466 pp.

Wilson, D., 1995. Endophyte—the evolution of term, a classification of its use and definition. Oikos 73 (2), 274–276. https://doi.org/10.2307/35.45919.

Wilson, R.A., Talbot, N.J., 2009. Fungal physiology—a future perspective. Microbiology 155 (12). https://doi.org/10.1099/mic.0.035436-0.

Woese, C.R., Kandler, O., Wheelis, M.L., 1990. Towards a natural system of organisms: proposal for the domains Archaea, Bacteria, and Eucarya. Proc. Natl. Acad. Sci. U. S. A. 87 (12), 4576–4579.

Zillig, W., 1991. Comparative biochemistry of archaea and Bacteria. Curr. Opin. Genet. Dev. 1 (4), 544–551.

Further reading

Barton, L.L., Northup, D.E., 2011. Microbial Ecology. Wiley-Blackwell. 407 pp.

Chapter 2

Wood as an ecological niche for microorganisms: Wood formation, structure, and cell wall composition

Uwe Schmitt[a], Adya P. Singh[b], and Yoon Soo Kim[c]

[a]Thünen Institute of Wood Research, Hamburg, Germany, [b]Scion, Rotorua, New Zealand, [c]Department of Wood Science and Engineering, Chonnam National University, Gwangju, South Korea

Chapter Outline

1. Introduction

Wood is a natural material with many desirable properties which enable this important renewable resource to be cost effectively utilized for a variety of useful products, ranging from building construction and pulp and paper production to generating products of even greater value, such as useful and novel chemicals and biofuels. Technologies are advancing rapidly for devising cost-effective methods for also nano-biotechnological applications, such as production of nano-cellulose which can be used in combination with other biodegradable materials to produce light weight, high strength materials for use in biomedical areas and as panels for automobiles and airplanes.

As an important sink for the atmospheric carbon dioxide, wood makes important contributions to the health of our environment which is at the risk of rapid deterioration from the factors causing climate change, such as global warming. Unfortunately, wood is becoming a dwindling natural resource because of excessive and illegal logging of trees that far exceeds the plantation of replacement forest trees. One way forward for tipping the balance in favor of greater production of wood than extraction is to manage forests more efficiently while generating faster growing and high yielding trees, taking advantage of the biotechnological knowledge available, particularly in the areas of genetic engineering and in vitro propagation. However, our understanding of the processes, particularly at the level of genetic and molecular controls, which regulate wood formation and the effect of environmental factors on these processes is not complete, and progress on these fronts has to be made for producing high growth forests that can not only mitigate the detrimental effects of environmental change but can produce wood biomass superior in quality and quantity.

Wood tissues also serve as crucial niches for inhabitation by diverse microbiomes. Some of them are capable of degrading wood cell walls (primary degraders) to obtain nutrients for their growth. Thus wood serves as an important source of nutrients for these microorganisms. Others which are incapable of depolymerizing lignocellulosic components of wood cell walls depend on the simple organic residues leftover from primary degradation. To understand wood decay mechanisms by microbiomes, it is indispensable to know details of the formation and structure of wood as well as its chemical composition.

The information of the current paper is therefore presented under following major headings: wood formation, principles of wood anatomy (Fig. 2.1), ultrastructure of wood cell walls, chemical characteristics of wood cell walls.

Forest Microbiology. https://doi.org/10.1016/B978-0-12-822542-4.00010-3

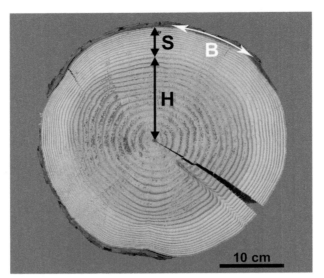

FIG. 2.1 Stem disc of Douglas fir (*Pseudotsuga menziesii*) showing the principal arrangement of tissues with the inner mostly darker colored heartwood (H) and outer sapwood (S). Not all tree species are able to form heartwood, birch is one example. Attached to wood is the cambium (only a few cell layers thick and not visible in the present picture) and finally the bark (B). *(With personal permission of C. Waitkus.)*

2. Wood formation

2.1 Cambium and wood formation

Annual plants rely largely on primary tissues for mechanical support and the transport of water and nutrients. Perennial plants also develop secondary tissues which, when fully differentiated, take over these functions from the primary tissues. The secondary xylem (wood) and phloem develop from a secondary meristem called cambium (technically vascular cambium). Under genetic control (Robischon et al., 2011), the cambium has its origin from procambium in apical regions of plant organs (Esau, 1965, 1977; Larson, 1994), and once developed as a continuous concentric layer it begins to form secondary xylem inwards and the secondary phloem outwards in shoots and roots. Thus the cambium plays a crucial role in the diameter growth of perennial plants by continually adding secondary xylem and phloem tissues year after year, which ensures that even the tallest trees are well supported mechanically as well as physiologically through the transport of water and nutrients throughout their life. The economic benefits from cambium lie in the utilization of wood biomass generated in diverse and important ways, which enhances the quality of human life while enriching environmental values.

Generation of secondary tissues from the cambium involves a complex set of genetically regulated processes, initiated with cell division in cambium and its recent derivatives (xylem and phloem mother cells). The cambium is generally regarded as a one-cell layer thick lateral meristem, which consists of cells that remain meristematic and are referred to as cambial initials. However, it is difficult to precisely identify the cambial layer within the cambial zone consisting of morphologically and structurally similar cells, with narrow radial diameter and thin cell walls as observable in transversely cut sections. The cambium consists of two types of initials, fusiform and ray initials. The products of cell division in fusiform initials are xylem and phloem mother cells, which divide to produce a series of thin walled cells in the radial direction. Because of greater cell divisions in xylem mother cells compared to phloem mother cells proportionately secondary xylem tissues are significantly more abundant. The majority cells (axial tracheids) of the secondary xylem (wood) produced in gymnosperms are morphologically similar, and therefore gymnospermous wood is regarded as a homogenous structure. In comparison, the secondary xylem produced in angiosperms consists of several cell types, namely vessels, fibers, fiber-tracheids, and axial parenchyma, which vary in their dimension and also in cell wall structure and composition.

In addition to contributing to the diameter growth, the cambium is specialized for generating a system of primarily nutrient transporting tissues, which constitutes rays. The ray initials produce rays, which form a radial system spanning across secondary xylem and secondary phloem. The main function of this system of homocellular or heterocellular tissues is transport of nutrients derived from leaves, where photosynthesis takes place, necessary for the life and activity of cambium and for the development of secondary tissues. Together, the axial (secondary xylem and phloem) and radial (rays) systems establish a three-dimensional interconnecting network that is necessary for efficient transport of water and nutrients as well as maintenance of tissue integrity. Radial cell division in the cambium ensures that this lateral meristem keeps pace with the increase in diameter growth of stems and roots in perennial plants and thus maintains its integrity.

The products of divisions in secondary xylem mother cells undergo developmental processes involving cell expansion, cell wall thickening and lignification (Wardrop, 1965; Grits and Ifju, 1984), and finally programmed cell death (except for parenchyma), enabling matured tissues to perform efficient water conduction (vessels and tracheids) and mechanical support functions (tracheids and fibers). These processes are interrelated, coordinated, and regulated involving intrinsic (gene expression, hormonal signals, enzymes) and environmental factors (temperature, water) (Plomion et al., 2001; Schrader et al., 2003, 2004; Deslauriers and Morin, 2005; Gričar et al., 2007; Begum et al., 2013; Prislan et al., 2013).

2.2   Seasonality of wood formation

Wood formation in most cases shows a seasonal rhythm with alternating dormant and active stages of the cambium, which largely depend on the climate in which trees are growing. Temperature and precipitation seem to play the most important role in regulating cambial activity. During dormant periods, fusiform and ray initials in the cambium do not undergo cell divisions. When viewed with a transmission electron microscope in the transverse direction, the inactive fusiform initials are regularly characterized by round to oval nuclei. Axially, the nuclei are elongated. Numerous small vacuoles, a high number of lipid droplets, as well as free ribosomes appear in the cytoplasm, and the endoplasmic reticulum is smooth. The plasma membrane, closely appressed to the cell wall, is characterized by a smooth outline. Dictyosomes are without or with only a few secretory vesicles, and plastids do not contain starch grains. Radial cell walls of dormant cambial cells are distinctly thicker than their tangential walls, a phenomenon, which can routinely be observed and measured with an electron microscope, and even a light microscope can show this feature in principle (Figs. 2.2 and 2.3). Active fusiform initials characteristically show a large vacuole with narrow surrounding cytoplasm closely appressed to the cell wall. Both radial and tangential walls are distinctly thinner than in dormant cells. Stages of cell division are regularly found, the endoplasmic reticulum is now rough, and the plastids regularly contain starch grains (Fig. 2.2).

When comparing dormant and active cambium it also becomes obvious that the number of cell rows is generally higher in an active cambium (Fig. 2.3). Vaganov et al. (2006) provide data for several wood species with only a small increase in cell rows like in *Abies alba* and with a huge increase like in *Pinus strobus*. Fig. 2.3 shows the cambium of *Robinia pseudoacacia* with a small increase of its cambial cell rows between dormancy and activity.

As mentioned earlier, cambial cells contribute to the secondary thickening of trees by the formation of new cell layers which are added in the radial direction. The corresponding cell divisions with the new walls oriented tangentially are called periclinal divisions. Anticlinal divisions with the new walls oriented in the radial direction are necessary for compensation of the increasing girth of a tree.

Annual rhythms of wood formation with one dormancy period during cold months and one activity period during warm months are regularly found in temperate and boreal climate zones. Distinct growth rings with earlywood and latewood are then formed. There are differences in the pattern of cells across a growth ring, with three patterns recognizable. (1) no change in the cell pattern across the growth ring; (2) gradual reduction in the size of cells from earlywood to latewood;

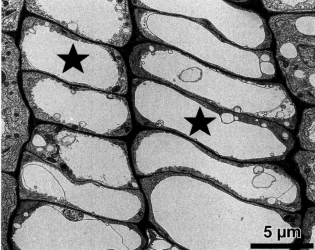

FIG. 2.2   Electron micrographs of the cambium in *Populus* sp.: Dormant cambium cells (left) with numerous smaller vacuoles (★) and active cambium (right) with a single large and central vacuole (★), cytoplasm closely appressed to the cell wall.

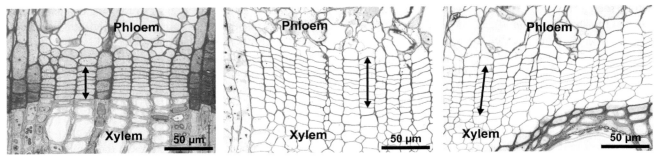

FIG. 2.3 Light micrographs of *Robinia pseudoacacia* cambium (between inner wood, synonymously called xylem and outer phloem which belongs to the bark) at different months in the year showing a slightly increased number of cambial cell rows during the vegetation period in May (middle) and June (right) when compared to dormancy in December (left). Double arrows indicate the thickness of the cambial zone.

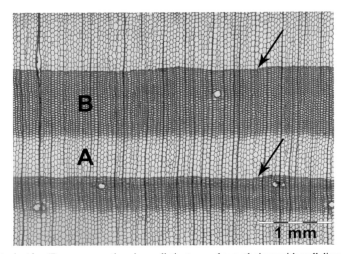

FIG. 2.4 Light micrograph of *Larix decidua*: Transverse section shows distinct annual growth rings with well discernable earlywood (A) (tracheids with wide lumens and thin walls) and latewood portions (B) (tracheids with narrow lumens and thick walls). Arrows indicate growth ring boundaries. (With personal permission of HG Richter).

(3) abrupt change in the diameter from earlywood to latewood, rendering a clear distinction between the earlywood and latewood boundary in softwoods (Fig. 2.4) and hardwoods, but because of distinct differences in anatomical features of the cell types (softwoods with tracheids, and hardwoods with vessels, fibers and fiber-tracheids in the axial system) appearances in the pattern vary and have been used as a diagnostic feature in wood identification. In trees with clear annual growth rings, wood formation follows a more or less sigmoid curve with a slow start after dormancy, an intense stage during warm months, and a slow decrease of wood formation at the end of the vegetation period.

Trees growing in subtropical zones with winter rain might have wood formation during the entire year with only variations in the intensity of cambial activity (Barnett, 1971; Prislan et al., 2016). Growth rings are sometimes less distinct because of a less pronounced border between two growth rings. For tropical regions, an annual rhythm of wood formation in trees was variously described; however, there are also reports about tropical trees with less distinct or even lacking growth zones (overview in Worbes, 2002).

It is well known that climate change also affects wood formation and even wood quality. Rise in the temperature causes drought, which can adversely affect tree growth and productivity. The information available on the effect of drought on wood forming processes varies, ranging from strong influence on cell production but little effect on wood anatomy (Balducci et al., 2016), warming together with water restriction causing *Picea mariana* sapling mortality (Balducci et al., 2015), drought resulting in shorter period of wood formation and lower wood increment but significantly larger tracheids with thinner cell walls (Eilmann et al., 2011), to reduction in cambial activity and fiber lumina but increased saccharification potential (relevant to the production of second generation bioenergy) (Wildhagen et al., 2018). The work of Wildhagen and coworkers focused on the relationship of drought to gene expression, demonstrating that some genes were differentially expressed in response to drought. Variable results reported may be attributed to differences in tree species and age, magnitude of stress (the degree and length of plant exposure to water-limiting conditions), and interaction of drought with other factors, such as the temperature where plants were also subjected to higher than normal temperatures. The overall impression gained from these and other studies is that plants may develop adaptations to cope particularly with low to moderate environmental stress conditions, but not from severe stresses.

3.   Principles of wood anatomy

Wood cells produced from the cambium are organized in a way that gives the wood its distinctive character. There are basic differences between angiosperms and gymnosperms with respect to cell types, cell size, cell wall thickness, cell wall ultrastructure, and in some cases cell wall composition. Both in angiosperms and gymnosperms secondary xylem tissues are organized into two contrasting but functionally supporting systems: axial and radial. Axial tissues perform upward transport of water and nutrients derived from the soil by roots (root hairs), crucial for the development, growth, and function of aerial parts, including stomata, which are responsible for such vital functions as transpiration and intake of atmospheric CO_2. Some axial tissues are designed to also perform mechanical functions needed to support the weight of the crown on the stem, which can be overwhelming, particularly for the species endowed with massive leafy biomass of the crown. Radial tissues (rays) mainly function in the transport of nutrients produced during photosynthesis in leaves and other green structures of plants and translocated by the elements of the phloem tissues of the physiologically functional bark. Continuity of the rays across the secondary xylem and phloem ensures nutrient supply to actively functioning cambium and developing xylem cells. Reticulated interconnecting axial and radial systems thus supply water and nutrients to the entire plant body. For anatomical studies wood tissue is routinely viewed when sectioned in three different planes.

3.1   Sectioning planes

There are three different planes, i.e., transverse, radial longitudinal, and tangential longitudinal, in which wood can be cut to provide anatomical information.

3.1.1   *Transverse section* (Fig. 2.5 top)

Cutting wood pieces transversely to the direction of tree growth often reveals concentric arrangement of the xylem in rings. If annual growth rings are formed, counting of rings gives a clue to the age of a tree (see also Fig. 2.4). Information on climatic conditions of the past (i.e., past history of weather patterns) can be obtained by examining the features of growth rings. Slower growth produces narrower rings likely affected by adverse climatic conditions during the period of xylem production. Dendrology is the study of tree ring. Also, transverse face produces information on the extent and distribution of earlywood and latewood as well as pore size and distribution, a feature of diagnostic value in the identification of hardwood species. The widths of rays are clearly revealed (one cell wide or multicell wide), and where present the size and distribution of resin canals. Trees without annual growth rings like in many trees growing in the tropics may show well visible growth increments which are not necessarily formed during one year or they even lack distinguishable growth increments.

3.1.2   *Radial longitudinal section* (Fig. 2.5 middle)

Radial longitudinal faces of axial tracheids, fibers, and vessels are exposed by cutting wood longitudinally (lengthwise) along the rays and in the direction parallel to the tree axis. In hardwoods, structural characteristics of vessels, such as wall thickenings, type of perforation plate, or pitting between vessels and neighboring cells, can be well recognized. In many conifers, this is the face of the cell wall that bears bordered pits. Therefore radial longitudinal section is the ideal plane to study the number, distribution, and characteristics of bordered pits. Rays, which appear uniseriate (single cell wide) in the transversely cut face appear stripy, and in favorable sections the entire depth of rays (i.e., how many cell layers they consist of) is revealed. In addition to ray parenchyma, many conifer rays consist of tracheids that are specific to rays (ray tracheids). Radial faces reveal the distribution of ray tracheids and their pitting and cell wall characteristics, such as ray dentation (dentate appearance of cell walls).

3.1.3   *Tangential longitudinal section* (Fig. 2.5 bottom)

Tangential longitudinal sections are best suited to revealing the ends of rays, which can provide information on ray height, composition of rays (in softwoods, number and distribution of tracheids relative to parenchyma in individual rays; in hardwoods, composition of rays). This information is vital for understanding the role of ray tracheids in wood water relations as ray tracheids facilitate lateral flow of water, which is relevant to many of the industrial processes involving impregnation of wood with preservatives, coatings, etc. In conifers, wood rays are the main pathway for liquid movement, and ray tracheids

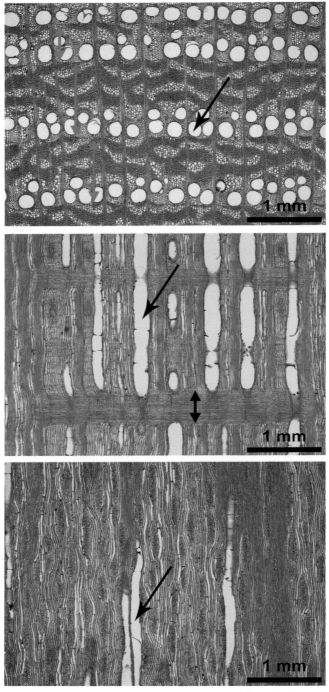

FIG. 2.5 Light micrographs of transverse (top), radial longitudinal (middle), and tangential longitudinal (bottom) sections of *Zelkova carpinifolia* (Ulmaceae). Transverse sections are well demonstrating growth increments (red arrows), vessel diameter (black arrow points to a vessel), and distribution of other elements, radial longitudinal is especially suited for showing cell lengths (arrow points to a vessel) and height of rays (black double arrow), tangential longitudinal section is showing cell lengths (black arrow points to a vessel) and preferably distribution of rays as well as their heights and widths (three rays are surrounded by a red line). *(With personal permission of H.G. Richter.)*

with their connection to axial tracheids through bordered pitting can greatly facilitate liquid movement within wood. Together with ray parenchyma, ray tracheids play a role in the radial movement of water which can enter into axial tracheids via ray pit connections, and this mechanism may enable water to be transported axially in the deeper layers of sapwood that may lack connection to transpiring leaves. Also vessel features in hardwoods can be observed, similar to radial longitudinal sections. In both soft- and hardwoods, radial sections also provide details on cell lengths and widths.

3.2 Softwoods

Softwoods systematically belong to the gymnosperms. With regard to evolution, hardwoods are more advanced than softwoods (e.g., Carlquist, 1975). Softwoods mainly consist of one type of cells in the axial system, the tracheids, which form the bulk of wood. Thus the wood appears rather homogenous (Figs. 2.1 and 2.4). Tracheids in some softwoods are also present in the radial system together with rays. Axial tracheids are many times longer (about 100 times) than wide and perform both water transport and mechanical functions. Usually, tracheids are 2–4 mm long, whereby especially for *Pseudotsuga* spec. Longer values have been recorded (Blohm, 2015). In transversely cut faces they appear rectangular shaped, and within a growth ring are differentiated as earlywood and latewood, which vary in size and cell wall thickness (Fig. 2.4). Axial tracheids are joined in a manner where their ends overlap (as opposed to end to end). Bordered pits are present on both nonoverlapping and overlapping parts. Thus water flow through tracheids is not straight but takes a slightly zigzag path. The design of pits (narrow diameter and presence of pits on overlapping ends) poses significant resistance to water movement, unlike the conducting cells (vessel elements) of hardwoods which are much larger in diameter and are joined end to end (Wiedenhoeft and Miller, 2005). As already mentioned before, tracheids in some softwoods are also present in the radial system within the ray.

Resin canals are present in some softwoods in both axial (axial resin canals, see also Fig. 2.6) and radial (radial resin canals) systems. The rays are called fusiform rays when canals are present in them. The canal itself is a cavity (appearing like a canal in longitudinal views) into which surrounding living cells (epithelial cells) secrete resins as a course of natural physiological process. The secretion of resins, which is a defense strategy, increases under biotic and abiotic stresses. For example, trees produce traumatic resin canals in response to injury. The composition of substances produced by epithelial cells varies (oleoresins, gums) and the biosynthesis of the secretory products is under genetic control. In wood species where resin canals can be detected visually or using a hand lens, a rapid assessment of the effect of stressful environmental conditions, such as drought, salinity, can be made by looking into the number, size, and distribution of resin canals.

When viewed in transversely cut sections, rays in softwoods appear as a dark line running in the radial direction (Fig. 2.4). Such rays are only one cell wide and consist largely of parenchyma cells, but in some wood species also of tracheids as mentioned earlier. The height of rays is revealed in longitudinal sections. The appearance of rays varies in longitudinal views: a stripy pattern is visible in radial longitudinal sections, and in tangential longitudinal sections end views of rays are visible, features useful in recognizing radial planes from tangential planes. These planes also serve to assess the pattern of distribution of ray tracheids relative to parenchyma. Whereas ray parenchyma mainly function in the storage and transport of photoassimilates (supply particularly critical for cambium and developing xylem), ray tracheids function in water movement. Ray tracheids are shorter and narrower than axial tracheids, but contain bordered pits typical of axial tracheids, although bordered pits in ray tracheids are distinctly smaller than those in axial tracheids. The ray tracheid cell walls are variously sculptured (dentate being a common pattern) which is a useful diagnostic feature in wood identification. Cross-field pitting is also an important diagnostic feature. Cross-field is the region of intersect between ray parenchyma and axial tracheids. The number, shape, and size of pits in this region serve to distinguish wood species.

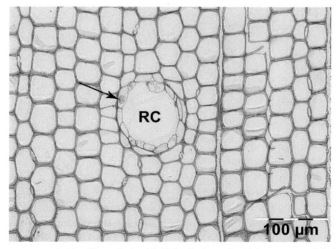

FIG. 2.6 Resin canal (RC) in *Pinus* sp. surrounded by thin-walled epithelial cells (arrow) which are responsible for resin synthesis.

3.3 Hardwoods

Hardwoods systematically belong to the angiosperms. They consist of several types of cells in their axial system, i.e., vessels, fibers, fiber-tracheids, and ray cells. These cells vary in length and diameter. When viewed on a transverse section, vessels in hardwoods are distinctly wider than fibers, particularly the earlywood vessels, and this feature has been taken advantage of by many wood anatomists to readily recognize vessels using a hand lens, making wood identification based on vessel distribution convenient, on the spot and rather rapid (Fig. 2.7). Synonymously, vessels are also called pores and the varying patterns of their size and arrangement within a growth increment are called porosity. Based on these features three porosity types are recognizable. In ring-porous wood, vessels are arranged in rings. The transition from early to latewood in this type is abrupt with a clear distinction between the wider earlywood vessels and much narrower latewood vessels (Fig. 2.7). This gives an appearance of vessels being positioned in a ring. In diffuse-porous wood, vessels are more or less uniformly distributed, and a clear distinction between early and latewood is not present. Semi-ring porous pattern defines an intermediate situation between ring-porous and diffuse-porous showing a gradual transition from larger earlywood to smaller latewood vessels.

Rays in angiosperm woods are usually larger, with some wood species displaying massive rays. In addition to transport function, these rays may have a mechanical role (Reiterer et al., 2002).

3.3.1 Vessels

The most distinctive feature of hardwoods that separates them from softwoods is the presence of vessels, which can be exceedingly large in their diameter (200 μm), visible even with the unaided eye. Although vessels are shorter than tracheids, their end-to-end axial arrangement, larger diameter, and relatively unobstructed end walls where they join make them the most efficient water transporting elements.

Technically, individual vessels are called vessel elements or vessel members, which join at their ends to form a vessel. Longitudinal walls of vessel elements contain numerous pits, and the type and arrangement of pits (pitting pattern) varies depending upon whether vessel elements are in contact with one another or they join to a different cell type, e.g., fibers and parenchyma. The arrangement of intervessel pits varies, with opposite, alternate, and scalariform types as the common patterns. Although pit borders can be present in pit pairs common to vessel elements and also between a vessel element and a fiber, pit membranes lack a torus, which is typically present in intertracheid bordered pits in conifers. Another distinctive feature of vessel elements is that where vessel elements join end to end there are no true pits in this region; instead, the cell wall is perforated and is called perforation plate. When only a large single perforation develops it is referred to as simple perforation plate. When multiple perforations are partitioned by bars, the pattern is described as scalariform perforation. Thus simple, multiple, and combination perforation plates can be present (Meylan and Butterfield, 1975). The development of perforation plates has been extensively investigated. The common view is that when developed (from enzymatic action on the end wall) a membranous structure is initially present in the regions destined to become perforated, and then disappear partially or entirely with the force of moving water within the vessel.

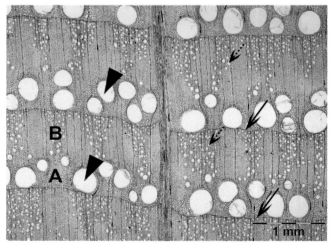

FIG. 2.7 Light micrograph of *Quercus petraea*: Transverse section showing distinct annual growth rings (arrows indicate growth ring borders). Note the ring-porous vessel arrangement with distinctly larger earlywood vessels (arrowheads) and much smaller latewood vessels (dotted arrows). Abrupt transition between earlywood (A) with its large vessels and latewood (B) with small vessels. (With personal permission of HG Richter).

3.3.2   Fibers

The main function of fibers, which are extremely thick walled, is support. In addition to being thick, fiber cell walls may become multilamellar (Singh et al., 2018), normally consisting of alternating thick and thin lamellae with differing orientations of microfibrils, much like in the bamboo (Parameswaran and Liese, 1976), a cell wall design that maximizes strength. The lamellae may also vary in lignin concentration. Fiber cell walls can be extremely thick, and thus wood density increases. Fiber secondary wall may consist of the usual three layered (S1, S2, S3) structure, and in some hardwoods more than three lamellae may be present. Woods containing thin walled fibers, such as balsa (*Ochroma pyramidale*), are soft in texture, and those with extremely thick walled fibers, such as belian, the wood density and strength is high. Fibers are usually shorter than softwood tracheids and narrower, but are longer than vessel elements. Pits are also present on fiber walls, but are not as numerous as in vessel elements. Fiber pits can be simple or bordered. Thickness variability is an interesting feature of fiber pit membranes which can be related to impregnability of woods with wood property modifying agents. For example, an investigation employing TEM showed differences between alder and eucalypt in wood impregnability with a wood-hardening formulation (Singh et al., 1999). The impregnability of alder fibers, which had thin membranes, was considerably greater than that of eucalypt fibers, which had much thicker pit membranes.

3.3.3   Axial parenchyma

Axial parenchyma are commonly present in hardwoods and are distributed in a specific pattern which is characteristic of a wood species. Therefore the pattern of axial parenchyma distribution serves as a useful diagnostic feature in wood identification. The proposed terminology to describe the distribution of axial parenchyma defines the arrangement of parenchyma relative to vessels. Two main patterns have been proposed: apotracheal and paratracheal. In the apotracheal type, parenchyma are not associated with vessels. Parenchyma are associated with vessels in the paratracheal type. Depending on the arrangement and formations, paratracheal type is classified as vasicentric (around the vessels), aliform (wing-like formations around vessels), and confluent (parenchyma often forming a band). Several classes have been proposed also for apotracheal parenchyma. In the diffuse type, parenchyma are scattered; in diffuse-aggregate type, parenchyma are arranged in small groups, which have a diffuse pattern. In banded type, parenchyma appear arranged in bands. Regardless of their arrangement, all parenchyma have similar function, i.e., they store substances.

3.4   Pits

Pits are openings in secondary cell walls partitioned by a thin structure called membrane. They are major pathways for the movement of water and minerals in wood. Pits have their origin at the primary cell wall stage. It is thought that in anticipation of pit development certain regions of the primary cell wall, which contain cytoplasmic connections between neighboring cells, become defined. These thinner primary cell wall areas are prevented from the deposition of secondary cell wall material when the secondary wall is laid down. The pit membrane is essentially a primary wall (middle lamella plus primary wall to be more accurate), which had undergone partial dissolution or remodeling. In wood tissues pits develop as pairs in exactly the same location within the neighboring cell, and suggestions are that perfect pairing of pits may be an outcome of intercommunication among neighboring cells leading to developmental processes that are regulated and coordinated.

Occasionally a pit is not paired, and this type of pit is called blind pit. Pits occur in several formations: bordered, half-bordered, and simple (Esau, 1977; Raven et al., 1999). During the development of bordered pits, which are typically present in conifer wood, the secondary wall is laid down overarching the pit membrane resulting in circular-oval configuration. Bordered pits typically consist of a membrane that is differentiated into a central disc like structure called torus (usually thickened), and a peripheral membranous structure called margo (Fig. 2.8).

Water movement occurs preferentially through the margo, which is a highly porous part of the membrane. Water movement through torus is restricted because of its compact, dense structure. Bordered pits are best viewed when face views are combined with sectional views. In face view, they are of circular to oval forms with an opening in the center (pit aperture), through which water gains entry from the lumen and passes across the margo web to enter into the neighboring cell. The space within the confines of the borders is referred to as pit cavity or pit chamber. Angiosperm woods may or may not develop borders in their pits. The main difference however between softwood and hardwood pits is that whereas softwood pits develop a complex membrane differentiated into torus and margo, the membranes in hardwood pits are more homogenous, with nanometer size pores present. Functionally, bordered pits of conifers have attracted much attention (Pittermann et al., 2005; Choat et al., 2008), as the pit membrane design is well suited to regulating hydraulic efficiency of trees, particularly when trees are exposed to stressful environmental conditions, such as drought. The membranes are highly effective in preventing passage of air bubbles (air seeding) from dysfunctional to functional tracheids by closing the aperture via a process called aspiration, where the torus moves with the flexible membrane to one side resulting in the closure of aperture. Pits

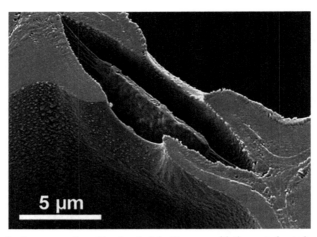

FIG. 2.8 Scanning electron micrograph of a cross sectioned bordered pit in *Pinus* sp. with the central membrane differentiated into an inner disc-like torus and an outer margo with fibrillar structure. *(From Sano, Y., 2016. Bordered pit structure and cavitation resistance in woody plants. In: Kim, Y., Funada, R., Singh, A. (Eds.), Secondary Xylem Biology: Origins, Functions, and Applications, first ed., Elsevier, Amsterdam, pp. 113–130 with permission of Elsevier.)*

are also important from the perspective of wood behavior in processes leading to improvements in practical applications of wood products, for example treatment of wood with preservatives, wood property modifying substances, and application of protective coatings. In these processes pit behavior (aspiration, partial aspiration, or no aspiration) determines the movement of such substances. These substances cannot directly pass through cell walls and therefore pit membrane condition determines impregnability of wood (De Meijer et al., 1998; Singh et al., 1999; Rijkaart et al., 2001; Singh and Dawson, 2004; Singh et al., 2010).

Vestured pits are a special type of pits typically present in hardwoods. Functional characteristics of these pits are less well understood than torus-margo type pits or simple pits. Vestured pits derive their name from their sieve-like appearance in face views. Again, the pitting is complementary, occurring in neighboring cells as pit pairs. Vestures originate as an outgrowth from pit chamber wall (Choat et al., 2008). Connection between vestures and the cell wall overarching pit chamber is clearly resolved in TEM images of ultrathin sections (Singh et al., 1999), even in wood species where vestures are tiny. Because of the inferior resolution, vesture-pit chamber wall connections cannot be resolved by light microscopy. Scanning electron microscopy provides most spectacular views, particularly when examined in the face view. TEM images of vestured pit membranes show straight to slightly deflected profiles of the membranes, and it is thought that vestures may support pit membranes against deflection. Based on the pit design, it is likely that water can move across the pit membrane rather freely, although the pathway can be tortuous, particularly where the density of vestures is high. Opinions vary regarding the composition of vestures. In TEM views of ultrathin sections stained with potassium permanganate, a reagent used to enhance the contrast of lignin in plant cell walls, vestures in some wood species appear denser than the parent wall, which suggests that vestures may be more highly lignified than the parent wall (Singh et al., 1993).

4. Ultrastructure of wood cell walls

Early studies on cell wall ultrastructure involved use of TEM as a high resolution tool (Côté, 1965). While TEM continued to be widely applied, industrial use of wood, such as for fiber production for paper and composites, and as biomass for bioethanol production, has prompted probing into the distribution of cell wall components across the cell wall and on fiber surfaces using a range of complementary tools and techniques, such as specific staining and immunolabeling in combination with TEM and light microscopy, and other high resolution tools, namely field emission scanning electron microscope (FE-SEM) and atomic force microscope (AFM) (Maurer and Fengel, 1990; Fromm et al., 2003; Fahlén and Salmén, 2005; Kim et al., 2010, 2012; Donaldson and Knox, 2012). More recently, high resolution chemical microscopy, such as confocal Raman microscopy (CRM) and spectroscopy, has yielded information on the composition of cell walls and distribution of cell wall components at submicrometer and nanometer scales, with the additional advantage being that CRM permits probing into cell walls in their native state (Gierlinger et al., 2006; Gierlinger, 2018). For example, the unique capabilities of CRM proved valuable in combining CRM with AFM in a recent study to sequentially examine same tissues and cell walls with these tools, which revealed that compression of wood under ambient conditions leads to molecular-level changes within the cell wall (Felhofer et al., 2020).

Structurally wood cell walls display a lamellar organization, and compositionally it consists of three main components (cellulose, hemicellulose, and lignin), as described in greater detail under "Chemical Characteristics." Within the cell wall these components are distributed and organized in a fashion that forms a complex network of interconnecting molecules, intricacies of which are still being uncovered (Kang et al., 2019). A consistent view has emerged that rigid, parallel ordered cellulosic microfibrils are embedded within a more flexible matrix of hemicellulose and lignin.

The lamellar cell wall consists of a compound middle lamella (middle lamella proper plus primary cell wall) between adjoining cells. Pectic components are confined to this region of the cell wall, where they establish a unique nano-architecture that is different from the architecture of the rest of the cell wall. The secondary cell wall which develops after the completion of cell expansion is differentiated into three layers, S1, S2, and S3 (S referring to secondary wall), which differ in their thickness and orientation of microfibrils (Donaldson, 2008) (Fig. 2.9). Being thickest, particularly prominent in the latewood, the S2 layer is considered to make greatest contribution to the strength and stiffness of wood. Microfibrils in this layer are oriented in the long direction of cells whereas in S1 and S3 layers they are oriented in a direction perpendicular to the cell axis. At the S1-S2 and S2-S3 interface a shift in the direction of microfibrils has been observed, and some workers have described the interface regions as transitional layers (Abe et al., 1991), which have also been incorporated into more recent cell wall models. Indications are that growth stresses generated during wood formation may influence the thickness of cell wall layers, particularly S1 and S3 layers. Considerable variability in the thickness of the S3 layer in adjoining tracheids (Singh et al., 2002) argues in favor of this view. While the S2 layer is important for the strength and stiffness of wood, the characteristics of S1 and S3 layers are relevant to tree's physiological and growth processes as well as some important industrial processes and products. The S1 layer, when exposed, determines the chemical, physical, and ultrastructural characteristics of fiber surfaces (Duchesne and Daniel, 1999) while the S3 layer influences fiber collapse. In many wood species the S3 layer is covered with a warty layer (Liese and Ledbetter, 1963), and albeit extremely thin, this layer may constitute a barrier to agents with which wood is impregnated for its protection and enhancement of its properties (Fig. 2.9).

The concentration and distribution of polymers has also attracted much attention from both fundamental and industrial perspectives. For example, lignin in cell walls has been of much concern to those engaged in pulping research as well as to scientists and technologists who utilize woody biomass to generate biofuels. During chemical pulping, lignin is removed from cell walls, and it forms a barrier to the entry of enzymes into cell walls during enzymatic pretreatment of lignocellulosic biomass for efficient production of ethanol. Middle lamella is the most highly lignified part of the cell wall in both soft and hardwoods (Donaldson, 2001; Singh and Daniel, 2001), particularly in softwoods. Secondary cell wall layers

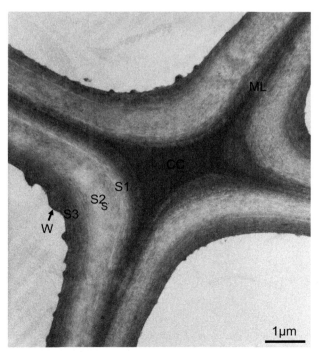

FIG. 2.9   TEM micrograph of a cell corner region of *Pinus sylvestris* showing the lamellar structure of a woody cell wall. ML middle lamella, CC cell corner, S1, S2, S3 secondary wall layers, W warty layer.

are similar in lignin concentration with some variability, with the S3 layer containing higher concentration of lignin than other secondary wall layers in some wood species. Lignin concentration in the S3 layer of *P. radiata* reaches around 50% (Donaldson, 1987). The concentration and distribution of lignin in cell walls appears to be developmentally regulated and is also influenced by biotic and abiotic stresses. For example, the concentration of lignin is shown to be variable in *P. radiata* trees severely affected by drought (Donaldson, 2002), and lignin concentration and type is also affected by mechanical stress (Schmitt et al., 2006). Displacement of lignin within cell walls has been reported in wood tissues subjected to compressive forces (Felhofer et al., 2020).

5. Chemical characteristics of wood cell walls

It has taken 200 million years for woody plants to achieve the optimization of their structures through the constant cross-talking to the surrounding environments. Chemically, only three polymers construct the wood cell walls; nonetheless, a number of adaptation strategies are embedded in the cell walls for the structural integrity and protection against biotic and abiotic agents. The secondary cell wall of woody plants is comprised of cellulose (~35%–54%), hemicelluloses (~20%–30%), and lignin (~20%–35%). However, the composition of chemical components varies with taxa (softwoods, hardwoods), cell types (tracheid, fiber, vessel, and parenchyma), and cell wall layers (secondary wall layers, middle lamella). Chemical composition also differs between early- and latewood and sap- and heartwood (Bertaud and Holmbom, 2004; Xiao et al., 2019) (Table 2.1).

5.1 Major cell wall components

5.1.1 Cellulose

Presence of cellulose is not confined to the plant kingdom, but it is also present in bacteria and even in tunicates. The amount of cellulose varies with plant species; relatively higher content in cotton (over 95%), medium in bamboo and wood (40%–50%), and lower in horse-tail (20%–30%). Cellulose is a linear polymer consisting exclusively of (1,4)-linked β-D-glucose. Three hydroxyl (OH) groups in each glucose unit are present on the surface of cellulose molecules, which are responsible for the chemical and physical behavior of cellulose (Fig. 2.10). The OH-group at C_1-end has reducing properties whereas the OH-group at C_4-end nonreducing properties. OH groups in cellulose form two types of hydrogen bonds:

TABLE 2.1 Chemical composition of Eucalyptus sp. (%).

	Heartwood	Sapwood
Cellulose	35.8	46.3
Hemicellulose (xylose)	27.3 19.9	19.9 11.6
Klason lignin	26.9	21.5
Acid-soluble lignin	6.9	4.5

Data from Xiao, M., Chen, W., Hong, S., Pang, B., Cao, X., Wang, Y., Yuan, T., Sun, R., 2019. Structural characterisation of lignin in heartwood, sapwood and bark of eucalyptus. Int. J. Biol. Macromol. 138, 519–527, with permission of Elsevier.

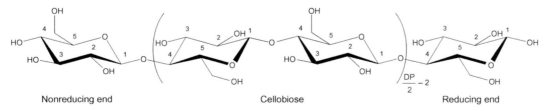

FIG. 2.10 Molecular structure of cellulose; the β-1,4 glucan chains make up the cellulose microfibrils with the alternating glucose residue rotated 180° to one another. *(From Klemm, D., Cranston, E., Fischer, D., Gama, M., Kedzior, S., Kralisch, D., Kramer, F., Kondo, T., Lindström, T., Nietzsche, S., Petzold-Welcke, K., Rauchfuß, F., 2018. Nanocellulose as a natural source for groundbreaking applications in materials science, Today's state, Mater. Today, 21(7), 720–748 with permission of Elsevier.)*

intramolecular (within the same cellulose molecule) and intermolecular hydrogen bonds (neighboring cellulose molecules) which are responsible for the formation of supramolecular structures of cellulose. Owing to inter- and intra-chain hydrogen bond, cellulose chains pack together to form cellulose microfibrils, which have an ordered arrangement, and contribute to load-bearing function.

The majority cellulose in nature contains both crystalline (ordered) and amorphous (less ordered) regions with different ratios. Crystalline index (CI) is used to describe the relative amount of crystalline material in cellulose. The value of CI (about 70% in wood cellulose) varies significantly depending on the measurement methods. The simplest and most widely used method is X-ray diffraction (XRD), but the crystallinity value from XRD was shown to be usually higher than other assessment methods (Park et al., 2010). Crystallinity affects the accessibility of cellulose to enzymes and chemicals. Two distinct crystalline forms, triclinic (Iα) and monoclinic (Iβ), are present in cellulose. Iβ is the predominant form in softwood (Atalla and Vanderhart, 1984). Cellulose microfibrils are also classified as interior crystalline cellulose and surface cellulose molecules based on solid-state nuclear magnetic resonance (NMR) spectra by different hydroxymethyl conformation of the glucose units (Dupree et al., 2015). A diameter of microfibril in spruce wood was estimated to be 3–4 nm with the models of 18-chain (Terrett et al., 2019), 24-chain (Fernandes et al., 2011; Newman et al., 2013) and 36-chain microfibril (Himmel et al., 2007).

Due to structural characteristics of cellulose, diverse enzymes are needed to dissolve the cellulose enzymatically. At least three types of carbohydrate active enzymes (CAZs), such as cellobiohydrolase (CBH), glucanohydrolase (EG), and cellobiase, are involved in the degradation of cellulose (Lombard et al., 2014). The crystalline microfibril is responsible for the tensile strength in woody plants. Cellulose nanofibrils in the form of cellulose nanocrystals extracted from the lignocelluloses have recently emerged as the key components for advanced materials, such as food packaging and conducting materials, biosensors, electronics templates, drug delivery, and diverse functional fibers (Isogai et al., 2011; Klemm et al., 2018).

5.1.2 Hemicelluloses

Unlike cellulose, which consist entirely of β-(1–4) D-glucan chains, hemicelluloses consist of multiple types of sugar units with some uronic acids. Hemicelluloses also display a chain-like structure, but the chains are much shorter than cellulose and often branched. Hemicelluloses contribute to strengthening the cell wall by interaction with cellulose and lignin. They coat and tether cellulose microfibrils, enhancing the mechanical strength of the secondary wall. When compared to cellulose, hemicelluloses are more readily hydrolyzed by acids and are more soluble in dilute alkali. The composition of sugar units in hemicelluloses varies with wood and plant species (Table 2.2). Also, hardwoods contain in general higher amounts of hemicelluloses than softwoods. The major component of hemicelluloses in hardwoods is xyloglucan (XG), accounting for 20%–30%. XG has a backbone similar to cellulose, but it is decorated with xylose branches on 3 out of 4 glucose residues. The xylose is also appended with galactose and fucose residues (Scheller and Ulvskov, 2010).

The dominant hemicellulose in softwoods is glucomannan (GM) (10%–30%), which contains an unbranched chain containing one glucose residue for every 4 mannose unit. GM has small amount of acetyl groups and galactose residues. GM is also a dominant hemicellulose in *Equisetum* and *Psilotum nudum* (Timell, 1982). The second major hemicellulose of softwoods is glucuronoarabinoxylan (GAX) (5%–10%). In GAX, for every ten xylose units, one arabinose and two 4-*O*-methylglucuronic acid substitutions (10:1.3:2) are attached. The side chain substituents make GAX vulnerable to acid hydrolysis but resistant to alkali-catalyzed degradation (Sjoström, 1981). Softwoods also contain galactoglucomannan (GGM) in the ratio of 1:1:3 and its amount is in the range of 5%–8%. In ferns and *Lycopodium*, glucomannan and xylan are present in equal amounts (Timell, 1982).

TABLE 2.2 Occurrence of hemicelluloses in secondary walls of plants.

Amount of polysaccharide in wall (%, *w*/w)			
Polysaccharide	Dicots	Conifers	Grasses
Xyloglucan (XG)	20–30	–	–
Glucuronoarabinoxylan(GAX)	–	5–10	40–50
(Gluco)mannan (GM)	2–5	10–30	0–5
Galactoglucomannan(GGM)	0–3	5–8	–

Data from Timell, T., 1982. Recent progress in the chemistry and topochemistry of compression wood. Wood Sci. Technol. 16(2), 83–122; Fengel, D., Wegener, G., 1984. Wood Chemistry, Ultrastructure, Reactions. first ed., De Gruyter, Berlin; Scheller, H., Ulvskov, P., 2010. Hemicelluloses. Annu. Rev. Plant Biol. 61, 263–289..

5.1.3 *Lignin*

Lignin is the second most abundant biopolymer in woody plants after cellulose, accounting for about 30% in softwoods and 20%–25% in hardwoods, respectively. Lignin is a complex aromatic heteropolymer, which renders structural rigidity to the cell wall and waterproofs it, enabling efficient transport of solutes and water through the vascular system. Lignin also plays an important role in protecting cell wall polysaccharides from microbial degradation, thus imparting decay resistance. It is thought that lignin evolved in terrestrial plants, providing the structural support necessary for an erect growth (Frey-Wyssling, 1976). Recent genomic studies on the presence of lignin in the marine red alga (*Calliarthron*) implied that lignin in the secondary walls of red alga may have also evolved to resist bending stresses imposed by strong waves, similar to its biomechanical support function in vascular plants (Martone et al., 2009).

Unlike cellulose and hemicellulose, lignin lacks a repeat structure and consists of many building blocks of hydroxyl cinnamyl alcohol monomers (monolignols). Depending on the degree of methoxylaton in monolignols, three types of monolignols are classified, such as coniferyl alcohol (one methoxy group), sinapyl alcohol (two methoxy groups), and *p*-coumaryl alcohol (no methoxy group). These monolignols produce guaiacyl (G), syringyl (S), and *p*-hydroxy phenyl propanoid units (H), respectively (Fig. 2.11). Monolignol radicals favor coupling at their β position, resulting in the β-β, β-O-4, and β-5 dimers in lignin. These bonds are susceptible to attacks by chemicals and enzymes (Ralph et al., 2004).

The composition of lignin varies with wood species, cell types, and individual cell wall layers. G-unit is the main lignin in gymnosperm with low amount of H units, whereas angiosperm lignins consist of G- and S-units, with traces of H units. Monocotyledon lignin contains more H units than hardwoods, but comparable levels of G and S units. In birch wood, the compound middle lamella contains both G- and S unit, whereas secondary wall layers contain mainly S units. Interestingly, both vessel walls and compound middle lamella are rich in G-units (Timell, 1982).

In wood cell walls lignin is the most recalcitrant polymer due to its resistance to delignification by chemicals and enzymes. Diverse oxidoreductases are involved in the enzymatic degradation of lignin such as lignin peroxidase (LiP), Mn-dependent peroxidase (MnP) and laccase (Janusz et al., 2017). Genetic engineering has been applied to produce lignin-less trees (or trees with lower amount of lignin) for saving energy for pulping. Specific enzymes (COMT, CAD) involved in biosynthesis of lignin have been down-regulated or up-regulated to modify lignin structures. Such trees showed a shift in G and S levels and more aldehyde but the effect on the development and strength of trees was minor.

5.2 Molecular architecture and distribution of chemical components

Spatial relationship between lignin and polysaccharides at subnanometer resolution is crucial to understand the interaction of polymers in the secondary cell wall (Salmén et al., 2012). In the molecular architecture of secondary cell wall, hemicellulose plays an important role by close association with cellulose and lignin. Hemicellulose functions as a matrix component, which tethers cellulose fibrils and also binds to lignin, mainly at Cα of β-*O*-4 position (Dammstrom et al., 2009; Kang et al., 2019). Recent multidimensional solid-state NMR works showed that hemicelluloses, such as galactoglucomannan

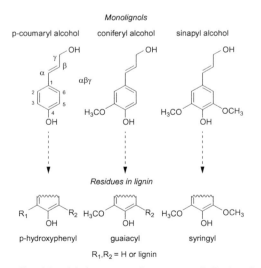

FIG. 2.11 Three main precursors of lignin (monolignols) and their corresponding structures in lignin polymers. (*From Laurichess, S., Averous, L., 2014. Chemical modification of Lignins: towards biobased polymers. Prog. Polym. Sci. 39(7), 1266–1290 with permission of Elsevier.*)

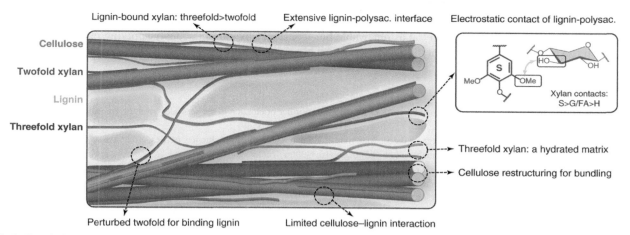

FIG. 2.12 Lignin-polysaccharide packing in secondary cell wall. Cellulose microfibrils, two- and threefold xylan, and lignin are depicted in red, purple, blue and yellow, respectively. The two hydrophobic cores of lignin and cellulose are bridged by xylans. The lignin-xylan interactions are boxed in blue. Polysaccharides are abbreviated. *(From Kang, X., Kirui, A., Dickwella Widanage, M., Mentink-Vigier, F., Cosgrove, D., Wang, T., 2019. Lignin-polysaccharide interactions in plant secondary cell walls revealed by solid-state NMR. Nat. Commun. [online] 10, 347. Available at: doi:10.1038/s41467-018-08252-0 [Accessed 7 Nov. 2020], Creative Commons CC BY license, Springer Nature.)*

and xylan, coat the surface of same cellulose microfibrils. In particular, most of the xylan binds to the hydrophilic faces of cellulose microfibrils whereas some xylan with threefold conformation binds to lignin with noncovalent interactions (Simmons et al., 2016; Kang et al., 2019; Terrett et al., 2019) (Fig. 2.12). The hemicellulose-coated elementary cellulose fibrils and lignin assemble into larger order structures in woody plants and even in grasses. Thus lignin is mainly localized on the surface of a polysaccharide macrofibril core. Owing to the close interactions and sophisticated arrangement of secondary cell wall components within a dense and hydrophobic matrix, chemical disintegration of the cell wall components usually brings about the loss and modification of their original structure.

Major components in the secondary cell wall are heterogeneously distributed in the individual cell wall layers. The highest concentration of lignin is found in the cell corner, reaching more than 80%. However, total lignin percentage of cell corner covers only 12% because of the small volume of cell corner regions (4%). Consequently, over 70% of total lignin is distributed in the S2 layer which covers nearly 90% of the secondary cell wall. In secondary cell wall, noncondensed G- and S-lignin is deposited preferentially in S1 and S3 layers in concomitant incorporation of condensed G and S-subunits within the entire S2 layer (Ruel et al., 2006; Terashima et al., 2012). Maximum amount of galactose, rhamnose, arabinose, and uronic acids are present mainly at the interface of primary and secondary cell walls. In contrast, glucomannan shows much higher distribution in the outer part of S2 and S3 layers compared to S1 and S2 layers and arabinoglucuronoxylan in softwoods is distributed mainly in the S1 layer (Meier, 1961). Absolute amount of xylose and cellulose increased until the end of fiber development whereas arabinose residues decreased (Huwyler et al., 1979). Noninvasive techniques, such as confocal Raman microscopy with an imaging program, contribute to enhancing our understanding of the molecular architecture of secondary cell wall layers in situ with a high spatial resolution (0.5 μm) (Zhang et al., 2017; Gierlinger, 2018; Agarwal, 2019).

6. Future perspectives

Wood serves as a habitat for a wide range of microorganisms, some of which can degrade it to obtain nutrients. In this context, a thorough understanding of wood is important covering all aspects related to wood formation, structure, and composition. Challenges lie ahead in more precisely understanding genetic regulation of wood formation from tissue to cell wall composition level. The emerging information will be of value in a number of ways. For example, with increasing environmental pressure arising from global warming on tree growth and wood formation probing into gene expression related to the effect of environmental factors, such as drought, would be helpful in understanding adaptation strategies of trees. This will help forestry scientists to genetically engineer trees or identify tree species best suited to particular environments. Benefits will also accrue in producing wood with the characteristics or qualities fit for certain industrial applications of wood and wood-derived materials, such as bioenergy (Malico et al., 2019), solid wood conversion into smart products (reviewed in Burgert et al., 2016), and nano-cellulose (Trache et al., 2020) based products which are proving to have wide-ranging applications from biomedical areas to strong light weight panels especially for automobiles and airplanes. Finally, so-called second generation or new generation wood products are using lignocellulosic biomass from forest plantations with improved or modified genetic material as renewable bio-feedstocks by simultaneously applying innovative technologies which can support the development of knowledge-based bio-economies.

References

Abe, H., Ohtani, J., Fukazawa, K., 1991. FE-SEM observations on the Microfibrillar orientation in the secondary wall of Tracheids. IAWA Bull. 12 (4), 431–438.

Agarwal, U. (2019). Analysis of Cellulose and Lignocellulose Materials by Raman Spectroscopy: A Review of the Current Status. Molecules, [online] 24 (9), 1659. [Accessed 5 November 2020] Available at: https://doi.org/10.3390/molecules24091659.

Atalla, R., Vanderhart, D., 1984. Native cellulose: a composite of 2 distinct crystalline forms. Science 233 (4633), 283–285.

Balducci, L., Deslauriers, A., Giovannelli, A., Beaulieu, M., Delzon, S., Rossi, S., Rathgeber, C., 2015. How do drought and warming influence wood traits of *Picea mariana* saplings? J. Exp. Bot. 66 (1), 377–389.

Balducci, L., Cuny, H., Rathgeber, C., Deslauriers, A., Giovannelli, A., Rossi, S., 2016. Compensatory mechanisms mitigate the effect of warming and drought on wood formation. Plant Cell Environ. 39 (6), 1338–1352.

Barnett, J., 1971. Winter activity in the cambium of *Pinus radiata*. N. Z. J. For. Sci. 1 (2), 208–222.

Begum, S., Nakaba, S., Yamagishi, Y., Oribe, Y., Funada, R., 2013. Regulation of cambial activity in relation to environmental conditions: understanding the role of temperature in wood formation of trees. Physiol. Plant. 147 (1), 46–54.

Bertaud, F., Holmbom, B., 2004. Chemical composition of Earlywood and latewood in Norway spruce heartwood, sapwood, and transition zone wood. Wood Sci. Technol. 38 (4), 245–256.

Blohm, J., 2015. Holzqualität und Eigenschaften des juvenilen und adulten Holzes der Douglasie (*Pseudotsuga menziesii* (Mirb.) Franco) aus süddeutschen Anbaugebieten. PhD Thesis, University of Hamburg.

Burgert, I., Keplinger, T., Cabane, E., Merk, V., Rüggeberg, M., 2016. Biomaterial wood: wood-based and bioinspired materials. In: Kim, Y., Funada, R., Singh, A. (Eds.), Secondary Xylem Biology: Origins, Functions, and Applications, first ed. Academic Press/Elsevier, London, pp. 259–281.

Carlquist, S., 1975. Ecological Strategies of Xylem Evolution, first ed. University of California Press, Berkeley.

Choat, B., Cobb, A., Jansen, S., 2008. Structure and function of bordered pits: new discoveries and impacts on whole-plant hydraulic function. New Phytol. 177 (3), 608–626.

Côté, W., 1965. Cellular ultrastructure of woody plants. In: Proceedings of the Advanced Science Seminar. Syracuse University Press, Syracuse, pp. 61–97.

Dammstrom, S., Salmén, L., Gatenholm, P., 2009. On the interactions between cellulose and xylan, a biomimetic simulation of the hardwood cell wall. Bioresources 4 (1), 3–14.

De Meijer, K., Thurich, K., Militz, H., 1998. Comparative study on penetration characteristics of modern wood coatings. Wood Sci. Technol. 32 (5), 347–365.

Deslauriers, A., Morin, H., 2005. Intra-annual tracheid production in balsam fir stems and the effect of meteorological variables. Trees 19 (4), 402–408.

Donaldson, L., 1987. S3 lignin concentration in radiata pine tracheids. Wood Sci. Technol. 21 (3), 227–234.

Donaldson, L., 2001. Lignification and lignin topochemistry: an ultrastructural view. Phytochemistry 57 (6), 859–873.

Donaldson, L., 2002. Abnormal lignin distribution in wood from severely stressed *Pinus radiata* trees. IAWA J. 23 (2), 161–178.

Donaldson, L., 2008. Microfibril angle: measurement, variation and relationships—a review. IAWA J. 29 (4), 345–386.

Donaldson, L., Knox, J., 2012. Localisation of cell wall polysaccharides in normal and compression wood of radiata pine: relationship with lignification and microfibril orientation. Plant Physiol. 158 (2), 642–653.

Duchesne, I., Daniel, G., 1999. The ultrastructure of wood fibre surfaces as shown by a variety of microscopical methods—a review. Nord. Pulp Pap. Res. J. 14 (2), 129–139.

Dupree, R., Simmons, T., Mortimer, J., Patel, D., Iuga, D., Brown, S., Dupree, P., 2015. Probing the molecular architecture of *Arabidopsis thaliana* secondary cell walls using two- and three-dimensional ^{13}C solid state nuclear magnetic resonance spectroscopy. Biochemistry 54 (14), 2335–2345.

Eilmann, B., Zweifel, R., Buchmann, N., Graf Pannatier, E., Rigling, A., 2011. Drought alters timing, quantity and quality of wood formation in scots pine. J. Exp. Bot. 62 (8), 2763–2771.

Esau, K., 1965. Plant Anatomy. Wiley, New York.

Esau, K., 1977. Anatomy of Seed Plants, second ed. Wiley, New York.

Fahlén, J., Salmén, L., 2005. Pore and matrix distribution in the fibre wall revealed by atomic force microscopy and image analysis. Biomacromolecules 6 (1), 433–438.

Felhofer, M., Bock, P., Singh, A., Prats-Mateu, B., Zirbs, R., Gierlinger, N., 2020. Wood deformation leads to rearrangement of molecules at the nanoscale. Nano Lett. 20 (4), 2647–2653.

Fernandes, A., Thomas, L., Altaner, C., Callow, P., Forsyth, V., Apperley, D., Kennedy, C., Jarvis, M., 2011. Nanostructure of cellulose microfibrils in spruce wood. Proc. Natl. Acad. Sci. U. S. A. 108 (47), E1195–E1203.

Frey-Wyssling, A., 1976. The Plant Cell Wall, first ed. Gebr. Borntraeger, Berlin.

Fromm, J., Rockel, B., Lautner, S., Windeisen, E., Wanner, G., 2003. Lignin distribution in wood cell walls determined by TEM and backscattered SEM techniques. J. Struct. Biol. 143 (1), 77–84.

Gierlinger, N., 2018. New insights into plant cell walls by vibrational microspectroscopy. Appl. Spectrosc. Rev. 53 (7), 517–551.

Gierlinger, N., Schwanninger, M., Reinecke, A., Burgert, I., 2006. Molecular changes during tensile deformation of single wood fibres followed by Raman microscopy. Biomacromolecules 7 (7), 2077–2081.

Gričar, J., Zampančič, M., Čufar, K., Oven, P., 2007. Regular cambial activity and xylem and phloem formation in locally heated and cooled stem portions of Norway spruce. Wood Sci. Technol. 41 (6), 463–475.

Grits, G., Ifju, G., 1984. Differentiation of tracheids in developing secondary xylem of *Tsuga canadiensis* L Carr. Changes in morphology and cell-wall structure. Wood Fiber Sci. 16 (1), 20–36.

Himmel, M., Ding, S., Johnson, D., Adney, W., Nimlos, M., Brady, J., Foust, T., 2007. Biomass recalcitrance: engineering plants and enzymes for biofuels production. Science 315 (5813), 804–807.

Huwyler, H., Franz, G., Meier, H., 1979. Changes in the composition of cotton fibre cell walls during development. Planta 146 (5), 635–642.

Isogai, A., Saito, T., Fukuzumi, H., 2011. TEMPO-oxidized cellulose nanofibres. Nanoscale 3 (1), 71–85.

Janusz, G., Pawlik, A., Sulej, J., Swiderska-Bulek, U., Jarosz-Wilkolazka, U., Paszczynski, A., 2017. Lignin degradation: microorganisms, enzymes involved, genomes analyses and evolution. FEMS Microbiol. Rev. 41 (6), 941–962.

Kang, X., Kirui, A., Dickwella Widanage, M., Mentink-Vigier, F., Cosgrove, D. and Wang, T. (2019). Lignin-polysaccharide interactions in plant secondary cell walls revealed by solid-state NMR. Nat. Commun., [online] 10, 347. [Accessed 7 November 2020]. Available at: https://doi.org/10.1038/s41467-018-08252-0.

Kim, J., Awano, T., Yoshinaga, A., Takabe, K., 2010. Immunolocalisation and structural variations of Xylan in differentiating earlywood cell walls of *Cryptomeria japonica*. Planta 232 (4), 817–824.

Kim, J., Sandquist, D., Sundberg, B., Daniel, G., 2012. Spatial and temporal variability of xylan distribution in differentiating secondary xylem of hybrid Aspen. Planta 235 (6), 1315–1330.

Klemm, D., Cranston, E., Fischer, D., Gama, M., Kedzior, S., Kralisch, D., Kramer, F., Kondo, T., Lindström, T., Nietzsche, S., Petzold-Welcke, K., Rauchfuß, F., 2018. Nanocellulose as a natural source for groundbreaking applications in materials science, Today's state. Mater. Today 21 (7), 720–748.

Larson, P., 1994. The Vascular Cambium: Development and Structure, first ed. Springer, Berlin.

Liese, W., Ledbetter, M., 1963. Occurrence of a warty layer in vascular cells of plants. Nature 197 (4863), 201–202.

Lombard, V., Golaconda-Ramulu, H., Drula, E., Coutinho, P., Henrissat, B., 2014. The carbohydrate-active enzymes database (CAXy) in 2013. Nucleic Acids Res. 42 (D1), D490–D495.

Malico, I., Pereira, R., Goncalves, A. and Sousa, A. (2019). Current status and future perspectives for energy production from solid biomass in the European industry. Renew. Sustain. Energy Rev., [online] 112, pp. 960-977. [Accessed 10 November 2020] Available at: https://doi.org/10.1016/j.rser.2019.06.022.

Martone, P., Estevez, J., Lu, F., Ruel, K., Denny, M., Somerville, C., Ralph, J., 2009. Discovery of lignin in seaweed reveals convergent evolution of cell-wall architecture. Curr. Biol. 19 (2), 169–175.

Maurer, A., Fengel, D., 1990. A process for improving the quality and lignin staining of ultrathin sections from wood tissues. Holzforschung 44 (6), 453–460.

Meier, H., 1961. The distribution of morphological aspects of the fine structure of wood. Pure Appl. Chem. 5 (1), 37–52.

Meylan, B., Butterfield, B., 1975. Occurrence of simple, multiple, and combination perforation plates in the vessels of New Zealand woods. N. Z. J. Bot. 13 (1), 1–18.

Newman, R., Hill, S., Harris, P., 2013. Wide-angle X-ray scattering and solid-state nuclear magnetic resonance data combined to test models for cellulose microfibrils in mung bean cell walls. Plant Physiol. 163 (4), 1558–1567.

Parameswaran, N., Liese, W., 1976. On the fine structure of bamboo fibres. Wood Sci. Technol. 10 (4), 231–246.

Park, S., Baker, J., Himmel, M., Parilla, P. and Johnson, D. (2010). Cellulose crystallinity index: measurement techniques and their impact on interpreting cellulose performance. Biotechnol. Biofuels, [online] 3, 10. [Accessed 10 November 2020] Available at: http://www.biotechnologyforbiofuels.com/content/3/1/10.

Pittermann, J., Sperry, J., Hacke, U., Wheeler, J., Sikkema, E., 2005. Torus-margo pits help conifers compete with angiosperms. Science 310 (5756), 1924.

Plomion, C., Leprovost, G., Stokes, A., 2001. Wood formation in trees. Plant Physiol. 127 (4), 1513–1523.

Prislan, P., Čufar, K., Koch, G., Schmitt, U., Gričar, J., 2013. Review of cellular and subcellular changes in the cambium. IAWA J. 34 (4), 391–407.

Prislan, P., Gričar, J., de Luis, M., Novak, K., Martinez del Castillo, E., Schmitt, U., Koch, G., Štrus, J., Mrak, P., Žnidarič, M. and Čufar, K. (2016). Annual cambial rhythm in *Pinus halepensis* and *Pinus sylvestris* as indicator for climate adaptation. Front. Plant Sci., [online] 7, 1923. [Accessed 10 November 2020] Available at: https://doi.org/10.3389/fpls.2016.01923.

Ralph, J., Lundquist, K., Brunow, G., Lu, F., Kim, H., Schatz, P., Marita, J., Hatfield, R., Ralph, S., Christensen, J., Boerjan, W., 2004. Lignins: natural polymers from oxidative coupling of 4-hydroxylphenyl-propanoids. Phytochem. Rev. 3 (1–2), 29–60.

Raven, P., Evert, R., Eichhorn, S., 1999. Biology of Plants, sixth ed. W.H. Freeman, New York.

Reiterer, A., Burgert, I., Sinn, G., Tschegg, S., 2002. The radial reinforcement of the wood structure and its implication on mechanical and fracture mechanical properties: a comparison between two species. J. Mater. Sci. 37 (5), 935–940.

Rijkaart, V., Stevens, M., de Meijer, M., Militz, H., 2001. Quantitative assessment of the penetration of water-borne and solvent-borne wood coatings in scots pine sapwood. Holz Roh Werkst. 59 (4), 278–287.

Robischon, M., Du, J., Miura, E., Groover, A., 2011. The *Populus* class III HD ZIP, *pop* REVOLUTA, influences cambium initiation and patterning of woody stems. Plant Physiol. 155 (3), 1214–1225.

Ruel, K., Chevalier-Billosta, V., Guillemin, F., Berrio Sierra, J., Joseleau, J., 2006. The wood cell wall at the ultrastructural scale-formation and topochemical organization. Maderas Cienc. Tecnol. 8 (2), 107–116.

Salmén, L., Olsson, A., Stevanic, J., Simonovič, J., Radotič, K., 2012. Structural organisation of the wood polymers in the wood fibre structure. Bioresources 7 (1), 521–532.

Scheller, H., Ulvskov, P., 2010. Hemicelluloses. Annu. Rev. Plant Biol. 61, 263–289.

Schmitt, U., Singh, A., Frankenstein, C., Möller, R., 2006. Cell wall modifications in woody stems induced by mechanical stress. N. Z. J. For. Sci. 36 (1), 72–86.

Schrader, J., Baba, K., May, S., Palme, K., Bennett, M., Bhalerao, R., Sandberg, G., 2003. Polar auxin transport in the wood-forming tissues of hybrid aspen is under simultaneous control of developmental and environmental signals. Proc. Natl. Acad. Sci. U. S. A. 100 (17), 10096–10101.

Schrader, J., Nilsson, J., Mellerowics, E., Berglund, A., Nilsson, P., Hertzberg, M., Sandberg, G., 2004. A high-resolution transcript profile across the wood forming meristem of poplar identifies potential regulators of cambial stem cell identity. Plant Cell 16 (9), 2278–2292.

Simmons, T., Mortimer, J., Bernardinelli, O., Pöppler, A., Brown, S., de Azevedo, E., Dupree, R. and Dupree, P. (2016). Folding of xylans onto cellulose fibrils in plant cell walls revealed by solid-state NMR. Nat. Commun., [online] 7, 13902. [Accessed 10 November 2020] Available at: https://www.nature.com/articles/ncomms13902.

Singh, A., Daniel, G., 2001. The S2 layer in the tracheid walls of *Picea abies* wood: inhomogeneity in lignin distribution and cell wall microstructure. Holzforschung 55 (4), 373–378.

Singh, A., Dawson, B., 2004. Confocal microscope—a valuable tool for examining wood-coating Interface. J. Coat. Technol. Res. 1 (3), 235–237.

Singh, A., Nilsson, T., Daniel, G., 1993. *Alstonia scholaris* vestures are resistant to degradation by tunnelling bacteria. IAWA J. 14 (2), 119–126.

Singh, A., Dawson, B., Franich, R., Cowan, F., Warnes, J., 1999. The relationship between pit membrane ultrastructure and chemical impregnability of wood. Holzforschung 53 (4), 341–346.

Singh, A., Daniel, G., Nilsson, T., 2002. High variability in the thickness of the S3 layer in *Pinus radiata* tracheids. Holzforschung 56 (2), 111–116.

Singh, A., Singh, T., Rickard, C., 2010. Visualising impregnated chitosan in *Pinus radiata* earlywood cells using light and scanning electron microscopy. Micron 41 (3), 263–267.

Singh, A., Wong, A., Kim, Y., Wi, S., 2018. Resistance of the S1 layer in kempas heartwood fibres to soft rot decay. IAWA J. 39 (1), 37–42.

Sjöström, E., 1981. Wood Chemistry, Fundamentals and Applications. Academic Press, New York.

Terashima, N., Yoshida, M., Hafrén, J., Fukushima, K., Westermark, U., 2012. Proposed supramolecular structure of lignin in softwood tracheid compound middle lamella regions. Holzforschung 66 (8), 907–915.

Terrett, O., Lyczakowski, J., Yu, L., Iuga, D., Trent Franks, O., Brown, S., Dupree, R. and Dupree, P. (2019). Molecular architecture of softwood revealed by solid-state NMR. Nat. Commun., [online] 10, 4987. [Accessed 10 November 2020] Available at: https://www.nature.com/articles/s41467-019-12979-9.

Timell, T., 1982. Recent progress in the chemistry and topochemistry of compression wood. Wood Sci. Technol. 16 (2), 83–122.

Trache, D., Tarchoun, A., Derradji, M., Hamidon, T., Masruchin, N., Brosse, N. and Hussin, M. (2020). Nanocellulose: from fundamentals to advanced applications. Front. Chem., [online] 8, 392. [Accessed 10 November 2020] Available at: https://www.frontiersin.org/articles/10.3389/fchem.2020.00392/full.

Vaganov, E., Hughes, M., Shashkin, A., 2006. Growth Dynamics of Conifer Tree Rings: Images of Past and Future Environments, first ed. Springer, Berlin-Heidelberg.

Wardrop, A., 1965. Cellular differentiation in xylem. In: Côté, W. (Ed.), Cellular Ultrastructure of Woody Plants, first ed. Syracuse University Press, Syracuse, pp. 61–97.

Wiedenhoeft, A., Miller, R., 2005. Structure and function of wood. In: Rowell, R. (Ed.), Handbook of Wood Chemistry and Wood Composites. CRC Press, Boca Raton, pp. 9–33.

Wildhagen, H., Paul, S., Allwright, M., Smith, H., Malinowska, M., Schnabel, S., Paulo, M., Cattonaro, F., Vendramin, V., Scalabrn-in, S., Janz, D., Douthe, C., Brendel, O., Buré, C., Cohen, D., Hummel, I., Le Thiec, D., Van Eeuwijk, F., Keurentjes, J., Flexas, J., Morgante, M., Robson, P., Bogeat-Triboulot, M., Taylor, G., Polle, A., 2018. Genes and gene clusters related to genotype and drought-induced variation in saccharification potential, lignin content and wood anatomical traits in *Populus nigra*. Tree Physiol. 38 (3), 320–339.

Worbes, M., 2002. One hundred years of tree-ring research in the tropics: a brief history and an outlook to future challenges. Dendrochronologia 20 (1–2), 217–231.

Xiao, M., Chen, W., Hong, S., Pang, B., Cao, X., Wang, Y., Yuan, T., Sun, R., 2019. Structural characterisation of lignin in heartwood, sapwood and bark of eucalyptus. Int. J. Biol. Macromol. 138, 519–527.

Zhang, X., Chen, S., Ramaswamy, S., Kim, Y., Xu, F., 2017. Obtaining pure spectra of hemicellulose and cellulose from poplar cell wall Raman imaging data. Cellul. 24 (11), 4671–4682.

Chapter 3

Methods for studying the forest tree microbiome

Kathrin Blumenstein[a], Eeva Terhonen[a], Hui Sun[b], and Fred O. Asiegbu[c]

[a]*Forest Pathology Research Group, Department of Forest Botany and Tree Physiology, Faculty of Forest Sciences and Forest Ecology, University of Göttingen, Göttingen, Germany,* [b]*Department of Forest Protection, College of Forestry, Nanjing Forestry University, Nanjing, China,* [c]*Department of Forest Sciences, Faculty of Agriculture and Forestry, University of Helsinki, Helsinki, Finland*

Chapter Outline

1. Introduction

The research fields of "microbiome" and "genomics" are expanding rapidly as new microbial genome sequences are currently being produced at an exponential rate. Traditionally, identifying all forest tree-associated microbes using conventional isolation methods has been demanding as many are difficult to cultivate outside their hosts. The recent advances in molecular approaches, including high-throughput sequencing (HTS) or next-generation sequencing (NGS) and single cell genomic approaches, have facilitated overcoming these obstacles (Guttman et al., 2014; Siegl et al., 2011; Woyke et al., 2006). The application of metagenomics (HTS), including metatranscriptomics, has equally facilitated the accumulation of a vast amount of data on microbiomes colonizing different plant tissues (endosphere, phyllosphere, rhizosphere) and is now being extended to forest ecosystem and soils. These novel technological advances in "omics" and bioinformatics have contributed considerably to furthering our understanding of molecular interaction between microbiomes and their host trees or within a particular ecosystem. In addition to the genomic techniques, nonnucleic acid approaches such as metabolomics and metaproteomics are increasingly being used, further contributing to our understanding of functional dynamics of the microbiome community (Bashiardes et al., 2016). Metagenomics provides information on the genetic material of the microbial community in an environmental sample, whereas metatranscriptomics allows us to gain insight on the transcript abundance or information on the gene expression profile of a transcriptionally active microbial community (Cox et al., 2017).

Forest Microbiology. https://doi.org/10.1016/B978-0-12-822542-4.00016-4

Metaproteomics, on the other hand, gives insight about all the proteins produced by the microbiome community while metabolomics provides information on the complete set of the metabolite composition of the microbiota community. The integration of these functional genomic approaches together with complementary nonnucleic acid studies on the forest microbiome is expected to provide insights as well as a mechanistic understanding of lifestyles of forest microbes and their interactions with direct relevance to forest biomes.

2. Traditional methods for studying phyllosphere and endophytic microbiota

All plant tissues host microbial communities, including bacterial, archaeal, fungal (Compant et al., 2019), and protist taxa. For many years, the traditional nonmolecular methods have been used for the study of microbiota in the layers of the phyllosphere (the plant aerial surfaces) and the endosphere (internal tissues). The phyllosphere is nutrient poor and microbes on the plant tissues defined as epiphytes are exposed to extremes of temperature, radiation, and moisture (Vorholt, 2012). Microbes inside the plant tissues, such as leaves, roots, or stems (the endosphere), are defined as endophytes and can establish beneficial, neutral, or detrimental symbiotic associations with their host plants (Turner et al., 2013). The above-ground plant microbiota mainly originates from the soil, the rhizosphere environment, the seed and air, and resides on or inside the plant tissue (see Fig. 3.1A–C). The composition of the microbiota in above- and belowground plant parts is influenced by biotic and abiotic factors, including external environmental conditions such as soil, climate, pathogen presence, and human practices (Compant et al., 2019; Hardoim et al., 2015). Endophytes usually spread systemically via the xylem to distinct areas of the plant like stem and leaves (Fig. 3.1D) (Compant et al., 2010). Endophytes can also enter plant tissues through aerial parts of the plant such as flowers and fruits (Fig. 3.1C) (Compant et al., 2011). Stem and leaf

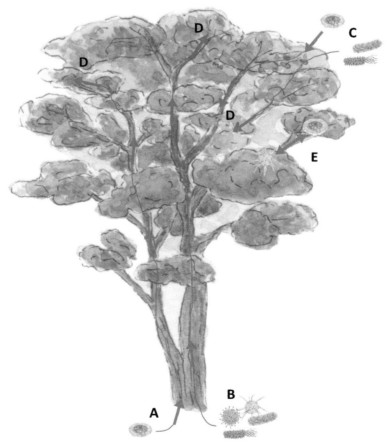

FIG. 3.1 Diverse microbiota associated with a tree. (A) Aggregate mycelia of filamentous fungi, (B) bacteria and protists can enter the tree through roots. (C) Airborne dispersed spores enter the tree through stomata or wounds. Inside their host, (D) microbiota interacts with each other and competes for nutrients and space. (E) Microbiota spreads inside the host through the xylem vessels from the roots to the leaves or through the sap flow from the leaves to twigs and branches. *(Artistic drawing by Dr. Marta Agostinelli)*

surfaces also produce exudates that attract microorganisms (Compant et al., 2010). But only adapted bacteria can survive and enter the plant via stomata, wounds, and hydathodes (Fig. 3.1C) (Compant et al., 2010; Hallmann, 2001). Interactions between and also within microbial communities and their host (Fig. 3.1E) reveal important information about the plant health and productivity (Berendsen et al., 2012). Though bacterial and fungal endophytic communities are usually studied separately, an investigation on how the interactions are regulated within the plant is a fascinating new field in endophyte research (Frey-Klett et al., 2011). Therefore, an in-depth study of the entire diversity of a microbiome is crucial (Turner et al., 2013). The term "microbiome" refers to the collective genomes of microorganisms living in association with the host plant (Hardoim et al., 2015). Discovering the microbiome functionality in relation with its host could provide useful information on how trees benefit from these associations (Hacquard, 2016; Hardoim et al., 2015). The core microbiome consists of microorganisms that are closely linked with a certain plant species independent of environmental conditions (Toju et al., 2018). So-called satellite taxa appear only rarely in few sites (Hanski, 1982; Magurran and Henderson, 2003). The importance of satellite taxa is increasingly being recognized as drivers of key functions for the ecosystem (Compant et al., 2019).

The first challenge to overcome in tree microbiota study is determining the appropriate method to be used for the visualization, isolation, and identification (McCully, 2001). The use of direct culture-dependent methods is an efficient initial approach for the isolation of phyllosphere and endophytic microbiota. Depending on the organism to be isolated, culture media and growth conditions need to meet the species' preferences. The aim is to obtain pure cultures of the organisms. However, vast majority of microbial diversity in an environment is often missed using the culture-dependent techniques (Turner et al., 2013). This obstacle is now partly overcome by the use of HTS technique (see Section 4).

2.1   Culture dependent-based isolation methods

Culture-dependent isolation method is useful for studying the characteristics of a microorganism. The microbe needs to be isolated from the host material and cultured on artificial media. The number of species to be isolated depends on factors such as the biotic and abiotic conditions, the isolation procedure, the seasonal and climatic conditions, and the amount of the host sample (Sieber, 2007).

2.1.1   Isolation of endophytic or phyllosphere fungi

The plant material (e.g., needles, leaves, twigs, or roots) is cut up into small (few centimeters) pieces. The samples are surface sterilized (Fig. 3.2A) in a laminar flow chamber under axenic conditions. This is to guarantee that only fungal organisms residing inside the host tissues are isolated. Firstly, the plant material is immersed in 70% ethanol. The duration of the immersion depends on the type of the individual plant material. The purpose is to sterilize only the surface but to prevent the plant material from absorbing the chemical and kill internal fungal organisms. For twigs and leaves, 30–60 s are recommended. The next immersion step is in 0.5%–10% sodium hypochlorite (NaOCl) (twigs: 5 min; leaves: 1 min). This is followed by further dipping in 70% ethanol (twigs and leaves: 15–30 s). Thereafter, the sample is thoroughly washed with sterile water (one or two times) for about 30 s and allowed to dry. Several protocols exist for the surface sterilization

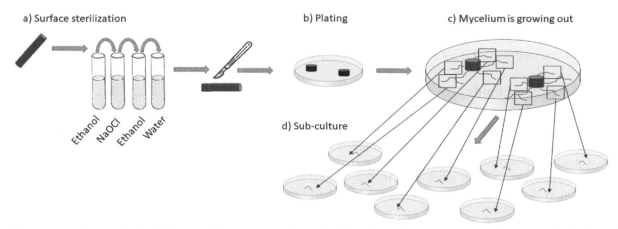

FIG. 3.2   Isolation of endophytic fungi. Plant material (twigs, roots, or leaves) is (A) surface sterilized and cut into smaller pieces. (B) The slices, chips, or disks of plant material are placed onto Petri dishes filled with artificial growth media. (C) Starting after one day, mycelium of the fast-growing fungal species grows out of the plant material and needs to be (D) subcultured immediately.

and differ by the type of host plant material (see Arnold et al., 2003; Helander et al., 2007; Martín et al., 2013, 2015; Stone et al., 2004; Terhonen et al., 2014). Successful surface sterilization can be verified by pressing the sterilized plant tissue on to a suitable agar plate as described by Schulz et al. (1999), or by incubating the last rinsing water on an agar plate.

For woody samples, they can be separated into bark and xylem tissues and cut into smaller disks. Leaves can be cut into 1 cm^2 discs and placed on artificial growth media in Petri dishes (Fig. 3.2B). The dishes are incubated under the preferred growth conditions of the targeted endophytic fungi. This can be in darkness or under a low light intensity, varying air moisture and temperature range. Usually, these settings reflect the host plants' natural conditions. After one day, fast-growing species might start to grow out of the pieces of wood or plant material in the agar medium (Fig. 3.2C). For the next days and weeks, careful observation is necessary, as outgrowing mycelium needs to be subcultured immediately onto new Petri dishes (Fig. 3.2D) in order not to miss rare or slow-growing species. Surface sterilization is not required for the isolation of phyllosphere epiphytic microbiota (i.e., step "A" in Fig. 3.2 can be omitted).

The choice of the growth medium determines the targeted fungal species. Malt extract agar is suitable for most fungal species. The initial growth and isolation of slow-growing fungi can be inhibited by rapidly growing fungi (Martín et al., 2013). The isolation of basidiomycetes might require the use of special media (see also Bußkamp, 2018). To facilitate the formation of distinctive morphology of filamentous fungi, Malt-Yeast-Peptone (MYP) medium (7 g malt extract, 1 g peptone, 0.5 g yeast extract, and 15 g agar/L water) can be used (Langer, 1994). If a high nutrient medium such as Potato Dextrose Agar (PDA) is used as culture medium, fungi can lose their ability to produce enzymes or metabolites, or perform other functions (Smith and Onions, 1994). Sabouraud's dextrose agar (SDA) can be used for the cultivation of fungi, but since Gram-positive and Gram-negative bacteria prefer that medium as well, the selective isolation of fungal species can become challenging (Littman, 1947).

2.1.2 Cultivation of endophytes using the dilution-to-extinction method

A method described by Solis et al. (2016) can also be applied for the isolation of endophytic fungi from plant material. The plant material should be grinded into small particles for uniform homogenization. The smallest hyphal particles (ø < 0.2 mm) are diluted and plated onto malt extract agar (MEA, 1.5%). For high-throughput screening, by using multiwell plates (e.g., 48-well), a separated inspection for emerging colonies is feasible. Single colonies can be transferred to Petri dishes containing MEA.

2.1.3 Isolation of endophytic bacteria

The surface sterilization procedure is similar as described before for endophytic fungi. To isolate bacteria from internal tissues, host tree samples (roots or shoots) are washed in sterile distilled water for 5 min, followed by immersion in a solution containing 1% NaOCl for 5 min with an addition of a droplet of Tween 80 per 100 mL solution. Thereafter, the plant material is rinsed in sterile water up to three times. Finally, the plant material is macerated in 10 mL of 10 mM MgSO$_4$. Dilution series (100:1) are plated on nonselective media (Taghavi et al., 2009). The surface sterilization for the isolation of endophytic bacteria requires extra care because of the diverse niches on the plants surface. Bacteria might be unreachable to the chemicals used for surface sterilization. Alternatively, the studies by Bulgarelli et al. (2012) and Lundberg et al. (2012) describe how sonication was applied to remove the surface layer of plant tissue. Organisms isolated from beneath the surface layer could then be defined as the endophytic microbiota (Bulgarelli et al., 2012; Lundberg et al., 2012). Bacterial endophytes can be isolated using culture media such as Yeast Peptone Dextrose Agar (YPDA), Brain Heart infusion medium (BHI), Luria agar (LA), King's B agar (KBA), or Tryptic soy agar (TSA) (e.g., Rashid et al., 2012).

2.2 Histological methods

Culture-dependent methods could be complemented with other approaches, such as microscopy, image visualization, as well as immunological assays. The initial challenge to overcome is the localization of the microorganisms (McCully, 2001). Image analyses make the exact location of the invasive microbe within plant tissues and the physical contacts between microbial groups visible (Cardinale, 2014; Compant et al., 2011). High-quality light microscopy, scanning electron microscopy (SEM), and/or transmission electron microscopy (TEM) of fixed and resin-embedded samples (Asiegbu et al., 1993, 1994; Compant et al., 2010; Monteiro et al., 2012) gives reliable visual results. Light microscopy can be useful for screening the host plant tissue for fungal hyphae (e.g., Cabral et al., 1993). After vital staining, living fungal species can be detected (Schulz and Boyle, 2005). SEM and TEM are recommended for the visualization of the fungal structures (Christensen et al., 2002; Sequerra et al., 1995). Confocal laser scanning microscopy (CLSM) is applied for the identification and localization of microbial associations in their natural habitat. For the quantification, a combination of hemocytometer, flow cytometry, and immunological techniques (enzyme linked immunosorbent assays, ELISA) could be used (Asiegbu et al., 1995; Hartmann et al., 2019). Alternatively, to localize and identify single cells, the Fluorescence In Situ Hybridization (FISH) technique can be applied (Schmid et al., 2007). The quantitative presence of bacterial endophytes can be estimated by the ELISA technique (Schloter and Hartmann, 1998). Many of those studies revealed that endophytic bacteria mostly reside in the intercellular apoplast, in dead or dying cells and also in xylem vessels (Turner et al., 2013). Additionally, epiphytic bacteria appear in bigger amounts than endophytic ones, and younger plants have higher bacterial concentrations than mature ones (Turner et al., 2013).

Microscopy visualization of fungal hyphae inside their host tissue neither allow the discrimination of fungal mycelium due to their very localized appearance nor the taxonomic identification (Stone et al., 1994; Sieber, 2002). For determining a fungal species' taxonomy, morphological characteristics of the hyphae, spores, or fruiting bodies could be useful. Using microscope and isolation of the fungus onto pure culture might further facilitate taxonomic identification (see also Sieber, 2002).

3. Biochemical methods (microbiota—bacteria and fungi)

3.1 Metabolic activity

Living inside their host tree, bacterial and fungal microorganisms have to adapt their metabolism to their internal environment (de Santi Ferrara et al., 2012; Monteiro et al., 2012) in order to guarantee their survival. The microbiota can interact with their host plant and vice versa in different ways: (a) the host tree responds with the production of novel metabolites due to the interactions involved in the microorganism-host relationship (Kusari et al., 2012). (b) The microbiota is also able to influence the secondary metabolism of their host plant (Zhang et al., 2006). (c) In order to avoid adverse effect on the host, which might reduce its ecological fitness, most endophytes tend not to provoke host defense reactions (Hardoim et al., 2015). This kind of association is influenced and regulated by the nature and secreted secondary metabolites that are involved in signaling and host defenses (Schulz and Boyle, 2005). The metabolites produced by the microbes can influence the dynamics in their community or effect the host physiology. Endophytic microorganisms have the ability to produce a diverse range of bioactive secondary metabolites (Schulz et al., 2002; Strobel, 2003; Tan and Zou, 2001). The compounds produced by endophytes belong to diverse structural groups such as terpenoids, steroids, xanthones, chinones, phenols, isocoumarins, benzopyranones, tetralones, cytochalasines, and enniatines (Schulz et al., 2002). Often, these compounds have beneficial or antimicrobial activities, such as the protection against pests. It is also important for the endophyte to produce defensive compounds when it is in competition with other organisms (Schulz et al., 1999). In some cases, the antimicrobial and herbicidal compounds were directed against the host plant (Schulz et al., 1999) and had detrimental effects. The identification of metabolites produced by microorganisms can be challenging, as well as identifying which particular microorganism synthesized a certain compound. (d) Microbiota and host might also share parts of a specific metabolic pathway and both contribute to the final product. (e) Or microbes can share compounds with their host, for example, endophytic bacteria from the genus *Bacillus* produce antimicrobial compounds (oxylipin family) that are also secreted within the stems of their host plant (Amaranthaceae) (Trapp et al., 2015). (f) The microbiota and the host plant can metabolize products produced by the other, e.g., facultative endophytes consume nutrients provided by their host (Knight et al., 2018). Many of the metabolic activities outlined before can be determined by use of enzymatic assays accompanied by spectrophotometric quantification (Tienaho et al., 2019).

3.2 Methods to investigate fungal secondary metabolites

Many endophytic fungi produce secondary metabolites which can antagonize competing fungi to occupy the same ecological niche in the host plant (e.g., Mandyam and Jumpponen, 2005; Miller et al., 2002; Schulz et al., 1999; Sumarah et al., 2010, 2011; Tellenbach et al., 2013). Characterizing the bioactive properties of metabolites secreted by endophytic fungi enables their potential use as biocontrol agents (Blumenstein, 2015; Terhonen et al., 2016). In order to conduct a metabolic profiling, the extracellular compounds need to be extracted from the fungus or environmental sample. Typical approaches for an extraction with LC-MS methods are described in Terhonen et al. (2016) and Tienaho et al. (2019). Pure fungal cultures are incubated in a suitable growth medium, as for example 2% malt extract, with mild shaking. The duration of the incubation depends on the general growth rate of the fungal colonies. Fungal material is removed by filtering through miracloth or filter paper. Further, the filtrate is extracted with two equal volumes of the ethyl acetate, which is evaporated in a vacuum dryer. The extract is weighed and resuspended to a smaller, concentrated volume. In order to identify the chemical constituents of the extracts, they are resuspended in acetonitrile and dried under nitrogen. Next, the extracts are filtered and redissolved in acetonitrile and screened by UPLC-QTOF/MS using electrospray ionization in both positive and negative ion mode (Terhonen et al., 2016). The UPLCMS data of metabolites can be analyzed using the two pattern recognition methods, PCA and OPLS-DA, to distinguish differences between groups, identify possible outliers, and identify metabolites responsible for differences between noninhibitory endophytes. However, for a high-throughput community profiling of microbial metabolome, novel computational algorithms are being developed, which are able to predict diverse range of unobserved metabolites (Mallick et al., 2019).

3.3 "Phenomics" methods

To link ecological habit of microorganisms to their functional traits requires reliable phenotyping (Blumenstein et al., 2015b). The observable characteristics of cells, including all types of cell properties, are the phenotype, which complements the characteristics gained from biochemical and molecular studies (Bochner, 2008). The taxonomic and morphological diversity and the physiological adaptability of microorganisms make phenotyping a challenging task for achieving biologically relevant information (Blumenstein et al., 2015b). All essential cellular physiological processes of a microbe need to be taken into consideration, such as sufficient regulation of the water activity, internal pH balance, energy production, cell division, biosynthesis of the required biochemical components, and acquisition of vital nutrients (e.g., C, N, P, S) (Bochner, 2008). Fungal hyphae could respond more effectively to the conditions across the whole colony by acting as a collective mass in comparison to isolated individuals (Falconer et al., 2005). By contrast, the conditions are different when investigating cells of bacteria biota. Almost 100 years ago, it was demonstrated that bacteria could be distinguished on agar media with different C- and N-sources through compiling growth assays (Den Dooren de Jong, 1926). The use of multiwell plates to cultivate different microorganisms for phenotyping in small volumes is further explained later. The methods for high-throughput phenotyping of microorganisms are referred to as "phenomics" and comprise among others the computer visualization, such as the imaging of cell traits with imaging technologies, such as NIR and fluorescence, as well as reporter gene expression (Houle et al., 2010).

3.3.1 Phenotype Microarrays

Phenotype Microarrays (PM) enable the culturing of microorganisms in microtiter plates for the investigation of phenotypic changes in a cell caused by different environmental conditions (Decorosi et al., 2011; Line et al., 2010; Stolyar et al., 2007). The method provides (semi-) high-throughput assays information about phenotypes of microorganism like bacteria or filamentous fungi (Blumenstein et al., 2015b). Phenotypic microarray plates are commercially available through Biolog, Inc. (Hayward, CA), which provides a suite of twenty 96-well plates with each containing a diverse set of substrates, stressors, or nutrients (Bochner, 2003; Bochner et al., 2001). This method allows 1000 assays of chemical sensitivity, 100 assays of ion effects and osmolarity, 200 assays of C-source metabolism, 100 assays of S-source and P-source metabolism, 400 assays of N-source metabolism, 100 assays of biosynthetic pathways, and 100 assays of pH control and pH effects with deaminases and decarboxylases (Bochner, 2008). Every microarray plate has 96 micro-wells. One micro-well contains no substrate and serves as a control, while 95 are filled with a predetermined set of different C-sources. For example, two plate types including carbon sources provide sources, including tetra-, tri-, di-, mono-, and methyl saccharides, polysaccharides, saccharide phosphates, amino acids, alcohols, carboxylic acids, nucleosides, phenolic compounds, and surfactants. There is a redox dye (tetrazolium violet) as an indicator in each well on a plate. A target cell suspension with a standardized cell density is added with an inoculation fluid containing, e.g., Fe, Ca, Na, Mg, K, P, N, C, pyrimidines, purines, amino acids, and vitamins at sufficient levels to maintain cell viability. A unique culture condition is created. The tetrazolium violet is reduced to formazan if the substrate in a particular well is metabolized by the microorganism, resulting in a change in that well's absorbance (Blumenstein et al., 2015a; Bochner, 2008). The color change and the optical density (OD) in each well of the microtiter plates can be recorded spectrophotometrically at user-specified time intervals (Atanasova and Druzhinina, 2010). The color reaction is mainly indicative for bacteria. The growth response of filamentous fungi can be recorded as change in the optical density at 750 nm (OD750) (Tanzer et al., 2003). Measurements of growth can also be conducted at 590 nm (Blumenstein et al., 2015a).

The method offers optimal application possibilities for bacterial cells as it was initially designed for their study. For example, a metabolic fingerprint can be created, which can also be used for the identification of bacteria (Bochner et al., 2001). The method has also been adapted for filamentous fungi (Blumenstein et al., 2015a; Singh, 2009). The substrate utilization patterns of fungi can be compared, providing valuable information about nutrient preferences when investigating competitive interactions between pathogens and endophytes, coexisting in their host (Blumenstein et al., 2015b). In order to optimize growth media for the production of secondary metabolites produced by filamentous fungi, PM technology has also been used (Singh, 2009). Another example for the application is to study interactions with other organisms *in planta* by investigating niche partitioning (Blumenstein et al., 2015a). Inside their host plants, microorganisms share nutrients and space. In this regard, according to the niche concept, niches are defined by the needs and impacts of the species that are present, which determines whether a given community of species can coexist in a common ecological community (Chase and Leibold, 2003).

3.3.2 Enrichment cultures

The coexistence of a microorganism community is possible due to the role of each individual microorganism in that heterogeneous ecosystem, which provides diverse conditions for different types of organisms (Veldkamp, 1970). To investigate single species of a microbiota in pure culture, the individuals need to be separated to enable their isolation. Usually, the streak plate technique is used for the separation of bacteria. The environmental sample is spread out onto agar plate. This allows individual cells to multiply separately by forming colonies, which can be picked and subcultured for further purification into pure culture. The enrichment technique provides possibilities to selectively change physicochemical conditions that favor only certain microorganisms to survive. Alternatively, the nutrient content of a culture medium can be modified, providing only selected conditions for a certain organism to grow preferably in the medium. Thus the growth of dominant or rare species is enriched and can be isolated (Veldkamp, 1970). Beijerinck (1901) and Winogradsky (1949) were the first to describe enrichment techniques in the beginning of the twentieth century, which was an important methodological step for studying natural microbial communities (Dworkin, 2006; Schlegel and Jannasch, 1967). In molecular genetics, the enrichment technique is used to select bacteria that carry a plasmid with an antibiotic-resistant gene. The bacteria are incubated in a medium that contains the respective antibiotic. Only bacteria cells harboring the plasmid with the antibiotic resistance gene will grow and survive. This concept can be applied with any kind of inhibitor or toxin, as it facilitates specific isolation of target microbes of interest (Schlegel and Jannasch, 1967).

3.3.3 Phospholipid fatty acid analysis

Phospholipid fatty acid analyses (PLFAs) are essential components of every living cell. Due to their wide structural diversity, they can be used as quantitative biomarkers (Tornberg et al., 2003; Zelles, 1997). PLFA profiles are also used to characterize microbial communities, because they represent only the viable microbial community. This is because they are not found in storage products or in dead cells because the phosphate group is quickly hydrolyzed to diglycerides after the death of the cell (White et al., 1979; Zelles, 1999). Using PLFAs as biomarkers is advantageous to track dynamic changes in microbial communities over seasons (Moore-Kucera and Dick, 2008).

3.4 Use of nonnucleic acid approaches (metaproteomics and metabolomics) in microbial community profiling of environmental samples

The "omics" technologies, including metaproteomics and metabolomics, have recently widely been used to investigate genetic and metabolic activities of microbes in environmental samples. Both metaproteomics and metabolomics are important tools for large-scale profiling of microbial communities in an ecosystem (Dubey et al., 2020). As previously highlighted, the application of metaproteomics will facilitate connecting proteins to their corresponding gene sequences and biological functions (bioindicators) (Maron et al., 2007). It will also provide insights into bacterial responses to different growth conditions (Pedersen et al., 1978) as well as facilitate the identification of stress-responsive proteins (Maron et al., 2007). Metabolomics, on the other hand, will provide valuable information about the metabolites secreted by the microbial community and their potential metabolic functions and activities (Hettich et al., 2013).

3.4.1 Metaproteomics

The metaproteomics analysis consists of three major steps, which are explained as follows: (1) the protein extraction, (2) separation, and (3) identification.

(1) Protein extraction

Successful extraction of microbial protein from environmental sample is essential for high-quality analysis of proteome data. The proteome can be difficult to isolate due to the heterogeneity and complexity of the environment, where the sample originates from (Maron et al., 2007). Depending on the selected proteins, whether from prokaryotes or eukaryotes, cellular or extracellular, the method for the extraction is to be adapted and optimized (Maron et al., 2007). Compounds that can disturb the digestion, chromatographic separation, or mass spectrometric studies need to be removed (Dubey et al., 2020). Also, the quality and quantity of the extracts should be representative of the sample (Maron et al., 2007).

(2)   Protein separation

For the protein analysis, sodium dodecyl sulfate polyacrylamide gel electrophoresis (SDS-PAGE) or native PAGE or two-dimensional gel electrophoresis (2-DE) or high-performance liquid chromatography (HPLC) can be used to separate proteins (O'Farrell, 1975). Proteins separated by 2-DE gels can be used to create a comparative protein map (proteotyping). The separation procedure could be accompanied by in situ spot protein digestion. Specific polypeptides or enzymatic activities are measured in order to create a "proteofingerprint" (Maron et al., 2007). Data analysis of metaproteomic studies includes the use of large protein databases obtained from hundreds or thousands of organisms, as well as numerous processing steps to ensure high data quality (Jagtap et al., 2015).

(3)   Identification

The individual separated protein spots could be subjected to enzymatic proteolytic digestion and analyzed with the aid of mass spectrometry (MS; MALDI-MS) (Cristea et al., 2004; Pandey and Lewitter, 1999). Alternatively, the peptides can also be separated using liquid chromatography (LC) together with tandem mass spectrometry (MS/MS) (LC-MS/MS) (Cristea et al., 2004). The proteome is further linked to their corresponding genes via genomic database searching (Cristea et al., 2004; Dubey et al., 2020; Mann and Pandey, 2001). It is essential to facilitate identification of the proteins, although it is not critical to have complete genome sequences available for the samples (Hettich et al., 2013). Short peptide sequences could also be searched against protein databases (https://www.uniprot.org/). Proteins can also be detected by isotopic labeling (Goodlett and Yi, 2003; Ogunseitan, 1996) and visualized by autoradiography after separation on acrylamide gels (Ogunseitan, 1996).

3.4.2   Metabolomics

Metabolomics provides insights that facilitate the understanding of the regulation of the entire metabolite diversity from the microbiome in one ecosystem, the "metabolome" (Goodacre et al., 2004). The microbial activity reveals the actual information where a substrate becomes a product, leading to the metabolome. Metabolomics enables the detection and quantification of produced metabolites and thus provides information about how microbes may react to diverse habitats by altering their metabolite production which functions as signature molecules for diverse cellular processes (Hettich et al., 2013). Networks of metabolic activity can be used to estimate microbial community structure and its functions through modeling the molecular mechanisms of certain organisms (Knight et al., 2018). The method finds practical applications in the comparison of metabolites from natural sources to those from isolates in culture. By matching tandem mass spectrometry data, a particular metabolite signature might be typical for either the cultured microorganisms or the natural ones (Knight et al., 2018; Quinn et al., 2016). Changes in the metabolite profile of a microbial community reflect changes in its biosynthetic activity, mRNA and protein expression, and protein activity (Roume et al., 2013).

Knight et al. (2018) stated that "multi-omics analysis integrates chemical and biological knowledge to provide a more complete picture of a biological system and is an active area of research with largely untested methods." Thus the potential for omics integration is high. Several studies show the integration of metagenome, metatranscriptome, and metabolome data and correlating microorganisms with metabolites (Whiteson et al., 2014). Mallick et al. (2019) reported the development of a computational model (MelonnPan) for predicting metabolic profiles of microbiome community based on genome information. The success of this predictive model in human microbiome project suggests it can be extended to obtain functional insights on the metabolome of tree microbiota.

4.   High-throughput or next-generation sequencing: Principles, concept, and applications

DNA sequencing methods have been facilitated by three technological revolutions: first-generation sequencing (whole genome shotgun sequencing), next-generation sequencing (NGS or HTS), and the third-generation sequencing (single molecule long read sequencing). Three technologies are currently available for generation of high-throughput data: (a) high-throughput pyrosequencing on beads, (b) sequencing by ligation on beads, and (c) sequencing by synthesis on a glass substrate (Cao et al., 2017). However, each of them has unique advantages and disadvantages (Reuter et al., 2015). The next-generation sequencing (NGS), also called "high-throughput technology (HTS)," allowed the generation of thousands to millions of short sequencing reads in a single machine run. The advantages of NGS over the Sanger sequencing include in vitro construction of the sequencing library, in vitro clonal amplification of DNA fragments, array-based sequencing enables DNA fragments to be multiplexed, and solid-phase immobilization of DNA (Cao et al., 2017).

NGS technologies represent high-throughput and cost-effective method for sequencing and offer the possibility of massive parallel multigene analysis (Serratì et al., 2016). They can be used to sequence the entire exomes, DNA protein-coding regions, transcriptomes, or genomes (Horak et al., 2016; Wang and Xu, 2017). NGS are increasingly applied to microbial research, including de novo sequencing of bacterial and viral genomes (Chaisson and Pevzner, 2008), and characterizing the transcriptomes (RNA-seq) of cells and organisms (Hiller et al., 2009).

4.1 Amplicon sequencing

Amplicon sequencing is a highly targeted NGS for analysis of genetic variation, identification, and characterization of specific genomic regions. The method involves PCR amplification of targeted region of interest followed by next-generation sequencing (NGS, HTS). The PCR amplicons from different samples can be pooled with a barcode (index) added to each sample to label the identity. Sequence adapters are added to individual sample for amplicon sequencing, which allows formation of barcoded amplicons and also for the amplicons to adhere to the flow cell for sequencing.

Amplicon sequencing can detect known and novel variants within the region of interest (Figs. 3.3 and 3.4). The most common applications are amplicon sequencing of 16S rRNA (for bacteria), 18S rRNA (for microeukaryotes and unicellular eukaryotes), or internal transcribed spacers (ITS—for fungal communities) across multiple species (Kovalchuk et al., 2018; Ren et al., 2018). It is a widely used method to study the phylogeny and taxonomy in diverse metagenomic samples (Vincent et al., 2017). The most common case of NGS amplicon sequencing for microbial identification is the 16S rRNA gene sequencing for bacteria (Engelbrektson et al., 2010). Bacteria contain 16S rRNA gene covering nine hypervariable regions flanked with conserved sequences (Fig. 3.4) (Neefs et al., 1993). The conserved region could be used to design PCR primers to amplify and sequence these hypervariable regions to characterize the bacterial taxonomy. Similarly, the 18S rRNA gene or ITS can be used to identify fungi (Fig. 3.3). Amplicon-based approaches targeting variable regions of specific markers can also be used for functional studies, e.g., targeting enzyme-coding genes catalyzing C, N, and P cycles, for example, β-glucosidases (Pathan et al., 2015), protease genes (Baraniya et al., 2016), or alkaline phosphatases (Bergkemper et al., 2016).

The characteristics of amplicon sequencing include: (a) Discovery, validation, and screening genetic variants using a highly targeted approach; (b) high coverage of multiplexing of hundreds to thousands of amplicons in a single run; (c) obtaining highly targeted resequencing even in difficult-to-sequence areas, such as GC-rich regions; (d) flexibility for a wide range of experimental designs; (e) reduced sequencing costs and turnaround time compared to broader approaches such as whole-genome sequencing; and (f) microbial culture free (Raza and Ahmad, 2016). However, the amplicon sequencing can lead to individual targeted region polymorphism and ultimately an overestimation of community diversity due to genetic exchange between closely and distantly related taxa. One example is that 16S rRNA gene can be transferred between bacterial genotypes (Acinas et al., 2004).

The eukaryotic rRNA cistron consists of the 18S, 5.8S, and 28S rRNA genes transcribed as a unit by the RNA polymerase I. During posttranscriptional processes, the cistron is split and the two internal transcribed spacers (ITS) are removed. These two spacers, including the 5.8S gene and the ITS region (ITS1-5.8S-ITS2) (Fig. 3.3), constitute the official fungal "barcode of life" (Schoch et al., 2012). It has been used in fungal ecology for 30 years (Gardes et al., 1991). As of 2012, ~ 172,000 full-length fungal ITS sequences were deposited in GenBank (Schoch et al., 2012).

Depending on the target fungal isolates, the choice of the right primers is a crucial step in amplicon study. Primers mismatching other lineages of eukaryotes in the environmental sample can be used in order to reduce their amplification. Alternatively, primers with a broad spectrum which amplify nonfungal lineages can also be used. The nonfungal lineages can be removed during the subsequent stages of the analysis (Taylor et al., 2016). Only a few primers are able to

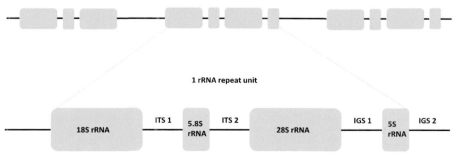

FIG. 3.3 1 rRNA gene repeat unit, including 18S rRNA, 5.8S rRNA, and 28S rRNA.

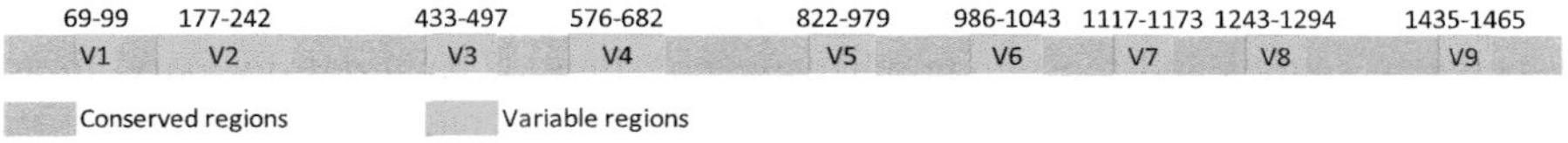

69-99	177-242		433-497	576-682		822-979	986-1043	1117-1173	1243-1294		1435-1465
V1	V2		V3	V4		V5	V6	V7	V8		V9

▢ Conserved regions   ▢ Variable regions

FIG. 3.4 The 16S ribosomal RNA gene conserved and variable regions. Numbers in the figure show the position of variable regions in *E. coli* 16S rRNA gene.

amplify > 90% of all fungal groups (Tedersoo et al., 2015). Using the classic ITS primers, biased amplification of basidiomycetes can happen due to mismatches to ascomycete sequences. Such biases can lead to false results in amplicon metagenomic studies on fungal diversity and community structures. The coverage of primers influences to a great extent the reliability of the data produced by HTS studies (Toju et al., 2012). For ITS amplicon sequencing on the Illumina platform, several primer attributes are important: PCR efficiency, coverage, selectivity, and variation in amplicon size (Taylor et al., 2016). Optimal primer selection for short-amplicon HTS applications relies on the following criteria: high coverage, taxonomic resolution, and accuracy and short amplicon length (Bokulich and Mills, 2012). Therefore primers for HTS studies need to be able to amplify the sequences of diverse Dikarya fungi (i.e., a subkingdom of fungi consisting of the two phyla Ascomycota and Basidiomycota) without significant taxonomic biases (Toju et al., 2012). Suitable primers lay in various parts of the ITS and surrounding ribosomal coding regions (Bokulich and Mills, 2013; Gardes and Bruns, 1996; Ihrmark et al., 2012; Taylor and McCormick, 2008; Toju et al., 2012). Most HTS-based studies focus on either the ITS1 or ITS2 subregion of typically 250–400 bases. While the ITS2 subregion includes lower length variation and more universal primer sites compared to ITS1, which results in less taxonomic bias than with ITS1 (Tedersoo et al., 2015). Taylor et al. (2016) identified one promising primer in the 5.8S, named 5.8S-Fun, and one in the LSU, named ITS4-Fun. Most recent advances in HTS studies allow the use of the entire ITS region and flanking rRNA genes for third-generation techniques with platforms such as Pacific Biosciences (PacBio) and Oxford Nanopore, which has the advantage of a higher taxonomic resolution. Tedersoo et al. (2015) suggest the use of ITS1 and ITS2 subregions in HTS for optimal functional and taxonomic resolution.

4.2 16S sequencing principle and bias

The 16S rRNA gene is a housekeeping gene of about 1550 bp long with both conserved and variable regions. The conserved regions provide possibility to design universal primers to amplify and sequence the variable regions in a wide range of different bacteria from a single sample. The 16S rRNA gene amplicon sequencing refers to the amplification and sequencing of the variable regions in 16S ribosomal RNA genes (Fig. 3.4). The 16S rRNA sequencing, metagenomics, and metatranscriptomics are the three basic sequencing strategies used in the taxonomic identification and characterization of bacterial biome. These sequencing strategies have used different HTS platforms for DNA and RNA sequence identification, which can provide deeper taxonomic identification of complex microbiome from environmental sample (Cao et al., 2017; Sun et al., 2013). The 16S rRNA sequencing has been applied in many research fields such as environmental conservation, agricultural production, petroleum exploration, and industrial manufacturing. It also has been used for enumerating the global bacteria communities in both symptomatic and asymptomatic groups as well as disease diagnosis, biomarker discovery, and forest management (Ma et al., 2020; Qu et al., 2020; Ren et al., 2018). The 16S rRNA sequencing bypasses the conventional bacterial culture-dependent method and facilitates the analysis of the entire microbial community. This approach can be used to profile thousands of species simultaneously from a single sample. It offers the sensitivity needed to detect environmental DNA (eDNA) present at low levels in the environment and a cost-effective technique for the identification of isolates that may not be found using culture-dependent methods. The 16S rRNA sequencing generally involves the PCR targeted amplification of different variable regions followed by next/third-generation sequencing. The 16S rRNA analysis pipeline for phylogenetic assignment uses three popular databases: Silva, Green Genes, and Ribosomal Database Project (RDP). The commonly used software to analyze 16S rDNA data from environmental samples include QIIME (Quantitative Insights Into Microbial Ecology) (Caporaso et al., 2010), Mothur (Schloss et al., 2009) and USEARCH (ultra-fast sequence analysis) (Edgar, 2010, 2013; Edgar et al., 2011). The bacterial 16S rRNA sequencing workflow includes several steps: (1) DNA extraction and library preparation with PCR for targeted 16S rRNA regions; (2) sequencing by the next/third-generation sequencing technology; (3) data processing by bioinformatics tools, including raw data denoising, sequence quality control (PCR error, chimera checking, ambiguous bases), alpha(α)- and beta(β)-diversity analysis, and taxonomic assignment.

The amplification bias in 16S rRNA sequencing has to be considered when assessing the microbial diversity. The primer choice is one of the most important factors, which impacts the sequencing accuracy in terms of coverage of targeted region, specificity of primer-template binding, and amplification efficiency, resulting in diversity bias (Ibarbalz et al., 2014). One example is the primer pairs targeting V1-V2 and V3-V4 regions of 16S gene. The flanking region of v1-v2 is less conserved than that of v3-v4, which causes more biases in the diversity and evenness due to primer mismatches (Klindworth et al., 2013). Moreover, the lack of consensus for the commonly used universal primers could prevent comparison between studies and limit comprehensive coverage of bacterial diversity. In addition, 16S rRNA sequencing can only provide the information on the bacteria identification, but not the potential metabolic functions of the community.

5. Data analysis: Clustering, sequence identification, and operational taxonomic units

The main steps in data analysis are as follows: PCR error checking, quality control, chimera checking, clustering, taxonomic assignment, and finally generating Operational Taxonomic Unit (OTU) abundance tables. These tables can be used further to compare the diversity and composition of fungal and bacterial taxa between samples. Apart from listing the species within ecological samples, recent research is moving toward defining the functional groups of observed microbes (Cleary et al., 2019; Nguyen et al., 2016). A number of online bioinformatics tools are available for downstream sequence processing and analysis. Among them, QIIME2 (v2018.2; Bolyen et al., 2019; Caporaso et al., 2010) is an open-source next-generation microbiome bioinformatics platform that is extensible and free.

To take full advantage of the large amount of data, improved integrated analysis tools and comprehensive databases are needed. The major pitfalls are in taxonomic classification as the metagenomic datasets rely on the use of reference databases, which are biased toward model organisms or readily culturable microorganisms (Nilsson et al., 2019, 2012). This is a major limitation for taxonomic classification of microbial communities in the ecosystems, as up to 90% of the sequences of a metagenomic dataset may remain unidentified due to the lack of a reference sequence (Nilsson et al., 2012; Sun et al., 2013; Terhonen et al., 2013).

5.1 Data quality filtering

After receiving the sequence data (reads), the quality control analysis starts with merging paired-end sequences (usually Illumina platform), followed by removing mismatching primers and barcodes. Furthermore, primers and possible tag sequences are removed (Terhonen et al., 2013). Quality trimming also includes removing sequences that have ambiguous bases, homopolymers, and/or average quality score lower than 25 (Sun et al., 2013, 2016; Terhonen et al., 2013). The minimum length and maximum length of reads have to be carefully decided. The length of the ITS region is variable, and these regions (ITS1 and/or ITS2) can be extremely short (50 bases) (Nilsson et al., 2019). This is one of the reasons to use paired sequences (same region sequenced from reverse and forward directions and later on combined as one sequence) (Nilsson et al., 2019). When analyzing ITS regions, it is important to remove bordering subregions (SSU, 5.8S, and LSU) in order to increase taxonomic resolution (Nilsson et al., 2012). Using ITSx, a software tool for detecting and extracting ITS1 and ITS2 sequences (Bengtsson-Palme et al., 2013) with threshold values as well as for removal of flanking gene regions is highly recommended (Nilsson et al., 2019).

The following programs could be used for quality filtering including ITS extraction using ITSx (Bengtsson-Palme et al., 2013): LotuS (v1.59; Hildebrand et al., 2014); PipeCraft (v1.0; Anslan et al., 2017); PIPITS (v2.0; Gweon et al., 2015). A further quality filtering programs plus additional ITS extraction using ITSx (Table 3.1): trimmomatic (Bolger et al., 2014) can also be used: DADA2 (Callahan et al., 2016) sdm (Hildebrand et al., 2014); fastx (http://hannonlab.cshl.edu/fastx_toolkit); Galaxy (v.2.1.1; Afgan et al., 2018).

5.2 Chimera filtering

Chimeras are artificially formatted sequences that are incorrectly joined together during PCR amplification when using environmental samples as templates (Ashelford et al., 2005, 2006; Huber et al., 2004; Quince et al., 2009). There are several programs that can be used for chimera detection (Table 3.1). Most used are USEARCH (Edgar, 2010, 2013; Edgar et al., 2011), VSEARCH (Rognes et al., 2016), UCHIME2 (Edgar, 2016a), and UNOISE2 (Edgar, 2016b).

5.3 Clustering

Programs that can be used for clustering include USEARCH (Edgar, 2010, 2013; Edgar et al., 2011) and VSEARCH (Rognes et al., 2016). A crucial step in the taxonomic analysis of large metagenomic datasets is the clustering. Within this step, the sequences derived from a mixture of different organisms are assigned to phylogenetic groups (species-level OTUs) according to their taxonomic origins. Typically, sequence (ITS) similarity thresholds of $\geq$ 97.0% between species-level taxonomic resolution are used (Nilsson et al., 2019; Tedersoo et al., 2014). For fungal (ITS1 or ITS2) datasets, de novo single linkage clustering methods perform the best (Nilsson et al., 2019). For bacterial (16S rRNA) sequences, the gaps in the alignment are less frequent than in ITS alignments. In that sense, using the default settings of clustering tools is often sufficient. The commonly used algorithms for 16S rRNA are implemented in USEARCH, VSEARCH, and mothur software. It is important to remember that well-documented clustering approaches and deposition of metadata as open access will increase scientific reproducibility and comparability across studies.

TABLE 3.1 Software tools available for analysis and assembly of metagenomics data.

	Software/program/algorithm	Note	Reference
Quality filtering	LotuS v1.59	ITSx included	Hildebrand et al. (2014)
	PipeCraft v1.0	ITSx included	Anslan et al. (2017)
	PIPITS v2.0	ITSx included	Gweon et al. (2015)
	Trimmomatic		Bolger et al. (2014)
	DADA2		Callahan et al. (2016)
	fastx		http://hannonlab.cshl.edu/fastx_toolkit
	Galaxy v.2.1.1		Afgan et al. (2016)
	sdm		Hildebrand et al. (2014)
Chimera	USEARCH		Edgar (2010, 2013), Edgar et al. (2011)
	VSEARCH		Rognes et al. (2016)
ITS extraction	ITSx		Bengtsson-Palme et al. (2013)
Clustering	USEARCH		Edgar (2010, 2013), Edgar et al. (2011)
	VSEARCH		Rognes et al. (2016)
Mapping/alignment	USEARCH		Edgar (2010, 2013), Edgar et al. (2011)
	VSEARCH		Rognes et al. (2016)

5.4 Taxonomic identification

The taxonomic identification of OTUs is based on comparison of sequences in reference databases. One of the most used tool for comparison is BLAST (Altschul et al., 1997) searches against sequence archive in database of National Center for Biotechnology information (NCBI, https://www.ncbi.nlm.nih.gov/) (Sayers et al., 2011). Similarity $\geq$ 97% across the entire length of the pairwise alignments is usually taken to indicate conspecificity (Arnold and Lutzoni, 2007). For ITS reference database of the fungal kingdom, the UNITE database (https://unite.ut.ee/) is recommended (Nilsson et al., 2019). However, besides relying only algorithms, also manual data curation and searches done in reference databases could improve taxonomic identifications. Therefore manual organization and confirmation of OTUs is an important step in obtaining robust datasets (Anslan et al., 2018). Interpreting the correct taxonomic level based only on ITS sequence is challenging. Nilsson et al. (2019) provided some guidelines that are commonly recommended to be used: (a) $\geq$ 70% similarity corresponds to the phylum level, (b) $\geq$ 75% corresponds to the class level, (c) $\geq$ 80% corresponds to the order level, (d) $\geq$ 85% corresponds to the family level, (e) $\geq$ 90% similarity corresponds to the genus level, and (f) $\geq$ 97.0%–98.5% similarity (in a pairwise alignment covering $\geq$ 90% of the query sequence) corresponds to the species level (Tedersoo et al., 2014; Nilsson et al., 2019). The sequences for bacterial datasets are usually aligned against databases. One of the recommended databases is the SILVA alignment database (https://www.arb-silva.de/). Thereafter, the sequences can be clustered into OTUs defined by a 97% similarity using the average neighbor algorithm (Yilmaz et al., 2014).

5.5 Fungal guild

FUNGuild is an open annotation tool for taxonomically parsing fungal OTUs by ecological guild independent of analysis pipeline or sequencing platform (Nguyen et al., 2016). FUNGuild can assign fungal taxa to three trophic modes: pathotrophic (pathogens), saprotrophs, and symbiotrophic (mutualists), which are further annotated to different guilds. However, many genera contain more than one trophic strategy, which also highlights the need for manual evaluation. Fungal taxa categorized into combined pathotrophic-saprotroph or pathotrophic-mutualist levels include those that may exhibit different lifestyles depending on life cycle stage and/or environmental conditions.

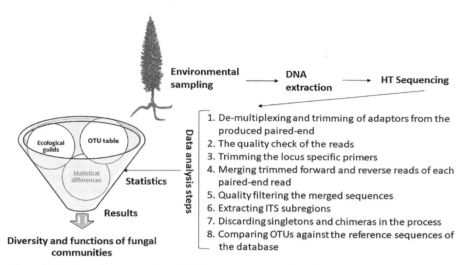

Environmental sampling → **DNA extraction** → **HT Sequencing**

Ecological guilds

OTU table

Statistical differences

Statistics

Results

Diversity and functions of fungal communities

Data analysis steps

1. De-multiplexing and trimming of adaptors from the produced paired-end
2. The quality check of the reads
3. Trimming the locus specific primers
4. Merging trimmed forward and reverse reads of each paired-end read
5. Quality filtering the merged sequences
6. Extracting ITS subregions
7. Discarding singletons and chimeras in the process
8. Comparing OTUs against the reference sequences of the database

FIG. 3.5   Analysis workflow of high-throughput sequencing (HTS) based on metabarcoding ITS datasets.

5.6   Data analysis

Usually diversity indices are calculated from the number of reads of each OTU observed in one sample/treatment. The most common ones are the Shannon-Wiener index (Shannon, 1948) and the Simpson index (Simpson, 1949) and Chao 1 (Chao, 1984). When comparing the diversity in each habitat the Permutational Multivariate Analysis of Variance (PERMANOVA) in VEGAN package version 2.4 (Oksanen et al., 2016) can be used to test the differences/similarities between samples (factors: treatment/site/sample/niche, etc.). Nonmetric Multidimensional Scaling (NMDS) in VEGAN package that uses a Bray-Curtis distance matrix can be used to visualize the community structure based on the OTU abundance from each sample.

A rarefaction curve plot is generated to show the number of OTUs versus the number of sequences (reads). If the rarefaction curves do not reach the plateau, it indicates that a higher diversity of OTUs could still be detected by increasing the sequencing depth. Reaching the plateau indicates that the sequence depth is sufficient. A flow chart of the steps in the analysis of HTS metabarcoding datasets is shown in Fig. 3.5.

6.   RNA-seq and DNA GeoChip for microbiome analyses

6.1   RNA-seq for microbiome study

RNA sequencing (RNA-Seq) is a transcriptome profiling approach and involves a wide variety of applications, ranging from simple mRNA profiling to discovery of the entire transcriptome (Fig. 3.6). It was first developed in the mid-2000s with the advent of NGS technology (Weber, 2015). The advantage of RNA-seq and associated transcriptomic information is that prior data or knowledge of genomic sequence of the target organism is not required (Grada and Weinbrecht, 2013). The key features of RNA-seq are detection of transcripts with low expression levels, transcript analysis with or without reference sequence, and characterization of alternative splicing and polyadenylation (Raza and Ahmad, 2016).

Many available approaches can be used to analyze RNA-seq data. Despite of the different analysis methods, the common steps for RNA-seq analysis usually require initial filtering of raw sequence reads, assembling reads into transcripts or aligning reads to reference sequences, annotating putative transcripts, and comparison of transcript abundance across samples. The genome-guided and de novo assembly are the two methods to assemble the transcriptome. The genome-guided method is similar to DNA alignment with the additional complexity of aligning reads that cover noncontinuous portions of the reference genome (www.illumina.com). De novo assembly refers to reconstruction of the sequences and transcriptome without a reference genome particularly when the information on the genome is either unknown, incomplete, or substantially altered (Grabherr et al., 2011). A detailed analysis protocol of RNA-seq can be found at: https://rnaseq.uoregon.edu/index.html#analysis.

Plant RNA-seq datasets, which were originally generated to study the host transcriptome, can now be used as a novel dataset to explore plant-associated microbiota (Cox et al., 2017). RNA-seq dataset has recently been used to detect the taxonomic and functional diversity of root microbiota in tomato (Chialva et al., 2019). The RNA-seq analysis was able to

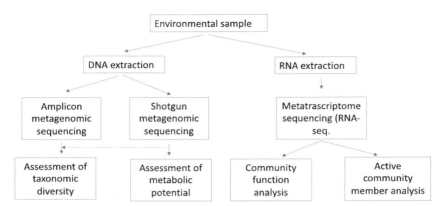

FIG. 3.6 Simplified diagram of metagenomic and metatranscriptomic sequencing workflow. *(Modified from Simon, C., Daniel, R., 2011. Metagenomic analyses: past and future trends. Appl. Environ. Microbiol. 77, 1153–1161.)*

uncover the composition and the metabolic activities of the microbiota shaping the tomato plant responses (Chialva et al., 2019). RNA-seq has also been used to study the active fungal communities of susceptible and resistant clones of *Eucalyptus grandis* (Messal et al., 2019). These studies demonstrate that RNA-seq analysis is a useful and novel resource to study microbe-host and microbe-microbe interactions.

6.2   DNA GeoChip for microbiome analysis

The GeoChip is a DNA microarray designed to identify the functional genes involved in different biogeochemical processes such as C, N, S, and P cycling. The GeoChip array contains probes of gene markers covering microorganisms from archaea, bacteria, and fungi. It is currently manufactured by Agilent Technologies (Santa Clara, CA, United States) and has been widely used as a high-throughput metagenomic tool for profiling environmental microbial community. It is particularly useful in terms of community metabolic potential, functional structure and diversity, and correlation of microbial community structure to ecosystem functioning (Sun et al., 2015, 2016; Zhou et al., 2010).

The GeoChip was initially designed to address two major challenges for studying functional genes in microbial community. The first is the low probe specificity due to presence of some genes with high homologies across species and the second one is the limited number of comprehensive probe sets (He et al., 2007). Several generations of GeoChips have been developed since the first version was designed in 2004 (Rhee et al., 2004). The GeoChip 5.0 is the current version in use and contains 167,044 distinct probes covering 395,894 coding sequences (CDS) from ~ 1500 functional gene families involved in microbial carbon (degradation, fixation, methane), nitrogen, sulfur, and phosphorus cycling, energy metabolism, metal homeostasis, organic remediation, "Other" (phylogenetic genes and CRISPR system), secondary metabolism (e.g., antibiotic metabolism, pigments), stress responses, viruses (both bacteriophages and eukaryotic viruses), and virulence. It has 3 formats, including small (60 K × 8, 8 arrays with 60,000 probes each on 1 slide), medium (180 K × 4), large (400 K × 2), and extra-large (1.0 M × 1) formats (https://www.glomics.com/gch-tech.html).

6.3   GeoChip: Data generation, normalization, and analysis

The data generation process includes the DNA extraction and hybridization (He et al., 2007). The total genomic DNA of samples from the environmental sample is extracted, labeled with a cyanine dye, and hybridized to the GeoChip array. Any unhybridized DNA is washed off and the array is imaged. The high quality of community DNAs is critical to minimize experimental variations for improving microarray-based quantitative accuracy. If the yield of DNA is insufficient, whole community genome amplification can be performed to increase the quantity of DNA (He et al., 2007).

The GeoChip microarrays have an in-house data analysis pipeline (http://ieg.ou.edu/microarray/), which allows the user to set up signal cutoff and select normalization protocols (Van Nostrand et al., 2016). Many other microarray software can also be used for GeoChip data analysis. The data normalization and quality filtering are the two crucial steps for downstream analysis (Deng and He, 2014; Liang et al., 2010). Generally, the major steps for GeoChip data normalization and filtering microarray data include poor-quality spots removal, normalization of signal intensity of each spot by mean and removal of outliers. Detailed protocol can be found on the website of the Institute for Environmental Genomics, University of Oklahoma (United States) (http://ieg.ou.edu/entrance.html). Briefly, the criteria for poor-quality spot are spots with signal-to-noise ratio [SNR = (signal mean − background mean)/background standard deviation] less than 2.0, the coefficient of variation (CV) of the background more than 0.8, and the signal intensity at least 1.3 times the background (Wu et al., 2006). Spots with signal intensities less than ~ 200, signal values less than twice of the average background signal, and spots detected only in one sample are also poor quality and should be removed (Van Nostrand et al., 2016). For normalization, the average signal intensity of common oligo reference standard is calculated, and the maximum average value is applied to normalize the signal intensity. The sum of the signal intensity of samples is calculated, and the maximum sum value is applied to normalize the signal intensity of all spots in an array, producing a normalized value for each spot.

After data normalization, further analysis can be done using statistical methods. The signals of all spots are normally logarithmic transformed (log10) to relative abundance before statistical analysis. The GeoChip data analysis pipeline has a variety of analysis tools for microarray data (http://ieg.ou.edu/microarray/). These include calculating the relative abundance of genes or gene categories or subcategories, richness and α and β diversity of functional genes, and gene overlap between individual samples or sample groups. The response ratios can be used to compare gene levels or signal intensity between conditions (e.g., treatment versus control, contaminated versus uncontaminated) (Luo et al., 2006).

To visualize the gene structure differences, the ordination methods, such as principal component analysis (PCA), correspondence analysis (CA), canonical correspondence analysis (CCA), detrended correspondence analysis (DCA) can be used. Nonmetric multidimensional scaling (NMDS), which represents the relative interrelatedness of samples on a priori dimensions, could also be used (Van Nostrand et al., 2016). The hierarchical cluster analysis, T-tests, analysis of variance (ANOVA), and dissimilarity tests could be used to investigate the differences between conditions. The nonparametric multivariate statistical tests, including nonparametric multivariate analysis of variance (ADONIS), analysis of similarity (ANOSIM), and Permutational multivariate analysis of variance (PERMANOVA), can be used to test the dissimilarities between treatments. To illustrate relationships between functional genes structure and other abiotic or biotic factors, different constrained ordination programs can be used, such as canonical correspondence analysis (CCA) (Ter Braak, 1987), distance-based redundancy analysis (db-RDA) (Legendre and Anderson, 1999), and variation partitioning analysis (VPA) (Økland and Eilertsen, 1994; Ramette and Tiedje, 2007).

7. Metagenome and metagenomics

Metagenomics is the study of the metagenome or the collective genome of microorganisms from an environmental sample (Fig. 3.6). This definition in a broader context also includes eco-genomics, community genomics, or environmental genomics. It is a nontargeted approach that results in the description and quantification of the copy number and allelic variants of genes of the microbial community (Vincent et al., 2017). The metagenomics can provide information on both microbial phylogenies and community metabolic functions. It can provide insight into the evolutionary linkages between community phylogeny and function, as well as the information on the potential novel metabolites or enzymes in the environment (Chistoserdova, 2009).

The shotgun metagenomic sequencing refers to the adaptation of shotgun sequencing to metagenomic samples, in which the small fragments from the sheared DNA extracted from the environmental sample are sequenced (Thomas et al., 2012). Shotgun sequencing initially started with the cloning of environmental DNA, followed by functional expression screening (Handelsman et al., 1998). It was quickly complemented by direct random shotgun sequencing of environmental DNA bypassing the cloning steps (Tyson et al., 2004; Venter et al., 2004). Therefore the shotgun metagenomics analysis became a powerful and popular tool for researchers to investigate the community-level attributes from a pool of whole-community DNA extracted from environmental samples (Handelsman, 2004).

De novo assembly and genome-guided assembly are the two commonly used methods to assign metagenomic raw sequences to genomic feature (Grabherr et al., 2011). De novo assembly of metagenome and metatranscriptome is typically used when there is no known reference genome information available to reconstruct the genome/transcriptome (Grabherr et al., 2011). It needs specific bioinformatic tools for assembly and requires large computational capacity (Miller et al., 2010; Pevzner et al., 2001). The genome-guided assembly relies on the reference genome with the same methods used for DNA and RNA-seq alignment by additional complexity of aligning reads that cover noncontinuous portions of the reference genome (Dobin et al., 2013). The common software packages such as Newbler (Roche), AMOS (http://sourceforge.net/projects/amos/), or MIRA (Chevreux et al., 1999) can be used for genome-guided assembly analysis.

After assembly, the metagenomic sequences can be annotated in two ways based on the length of assembly contigs. If the contig length is long enough (over 30,000 bp), the existing pipeline for annotation can be used, such as RAST (Aziz et al., 2008) or IMG (Markowitz et al., 2009). If there are unassembled reads or short contigs, the annotation can be carried out on the entire community. The feature prediction and functional annotation are the two steps of annotation process of metagenomic data. The feature prediction is to identify and label sequences as genes or genomic elements and functional annotation is to assign these genes to putative functions and taxonomic classification (Thomas et al., 2012). Currently, MG-RAST, IMG/M, and CAMERA are three popular large-scale databases to handle and deposit metagenomic datasets (Glass et al., 2010; Markowitz et al., 2007; Sun et al., 2010). MG-RAST is a data repository and can also provide an analysis pipeline and a comparative genomics platform.

7.1 Metagenome-wide association study

Metagenome-wide association study (MWAS) refers to the use of microbiome DNA sequence data to identify genetic risk factors for diseases and has successfully been used in human microbiome project (Bush and Moore, 2012). MWAS provides new insights into microbial functions that are perturbed and associated with a number of important human diseases and potentially allow the development of improved tools for identifying and monitoring disease states (Flintoft, 2012). The ultimate goal of MWAS is to use genetic risk factors to make predictions about who is at risk and to identify the biological underpinnings of disease susceptibility for developing new prevention and treatment strategies (Bush and Moore, 2012). MWAS have largely been modeled on a genome scale to identify genetic variants in the human population that are associated with a disease. It typically focuses on associations between single-nucleotide polymorphisms (SNPs) and traits and can equally be applied to any other genetic variants and any other organisms. One of the most successful applications of MWAS has been in the area of pharmacology to identify DNA sequence variations associated with drug metabolism, efficacy, and adverse effects (Bush and Moore, 2012). MWAS can identify the associations for both microbial taxa that are more or less abundant and the functions that are enriched or depleted (Wang and Jia, 2016).

MWAS has been successfully used to study human diseases, such as type 2 diabetes (Karlsson et al., 2013) in which the single-nucleotide polymorphisms (SNPs) are used as explanatory variables for association analysis. Recently, MWAS has been applied on plant and soil metagenomic data and adopts the metagenomic species or gene cluster as the explanatory variables to determine the association between soil microbial taxa and crop productivity (Chang et al., 2017). The successful application of MWAS in human microbiome project suggests a potential merit for its use in forest microbiome study.

8. Choice of methods for microbiome studies: Marker gene, whole metagenome, or metatranscriptomic analysis

The choice of method used for microbiome studies determines the complexity and depth of the results. For example, whether it is for an overview of the species composition or if molecular traits at gene level are to be investigated. Cost-effectiveness, quality, and purity of the environmental sample are also determining factors to be considered. A combination of different methods would provide the full spectrum of results. Many studies often do not require all the methods, sometimes the budget does not allow access to all methods or the quality of the sample limits the use of a certain method. The three commonly used methods are further evaluated in the following sections (see also Knight et al., 2018).

8.1 Marker gene analysis

A well-established method, which provides a fast overview of a microbial community composition is the marker gene analysis. This method gives good results even if the environmental sample is contaminated by host DNA or has low biomass (Knight et al., 2018). The regions for amplification and sequencing are the 16S rRNA for bacteria and the ITS region for fungi. The chosen primers target a certain gene region whereas the resolution usually does not exceed the genus level (Knight et al., 2018). A critical step is deciding the target region of the primer because of the possibility for biases (Liu et al., 2007; Soergel et al., 2012). Marker gene studies allow an inexpensive sample preparation and analysis with good access to existing datasets (Caporaso et al., 2010; McDonald et al., 2015; Thompson et al., 2017).

8.2 Whole metagenome analysis

When sequencing the entire genomes that exist in an environmental sample, all DNA including that of eukaryotes or viruses is isolated. More detailed information on genomic and taxonomic level can be obtained and resolution to species level is possible. Even whole microbial genomes from short DNA sequences can be associated (Mukherjee et al., 2017). Though the method is cost intensive (Scholz et al., 2016), contamination from host-derived DNA may occur and the method does not discriminate between live, dead, or active cells (Knight et al., 2018).

8.3 Metatranscriptome analysis

The smallest metabolic changes in environmental conditions can be discovered in metatranscriptomics analyses (Bashiardes et al., 2016; Knight et al., 2018). RNA sequencing provides the profile transcriptions of microbiomes and information of the functional gene expression. Only viable and active cells are represented (Knight et al., 2018). This method is the most complex in terms of sample preparation, processing, and storage (Bikel et al., 2015).

9. Technical considerations and constraints

A major limitation and constraint in the use of traditional culture-dependent method for microbiome study is that it grossly underestimates the total number of species within the environmental sample (Turner et al., 2013). Additionally, unculturable fungi are unable to grow on any provided artificial growth medium, slow-growing fungi can easily be overgrown by fast-growing ones, or the chosen media was too selective (Siddique et al., 2017; Unterseher and Schnittler, 2009). Also obligate biotrophs can be undetected or underrepresented (Stone et al., 2004). Furthermore, serial dissection and plating of host material provides only a selection of the resident microbiota. It might not be possible to differentiate between a systemic colonization in the host tissue or several infections of the same species (Carroll, 1995; Stone, 1987). The use of culture-independent approach such as amplicon next-generation sequence (NGS or HTS) has helped to overcome many of these challenges and limitations. A major advantage of amplicon-NGS culture-independent approach is the ability to detect large number of microbial taxa and diversity compared to culture-based method (Zapka et al., 2017). It is equally useful for detecting slow-growing microbes and those microbes where growth conditions are not optimal as well as microbes that cannot grow on artificial culture media. Additionally, culture-independent method facilitates quick and speedy surveys of microbial species from extreme and harsh ecosystem as well as from large communities. However, there are some notable limitations with this approach, such as biases in PCR amplification and sequencing, tendency for overestimation of species, and lack or limited number of reference sequences. Consequently, many of the OTUs may not be fully identified. Furthermore, another constraint is that without live culture of the identified isolate, it may be difficult to expedite follow-up characterization study of a potential useful species in the laboratory. This can hamper or constrain the potential use of such isolates for industrial and biotechnological applications. Overall, HTS is considered superior approach over culture-dependent methods (Al-Sadi et al., 2015; Oono et al., 2015). However, the simultaneous use of several isolation methods is indispensable to obtain a complete picture of the entire microbial diversity.

References

Acinas, S.G., Marcelino, L.A., Klepac-Ceraj, V., Polz, M.F., 2004. Divergence and redundancy of 16S rRNA sequences in genomes with multiple rrn operons. J. Bacteriol. 186, 2629–2635.

Afgan, E., Baker, D., Van den Beek, M., Blankenberg, D., Bouvier, D., Čech, M., Chilton, J., Clements, D., Coraor, N., Eberhard, C., 2016. The Galaxy platform for accessible, reproducible and collaborative biomedical analyses. Nucleic Acids Res. 44, 3–10. https://doi.org/10.1093/nar/gkw343.

Afgan, E., Baker, D., Batut, B., Van Den Beek, M., Bouvier, D., Čech, M., Chilton, J., Clements, D., Coraor, N., Grüning, B.A., 2018. The Galaxy platform for accessible, reproducible and collaborative biomedical analyses: 2018 update. Nucleic Acids Res. 46, W537–W544.

Altschul, S.F., Madden, T.L., Schäffer, A.A., Zhang, J., Zhang, Z., Miller, W., Lipman, D.J., 1997. Gapped BLAST and PSI-BLAST: a new generation of protein database search programs. Nucleic Acids Res. 25, 3389–3402.

Al-Sadi, A.M., Al-Mazroui, S.S., Phillips, A.J.L., 2015. Evaluation of culture-based techniques and 454 pyrosequencing for the analysis of fungal diversity in potting media and organic fertilizers. J. Appl. Microbiol. 119, 500–509.

Anslan, S., Bahram, M., Hiiesalu, I., Tedersoo, L., 2017. PipeCraft: flexible open-source toolkit for bioinformatics analysis of custom high-throughput amplicon sequencing data. Mol. Ecol. Resour. 17, e234–e240.

Anslan, S., Nilsson, R.H., Wurzbacher, C., Baldrian, P., Tedersoo, L., Bahram, M., 2018. Great differences in performance and outcome of high-throughput sequencing data analysis platforms for fungal metabarcoding. MycoKeys, 29–40. https://doi.org/10.3897/mycokeys.39.28109.

Arnold, A.E., Lutzoni, F., 2007. Diversity and host range of foliar fungal endophytes: are tropical leaves biodiversity hotspots? Ecology 88, 541–549.

Arnold, A.E., Mejía, L.C., Kyllo, D., Rojas, E.I., Maynard, Z., Robbins, N., Herre, E.A., 2003. Fungal endophytes limit pathogen damage in a tropical tree. Proc. Natl. Acad. Sci. 100, 15649. https://doi.org/10.1073/pnas.2533483100.

Ashelford, K.E., Chuzhanova, N.A., Fry, J.C., Jones, A.J., Weightman, A.J., 2005. At least 1 in 20 16S rRNA sequence records currently held in public repositories is estimated to contain substantial anomalies. Appl. Environ. Microbiol. 71, 7724–7736.

Ashelford, K.E., Chuzhanova, N.A., Fry, J.C., Jones, A.J., Weightman, A.J., 2006. New screening software shows that most recent large 16S rRNA gene clone libraries contain chimeras. Appl. Environ. Microbiol. 72, 5734–5741.

Asiegbu, F., Daniel, G., Johansson, M., 1993. Studies on the infection of Norway spruce roots by *Heterobasidion annosum*. Can. J. Bot. 71, 1552–1561.

Asiegbu, F.O., Daniel, G., Johansson, M., 1994. Defence related reactions of seedling roots of Norway spruce to infection by *Heterobasidion annosum* (Fr.) Bref. Physiol. Mol. Plant Pathol. 45, 1–19.

Asiegbu, F.O., Daniel, G., Johansson, M., 1995. Infection and disintegration of vascular tissues of non-suberized roots of spruce by *Heterobasidion annosum* and use of antibodies for characterizing infection. Mycopathologia 129, 91–101.

Atanasova, L., Druzhinina, I.S., 2010. Global nutrient profiling by Phenotype MicroArrays: a tool complementing genomic and proteomic studies in conidial fungi. J. Zhejiang Univ. Sci. B 11, 151–168.

Aziz, R.K., Bartels, D., Best, A.A., DeJongh, M., Disz, T., Edwards, R.A., Formsma, K., Gerdes, S., Glass, E.M., Kubal, M., 2008. The RAST Server: rapid annotations using subsystems technology. BMC Genomics 9, 75.

Baraniya, D., Puglisi, E., Ceccherini, M.T., Pietramellara, G., Giagnoni, L., Arenella, M., Nannipieri, P., Renella, G., 2016. Protease encoding microbial communities and protease activity of the rhizosphere and bulk soils of two maize lines with different N uptake efficiency. Soil Biol. Biochem. 96, 176–179.

Bashiardes, S., Zilberman-Schapira, G., Elinav, E., 2016. Use of metatranscriptomics in microbiome research. Bioinform. Biol. Insights 10, BBI.S34610. https://doi.org/10.4137/BBI.S34610.

Beijerinck, M.W., 1901. Anhäufungsversuche mit Ureumbakterien. Zentralbl. Bakteriol. Parasitenkd. Infektionskr. Hyg. II7, 33–61.

Bengtsson-Palme, J., Ryberg, M., Hartmann, M., Branco, S., Wang, Z., Godhe, A., De Wit, P., Sánchez-García, M., Ebersberger, I., de Sousa, F., 2013. Improved software detection and extraction of ITS1 and ITS 2 from ribosomal ITS sequences of fungi and other eukaryotes for analysis of environmental sequencing data. Methods Ecol. Evol. 4, 914–919.

Berendsen, R.L., Pieterse, C.M., Bakker, P.A., 2012. The rhizosphere microbiome and plant health. Trends Plant Sci. 17, 478–486.

Bergkemper, F., Kublik, S., Lang, F., Krüger, J., Vestergaard, G., Schloter, M., Schulz, S., 2016. Novel oligonucleotide primers reveal a high diversity of microbes which drive phosphorous turnover in soil. J. Microbiol. Methods 125, 91–97.

Bikel, S., Valdez-Lara, A., Cornejo-Granados, F., Rico, K., Canizales-Quinteros, S., Soberon, X., Del Pozo-Yauner, L., Ochoa-Leyva, A., 2015. Combining metagenomics, metatranscriptomics and viromics to explore novel microbial interactions: towards a systems-level understanding of human microbiome. Comput. Struct. Biotechnol. J. 13, 390–401.

Blumenstein, K., 2015. Endophytic fungi in elms: implications for the integrated management of Dutch elm disease. Swedish University of Agricultural Sciences, Alnarp.

Blumenstein, K., Albrectsen, B.R., Martín, J.A., Hultberg, M., Sieber, T.N., Helander, M., Witzell, J., 2015a. Nutritional niche overlap potentiates the use of endophytes in biocontrol of a tree disease. BioControl 60, 655–667. https://doi.org/10.1007/s10526-015-9668-1.

Blumenstein, K., Macaya-Sanz, D., Martín, J.A., Albrectsen, B.R., Witzell, J., 2015b. Phenotype MicroArrays as a complementary tool to next generation sequencing for characterization of tree endophytes. Front. Microbiol. 6, 1033.

Bochner, B.R., 2003. New technologies to assess genotype–phenotype relationships. Nat. Rev. Genet. 4, 309–314.

Bochner, B.R., 2008. Global phenotypic characterization of bacteria. FEMS Microbiol. Rev. 33, 191–205.

Bochner, B.R., Gadzinski, P., Panomitros, E., 2001. Phenotype microarrays for high-throughput phenotypic testing and assay of gene function. Genome Res. 11, 1246–1255.

Bokulich, N.A., Mills, D.A., 2012. Next-generation approaches to the microbial ecology of food fermentations. BMB Rep. 45, 377–389.

Bokulich, N.A., Mills, D.A., 2013. Improved selection of internal transcribed spacer-specific primers enables quantitative, ultra-high-throughput profiling of fungal communities. Appl. Environ. Microbiol. 79, 2519–2526. https://doi.org/10.1128/AEM.03870-12.

Bolger, A.M., Lohse, M., Usadel, B., 2014. Trimmomatic: a flexible trimmer for Illumina sequence data. Bioinformatics (Oxf. Engl.) 30, 2114–2120. https://doi.org/10.1093/bioinformatics/btu170.

Bolyen, E., Rideout, J.R., Dillon, M.R., Bokulich, N.A., Abnet, C.C., Al-Ghalith, G.A., Alexander, H., Alm, E.J., Arumugam, M., Asnicar, F., 2019. Reproducible, interactive, scalable and extensible microbiome data science using QIIME 2. Nat. Biotechnol. 37, 852–857.

Bulgarelli, D., Rott, M., Schlaeppi, K., van Themaat, E.V.L., Ahmadinejad, N., Assenza, F., Rauf, P., Huettel, B., Reinhardt, R., Schmelzer, E., 2012. Revealing structure and assembly cues for *Arabidopsis* root-inhabiting bacterial microbiota. Nature 488, 91–95.

Bush, W.S., Moore, J.H., 2012. Chapter 11: genome-wide association studies. PLoS Comput. Biol. 8 (12), e1002822.

Bußkamp, J., 2018. Schadenserhebung, Kartierung und Charakterisierung des "Diplodia-Triebsterbens" der Kiefer, insbesondere des endophytischen Vorkommens in den klimasensiblen Räumen und Identifikation von den in Kiefer (*Pinus sylvestris*) vorkommenden Endophyten. Universität Kassel, Kassel.

Cabral, D., Stone, J.K., Carroll, G.C., 1993. The internal mycobiota of Juncus spp.: microscopic and cultural observations of infection patterns. Mycol. Res. 97, 367–376.

Callahan, B.J., McMurdie, P.J., Rosen, M.J., Han, A.W., Johnson, A.J.A., Holmes, S.P., 2016. DADA2: high-resolution sample inference from Illumina amplicon data. Nat. Methods 13, 581–583. https://doi.org/10.1038/nmeth.3869.

Cao, Y., Fanning, S., Proos, S., Jordan, K., Srikumar, S., 2017. A review on the applications of next generation sequencing technologies as applied to food-related microbiome studies. Front. Microbiol. 8, 1829.

Caporaso, J.G., Kuczynski, J., Stombaugh, J., Bittinger, K., Bushman, F.D., Costello, E.K., Fierer, N., Pena, A.G., Goodrich, J.K., Gordon, J.I., 2010. QIIME allows analysis of high-throughput community sequencing data. Nat. Methods 7, 335.

Cardinale, M., 2014. Scanning a microhabitat: plant-microbe interactions revealed by confocal laser microscopy. Front. Microbiol. 5. https://doi.org/10.3389/fmicb.2014.00094.

Carroll, G., 1995. Forest endophytes: pattern and process. Can. J. Bot. 73, 1316–1324.

Chaisson, M.J., Pevzner, P.A., 2008. Short read fragment assembly of bacterial genomes. Genome Res. 18, 324–330.

Chang, H.X., Haudenshield, J.S., Bowen, C.R., Hartman, G.L., 2017. Metagenome-wide association study and machine learning prediction of bulk soil microbiome and crop productivity. Front. Microbiol. 8, 519. https://doi.org/10.3389/fmicb.2017.00519.

Chao, A., 1984. Nonparametric estimation of the number of classes in a population. Scand. J. Stat. 11, 265–270.

Chase, J.M., Leibold, M.A., 2003. Ecological Niches: Linking Classical and Contemporary Approaches. University of Chicago Press.

Chevreux, B., Wetter, T., Suhai, S., 1999. Genome Sequence Assembly Using Trace Signals and Additional Sequence Information. Presented at the German conference on bioinformatics, Citeseer, pp. 45–56.

Chialva, M., Ghignone, S., Novero, M., Hozzein, W.N., Lanfranco, L., Bonfante, P., 2019. Tomato RNA-seq data mining reveals the taxonomic and functional diversity of root-associated microbiota. Microorganisms 8 (1).

Chistoserdova, L., 2009. Functional metagenomics: recent advances and future challenges. Biotechnol. Genet. Eng. Rev. 26, 335–352.

Christensen, M.J., Bennett, R.J., Schmid, J., 2002. Growth of Epichloë/Neotyphodium and p-endophytes in leaves of Lolium and Festuca grasses. Mycol. Res. 106, 93–106.

Cleary, M., Oskay, F., Doğmuş, H.T., Lehtijärvi, A., Woodward, S., Vettraino, A.M., 2019. Cryptic risks to forest biosecurity associated with the global movement of commercial seed. Forests 10, 459.

Compant, S., Clément, C., Sessitsch, A., 2010. Plant growth-promoting bacteria in the rhizo-and endosphere of plants: their role, colonization, mechanisms involved and prospects for utilization. Soil Biol. Biochem. 42, 669–678.

Compant, S., Mitter, B., Colli-Mull, J.G., Gangl, H., Sessitsch, A., 2011. Endophytes of grapevine flowers, berries, and seeds: identification of cultivable bacteria, comparison with other plant parts, and visualization of niches of colonization. Microb. Ecol. 62, 188–197.

Compant, S., Samad, A., Faist, H., Sessitsch, A., 2019. A review on the plant microbiome: ecology, functions, and emerging trends in microbial application. J. Adv. Res. 19, 29–37. https://doi.org/10.1016/j.jare.2019.03.004. Special Issue on Plant Microbiome.

Cox, J.W., Ballweg, R.A., Taft, D.H., Velayutham, P., Haslam, D.B., Porollo, A., 2017. A fast and robust protocol for metataxonomic analysis using RNAseq data. Microbiome 5 (1), 1–13.

Cristea, I.M., Gaskell, S.J., Whetton, A.D., 2004. Proteomics techniques and their application to hematology. Blood 103, 3624–3634. https://doi.org/10.1182/blood-2003-09-3295.

de Santi Ferrara, F.I., Oliveira, Z.M., Gonzales, H.H.S., Floh, E.I.S., Barbosa, H.R., 2012. Endophytic and rhizospheric enterobacteria isolated from sugar cane have different potentials for producing plant growth-promoting substances. Plant Soil 353, 409–417.

Decorosi, F., Santopolo, L., Mora, D., Viti, C., Giovannetti, L., 2011. The improvement of a phenotype microarray protocol for the chemical sensitivity analysis of *Streptococcus thermophilus*. J. Microbiol. Methods 86, 258–261.

Den Dooren de Jong, L.E., 1926. Bijdrage tot de kennis van het mineralisatieproces. Nijgh and van Ditmar Uitgevers-Mij, Rotterdam, pp. 1–200.

Deng, Y., He, Z., 2014. Microarray data analysis. In: He, Z. (Ed.), Microarrays: Current Technology, Innovations and Applications. Caister Academic Press, Norfolk, UK.

Dobin, A., Davis, C.A., Schlesinger, F., Drenkow, J., Zaleski, C., Jha, S., Batut, P., Chaisson, M., Gingeras, T.R., 2013. STAR: ultrafast universal RNA-seq aligner. Bioinformatics 29, 15–21.

Dubey, R.K., Tripathi, V., Prabha, R., Chaurasia, R., Singh, D.P., Rao, C.S., El-Keblawy, A., Abhilash, P.C., Rao, C.S., 2020. Metatranscriptomics and metaproteomics for microbial communities profiling. In: Dubey, R.K., Tripathi, V., Prabha, R., Chaurasia, R., Singh, D.P., El-Keblawy, A., Abhilash, P.C. (Eds.), Unravelling the Soil Microbiome: Perspectives for Environmental Sustainability. SpringerBriefs in Environmental Science, Springer International Publishing, Cham, pp. 51–60, https://doi.org/10.1007/978-3-030-15516-2_5.

Dworkin, M., 2006. The Prokaryotes. Symbiotic Associations, Biotechnology, Applied Microbiology, vol. 1 Springer Science & Business Media.

Edgar, R.C., 2010. Search and clustering orders of magnitude faster than BLAST. Bioinformatics 26, 2460–2461.

Edgar, R.C., 2013. UPARSE: highly accurate OTU sequences from microbial amplicon reads. Nat. Methods 10, 996.

Edgar, R., 2016a. UCHIME2: Improved chimera prediction for amplicon sequencing. BioRxiv 074252.

Edgar, R.C., 2016b. UNOISE2: Improved error-correction for Illumina 16S and ITS amplicon sequencing. BioRxiv 081257.

Edgar, R.C., Haas, B.J., Clemente, J.C., Quince, C., Knight, R., 2011. UCHIME improves sensitivity and speed of chimera detection. Bioinformatics 27, 2194–2200. https://doi.org/10.1093/bioinformatics/btr381.

Engelbrektson, A., Kunin, V., Wrighton, K.C., Zvenigorodsky, N., Chen, F., Ochman, H., Hugenholtz, P., 2010. Experimental factors affecting PCR-based estimates of microbial species richness and evenness. ISME J. 4, 642.

Falconer, R.E., Bown, J.L., White, N.A., Crawford, J.W., 2005. Biomass recycling and the origin of phenotype in fungal mycelia. Proc. R. Soc. B Biol. Sci. 272, 1727–1734.

Flintoft, L., 2012. Associations go metagenome-wide. Nat. Rev. Genet. 13, 756–757.

Frey-Klett, P., Burlinson, P., Deveau, A., Barret, M., Tarkka, M., Sarniguet, A., 2011. Bacterial-fungal interactions: hyphens between agricultural, clinical, environmental, and food microbiologists. Microbiol. Mol. Biol. Rev. 75, 583–609. https://doi.org/10.1128/MMBR.00020-11.

Gardes, M., Bruns, T.D., 1996. ITS-RFLP matching for identification of fungi. In: Species Diagnostics Protocols. Springer, pp. 177–186.

Gardes, M., White, T.J., Fortin, J.A., Bruns, T.D., Taylor, J.W., 1991. Identification of indigenous and introduced symbiotic fungi in ectomycorrhizae by amplification of nuclear and mitochondrial ribosomal DNA. Can. J. Bot. 69, 180–190.

Glass, E.M., Wilkening, J., Wilke, A., Antonopoulos, D., Meyer, F., 2010. Using the metagenomics RAST server (MG-RAST) for analyzing shotgun metagenomes. Cold Spring Harb. Protoc. 2010 (1). https://doi.org/10.1101/pdb.prot5368. pdb.prot5368.

Goodacre, R., Vaidyanathan, S., Dunn, W.B., Harrigan, G.G., Kell, D.B., 2004. Metabolomics by numbers: acquiring and understanding global metabolite data. Trends Biotechnol. 22, 245–252. https://doi.org/10.1016/j.tibtech.2004.03.007.

Goodlett, D.R., Yi, E.C., 2003. Stable isotopic labeling and mass spectrometry as a means to determine differences in protein expression. TrAC Trends Anal. Chem. 22, 282–290. https://doi.org/10.1016/S0165-9936(03)00505-3.

Grabherr, M.G., Haas, B.J., Yassour, M., Levin, J.Z., Thompson, D.A., Amit, I., Adiconis, X., Fan, L., Raychowdhury, R., Zeng, Q., 2011. Full-length transcriptome assembly from RNA-Seq data without a reference genome. Nat. Biotechnol. 29, 644.

Grada, A., Weinbrecht, K., 2013. Next-generation sequencing: methodology and application. J. Invest. Dermatol. 133, e11.

Guttman, D.S., McHardy, A.C., Schulze-Lefert, P., 2014. Microbial genome-enabled insights into plant–microorganism interactions. Nat. Rev. Genet. 15, 797–813.

Gweon, H.S., Oliver, A., Taylor, J., Booth, T., Gibbs, M., Read, D.S., Griffiths, R.I., Schonrogge, K., 2015. PIPITS: an automated pipeline for analyses of fungal internal transcribed spacer sequences from the I llumina sequencing platform. Methods Ecol. Evol. 6, 973–980.

Hacquard, S., 2016. Disentangling the factors shaping microbiota composition across the plant holobiont. New Phytol. 209, 454–457.

Hallmann, J., 2001. Plant Interactions With Endophytic Bacteria. CABI Publishing, New York.

Handelsman, J., 2004. Metagenomics: application of genomics to uncultured microorganisms. Microbiol. Mol. Biol. Rev. 68, 669–685.

Handelsman, J., Rondon, M.R., Brady, S.F., Clardy, J., Goodman, R.M., 1998. Molecular biological access to the chemistry of unknown soil microbes: a new frontier for natural products. Chem. Biol. 5, R245–R249.

Hanski, I., 1982. Dynamics of regional distribution: the core and satellite species hypothesis. Oikos, 210–221.

Hardoim, P.R., van Overbeek, L.S., Berg, G., Pirttilä, A.M., Compant, S., Campisano, A., Döring, M., Sessitsch, A., 2015. The hidden world within plants: ecological and evolutionary considerations for defining functioning of microbial endophytes. Microbiol. Mol. Biol. Rev. 79, 293–320. https://doi.org/10.1128/MMBR.00050-14.

Hartmann, A., Fischer, D., Kinzel, L., Chowdhury, S.P., Hofmann, A., Baldani, J.I., Rothballer, M., 2019. Assessment of the structural and functional diversities of plant microbiota: achievements and challenges—a review. J. Adv. Res. 19, 3–13. https://doi.org/10.1016/j.jare.2019.04.007. Special Issue on Plant Microbiome.

He, Z., Gentry, T.J., Schadt, C.W., Wu, L., Liebich, J., Chong, S.C., Huang, Z., Wu, W., Gu, B., Jardine, P., 2007. GeoChip: a comprehensive microarray for investigating biogeochemical, ecological and environmental processes. ISME J. 1, 67–77.

Helander, M., Ahlholm, J., Sieber, T.N., Hinneri, S., Saikkonen, K., 2007. Fragmented environment affects birch leaf endophytes. New Phytol. 175, 547–553.

Hettich, R.L., Pan, C., Chourey, K., Giannone, R.J., 2013. Metaproteomics: harnessing the power of high performance mass spectrometry to identify the suite of proteins that control metabolic activities in microbial communities. Anal. Chem. 85, 4203–4214. https://doi.org/10.1021/ac303053e.

Hildebrand, F., Tadeo, R., Voigt, A.Y., Bork, P., Raes, J., 2014. LotuS: an efficient and user-friendly OTU processing pipeline. Microbiome 2, 30. https://doi.org/10.1186/2049-2618-2-30.

Hiller, D., Jiang, H., Xu, W., Wong, W.H., 2009. Identifiability of isoform deconvolution from junction arrays and RNA-Seq. Bioinformatics 25, 3056–3059.

Horak, P., Fröhling, S., Glimm, H., 2016. Integrating next-generation sequencing into clinical oncology: strategies, promises and pitfalls. ESMO Open 1, e000094.

Houle, D., Govindaraju, D.R., Omholt, S., 2010. Phenomics: the next challenge. Nat. Rev. Genet. 11, 855–866.

Huber, T., Faulkner, G., Hugenholtz, P., 2004. Bellerophon: a program to detect chimeric sequences in multiple sequence alignments. Bioinformatics (Oxf. Engl.) 20, 2317–2319. https://doi.org/10.1093/bioinformatics/bth226.

Ibarbalz, F.M., Pérez, M.V., Figuerola, E.L.M., Erijman, L., 2014. The bias associated with amplicon sequencing does not affect the quantitative assessment of bacterial community dynamics. PLoS One 9 (6), e99722.

Ihrmark, K., Bödeker, I.T.M., Cruz-Martinez, K., Friberg, H., Kubartova, A., Schenck, J., Strid, Y., Stenlid, J., Brandström-Durling, M., Clemmensen, K.E., Lindahl, B.D., 2012. New primers to amplify the fungal ITS2 region—evaluation by 454-sequencing of artificial and natural communities. FEMS Microbiol. Ecol. 82, 666–677. https://doi.org/10.1111/j.1574-6941.2012.01437.x.

Jagtap, P.D., Blakely, A., Murray, K., Stewart, S., Kooren, J., Johnson, J.E., Rhodus, N.L., Rudney, J., Griffin, T.J., 2015. Metaproteomic analysis using the Galaxy framework. Proteomics 15, 3553–3565. https://doi.org/10.1002/pmic.201500074.

Karlsson, F.H., Tremaroli, V., Nookaew, I., Bergström, G., Behre, C.J., Fagerberg, B., Nielsen, J., Bäckhed, F., 2013. Gut metagenome in European women with normal, impaired and diabetic glucose control. Nature 498, 99–103.

Klindworth, A., Pruesse, E., Schweer, T., Peplies, J., Quast, C., Horn, M., Glöckner, F.O., 2013. Evaluation of general 16S ribosomal RNA gene PCR primers for classical and next-generation sequencing-based diversity studies. Nucleic Acids Res. 41 (1), e1. https://doi.org/10.1093/nar/gks808.

Knight, R., Vrbanac, A., Taylor, B.C., Aksenov, A., Callewaert, C., Debelius, J., Gonzalez, A., Kosciolek, T., McCall, L.-I., McDonald, D., Melnik, A.V., Morton, J.T., Navas, J., Quinn, R.A., Sanders, J.G., Swafford, A.D., Thompson, L.R., Tripathi, A., Xu, Z.Z., Zaneveld, J.R., Zhu, Q., Caporaso, J.G., Dorrestein, P.C., 2018. Best practices for analysing microbiomes. Nat. Rev. Microbiol. 16, 410–422. https://doi.org/10.1038/s41579-018-0029-9.

Kovalchuk, A., Mukrimin, M., Zeng, Z., Raffaello, T., Liu, M., Kasanen, R., Sun, H., Asiegbu, F.O., 2018. Mycobiome analysis of asymptomatic and symptomatic Norway spruce trees naturally infected by the conifer pathogens *Heterobasidion* spp. Environ. Microbiol. Rep. 10, 532–541.

Kusari, S., Hertweck, C., Spiteller, M., 2012. Chemical ecology of endophytic fungi: origins of secondary metabolites. Chem. Biol. 19, 792–798. https://doi.org/10.1016/j.chembiol.2012.06.004.

Langer, G., 1994. Die Gattung *Botryobasidium* DONK (Corticiaceae, Basidiomycetes). Schweizerbart Science Publishers, Stuttgart, Germany.

Legendre, P., Anderson, M.J., 1999. Distance-based redundancy analysis: testing multispecies responses in multifactorial ecological experiments. Ecol. Monogr. 69, 1–24.

Liang, Y., He, Z., Wu, L., Deng, Y., Li, G., Zhou, J., 2010. Development of a common oligonucleotide reference standard for microarray data normalization and comparison across different microbial communities. Appl. Environ. Microbiol. 76, 1088–1094.

Line, J.E., Hiett, K.L., Guard-Bouldin, J., Seal, B.S., 2010. Differential carbon source utilization by Campylobacter jejuni 11168 in response to growth temperature variation. J. Microbiol. Methods 80, 198–202.

Littman, M.L., 1947. A culture medium for the primary isolation of fungi. Science 106, 109–111.

Liu, Z., Lozupone, C., Hamady, M., Bushman, F.D., Knight, R., 2007. Short pyrosequencing reads suffice for accurate microbial community analysis. Nucleic Acids Res. 35, e120.

Lundberg, D.S., Lebeis, S.L., Paredes, S.H., Yourstone, S., Gehring, J., Malfatti, S., Tremblay, J., Engelbrektson, A., Kunin, V., Del Rio, T.G., 2012. Defining the core *Arabidopsis thaliana* root microbiome. Nature 488, 86–90.

Luo, F., Zhong, J., Yang, Y., Zhou, J., 2006. Application of random matrix theory to microarray data for discovering functional gene modules. Phys. Rev. E 73, 031924.

Ma, Y., Qu, Z.-L., Liu, B., Tan, J.-J., Asiegbu, F.O., Sun, H., 2020. Bacterial community structure of *Pinus thunbergii* naturally infected by the nematode *Bursaphelenchus xylophilus*. Microorganisms 8, 307.

Magurran, A.E., Henderson, P.A., 2003. Explaining the excess of rare species in natural species abundance distributions. Nature 422, 714–716.

Mallick, H., Franzosa, E.A., Mclver, L.J., Banerjee, S., Sirota-Madi, A., Kostic, A.D., Clish, C.B., Vlamakis, H., Xavier, R.J., Huttenhower, C., 2019. Predictive metabolomic profiling of microbial communities using amplicon or metagenomic sequences. Nat. Commun. 10, 1–11.

Mandyam, K., Jumpponen, A., 2005. Seeking the elusive function of the root-colonising dark septate endophytic fungi. Stud. Mycol. 53, 173–189. https://doi.org/10.3114/sim.53.1.173. The missing lineages: Phylogeny and ecology of endophytic and other enigmatic root-associated fungi.

Mann, M., Pandey, A., 2001. Use of mass spectrometry-derived data to annotate nucleotide and protein sequence databases. Trends Biochem. Sci. 26, 54–61. https://doi.org/10.1016/s0968-0004(00)01726-6.

Markowitz, V.M., Ivanova, N.N., Szeto, E., Palaniappan, K., Chu, K., Dalevi, D., Chen, I.-M.A., Grechkin, Y., Dubchak, I., Anderson, I., 2007. IMG/M: a data management and analysis system for metagenomes. Nucleic Acids Res. 36, D534–D538.

Markowitz, V.M., Mavromatis, K., Ivanova, N.N., Chen, I.-M.A., Chu, K., Kyrpides, N.C., 2009. IMG ER: a system for microbial genome annotation expert review and curation. Bioinformatics 25, 2271–2278.

Maron, P.-A., Ranjard, L., Mougel, C., Lemanceau, P., 2007. Metaproteomics: a new approach for studying functional microbial ecology. Microb. Ecol. 53, 486–493. https://doi.org/10.1007/s00248-006-9196-8.

Martín, J.A., Witzell, J., Blumenstein, K., Rozpedowska, E., Helander, M., Sieber, T.N., Gil, L., 2013. Resistance to Dutch elm disease reduces presence of xylem endophytic fungi in elms (*Ulmus* spp.). PLoS One 8, e56987.

Martín, J.A., Macaya-Sanz, D., Witzell, J., Blumenstein, K., Gil, L., 2015. Strong *in vitro* antagonism by elm xylem endophytes is not accompanied by temporally stable *in planta* protection against a vascular pathogen under field conditions. Eur. J. Plant Pathol. 142, 185–196. https://doi.org/10.1007/s10658-015-0602-2.

McCully, M.E., 2001. Niches for bacterial endophytes in crop plants: a plant biologist's view. Funct. Plant Biol. 28, 983–990. https://doi.org/10.1071/pp01101.

McDonald, D., Birmingham, A., Knight, R., 2015. Context and the human microbiome. Microbiome 3, 52.

Messal, M., Slippers, B., Naidoo, S., Bezuidt, O., Kemler, M., 2019. Active fungal communities in asymptomatic *Eucalyptus grandis* stems differ between a susceptible and resistant clone. Microorganisms 7. https://doi.org/10.3390/microorganisms7100375.

Miller, J.D., Mackenzie, S., Foto, M., Adams, G.W., Findlay, J.A., 2002. Needles of white spruce inoculated with rugulosin-producing endophytes contain rugulosin reducing spruce budworm growth rate. Mycol. Res. 106, 471–479. https://doi.org/10.1017/S0953756202005671.

Miller, J.R., Koren, S., Sutton, G., 2010. Assembly algorithms for next-generation sequencing data. Genomics 95, 315–327.

Monteiro, R.A., Balsanelli, E., Wassem, R., Marin, A.M., Brusamarello-Santos, L.C., Schmidt, M.A., Tadra-Sfeir, M.Z., Pankievicz, V.C., Cruz, L.M., Chubatsu, L.S., 2012. Herbaspirillum-plant interactions: microscopical, histological and molecular aspects. Plant Soil 356, 175–196.

Moore-Kucera, J., Dick, R.P., 2008. PLFA Profiling of microbial community structure and seasonal shifts in soils of a Douglas-fir chronosequence. Microb. Ecol. 55, 500–511. https://doi.org/10.1007/s00248-007-9295-1.

Mukherjee, S., Seshadri, R., Varghese, N.J., Eloe-Fadrosh, E.A., Meier-Kolthoff, J.P., Göker, M., Coates, R.C., Hadjithomas, M., Pavlopoulos, G.A., Paez-Espino, D., Yoshikuni, Y., Visel, A., Whitman, W.B., Garrity, G.M., Eisen, J.A., Hugenholtz, P., Pati, A., Ivanova, N.N., Woyke, T., Klenk, H.-P., Kyrpides, N.C., 2017. 1,003 reference genomes of bacterial and archaeal isolates expand coverage of the tree of life. Nat. Biotechnol. 35, 676–683. https://doi.org/10.1038/nbt.3886.

Neefs, J.-M., Van de Peer, Y., De Rijk, P., Chapelle, S., De Wachter, R., 1993. Compilation of small ribosomal subunit RNA structures. Nucleic Acids Res. 21, 3025–3049.

Nguyen, N.H., Song, Z., Bates, S.T., Branco, S., Tedersoo, L., Menke, J., Schilling, J.S., Kennedy, P.G., 2016. FUNGuild: an open annotation tool for parsing fungal community datasets by ecological guild. Fungal Ecol. 20, 241–248. https://doi.org/10.1016/j.funeco.2015.06.006.

Nilsson, R.H., Tedersoo, L., Abarenkov, K., Ryberg, M., Kristiansson, E., Hartmann, M., Schoch, C.L., Nylander, J.A.A., Bergsten, J., Porter, T.M., Jumpponen, A., Vasihampayan, P., Ovaskainen, O., Hallenberg, N., Bengtsson-Palme, J., Eriksson, K.M., Larsson, K.-H., Larsson, E., Kõljalg, U., 2012. Five simple guidelines for establishing basic authenticity and reliability of newly generated fungal ITS sequences. MycoKeys 4, 37–63. https://doi.org/10.3897/mycokeys.4.3606.

Nilsson, R.H., Larsson, K.-H., Taylor, A.F.S., Bengtsson-Palme, J., Jeppesen, T.S., Schigel, D., Kennedy, P., Picard, K., Glöckner, F.O., Tedersoo, L., Saar, I., Kõljalg, U., Abarenkov, K., 2019. The UNITE database for molecular identification of fungi: handling dark taxa and parallel taxonomic classifications. Nucleic Acids Res. 47, D259–D264. https://doi.org/10.1093/nar/gky1022.

Ogunseitan, O.A., 1996. Protein profile variation in cultivated and native freshwater microorganisms exposed to chemical environmental pollutants. Microb. Ecol. 31, 291–304.

Oksanen, J., Blanchet, F.G., Friendly, M., Kindt, R., Legendre, P., McGlinn, D., Minchin, P.R., O'Hara, R.B., Simpson, G.L., Solymos, P., 2016. vegan: Community Ecology Package. R package version 2.4-1.

Oono, R., Lefèvre, E., Simha, A., Lutzoni, F., 2015. A comparison of the community diversity of foliar fungal endophytes between seedling and adult loblolly pines (*Pinus taeda*). Fungal Biol. 119, 917–928.

O'Farrell, P.H., 1975. High resolution two-dimensional electrophoresis of proteins. J. Biol. Chem. 250, 4007–4021.

Pandey, A., Lewitter, F., 1999. Nucleotide sequence databases: a gold mine for biologists. Trends Biochem. Sci. 24, 276–280. https://doi.org/10.1016/S0968-0004(99)01400-0.

Pathan, S.I., Ceccherini, M.T., Hansen, M.A., Giagnoni, L., Ascher, J., Arenella, M., Sørensen, S.J., Pietramellara, G., Nannipieri, P., Renella, G., 2015. Maize lines with different nitrogen use efficiency select bacterial communities with different β-glucosidase-encoding genes and glucosidase activity in the rhizosphere. Biol. Fertil. Soils 51, 995–1004.

Pedersen, S., Bloch, P.L., Reeh, S., Neidhardt, F.C., 1978. Patterns of protein synthesis in *E. coli*: a catalog of the amount of 140 individual proteins at different growth rates. Cell 14, 179–190. https://doi.org/10.1016/0092-8674(78)90312-4.

Pevzner, P.A., Tang, H., Waterman, M.S., 2001. An Eulerian path approach to DNA fragment assembly. Proc. Natl. Acad. Sci. 98, 9748–9753.

Qu, Z., Liu, B., Ma, Y., Sun, H., 2020. Differences in bacterial community structure and potential functions among Eucalyptus plantations with different ages and species of trees. Appl. Soil Ecol. 149, 103515.

Quince, C., Lanzén, A., Curtis, T.P., Davenport, R.J., Hall, N., Head, I.M., Read, L.F., Sloan, W.T., 2009. Accurate determination of microbial diversity from 454 pyrosequencing data. Nat. Methods 6, 639–641. https://doi.org/10.1038/nmeth.1361.

Quinn, R.A., Phelan, V.V., Whiteson, K.L., Garg, N., Bailey, B.A., Lim, Y.W., Conrad, D.J., Dorrestein, P.C., Rohwer, F.L., 2016. Microbial, host and xenobiotic diversity in the cystic fibrosis sputum metabolome. ISME J. 10, 1483–1498.

Ramette, A., Tiedje, J.M., 2007. Multiscale responses of microbial life to spatial distance and environmental heterogeneity in a patchy ecosystem. Proc. Natl. Acad. Sci. 104, 2761–2766.

Rashid, S., Charles, T.C., Glick, B.R., 2012. Isolation and characterization of new plant growth-promoting bacterial endophytes. Appl. Soil Ecol. 61, 217–224.

Raza, K., Ahmad, S., 2016. Principle, analysis, application and challenges of next-generation sequencing: A review. ArXiv Prepr. ArXiv160605254.

Ren, F., Kovalchuk, A., Mukrimin, M., Liu, M., Zeng, Z., Ghimire, R.P., Kivimäenpää, M., Holopainen, J.K., Sun, H., Asiegbu, F.O., 2018. Tissue microbiome of Norway spruce affected by Heterobasidion-induced wood decay. Microb. Ecol. 77, 640–650.

Reuter, J.A., Spacek, D.V., Snyder, M.P., 2015. High-throughput sequencing technologies. Mol. Cell 58, 586–597.

Rhee, S.-K., Liu, X., Wu, L., Chong, S.C., Wan, X., Zhou, J., 2004. Detection of genes involved in biodegradation and biotransformation in microbial communities by using 50-mer oligonucleotide microarrays. Appl. Environ. Microbiol. 70, 4303–4317.

Rognes, T., Flouri, T., Nichols, B., Quince, C., Mahé, F., 2016. VSEARCH: a versatile open source tool for metagenomics. PeerJ 4, e2584. https://doi.org/10.7717/peerj.2584.

Roume, H., Muller, E.E., Cordes, T., Renaut, J., Hiller, K., Wilmes, P., 2013. A biomolecular isolation framework for eco-systems biology. ISME J. 7, 110–121.

Sayers, E.W., Barrett, T., Benson, D.A., Bolton, E., Bryant, S.H., Canese, K., Chetvernin, V., Church, D.M., DiCuccio, M., Federhen, S., Feolo, M., Fingerman, I.M., Geer, L.Y., Helmberg, W., Kapustin, Y., Landsman, D., Lipman, D.J., Lu, Z., Madden, T.L., Madej, T., Maglott, D.R., Marchler-Bauer, A., Miller, V., Mizrachi, I., Ostell, J., Panchenko, A., Phan, L., Pruitt, K.D., Schuler, G.D., Sequeira, E., Sherry, S.T., Shumway, M., Sirotkin, K., Slotta, D., Souvorov, A., Starchenko, G., Tatusova, T.A., Wagner, L., Wang, Y., Wilbur, W.J., Yaschenko, E., Ye, J., 2011. Database resources of the National Center for Biotechnology Information. Nucleic Acids Res. 39, D38–D51. https://doi.org/10.1093/nar/gkq1172.

Schlegel, H.G., Jannasch, H.W., 1967. Enrichment cultures. Annu. Rev. Microbiol. 21, 49–70.

Schloss, P.D., Westcott, S.L., Ryabin, T., Hall, J.R., Hartmann, M., Hollister, E.B., Lesniewski, R.A., Oakley, B.B., Parks, D.H., Robinson, C.J., Sahl, J.W., Stres, B., Thallinger, G.G., Van Horn, D.J., Weber, C.F., 2009. Introducing mothur: open-source, platform-independent, community-supported software for describing and comparing microbial communities. Appl. Environ. Microbiol. 75, 7537–7541. https://doi.org/10.1128/AEM.01541-09.

Schloter, M., Hartmann, A., 1998. Endophytic and surface colonization of wheat roots (*Triticum aestivum*) by different *Azospirillum brasilense* strains studied with strain-specific monoclonal antibodies. Symbiosis 25 (1), 159–179.

Schmid, M., Rothballer, M., Hartmann, A., 2007. Analysis of microbial communities in soil microhabitats using fluorescence in situ hybridization. In: Modern Soil Microbiology. II. CRC Press, Boca Raton, FL, pp. 317–335.

Schoch, C.L., Seifert, K.A., Huhndorf, S., Robert, V., Spouge, J.L., Levesque, C.A., Chen, W., Fungal Barcoding Consortium, 2012. Nuclear ribosomal internal transcribed spacer (ITS) region as a universal DNA barcode marker for fungi. Proc. Natl. Acad. Sci. 109, 6241–6246. https://doi.org/10.1073/pnas.1117018109.

Scholz, M., Ward, D.V., Pasolli, E., Tolio, T., Zolfo, M., Asnicar, F., Truong, D.T., Tett, A., Morrow, A.L., Segata, N., 2016. Strain-level microbial epidemiology and population genomics from shotgun metagenomics. Nat. Methods 13, 435–438. https://doi.org/10.1038/nmeth.3802.

Schulz, B., Boyle, C., 2005. The endophytic continuum. Mycol. Res. 109, 661–686. https://doi.org/10.1017/S095375620500273X.

Schulz, B., Römmert, A.-K., Dammann, U., Aust, H.-J., Strack, D., 1999. The endophyte-host interaction: a balanced antagonism? Mycol. Res. 103, 1275–1283. https://doi.org/10.1017/S0953756299008540.

Schulz, B., Boyle, C., Draeger, S., Römmert, A.-K., Krohn, K., 2002. Endophytic fungi: a source of novel biologically active secondary metabolites. Mycol. Res. 106, 996–1004. https://doi.org/10.1017/S0953756202006342.

Sequerra, J., Capellano, A., Gianinazzi-Pearson, V., Moiroud, A., 1995. Ultrastructure of cortical root cells of Alnus incana infected by *Penicillium nodositatum*. New Phytol. 130, 545–555.

Serratì, S., De Summa, S., Pilato, B., Petriella, D., Lacalamita, R., Tommasi, S., Pinto, R., 2016. Next-generation sequencing: advances and applications in cancer diagnosis. OncoTargets Ther. 9, 7355.

Shannon, C.E., 1948. A mathematical theory of communication. Bell Syst. Tech. J. 27, 379–423. https://doi.org/10.1002/j.1538-7305.1948.tb01338.x.

Siddique, A.B., Khokon, A.M., Unterseher, M., 2017. What do we learn from cultures in the omics age? High-throughput sequencing and cultivation of leaf-inhabiting endophytes from beech (*Fagus sylvatica* L.) revealed complementary community composition but similar correlations with local habitat conditions. MycoKeys 20, 1–16. https://doi.org/10.3897/mycokeys.20.11265.

Sieber, T.N., 2002. Fungal root endophytes. In: Waisel, Y., Eshel, A., Kafkafi, U. (Eds.), The Hidden Half. Dekker, New York.

Sieber, T.N., 2007. Endophytic fungi in forest trees: are they mutualists? Fungal Biol. Rev. 21, 75–89.

Siegl, A., Kamke, J., Hochmuth, T., Piel, J., Richter, M., Liang, C., Dandekar, T., Hentschel, U., 2011. Single-cell genomics reveals the lifestyle of Poribacteria, a candidate phylum symbiotically associated with marine sponges. ISME J. 5, 61–70.

Simpson, E.H., 1949. Measurement of diversity. Nature 163, 688. https://doi.org/10.1038/163688a0.

Singh, M.P., 2009. Application of Biolog FF MicroPlate for substrate utilization and metabolite profiling of closely related fungi. J. Microbiol. Methods 77, 102–108.

Smith, D., Onions, A.H., 1994. The Preservation and Maintenance of Living Fungi. CAB International.

Soergel, D.A.W., Dey, N., Knight, R., Brenner, S.E., 2012. Selection of primers for optimal taxonomic classification of environmental 16S rRNA gene sequences. ISME J. 6, 1440–1444. https://doi.org/10.1038/ismej.2011.208.

Solis, M.J.L., Cruz, T.E.D., Schnittler, M., Unterseher, M., 2016. The diverse community of leaf-inhabiting fungal endophytes from Philippine natural forests reflects phylogenetic patterns of their host plant species *Ficus benjamina, F. elastica* and *F. religiosa*. Mycoscience 57, 96–106.

Stolyar, S., He, Q., Joachimiak, M.P., He, Z., Yang, Z.K., Borglin, S.E., Joyner, D.C., Huang, K., Alm, E., Hazen, T.C., 2007. Response of *Desulfovibrio vulgaris* to alkaline stress. J. Bacteriol. 189, 8944–8952.

Stone, J.K., 1987. Initiation and development of latent infections by *Rhabdocline parkeri* on Douglas-fir. Can. J. Bot. 65, 2614–2621.

Stone, J.K., Viret, O., Petrini, O., Chapela, I.H., Ouellette, G.B., Petrini, O., 1994. Histological studies of host penetration and colonization by endophytic fungi. In: Host Wall Alterations by Parasitic Fungi. APS Press, St Paul, MN, pp. 115–126.

Stone, J.K., Polishook, J.D., White, J.F., 2004. Endophytic fungi. In: Biodiversity of Fungi. Elsevier Academic Press, Burlington, MA, pp. 241–270.

Strobel, G.A., 2003. Endophytes as sources of bioactive products. Microbes Infect. 5, 535–544.

Sumarah, M.W., Puniani, E., Sørensen, D., Blackwell, B.A., Miller, J.D., 2010. Secondary metabolites from anti-insect extracts of endophytic fungi isolated from *Picea rubens*. Phytochemistry 71, 760–765. https://doi.org/10.1016/j.phytochem.2010.01.015.

Sumarah, M.W., Kesting, J.R., Sørensen, D., Miller, J.D., 2011. Antifungal metabolites from fungal endophytes of *Pinus strobus*. Phytochemistry 72, 1833–1837. https://doi.org/10.1016/j.phytochem.2011.05.003.

Sun, S., Chen, J., Li, W., Altintas, I., Lin, A., Peltier, S., Stocks, K., Allen, E.E., Ellisman, M., Grethe, J., 2010. Community cyberinfrastructure for advanced microbial ecology research and analysis: the CAMERA resource. Nucleic Acids Res. 39, D546–D551.

Sun, H., Terhonen, E., Koskinen, K., Paulin, L., Kasanen, R., Asiegbu, F.O., 2013. The impacts of treatment with biocontrol fungus (*Phlebiopsis gigantea*) on bacterial diversity in Norway spruce stumps. Biol. Control 64, 238–246.

Sun, H., Santalahti, M., Pumpanen, J., Köster, K., Berninger, F., Raffaello, T., Jumpponen, A., Asiegbu, F.O., Heinonsalo, J., 2015. Fungal community shifts in structure and function across a boreal forest fire chronosequence. Appl. Environ. Microbiol. 81, 7869–7880.

Sun, H., Terhonen, E., Kovalchuk, A., Tuovila, H., Chen, H., Oghenekaro, A.O., Heinonsalo, J., Kohler, A., Kasanen, R., Vasander, H., Asiegbu, F.O., 2016. Dominant tree species and soil type affect the fungal community structure in a boreal peatland forest. Appl. Environ. Microbiol. 82, 2632–2643. https://doi.org/10.1128/AEM.03858-15.

Taghavi, S., Garafola, C., Monchy, S., Newman, L., Hoffman, A., Weyens, N., Barac, T., Vangronsveld, J., van der Lelie, D., 2009. Genome survey and characterization of endophytic bacteria exhibiting a beneficial effect on growth and development of poplar trees. Appl. Environ. Microbiol. 75, 748–757.

Tan, R.X., Zou, W.X., 2001. Endophytes: a rich source of functional metabolites. Nat. Prod. Rep. 18, 448–459.

Tanzer, M.M., Arst, H.N., Skalchunes, A.R., Coffin, M., Darveaux, B.A., Heiniger, R.W., Shuster, J.R., 2003. Global nutritional profiling for mutant and chemical mode-of-action analysis in filamentous fungi. Funct. Integr. Genomics 3, 160–170.

Taylor, D.L., McCormick, M.K., 2008. Internal transcribed spacer primers and sequences for improved characterization of basidiomycetous orchid mycorrhizas. New Phytol. 177, 1020–1033. https://doi.org/10.1111/j.1469-8137.2007.02320.x.

Taylor, D.L., Walters, W.A., Lennon, N.J., Bochicchio, J., Krohn, A., Caporaso, J.G., Pennanen, T., 2016. Accurate estimation of fungal diversity and abundance through improved lineage-specific primers optimized for Illumina Amplicon Sequencing. Appl. Environ. Microbiol. 82, 7217–7226. https://doi.org/10.1128/AEM.02576-16.

Tedersoo, L., Bahram, M., Põlme, S., Kõljalg, U., Yorou, N.S., Wijesundera, R., Ruiz, L.V., Vasco-Palacios, A.M., Thu, P.Q., Suija, A., Smith, M.E., Sharp, C., Saluveer, E., Saitta, A., Rosas, M., Riit, T., Ratkowsky, D., Pritsch, K., Põldmaa, K., Piepenbring, M., Phosri, C., Peterson, M., Parts, K., Pärtel, K., Otsing, E., Nouhra, E., Njouonkou, A.L., Nilsson, R.H., Morgado, L.N., Mayor, J., May, T.W., Majuakim, L., Lodge, D.J., Lee, S.S., Larsson, K.-H., Kohout, P., Hosaka, K., Hiiesalu, I., Henkel, T.W., Harend, H., Guo, L., Greslebin, A., Grelet, G., Geml, J., Gates, G., Dunstan, W., Dunk, C., Drenkhan, R., Dearnaley, J., Kesel, A.D., Dang, T., Chen, X., Buegger, F., Brearley, F.Q., Bonito, G., Anslan, S., Abell, S., Abarenkov, K., 2014. Global diversity and geography of soil fungi. Science 346. https://doi.org/10.1126/science.1256688.

Tedersoo, L., Anslan, S., Bahram, M., Põlme, S., Riit, T., Liiv, I., Kõljalg, U., Kisand, V., Nilsson, H., Hildebrand, F., Bork, P., Abarenkov, K., 2015. Shotgun metagenomes and multiple primer pair-barcode combinations of amplicons reveal biases in metabarcoding analyses of fungi. MycoKeys 10, 1–43. https://doi.org/10.3897/mycokeys.10.4852.

Tellenbach, C., Sumarah, M.W., Grünig, C.R., Miller, J.D., 2013. Inhibition of *Phytophthora* species by secondary metabolites produced by the dark septate endophyte *Phialocephala europaea*. Fungal Ecol. 6, 12–18. https://doi.org/10.1016/j.funeco.2012.10.003.

Ter Braak, C.J., 1987. The analysis of vegetation-environment relationships by canonical correspondence analysis. Vegetatio 69, 69–77.

Terhonen, E., Sun, H., Buée, M., Kasanen, R., Paulin, L., Asiegbu, F.O., 2013. Effects of the use of biocontrol agent (*Phlebiopsis gigantea*) on fungal communities on the surface of *Picea abies* stumps. For. Ecol. Manag. 310, 428–433. https://doi.org/10.1016/j.foreco.2013.08.044.

Terhonen, E., Keriö, S., Sun, H., Asiegbu, F.O., 2014. Endophytic fungi of Norway spruce roots in boreal pristine mire, drained peatland and mineral soil and their inhibitory effect on *Heterobasidion parviporum* in vitro. Fungal Ecol. 9, 17–26.

Terhonen, E., Sipari, N., Asiegbu, F.O., 2016. Inhibition of phytopathogens by fungal root endophytes of Norway spruce. Biol. Control 99, 53–63. https://doi.org/10.1016/j.biocontrol.2016.04.006.

Thomas, T., Gilbert, J., Meyer, F., 2012. Metagenomics—a guide from sampling to data analysis. Microb. Inform. Exp. 2, 3.

Thompson, L.R., Sanders, J.G., McDonald, D., Amir, A., Ladau, J., Locey, K.J., Prill, R.J., Tripathi, A., Gibbons, S.M., Ackermann, G., 2017. A communal catalogue reveals Earth's multiscale microbial diversity. Nature 551, 457–463.

Tienaho, J., Karonen, M., Muilu-Mäkelä, R., Wähälä, K., Leon Denegri, E., Franzén, R., Karp, M., Santala, V., Sarjala, T., 2019. Metabolic profiling of water-soluble compounds from the extracts of dark septate endophytic fungi (DSE) isolated from scots pine (*Pinus sylvestris* L.) seedlings using UPLC–orbitrap–MS. Molecules 24, 2330.

Toju, H., Tanabe, A.S., Yamamoto, S., Sato, H., 2012. High-coverage ITS primers for the DNA-based identification of Ascomycetes and Basidiomycetes in environmental samples. PLoS One 7, e40863. https://doi.org/10.1371/journal.pone.0040863.

Toju, H., Peay, K.G., Yamamichi, M., Narisawa, K., Hiruma, K., Naito, K., Fukuda, S., Ushio, M., Nakaoka, S., Onoda, Y., Yoshida, K., Schlaeppi, K., Bai, Y., Sugiura, R., Ichihashi, Y., Minamisawa, K., Kiers, E.T., 2018. Core microbiomes for sustainable agroecosystems. Nat. Plants 4, 247–257. https://doi.org/10.1038/s41477-018-0139-4.

Tornberg, K., Bååth, E., Olsson, S., 2003. Fungal growth and effects of different wood decomposing fungi on the indigenous bacterial community of polluted and unpolluted soils. Biol. Fertil. Soils 37, 190–197.

Trapp, M.A., Kai, M., Mithöfer, A., Rodrigues-Filho, E., 2015. Antibiotic oxylipins from *Alternanthera brasiliana* and its endophytic bacteria. Phytochemistry 110, 72–82. https://doi.org/10.1016/j.phytochem.2014.11.005.

Turner, T.R., James, E.K., Poole, P.S., 2013. The plant microbiome. Genome Biol. 14, 209. https://doi.org/10.1186/gb-2013-14-6-209.

Tyson, G.W., Chapman, J., Hugenholtz, P., Allen, E.E., Ram, R.J., Richardson, P.M., Solovyev, V.V., Rubin, E.M., Rokhsar, D.S., Banfield, J.F., 2004. Community structure and metabolism through reconstruction of microbial genomes from the environment. Nature 428, 37–43.

Unterseher, M., Schnittler, M., 2009. Dilution-to-extinction cultivation of leaf-inhabiting endophytic fungi in beech (*Fagus sylvatica* L.)–different cultivation techniques influence fungal biodiversity assessment. Mycol. Res. 113, 645–654.

Van Nostrand, J.D., Yin, H., Wu, L., Yuan, T., Zhou, J., 2016. Hybridization of environmental microbial community nucleic acids by GeoChip. In: Microbial Environmental Genomics (MEG). Springer, pp. 183–196.

Veldkamp, H., 1970. Enrichment cultures of prokaryotic organisms. In: Norris, J.R., Ribbons, D.W. (Eds.), Methods in Microbiology. Academic Press, pp. 305–361 (Chapter 5) https://doi.org/10.1016/S0580-9517(08)70543-9.

Venter, J.C., Zhang, J.N., Liu, X., Rowe, W., Cravchik, A., Kalush, F., Naik, A., Subramanian, G., Woodage, T., 2004. Polymorphisms in known genes associated with human disease, methods of detection and uses thereof.

Vincent, A.T., Derome, N., Boyle, B., Culley, A.I., Charette, S.J., 2017. Next-generation sequencing (NGS) in the microbiological world: how to make the most of your money. J. Microbiol. Methods 138, 60–71.

Vorholt, J.A., 2012. Microbial life in the phyllosphere. Nat. Publ. Group 10, 828–840.

Wang, J., Jia, H., 2016. Metagenome-wide association studies: fine-mining the microbiome. Nat. Rev. Microbiol. 14, 508.

Wang, K., Xu, C., 2017. Applications of next-generation sequencing in cancer research and molecular diagnosis. J. Clin. Med. Genom. 5, 147.

Weber, A.P., 2015. Discovering new biology through sequencing of RNA. Plant Physiol. 169, 1524–1531.

White, D.C., Davis, W.M., Nickels, J.S., King, J.D., Bobbie, R.J., 1979. Determination of the sedimentary microbial biomass by extractible lipid phosphate. Oecologia 40, 51–62.

Whiteson, K.L., Meinardi, S., Lim, Y.W., Schmieder, R., Maughan, H., Quinn, R., Blake, D.R., Conrad, D., Rohwer, F., 2014. Breath gas metabolites and bacterial metagenomes from cystic fibrosis airways indicate active pH neutral 2,3-butanedione fermentation. ISME J. 8, 1247–1258. https://doi.org/10.1038/ismej.2013.229.

Winogradsky, S., 1949. Microbiologie du sol. Oeuvres completes. Marson, Paris.

Woyke, T., Teeling, H., Ivanova, N.N., Huntemann, M., Richter, M., Gloeckner, F.O., Boffelli, D., Anderson, I.J., Barry, K.W., Shapiro, H.J., 2006. Symbiosis insights through metagenomic analysis of a microbial consortium. Nature 443, 950–955.

Wu, L., Liu, X., Schadt, C.W., Zhou, J., 2006. Microarray-based analysis of subnanogram quantities of microbial community DNAs by using whole-community genome amplification. Appl. Environ. Microbiol. 72, 4931–4941.

Yilmaz, P., Parfrey, L.W., Yarza, P., Gerken, J., Pruesse, E., Quast, C., Schweer, T., Peplies, J., Ludwig, W., Glöckner, F.O., 2014. The SILVA and "All-species Living Tree Project (LTP)" taxonomic frameworks. Nucleic Acids Res. 42, D643–D648. https://doi.org/10.1093/nar/gkt1209.

Zapka, C., Leff, J., Henley, J., Tittl, J., De Nardo, E., Butler, M., Griggs, R., Fierer, N., Edmonds-Wilson, S., 2017. Comparison of standard culture-based method to culture-independent method for evaluation of hygiene effects on the hand microbiome. mBio 8 (2). e00093-17.

Zelles, L., 1997. Phospholipid fatty acid profiles in selected members of soil microbial communities. Chemosphere 35, 275–294.

Zelles, L., 1999. Fatty acid patterns of phospholipids and lipopolysaccharides in the characterisation of microbial communities in soil: a review. Biol. Fertil. Soils 29, 111–129.

Zhang, H.W., Song, Y.C., Tan, R.X., 2006. Biology and chemistry of endophytes. Nat. Prod. Rep. 23, 753–771. https://doi.org/10.1039/B609472B.

Zhou, J., He, Z., Van Nostrand, J.D., Wu, L., Deng, Y., 2010. Applying GeoChip analysis to disparate microbial communities. Microbe 5, 60–65.

Økland, R.H., Eilertsen, O., 1994. Canonical correspondence analysis with variation partitioning: some comments and an application. J. Veg. Sci. 5, 117–126.

Chapter 4

Abiotic factors affecting the composition of forest tree microbiomes

Risto Kasanen

Department of Forest Sciences, Faculty of Agriculture and Forestry, University of Helsinki, Helsinki, Finland

Chapter Outline

1. Introduction

Currently, it is widely recognized that all living organisms are functionally microbe-associated holobionts instead of single biological entities. Accumulating evidence suggests that microbiomes contribute to ecological adaptation, as plants may be able to modulate their microbiota dynamically (Vandenkoornhuyse et al., 2015). The ability of organisms to adapt to environmental change with microbial assistance is called microbiome flexibility (Voolstra and Ziegler, 2020). The tolerance toward abiotic stress is an important component of fitness. Separating the plant fitness component from microbiome fitness is challenging. To be able to grow a plant without a microbiome, the growth conditions would need to be disconnected from the environment, which would make any definition of fitness irrelevant. The interactions of plants, microbiomes, and abiotic factors of the environment are complex (Fig. 4.1). In addition to a plant's own modulation of its microbiome, plant symbionts in roots also have a role in modifying the microbiome (Uroz et al., 2019). The roles of a beneficial mycobiome in plant health (Mendes et al., 2013) and in the induction of systemic resistance against pathogens have been demonstrated (Pieterse et al., 2014). Carrell and Frank (2015) observed that giant coast redwood trees (*Sequoia sempervirens*) and giant sequoia (*Sequoiadendron giganteum*) host endophyte microbiomes that are known to be related in the protection of the host against biotic and abiotic stress as well as nitrogen fixation. The authors hypothesize that the endophytic microbiome might be a key for the success of these giants.

The role of abiotic factors in the microbiome structure of forest trees is not well known. Multiscale habitats and interactions shape the microbiome of trees. Therefore, it is difficult to predict the structure and composition of forest microbiota, including the effects of disturbances, soil type, climate, and management. Most published forest microbiome studies focus independently on either bacterial or fungal communities, but rarely on both (Uroz et al., 2016).

Forest Microbiology. **https://doi.org/10.1016/B978-0-12-822542-4.00011-5**

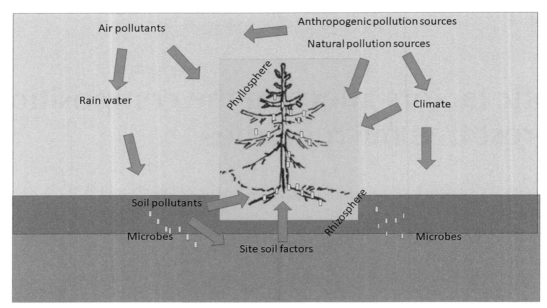

FIG. 4.1 Abiotic factors affecting the host plant and its microbiome.

The impacts of water flooding, drought, and N and P deficiency stress and shape the root microbiome. Again, the majority of the research available focuses on the responses of root bacteria communities, and the impacts on the mycobiome are poorly known (Hartman and Tringe, 2019).

2. The impacts of water: Flooding and drought

2.1 Phyllosphere microbiota and drought

Trees respond to abiotic stress factors by adjusting the stomata of the leaves, which modulates water use and photosynthesis. In addition, the growth rate and utilization of carbohydrate storage are altered. Water stress might increase the susceptibility of trees to pests and pathogenic microbes (Netherer et al., 2015; Hossain et al., 2019; Matthews et al., 2018). Therefore, indirect effects of abiotic water stress on the tree microbiome caused by invading harmful organisms and consequent microbial succession are likely (Table 4.1).

Drought increases the hydraulic soil-leaf tension in the xylem, which increases the risk of xylem cavitation in trees. The ability of xylem conduits (vessels and tracheids) to maintain water transport by water tension is limited by their ability to withstand embolism. Embolism by water stress is caused by air seeding at the pit membranes between xylem conduits in the vessels and tracheids. Herbivory or mechanical damage to stems and leaves commonly causes embolism (Tyree and Sperry, 1989). The results by Paljakka et al. (2020) suggest an additional strategy that pathogens might utilize to disturb water transport: the surface tension of xylem sap decreases and the hydraulic conductivity of trees deteriorates.

Aghai et al. (2019) provided specific evidence of a beneficial tree mycobiome and demonstrated the transferability of microbial-driven drought resistance. Microbial endophytes from the phyllosphere of drought-tolerant Salicaceae were introduced to Douglas firs (*Pseudotsuga menziesii*) and *Thuja plicata*. Tree inoculation with drought tolerance-related endophytes increased survival during drought stress. The inoculated, unfertilized *Thuja plicata* seedlings were as green as the fertilized control at 420 days after inoculation (Aghai et al., 2019).

The effects of drought stress of plants on fungi have been mostly studied at the level of individual species or strains, especially pathogenic fungi (Table 4.1). Drought stress seems to reduce the species diversity of the endophytic mycobiota in cork oak (*Quercus suber*) and to promote the proliferation of some potentially pathogenic endophytes (Linaldeddu et al., 2011). Terhonen et al. (2019) found that the colonization success of the *Heterobasidion* species in stem wood and bark tissues was higher when *Picea abies* seedlings were drought-stressed experimentally. Linnakoski et al. (2017) demonstrated that the variation in the ability of *Endoconidiophora polonica* to colonize drought-stressed seedlings of *Picea abies* was strain-specific. It is known that the mycobiome structure of symptomatic *Heterobasidion*-infected trees is different from asymptomatic trees (Kovalchuk et al., 2018). Therefore, drought events are likely to cause indirect shifts in tree microbiomes due to the enhanced colonization success of pathogens.

TABLE 4.1 Impacts of water stress on microbiomes.

Phyllosphere	Factor	Tree host	Effect	References
	Drought	Norway spruce *Picea abies*	Increased susceptibility to pests and pathogens	Netherer et al. (2015) Hossain et al. (2019) Matthews et al. (2018) Terhonen et al. (2019) Linnakoski et al. (2017)
		Holm oak *Quercus ilex* Scots pine *Pinus sylvestris*	Decrease of microbial diversity	Preece et al. (2020)
		Cork oak *Qurcus suber*	Reduced endophyte diversity	Linaldeddu et al. (2011)
		Douglas fir *Pseudotsuga menziesii* Red cedar *Thuja plicata*	Microbes derived from Salicaceae increase drought tolerance	Aghai et al. (2019)
	Flooding	Black alder *Alnus glutinosa* Chinese willow *Salix matsudana* Horse chestnut *Aesculus hippocastanum*	Increased susceptibility to *Phytophthora* sp.	Strnadová et al. (2010) Brasier et al. (2003)
		Sycamore *Acer pseudoplatanus* Linden *Tilia cordata* Common oak *Quercus robur*	Specialization of bacteria in height layers of canopy	Herrmann et al. (2020)
Rhizosphere	Drought	Holm oak *Quercus ilex*	Tree decline in the absence of EM fungi, *Trichoderma* sp. excluding *Phytophthora* sp.	Ruiz-Gomez et al. (2019)
		Eastern cottonwood *Populus deltoides*	Rhizosphere core set of bacteria responding to host stress	Timm et al. (2018)
	Flooding	Hybrid poplar *Populus tremula* × *Populus alba*	Increase of anaerobic species	Graff and Conrad (2005)
	Irrigation	Pine forest	Shift of microbiome from oligotrophic to copiotrophic	Hartmann et al. (2017)

2.2   Effects of flooding on the phyllosphere microbiome

Flooding events have been demonstrated to especially predispose trees to diseases caused by plant pathogens of the genus *Phytophthora* (Strnadová et al., 2010; Brasier et al., 2003). Also, it is probable that rapid changes in water availability and oxygen stress caused by flooding and the subsequent decline of affected trees will cause changes in the tree microbiome (Table 4.1). Herrmann et al. (2020) studied the effect of the canopy position of leaves (top, mid, and bottom) and tree species (*Acer pseudoplatanus*, *Quercus robur*, and *Tilia cordata*) for the phyllosphere microbiome in a floodplain hardwood forest. The richness and abundance of bacterial species increased toward the mid and bottom position of the canopy. Thirty core OTUs composed up to 66% of Actinobacteria, likely reflecting ecological preferences within the canopy. Tree species OTU networks were observed and 55%–57% of the OTUs were unique to each tree species.

2.3   Drought tolerance requires a beneficial microbiome in the rhizosphere

Naylor and Coleman-Derr (2018) reviewed the effect of drought on root-associated bacterial flora. In general, drought decreases the total bacterial biomass, but appears to have a minor impact on bacterial phylogenetic diversity. However, the community composition is significantly affected by drought. Again, the majority of the studies have been conducted with perennial plants and agricultural crops. Xu et al. (2018) showed that drought stress caused a delay in the development of the sorghum root microbiome and modified the bacterial flora toward drought-resistant bacteria. de Vries et al. (2020) concluded that during short drought events, plant growth-promoting rhizobacteria (PGPR) and arbuscular mycorrhizal fungi (AMF) increase plant drought tolerance, but finally plants and microbes perish if the drought continues.

It appears that the mycobiome facilitates tree survival in abiotic stress and also counteracts the decrease of resistance due to stress. Extreme drought episodes and decreasing precipitation associated with global warming have been connected to the decline of Mediterranean forest ecosystems. Ruiz-Gomez et al. (2019) found a high number of pathogens in the soil mycobiome of declining forests. The presence of ectomycorrhizal fungi improved the canopy condition expressed as a low defoliation degree. The suspected causal agents of tree decline, *Phytophthora cinnamomi* and *Pythium spiculum*, were not the most common oomycetes and were not related directly to the tree condition. The presence of the *Trichoderma* species excluded the pathogenic *Phytophthora* spp. in the soil mycobiome. In addition, in the rhizosphere of *Populus deltoides*, Timm et al. (2018) identified a core set of bacteria that responded to tree stress caused by drought, shading, and metal toxicity.

Preece et al. (2020) found evidence that drought might have a stronger overall effect than nitrogen or phosphorus deposition changes on the soil microbiomes of Mediterranean *Quercus ilex* and *Pinus sylvestris* forests. Drought decreased microbial diversity and evenness. The activity of soil microbes measured as respiration decreased for *Quercus ilex* but increased for *Pinus sylvestris* when nitrogen was added. The addition of phosphorus during the dry period did not change soil respiration. The microbial community evenness for *Pinus sylvestris* changed due to the interaction of soil-water content. Soil respiration indicated that the activity of microbial associates of *Quercus ilex* resumed after the drought stopped. Although soil respiration for *Pinus sylvestris* decreased, the soil microbiome of *Pinus sylvestris* showed more resilience than *Quercus ilex*.

2.4 Increasing water content in the rhizosphere benefits anaerobic and copiotrophic microbes

The intake of macronutrients is disturbed in flooded plants due to root mortality, disruption of mycorrhizae, and decreased root metabolism (Kozlowski, 1997). Limitations in the oxygen availability of plant roots due to flooding may theoretically affect the rhizosphere microbiome (Hartman and Tringe, 2019). However, information on the effects of flooding is sparse. Cui et al. (2019) studied the effect of flooding on the microbiome of rice culms and observed that stalks growing in flooded and nonflooded soils had strikingly different bacteria flora. Sjogaard et al. (2018) examined coastal flooding with seawater and documented accelerated heterotrophic activity of the soil microbiome followed by decreased microbial activity after 4 months. The effects on the forest tree microbiome are difficult to extrapolate based on these results.

Graff and Conrad (2005) used the terminal restriction fragment length polymorphism methodology to study poplar (*Populus tremula* × *Populus alba*) tree microcosms and found that bacterial communities were different in bulk soil, the rhizosphere, and the rhizoplane. *Bacillus* sp. were highest in the unflooded bulk soil and the rhizosphere. Flooding modified the bacterial communities, especially *Aquaspirillum* sp., a genus containing anaerobic species; it was rare elsewhere but dominated in the rhizosphere of flooded microcosms.

Even without flooding events, increasing rainfall might cause substantial changes in a water-limited pine forest ecosystem. Hartmann et al. (2017) reported that a decade of irrigation increased tree growth, litter fall, and root biomass. The enhanced primary production stimulated soil microbial activity and shifted the microbiome from oligotrophic to more copiotrophic.

3. Impact of site factors

3.1 Nutrient availability and pH characterize rhizosphere communities

Nutrient-poor, acidic forest soils differentiated into organic and mineral soil horizons are common in temperate areas. Uroz et al. (2013) studied the effect of the variation in nutrient availability and pH between the horizons on the soil microbiome and its function. The high-throughput sequencing, functional assays, and functional metagenomics of two soil horizons revealed complex, differentiated microbial communities that appeared to specialize functionally. The organic soil horizon harbored Bacteria and Ascomycota, whereas in the mineral horizon, the proportion of Archaea increased and eukaryotes decreased. The microbiome of the organic horizon was rich in genes related to the degradation of soluble carbohydrates and polysaccharides, whereas the genes involved in access to amino acids were more abundant in the mineral horizon. In conclusion, the soil stratification and soil resource availability have clear effects on the functional diversity and modify the taxonomic diversity of the bacterial communities.

The results of an impressive survey of 1251 field plots over several years suggest that soil pH is the major driver of soil fungal diversity (Tedersoo et al., 2020). The effect of tree species (*Pinus sylvestris*, *Picea abies*, and *Populus × wettsteinii*) and ectomycorrhizal presence were higher on biotrophic than saprotrophic fungi. The fungi richness was highest in woodlands and nearby ruins of buildings. The diversity of soil fungi in forest island habitats was probably an edge effect, as forest fragmentation decreased fungal richness. The fungal diversity of pristine forests was higher than old nonpristine forests. Tree diversity and soil properties increased the overall fungal richness. The diversity of most fungal groups suffered from management, but especially for forests, the results depended on the fungal group and the time since partial harvesting.

In addition to the findings of the beneficial core microbiome described for the *Populus* species (Firrincieli et al., 2020; Timm et al., 2018), Colin et al. (2017) identified a core beech (*Fagus sylvatica*) rhizosphere microbiome. They also studied whether beech trees select bacteria in the rhizosphere according to the growth site properties such as pH, content of organic matter, carbon, nitrogen, limestone (calcium), and exchangeable nutritive cations. It was evident that the soil type influenced the bacteria colonizing the beech rhizosphere.

The comparison of plant phenotypic vs genotypic variation of microbiomes is poorly known.

Gallart et al. (2018) studied the microbiomes of the rhizosphere of two genotypes of *Pinus radiata*, which had different responses to organic or inorganic nitrogen. They analyzed the response of root microbial assemblages to the host genotype, the form of nitrogen, and the interaction of the tree genotype and the environment. The tree genotype modified the root microbiome. The form of nitrogen and tree genotype had strong effects on the abundance of taxa. The study highlighted the importance of genotype-by-environment interactions in the microbiome structure. The effect of the mycobiome in the efficiency of fertilizer use was observed in a study, which followed the performance of seedlings inoculated with conifer endophyte strains with symbiotic potential (Aghai et al., 2019).

Also, Karlinski et al. (2020) concluded that the tree genotype and soil environment determined the biomass of microorganisms and their contributions to the rhizosphere community. They examined the differences in biomass and community composition of the soil microbiome of four poplar genotypes grown in different soil conditions and depths. The soil environments were highly different: one site was located in a polluted area near a copper smelter. In general, the site affected the microbial biomass (excluding Actinobacteria), whereas the amount of fungal and bacterial groups in the microbiome and the abundance of arbuscular mycorrhizae in the fungal community depended on the tree genotype. Site factors and tree genotype had effects on the bacterial biomass and fungal biomass, respectively. Surprisingly, the role of tree genotype had a strong effect on the microbiome despite obvious site effects.

Interestingly, Haas et al. (2018) suggested that long-term anthropogenic nutrient input can have positive effects on belowground biodiversity, which would enhance ecosystem function. Ectomycorrhizal fungi (EMF) and bacterial communities increased in the rhizosphere of Norway spruce after long-term (25 years) nutrient optimization. Over time, the abundance of nitrophilic EMF and bacterial species as well as the richness and diversity of the rhizosphere microbiota increased.

3.2 Site-specific environment defines phyllosphere microbiome

Firrincieli et al. (2020) suggested that only a minor fraction of the phyllosphere bacterial microbiome diversity correlated with climate variables. The sampling site had a major effect on the microbiome of wild *Populus trichocarpa* plants from nutrient-poor environments: hot-dry (xeric) riparian zones, riparian zones with mid hot-dry, and moist (mesic) climates. The study also described the core bacterial microbiome in the phyllosphere of *Populus trichocarpa*.

Barge et al. (2019) used DNA metabarcoding to characterize the foliar mycobiomes of *Populus trichocarpa* across its geographic range with sharp climatic transition. The foliar microbiome varied among sites and between regions. Climate appeared as a stronger driver of community composition than geographic distance. Therefore, the environmental effects of sites are more important in shaping foliar mycobiomes than dispersal limitation.

Haas et al. (2018) observed no clear environmental effect when they analyzed the response of the microbial community in the needles of Norway spruce to the growing season and long-term (25 years) nutrient addition with the NGS of the bacterial 16S rRNA gene and the fungal ITS1 region. The phyllosphere diversity increased significantly over the growing season but was not influenced by the improved nutrient status of the trees.

4. The effects of pollution on a microbiome

4.1 Air pollution

Air pollution is released in the atmosphere from numerous natural (biogenic) and human-created (anthropogenic) sources (Fig. 4.1). So-called criteria pollutants include carbon monoxide, nitrogen and sulfur dioxide, ground-level ozone, lead

(Pb), and particulate matter. Directly phytotoxic pollutants include ozone, nitrogen and sulfur oxides, fluorides, and peroxyacetyl nitrate (Vallero, 2014). Acid rain is a general term related to the introduction of acidic substances via low pH precipitation to flora, soil, and surface waters. In practice, deposition is considered acidic when pH < 5.7, which is about the same as the natural pH of rainfall (Vallero, 2014). Volatile organic compounds (VOCs) are a wide array of gaseous substances that might have anthropogenic or biogenic origin (Vallero, 2014).

The effects of pollutants on the ecosystem were intensively studied in the 1980s and 1990s, and several studies revealed that plant-associated epiphytic fungi on the phyllosphere were sensitive to air pollutants. Also, endophytic fungi responded negatively to sulfuric acid and heavy metal deposition (Vesterlund and Saikkonen, 2011). In addition, the sensitivity of the pathogenic fungus *Gremmeniella abietina* to acidic rain and endophyte competition was demonstrated (Ranta and Neuvonen, 1994). It is noteworthy that most of the studies of acid precipitation at the time were performed with a culture-dependent methodology, and a full resolution of microbiomes might require NGS technologies. However, the information is still sparse (Table 4.2). Current knowledge of the impacts of ozone on the plant microbiota focuses on rhizospheric processes. Increases in ground-level ozone might reduce the allocation of C derived from the soil (Agathokleous et al., 2020).

4.2 Nitrogen and phosphorus deposition

Nitrogen and phosphorus are among the key resources for plants. However, nitrogen oxides (NO_x) are especially important air pollutants that are chronically deposited in ecosystems (Vallero, 2014). The models developed by Averill et al. (2018) suggest that nitrogen deposition may change interactions in the forest microbiome, followed by a decrease of the capacity of forests to sequester carbon. Nitrogen deposition favors arbuscular mycorrhizal trees more than ectomycorrhizal trees, and thus decreases soil carbon stocks (Averill et al., 2018). Experimentally, Chen et al. (2019) found evidence that the root-associated microbiomes of *Broussonetia papyrifera* were differentiated due to different climate types, with the only exception being the bacterial assemblage of the rhizosphere. Especially, the concentration of soil phosphate and nitrate had a key role in the studied environmental factors. The root-associated microbiome contained Proteobacteria, especially *Pseudomonas*, *Rhizobium*, and Basidiomycota.

4.3 Particulate matter

Currently, an increasing amount of anthropogenic nanoparticles from various activities is released into the environment (Table 4.2). The excess amount of ultrafine particulate matter (UFP) and black carbon on leaves had effect on the microbial communities (Espenshade et al., 2019). The study compared the effects of seasons, sites, and air pollutants on the leaf bacterial communities of *Platanus × hispanica* trees in a city center and a nearby forest area. Sequencing of the 16S ribosomal genes of the microbiome revealed large annual variation. Trees growing in the city center had different bacterial community composition compared to the natural site. In addition, human-associated bacteria were associated with the leaves in the city center. Vitali et al. (2019) introduced silver nanoparticles at the leaf and root level of poplar plants to study the effects on plant microbiota. The application of particles on leaves increased the evenness by reducing bacteria; it also decreased the bacterial and fungal biodiversity in the roots. The treatment caused the bacterial community to shift from aerobic to facultative anaerobic and oxidative stress-tolerant bacteria.

4.4 Soil pollution

Major types of soil pollutants are heavy metals and their salts as well as other inorganic pollutants such as aluminum, beryllium, fluorine, and radionuclides (Mirsal, 2008). PAHs are toxic compounds mainly produced in various combustion processes as well as volcanic activity and natural fires. The major PAH source is anthropogenic-currently, even the lowest concentrations of PAHs in soil are frequently 10 times higher than natural. PAHs are mainly associated with soil organic matter and soot-like C. The PAHs are lowly soluble to water and thus PAH uptake by plants is low. Most PAHs detected in plant tissue are from atmospheric deposition (Wilcke, 2000).

The approach of using trees for phytoremediation has emphasized the role of microbiomes in the tolerance of plants to toxic compounds (Domka et al., 2019). Indeed, trees in urban parks provide important ecosystem services as the soil has the capacity to store nutrients and metals (Setala et al., 2017). Microbiomes of trees have been studied more extensively in natural ecosystems but less in urban and built environments.

Roslund et al. (2018) found that PAHs affected the microbiome structure when they studied the half-lives of PAHs during 12 weeks of a degradation experiment of landscaping materials. Bacterial assemblages were subject to Illumina MiSeq 16S rRNA gene metabarcoding. If the landscaping materials contained 1%–2% organic matter, the half-lives of PAHs with

TABLE 4.2 Impacts of pollution on tree microbiomes.

Pollutant	Host	Affected part	Effect	References
Sulfuric acid deposition	Scots pine *Pinus sylvestris*	Needles	Decrease of endophytic fungi and *Gremmeniella abietina* infections	Vesterlund and Saikkonen (2011) Ranta and Neuvonen (1994)
Nitrogen deposition	Model, ecosystem level	Rhizosphere mycorrhiza	Decrease of soil carbon stocks due to microbiome shifts	Averill et al. (2018)
Nitrate and phosphate	Paper mulberry *Broussonetia papyrifera*	Rhizosphere	Differentiation to climate type	Chen et al. (2019)
Particulate matter	London plane *Platanus × hispanica*	Phyllosphere	Differentiation of bacterial communities in city center vs natural sites	Espenshade et al. (2019)
Black carbon	London plane	Phyllosphere	Differentiation of bacterial communities in city center vs natural sites	Espenshade et al. (2019)
Silver nanoparticles	Black poplar *Populus nigra*	Leaves Roots	Reducing bacteria Reducing bacterial and fungal biodiversity	Vitali et al. (2019)
Combined pollutants in urban environment	Burr oak *Quercus macrocarpa*	Phyllosphere	Differences in fungal richness, diversity, and community composition	Jumpponen and Jones (2010)
	Acer platanoides, Acer rubrum, Acer saccharum, Celtis occidentalis, Fraxinus americana, Fraxinus pennsylvanica, Picea glauca	Phyllosphere	Urban intensity increased the diversity of phyllosphere microbial communities and number of taxa	Laforest-Lapointe et al. (2017)
	Alnus rubra	Phyllosphere	Differences in endophyte communities	Wolfe et al. (2018)
	Various evergreen and deciduous	Rhizosphere ECM	Host tree effect in parks exceeds site effects	Hui et al. (2017)
	Herbaceous plants	Rhizosphere	Roadside effect (Cd) decreased AM species richness and evenness	Lin et al. (2020)
	Caprinus betulus	Phyllosphere	The structure and function of microbiomes was different in the forests and city center	Imperato et al. (2019)

low molecular weights (phenanthrene, fluoranthene, and pyrene) were 1.5–4.4 weeks. Increasing the proportion of organic matter to 13% resulted in a similar half-life of 2.5 weeks, but a high content of 56% dramatically extended the half-lives up to 52 weeks. A low content of organic matter shortened the half-lives of phenanthrene and fluoranthene. The presence of Beta-, Delta-, and Gammaproteobacteria as well as the diversity of Bacteroidetes and Betaprotebacteria inversely affected the half-life of pyrene. The high molecular weight compound benzo(b)fluoranthene was not degradable and an abundance of Betaproteobacteria decreased the half-life of chrysene. The organic content and bacterial community of landscaping materials affected the breakdown of PAHs.

4.5 Combined effects of pollutants in urban areas

4.5.1 Phyllosphere microbiome in urban areas

Jumpponen and Jones (2010) observed differences between urban and nonurban stands of *Quercus macrocarpa* in fungal richness, diversity, and community composition over a growing season. The authors proposed that these differences are due to stand management, size, and isolation as well as the accumulation of nutrients and pollutants. The temporal dynamics of diverse mycobiomes were probably due to fungal life cycles or environmental tolerances. The concentrations of macronutrients (N, K, S), micronutrients (B, Mn, Se, Zn), and trace elements (Cd, Pb) were higher in the urban stands.

Laforest-Lapointe et al. (2017) examined the bacterial communities of phyllospheres of seven tree species in urban areas and concluded that anthropogenic activity and plant microbiomes are drivers of urban microbiomes. They also followed the changes in structure and diversity of bacterial communities over a gradient of urban intensity and tree isolation. At higher urban intensity, the diversity of phyllosphere microbial communities increased while a higher number of taxa were also found. Wolfe et al. (2018) isolated the foliar microbiomes of *Alnus rubra* from the metropolitan area. The results suggested the combined effects of local air pollution sources and other site characteristics on fungal endophyte community composition.

4.5.2 Urban areas share similarities with forest rhizospheres

Hui et al. (2017) showed that the richness of ECM fungi was only slightly higher and highly similar in forests compared to city parks, which hosted rich and diverse ECM fungal communities. The results imply that the presence of host trees shapes the structure and diversity of the ECM fungal community more than the characteristics of the soil or site disturbances.

Francini et al. (2018) studied the response of soil microbes to different plant assemblages (lawns only and lawns with deciduous or evergreen trees) and park age. Old deciduous trees were accompanied by the highest fungal abundances and fungal-feeding nematodes in the soil. The biomass of microbes in urban parks increased over time, but fungal-feeding nematodes decreased. Minimally managed natural forests had a threefold microbial biomass compared to the oldest parks. Urban parks are different from natural forests regarding community composition and biomass, although parks harbor diverse soil microbial diversity and biomass.

Lin et al. (2020) explored the arbuscular mycorrhizal (AM) fungal communities in urban ecosystems using 454 pyrosequencing. Urbanization had no effect on diversity or the community composition of the AM fungi. Within urban areas, the roadside effect explained by the soil cadmium content decreased AM species richness and evenness. The richness of herbaceous plants increased the richness of AM fungi.

4.6   Bidirectional effects of plant or microbiota VOCs

Biogenic VOCs have antimicrobial effects and are potential carbon sources; therefore, emissions potentially shape the microbiomes on plant surfaces. Epiphytic microbiomes may produce their own VOCs, which may modify the function of plant VOCs.

Microbes affect the production and emission of VOCs via effects in plant physiology. Microbes also metabolize the VOCs emitted by the plant (Farré-Armengol et al., 2016).

Imperato et al. (2019) described bacterial assemblages of *Carpinus betulus* phyllosphere in three locations: Warsaw (Poland), a protected forest (Białowieża), and a forest in an operational oil field (Bóbrka). *Carpinus betulus* leaves in the city center had increased concentrations of particulate matter (PM), with more palladium and radon than in the forests. Air VOCs of the oil field had more butanone methyl propanal, butylbenzene, and cyclohexane than in the city and protected forest. In the city, more xylene and toluene were measured. Sequencing revealed a high abundance of Gammaproteobacteria (71%), mainly *Pseudomonas* spp., Actinobacteria, Alpha- and Beta-proteobacteria, and Firmicutes. The structure and function of the microbiomes were different in the forests and city center. Statistically, more genes of hydrocarbon degradation were detected in the protected forest in comparison to the city and oil field. The protected forest had the most beneficial bacteria with the ability to degrade diesel.

5.   Global warming and elevated CO_2

The effects of predicted climatic change are increasing the atmospheric CO_2 and temperature; episodic drought caused by heat waves might be more common, although overall precipitation may increase in the Boreal zone. In general, forests are climate-sensitive ecosystems. While increasing temperature has increased global plant growth (Nemani et al., 2003), the effects of climate extremes on forest productivity and health are difficult to forecast (Itter et al., 2017). The climate extremes will result in a combination of impacts such as heat waves causing drought and increased soil temperature. Changes in climate will affect the microbiomes in forest ecosystems, but the types of microorganisms and effects are highly variable (Jansson and Hofmockel, 2020).

5.1.   Rhizosphere

Frey et al. (2008) studied the effect of chronic soil warming on the microbiome, its biomass, and function. Twelve years of soil warming at a constant 5°C above ambient resulted in a significant reduction in microbial biomass and the utilization of carbon. Heating significantly reduced the abundance of fungi. Melillo et al. (2017) analyzed the same experiment and reported the effects of long-term warming by heating the soil by 5°C for 26 years. A four-phase pattern was observed. First, the increasing microbiome respiration released more carbon rapidly. Second, the labile C pool was depleted, indicating a reduced microbial biomass. Next, the microbiome was changed to a more diverse, oligotrophic microbial assemblage with an increase in soil respiration toward the final phase of reducing the microbial biomass. Overall, the results suggest a further release of carbon from temperate forests.

Based on the modeling of 10 individual studies by Veresoglou et al. (2016), the accumulation of atmospheric CO_2 would increase fungal diversity. The experiment duration and natural processes could explain this; some species either fail to adapt or are outcompeted due to short-term changes. In the end, new species might colonize the released resources or evolve to new conditions.

However, the microbiomes also face extreme droughts and flooding effects, which are described earlier in this book chapter. Therefore, the outcome of increasing CO_2 on microbiomes globally is difficult to conclude due to indirect, local effects.

Forest fires are a consequence of extended drought. Sun et al. (2016a) studied the effect of fire on the soil microbiome and found that soil temperature, pH, and water contents were more important factors than fire on bacterial assemblages. The results revealed the similarity of the bacterial diversity between the burned areas of different ages, which indicates that the bacterial diversity recovers rapidly after fire. The most common taxa were Proteobacteria (39%), Acidobacteria (34%), and Actinobacteria (17%). Genes related to C and N cycling were present in all sites. It should be noted that sites with different fire histories formed separate clusters in gene signal intensity analysis, which suggests potential differences in biogeochemical soil processes.

Bond-Lamberty et al. (2018) recognized the evidence of increasing global heterotrophic respiration due to microbial mineralization, which probably is caused by environmental changes. Numerous metaanalyses and surveys support these observations. Climate change is therefore causing an observable and consistent global loss of soil carbon.

Global warming will cause shifts in the distribution of any organism, which are currently limited by temperature regimes. These include plant species distribution, which has direct and indirect effects on plant-microbe interactions. However, the causes and modes of these interactions are poorly understood. Plants will spread to higher latitudes with different light environments. This may also indirectly influence the rhizosphere and its mycobiome. Saravesi et al. (2019) studied the effects of temperature and light on the root mycobiome of *Pinus sylvestris*. Tree seedlings from different geographical origins were experimentally grown under similar temperatures in southern (60°N) and northern (69°N) Finland. The Basidiomycota, especially ECM fungi, were more abundant in the roots of native pines than pines from different origins. Population origin had an effect on the biomass increase of seedlings. The results indicate that the root mycobiome may be modified to respond differently by native vs nonnative light environments.

5.2   Phyllosphere

Khan et al. (2016) inoculated a *Populus deltoides × Populus nigra* clone OP-367 with a consortium of endophytes. Poplar plants inoculated with endophytes tolerated drought stress and their leaf physiology was improved. In addition, reduced damage by ROS was observed. The endophytes produced important phytohormones and expressed putative microbial genes, which indicate a possible mechanism for drought tolerance provoked by symbiosis.

The responses of the microbiomes of particular tree species to variable climatic conditions have been used in assessing the effects of environmental change. Wallace et al. (2018) studied the microbiomes of the leaves and roots of seedlings of the sugar maple (*Acer saccharum*) at the species' elevational range limit in Quebec. The microbiomes of tree seedlings were different at different elevations. Characteristic microbiomes were found in plant compartments, although the endophytes were found to be a subset of the epiphytes. It can be concluded that microbiomes of the sugar maple are different at the edge of the species' elevational limit from the natural range. Also, Lazarević and Menkis (2020) studied the needle mycobiome of *Pinus heldreichii* and found that the fungal communities of trees growing in harsh conditions were different from three sites at lower altitudes (milder growing conditions), suggesting that environmental conditions were among the major determinants of fungal communities associated with the needles of *Pinus heldreichii*. Needle pathogens appeared to be common in fungal communities in the phyllosphere of *Pinus heldreichii*, especially in those trees under strong abiotic and insect herbivory stress.

6. Effects of genetic modification of trees

The genetic modification of trees that started in the 1980s has resulted in a large amount of field-and laboratory-based performance and safety data. Walter et al. (2010) found no evidence of any substantive harm to biodiversity, human health, or the environment. It should be noted that information on the effect of the genetic transfer on the microbiome of woody plants is sparse.

Hur et al. (2011) compared the bacterial and archaeal communities of the rhizosphere of genetically modified and wild-type poplars, which were grown on contaminated soils over a period of 3 years. DNA pyrosequencing revealed that the poplar type and growth stages were causing shifts in the microbial community. The microbial communities were statistically identical at the tree age of 3 years, but the rate of change toward the final community structure was faster in poplars, which were genetically modified for phytoremediation.

Wang et al. (2019) analyzed the effects of environmental conditions, ecological niches, and plant stress resistance-related transcription factors on the microbiome composition of transgenic *Populus alba × Populus berolinensis*. The results suggested that genetic transformation did not affect the endophytic microbiome diversity. Actinobacteria, Proteobacteria, Bacteroidetes, and Firmicutes were the most common bacteria, and Dothideomycetes, Agaricomycetes, Leotiomycetes, and Sordariomycetes the most abundant fungi. The microbiome structure depended on the pH and the soil organic matter content. Each type of plant tissue as an ecological niche harbored unique microbial communities. The study highlighted the importance of the environment on the microbiome.

Beckers et al. (2016) reported contrasting results of a study on the effect of Cinnamoyl-CoA reductase (CCR) on the rhizosphere (roots) and endosphere (shoots and leaves) microbiomes. CCR is a key enzyme in lignin synthesis and therefore a potential target of genetic modification to improve the quality of lignocellulose for industrial purposes. However, silencing the CCR gene affects the general and monolignol-specific lignin pathways, and phenolic compounds accumulate in the xylem. The metabolic behavior and structures of bacterial communities in the endosphere responded highly to the poplar genotype. Contrastingly, the rhizosphere microbiomes of CCR-deficient and WT poplars were highly similar. Also, evidence that the plant microbiome might modulate the function of the plant genome was obtained, as the compounds that accumulate due to changes in the CCR gene expression are apparently further metabolized by the endophytic community. These new metabolites may interfere with the intended plant phenotype.

7. The effects of forest management on a tree microbiome

The majority of the current temperate and boreal forests are nutrient-poor and grow on rocky soils due to the historical development of agriculture on fertile soil types. Therefore, forest ecosystems strongly differ from other terrestrial environments (Uroz et al., 2016). Studies on the effect of forest management on the forest microbiome are few. The soil fungi provide fungal inocula for the rhizosphere, and therefore information on the surrounding soils is important. Castano et al. (2018) found no significant effect of thinning of a *Pinus pinaster* forest on the soil fungal community. They observed annual variation in the mycobiome driven by precipitation and temperature. The result indicates that the soil fungal communities of Mediterranean forests are resistant to forest-thinning operations.

A comparison of abiotic (drought) and anthropogenic disturbance (shrub removal) was provided by Gehring et al. (2014), who found evidence of complex, genetic-based interactions among species. They studied the influence of plant genetic variation, drought, shrub removal, and insect herbivory on the ectomycorrhizal (EM) fungal community of *Pinus edulis* in a 16-year field survey.

The EM of insect-susceptible trees was not affected by drought. On the other hand, evidence of an interaction of the genotype and environment was provided by the observation that the EM community of insect-resistant trees shifted significantly. The management activity of shrub removal was beneficial for insect-susceptible trees and altered the EM assemblage of insect-resistant trees. In addition, the shoot growth of insect-susceptible trees responded with a sevenfold effect compared to insect-resistant trees. The shift in the EM community of insect-resistant trees triggered by shrub removal was associated with greater shoot growth, which might be the result of competitive release. Interestingly, insect-susceptible trees grew faster than resistant trees, suggesting that the EM fungi associated with susceptible trees were overcompensating for the effects of pests.

7.1 Case studies of the effect of forest management on microbiomes

7.1.1 Case: Effects of drainage on microbiomes of boreal peatland

Boreal peatlands and associated microbiota are important carbon reservoirs, and their utilization plays a crucial role in the global carbon cycle. The peatlands have been subject to strong management, such as the drainage of the majority of the fertile spruce-dominated pristine mires in South Finland to increase timber production. The effects of natural differences or management such as drainage on peatland microbiomes are not well known.

Sun et al. (2016a) examined the diversity and structure of mycobiomes in the organic soil layer in boreal forest soils. They also compared the primary colonization of buried wood with the surrounding fungal community. High-throughput sequencing of the internal transcribed spacer (ITS) region revealed that the soil types and tree species had an impact on the composition of the fungal community. The environmental variables such as the availability of nutrients (Ca, Fe, and P) within the site determined the fungal community composition.

Terhonen et al. (2014) investigated the effect of drainage on the species composition of *Picea abies* root endophytes by comparing the drained mire, pristine mire, and mineral soil. A total of 113 isolates resulting in 15 different OTUs of fungal root endophytes were isolated from nonmycorrhizal roots. Dark septate endophytes (77%) were the most common, with the *Phialocephala fortinii* s.l.-*Acephala applanata* species covering 52% of all the isolates. Interestingly, 19 (17%) of the fungal isolates inhibited the growth of *Heterobasidion annosum*, a root pathogen of conifers, in vitro. The mineral soil sites had the lowest endophyte diversity whereas a man-made ecosystem, the drained peatland, had the highest microbial diversity.

Sun et al. (2014a,b) used the 16S rRNA sequencing approach to analyze changes in the bacterial community and during the early colonization of wood. The forest soil types had different communities of the primary wood-inhabiting bacteria and the bacteria community remained unchanged over a prolonged incubation time. The dominant phyla were Proteobacteria, followed by Bacteroidetes, Acidobacteria, Actinobacteria, Amatimonadetes, Planctomycetes, and the TM7 group. The results suggest that variations in soil bacterial community composition are driving the wood-inhabiting bacterial structure.

The distribution of bacterial communities in different peat soils and the effect of forest management practices are still poorly known. Sun et al. (2014b) explored the bacterial diversity and community structure of eight types of peat soils (pristine and drained) and two mineral forest soil types. The main tree species was either spruce or pine. The most abundant microbes were Proteobacteria, Acidobacteria, Actinobacteria, Bacteroidetes, Planctomycetes, and Verrucomicrobia. The soil types and vegetation determined the relative abundance of the bacterial phyla and genera. The only significant differences were found for Gemmatimonadetes and Cyanobacteria between the pristine and the drained soils. The relative abundance of genera including *Burkholderia, Caulobacter, Opitutus, Mucilaginibacter, Acidocella, Mycobacterium, Bradyrhizobium, Dyella*, and *Rhodanobacter* changed significantly between the peat soils with the same or different tree species. The peat forest soils harbored high bacterial diversity and richness. The tree species had a stronger effect on the bacterial diversity than the type of peat soil, which determined the community structure.

7.1.2   Case: Effects of stump treatment for the control of forest disease on the wood microbiome

The biocontrol agent *Phlebiopsis gigantea* has been used as a control on the surface of *Picea abies* against *Heterobasidion* root rot for three decades. This pathogen is the major cause of root rot disease in conifers, resulting in severe economic losses (Asiegbu et al., 2005). The goal of the treatment is to prevent the primary infection of the pathogen on the freshly cut stumps and subsequent secondary infections to the root system and disease transfer to the next tree generation. Although the performance of *Phlebiopsis gigantea* in biological control is empirically shown, the impacts on the microbiomes of the stumps have been poorly understood. The stumps of harvesting operations appear to constitute a low-quality substrate for maintaining fungal biodiversity (Vasaitis et al., 2016), but they are an increasing wood resource and therefore important for the microbial diversity of managed forests.

To obtain information on the long-term effects of the treatment, high-throughput sequencing was used to characterize the diversity of bacteria (Sun et al., 2013) and fungi (Terhonen et al., 2013) inhabiting the stumps of *Picea abies*. The 1-, 6-, and 13-year-old stumps were previously treated with *Phlebiopsis gigantea* directly after tree felling. Also, nontreated stumps in the areas were sampled for control.

Biological control had a significant decreasing effect on the bacterial richness, but the bacterial communities of stumps gradually recovered (Sun et al., 2013). One year after treatment, a significant increase of the phylum Acidobacteria was observed. During the early stages of decomposition, Proteobacteria were the most abundant, whereas Acidobacteria increased and were the most common in older stumps.

Regarding fungal diversity, more OTUs were detected in the control stumps, although there was no statistical difference between treated and nontreated stumps (Terhonen et al., 2013). The biocontrol fungus *Phlebiopsis gigantea* was found only from stumps 1 year posttreatment. The target of control, pathogenic *Heterobasidion annosum* s.l., was not detected. *Phlebiopsis gigantea* has been found in other studies at 6 years after application, but not detected later (Vainio et al., 2001; Vasiliauskas et al., 2004). Vasiliauskas et al. (2005) reported that the fungal richness was lower in *Picea abies* stumps that had been treated with *Phlebiopsis gigantea*.

The results implied that the biological control of *Heterobasidion* sp. by stump treatment caused no clear negative effects on the stump microbiome. It can be considered that the long-term environmental impacts of the use of *Phlebiopsis gigantea* in the control of *Heterobasidion* root rot are negligible.

8. Concluding remarks

Abiotic stress, which is increasingly connected to consequences of global change, will affect the microbial partners of plant holobionts. Most of the information, however, is gathered from perennial plants and agricultural crops, not forest trees. Increasing precipitation, especially events of flooding, will change microbiomes toward anaerobic and copiotrophic behavior. Increased drought periods will decrease the total biomass of the rhizosphere microbiome and especially the mycorrhizae. On the other hand, the infection success of pathogenic fungi will increase. The effects of forest management are not well known, but it appears that intensive operations such as drainage have direct effects on microbial flora, but moderate-scale operations such as thinning and biological control of disease produce only minor shifts in microbiomes.

In conclusion, the environmental change will affect the composition and activity of microbiomes. Excess nitrogen deposition might modify microbiomes toward more carbon-releasing assemblages. Accumulated evidence suggests that global warming and elevated CO_2 cause shifts in the microbial activity of soils, and it is likely that changes will also occur in the microbiome structure. Although the total microbial activity decreases, a net increase of carbon release has been observed. Most of the literature available focuses on total microbial activity, bacteria, or fungi separately or specific functional groups such as mycorrhiza. Therefore, full microbiome scale approaches would shed more light on the effect of abiotic factors on tree holobionts.

References

Agathokleous, E., Feng, Z., Oksanen, E., Sicard, P., Wang, Q., Saitanis, C.J., Araminiene, V., Blande, J.D., Hayes, F., Calatayud, V., 2020. Ozone affects plant, insect, and soil microbial communities: a threat to terrestrial ecosystems and biodiversity. Sci. Adv. 6, 1–17.

Aghai, M.M., Khan, Z., Joseph, M.R., Stoda, A.M., Sher, A.W., Ettl, G.J., Doty, S.L., 2019. The effect of microbial endophyte consortia on *Pseudotsuga menziesii* and *Thuja plicata* survival, growth, and physiology across edaphic gradients. Front. Microbiol. 10, 1353.

Asiegbu, F.O., Adomas, A., Stenlid, J., 2005. Conifer root and butt rot caused by Heterobasidion annosum (Fr.) Bref. sl. Mol. Plant Pathol. 6, 395–409.

Averill, C., Dietze, M.C., Bhatnagar, J.M., 2018. Continental-scale nitrogen pollution is shifting forest mycorrhizal associations and soil carbon stocks. Glob. Chang. Biol. 24, 4544–4553.

Barge, E.G., Leopold, D.R., Peay, K.G., Newcombe, G., Busby, P.E., 2019. Differentiating spatial from environmental effects on foliar fungal communities of *Populus trichocarpa*. J. Biogeogr. 46, 2001–2011.

Beckers, B., De Beeck, M.O., Weyens, N., Van Acker, R., Van Montagu, M., Boerjan, W., Vangronsveld, J., 2016. Lignin engineering in field-grown poplar trees affects the endosphere bacterial microbiome. Proc. Natl. Acad. Sci. USA 113, 2312–2317.

Bond-Lamberty, B., Bailey, V.L., Chen, M., Gough, C.M., Vargas, R., 2018. Globally rising soil heterotrophic respiration over recent decades. Nature 560, 80–83.

Brasier, C.M., Sanchez-Hernandez, E., Kirk, S.A., 2003. Phytophthora inundata sp. nov., a part heterothallic pathogen of trees and shrubs in wet or flooded soils. Mycol. Res. 107, 477–484.

Carrell, A.A., Frank, A.C., 2015. Bacterial endophyte communities in the foliage of coast redwood and giant sequoia. Front. Microbiol. 6, 1008.

Castano, C., Alday, J.G., Lindahl, B.D., Martinez de Aragon, J., de Miguel, S., Colinas, C., Parlade, J., Pera, J., Antonio Bonet, J., 2018. Lack of thinning effects over inter-annual changes in soil fungal community and diversity in a Mediterranean pine forest. For. Ecol. Manag. 424, 420–427.

Chen, P., Zhao, M., Tang, F., Hu, Y., Peng, X., Shen, S., 2019. The effect of environment on the microbiome associated with the roots of a native woody plant under different climate types in China. Appl. Microbiol. Biotechnol. 103, 3899–3913.

Colin, Y., Nicolitch, O., Van Nostrand, J.D., Zhou, J.Z., Turpault, M.-P., Uroz, S., 2017. Taxonomic and functional shifts in the beech rhizosphere microbiome across a natural soil toposequence. Sci. Rep. 7, 9604.

Cui, H.-L., Duan, G.-L., Zhang, H., Cheng, W., Zhu, Y.-G., 2019. Microbiota in non-flooded and flooded rice culms. FEMS Microbiol. Ecol. 95, fiz036.

de Vries, F.T., Griffiths, R.I., Knight, C.G., Nicolitch, O., Williams, A., 2020. Harnessing rhizosphere microbiomes for drought-resilient crop production. Science 368, 270–274.

Domka, A.M., Rozpadek, P., Turnau, K., 2019. Are fungal endophytes merely mycorrhizal copycats? The role of fungal endophytes in the adaptation of plants to metal toxicity. Front. Microbiol. 10, 371.

Espenshade, J., Thijs, S., Gawronski, S., Boye, H., Weyens, N., Vangronsveld, J., 2019. Influence of urbanization on epiphytic bacterial communities of the Platanus x hispanica tree leaves in a biennial study. Front. Microbiol. 10, 675.

Farré-Armengol, G., Filella, I., Llusia, J., Peñuelas, J., 2016. Bidirectional interaction between phyllospheric microbiotas and plant volatile emissions. Trends Plant Sci. 21, 854–860.

Firrincieli, A., Khorasani, M., Frank, A.C., Doty, S.L., 2020. Influences of climate on phyllosphere endophytic bacterial communities of wild poplar. Front. Plant Sci. 11, 203.

Francini, G., Hui, N., Jumpponen, A., Kotze, D.J., Romantschuk, M., Allen, J.A., Setala, H., 2018. Soil biota in boreal urban greenspace: responses to plant type and age. Soil Biol. Biochem. 118, 145–155.

Frey, S.D., Drijber, R., Smith, H., Melillo, J., 2008. Microbial biomass, functional capacity, and community structure after 12 years of soil warming. Soil Biol. Biochem. 40, 2904–2907.

Gallart, M., Adair, K.L., Love, J., Meason, D.F., Clinton, P.W., Xue, J., Turnbull, M.H., 2018. Host genotype and nitrogen form shape the root microbiome of *Pinus radiata*. Microb. Ecol. 75, 419–433.

Gehring, C., Flores-Renteria, D., Sthultz, C.M., Leonard, T.M., Flores-Renteria, L., Whipple, A.V., Whitham, T.G., 2014. Plant genetics and interspecific competitive interactions determine ectomycorrhizal fungal community responses to climate change. Mol. Ecol. 23, 1379–1391.

Graff, A., Conrad, R., 2005. Impact of flooding on soil bacterial communities associated with poplar (*Populus* sp.) trees. FEMS Microbiol. Ecol. 53, 401–415.

Haas, J.C., Street, N.R., Sjodin, A., Lee, N.M., Hogberg, M.N., Nasholm, T., Hurry, V., 2018. Microbial community response to growing season and plant nutrient optimisation in a boreal Norway spruce forest. Soil Biol. Biochem. 125, 197–209.

Hartman, K., Tringe, S.G., 2019. Interactions between plants and soil shaping the root microbiome under abiotic stress. Biochem. J. 476, 2705–2724.

Hartmann, M., Brunner, I., Hagedorn, F., Bardgett, R.D., Stierli, B., Herzog, C., Chen, X., Zingg, A., Graf-Pannatier, E., Rigling, A., Frey, B., 2017. A decade of irrigation transforms the soil microbiome of a semi-arid pine forest. Mol. Ecol. 26, 1190–1206.

Herrmann, M., Geesink, P., Richter, R., Kuesel, K., 2020. Canopy position has a stronger effect than tree species identity on phyllosphere bacterial diversity in a floodplain hardwood forest. Microb. Ecol. 81, 157–168.

Hossain, M., Veneklaas, E.J., Hardy, G.E.S.J., Poot, P., 2019. Tree host-pathogen interactions as influenced by drought timing: linking physiological performance, biochemical defence and disease severity. Tree Physiol. 39, 6–18.

Hui, N., Liu, X., Kotze, D.J., Jumpponen, A., Francini, G., Setälä, H., 2017. Ectomycorrhizal fungal communities in urban parks are similar to those in natural forests but shaped by vegetation and park age. Appl. Environ. Microbiol. 83, 1281–1295.

Hur, M., Kim, Y., Song, H.-R., Kim, J.M., Choi, Y.I., Yi, H., 2011. Effect of genetically modified poplars on soil microbial communities during the phytoremediation of waste mine tailings. Appl. Environ. Microbiol. 77, 7611–7619.

Imperato, V., Kowalkowski, L., Portillo-Estrada, M., Gawronski, S.W., Vangronsveld, J., Thijs, S., 2019. Characterisation of the *Carpinus betulus* L. phyllomicrobiome in urban and forest areas. Front. Microbiol. 10, 1110.

Itter, M.S., Finley, A.O., D'Amato, A.W., Foster, J.R., Bradford, J.B., 2017. Variable effects of climate on forest growth in relation to climate extremes, disturbance, and forest dynamics. Ecol. Appl. 27, 1082–1095.

Jansson, J.K., Hofmockel, K.S., 2020. Soil microbiomes and climate change. Nat. Rev. Microbiol. 18, 35–46.

Jumpponen, A., Jones, K.L., 2010. Seasonally dynamic fungal communities in the *Quercus macrocarpa* phyllosphere differ between urban and nonurban environments. New Phytol. 186, 496–513.

Karlinski, L., Ravnskov, S., Rudawska, M., 2020. Soil microbial biomass and community composition relates to poplar genotypes and environmental conditions. Forests 11, 262.

Khan, Z., Rho, H., Firrincieli, A., Hung, S.H., Luna, V., Masciarelli, O., Kim, S.-H., Doty, S.L., 2016. Growth enhancement and drought tolerance of hybrid poplar upon inoculation with endophyte consortia. Curr. Plant Biol. 6, 38–47.

Kovalchuk, A., Mukrimin, M., Zeng, Z., Raffaello, T., Liu, M., Kasanen, R., Sun, H., Asiegbu, F.O., 2018. Mycobiome analysis of asymptomatic and symptomatic Norway spruce trees naturally infected by the conifer pathogens Heterobasidion spp. Environ. Microbiol. Rep. 10, 532–541.

Kozlowski, T.T., 1997. Responses of woody plants to flooding and salinity. Tree Physiol. 17, 490.

Laforest-Lapointe, I., Messier, C., Kembel, S.W., 2017. Tree leaf bacterial community structure and diversity differ along a gradient of urban intensity. Msystems 2. UNSP e00087-17.

Lazarević, J., Menkis, A., 2020. Fungal diversity in the phyllosphere of *Pinus heldreichii* H. Christ—an endemic and high-altitude pine of the mediterranean region. Diversity 12, 172.

Lin, L., Chen, Y., Qu, L., Zhang, Y., Ma, K., 2020. Cd heavy metal and plants, rather than soil nutrient conditions, affect soil arbuscular mycorrhizal fungal diversity in green spaces during urbanization. Sci. Total Environ. 726, 138594.

Linaldeddu, B.T., Sirca, C., Spano, D., Franceschini, A., 2011. Variation of endophytic cork oak-associated fungal communities in relation to plant health and water stress. For. Pathol. 41, 193–201.

Linnakoski, R., Sugano, J., Junttila, S., Pulkkinen, P., Asiegbu, F.O., Forbes, K.M., 2017. Effects of water availability on a forestry pathosystem: fungal strain-specific variation in disease severity. Sci. Rep. 7, 13501.

Matthews, B., Netherer, S., Katzensteiner, K., Pennerstorfer, J., Blackwell, E., Henschke, P., Hietz, P., Rosner, S., Jansson, P.-E., Schume, H., Schopf, A., 2018. Transpiration deficits increase host susceptibility to bark beetle attack: experimental observations and practical outcomes for Ips typographus hazard assessment. Agric. For. Meteorol. 263, 69–89.

Melillo, J.M., Frey, S.D., DeAngelis, K.M., Werner, W.J., Bernard, M.J., Bowles, F.P., Pold, G., Knorr, M.A., Grandy, A.S., 2017. Long-term pattern and magnitude of soil carbon feedback to the climate system in a warming world. Science 358, 101–105.

Mendes, R., Garbeva, P., Raaijmakers, J.M., 2013. The rhizosphere microbiome: significance of plant beneficial, plant pathogenic, and human pathogenic microorganisms. FEMS Microbiol.Rev. 37 (5), 634–663.

Mirsal, I.A., 2008. Soil Pollution. Springer, Berlin.

Naylor, D., Coleman-Derr, D., 2018. Drought stress and root-associated bacterial communities. Front. Plant Sci. 8, 2223.

Nemani, R.R., Keeling, C.D., Hashimoto, H., Jolly, W.M., Piper, S.C., Tucker, C.J., Myneni, R.B., Running, S.W., 2003. Climate-driven increases in global terrestrial net primary production from 1982 to 1999. Science 300, 1560–1563.

Netherer, S., Matthews, B., Katzensteiner, K., Blackwell, E., Henschke, P., Hietz, P., Pennerstorfer, J., Rosner, S., Kikuta, S., Schume, H., Schopf, A., 2015. Do water-limiting conditions predispose Norway spruce to bark beetle attack? New Phytol. 205, 1128–1141.

Paljakka, T., Rissanen, K., Vanhatalo, A., Salmon, Y., Jyske, T., Prisle, N.L., Linnakoski, R., Lin, J.J., Laakso, T., Kasanen, R., 2020. Is decreased xylem sap surface tension associated with embolism and loss of xylem hydraulic conductivity in pathogen-infected Norway spruce saplings? Front. Plant Sci. 11, 1–12.

Pieterse, C.M.J., Zamioudis, C., Berendsen, R.L., Weller, D.M., Van Wees, S.C.M., Bakker, P.A.H.M., 2014. Induced systemic resistance by beneficial microbes. Annu. Rev. Phytopathol. 52 (52), 347–375.

Preece, C., Farre-Armengol, G., Penuelas, J., 2020. Drought is a stronger driver of soil respiration and microbial communities than nitrogen or phosphorus addition in two Mediterranean tree species. Sci. Total Environ. 735, 139554.

Ranta, H., Neuvonen, S., 1994. The host-pathogen system of *Gremmeniella abietina* (Lagerb) Morelet and Scots pine—effects of nonpathogenic phyllosphere fungi, acid-rain and environmental-factors. New Phytol. 128, 63–69.

Roslund, M.I., Grönroos, M., Rantalainen, A.-L., Jumpponen, A., Romantschuk, M., Parajuli, A., Hyöty, H., Laitinen, O., Sinkkonen, A., 2018. Half-lives of PAHs and temporal microbiota changes in commonly used urban landscaping materials. PeerJ 6, e4508.

Ruiz-Gomez, F.J., Perez-de-Luque, A., Maria Navarro-Cerrillo, R., 2019. The involvement of Phytophthora root rot and drought stress in holm oak decline: from ecophysiology to microbiome influence. Curr. For. Rep. 5, 251–266.

Saravesi, K., Markkola, A., Taulavuori, E., Syvanpera, I., Suominen, O., Suokas, M., Saikkonen, K., Taulavuori, K., 2019. Impacts of experimental warming and northern light climate on growth and root fungal communities of Scots pine populations. Fungal Ecol. 40, 43–49.

Setala, H., Francini, G., Allen, J.A., Jumpponen, A., Hui, N., Kotze, D.J., 2017. Urban parks provide ecosystem services by retaining metals and nutrients in soils. Environ. Pollut. 231, 451–461.

Sjogaard, K.S., Valdemarsen, T.B., Treusch, A.H., 2018. Responses of an agricultural soil microbiome to flooding with seawater after managed coastal realignment. Microorganisms 6. UNSP 12.

Strnadová, V., Černý, K., Holub, V., Gregorová, B., 2010. The effects of flooding and Phytophthora alni infection on black alder. J. For. Sci. 56, 41–46.

Sun, H., Terhonen, E., Koskinen, K., Paulin, L., Kasanen, R., Asiegbu, F.O., 2013. The impacts of treatment with biocontrol fungus (*Phlebiopsis gigantea*) on bacterial diversity in Norway spruce stumps. Biol. Control 64, 238–246.

Sun, H., Terhonen, E., Kasanen, R., Asiegbu, F.O., 2014a. Diversity and community structure of primary wood-inhabiting bacteria in boreal forest. Geomicrobiol J. 31, 315–324.

Sun, H., Terhonen, E., Koskinen, K., Paulin, L., Kasanen, R., Asiegbu, F.O., 2014b. Bacterial diversity and community structure along different peat soils in boreal forest. Appl. Soil Ecol. 74, 37–45.

Sun, H., Terhonen, E., Kovalchuk, A., Tuovila, H., Chen, H., Oghenekaro, A.O., Heinonsalo, J., Kohler, A., Kasanen, R., Vasander, H., Asiegbu, F.O., 2016a. Dominant tree species and soil type affect the fungal community structure in a boreal peatland forest. Appl. Environ. Microbiol. 82, 2632–2643.

Tedersoo, L., Anslan, S., Bahram, M., Drenkhan, R., Pritsch, K., Buegger, F., Padari, A., Hagh-Doust, N., Mikryukov, V., Gohar, D., 2020. Regional-scale in-depth analysis of soil fungal diversity reveals strong pH and plant species effects in Northern Europe. Front. Microbiol. 11, 1953. https://www.frontiersin.org/articles/10.3389/fmicb.2020.01953/full.

Terhonen, E., Sun, H., Buee, M., Kasanen, R., Paulin, L., Asiegbu, F.O., 2013. Effects of the use of biocontrol agent (*Phlebiopsis gigantea*) on fungal communities on the surface of *Picea abies* stumps. For. Ecol. Manag. 310, 428–433.

Terhonen, E., Keriö, S., Sun, H., Asiegbu, F.O., 2014. Endophytic fungi of Norway spruce roots in boreal pristine mire, drained peatland and mineral soil and their inhibitory effect on Heterobasidion parviporum in vitro. Fungal Ecol. 9, 17–26.

Terhonen, E., Langer, G.J., Busskamp, J., Rascutoi, D.R., Blumenstein, K., 2019. Low water availability increases necrosis in *Picea abies* after artificial inoculation with fungal root rot pathogens Heterobasidion parviporum and Heterobasidion annosum. Forests 10, 55.

Timm, C.M., Carter, K.R., Carrell, A.A., Jun, S.-R., Jawdy, S.S., Velez, J.M., Gunter, L.E., Yang, Z., Nookaew, I., Engle, N.L., Lu, T.-Y.S., Schadt, C.W., Tschaplinski, T.J., Doktycz, M.J., Tuskan, G.A., Pelletier, D.A., Weston, D.J., 2018. Abiotic stresses shift belowground Populus-associated bacteria toward a core stress microbiome. Msystems 3, e00070.

Tyree, M.T., Sperry, J.S., 1989. Vulnerability of xylem to cavitation and embolism. Annu. Rev. Plant Biol. 40, 19–36.

Uroz, S., Ioannidis, P., Lengelle, J., Cébron, A., Morin, E., Buée, M., Martin, F., 2013. Functional assays and metagenomic analyses reveals differences between the microbial communities inhabiting the soil horizons of a Norway spruce plantation. PLoS One 8, e55929.

Uroz, S., Buee, M., Deveau, A., Mieszkin, S., Martin, F., 2016. Ecology of the forest microbiome: highlights of temperate and boreal ecosystems. Soil Biol. Biochem. 103, 471–488.

Uroz, S., Courty, P.E., Oger, P., 2019. Plant symbionts are engineers of the plant-associated microbiome. Trends Plant Sci. 24, 905–916.

Vainio, E.J., Lipponen, K., Hantula, J., 2001. Persistence of a biocontrol strain of *Phlebiopsis gigantea* in conifer stumps and its effects on within-species genetic diversity. For. Pathol. 31, 285–295.

Vallero, D.A., 2014. Fundamentals of Air Pollution. Academic Press, San Diego.

Vandenkoornhuyse, P., Quaiser, A., Duhamel, M., Le Van, A., Dufresne, A., 2015. The importance of the microbiome of the plant holobiont. New Phytol. 206, 1196–1206.

Vasaitis, R., Burnevica, N., Uotila, A., Dahlberg, A., Kasanen, R., 2016. Cut *Picea abies* stumps constitute low quality substrate for sustaining biodiversity in fungal communities. Balt. For. 22, 239–245.

Vasiliauskas, R., Lygis, V., Thor, M., Stenlid, J., 2004. Impact of biological (Rotstop) and chemical (urea) treatments on fungal community structure in freshly cut *Picea abies* stumps. Biol. Control 31, 405–413.

Vasiliauskas, R., Larsson, E., Larsson, K.-H., Stenlid, J., 2005. Persistence and long-term impact of Rotstop biological control agent on mycodiversity in *Picea abies* stumps. Biol. Control 32, 295–304.

Veresoglou, S.D., Anderson, I.C., de Sousa, N.M.F., Hempel, S., Rillig, M.C., 2016. Resilience of fungal communities to elevated CO2. Microb. Ecol. 72, 493–495.

Vesterlund, S.-R., Saikkonen, K., 2011. Responses of Foliar Endophytes to Pollution, Endophytes of Forest Trees. Springer, Dordrecht.

Vitali, F., Raio, A., Sebastiani, F., Cherubini, P., Cavalieri, D., Cocozza, C., 2019. Environmental pollution effects on plant microbiota: the case study of poplar bacterial-fungal response to silver nanoparticles. Appl. Microbiol. Biotechnol. 103, 8215–8227.

Voolstra, C.R., Ziegler, M., 2020. Adapting with microbial help: microbiome flexibility facilitates rapid responses to environmental change. BioEssays 42 (7), 1–9. 2000004.

Wallace, J., Laforest-Lapointe, I., Kembel, S.W., 2018. Variation in the leaf and root microbiome of sugar maple (*Acer saccharum*) at an elevational range limit. Peerj 6, e5293.

Walter, C., Fladung, M., Boerjan, W., 2010. The 20-year environmental safety record of GM trees. Nat. Biotechnol. 28, 656+.

Wang, Y., Zhang, W., Ding, C., Zhang, B., Huang, Q., Huang, R., Su, X., 2019. Endophytic communities of transgenic poplar were determined by the environment and niche rather than by transgenic events. Front. Microbiol. 10, 588.

Wilcke, W., 2000. Synopsis polycyclic aromatic hydrocarbons (PAHs) in soil—a review. J. Plant Nutr. Soil Sci. 163, 229–248.

Wolfe, E.R., Kautz, S., Singleton, S.L., Ballhorn, D.J., 2018. Differences in foliar endophyte communities of red alder (*Alnus rubra*) exposed to varying air pollutant levels. Botany 96, 825–835.

Xu, L., Naylor, D., Dong, Z., Simmons, T., Pierroz, G., Hixson, K.K., Kim, Y.-M., Zink, E.M., Engbrecht, K.M., Wang, Y., Gao, C., DeGraaf, S., Madera, M.A., Sievert, J.A., Hollingsworth, J., Birdseye, D., Scheller, H.V., Hutmacher, R., Dahlberg, J., Jansson, C., Taylor, J.W., Lemaux, P.G., Coleman-Derr, D., 2018. Drought delays development of the sorghum root microbiome and enriches for monoderm bacteria. Proc. Natl. Acad. Sci. USA 115, E4284–E4293.

Chapter 5

Interspecific interactions within fungal communities associated with wood decay and forest trees

H. Umair Masood Awan[a,b] and Fred O. Asiegbu[a]

[a]Department of Forest Sciences, Faculty of Agriculture and Forestry, University of Helsinki, Helsinki, Finland, [b]Helclean Consultancy Services, Helsinki, Finland

Chapter Outline

1. Introduction

Interspecific fungal interaction is a dynamic process requiring the participation of two or more fungal species in their natural substrate. For many years, the nature of the community dynamics of interspecific fungal interactions has been studied, either in their terrestrial environment, under in vitro laboratory microcosm conditions, or by the agar model system (Toljander et al., 2006; White et al., 1998). The complex nature of the results from these community dynamic studies in either a microcosm or a natural ecosystem further underlies the inherent difficulty in the interpretation of data from both systems (Rayner and Boddy, 1988). This could be partly due to the diverse nature of fungal responses when grown under similar or the same experimental conditions (Halley et al., 1996). Many authors have observed that the development, colonization, and morphological growth of an individual mycelium within a heterogenous terrestrial ecosystem is governed by substrate availability, the nature of the abiotic microclimatic conditions, and biotic factors (e.g., competing organisms) (Rayner and Boddy, 1988; Ritz et al., 1996; White et al., 1998). According to White et al. (1998), the nature of the interaction between individual fungal mycelia on a terrestrial substrate could impact the compatibility or incompatibility of the species and the survival of the population as well as the subsequent development of the community. These interactions between the diverse species impact the assemblage of fungal communities on woody materials and other terrestrial substrates. A notable outcome in many interspecific fungal interactions is changes in the mycelia structure, particularly for basidiomycetes, which results in the formation of cords or barrage zones (Rayner et al., 1994). The hyphal interaction could also lead to protoplasmic fusion or anastomosis between the hyphae of closely related species of basidiomycetes or ascomycetes, further enhancing the survival of the populations (Agrios, 2005; Carlile, 1995). Largely, incompatible interspecific fungal interactions most often manifest in the form of combative interactions resulting in direct antagonism (hyphal interference, mycoparasitism) and indirect antagonism (formation of barrage zones and antibiosis, competition), which ultimately lead to fungal displacement and succession (Dix and Webster, 1995). In a forest ecosystem, the interspecific interaction that facilitates the decomposition of woody substrates and debris is crucial for nutrient and carbon cycling (Bellassen and Luyssaert, 2014;

Luyssaert et al., 2010; Purahong et al., 2018). Although the phenomenon of fungal interspecific interaction has been studied for a long time, not much is known about the impact on species diversity, species richness, or abundance. Furthermore, very little is known about the mechanisms underlying competitive fungal interactions that facilitate the wood decay process (Hiscox et al., 2018). Apart from ecological roles in wood decomposition and nutrient cycling, competitive combative fungal interactions could also have industrial applications in the biological control of pathogenic microbes (Hiscox et al., 2018).

1.1 Fungal community species assemblage, diversity, and abundance

In ecology, a community refers to interactions among a group of species present at a particular location or substrate at a specific time period (Pyron, 2010). Understanding the factors driving community development, particularly as it relates to species diversity and abundance, is of significant importance. In nature, multiple communities are known to exist together. Due to interspecific interactions (competition or cooperation), many of these communities have a disparate number of species or different abundances of species. Comparisons can be made among communities using attributes such as species diversity, species evenness, and abundance (Pyron, 2010). Species diversity refers to the abundance or number of individuals of different species in a given community, ecosystem, or region (Pyron, 2010). Species richness, on the other hand, is the number of different species within a biological community; however, it does not take into account the relative abundance of individuals per species in that community (Pyron, 2010). Species evenness is a reflection of the relative abundance of the different species in a given community or how close the numbers of each species are within an area. Not much is known about the ecological impact of interspecific fungal interactions on species diversity indices.

Fungal species diversity is ecologically crucial for forest health, particularly as primary decomposers. They contribute to nutrient and carbon cycling that facilitates the regeneration and establishment of young trees. Differences in species abundance and the diversity of wood decay fungi have been reported between natural and managed forests in Fennoscandia (see Niemelä et al., 2002; Siitonen, 2001). The abundance of wood-decaying fungi was higher in old natural spruce forests than in the adjacent managed forests (Siitonen, 2001). Several authors have noted that intensively managed forestry practices are unlikely to maintain the requirements of fungal decomposers comparable to the myco-decayers of old growth forests (Jonsson, 2000; Niemelä et al., 2002). Siitonen (2001) attributed the species composition in the natural forest stands to the high amount of late decay stages of logs, older forest age, low human impact, and high volumes of coarse woody debris. Ottosson (2013) found that highly decayed wood harbored a higher number of ascomycetes than basidiomycetes, which were predominantly observed in less decayed wood.

A spatiotemporal comparison further reveals that the fungal species assemblage of northern hemispheric forests is different than southern hemispheric forests of the world. Partly due to varied habitat conditions (Merckx, 2012). The composition of species and their assemblage changes with time. There are different factors that can bring about changes in temporal terms such as biotic or abiotic disturbances, including stochasticity of assemblages. However, according to a few studies, the concept of succession (a timely ordered sequence, which will be discussed later) does not support the idea of stochasticity (random colonization model) in nature (Boddy, 2001; Fukami et al., 2010). A recent study demonstrates that the fungal species richness of saprotrophs and ectomycorrhiza symbionts is strongly correlated with land use and climate conditions, especially concerning seasonality. Other ongoing global change processes will also affect fungal species richness patterns at large scales (Andrew et al., 2019).

Species diversity and richness are intricately linked to the coexistence or the lack of coexistence, which is impacted by interspecific interactions. The coexistence and noncoexistence of species in a single community are therefore potentially linked to cooperation and competition mechanisms. Toljander et al. (2006) reported that under an in vitro microcosm with 16 wood decay fungal species (eight white rot and eight brown rot fungi), the highest wood decomposition rates were observed at intermediate levels of fungal diversity. They noted that niche partitioning could be critical in sustaining fungal species diversity in a wood decay community. Without niche partitioning in the woody substrate, highly competitive species would dominate with the exclusion of weak or less combative species (Toljander et al., 2006). Fig. 5.1 depicts fungal colonization and niche partitioning on fallen wood.

1.2 Ecological aspects of fungal interactions in dead wood

Fungi rarely exist alone; as a result, mycelial interactions and competition for resources are continuous processes. These dynamic processes dictate fungal community development as well as the outcome of interactions. The interaction outcomes are directly affected by these processes because different species influence mineralization, decomposition, and nutrient translocation to a great extent. The decay rate measured as CO_2 release has been observed to increase during a deadlock

FIG. 5.1 Fungal colonization and niche partitioning on fallen wood. The red arrows show a few distinct areas of niche partitions.

between *Flammulina velutipes* and *Lenzites betulina* as well as when *Bjerkandera adusta* replaces *Stereum gausapatum* during interaction in wood (Boddy, 2001). A decrease in CO_2 production was observed during interactions where *Stereum hirsutum* was replaced by *Chondrostereum purpureum* (Woodward and Boddy, 2008). The movement and partitioning of carbon within fungal threads is affected by interactions. It has been noted that *Phanerochaete velutina* opportunistically switched carbon dependence from the wood it was growing on to the wood that was captured and mobilized earlier by its competitor (Woodward and Boddy, 2008). Evidence of interspecific carbon exchange has also been observed in an interaction deadlock outcome in soil, perhaps due to the leakage of carbon from the damaged hyphae into the soil (Woodward and Boddy, 2008).

Interspecific fungal interactions also affect mineral nutrient uptake, including partitioning as well as movement and evolution. It has been noted that the presence of another combative mycelium affects the uptake kinetics of ^{32}P of filamentous basidiomycetes. The outcome of the interactions was, however, not related to variations in uptake capacity. It is possible that species divert themselves from phosphorus uptake in order to defend their territory (Wells and Boddy, 2002). It has also been hypothesized that in the presence of grazing invertebrates, nutrients are released into the soil during interspecific fungal interactions (Woodward and Boddy, 2008). Higher losses of ^{32}P to soil were observed over self-pairings when *Phanerochaete velutina* and *Hypholoma fasciculare* interacted in a soil microcosm under laboratory conditions (Wells and Boddy, 2002). Nutrients were lost among nonself pairings in the soil. Both cases revealed that the leakage occurred in the interaction zone as well as in other places.

In nature, there may be niche partitioning in boreal forests that contributes to reducing interactions between saprotrophs and ectomycorrhiza (EM) fungi. Several authors have noted that saprotrophs dominate the upper organic soil horizons while EM species appear to dominate the lower mineral soil zones (Woodward and Boddy, 2008). These observations suggest that saprotrophic fungi play a role in the release of carbon from soil organic matter, whereas ectomycorrhizal symbionts facilitate the mobilization of organic nitrogen (Lindahl et al., 2007). Apart from nutrient recycling, several basidiomycetes cause economic losses to forest trees in the form of diseases. Many of these fungi (e.g., *Heterobasidion*, *Armillaria*, and *Phellinus*) often spend a considerable part of their life cycle living as saprotrophs. A mechanistic understanding of interspecific fungal interactions could provide novel insights on the identification of potential antagonists to reduce the threat posed by these pathogens. The discovery of *Phlebiopsis gigantea* as a biocontrol agent against *Heterobasidion* spp. is one such example. A number of other fungi (e.g., *Bjerkandera adusta*, *Resinicium bicolor*, *Trichoderma* spp.) have been tested through interspecific interaction as potential biocontrol agents against these pathogens (Holdenrieder and Greig, 1998; Holmer and Stenlid, 1997a; Nicolotti et al., 1999). Only *P. gigantea* is currently commercially used as a successful antagonist against *Heterobasidion* spp. (Pratt et al., 2000).

Combative interactions govern community development during wood decay. The community development and composition are dependent on the assemblages of fungi. The fungal territories are modified chemically and physically by changing water content, pH, and secreted secondary metabolites. This transformation of the original niche may serve as a constitutive defense of certain species while leaving others vulnerable to exclusion. The effect where early arrivers exert changes on the late-coming species is commonly known as the **priority effect** (Fukami, 2015; Ottosson et al., 2014). These sorts of priority effects are common in wood decomposition communities, as suggested by a few authors (Fukami, 2015; Fukami et al., 2010; Hiscox et al., 2015a). Different authors have tried to show examples of predecessor and successor relationships where certain species completely replaced other species (Heilmann-Clausen and Christensen, 2004; Rayner et al., 1987).

Fungal species composition within a resource defines the rate of decay because diverse species decay wood at different rates (Van Der Wal et al., 2015). Additionally, interspecific interactions indirectly govern decomposition rates by utilizing resources at varying levels. For instance, Hiscox et al. (2015a) observed that a considerable amount of carbon dioxide was released by interactive fungi compared to noninteractive fungi in a microcosm experiment. Recently, due to climate change, it has been reported that the interaction outcome is sensitive to even minute changes in abiotic conditions, as these changes may pose serious effects on decay processes (Hiscox et al., 2016). Also Hiscox et al. (2016) noted that the amount of carbon dioxide released to the atmosphere by wood decay fungi is partly affected by fluctuating environmental conditions as well as its carbon use efficiency (CUE). It has been observed that the CUE of artificial wood decomposition communities was decreased when community complexity and fluctuating temperature were increased (Toljander et al., 2006). Many authors have forecast that disruptive environmental conditions are needed to transform the ecosystem functions of wood decay communities, and combative hierarchies are likely to play important roles in this process (Hiscox and Boddy, 2017).

1.3 Cofungal interactions within living trees

The cooccurrence of different fungal species is common in healthy and standing living trees (Arhipova et al., 2011; Boddy, 2001). To facilitate fungal interaction within living wood, the individual fungal propagule has to enter in any of the following three ways: (a) wounded bark, (b) via roots, or (c) via vectors. The bark and the limited aeration due to the high moisture content of the sapwood of healthy living trees naturally help to protect it from fungal infection. Although the heartwood of trees contains toxic substances, because of its limited active defenses, heartwood is more susceptible to invasion by many necrotrophic pathogens. A few species such as *Heterobasidion* spp. and *Phellinus pini* are also able to enter into heartwood through wounds or root contacts (Asiegbu et al., 2005; Brazee and Lindner, 2013). The *Dalbergia* dieback of the multipurpose economically important tree species Shisham (*Dalbergia sissoo*) in Southeast Asia is most likely a result of several cofungal interactions in roots (Shakya and Lakhey, 2007). Among them, *Fusarium solani* and *Botryodiplodia theobromae* stand out as important players in the young seedlings of *D. sissoo* (Ahmad et al., 2013). The basal stem rot (BSR) and upper stem rot (USR) diseases of the oil palm caused by *Ganoderma boninense* enter primarily via roots (Naidu et al., 2017). Another way for fungal entrance into a standing tree is through insect vectors. This may also serve as an opportunity for saprotrophic species (*Bjerkandera adusta*, *Fomitopsis pinicola*, *Phlebiopsis gigantea*, *Stereum sanguinolentum*, *Trichaptum abietinum*) to be introduced into a living tree system (Vasaitis, 2013). Similarly, opportunistic fungal species cause diverse diseases in healthy trees such as *Sphaeropsis* blight in economically important pine trees (Awan and Pettenella, 2017). Many necrotrophic pathogens (e.g., *Heterobasidion annosum*) possess a dual nutritional lifestyle that enables them to live as a saprotroph on dead tissues after tree mortality. The subsequent colonization of the dead tree can facilitate naturally programmed successional trajectories (Ottosson, 2013). Kovalchuk et al. (2018) reported that *Heterobasidion*-rotted trees were more susceptible to coinfection by other wood-degrading fungi compared to nonrotted trees.

2. Wood decay, colonization, and methods for classifying interspecific fungal interaction

Wood decay patterns are defined by several parameters, of which the nature of fungal organisms that feed on the wood is significantly important. Until now, three (+1 undefined) different fungal attacking ways on wood that lead toward different wood decay patterns have been identified in the literature. A recent paper provides interesting insight into how wood chemical residues can be used as a tool to determine the fungal rot type and link the information to a measurable outcome for an undefined decay form (grey rot) (Schilling et al., 2020). Table 5.1 summarizes the patterns of fungal wood decay and their causal agents, whereas Table 5.2 summarizes diverse laboratory experiments on quantifying wood decay patterns caused by fungi.

The first three patterns of fungal wood decay are economically important, that is, white rot, brown rot, and soft rot (Goodell et al., 2008; Schwarze et al., 2000; Viitanen et al., 2010). However, comparative genomics have revealed that the mechanisms of wood decay are diverse in Agaricomycotina and the separation into white rot or brown rot should be based on experimental evidence (Floudas et al., 2015). For instance, *Cylindrobasidium torrendii* and *Schizophyllum commune* exhibit features similar to soft rot decay, but they are considered intermediates between brown rot and white rot fungi. *Heterobasidion annosum* causes white rot in spruce, pine, and fir (Asiegbu et al., 2005; Schwarze et al., 2000) and *Fomitopsis pinicola* causes brown rot damage in both conifers and broadleaf trees (Schwarze et al., 2000). White rot is characterized by the decomposition of all wood components: cellulose, hemicellulose, and lignin (Song et al., 2012). A white color is left over as a result of white rot fungal activity (Hatakka and Hammel, 2011; Song et al., 2012). The wood is usually bleached in appearance with a fibrous structure. Unlike most white rot fungi, many brown rot fungi only utilize cellulose and hemicellulose components, leaving behind modified lignin that leaves the wood brownish (Filley et al., 2002; Goodell, 2003; Hatakka and Hammel, 2011; Niemenmaa et al., 2008; Song et al., 2012). The evolution of brown rotters as compared to white rot fungi was accompanied by reductions in several proteins important for the effective degradation of lignin (Eastwood et al., 2011; Floudas et al., 2012; Mäkelä et al., 2014). In soft rot decay, mostly cellulose fibers are degraded, usually by members of the fungal group ascomycetes (Hatakka and Hammel, 2011).

Żółciak et al. (2016) showed that the rate of wood (Norway spruce) decay is dependent on its density (high or low) and most likely related to the enzymatic preferences of the fungi (*Antrodia gossypium*, *Phlebiopsis gigantea* strains from Finland and the United Kingdom (UK), *Heterobasidion parviporum* strains). P. gigantea strains (PgFI, PgGB) caused almost double the weight loss (28%, 28.1%) over that caused by *A. gossypium* (15.2%) after 6 months incubation on a spruce wood block (Table 5.2). The three isolates of *H. parviporum* (Hp1, Hp2, Hp3) showed 24.7%, 19.6%, and 21.6% weight loss, respectively. Unlike P. gigantea isolates (PgFI, PgGB), which exhibited almost similar preferences for wood substrate density, *H. parviporum* isolates showed less decomposition on high-density wood blocks. Similar decay rates of *H. parviporum* isolates (Hp1, Hp2, Hp3) and its competitor P. gigantea (PgFI, PgGB) were also observed. The different preferences for colonized wood with respect to density suggest that the effective protection of stands against root rot requires simultaneous treatment with different biological control agents (Żółciak et al., 2016).

Mgbeahuruike et al. (2011) examined the patterns of wood decay caused by 64 wild heterokaryotic isolates of P. gigantea. Among them, isolate number 4143 caused the maximum weight loss of 40.7% in Norway spruce whereas isolate 4146 caused a 40.2% weight loss in pine wood blocks (Table 5.2). Microscopic observation revealed that the degradation pattern in both wood species had typical features of white rot decay. Unlike spruce, which was selectively decayed, there was a simultaneous decay noticed in pine wood blocks (Mgbeahuruike et al., 2011). Other authors have previously reported simultaneous wood decay in other basidiomycete wood decay fungi such as *H. annosum* s.s. (Daniel, 2003; Daniel et al., 1998). Another wood decay fungus *Stereum sanguinolentum* was found to be able to penetrate early wood tracheids of the Norway spruce, but further invasive growth within the tracheids of latewood was restricted due to an increased number of bordered pits (Kleist and Seehann, 1997; Schwarze et al., 2000).

The type and chemistry of wood components play an important role in the competitive advantage of an individual fungus inside a diverse wood-inhabiting fungal community (Song et al., 2012). Asiegbu et al. (1996) reported a significant loss of native wood lignin in paired or multiple cultures of three white rot fungi (*Pleurotus sajor-caju*, *Phanerochaete chrysosporium*, *Trametes versicolor*) (see Table 5.2). Similar synergistic effects were also documented on polysaccharide degradation in cocultures of the fungi. By contrast, Toljander et al. (2006) did not observe any synergistic effect in wood decomposition under constant temperature regimes in cocultures of the two functional groups, white and brown rot fungi. The weight loss was dictated by the temperature fluctuations. In a 16 community's assembly of four and eight species (each separately), the gross decomposition rates were found to be higher with fluctuating temperatures than a constant temperature regime. However, this effect was not observed in the presence of the highest amount of species richness.

A summary of different studies involving fungal monocultures and mixed cultures is presented in Table 5.2.

3. Mechanisms of combative interactions (mycoparasitism, competition, hyphal interference, antibiosis)

In nature, fungi are known to interact with other organisms: bacteria, viruses, plants, and animals (A'Bear et al., 2013; de Boer et al., 2005; De Boer et al., 2010). Many of these fungal interactions are mainly impacted by both biotic (host, competitors) and abiotic agents (temperature, humidity, soil) in the environment. The outcome of the interactions could be of benefit to all parties involved (synergistic/mutualistic/complementaristic/additive interactions). In many others, they are detrimental for at least one partner (antagonistic/competitive/combative/extractive interactions). As many of these interactions are dynamic, invariably, a shift from synergism to antagonism may occur at both the ecological and evolutionary timescales, depending on the cost to benefit ratios in the natural continuum (Toby Kiers et al., 2010).

Antagonistic interaction refers to the activity of any one organism that adversely interferes or suppresses the normal growth of another organism in its vicinity or at a distance. In interspecific fungal interaction, antagonism could be indirect, whereby the progression of defensive hyphal barriers at the mycelial fronts is restricted by chemicals secreted by the opposing hyphae, leading to antibiosis (Fig. 5.2) (Dix and Webster, 1995). By contrast, in aggressive direct antagonistic interaction, the hyphae of the competitor advance into the mycelia of the other and destroy it, resulting in mycoparasitism

TABLE 5.1 Summary of important types of wood decay with respect to their causal agents.

Pattern type	Agent	Color	Texture	Chemistry	Examples	References
White rot	Basidiomycota	± bleached	fibrous	all components removed	*Heterobasidion* spp., *Gelatoporia subvermispora. Trametes versicolor, Pleurotus ostreatus, Pycnoporus sanguineus, Ganoderma lipsiense and lucidum, Phanerochaete chrysosporium, Mycena leaiana*	Asiegbu et al. (2005), Bari et al. (2015), Kim et al. (2019), Luna et al. (2004), Niemelä (1985), Skyba et al. (2016), and Worrall et al. (1997)
Brown rot	Basidiomycota	± Brown	Fibrous texture lost early, cross-checking	Primarily carbohydrates lost, lignin mostly remains	*Antrodia sinuosa, Coniophora puteana, Fomitopsis pinicola, Gloeophyllum trabeum, Serpula lacrymans, Postia placenta, Lenzites trabea*	Green and Highley (1997), Kim et al. (2019), Kirk (1975), Kwang et al. (2004), Mansfield et al. (1998), Nurika et al. (2020), Sugano et al. (2019), and Villavicencio et al. (2020)
Soft rot	Asco-, Deutero-mycota	Bleached or brown	Usually on surface, some fibrous texture lost, cross-checking in some cases	Carbohydrates preferred, but some lignin lost too	*Aspergillus niger, Penicillium chrysogenum, Phialophora sp., Scytalidium sp., Xylaria sp.*	Hamed (2013), Kim et al. (2019), and Worrall et al. (1997)
Stain and Mold (Sapstain)	Asco-, Deutero-mycota	Blue, greyish black	Nonfibrous	Discoloration of wood	*Aureobasidium pullulans, Leptographium* spp., *Ophiostoma* sp., *Stachybotrys chartarum, Sphaeropsis sapinea/Diplodia (sa)pinea, Cladosporium herbarum, Ceratocystis pilifera*	Kim et al. (2019), Kim (2005), and Swart (1991)

or hyphal interference (Fig. 5.3) (Dix and Webster, 1995). Most often, nonself interspecific interaction leads to the production of biologically active metabolites (BAM), either by the aggressive antagonistic species or the defending species. Many of the BAMs observed during nonself interspecific interaction include the production of extracellular enzymes, antibiotics, gas phase compounds/info-chemicals/volatile organic compounds (VOCs), and diffusible organic compounds (DOCs) (Boddy, 2000). This could also be accompanied by morphological changes and transformations in the hyphae.

In interspecific fungal interaction, competition among higher fungal species is generally negative (Boddy, 2000). Fungal competition can occur in a primary uninhabited organic substrate or in a secondary resource substrate that is already precolonized by other fungi. The colonization of the primary resource inhabited substrate is often facilitated by the ability to adapt or utilize the substrate as well as rapid dispersal and growth (Boddy, 2000). The success in establishing on the secondary resource substrate is partly dependent on the competitive ability and aggressiveness of the antagonist to capture occupied territory or to defend it against aggressors (Boddy, 2000; Hiscox and Boddy, 2017; Woodward and Boddy, 2008). *Xylaria hypoxylon* is a typical example of an early colonizing ascomycete able to extend its territory once no uncolonized resource remains (Boddy, 2000).

Antibiosis refers to the biological interaction between two organisms in which one of them is adversely affected partly due to metabolic products secreted by another. The concept of antibiosis was reinforced through the work of Alexander Fleming, which led to the accidental discovery of the antibiotic penicillin in 1928 (Szentivanyi and Herman Friedman, 1987). Antibiosis is an important antagonistic mechanism that could be of relevance for exploitation in biological control, where the antagonist releases substance(s) that could act as antibiotics, lytic enzymes, volatile compounds, or toxins against pathogens to disrupt them (Boddy, 2016; Woodward and Boddy, 2008). In an artificial agar culture with wood decay fungi, mutual inhibition at a distance of 15 mm or more was considered to be antibiosis (Boddy, 2000). However, Boddy (2000) noted that in natural ecosystems, these inhibitions can occur unnoticed or at shorter distances. The interaction between *Diplodia pinea* and *Botrytis cinerea* has been observed as inhibition at a distance by de Oliveira et al. (2018). Recently, antibiosis has been observed between *Heterobasidion parviporum* and *Mycena* sp. (Wen et al., 2019) possibly due in part to the effects of DOCs (Boddy, 2016).

Hyphal interference (HI), a form of direct antagonistic interaction, is a biological mechanism that triggers programmed hyphal death when two hyphae or spores from different species meet (Boddy, 2000; Silar, 2012). This is an ancient and conserved phenomenon and like many other traits, probably the result of convergent evolution in fungi (Silar, 2012). The biochemical and molecular basis of how fungi differentiate self from nonself, which promotes hyphal death, is still not fully understood. Although this mechanism is present in many fungal species, it is relatively less studied than mycoparasitism, which leaves us with many unanswered questions. Ikediugwu and Webster (1970) were the first to use the term HI to describe observations with interactions of *Coprinus heptemerus* and *Ascobolus crenulatus*, where *Ascobolus* hyphae were killed because they were vacuolated and lost turgor. The most studied HI example in the literature points toward squashing of the *H. annosum* hyphae by *P. gigantea* (Deacon, 2013; Ikediugwu, 1976a; Ikediugwu, 1976b; Rayner and Boddy, 1988). The combined dual immunofluorescence labelling showed that a potential mechanism for the superior antagonistic potential of *P. gigantea* over *H. annosum* is by hyphal thinning (Fig. 5.3) (Asiegbu, unpublished). Due to this mechanism, *Phlebiopsis gigantea* is commercially used as a biocontrol agent against *H. annosum* (Pratt et al., 2000). Interestingly, nonenzymatic diffusible metabolite has not been fully described in the antagonism between these two species. The absence or presence of an antagonist does not necessarily affect the release of nonenzymatic diffusible compounds in the case of *Hypomyces aurantius* (see Boddy, 2000). HI has also been observed among the interacting mycelia of other wood-degrading basidiomycetes (Deacon, 2013). For example, HI was observed between *Phanerochaete magnoliae* and *Datronia mollis* where the latter is killed in vitro (Ainsworth and Rayner, 1991). Like basidiomycetes, HI is also not uncommon among ascomycetes. *Podospora anserina* (ascomycete) and *Coprinopsis cinerea* (basidiomycete) showed HI where the hyphae of the former were killed in vitro (*Silar, 2005*). Woodward and Boddy (2008) reported that during HI, the affected cells are vacuolated with the loss of opacity and hydrostatic pressure accompanied by hyphal swelling and burst (Woodward and Boddy, 2008). HI is regarded as a less-aggressive strategy used by fungi than mycoparasitism. Usually, the phenomenon of mycoparasitism has been suggested as more favored and reliable for practical application in biocontrol than the HI mechanism (Silar, 2012).

Unlike HI, mycoparasitism is a foraging strategy that refers to a process in which a parasitic fungus (mycoparasite) parasitizes another fungus (mycohost) in order to obtain nutrients. Mycoparasitism has been widely used for the development of mitosporic fungi to be used in horticulture and agriculture as biocontrol agents (Whipps, 2001). However, this mechanism has been studied infrequently for saprotrophic basidiomycetes (Woodward and Boddy, 2008). In nature, the process of HI is different from mycoparasitism. In HI, fungal species appear to be more insensitive (tolerant) to the carbon/nitrogen ratio or the nature of the substrate than in mycoparasitism (Ikediugwu and Webster, 1970; Silar, 2012). Mycoparasitism could be mediated by active adhesives (e.g., lectin/agglutinin-carbohydrate binding) (Boddy, 2000). Subsequent invasive

TABLE 5.2 An overview of studies employing mono and mixed cultures of fungi in the degradation of woody substrates.

Wood species	Wood sample taken from	Fungal isolate no.	Fungal name
Picea abies (Norway Spruce)	Sawdust from 60 to 65 years old spruce and 20–30 years old birch	CMI210864	Trametes versicolor
Picea abies (Norway Spruce)	Sawdust from 60 to 65 years old spruce and 20–30 years old birch	CMI747691	Phanerochaete chrysosporium
Populus deltoides (Poplar)	Sound poplar wood	Starin163	Pycnoporus sanguineus**
Populus deltoides (Poplar)	Sound poplar wood	Strain 340	Ganoderma lucidum***+
Populus tremula (Aspen)	Sapwood of aspen stem	T24li	Physisporinus rivulosus**
Populus tremula (Aspen)	Sapwood of aspen stem	CZ-3	Ceriporiopsis subvermispora**
Populus tremula (Aspen)	Sapwood of aspen stem	ME446	Phanerochaete chrysosporium**
Populus tremula (Aspen)	Sapwood of aspen stem	B6	Pleurotus ostreatus**
Pinus densiflora (Pine)	20-year-old pine	KCCM34740	Phanerochaete chrysosporium**
Pinus densiflora (Pine)	20-year-old pine	KFRI21078	Ceriporiopsis subvermispora**
Pinus densiflora (Pine)	20-year-old pine	KCCM11258	Trametes versicolor**
Populus albaxglandulosa (Aspen)	20-year-old poplar	KCCM34740	Phanerochaete chrysosporium**
Populus albaxglandulosa (Aspen)	20-year-old poplar	KFRI21078	Ceriporiopsis subvermispora**
Populus albaxglandulosa (Aspen)	20-year-old poplar	KCCM11258	Trametes versicolor**
Eucalyptus grandis	12-year-old plantation	PPRI 6762	Pycnoporus sanguineus***
Azadirachta indica (Neem)	Healthy Neem wood	NG	Trichoderma harzianum**
Azadirachta indica (Neem)	Healthy Neem wood	NG	Chrysosporium asperatum***+
Picea abies (Norway Spruce)	Norway spruce	4143	Phlebiopsis gigantea***+
Pinus sylvestris (Pine)	Pinus sylvestris	4146	Phlebiopsis gigantea***
Picea abies (Norway Spruce)	Norway spruce	Pg 4, Pg 15	Phlebiopsis gigantea
Tectonia grandis (Teak)	Sound sapwood from stem	NG	Irpex lacteus***+
Tectonia grandis (Teak)	Sound sapwood from stem	NG	Phanerochaete chrysosporium***+
Populus tomentosa (Chinese white poplar)	Fresh poplar wood	C5617	Lenzites betulinus
Populus tomentosa (Chinese white poplar)	Fresh poplar wood	D10149	Trametes velutina
Populus tomentosa (Chinese white poplar)	Fresh poplar wood	C6320	Trametes orientalis
Picea abies (Norway Spruce)	Fresh Norway spruce	MSCL 1023	Heterobasidion parviporum
Picea abies (Norway Spruce)	Fresh Norway spruce	T. viride 969	Trichoderma viride
Picea abies (Norway Spruce)	Norway spruce	NG	Pycnoporus sanguineus**
Hevea brasiliensis (Rubber tree)	Tree clone NIG 801	MUCL45064	Rigidiporus microporus***
Hevea brasiliensis (Rubber tree)	Tree clone NIG 801	ED310, M13, MS564b	Rigidiporus microporus***
Ailanthus excelsa (Heaven tree)	Fallen tree branches	NG	Bjerkandera adusta***
Shorea gibbosa (Yellow meranti)	Uninfected heartwood of the meranti stem	YM3	Phlebia brevispora***

Material size (mm) or mass (g)	Incubation in months (weeks)	Material used	Max decay (%)	Decay pattern	References
20 g	1.5 (6)	Sawdust	5	WR	Asiegbu et al. (1996)
20 g	1.5 (6)	Sawdust	5.7	WR	Asiegbu et al. (1996)
20×20×20	5 (20)	Wood blocks	59.05	WR	Luna et al. (2004)
20×20×20	5 (20)	Wood blocks	52.09	WR	Luna et al. (2004)
16×16×7	1.5 (6)	Wood blocks	17	WR	Chi et al. (2007)
16×16×7	1.5 (6)	Wood blocks	21	WR	Chi et al. (2007)
16×16×7	1.5 (6)	Wood blocks	15	WR	Chi et al. (2007)
16×16×7	1.5 (6)	Wood blocks	4	WR	Chi et al. (2007)
20×20×20	2 (8)	Wood blocks	10.8	WR	Kang et al. (2007)
20×20×20	2 (8)	Wood blocks	10.9	WR	Kang et al. (2007)
20×20×20	2 (8)	Wood blocks	9	WR	Kang et al. (2007)
20×20×20	2 (8)	Wood blocks	10.8	WR	Kang et al. (2007)
20×20×20	2 (8)	Wood blocks	10.8	WR	Kang et al. (2007)
20×20×20	2 (8)	Wood blocks	13	WR	Kang et al. (2007)
6×9	0.5 (2)	Wood chips	28	WR	Van Heerden et al. (2008)
20×20×20	4 (16)	Wood blocks	43.3	WR	Koyani et al. (2011)
20×20×20	4 (16)	Wood blocks	46.7	WR	Koyani et al. (2011)
30×10	4 (16)	Wood chips	40.7	WR	Mgbeahuruike et al. (2011)
30×10	4 (16)	Wood chips	40.2	WR	Mgbeahuruike et al. (2011)
10×25×50	6 (24)	Wood blocks	44, 45	WR	Żółciak et al. (2012)
20×20×20	4 (16)	Wood blocks	27.97	WR	Koyani and Rajput (2014)
20×20×20	4 (16)	Wood blocks	30.05	WR	Koyani and Rajput (2014)
20 mesh	3 (12)	Wood residues	30	WR	Wang et al. (2014)
20 mesh	3 (12)	Wood residues	37.9	WR	Wang et al. (2014)
20 mesh	3 (12)	Wood residues	24	WR	Wang et al. (2014)
20×10×10	7.5 (30)	Wood blocks	23.6	WR	Alksne et al. (2015)
20×10×10	7.5 (30)	Wood blocks	27.9	WR	Alksne et al. (2015)
20×20×5	2.5 (10)	Soil block culture	31	WR	Kim et al. (2015)
30×10×5	6 (24)	Wood blocks	27.2	WR	Oghenekaro et al. (2015)
30×10×5	6 (24)	Wood blocks	21.2, 15.7, 4.3	WR	Oghenekaro et al. (2015)
20×20×20	4 (16)	Wood blocks	69	WR	Pramod et al. (2015)
20×20×10	3 (12)	Wood blocks	12.34	WR	Erwin (2016)

(Continued)

TABLE 5.2 An overview of studies employing mono and mixed cultures of fungi in the degradation of woody substrates—cont'd

Wood species	Wood sample taken from	Fungal isolate no.	Fungal name
Azadirachta indica (Neem)	Healthy sapwood from 8 to 10-year-old juvenile wood	NA	*Schizophyllum commune***+
Ailanthus excelsa (Heaven tree)	Healthy sapwood from 8 to 10-year-old juvenile wood	NA	*Schizophyllum commune***+
Tectonia grandis (Teak)	Healthy sapwood from 8 to 10-year-old juvenile wood	NA	*Schizophyllum commune***+
Eucalyptus sp	Healthy sapwood from 8 to 10-year-old juvenile wood	NA	*Schizophyllum commune***+
Leucaena leucocephala (Tamarind)	Healthy sapwood from 8 to 10-year-old juvenile wood	NA	*Schizophyllum commune***+
Picea abies (Norway Spruce)	30-year-old healthy Norway spruce tree		*Antrodia gossypium*
Picea abies (Norway Spruce)	30-year-old healthy Norway spruce tree	PgFl, PgGB	*Phlebiopsis gigantea*
Picea abies (Norway Spruce)	30-year-old healthy Norway spruce tree	Hp1, Hp2, Hp3	*Heterobasidion parviporum*
Elaeis guineensis (Oil Palm)	Healthy/diseased tissue portions	Strain ST2	*Grammothele fuligo****
Elaeis guineensis (Oil Palm)	Healthy/diseased tissue portions	Strain PTG 13	*Grammothele fuligo****
Elaeis guineensis (Oil Palm)	Healthy/diseased tissue portions	Strain PTG 12	*Grammothele fuligo****
Elaeis guineensis (Oil Palm)	Healthy/diseased tissue portions	Strain FBR	*Pycnoporus sanguineus***
Elaeis guineensis (Oil Palm)	Healthy/diseased tissue portions	PER 71	*Ganoderma boninense*
Erythrophleum fordii (Oliver wood)	Healthy heartwood parts	ATCC 90872	*Phanerochaete sordida***
Erythrophleum fordii (Oliver wood)	Healthy heartwood parts	ATCC 34541	*Phanerochaete chrysosporium****
Fagus crenata (Japanese beech)	Healthy heartwood parts	ATCC 90872	*Phanerochaete sordida***
Fagus crenata (Japanese beech)	Healthy heartwood parts	ATCC 34541	*Phanerochaete chrysosporium****
Ficus microcarpa (curtain fig)	NG	Pn (NTU5-7)	*Phellinus noxius*
Ficus microcarpa (curtain fig)	NG	Tk (ML56)	*Trichoderma koningiopsis*
Picea abies (Norway Spruce)	Single ~70-year aged individual tree	FBCC0110	*Trichaptum abietinum*
Picea abies (Norway Spruce)	Single ~70-year aged individual tree	FBCC0043	*Phlebia radiata*
Picea abies (Norway Spruce)	Single ~70-year aged individual tree	FBCC1181	*Fomitopsis pinicola*
Populus tremuloides (Aspen)	NG	202A	*Antrodia* sp.
Populus tremuloides (Aspen)	NG	753FPL	*Antrodia carbonica*
Populus tremuloides (Aspen)	NG	ChBrnRt1	*Laetiporus squalidus*
Populus tremuloides (Aspen)	NG	PC2-2	*Postia* sp.
Populus tremuloides (Aspen)	NG	FP-103444-T	*Fistulina hepatica*
Populus tremuloides (Aspen)	NG	33A	*Fomitopsis cajanderi*
Populus tremuloides (Aspen)	NG	206A	*Gloeophyllum sepiarium*

Material size (mm) or mass (g)	Incubation in months (weeks)	Material used	Max decay (%)	Decay pattern	References
20×20×20	4 (16)	Wood blocks	30.12	WR	Koyani et al. (2016)
20×20×20	4 (16)	Wood blocks	34.44	WR	Koyani et al. (2016)
20×20×20	4 (16)	Wood blocks	24.05	WR	Koyani et al. (2016)
20×20×20	4 (16)	Wood blocks	29.68	WR	Koyani et al. (2016)
20×20×20	4 (16)	Wood blocks	30.82	WR	Koyani et al. (2016)
10×25×50	6 (24)	Wood blocks	15.2	BR	Żółciak et al. (2016)
10×25×50	6 (24)	Wood blocks	28, 28.1	WR	Żółciak et al. (2016)
10×25×50	6 (24)	Wood blocks	24.7, 19.6, 21.6	WR	Żółciak et al. (2016)
20×20×40	4 (16)	Wood blocks	58/79****	WR	Naidu et al. (2017)
20×20×40	4 (16)	Wood blocks	36/67****	WR	Naidu et al. (2017)
20×20×40	4 (16)	Wood blocks	29/58****	WR	Naidu et al. (2017)
20×20×40	4 (16)	Wood blocks	40/77****	WR	Naidu et al. (2017)
20×20×40	4 (16)	Wood blocks	19/48****	WR	Naidu et al. (2017)
20×20×5	1 (4)	Wood blocks	2	WR	Nguyen et al. (2018)
20×20×5	1 (4)	Wood blocks	2	WR	Nguyen et al. (2018)
20×20×5	1 (4)	Wood blocks	14	WR	Nguyen et al. (2018)
20×20×5	1 (4)	Wood blocks	12	WR	Nguyen et al. (2018)
~25×25×7	3 (12)	Wood blocks	11.5	WR	Chou et al. (2019)
~25×25×7	3 (12)	Wood blocks	1.8	WR	Chou et al. (2019)
10 g	3 (12)	Sawdust	7	WR	Mali et al. (2019)
10 g	3 (12)	Sawdust	6	WR	Mali et al. (2019)
10 g	3 (12)	Sawdust	22.5	BR	Mali et al. (2019)
19×19×19	1 (4)	Soil block microcosm	0.97	BR	Schilling et al. (2020)*
19×19×19	1 (4)	Soil block microcosm	0.15	BR	Schilling et al. (2020)*
19×19×19	1 (4)	Soil block microcosm	7.64	BR	Schilling et al. (2020)*
19×19×19	1 (4)	Soil block microcosm	0.09	BR	Schilling et al. (2020)*
19×19×19	1 (4)	Soil block microcosm	0.04	BR	Schilling et al. (2020)*
19×19×19	1 (4)	Soil block microcosm	0.31	BR	Schilling et al. (2020)*
19×19×19	1 (4)	Soil block microcosm	6.05	BR	Schilling et al. (2020)*

(Continued)

TABLE 5.2 An overview of studies employing mono and mixed cultures of fungi in the degradation of woody substrates—cont'd

Wood species	Wood sample taken from	Fungal isolate no.	Fungal name
Populus tremuloides (Aspen)	NG	751	*Neolentinus lepideus*
Populus tremuloides (Aspen)	NG	212A	*Oligoporus balsaminus*
Populus tremuloides (Aspen)	NG	209	*Phaeolus schweinitzii*
Populus tremuloides (Aspen)	NG	PJ-1	*Pyrofomes demidoffii*
Populus tremuloides (Aspen)	NG	BY1	*Conferticium ravum*
Populus tremuloides (Aspen)	NG	105725FPL	*Ceriporiopsis subvermispora***
Populus tremuloides (Aspen)	NG	Tyro292	*Aurantiporus* sp.
Populus tremuloides (Aspen)	NG	4C	*Dichomitus squalens***
Populus tremuloides (Aspen)	NG	GL-MN1	*Ganoderma sessile*
Populus tremuloides (Aspen)	NG	WI-7C	*Ganoderma tsugae***
Populus tremuloides (Aspen)	NG	H-2MN	*Hymenochaete corrugata*
Populus tremuloides (Aspen)	NG	ID1	*Inonotus dryophilus***
Populus tremuloides (Aspen)	NG	34A	*Irpex lacteus*
Populus tremuloides (Aspen)	NG	Ten. #74	*Peniophorella praetermissa*
Populus tremuloides (Aspen)	NG	11A	*Perenniporia subacida***
Populus tremuloides (Aspen)	NG	PM-1	*Phellinus arctostaphyli*
Populus tremuloides (Aspen)	NG	TAB19	*Phellinus pini***
Populus tremuloides (Aspen)	NG	64C	*Phlebia brevispora*
Populus tremuloides (Aspen)	NG	604	*Phlebia chrysocreas*
Populus tremuloides (Aspen)	NG	Park#82	*Phlebia* sp.
Populus tremuloides (Aspen)	NG	PRL2845	*Phlebia tremellosa***
Populus tremuloides (Aspen)	NG	B360	*Baltazaria* sp.**

Material size (mm) or mass (g)	Incubation in months (weeks)	Material used	Max decay (%)	Decay pattern	References
19×19×19	1 (4)	Soil block microcosm	0.02	BR	Schilling et al. (2020)*
19×19×19	1 (4)	Soil block microcosm	0.14	BR	Schilling et al. (2020)*
19×19×19	1 (4)	Soil block microcosm	0.35	BR	Schilling et al. (2020)*
19×19×19	1 (4)	Soil block microcosm	0.01	BR	Schilling et al. (2020)*
19×19×19	1 (4)	Soil block microcosm	1.55	WR	Schilling et al. (2020)*
19×19×19	1 (4)	Soil block microcosm	4.55	WR	Schilling et al. (2020)*
19×19×19	1 (4)	Soil block microcosm	1.04	WR	Schilling et al. (2020)*
19×19×19	1 (4)	Soil block microcosm	3.85	WR	Schilling et al. (2020)*
19×19×19	1 (4)	Soil block microcosm	2.94	WR	Schilling et al. (2020)*
19×19×19	1 (4)	Soil block microcosm	3.35	WR	Schilling et al. (2020)*
19×19×19	1 (4)	Soil block microcosm	0.06	WR	Schilling et al. (2020)*
19×19×19	1 (4)	Soil block microcosm	0.46	WR	Schilling et al. (2020)*
19×19×19	1 (4)	Soil block microcosm	0.32	WR	Schilling et al. (2020)*
19×19×19	1 (4)	Soil block microcosm	0.08	WR	Schilling et al. (2020)*
19×19×19	1 (4)	Soil block microcosm	0.01	WR	Schilling et al. (2020)*
19×19×19	1 (4)	Soil block microcosm	0.47	WR	Schilling et al. (2020)*
19×19×19	1 (4)	Soil block microcosm	0.38	WR	Schilling et al. (2020)*
19×19×19	1 (4)	Soil block microcosm	0.96	WR	Schilling et al. (2020)*
19×19×19	1 (4)	Soil block microcosm	4.08	WR	Schilling et al. (2020)*
19×19×19	1 (4)	Soil block microcosm	0.96	WR	Schilling et al. (2020)*
19×19×19	1 (4)	Soil block microcosm	5.14	WR	Schilling et al. (2020)*
19×19×19	1 (4)	Soil block microcosm	1.52	WR	Schilling et al. (2020*

(Continued)

TABLE 5.2 An overview of studies employing mono and mixed cultures of fungi in the degradation of woody substrates—cont'd

Wood species	Wood sample taken from	Fungal isolate no.	Fungal name
Populus tremuloides (Aspen)	NG	Calem#67	*Stereum hirsutum*
Populus tremuloides (Aspen)	NG	611A	*Trametes betulina*
Populus tremuloides (Aspen)	NG	303B	*Kretzschmaria hedjaroudei*
Populus tremuloides (Aspen)	NG	Di44-5	*Jaapia argillacea*
Populus tremuloides (Aspen)	NG	Quin.25A	*Sistotrema brinkmannii*
Populus tremuloides (Aspen)	NG	WBR-1	*Sistotrema coronilla*
Beech wood	NG	HfGTWV2	*Hypholoma fasciculare*
Beech wood	NG	TVWGP01	*Trametes versicolor*
Beech wood	NG	VcWVJH1	*Vuilleminia comedens*
Beech wood	NG	EMPA62	*Coniophora puteana*
Beech wood	NG	FPEF01	*Fomitopsis pinicola*
Beech wood	NG	LSPT01	*Laetiporus sulphureus*
Dual culture			
Picea abies (Norway Spruce)	Sawdust from 60 to 65 years old spruce and 20–30 years old birch	CMI210864 + CMI199761	*Trametes versicolor + Pleurotus sajor-caju*
Picea abies (Norway Spruce)	Sawdust from 60 to 65 years old spruce and 20–30 years old birch	CMI199761 + CMI747691	*Pleurotus sajor-caju + Phanerochaete chrysosporium*
Picea abies (Norway Spruce)	Sawdust from 60 to 65 years old spruce and 20–30 years old birch	CMI210864 + CMI747691	*Trametes versicolor + Phanerochaete chrysosporium*
Populus tremula (Aspen)	Sapwood of aspen stem	T24li + B6	*Physisporinus rivulosus + Pleurotus ostreatus*
Populus tremula (Aspen)	Sapwood of aspen stem	CZ-3 + B6	*Ceriporiopsis subvermispora + Pleurotus ostreatus*
Populus tremula (Aspen)	Sapwood of aspen stem	ME446 + B6	*Phanerochaete chrysosporium + Pleurotus ostreatus*
Populus tremula (Aspen)	Sapwood of aspen stem	T24li + CZ-3	*Physisporinus rivulosus + Cerporiopsis subvermispora*
Populus tremula (Aspen)	Sapwood of aspen stem	T24li + ME446	*Physisporinus rivulosus + Phanerochaete chrysosporium*
Eucalyptus grandis	12-year-old plantation	PPRI 6762 + ABA006	*Pycnoporus sanguineus + Pichia guilliermondii*
Eucalyptus grandis	12-year-old plantation	PPRI 6762 + J11904	*Pycnoporus sanguineus + Aspergillus flavipes*
Populus tomentosa (Chinese white poplar)	Fresh poplar wood	C5617 + C6320	*Lenzites betulinus + Trametes orientalis*
Populus tomentosa (Chinese white poplar)	Fresh poplar wood	C5617 + D10149	*Lenzites betulinus + Trametes velutina*

Material size (mm) or mass (g)	Incubation in months (weeks)	Material used	Max decay (%)	Decay pattern	References
19×19×19	1 (4)	Soil block microcosm	2.93	WR	Schilling et al. (2020)∗
19×19×19	1 (4)	Soil block microcosm	0.26	WR	Schilling et al. (2020)∗
19×19×19	1 (4)	soil block microcosm	0.28	WR	Schilling et al. (2020)∗
19×19×19	1 (4)	soil block microcosm	0.4	?	Schilling et al. (2020)∗
19×19×19	1 (4)	Soil block microcosm	0.24	?	Schilling et al. (2020)∗
19×19×19	1 (4)	Soil block microcosm	0.06	?	Schilling et al. (2020)∗
20×20×20	3 (12)	Wood blocks	~9	WR	Fukasawa et al. (2020)
20×20×20	3 (12)	Wood blocks	~24	WR	Fukasawa et al. (2020)
20×20×20	3 (12)	Wood blocks	~8	WR	Fukasawa et al. (2020)
20×20×20	3 (12)	Wood blocks	~4	BR	Fukasawa et al. (2020)
20×20×20	3 (12)	Wood blocks	~8	BR	Fukasawa et al. (2020)
20×20×20	3 (12)	Wood blocks	~12	BR	Fukasawa et al. (2020)
20 g	1.5 (6)	Sawdust	14.2	*Synergistic effect*	Asiegbu et al. (1996)
20 g	1.5 (6)	Sawdust	8	*Synergistic effect*	Asiegbu et al. (1996)
20 g	1.5 (6)	Sawdust	13.5	*Synergistic effect*	Asiegbu et al. (1996)
16×16×7	1.5 (6)	Wood blocks	19	*Synergistic effect*	Chi et al. (2007)
16×16×7	1.5 (6)	Wood blocks	26	*Synergistic effect*	Chi et al. (2007)
16×16×7	1.5 (6)	Wood blocks	13	*Antagonistic effect*	Chi et al. (2007)
16×16×7	1.5 (6)	Wood blocks	21	*Synergistic effect*	Chi et al. (2007)
16×16×7	1.5 (6)	Wood blocks	18	*Synergistic effect*	Chi et al. (2007)
6×9	0.5 (2)	Wood chips	~31	*Synergistic effect*	Van Heerden et al. (2008)
6×9	0.5 (2)	Wood chips	~27	*No synergistic effect*	Van Heerden et al. (2008)
80 mesh	3 (12)	Wood residues	36	*Synergistic effect*	Wang et al. (2014)
80 mesh	3 (12)	Wood residues	34	*Antagonistic effect*	Wang et al. (2014)

(Continued)

TABLE 5.2 An overview of studies employing mono and mixed cultures of fungi in the degradation of woody substrates—cont'd

Wood species	Wood sample taken from	Fungal isolate no.	Fungal name
Populus tomentosa (Chinese white poplar)	Fresh poplar wood	C6320 + D10149	*Trametes orientalis + Trametes velutina*
Picea abies (Norway Spruce)	Fresh Norway spruce	T. viride 945 + H. parviporum 1023	*Trichoderma viride + Heterobasidion parviporum*
Ficus microcarpa (curtain fig)	NG	Pn (A42) + Tk (ML56)	*Phellinus noxius + Trichoderma koningiopsis*
Ficus microcarpa (curtain fig)	NG	Tasp (TA) + Pn (A42)	*Trichoderma asperellum + Phellinus noxius*
Ficus microcarpa (curtain fig)	NG	Tk (ML56) + Pn (A42)	*Trichoderma koningiopsis + Phellinus noxius*
Picea abies (Norway Spruce)	Single ~70-year aged individual tree	FBCC1181 + FBCC0110	*Fomitopsis pinicola + Trichaptum abietinum*
Picea abies (Norway Spruce)	Single ~70-year aged individual tree	FBCC1181 + FBCC0043	*Fomitopsis pinicola + Phlebia radiata*
Picea abies (Norway Spruce)	Single ~70-year aged individual tree	FBCC0110 + FBCC0043	*Trichaptum abietinum + Phlebia radiata*
Beech wood	NG	HfGTWV2 + EMPA62	*Hypholoma fasciculare + Coniophora puteana*
Beech wood	NG	TVWGP01 + EMPA62	*Trametes versicolor + Coniophora puteana*
Beech wood	NG	TVWGP01 + LSPT01	*Trametes versicolor + Laetiporus sulphureus*
Beech wood	NG	TVWGP01 + FPEF01	*Trametes versicolor + Fomitopsis pinicola*
Beech wood	NG	VcWVJH1 + EMPA62	*Vuilleminia comedens + Coniophora puteana*
Beech wood	NG	VcWVJH1 + FPEF01	*Vuilleminia comedens + Fomitopsis pinicola*
Triple culture			
Picea abies (Norway Spruce)	Sawdust from 60 to 65 years old spruce and 20–30 years old birch	CMI210864 + CMI747691 + CMI199761	*Trametes versicolor + Phanerochaete chrysosporium + Pleurotus sajor-caju*
Eucalyptus grandis	12-year-old plantation	PPRI 6762 + ABA006 + ABA003	*Pycnoporus sanguineus + Pichia guilliermondii + Rhodotorula glutinis***
Eucalyptus grandis	12-year-old plantation	PPRI 6762 + ABA006 + ABA003	*Pycnoporus sanguineus + autoclaved Pichia guilliermondii + autoclaved Rhodotorula glutinis***
Picea abies (Norway Spruce)	Fresh Norway spruce	T. viride 945 + P. gigantea 702 + H. parviporum 1023	*Trichoderma viride + Phlebiopsis gigantea + Heterobasidion parviporum*
Picea abies (Norway Spruce)	Single ~70-year aged individual tree	FBCC1181 + FBCC0110 + FBCC0043	*Fomitopsis pinicola + Trichaptum abietinum + Phlebia radiata*

?, unknown; *WR*, white rot; *BR*, brown rot; *NA*, not applicable; *NG*, not given.

*The intent of this study was not to demonstrate mass loss potential, but to capture early decay stages and chemical changes in wood.

** Selective or preferential delignification

*** Simultaneous delignification

**** Decay of healthy and diseased wood samples separated by slash (/), respectively

***+ Both selective and simultaneous delignification

Material size (mm) or mass (g)	Incubation in months (weeks)	Material used	Max decay (%)	Decay pattern	References
80 mesh	3 (12)	Wood residues	24	*Antagonistic effect*	Wang et al. (2014)
20×10×10	7.5 (30)	Wood blocks	27.5	*Antagonistic effect*	Alksne et al. (2015)
~25×25×7	3 (12)	Wood blocks	4.3	*Antagonistic effect*	Chou et al. (2019)
~25×25×7	3 (12)	Wood blocks	3	*Antagonistic effect*	Chou et al. (2019)
~25×25×7	3 (12)	Wood blocks	3	*Antagonistic effect*	Chou et al. (2019)
10 g	3 (12)	Sawdust	15.9	*Antagonistic effect*	Mali et al. (2019)
10 g	3 (12)	Sawdust	16	*Antagonistic effect*	Mali et al. (2019)
10 g	3 (12)	Sawdust	6.5	*Antagonistic effect*	Mali et al. (2019)
20×20×20	3 (12)	Wood blocks	~11	*Antagonistic effect*	Fukasawa et al. (2020)
20×20×20	3 (12)	Wood blocks	~24	*No synergistic effect*	Fukasawa et al. (2020)
20×20×20	3 (12)	Wood blocks	~28	*No synergistic effect*	Fukasawa et al. (2020)
20×20×20	3 (12)	Wood blocks	~21	*Antagonistic effect*	Fukasawa et al. (2020)
20×20×20	3 (12)	Wood blocks	~10	*No synergistic effect*	Fukasawa et al. (2020)
20×20×20	3 (12)	Wood blocks	~16	*No synergistic effect*	Fukasawa et al. (2020)
20 g	1.5 (6)	Sawdust	16.1	*Synergistic effect*	Asiegbu et al. (1996)
6×9	0.5 (2)	Wood chips	~30	*No synergistic effect*	Van Heerden et al. (2008)
6×9	0.5 (2)	Wood chips	~30	*No synergistic effect*	Van Heerden et al. (2008)
20×10×10	7.5 (30)	Wood blocks	34.3	*Antagonistic effect*	Alksne et al. (2015)
10 g	3 (12)	Sawdust	18.5	*Antagonistic effect*	Mali et al. (2019)

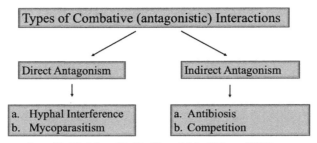

FIG. 5.2 Mechanisms of combative interactions. *Modified from Dix Neville and John Webster (1995).*

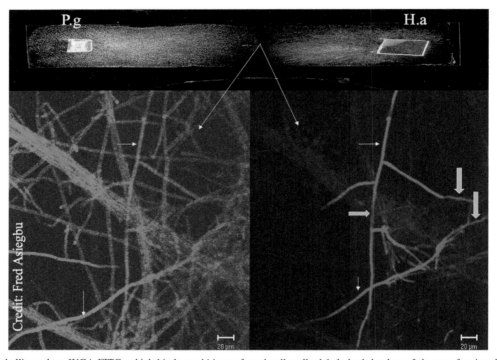

FIG. 5.3 Dual labelling where WGA-FITC, which binds to chitin on fungal cell walls, labels both hyphae of the two fungi and emits a greenish fluorescence. The goat antimouse IgG conjugated to Texas Red that binds to the monoclonal antibody (BC-AB9) raised for *Botrytis cineria* with high specificity and affinity to *H. annosum* (H.a) gives a reddish fluorescence. Hyphal interference interaction at the barrage or interaction zone between *Heterobasidion annosum* (thin arrows) and *Phlebiopsis gigantea* (P.g) as observed under confocal microscope. Thicker arrows indicate squashed *Heterobasidion hyphae.*

growth or penetration into a potential host may be accompanied by the mycoparasite growing along and around the host hyphae (Jeffries, 1995; Whipps, 2001). Several authors have observed that metabolites (enzymes and toxins) secreted by the parasite induce lysis in the cell walls of the host hyphae, thereby facilitating penetration (Deane et al., 1998; Howell, 1998; Vázquez-Garcidueñas et al., 1998). Some of the potential enzymes engaged by one of the commonly studied mycoparasites, *Trichoderma* spp., include β-glucanases, proteases, and chitinases. A number of *Trichoderma* spp. also secrete peptides (e.g., peptaibols) with antibiotic and antifungal properties, which disrupt cytoplasmic membranes and cause hyphal leakage and eventual cell death (Sharma et al., 2011; Shi et al., 2012). Jeffries (1995) noted that fungi do not always parasitize other mycelia, but some have been found to parasitize fruiting bodies, spores, and sclerotia (Maurice et al., 2021).

Two broad categories of mycoparasitic mechanisms and habits have been recognized in biotrophic and necrotrophic fungi. Several of the most studied mycoparasitic species belong to the division ascomycetes (e.g., *Trichoderma, Clonostachys rosea*). The mycoparasitic fungus, *Clonostachys rosea*, parasitizes phytopathogens such as *Botrytis cinerea* and *Fusarium graminearum* (Nygren et al., 2018). In wood systems, *L. betulina* parasitizes *Coriolus spp.*, and *Pseudotrametes gibbosa* feeds on *Bjerkandera* species. It has been suggested that the mycoparasitic mechanism plays a partial role in the establishment of mycoparasites so as to temporarily extract nutrients from the host mycelial surface (Rayner et al., 1987). Additional examples related to mycoparasitism are presented in Table 5.3. In principle, there is a large overlap between the different types of combative interactions in fungi. It is therefore possible that the majority of the strategies engaged during the various interactions stem from the same signaling pathways, possibly involving reactive oxygen species (ROS). This ultimately will lead to the production of diverse secondary metabolites used to kill the combatant hyphae, penetrate them, or provoke morphological differentiation and transformation. Overall, antibiosis, unlike other forms of antagonism (HI and mycoparasitism), does not necessarily require the fungi to invade the territory of others, therefore, Boddy (2000) suggested the use of the name "antagonism at a distance" to qualify it.

4. Types of response to competitive or combative interactions

4.1 Morphological responses to competitive interactions

Mycelial morphological changes are obvious in the areas that are in direct contact with their competitor (i.e., barrage or interaction zones). Hyphae may lump together to form barrage zones (Fig. 5.4), which physically block the entrance of the competitor or form linear aggregates of hyphae to invade or evade the defense system of the competitor. Microenvironmental conditions and local stimuli often dictate the morphological structures of interacting mycelia. In the mycelial front of two interacting fungi, the morphological features and structures may appear different depending on the stimulus (Rayner et al., 1994). In dual in vitro paired interactions, *H. parviporum* overgrew *Cortinarius gentilis* (ectomycorrhiza) and *Phialocephala sphaeroides* (endophyte) (Wen et al., 2019). The barrage zone formation was observed in paired cultures of saprotrophs (*Phlebiopsis gigantea* and *Phanerochaete chrysosporium*) with *H. parviporum* (Wen et al., 2019). A similar barrage zone was documented in cocultures of *Rigidoporus microporus* and *Phlebiopsis gigantea* (Oghenekaro et al., 2020). Using a confocal microscope, morphological distortions and the squashing of *Heterobasidion* interacting hyphae in contact with *P. gigantea* mycelia have been documented (Fig. 5.3, Asiegbu, unpublished). Qian and Chen (2012) observed *P. chrysosporium* and *T. versicolor* clearly change their hyphal morphology during interspecific interactions and deadlocked each other on an agar plate. Recently, interspecific interactions between three white rot fungi—*T. versicolor*, *T. maxima*, and *Ganoderma* sp.—were evaluated against *Aspergillus niger* (Lira-Pérez et al., 2020). A deadlock was observed when *T. versicolor* and *T. maxima* were cocultured separately with *A. niger*, whereas *Ganoderma* sp. partially replaced *A. niger* on the agar medium (Lira-Pérez et al., 2020). Another experiment in the lab on the agar medium showed that *F. velutipes* and *S. commune* deadlocked each other with a barrage zone whereas *S. commune* did replace *S. lacrymans*, and even fruiting bodies started to appear in the interaction of *G. lucidum* and *S. commune* after 1 month of cocultivation (Krause et al., 2020). An endophytic fungus, *Trichoderma gamsii*, was found to encircle and form coils around phytopathogens of the Chinese ginseng plant, including *Epicocum nigrum*, *Scytalidium lignicola*, *Phoma herbarum*, and *Fusarium flocciferum*, on agar plates (Chen et al., 2016).

4.2 Physicochemical responses including the production of volatile organic compounds (VOCs)

Deciphering the impact of metabolic products such as VOCs, enzymes, and other metabolites on interspecific interaction will be crucial in our understanding of their functional role as a driver of fungal community structure (O'Leary et al., 2019). There are more than 250 fungal gas phase infochemicals or VOCs that have been identified and characterized in fungi showing either antifungal or profungal (attractants) activities. In many cases, they exist as mixtures of simple hydrocarbons, heterocycles, aldehydes, ketones, alcohols, phenols, thioalcohols, thioesters, and their derivatives, including, among others, benzene derivatives and cyclohexanes (Korpi et al., 2009; Morath et al., 2012; Ortíz-Castro et al., 2009). El Ariebi et al., (2016) observed that VOCs have both inhibitory and stimulatory effects on interspecific fungal interaction. Similar observations on the inducible effects of VOCs on interspecific interactions were also made by other authors (Heilmann-Clausen and Boddy, 2005; Rayner et al., 1994). VOCs are likely to have a diverse neutral, negative, or positive impact on a wide range of fungi and other organisms in the wider biosphere (Humphris et al., 2002; Wheatley, 2002). Fungal infochemical or VOC production is highly dictated by the nature of the fungal species and pairs of species combinations (Mali et al., 2019). Several authors have recently reviewed various aspects of VOCs produced by fungi in self-pairing or during interactions with other organisms (Campos et al., 2010; de Boer et al., 2005; Morath et al., 2012; Wenke et al., 2010; Wheatley, 2002). One of the noted effects of VOCs is the impact on the rate of hyphal extension and mycelial growth (El Ariebi et al., 2016). The authors observed that the effects of VOCs differed depending on the substrate, fungal species, and status of the interaction (El Ariebi et al., 2016). A VOC lactone, (*rac*)-3,4-dimethylpentan-4-olide, produced in culture both by *Hymenoscyphus fraxineus* and *Hymenoscyphus albidus* inhibits germination of the seeds of *F. excelsior* and causes necroses in young seedlings (Citron et al., 2014; Halecker et al., 2020).

TABLE 5.3 Examples of mycoparasitism.

Parasitic types	Mycoparasite	Mycohost	Mycoparasite (sub) phylum	Mycohost (sub) phylum	Mechanism of parasitism	References
Intracellular biotroph	*Ampelomyces* spp.	*Arthrocladiella mougeotii, Blumeria graminis, Sawadaea bicornis* (all powdery mildews)	Ascomycota	Ascomycota	Entire thallus of the parasite enters the hypha of the host; host cell remains functional	Kiss (1997); Kiss (1998)
Haustorial biotroph	*Piptocephalis* spp. *Dimargaris* spp. *Filobasidiella depauperata*	At least 20 genera of Mucorales	Zoopagomycotina Kickxellomycotina Basidiomycota	Mucoromycotina Mucoromycotina Ascomycota	A short haustorial branch from a parasite hypha penetrates the host; host cell remains functional	Berry and Barnett (1957)
Fusion biotroph	*Gonatobotrys simplex Dicyma parasitica*	*Alternata alternata Physalospora obtusa*	Ascomycota Ascomycota	Ascomycota Ascomycota	Host and parasite are in intimate contact; micropore(s) form between the adpressed host and parasite hyphae, or from a short penetrative branch from the parasite hypha; host cell remains functional	Hoch (1977a); Hoch (1977b)
Contact necrotroph (hyphal intereference)	*Phlebiopsis gigantea*	*Heterobasidion annosum*	Basidiomycota	Basidiomycota	Parasite contacts but does not penetrate the host hyphae. Host cytoplasm degenerates and lysis may occur	Ikediugwu (1976a)
	Coprinellus heptemerus	*Ascobolus crenulates*	Basidiomycota	Ascomycota		Ikediugwu (1976b)
	Panaeolus sphinctrinus	*Bolbitius vitellins*	Basidiomycota	Basidiomycota		See Boddy (2016)
	Cladosporium sp	*Exobasidium camelliae*	Ascomycota	Basidiomycota		Mims et al. (2007)
	Trichoderma spp	*Phellinus noxius*	Ascomycota	Basidiomycota		Chou et al. (2019)
	Trichoderma gamsii	*Epicocum nigrum, Phoma herbarum, Fusarium flocciferum, Sytalidium lignicola*	Ascomycota	Ascomycota		Chen et al. (2016)
Invasive necrotroph	*Rozella* species	*Allomyces, Chytridium, Rhizophlyctis, Rhizophydium, Zygorrhizidium*	Cryptomycota	Chytridiomycota	Following contact, the parasite penetrates and enters the host; host cytoplasm rapidly degenerates and hyphal lysis often occurs	Bostick (1968)

Syncephalis californicus	Rhizopus oryzae	Zoopagomycotina	Mucoromycotina	Jeffries (1995)
Nectria inventa	Alternaria brassicae (hyphae and conidia)	Ascomycota	Ascomycota	Tsuneda et al. (1976)
Coniothyrium minitans, Talaromyces falvus	Sclerotinia sclerotiorum (sclerotia)	Ascomycota	Ascomycota	Turner and Tribe (1976)
Cladosporium uredinicola	Puccinia violae (uredospores)	Ascomycota	Basidiomycota	Traquair et al. (1984)
Fusarium merismoides	Pythium ultimum (oospores)	Ascomycota	Oomycota	Hoch and Abawi (1979)
Mycogone perniciosa	Rhopalomyces elegans (conidia)	Ascomycota	Mucoromycotina	Dayal and Barron (1970)
Mycogone perniciosa	Agaricus and Pluteus fruit bodies	Ascomycota	Basidiomycota	Gray and Morgan-Jones (1981)
Trichoderma spp.	Many such as Rhizoctonia solani (=Thanatephorus cucumeris), Corticium rolfsii	Ascomycota	Basidiomycota	Elad (1983) and Tseng et al. (2008)
Trichoderma harzianum	Botrytis cinerea	Ascomycota	Ascomycota	Elad and Kapat (1999)

Source: Adapted and modified form of Boddy, L., 2016. Interactions between fungi and other microbes, in: The Fungi, third ed. https://doi.org/10.1016/B978-0-12-382034-1.00010-4 and Jeffries, P., 1995. Biology and ecology of mycoparasitism. Can. J. Bot. https://doi.org/10.1139/b95-389.

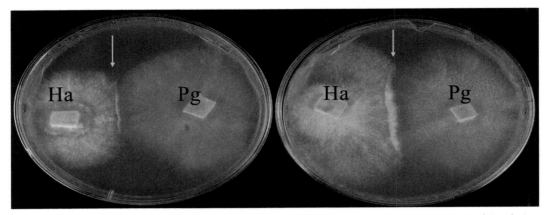

FIG. 5.4 Barrage zone formation (arrows) between *Heterobasidion annosum* and *Phlebiopsis gigantea*. *(Photo credit: Fred Asiegbu.)*

In nature, VOCs can pervade and disperse in soil pores for long distances (Aochi and Farmer, 2005). For instance, it has been argued that VOCs released by *Trichoderma* spp. in the soil are crucial in controlling plant pathogens (Dennis and Webster, 1971; Effmert et al., 2012), activating plant immunity, and enhancing plant growth (Contreras-Cornejo et al., 2014). Generally, *Muscodor* spp. isolates produce propanoic acid, 2-methyl-, and methyl esters (Kudalkar et al., 2012). However, for the first time, a few unusual compounds such as thujopsene, chamigrene, isocaryophyllene, and butanoic acid, 2-methyl were generated by *Muscodor sutura* (Kudalkar et al., 2012). After a 2-day exposure of VOCs produced by *M. sutura* to the 15 test fungi, two yeasts, and two bacterial species, it was found that 13 test fungi were completely inhibited (Kudalkar et al., 2012). The only two fungi that were not completely inhibited were *Trichoderma viridae* and *Fusarium solani*, showing about 42% and 58% growth inhibition, respectively (Kudalkar et al., 2012). Between two yeasts, *Candida albicans* and *Saccharomyces cerevisiae*, the former survived while the latter was inhibited and killed. The *M. sutura* VOCs were not inhibitory against *Escherichia coli* (Gram-) and *Bacillus subtilis* (Gram +) (Kudalkar et al., 2012). A recent report demonstrated that the production of VOCs by a strain (41E) of *Candida sake* could play a significant role in preventing blue mold on red delicious apples stored at low temperatures (Arrarte et al., 2017). Furthermore, several of the most studied mycoparasitic species of *Trichoderma* are very well known due to their special capability to produce VOCs and secondary metabolites (Schubert et al., 2008). It has also been observed that some classes of VOCs exert more profound effects than others. Aldehydes and ketones, including decanal, heptanal, 2-propanone, 2-methyl-1-butanol, and octanal, produced by *Trichoderma* spp. are inhibitory to the mycelia proliferation of wood-decaying basidiomycetes fungi (Boddy, 2016). VOCs have also been reported to have effects on gene transcript abundance (Boddy, 2016).

4.3 Biochemical responses

A key feature of antagonistic or interspecific fungal interactions is the production of secondary metabolites. Out of a total of 113 fungal isolates, Terhonen et al. (2014) found 19 isolates (17%) that inhibited the in vitro growth of the root rot pathogen *H. parviporum*. The root endophytes among them, that is, *Phialocephala sphaeroides*, *Cryptosporiopsis* sp., and their secondary metabolites, were observed to be promising antagonists of pathogenic *Heterobasidion parviporum*, *Phytophthora pini*, and *Botrytis cinerea* on the agar medium (Terhonen et al., 2016). Recently, secondary metabolites (postrediene A, B, and C) released during interspecific interactions of *P. ostreatus* and *Trametes robiniophila* showed great success in vitro as novel substances with high fungicidal activity against the human pathogenic fungi, *Candida albicans* and *Cryptococcus neoformans* (Shen et al., 2019). In vitro coculture of two citrus pathogens, *Penicillium digitatum* and *Penicillium citrinum*, that live on the same host plant revealed that secondary metabolites such as tryptoquialanines, citrinadins, chyrsogenamide A, and tetrapeptides have high antifungal activity and reduced 67% of *P. digitatum* radial growth. The sporulation of *P. citrinum* was inhibited by tryptoquialanines (Costa et al., 2019). The endophyte *Hypoxylon rubiginosum* can colonize axenically cultured *Fraxinus excelsior* (European ash) asymptomatically; it primarily produces phomopsidin, an antifungal metabolite, that is toxic to the ash dieback causal agent *Hymenoscyphus fraxineus* (Halecker et al., 2020).

The synergistic effect of unique metabolites such as sclerin and others found in the interaction zone on an artificial medium was most likely involved in inhibiting the growth of these pathogens (Terhonen et al., 2016). de Oliveira et al. (2018) observed the production of secondary metabolites (terpenoids, phenols, and alkaloids) in dual cultures between *D. pinea* and other phytopathogens (*Armillaria* sp., *B. adusta*, *B. cinerea*, *Rhizoctonia* sp.). The growth of mycelia may decrease or cease as these secondary metabolites diffuse in the agar (Schoeman et al., 1999) and may be correlated with the antagonistic activity depicted by *D. pinea* against phytopathogens (de Oliveira et al., 2018). A notable example is the increased enzymatic activity documented at the barrage or interaction zones (Eyre et al., 2010). Eyre et al. (2010) reported the upregulation of enzymes involved in the production of ROS (e.g., NADPH oxidases, phenol oxidases/laccases, peroxidase) at the interaction zone (Eyre et al., 2010). The production of ROS, phenol oxidases, and β-glucosidase was found to increase widely in mycelia, whereas the activities of laccases and manganese-dependent lignin peroxidase were enhanced at the contact or interaction zones (Eyre et al., 2010). Although the function of ROS in the interspecific interaction is not fully understood, their accumulation may likely have a toxic effect with potential impact on disruptions of cellular processes resulting in cell death (Silar, 2005; Tornberg and Olsson, 2002). Other authors reported that interacting fungi could engage their diverse biochemical and DNA repair machinery to resist or tolerate the combative damage caused by ROS toxicity (Eyre et al., 2010; Iakovlev et al., 2004). Zhao et al. (2015) reported that the production of a signaling compound such as nitric oxide (NO) during the interaction between *Inonotus obliquus* and *Phellinus morii* triggers the secretion of fungistatic substances.

Baldrian (2004) observed increased levels of phenol oxidase (laccase) and peroxidase activities in the interaction zones of *Trametes versicolor* and *Trichoderma harzianum*. The increased utilization or competition for nutrients during combative interaction between fungi could lead to the production of phenol oxidases and peroxidases and the subsequent generation of ROS between fungi. Laccase activities in mixed cultures were 2–25 times greater than in monocultures, depending on the species (Baldrian, 2004). For instance, cocultures of *T. versicolor* and *P. ostreatus* had significantly higher laccase activities than did monocultures of either fungus (Baldrian, 2004). The biological role of laccase is often associated with diverse physiological functions ranging from the polymerization of phenols to differentiation and fungal fruit body development (Mayer and Staples, 2002; Thurston, 1994).

Seven cell wall-degrading enzymes consisting of chitinase, cellulase, xylanase, β-1,3-glucanase, β-1,6-glucanase, mannanase, and protease were detected during interactions between *T. harzianum* and *R. solani* (Tseng et al., 2008). Mali et al. (2017) observed that *Phlebia radiata* and *Trichaptum abietinum* had the highest production of lignin-modifying oxidoreductases (manganese peroxidase, laccases) in cocultures or with *F. pinicola*. A major role of these enzymes might be the detoxification of the extracellular VOCs and DOCs of competitors (Baldrian, 2004; Hiscox et al., 2010). Melanin formation could also be linked to the enhanced laccase production, as it is demonstrated to be involved in the defense mechanisms during interspecific fungal interactions. The antimicrobial properties of the cell wall-bound melanin of *Phellinus weirii* have previously been demonstrated (Haars and Hüttermann, 1980). Melanin has also been reported to play a role in the protection of fungi against hydrolytic enzymes that degrade cell walls (Bloomfield and Alexander, 1967). Chitinases and *N*-acetylglucosaminidase (NAGase), which are induced during fungal-fungal interactions, might be involved in chitin degradation with consequent cell lysis and the release of sequestered nitrogen (Boddy, 2016).

Laccase production was enhanced in the interaction zone of *P. chrysosporium* and *T. versicolor* on the agar medium (Qian and Chen, 2012). It has been documented that *P. chrysoporium* does not produce laccase itself; perhaps the enhanced production was from *T. versicolor* (Qian and Chen, 2012). Other studies also reported that hyphal interactions between wood decay fungi are always associated with increased levels of laccase production (Kuhar et al., 2015; Zhong et al., 2019). Lira-Pérez et al. (2020) noted that interactions between *T. maxima* and *A. niger* result in high levels of H_2O_2 production and ligninolytic enzyme activities (laccases, MnP, and LiP). Nonetheless, the mechanistic understanding of laccase production during interspecific fungal interaction is not fully understood, and it appears to be dependent on the nature of the coculture species and their microenvironments (Hiscox et al., 2010; Qian and Chen, 2012; Zhong et al., 2019).

4.4 Transcriptional responses

In laboratory conditions, the activity of many microbial gene clusters was found to be cryptic or silent under controlled growth conditions (Harwani et al., 2018). Different tools and techniques that require prior comprehension of the gene sequences of the microbe have been developed. However, coculturing is one of the emerging approaches that mimics the natural environment in order to induce or trigger cryptic (orphan) genes during interspecific interactions (Harwani et al., 2018). The different antagonistic strategies employed by fungi against different combatants may lead to variable outcomes at the molecular level. The upregulation of 21% of gene transcripts was documented in the white rot fungus *Pycnoporus coccineus* when it interacted against the brown rot fungus *Coniophora puteana* and the grey mold fungus *Botrytis cinerea* (Arfi et al., 2013). This probably suggests the functional relevance of the abundant gene transcripts in the interspecific interaction. Morphological changes in mycelia observed during interspecific interaction are also linked to changes in gene expression when compared to noninteracting mycelia. Wen et al. (2019) reported a potential involvement in the interspecific fungal interaction of small secreted or effector-like proteins of *Heterobasidion parviporum* (HpSSPs). Several HpSSPs representatives were found to be up- or downregulated during the nonself interactions. HpSSPs that were commonly upregulated in combinations of *Phialocephala sphaeroides* (endophyte) and *Cortinarius gentilis* (ectomycorrhiza) appeared to be inactive with *Mycena* sp., which is a saprotroph (Wen et al., 2019). Another study reported that 1,3-beta glucan synthase and cytokinesis-related proteins were observed to be upregulated in *T. versicolor* during competition with *S. gausapatum* with evidence of cell wall alterations and cell divisions. The downregulation of genes encoding chitin synthase in *S. gausapatum* was observed at the same time (Eyre et al., 2010).

Melanin in the form of pigments at the interaction zones is often connected to morphological changes at the interacting mycelial fronts (Rayner et al., 1994). Melanins are involved in the hydrophobicity of hyphal surfaces and they reinforce the hyphal tensile strength by making cell-to-cell adhesion stronger (Bell and Wheeler, 1986). The significant upregulation of gene-encoding melanin was documented in *Neurospora crassa* when grown with a 72-h old *Penicillium chrysogenum* colony compared to expressions with the 24-h old *P. chrysogenum* (Villalta, 2011). Increased transcript levels of other genes associated with the cell wall, cell membrane, cross membrane transporters, Ca^{2+} dependent signaling, virulence, and transcriptional regulation were also observed (Villalta, 2011). Another metabolite with a putative functional role during interspecific fungal interaction is hydrophobin. Hydrophobin proteins play vital roles in the cell wall assembly and formation of aerial hyphae. The increased expression of genes encoding these proteins has been observed during interactions between *H. parviporum* and *P. gigantea* (Adomas et al., 2006). Hydrophobins may also prevent cytoplasmic loss during antagonistic fungal interactions by sealing off the damaged hyphae. Other gene transcripts encoding proteins with increased abundance documented in the barrage zone during the nonself hyphal interaction of *P. gigantea* and *H. annosum* include the ATP binding transporter, fructose-bisphosphate aldolase, endogalacturonase, glyceraldehyde-3-phosphate dehydrogenase, phosphoglucomutase, cyclophilin, and cytochrome P450 monooxygenase (Adomas et al., 2006). The same authors concluded that several of these genes encoding proteins important in the primary and secondary metabolisms could be relevant for nutrient acquisition and efficient substrate utilization, and might have a major influence on the outcome of the interaction. Another gene encoding hexagonal protein 1 (HEX-1) with a potential role in protecting against cytoplasmic loss was found to be upregulated during interaction between *Schizophyllum commune* and *Trichoderma viride* (Ujor et al., 2012).

Recently, genes encoding the carbohydrate active enzyme (CAZymes) as well as antifungal genes and oxidative stress-resistant genes were observed to be significantly upregulated in *Dichomitus squalens* (Ds) during interaction with *Trametes versicolor* (Tv) and *Pleurotus ostreatus* (Po) with the secretion of laccase (Zhong et al., 2019). The most significant upregulated genes during the interactions of DsTv or DsPo encoded terpenoid synthase, aldo/keto reductase, NAD(P)-binding protein, GroES-like protein, NADH:flavin oxidoreductase, laccase B, FAD-binding domain-containing protein, and glutathione S-transferase C-terminal-like protein. These proteins belong to diverse gene ontology (GO) categories, particularly to "monooxygenase activity," that may be linked to oxidative stress resistance, xenobiotic metabolism, and intracellular molecule binding, were found to be metabolically active in *Dichomitus squalens* (Zhong et al., 2019). Usually, xenobiotic stimulation, nutrient acquisition, and oxidative stress are the main reasons behind the promotion of laccase in cocultures. However, in a coculture system of *Pleurotus eryngii* var. *ferulae* versus *Rhodotorula mucilaginosa*, the genes related to oxidative stress were generally downregulated (Zhang et al., 2020). The comparison of the transcriptomic data of monoculture, coculture, and β-carotene culture confirmed that β-carotene was an essential substance for promoting laccase expression in this experiment (Zhang et al., 2020). Recently, *Trichoderma koningii* IABT1252 showed differential gene expressions in two stages of interaction (prior and after contact) with *Sclerotium rolfsii* (causes different forms of rot in many crops) in the agar medium. The prior contact interaction led to transcripts linked to signaling, secondary metabolite biosynthesis, hydrophobin, and transcription factors, whereas the after-contact condition harbors transcripts that were actively involved in mycoparasitism such as hydrolytic enzymes, proteases, signaling, transcription factor, and transporter proteins (Rabinal and Bhat, 2020).

5. Interaction outcome: Replacement, deadlock, metabiosis/antibiosis

The outcome of fungal interactions is crucial not only for the survival of the fungal population but also for community development and species diversity. Most often, the interaction outcome could be in the form of a replacement, deadlock, metabiosis, or antibiosis (Boddy, 2000). A deadlock or barrage zone between *Trametes versicolor* and *Phanerochaete chrysosporium* was observed by Qian and Chen (2012) on an artificial agar medium. Another deadlock was observed between *Diplodia pinea* and *Rhizoctonia* sp. on an agar plate (de Oliveira et al., 2018). Replacement, which is an overgrowth of the mycelia of a weak competitor by that of a strong combatant, was observed by the dominance of *P. velutina* over *Vuilleminia comedens* in just 1 week, whereas it took *Trametes versicolor* almost 4 weeks to replace *V. comedens* (Hiscox et al., 2015b). After initial inhibition, *Armillaria sp* was replaced by *D. pinea* when interacted on an agar medium (de Oliveira et al., 2018). Metabiosis is a form of commensalism where one organism creates a conducive environment for

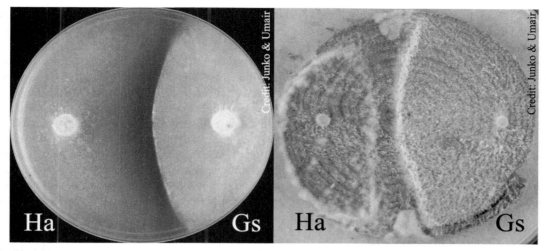

FIG. 5.5 Antibiosis phenomenon between white rot *Heterobasidion annosum* (Ha) and brown rot *Gloeophyllum sepiarium* (Gs) fungi on artificial agar media (left) and woody substrate (right).

the other to thrive. The phenomenon of metabiosis has been documented in interspecific interaction where secreted secondary metabolites stimulated the growth of the interactive species (Heilmann-Clausen and Boddy, 2005). The evidence of metabiosis could also be observed in successional changes among the wood decay community, which may be facilitated by the decomposition of the recalcitrant lignocellulose by white rot fungi that leaves behind easily degradable carbohydrates for consumption by ascomycetes. The in vitro interaction on agar between the phytopathogens *D. pinea* and *B. cinerea* suggested antibiosis between these species (de Oliveira et al., 2018). Antibiosis has been demonstrated in other systems such as interaction between *H. parviporum* and saprotrophic *Mycena* sp. (Wen et al., 2019). Antibiosis has been observed between *H. annosum* and *G. sepiarium* during interaction on agar and a woody substrate (Fig. 5.5). However, the weak correlation between behavior in the artificial culture and competitiveness in the natural conditions requires a cautious interpretation of antibiosis documented in laboratory studies.

6. Impact of biotic and abiotic factors on the outcome of interspecific fungal interactions

Fungal interaction outcomes are influenced by both biotic as well as abiotic factors. This makes predicting a winner of an interaction quite unpredictable and uncertain (Huisman and Weissing, 2001). Under an in vitro agar culture and ambient conditions, *H. fasciculare* outgrew and replaced *Steccherinum fimbriatum* mycelia but showed a deadlock in soil, and was itself overgrown in a woody substrate (Dowson et al., 1988). Mgbeahuruike et al. (2011) observed that the growth rate of *P. gigantea* was significantly affected by the substrate composition. They observed that 90% of isolates were able to displace *H. annosum* sensu stricto on a carbon-limited modified nutrient medium (NM) supplemented with ferulic acid, xylan, sawdust, and cellulose compared with only 4% in Hagem, a glucose-rich nutrient medium. This shows that the traits necessary for growth in wood are likely to be different from the traits needed in an artificial nutrient-rich medium. It was also noted that varying temperature conditions had an effect on the growth rate of the *P. gigantea* isolates. Over a 10-day incubation period, the highest average growth rate, 7.8 mm day^{-1} (with minimum 5.1 to maximum 8.4 mm day^{-1}), was observed at 20°C. The mean growth rate was 2.8 mm day^{-1} (0.4–3.6) at 10°C and 1.7 mm day^{-1} (0–5.7) at 30°C (Mgbeahuruike et al., 2011). Hiscox et al. (2016) also reported the impact of temperature during growth in wood. Higher temperatures favored mid-successional species whereas late or early successional species were found to be more successful at lower temperatures. The wood decay rates of eight brown rot and eight white rot fungi were found to be affected by temperature shifts (Toljander et al., 2006). Higher wood decay rates were observed with brown rot than the microcosm with white rot fungi under fluctuating temperatures (Toljander et al., 2006). By contrast, the interaction outcome between *P. velutina* and *R. bicolor* (both late successional species) was greatly altered when the temperature was increased in the soil microcosm from 15°C to 18°C (Crowther et al., 2012). In terms of fungal dominance, a 4°C temperature difference was found to alter the order of dominance (A'Bear et al., 2013). These authors observed that at 16°C, the dominance of the isolates was reflected in this order: *P. velutina* > *R. bicolor* > *H. fasciculare*. This, however, changed in ungrazed systems at an elevated temperature of 20°C: *R. bicolor* > *P. velutina* > *H. fasciculare*. Holmer and Stenlid (1993) observed that the success rate of competitive replacement was generally the lowest for mycelia that occupied 8% of the wood sectors and highest for those inhabiting 92% of the wood disc. They noted, however, that over the time course of the experiment, the competitive ability overrode the effects of inoculum size (Holmer and Stenlid, 1997b).

Furthermore, at a higher inoculum potential, *Gloeophyllum trabeum* outcompeted *Irpex lacteus* in two of the four wood types, but was shown to lose in an "equal-footing" interaction with *I. lacteus* (Song et al., 2015). The microclimatic factors such as moisture and gaseous regime were also found to effect the interactional outcomes. Boddy et al. (1985) reported that at a lower water potential, *Daldinia concentrica* was more combative, whereas other species were less combative in the agar medium. Recent findings suggest that the volume of natural substrate (wood) or territorial size, in lab conditions, affects the outcome of competitive interactions and the degradation rate of the substrate (Fukasawa et al., 2020). However, mycelia growing and occupying a larger substrate (wood) were more competitive than those occupying smaller ones and there was a negative relationship between the substrate (wood) volume and the percent mass loss of wood (Fukasawa et al., 2020). However, this experiment did not test monocultures on individual wood blocks of different volumes, so it is difficult to say whether this negative relationship between wood volume and decay was due to the competition or some other factor(s).

7. Fungal succession as an interaction outcome

Succession refers to the sequential occupation of a substrate or a change in species abundance and their types on a substrate after a disturbance. Another way of looking at succession is to observe biological changes that take place in an ecosystem after a disturbance, which often leads to predictable shifts in the sequential species composition (Odum, 2014). There are two types of succession: primary succession and secondary succession. Rayner and Boddy (1988) noted that pioneer fungi species are able to rapidly colonize the exposed fresh xylem wood surfaces but are later replaced by combative organisms of specific communities. The pioneer species usually lack detoxification capabilities against tree protective chemicals and are unable to degrade recalcitrant lignified wood components. Some members of the fungal group ascomycetes (e.g., *Hypoxylon* and *Xylaria* sp.) are pioneers that are able to colonize xylary tissues. Several members of basidiomycetes, *Inonotus obliquus,* and *Cerrena unicolor* causative agents of canker root rot as well as other root rot pathogens (e.g., *Armillaria, Heterobasidion,* and *Phellinus*) are usually not pioneers and are able to metabolize complex wood tissues (Blanchette, 1982a; Blanchette, 1982b; Enebak and Blanchette, 1989; Asiegbu et al., 2005; Rayner and Boddy, 1988).

Primary colonizers are generally regarded as weaker combatants than the secondary or tertiary colonizers (Hiscox et al., 2016). Generally, hierarchical (transitive) associations occur where early arrivers (primary colonizers) are the least combative whereas latecomers (secondary colonizers) are the most combative (Boddy, 2000; Hiscox et al., 2018). Though these hierarchies are often not strictly followed, nonhierarchical (intransitive) associations are usual between wood-degrading fungi (Boddy, 2000; Laird and Schamp, 2006). For this reason, the identity of the fungi and their interaction in primarily colonizing a substrate may influence the fate of newcomers. This outcome is known as the **priority effect** (Fukami, 2015; Fukami et al., 2010). The priority effects are starting to be included in studying fungal communities (Dickie et al., 2012; Kennedy et al., 2009; Peay et al., 2012). A few laboratory experiments have addressed questions related to primary colonizers in wood-inhabiting fungal communities. The identity of primary colonizers plays a significant role in the establishment of secondary colonizing species on decayed beech wood (Heilmann-Clausen and Boddy, 2005). Similar arguments have been presented by others regarding the priority effects of primary colonizers (Dickie et al., 2012; Fukami et al., 2010). The sequence of arrivers defines the composition of fungal communities in a habitat patch (Dickie et al., 2012; Kennedy et al., 2009; Peay et al., 2012). One mechanism may be where primary species use a preemptive competition strategy to exploit the resources in an area and establish themselves in a larger territory as compared to successive species in a habitat patch (Větrovský et al., 2011). Priority effects in wood-dwelling fungal communities often bring chemical changes in the wood in the form of toxins (Heilmann-Clausen and Boddy, 2005; Woodward and Boddy, 2008). In laboratory experiments, the predecomposition of wood by a primary species may exhibit inhibition but also facilitate the establishment of secondary or late colonizers, as suggested by Holmer et al. (1997). The same authors also observed that the species that were good successors under laboratory conditions also showed the same pattern in a natural environment when encountering early colonizers. Additionally, it has also been observed that the wood type, fungal species identity, and richness of species have some role in predetermining the wood degradation rates (Boddy, 2001; Van Der Wal et al., 2016). The wood decay was observed to be slower at the frontal parts of fungal interactions than the rear parts on wood blocks (Fukasawa et al., 2020), which potentially suggests that orientation is also important in decay patterns.

8. Challenges and constraints in interspecific interaction

The varying antagonistic activity of the same fungi in diverse or similar substrates and environments (Fig. 5.5; Table 5.2) requires cautious interpretation and extrapolation of in vitro results. The validity of the concept of competition for space and nutrients as well as models of fungal succession in a laboratory set-up may not fully correlate to observations in natural conditions. An additional limitation in studies of interspecific interaction is the inability to distinguish the hyphae of the interacting fungi under a microscope. This problem could be overcome with the use of specific dyes or probes visualized with the aid of a dual-channel confocal laser scanning microscope (see Fig. 5.3) and other novel imaging techniques. The advent of new technologies such as transcriptomic, metabolomic, and proteomic analysis as well as their increasing affordability would also be useful in the mechanistic understanding of various fungal interactions. The applications of these "omics" approaches will most likely provide further insights into different genes, metabolites, and proteins and how they are regulated during antagonistic and synergistic mycelial interactions. The signaling processes also vary where nonself mycelial interaction stirs changes in the profile of infochemicals or VOCs relative to self-pairings, both qualitatively and quantitatively (El Ariebi et al., 2016). Such morphological and metabolic changes might be linked with the modulation of gene expression and the secretion of effector proteins (Wen et al., 2019). Further studies on the regulatory patterns of the genes encoding the effector-like proteins during the interspecific interaction will help in our understanding of this complex ecological process. There is also a huge gap in the research conducted in the northern hemisphere and the southern hemisphere on interspecific fungal interactions, probably due to the allocation of research budgets and priorities. A recent joint research effort and the creation of a big database such as "ClimFun" across the pan-European level (Andrew et al., 2019) could be expanded to the pan-continental level for advancement of useful knowledge sharing and data acquisition.

References

A'Bear, A.D., Boddy, L., Jones, T.H., 2013. Bottom-up determination of soil collembola diversity and population dynamics in response to interactive climatic factors. Oecologia. https://doi.org/10.1007/s00442-013-2662-3.

Adomas, A., Eklund, M., Johansson, M., Asiegbu, F.O., 2006. Identification and analysis of differentially expressed cDNAs during nonself-competitive interaction between Phlebiopsis gigantea and Heterobasidion parviporum. FEMS Microbiol. Ecol. https://doi.org/10.1111/j.1574-6941.2006.00094.x.

Agrios, G., 2005. Plant Pathology, fifth ed. Elsevier, California, https://doi.org/10.1016/C2009-0-02037-6.

Ahmad, B.I., Khan, R.A., Siddiqui, M.T., 2013. Incidence of dieback disease following fungal inoculations of sexually and asexually propagated shisham (Dalbergia sissoo). For. Pathol. https://doi.org/10.1111/efp.12001.

Ainsworth, A.M., Rayner, A.D.M., 1991. Ontogenetic stages from coenocyte to basidiome and their relation to phenoloxidase activity and colonization processes in Phanerochaete magnoliae. Mycol. Res. https://doi.org/10.1016/S0953-7562(09)80395-X.

Alksne, L., Nikolajeva, V., Petriņa, Z., Eze, D., Gaitnieks, T., 2015. Influence of Trichoderma isolates and Phlebiopsis gigantea on the growth of Heterobasidion parviporum and wood decay of Norway spruce in controlled conditions. Environ. Exp. Biol., 159–168.

Andrew, C., Büntgen, U., Egli, S., Senn-Irlet, B., Grytnes, J.A., Heilmann-Clausen, J., Boddy, L., Bässler, C., Gange, A.C., Heegaard, E., Høiland, K., Kirk, P.M., Krisai-Greilhüber, I., Kuyper, T.W., Kauserud, H., 2019. Open-source data reveal how collections-based fungal diversity is sensitive to global change. Appl. Plant Sci. https://doi.org/10.1002/aps3.1227.

Aochi, Y.O., Farmer, W.J., 2005. Impact of soil microstructure on the molecular transport dynamics of 1,2-dichloroethane. Geoderma. https://doi.org/10.1016/j.geoderma.2004.11.024.

Arfi, Y., Levasseur, A., Record, E., 2013. Differential gene expression in pycnoporus coccineus during interspecific mycelial interactions with different competitors. Appl. Environ. Microbiol. https://doi.org/10.1128/AEM.02316-13.

Arhipova, N., Gaitnieks, T., Donis, J., Stenlid, J., Vasaitis, R., 2011. Butt rot incidence, causal fungi, and related yield loss in Picea abies stands of latvia. Can. J. For. Res. https://doi.org/10.1139/X11-141.

Arrarte, E., Garmendia, G., Rossini, C., Wisniewski, M., Vero, S., 2017. Volatile organic compounds produced by Antarctic strains of Candida sake play a role in the control of postharvest pathogens of apples. Biol. Control. https://doi.org/10.1016/j.biocontrol.2017.03.002.

Asiegbu, P.O., Paterson, A., Smith, J.E., 1996. The effects of co-fungal cultures and supplementation with carbohydrate adjuncts on lignin biodegradation and substrate digestibility. World J. Microbiol. Biotechnol. https://doi.org/10.1007/BF00360927.

Asiegbu, F.O., Adomas, A., Stenlid, J., 2005. Conifer root and butt rot caused by Heterobasidion annosum (Fr.) Bref. s.l. Mol. Plant Pathol. https://doi.org/10.1111/j.1364-3703.2005.00295.x.

Awan, H., Pettenella, D., 2017. Pine nuts: a review of recent sanitary conditions and market development. Forests 8, 367. https://doi.org/10.3390/f8100367.

Baldrian, P., 2004. Increase of laccase activity during interspecific interactions of white-rot fungi. FEMS Microbiol. Ecol. https://doi.org/10.1016/j.femsec.2004.07.005.

Bari, E., Nazarnezhad, N., Kazemi, S.M., Tajick Ghanbary, M.A., Mohebby, B., Schmidt, O., Clausen, C.A., 2015. Comparison between degradation capabilities of the white rot fungi Pleurotus ostreatus and Trametes versicolor in beech wood. Int. Biodeterior. Biodegrad. https://doi.org/10.1016/j.ibiod.2015.03.033.

Bell, A.A., Wheeler, M.H., 1986. Biosynthesis and functions of fungal Melanins. Annu. Rev. Phytopathol. https://doi.org/10.1146/annurev.py.24.090186.002211.

Bellassen, V., Luyssaert, S., 2014. Managing forests in uncertain times. Nature. https://doi.org/10.1038/506153a.

Berry, C.R., Barnett, H.L., 1957. Mode of parasitism and host range of *Piptocephalis virginiana*. Mycologia. https://doi.org/10.2307/3755686.

Blanchette, R.A., 1982a. Decay and canker formation by *Phellinus pini* in white and balsam fir (*Abies concolor*, *Abies balsamea*). Can. J. For. Res. https://doi.org/10.1139/x82-084.

Blanchette, R.A., 1982b. Progressive stages of discoloration and decay associated with the canker-rot fungus, *Inonotus obliquus*, in birch. Phytopathology. https://doi.org/10.1094/phyto-72-1272.

Bloomfield, B.J., Alexander, M., 1967. Melanins and resistance of fungi to lysis. J. Bacteriol. https://doi.org/10.1128/jb.93.4.1276-1280.1967.

Boddy, L., 2000. Interspecific combative interactions between wood-decaying basidiomycetes. FEMS Microbiol. Ecol. https://doi.org/10.1016/S0168-6496(99)00093-8.

Boddy, L., 2001. Fungal community ecology and wood decomposition processes in angiosperms: from standing tree to complete decay of coarse woody debris. Boreal For., 43–56. https://doi.org/10.2307/20113263.

Boddy, L., 2016. Interactions between fungi and other microbes. In: The Fungi, third ed., https://doi.org/10.1016/B978-0-12-382034-1.00010-4.

Boddy, L., Gibbon, O.M., Grundy, M.A., 1985. Ecology of *Daldinia concentrica*: effect of abiotic variables on mycelial extension and interspecific interactions. Trans.—Br. Mycol. Soc. https://doi.org/10.1016/S0007-1536(85)80183-2.

Bostick, L.R., 1968. Studies of the morphology of *Chytriomyces hyalinus*. J. Elisha Mitchell Sci. Soc., 94–99.

Brazee, N.J., Lindner, D.L., 2013. Unravelling the Phellinus pini s.l. complex in North America: a multilocus phylogeny and differentiation analysis of Porodaedalea. For. Pathol. https://doi.org/10.1111/efp.12008.

Campos, V.P., de Pinho, R.S.C., Freire, E.S., 2010. Volatiles produced by interacting microorganisms potentially useful for the control of plant pathogens. Cienc. Agrotecnol. https://doi.org/10.1590/s1413-70542010000300001.

Carlile, M.J., 1995. The success of the hypha and mycelium. In: Gow, N.A.R., Gadd, G.M. (Eds.), The Growing Fungus. Springer, Dordrecht, Netherlands, pp. 3–19.

Chen, J.L., Sun, S.Z., Miao, C.P., Wu, K., Chen, Y.W., Xu, L.H., Guan, H.L., Zhao, L.X., 2016. Endophytic *Trichoderma gamsii* YIM PH30019: a promising biocontrol agent with hyperosmolar, mycoparasitism, and antagonistic activities of induced volatile organic compounds on root-rot pathogenic fungi of *Panax notoginseng*. J. Ginseng Res. https://doi.org/10.1016/j.jgr.2015.09.006.

Chi, Y., Hatakka, A., Maijala, P., 2007. Can co-culturing of two white-rot fungi increase lignin degradation and the production of lignin-degrading enzymes? Int. Biodeterior. Biodegrad. https://doi.org/10.1016/j.ibiod.2006.06.025.

Chou, H., Xiao, Y.T., Tsai, J.N., Li, T.T., Wu, H.Y., Liu, L.Y.D., Tzeng, D.S., Chung, C.L., 2019. In vitro and in planta evaluation of *Trichoderma asperellum* TA as a biocontrol agent against *Phellinus noxius*, the cause of brown root rot disease of trees. Plant Dis. https://doi.org/10.1094/PDIS-01-19-0179-RE.

Citron, C.A., Junker, C., Schulz, B., Dickschat, J.S., 2014. A volatile lactone of *Hymenoscyphus pseudoalbidus*, pathogen of European ash dieback, inhibits host germination. Angew. Chemie – Int. Ed. https://doi.org/10.1002/anie.201402290.

Contreras-Cornejo, H.A., Macías-Rodríguez, L., Herrera-Estrella, A., López-Bucio, J., 2014. The 4-phosphopantetheinyl transferase of *Trichoderma virens* plays a role in plant protection against *Botrytis cinerea* through volatile organic compound emission. Plant Soil. https://doi.org/10.1007/s11104-014-2069-x.

Costa, J.H., Wassano, C.I., Angolini, C.F.F., Scherlach, K., Hertweck, C., Pacheco Fill, T., 2019. Antifungal potential of secondary metabolites involved in the interaction between citrus pathogens. Sci. Rep. https://doi.org/10.1038/s41598-019-55204-9.

Crowther, T.W., Littleboy, A., Jones, T.H., Boddy, L., 2012. Interactive effects of warming and invertebrate grazing on the outcomes of competitive fungal interactions. FEMS Microbiol. Ecol. https://doi.org/10.1111/j.1574-6941.2012.01364.x.

Daniel, G., 2003. Microview of wood under degradation by bacteria and fungi. In: Goodell, B., Nicholas, D., Schultz, T. (Eds.), ACS Symposium Series. American Chemical Society, Washington, DC, pp. 34–72, https://doi.org/10.1021/bk-2003-0845.ch004.

Daniel, G., Asiegbu, F., Johansson, M., 1998. The saprotrophic wood-degrading abilities of *Heterobasidium annosum* intersterility groups P and S. Mycol. Res. https://doi.org/10.1017/S0953756297005935.

Dayal, R., Barron, G.L., 1970. *Verticillium psalliotae* as a parasite of Rhopalomyces. Mycologia. https://doi.org/10.1080/00275514.1970.12019025.

de Boer, W., Folman, L.B., Summerbell, R.C., Boddy, L., 2005. Living in a fungal world: impact of fungi on soil bacterial niche development. FEMS Microbiol. Rev. 29, 795–811. https://doi.org/10.1016/j.femsre.2004.11.005.

De Boer, W., Folman, L.B., Klein Gunnewiek, P.J.A., Svensson, T., Bastviken, D., Öberg, G., Del Rio, J.C., Boddy, L., 2010. Mechanism of antibacterial activity of the white-rot fungus *Hypholoma fasciculare* colonizing wood. Can. J. Microbiol. https://doi.org/10.1139/W10-023.

de Oliveira, C.F., Moura, P.F., Rech, K.S., da Silva Paula de Oliveira, C., Hirota, B.C.K., de Oliveira, M., da Silva, C.B., de Souza, A.M., de Fátima Gaspari Dias, J., Miguel, O.G., Auer, C.G., Miguel, M.D., 2018. Antagonistic activity of *Diplodia pinea* against phytopathogenic fungi. Folia Microbiol. (Praha). https://doi.org/10.1007/s12223-018-00667-y.

Deacon, J., 2013. Fungal Biology, fourth ed. Blackwell Publishing Ltd., https://doi.org/10.1002/9781118685068.

Deane, E.E., Whipps, J.M., Lynch, J.M., Peberdy, J.F., 1998. The purification and characterization of a *Trichoderma harzianum* exochitinase. Biochim. Biophys. Acta – Protein Struct. Mol. Enzymol. https://doi.org/10.1016/S0167-4838(97)00183-0.

Dennis, C., Webster, J., 1971. Antagonistic properties of species-groups of trichoderma: II. Production of volatile antibiotics. Trans. Br. Mycol. Soc. https://doi.org/10.1016/S0007-1536(71)80078-5.

Dickie, I.A., Fukami, T., Wilkie, J.P., Allen, R.B., Buchanan, P.K., 2012. Do assembly history effects attenuate from species to ecosystem properties? A field test with wood-inhabiting fungi. Ecol. Lett. https://doi.org/10.1111/j.1461-0248.2011.01722.x.

Dix, N.J., Webster, J., 1995. Fungal ecology. Fungal Ecol. https://doi.org/10.1515/9781400846870.194.

Dowson, C.G., Rayner, A.D.M., Boddy, L., 1988. The form and outcome of mycelial interactions involving cord-forming decomposer basidiomycetes in homogeneous and heterogeneous environments. New Phytol. 109, 423–432. https://doi.org/10.1111/j.1469-8137.1988.tb03718.x.

Eastwood, D.C., Floudas, D., Binder, M., Majcherczyk, A., Schneider, P., Aerts, A., Asiegbu, F.O., Baker, S.E., Barry, K., Bendiksby, M., Blumentritt, M., Coutinho, P.M., Cullen, D., de Vries, R.P., Gathman, A., Goodell, B., Henrissat, B., Ihrmark, K., Kauserud, H., Kohler, A., LaButti, K., Lapidus, A., Lavin, J.L., Lee, Y.H., Lindquist, E., Lilly, W., Lucas, S., Morin, E., Murat, C., Oguiza, J.A., Park, J., Pisabarro, A.G., Riley, R., Rosling, A., Salamov, A., Schmidt, O., Schmutz, J., Skrede, I., Stenlid, J., Wiebenga, A., Xie, X., Kües, U., Hibbett, D.S., Hoffmeister, D., Högberg, N., Martin, F., Grigoriev, I.V., Watkinson, S.C., 2011. The plant cell wall-decomposing machinery underlies the functional diversity of forest fungi. Science 333, 762–765. https://doi.org/10.1126/science.1205411.

Effmert, U., Kalderás, J., Warnke, R., Piechulla, B., 2012. Volatile mediated interactions between bacteria and fungi in the soil. J. Chem. Ecol. https://doi.org/10.1007/s10886-012-0135-5.

El Ariebi, N., Hiscox, J., Scriven, S.A., Müller, C.T., Boddy, L., 2016. Production and effects of volatile organic compounds during interspecific interactions. Fungal Ecol. 20, 144–154. https://doi.org/10.1016/j.funeco.2015.12.013.

Elad, Y., 1983. Parasitism of *Trichoderma* spp. on *Rhizoctonia solani* and *Sclerotium rolfsii* – scanning electron microscopy and fluorescence microscopy. Phytopathology. https://doi.org/10.1094/phyto-73-85.

Elad, Y., Kapat, A., 1999. The role of *Trichoderma harzianum* protease in the biocontrol of *Botrytis cinerea*. Eur. J. Plant Pathol. https://doi.org/10.1023/A:1008753629207.

Enebak, S.A., Blanchette, R.A., 1989. Canker formation and decay in sugar maple and paper birch infected by Cerrena unicolor. Can. J. For. Res. https://doi.org/10.1139/x89-031.

Erwin, 2016. Short communication: Microscopic decay pattern of yellow meranti (*Shorea gibbosa*) wood caused by white-rot fungus *Phlebia brevispora*. Biodiversitas. https://doi.org/10.13057/biodiv/d170203.

Eyre, C., Muftah, W., Hiscox, J., Hunt, J., Kille, P., Boddy, L., Rogers, H.J., 2010. Microarray analysis of differential gene expression elicited in *Trametes versicolor* during interspecific mycelial interactions. Fungal Biol. https://doi.org/10.1016/j.funbio.2010.05.006.

Filley, T.R., Cody, G.D., Goodell, B., Jellison, J., Noser, C., Ostrofsky, A., 2002. Lignin demethylation and polysaccharide decomposition in spruce sapwood degraded by brown rot fungi. Org. Geochem. https://doi.org/10.1016/S0146-6380(01)00144-9.

Floudas, D., Binder, M., Riley, R., Barry, K., Blanchette, R.A., Henrissat, B., Martínez, A.T., Otillar, R., Spatafora, J.W., Yadav, J.S., Aerts, A., Benoit, I., Boyd, A., Carlson, A., Copeland, A., Coutinho, P.M., De Vries, R.P., Ferreira, P., Findley, K., Foster, B., Gaskell, J., Glotzer, D., Górecki, P., Heitman, J., Hesse, C., Hori, C., Igarashi, K., Jurgens, J.A., Kallen, N., Kersten, P., Kohler, A., Kües, U., Kumar, T.K.A., Kuo, A., LaButti, K., Larrondo, L.F., Lindquist, E., Ling, A., Lombard, V., Lucas, S., Lundell, T., Martin, R., McLaughlin, D.J., Morgenstern, I., Morin, E., Murat, C., Nagy, L.G., Nolan, M., Ohm, R.A., Patyshakuliyeva, A., Rokas, A., Ruiz-Dueñas, F.J., Sabat, G., Salamov, A., Samejima, M., Schmutz, J., Slot, J.C., John, F.S., Stenlid, J., Sun, H., Sun, S., Syed, K., Tsang, A., Wiebenga, A., Young, D., Pisabarro, A., Eastwood, D.C., Martin, F., Cullen, D., Grigoriev, I.V., Hibbett, D.S., 2012. The paleozoic origin of enzymatic lignin decomposition reconstructed from 31 fungal genomes. Science. https://doi.org/10.1126/science.1221748.

Floudas, D., Held, B.W., Riley, R., Nagya, L.G., Koehlerd, G., Ransdell, A.S., Younus, H., Chow, J., Chiniquy, J., Lipzen, A., Tritt, A., Sun, H., Haridas, S., LaButti, K., Ohm, K.A., Küese, U., Blanchette, R.A., Grigoriev, I.V., Mintod, R.E., Hibbett, D.S., 2015. Evolution of novel wood decay mechanisms in Agaricales revealed by the genome sequences of *Fistulina hepatica* and *Cylindrobasidium torrendii*. Fungal Genet. Biol. 76, 78–92.

Fukami, T., 2015. Historical contingency in community assembly: integrating niches, species pools, and priority effects. Annu. Rev. Ecol. Evol. Syst. https://doi.org/10.1146/annurev-ecolsys-110411-160340.

Fukami, T., Dickie, I.A., Paula Wilkie, J., Paulus, B.C., Park, D., Roberts, A., Buchanan, P.K., Allen, R.B., 2010. Assembly history dictates ecosystem functioning: evidence from wood decomposer communities. Ecol. Lett. 13, 675–684. https://doi.org/10.1111/j.1461-0248.2010.01465.x.

Fukasawa, Y., Gilmartin, E.C., Savoury, M., Boddy, L., 2020. Inoculum volume effects on competitive outcome and wood decay rate of brown- and white-rot basidiomycetes. Fungal Ecol. https://doi.org/10.1016/j.funeco.2020.100938.

Goodell, B., 2003. Brown-rot fungal degradation of wood: our evolving view (Chapter 6). In: Goodell, B., Nicholas, D.D., Schultz, T.P. (Eds.), Wood Deterioration and Preservation. ACS Symposium Series. 845. American Chemical Society, Washington DC, pp. 97–118.

Goodell, B., Qian, Y., Jellison, J., 2008. Fungal decay of wood: soft rot-brown rot-white rot. In: Schultz,, T.P., Militz, H., Freeman, M.H., Goodell, B., Nicholas, D.D. (Eds.), Development of Commercial Wood Preservatives. ACS Symposium Series. 982. American Chemical Society, Washington, DC, pp. 9–31.

Gray, D.J., Morgan-Jones, G., 1981. Host-parasite relationships of *Agaricus brunnescens* and a number of mycoparasitic hyphomycetes. Mycopathologia. https://doi.org/10.1007/BF00439068.

Green, F.I.I.I., Highley, T.L., 1997. Brown-rot wood decay-insights gained from a low-decay isolate of *Postia placenta*. Trends Plant Pathol. 1.

Haars, A., Hüttermann, A., 1980. Function of laccase in the white-rot fungus *Fomes annosus*. Arch. Microbiol. https://doi.org/10.1007/BF00446882.

Halecker, S., Wennrich, J.P., Rodrigo, S., Andrée, N., Rabsch, L., Baschien, C., Steinert, M., Stadler, M., Surup, F., Schulz, B., 2020. Fungal endophytes for biocontrol of ash dieback: the antagonistic potential of *Hypoxylon rubiginosum*. Fungal Ecol. https://doi.org/10.1016/j.funeco.2020.100918.

Halley, J.M., Robinson, C.H., Comins, H.N., Dighton, J., 1996. Predicting straw decomposition by a four-species fungal community: a cellular automaton model. J. Appl. Ecol. https://doi.org/10.2307/2404979.

Hamed, S.A.M., 2013. In-vitro studies on wood degradation in soil by soft-rot fungi: *Aspergillus niger* and *Penicillium chrysogenum*. Int. Biodeterior. Biodegrad. https://doi.org/10.1016/j.ibiod.2012.12.013.

Harwani, D., Begani, J., Lakhani, J., 2018. Co-cultivation strategies to induce de novo synthesis of novel chemical scaffolds from cryptic secondary metabolite gene clusters. In: Fungi and Their Role in Sustainable Development: Current Perspective., https://doi.org/10.1007/978-981-13-0393-7_33.

Hatakka, A., Hammel, K.E., 2011. Fungal biodegradation of lignocelluloses. In: Industrial Applications., https://doi.org/10.1007/978-3-642-11458-8_15.

Heilmann-Clausen, J., Boddy, L., 2005. Inhibition and stimulation effects in communities of wood decay fungi: exudates from colonized wood influence growth by other species. Microb. Ecol. 49, 399–406. https://doi.org/10.1007/s00248-004-0240-2.

Heilmann-Clausen, J., Christensen, M., 2004. Does size matter? On the importance of various dead wood fractions for fungal diversity in Danish beech forests. In: Forest Ecology and Management., https://doi.org/10.1016/j.foreco.2004.07.010.

Hiscox, J., Boddy, L., 2017. Armed and dangerous – chemical warfare in wood decay communities. Fungal Biol. Rev. https://doi.org/10.1016/j.fbr.2017.07.001.

Hiscox, J., Baldrian, P., Rogers, H.J., Boddy, L., 2010. Changes in oxidative enzyme activity during interspecific mycelial interactions involving the white-rot fungus *Trametes versicolor*. Fungal Genet. Biol. https://doi.org/10.1016/j.fgb.2010.03.007.

Hiscox, J., Savoury, M., Müller, C.T., Lindahl, B.D., Rogers, H.J., Boddy, L., 2015a. Priority effects during fungal community establishment in beech wood. ISME J. https://doi.org/10.1038/ismej.2015.38.

Hiscox, J., Savoury, M., Vaughan, I.P., Müller, C.T., Boddy, L., 2015b. Antagonistic fungal interactions influence carbon dioxide evolution from decomposing wood. Fungal Ecol. https://doi.org/10.1016/j.funeco.2014.11.001.

Hiscox, J., Clarkson, G., Savoury, M., Powell, G., Savva, I., Lloyd, M., Shipcott, J., Choimes, A., Amargant Cumbriu, X., Boddy, L., 2016. Effects of pre-colonisation and temperature on interspecific fungal interactions in wood. Fungal Ecol. https://doi.org/10.1016/j.funeco.2016.01.011.

Hiscox, J., O'Leary, J., Boddy, L., 2018. Fungus wars: basidiomycete battles in wood decay. Stud. Mycol. https://doi.org/10.1016/j.simyco.2018.02.003.

Hoch, H.C., 1977a. Mycoparasitic relationships: Gonatobotrys simplex parasitic on *Alternaria tenuis*. Phytopathology. https://doi.org/10.1094/phyto-67-309.

Hoch, H.C., 1977b. Mycoparasitic relationships. III. Parasitism of Physalospora obtuse by *Calcarisporium parasiticum*. Can. J. Bot. https://doi.org/10.1139/b77-027.

Hoch, H.C., Abawi, G.S., 1979. Mycoparasitism of oospores of *Pythium ultimum* by *Fusarium merismoides*. Mycologia. https://doi.org/10.2307/3759071.

Holdenrieder, O., Greig, B.J.W., 1998. Biological methods of control. In: Woodward, S., Stenlid, J., Karjalainen, R., Hüttermann, A. (Eds.), *Heterobasidion annosum*, Biology, Ecology, Impact and Control. CAB International, Wallingford, UK, pp. 235–258.

Holmer, L., Stenlid, J., 1993. The importance of inoculum size for the competitive ability of wood decomposing fungi. FEMS Microbiol. Ecol. https://doi.org/10.1016/0168-6496(93)90012-V.

Holmer, L., Stenlid, J., 1997a. Resinicium bicolor; a potential biological control agent for Heterobasidion annosum. Eur. J. For. Pathol. https://doi.org/10.1111/j.1439-0329.1997.tb00857.x.

Holmer, L., Stenlid, J., 1997b. Competitive hierarchies of wood decomposing Basidiomycetes in artificial systems based on variable inoculum sizes. Oikos. https://doi.org/10.2307/3546092.

Holmer, L., Renvall, P., Stenlid, J., 1997. Selective replacement between species of wood-rotting basidiomycetes, a laboratory study. Mycol. Res. https://doi.org/10.1017/S0953756296003243.

Howell, C., 1998. Trichoderma and Gliocladium, Volume 2: Enzymes, Biological Control and Commercial Applications, second ed. CRC Press.

Huisman, J., Weissing, F.J., 2001. Fundamental unpredictability in multispecies competition. Am. Nat. https://doi.org/10.1086/319929.

Humphris, S.N., Bruce, A., Buultjens, E., Wheatley, R.E., 2002. The effects of volatile microbial secondary metabolites on protein synthesis in *Serpula lacrymans*. FEMS Microbiol. Lett. https://doi.org/10.1016/S0378-1097(02)00604-3.

Iakovlev, A., Olson, Å., Elfstrand, M., Stenlid, J., 2004. Differential gene expression during interactions between *Heterobasidion annosum* and *Physisporinus sanguinolentus*. FEMS Microbiol. Lett. https://doi.org/10.1016/j.femsle.2004.10.007.

Ikediugwu, F.E.O., 1976a. Ultrastructure of hyphal interference between *Coprinus heptemerus* and *Ascobolus crenulatus*. Trans. Br. Mycol. Soc. https://doi.org/10.1016/s0007-1536(76)80054-x.

Ikediugwu, F.E.O., 1976b. The interface in hyphal interference by *Peniophora gigantea* against *Heterobasidion annosum*. Trans. Br. Mycol. Soc. https://doi.org/10.1016/s0007-1536(76)80055-1.

Ikediugwu, F.E.O., Webster, J., 1970. Antagonism between *Coprinus heptemerus* and other coprophilous fungi. Trans. Br. Mycol. Soc. https://doi.org/10.1016/s0007-1536(70)80031-6.

Jeffries, P., 1995. Biology and ecology of mycoparasitism. Can. J. Bot. https://doi.org/10.1139/b95-389.

Jonsson, B.G., 2000. Availability of coarse woody debris in a boreal old-growth *Picea abies* forest. J. Veg. Sci. https://doi.org/10.2307/3236775.

Kang, K.-Y., Sung, J.-S., Kim, D.-Y., 2007. Evaluation of white-rot fungi for biopulping of wood. Mycobiology. https://doi.org/10.4489/myco.2007.35.4.205.

Kennedy, P.G., Peay, K.G., Thomas, B., 2009. Root tip competition among ectomycorrhizal fungi: are priority effects a rule or an exception? Ecology 90 (8), 2098–2107. https://doi.org/10.1890/08-1291.1.

Kim, S.H., 2005. Recent researches on sapstaining fungi colonizing pines. Plant Pathol. J. https://doi.org/10.5423/PPJ.2005.21.1.001.

Kim, J.S., Gao, J., Daniel, G., 2015. Cytochemical and immunocytochemical characterization of wood decayed by the white rot fungus *Pycnoporus sanguineus* I. preferential lignin degradation prior to hemicelluloses in Norway spruce wood. Int. Biodeterior. Biodegrad. https://doi.org/10.1016/j.ibiod.2015.08.008.

Kim, J.H., Sutley, E.J., Martin, F., 2019. Review of modern wood fungal decay research for implementation into a building standard of practice. J. Mater. Civ. Eng. https://doi.org/10.1061/(ASCE)MT.1943-5533.0002998.

Kirk, T.K., 1975. Effects of a brown-rot fungus, *Lenzites trabea*, on lignin in spruce wood. Holzforschung. https://doi.org/10.1515/hfsg.1975.29.3.99.

Kiss, L., 1997. Graminicolous powdery mildew fungi as new natural hosts of *Ampelomyces mycoparasites*. Can. J. Bot. https://doi.org/10.1139/b97-076.

Kiss, L., 1998. Natural occurrence of ampelomyces intracellular mycoparasites in mycelia of powdery mildew fungi. New Phytol. https://doi.org/10.1046/j.1469-8137.1998.00316.x.

Kleist, G., Seehann, G., 1997. Colonization patterns and topochemical aspects of sap streak caused by *Stereum sanguinolentum* in Norway spruce. Eur. J. For. Pathol. https://doi.org/10.1111/j.1439-0329.1997.tb01450.x.

Korpi, A., Järnberg, J., Pasanen, A.L., 2009. Microbial volatile organic compounds. Crit. Rev. Toxicol. https://doi.org/10.1080/10408440802291497.

Kovalchuk, A., Mukrimin, M., Zeng, Z., Raffaello, T., Liu, M., Kasanen, R., Sun, H., Asiegbu, F.O., 2018. Mycobiome analysis of asymptomatic and symptomatic Norway spruce trees naturally infected by the conifer pathogens *Heterobasidion* spp. Environ. Microbiol. Rep. https://doi.org/10.1111/1758-2229.12654.

Koyani, R.D., Rajput, K.S., 2014. Light microscopic analysis of *Tectona grandis* L.f. wood inoculated with *Irpex lacteus* and *Phanerochaete chrysosporium*. Eur. J. Wood Wood Prod. https://doi.org/10.1007/s00107-013-0763-7.

Koyani, R.D., Sanghvi, G.V., Rajput, K.S., 2011. Comparative study on the delignification of *Azadirachta indica* (L) Del., wood by *Chrysosporium asperatum* and *Trichoderma harzianum*. Int. Biodeterior. Biodegrad. https://doi.org/10.1016/j.ibiod.2010.10.010.

Koyani, R.D., Bhatt, I.M., Patel, H.R., Vasava, A.M., Rajput, K.S., 2016. Evaluation of *Schizophyllum* commune Fr. potential for biodegradation of lignin: a light microscopic analysis. Wood Mater. Sci. Eng. https://doi.org/10.1080/17480272.2014.945957.

Krause, K., Jung, E.M., Lindner, J., Hardiman, I., Poetschner, J., Madhavan, S., Matthäus, C., Kai, M., Menezes, R.C., Popp, J., Svatoš, A., Kothe, E., 2020. Response of the wood-decay fungus *Schizophyllum* commune to co-occurring microorganisms. PLoS One. https://doi.org/10.1371/journal.pone.0232145.

Kudalkar, P., Strobel, G., Riyaz-Ul-Hassan, S., Geary, B., Sears, J., 2012. *Muscodor sutura*, a novel endophytic fungus with volatile antibiotic activities. Mycoscience. https://doi.org/10.1007/s10267-011-0165-9.

Kuhar, F., Castiglia, V., Levin, L., 2015. Enhancement of laccase production and malachite green decolorization by co-culturing *Ganoderma lucidum* and *Trametes versicolor* in solid-state fermentation. Int. Biodeterior. Biodegrad. https://doi.org/10.1016/j.ibiod.2015.06.017.

Kwang, H.L., Seung, G.W., Singh, A.P., Yoon, S.K., 2004. Micromorphological characteristics of decayed wood and laccase produced by the brown-rot fungus *Coniophora puteana*. J. Wood Sci. https://doi.org/10.1007/s10086-003-0558-2.

Laird, R.a., Schamp, B.S., 2006. Competitive intransitivity promotes species coexistence. Am. Nat. 168, 182–193. https://doi.org/10.1086/506259.

Lindahl, B.D., Ihrmark, K., Boberg, J., Trumbore, S.E., Högberg, P., Stenlid, J., Finlay, R.D., 2007. Spatial separation of litter decomposition and mycorrhizal nitrogen uptake in a boreal forest. New Phytol. https://doi.org/10.1111/j.1469-8137.2006.01936.x.

Lira-Pérez, J., Rodríguez-Vázquez, R., Chan-Cupul, W., 2020. Effect of fungal co-cultures on ligninolytic enzyme activities, H_2O_2 production, and orange G discoloration. Prep. Biochem. Biotechnol. https://doi.org/10.1080/10826068.2020.1721534.

Luna, M.L., Murace, M.A., Keil, G.D., Otaño, M.E., 2004. Patterns of decay caused by *Pycnoporus sanguineus* and *Ganoderma lucidum* (Aphyllophorales) in poplar wood. IAWA J. https://doi.org/10.1163/22941932-90000375.

Luyssaert, S., Ciais, P., Piao, S.L., Schulze, E.D., Jung, M., Zaehle, S., Schelhaas, M.J., Reichstein, M., Churkina, G., Papale, D., Abril, G., Beer, C., Grace, J., Loustau, D., Matteucci, G., Magnani, F., Nabuurs, G.J., Verbeeck, H., Sulkava, M., van der Werf, G.R., Janssens, I.A., 2010. The European carbon balance. Part 3: forests. Glob. Chang. Biol. https://doi.org/10.1111/j.1365-2486.2009.02056.x.

Mäkelä, M.R., Donofrio, N., De Vries, R.P., 2014. Plant biomass degradation by fungi. Fungal Genet. Biol. https://doi.org/10.1016/j.fgb.2014.08.010.

Mali, T., Kuuskeri, J., Shah, F., Lundell, T.K., 2017. Interactions affect hyphal growth and enzyme profiles in combinations of coniferous wood-decaying fungi of Agaricomycetes. PLoS ONE. https://doi.org/10.1371/journal.pone.0185171.

Mali, T., Mäki, M., Hellén, H., Heinonsalo, J., Bäck, J., Lundell, T., 2019. Decomposition of spruce wood and release of volatile organic compounds depend on decay type, fungal interactions and enzyme production patterns. FEMS Microbiol. Ecol. https://doi.org/10.1093/femsec/fiz13.

Mansfield, S.D., Saddler, J.N., Gübitz, G.M., 1998. Characterization of endoglucanases from the brown rot fungi *Gloeophyllum sepiarium* and *Gloeophyllum trabeum*. Enzym. Microb. Technol. https://doi.org/10.1016/S0141-0229(98)00033-7.

Maurice, S., Arnault, G., Nordén, J., Botnen, S.B., Miettinen, O., Kauserud, H., 2021. Fungal sporocarps house diverse and host-specific communities of fungicolous fungi. ISME J. https://doi.org/10.1038/s41396-020-00862-1.

Mayer, A.M., Staples, R.C., 2002. Laccase: new functions for an old enzyme. Phytochemistry. https://doi.org/10.1016/S0031-9422(02)00171-1.

Merckx, V.S.F.T., 2012. Mycoheterotrophy: the biology of plants living on fungi. In: Mycoheterotrophy: The Biology of Plants Living on Fungi., https://doi.org/10.1007/978-1-4614-5209-6.

Mgbeahuruike, A.C., Sun, H., Fransson, P., Kasanen, R., Daniel, G., Karlsson, M., Asiegbu, F.O., 2011. Screening of *Phlebiopsis gigantea* isolates for traits associated with biocontrol of the conifer pathogen *Heterobasidion annosum*. Biol. Control 57, 118–129. https://doi.org/10.1016/j.biocontrol.2011.01.007.

Mims, C.W., Hanlin, R.T., Richardson, E.A., 2007. Light- and electron-microscopic observations of *Cladosporium* sp. growing on basidia of *Exobasidium camelliae* var. *gracilis*. Can. J. Bot. https://doi.org/10.1139/B06-153.

Morath, S.U., Hung, R., Bennett, J.W., 2012. Fungal volatile organic compounds: a review with emphasis on their biotechnological potential. Fungal Biol. Rev. https://doi.org/10.1016/j.fbr.2012.07.001.

Naidu, Y., Siddiqui, Y., Rafii, M.Y., Saud, H.M., Idris, A.S., 2017. Investigating the effect of white-rot hymenomycetes biodegradation on basal stem rot infected oil palm wood blocks: biochemical and anatomical characterization. Ind. Crop. Prod. https://doi.org/10.1016/j.indcrop.2017.08.064.

Nguyen, T.D., Nishimura, H., Imai, T., Watanabe, T., Kohdzuma, Y., Sugiyama, J., 2018. Natural durability of the culturally and historically important timber: *Erythrophleum fordii* wood against white-rot fungi. J. Wood Sci. https://doi.org/10.1007/s10086-018-1704-1.

Nicolotti, G., Gonthier, P., Varese, G.C., 1999. Effectiveness of some biocontrol and chemical treatments against *Heterobasidion annosum* on Norway spruce stumps. Eur. J. For. Pathol. https://doi.org/10.1046/j.1439-0329.1999.00159.x.

Niemelä, T., 1985. On Fennoscandian polypores 9. Gelatoporia n. gen. and *Tyromyces canadensis*, plus notes on Skeletocutis and Antrodia. Karstenia. https://doi.org/10.29203/ka.1985.233.

Niemelä, T., Wallenius, T., Kotiranta, H., 2002. The kelo tree, a vanishing substrate of specified wood-inhabiting fungi. Polish Bot. J. 47 (2), 91–101.

Niemenmaa, O., Uusi-Rauva, A., Hatakka, A., 2008. Demethoxylation of [O14CH3]-labelled lignin model compounds by the brown-rot fungi *Gloeophyllum trabeum* and *Poria* (Postia) *placenta*. Biodegradation. https://doi.org/10.1007/s10532-007-9161-3.

Nurika, I., Eastwood, D.C., Bugg, T.D.H., Barker, G.C., 2020. Biochemical characterization of *Serpula lacrymans* iron-reductase enzymes in lignocellulose breakdown. J. Ind. Microbiol. Biotechnol. https://doi.org/10.1007/s10295-019-02238-7.

Nygren, K., Dubey, M., Zapparata, A., Iqbal, M., Tzelepis, G.D., Durling, M.B., Jensen, D.F., Karlsson, M., 2018. The mycoparasitic fungus *Clonostachys rosea* responds with both common and specific gene expression during interspecific interactions with fungal prey. Evol. Appl. https://doi.org/10.1111/eva.12609.

Odum, E.P., 2014. The strategy of ecosystem development. The Ecological Design and Planning Reader. Island Press, Washington, DC, pp. 203–216.

Oghenekaro, A.O., Daniel, G., Asiegbu, F.O., 2015. The saprotrophic wood-degrading abilities of rigidoporus microporus. Silva Fenn. https://doi.org/10.14214/sf.1320.

Oghenekaro, A.O., Kovalchuk, A., Raffaello, T., Camarero, S., Gressler, M., Henrissat, B., Lee, J., Liu, M., Martínez, A.T., Miettinen, O., Mihaltcheva, S., Pangilinan, J., Ren, F., Riley, R., Ruiz-Dueñas, F.J., Serrano, A., Thon, M.R., Wen, Z., Zeng, Z., Barry, K., Grigoriev, I.V.V., Martin, F., Asiegbu, F.O., 2020. Genome sequencing of *Rigidoporus microporus* provides insights on genes important for wood decay, latex tolerance and interspecific fungal interactions. Sci. Rep. https://doi.org/10.1038/s41598-020-62150-4.

O'Leary, J., Hiscox, J., Eastwood, D.C., Savoury, M., Langley, A., McDowell, S., Hillary, H.J., Boddy, L., Müller, C.T., 2019. The whiff of decay: linking volatile production and extracellular enzymes to outcomes of fungal interactions at different temperatures. Fungal Ecol. 39, 336–348. https://doi.org/10.1016/j.funeco.2019.03.006.

Ortíz-Castro, R., Contreras-Cornejo, H.A., Macías-Rodríguez, L., López-Bucio, J., 2009. The role of microbial signals in plant growth and development. Plant Signal. Behav. https://doi.org/10.4161/psb.4.8.9047.

Ottosson, E., 2013. Succession of wood-inhabiting fungal communities: diversity and species interactions during the decomposition of norway spruce. Dep. For. Mycol. Plant Pathol. Fac. Nat. Resour. Agric. Sci.

Ottosson, E., Nordén, J., Dahlberg, A., Edman, M., Jönsson, M., Larsson, K.H., Olsson, J., Penttilä, R., Stenlid, J., Ovaskainen, O., 2014. Species associations during the succession of wood-inhabiting fungal communities. Fungal Ecol. https://doi.org/10.1016/j.funeco.2014.03.003.

Peay, K.G., Belisle, M., Fukami, T., 2012. Phylogenetic relatedness predicts priority effects in nectar yeast communities. Proc. R. Soc. Biol. Sci. 279 (1729), 749–758. https://doi.org/10.1098/rspb.2011.1230.

Pramod, S., Koyani, R.D., Bhatt, I., Vasava, A.M., Rao, K.S., Rajput, K.S., 2015. Histological and ultrastructural alterations in the *Ailanthus excelsa* wood cell walls by Bjerkandera adusta (Willd.) P. Karst. Int. Biodeterior. Biodegrad. https://doi.org/10.1016/j.ibiod.2015.02.026.

Pratt, J.E., Niemi, M., Sierota, Z.H., 2000. Comparison of three products based on *Phlebiopsis gigantea* for the control of *Heterobasidion annosum* in Europe. Biocontrol Sci. Tech. https://doi.org/10.1080/09583150050115052.

Purahong, W., Wubet, T., Krüger, D., Buscot, F., 2018. Molecular evidence strongly supports deadwood-inhabiting fungi exhibiting unexpected tree species preferences in temperate forests. ISME J. https://doi.org/10.1038/ismej.2017.177.

Pyron, M., 2010. Pyron, Mark. "Characterizing communities". Nat. Educ. Knowl. 3, 39.

Qian, L., Chen, B., 2012. Enhanced oxidation of benzo[a]pyrene by crude enzyme extracts produced during interspecific fungal interaction of *Trametes versicolor* and *Phanerochaete chrysosporium*. J. Environ. Sci. (China). https://doi.org/10.1016/S1001-0742(11)61056-5.

Rabinal, C., Bhat, S., 2020. Identification of differentially expressed genes in *Trichoderma koningii* IABT1252 during its interaction with *Sclerotium rolfsii*. Curr. Microbiol. https://doi.org/10.1007/s00284-019-01838-x.

Rayner, A.D.M., Boddy, L., 1988. Fungal Decomposition of Wood: Its Biology and Ecology. John Wiley & Sons Ltd, Chichester, Sussex, UK.

Rayner, A.D.M., Boddy, L., Dowson, C.G., 1987. Temporary parasitism of *Coriolus* spp. by *Lenzites betulina*: a strategy for domain capture in wood decay fungi. FEMS Microbiol. Lett. https://doi.org/10.1016/0378-1097(87)90042-5.

Rayner, A.D.M., Griffith, G.S., Wildman, H.G., 1994. Induction of metabolic and morphogenetic changes during mycelial interactions among species of higher fungi. Biochem. Soc. Trans. https://doi.org/10.1042/bst0220389.

Ritz, K., Millar, S.M., Crawford, J.W., 1996. Detailed visualisation of hyphal distribution in fungal mycelia growing in heterogeneous nutritional environments. J. Microbiol. Methods. https://doi.org/10.1016/0167-7012(95)00077-1.

Schilling, J.S., Kaffenberger, J.T., Held, B.W., Ortiz, R., Blanchette, R.A., 2020. Using wood rot phenotypes to illuminate the "Gray" among decomposer fungi. Front. Microbiol. 11, 1288. https://doi.org/10.3389/fmicb.2020.01288.

Schoeman, M.W., Webber, J.F., Dickinson, D.J., 1999. The development of ideas in biological control applied to forest products. Int. Biodeterior. Biodegrad. https://doi.org/10.1016/S0964-8305(99)00037-2.

Schubert, M., Fink, S., Schwarze, F.W.M.R., 2008. In vitro screening of an antagonistic *Trichoderma* strain against wood decay fungi. Arboric. J. https://doi.org/10.1080/03071375.2008.9747541.

Schwarze, F.W.M.R., Engels, J., Mattheck, C., 2000. Fungal Strategies of Wood Decay in Trees, Fungal Strategies of Wood Decay in Trees., https://doi.org/10.1007/978-3-642-57302-6.

Shakya, D.D., Lakhey, P.B., 2007. Confirmation of Fusarium solani as the causal agent of die-back of *Dalbergia sissoo* in Nepal. Plant Pathol. https://doi.org/10.1111/j.1365-3059.2007.01637.x.

Sharma, P., Vignesh Kumar, P., Ramesh, R., Saravanan, K., Deep, S., Sharma, M., Mahesh, S., Dinesh, S., 2011. Biocontrol genes from *Trichoderma* species: a review. Afr. J. Biotechnol. https://doi.org/10.5897/AJBX11.041.

Shen, X.T., Mo, X.H., Zhu, L.P., Tan, L.L., Du, F.Y., Wang, Q.W., Zhou, Y.M., Yuan, X.J., Qiao, B., Yang, S., 2019. Unusual and highly bioactive *Sesterterpenes* synthesized by *Pleurotus ostreatus* during coculture with *Trametes robiniophila* Murr. Appl. Environ. Microbiol. https://doi.org/10.1128/AEM.00293-19.

Shi, M., Chen, L., Wang, X.W., Zhang, T., Zhao, P.B., Song, X.Y., Sun, C.Y., Chen, X.L., Zhou, B.C., Zhang, Y.Z., 2012. Antimicrobial peptaibols from *Trichoderma pseudokoningii* induce programmed cell death in plant fungal pathogens. Microbiology. https://doi.org/10.1099/mic.0.052670-0.

Siitonen, J., 2001. Forest management, coarse woody debris and saproxylic organisms: *Fennoscandian boreal* forests as an example. Ecol. Bull. https://doi.org/10.2307/20113262.

Silar, P., 2005. Peroxide accumulation and cell death in filamentous fungi induced by contact with a contestant. Mycol. Res. https://doi.org/10.1017/S0953756204002230.

Silar, P., 2012. Hyphal interference: self versus non-self fungal recognition and hyphal death. In: Biocommunication of Fungi., https://doi.org/10.1007/978-94-007-4264-2_10.

Skyba, O., Cullen, D., Douglas, C.J., Mansfield, S.D., 2016. Gene expression patterns of wood decay fungi *Postia placenta* and *Phanerochaete chrysosporium* are influenced by wood substrate composition during degradation. Appl. Environ. Microbiol. https://doi.org/10.1128/AEM.00134-16.

Song, Z., Vail, A., Sadowsky, M.J., Schilling, J.S., 2012. Competition between two wood-degrading fungi with distinct influences on residues. FEMS Microbiol. Ecol. https://doi.org/10.1111/j.1574-6941.2011.01201.x.

Song, Z., Vail, A., Sadowsky, M.J., Schilling, J.S., 2015. Influence of hyphal inoculum potential on the competitive success of fungi colonizing wood. Microb. Ecol. https://doi.org/10.1007/s00248-015-0588-5.

Sugano, J., Linnakoski, R., Huhtinen, S., Pappinen, A., Niemelä, P., Asiegbu, F.O., 2019. Cellulolytic activity of brown-rot *Antrodia sinuosa* at the initial stage of cellulose degradation. Holzforschung. https://doi.org/10.1515/hf-2018-0145.

Swart, W.J., 1991. Biology and control of *Sphaeropsis sapinea* on *Pinus* species in South Africa. Plant Dis. https://doi.org/10.1094/pd-75-0761.

Szentivanyi, A., Herman Friedman, G.G., 1987. Antibiosis and Host Immunity. Springer Plenum Press.

Terhonen, E., Keriö, S., Sun, H., Asiegbu, F.O., 2014. Endophytic fungi of Norway spruce roots in boreal pristine mire, drained peatland and mineral soil and their inhibitory effect on *Heterobasidion parviporum* in vitro. Fungal Ecol. https://doi.org/10.1016/j.funeco.2014.01.003.

Terhonen, E., Sipari, N., Asiegbu, F.O., 2016. Inhibition of phytopathogens by fungal root endophytes of Norway spruce. Biol. Control 99, 53–63. https://doi.org/10.1016/j.biocontrol.2016.04.006.

Thurston, C.F., 1994. The structure and function of fungal laccases. Microbiology. https://doi.org/10.1099/13500872-140-1-19.

Toby Kiers, E., Palmer, T.M., Ives, A.R., Bruno, J.F., Bronstein, J.L., 2010. Mutualisms in a changing world: an evolutionary perspective. Ecol. Lett. https://doi.org/10.1111/j.1461-0248.2010.01538.x.

Toljander, Y.K., Lindahl, B.D., Holmer, L., Högberg, N.O.S., 2006. Environmental fluctuations facilitate species co-existence and increase decomposition in communities of wood decay fungi. Oecologia. https://doi.org/10.1007/s00442-006-0406-3.

Tornberg, K., Olsson, S., 2002. Detection of hydroxyl radicals produced by wood-decomposing fungi. FEMS Microbiol. Ecol. https://doi.org/10.1016/S0168-6496(02)00200-3.

Traquair, J.A., Meloche, R.B., Jarvis, W.R., Baker, K.W., 1984. Hyperparasitism of *Puccinia violae* by *Cladosporium uredinicola*. Can. J. Bot. https://doi.org/10.1139/b84-030.

Tseng, S.C., Liu, S.Y., Yang, H.H., Lo, C.T., Peng, K.C., 2008. Proteomic study of biocontrol mechanisms of *Trichoderma harzianum* ETS 323 in response to *Rhizoctonia solani*. J. Agric. Food Chem. https://doi.org/10.1021/jf703626j.

Tsuneda, A., Skoropad, W.P., Tewari, J.P., 1976. Mode of parasitism of *Alternaria brassicae* by *Nectria inventa*. Phytopathology. https://doi.org/10.1094/phyto-66-1056.

Turner, G.J., Tribe, H.T., 1976. On *Coniothyrium minitans* and its parasitism of *Sclerotinia* species. Trans. Br. Mycol. Soc. https://doi.org/10.1016/s0007-1536(76)80098-8.

Ujor, V.C., Peiris, D.G., Monti, M., Kang, A.S., Clements, M.O., Hedger, J.N., 2012. Quantitative proteomic analysis of the response of the wood-rot fungus, *Schizophyllum commune*, to the biocontrol fungus, *Trichoderma viride*. Lett. Appl. Microbiol. https://doi.org/10.1111/j.1472-765X.2012.03215.x.

Van Der Wal, A., Ottosson, E., De Boer, W., 2015. Neglected role of fungal community composition in explaining variation in wood decay rates. Ecology. https://doi.org/10.1890/14-0242.1.

Van Der Wal, A., Gunnewiek, P.J.A.K., Cornelissen, J.H.C., Crowther, T.W., De Boer, W., 2016. Patterns of natural fungal community assembly during initial decay of coniferous and broadleaf tree logs. Ecosphere 7. https://doi.org/10.1002/ecs2.1393.

Van Heerden, A., Le Roux, N.J., Swart, J., Gardner-Lubbe, S., Botha, A., 2008. Assessment of wood degradation by *Pycnoporus sanguineus* when co-cultured with selected fungi. World J. Microbiol. Biotechnol. https://doi.org/10.1007/s11274-008-9773-8.

Vasaitis, R., 2013. Heart rots, sap rots and canker rots. In: Infectious Forest Diseases., https://doi.org/10.1079/9781780640402.0197.

Vázquez-Garcidueñas, S., Leal-Morales, C.A., Herrera-Estrella, A., 1998. Analysis of the β-1,3-glucanolytic system of the biocontrol agent *Trichoderma harzianum*. Appl. Environ. Microbiol. https://doi.org/10.1128/aem.64.4.1442-1446.1998.

Větrovský, T., Voříšková, J., Šnajdr, J., Gabriel, J., Baldrian, P., 2011. Ecology of coarse wood decomposition by the saprotrophic fungus *Fomes fomentarius*. Biodegradation 22 (4), 709–718. https://doi.org/10.1007/s10532-010-9390-8.

Viitanen, H., Toratti, T., Makkonen, L., Peuhkuri, R., Ojanen, T., Ruokolainen, L., Räisänen, J., 2010. Towards modelling of decay risk of wooden materials. Eur. J. Wood Wood Prod. https://doi.org/10.1007/s00107-010-0450-x.

Villalta, C.F., 2011. Investigations into the Gene Expression of Neurospora Crassa During Mycelial Contact With Fungi of Increasing Phylogenetic Distance (Doctoral dissertation). University of California, Berkeley, USA, pp. 1–194.

Villavicencio, E.V., Mali, T., Mattila, H.K., Lundell, T., 2020. Enzyme activity profiles produced on wood and straw by four fungi of different decay strategies. Microorganisms. https://doi.org/10.3390/microorganisms8010073.

Wang, W., Yuan, T., Cui, B., 2014. Biological pretreatment with white rot fungi and their co-culture to overcome lignocellulosic recalcitrance for improved enzymatic digestion. Bioresources. https://doi.org/10.15376/biores.9.3.3968-3976.

Wells, J.M., Boddy, L., 2002. Interspecific carbon exchange and cost of interactions between basidiomycete mycelia in soil and wood. Funct. Ecol. https://doi.org/10.1046/j.1365-2435.2002.00595.x.

Wen, Z., Zeng, Z., Ren, F., Asiegbu, F.O., 2019. The conifer root and stem rot pathogen (*Heterobasidion parviporum*): effectome analysis and roles in interspecific fungal interactions. Microorganisms. https://doi.org/10.3390/microorganisms7120658.

Wenke, K., Kai, M., Piechulla, B., 2010. Belowground volatiles facilitate interactions between plant roots and soil organisms. Planta. https://doi.org/10.1007/s00425-009-1076-2.

Wheatley, R.E., 2002. The consequences of volatile organic compound mediated bacterial and fungal interactions. Antonie van Leeuwenhoek. Int. J. Gen. Mol. Microbiol. https://doi.org/10.1023/A:1020592802234.

Whipps, J.M., 2001. Microbial interactions and biocontrol in the rhizosphere. J. Exp. Bot. https://doi.org/10.1093/jexbot/52.suppl_1.487.

White, N.A., Sturrock, C., Ritz, K., Samson, W.B., Bown, J., Staines, H.J., Palfreyman, J.W., Crawford, J., 1998. Interspecific fungal interactions in spatially heterogeneous systems. FEMS Microbiol. Ecol. 27, 21–32. https://doi.org/10.1016/S0168-6496(98)00052-X.

Woodward, S., Boddy, L., 2008. Chapter 7. Interactions between saprotrophic fungi. Br. Mycol. Soc. Symp. Ser. 28, 125–141. https://doi.org/10.1016/S0275-0287(08)80009-4.

Worrall, J.J., Anagnost, S.E., Zabel, R.A., 1997. Comparison of wood decay among diverse lignicolous fungi. Mycologia 89, 199–219. https://doi.org/10.2307/3761073.

Zhang, Q., Zhao, L., Li, Y.R., Wang, F., Li, S., Shi, G., Ding, Z., 2020. Comparative transcriptomics and transcriptional regulation analysis of enhanced laccase production induced by co-culture of *Pleurotus eryngii* var. *ferulae* with *Rhodotorula mucilaginosa*. Appl. Microbiol. Biotechnol. https://doi.org/10.1007/s00253-019-10228-z.

Zhao, Y., Xi, Q., Xu, Q., He, M., Ding, J., Dai, Y., Keller, N.P., Zheng, W., 2015. Correlation of nitric oxide produced by an inducible nitric oxide synthase-like protein with enhanced expression of the phenylpropanoid pathway in *Inonotus obliquus* cocultured with *Phellinus morii*. Appl. Microbiol. Biotechnol. https://doi.org/10.1007/s00253-014-6367-2.

Zhong, Z., Li, N., He, B., Igarashi, Y., Luo, F., 2019. Transcriptome analysis of differential gene expression in *Dichomitus squalens* during interspecific mycelial interactions and the potential link with laccase induction. J. Microbiol. https://doi.org/10.1007/s12275-019-8398-y.

Żółciak, A., Sierota, Z., Małecka, M., 2012. Characterisation of some *Phlebiopsis gigantea* isolates with respect to enzymatic activity and decay of Norway spruce wood. Biocontrol Sci. Technol. https://doi.org/10.1080/09583157.2012.691156.

Żółciak, A., Sikora, K., Nowakowska, J.A., Małecka, M., Borys, M., Tereba, A., Sierota, Z., 2016. Antrodia gossypium, *Phlebiopsis gigantea* and *Heterobasidion parviporum*: in vitro growth and Norway spruce wood block decay. Biocontrol Sci. Technol. https://doi.org/10.1080/09583157.2016.1236365.

Section B

Phyllosphere microbiome

Chapter 6

The phyllosphere mycobiome of woody plants

Thomas Niklaus Sieber

ETH Zurich, Department of Environmental Systems Science, Forest Pathology and Dendrology, Zurich, Switzerland

Chapter Outline

1. Introduction

Microorganisms colonize the surfaces and interior of aerial and hypogean plant tissues. Seen worldwide, the total leaf surface represents a huge habitat for microorganisms (Fig. 6.1). The global leaf surface (adaxial and abaxial surface) was estimated at 1,017,260,200 km^2 or just under twice the Earth's surface (Vorholt, 2012). The exact proportion of woody plants on this surface is unknown. It can only be estimated.

Currently, about one-third of the land area on Earth is forested (Gilani and Innes, 2020). This area is home to an estimated 3 trillion trees (Crowther et al., 2015). Assuming that all these trees are 2 m Norway spruce trees (*Picea abies*) with 2.8 million needles each (Habermann, 2004), this would result in 8.5×10^{18} needles worldwide. The mean surface area of a Norway spruce needle varies between 0.15 and 0.5 cm^2 depending on the latitude and/or altitude (Rajala et al., 2013, Sieber, unpublished data). Assuming 0.5 cm^2 per needle, the surface area of all needles would account for 418,809,844 km^2, or almost three times the world's land area. This estimate is at the lower end of the true actual global leaf surface of all woody plants, as very young trees were used for estimation. This huge tree leaf habitat is subdivided into thousands of different microhabitats that are specific to the plant species and location, and thus are home to thousands of different biocoenoses of microorganisms.

The number of microbial propagules per surface unit can vary greatly among plant species, locations, seasons, and years. Even on the same tree, it can vary greatly among years. The median number of colony-forming fungal units per cm^2 of a Norway spruce needle surface (CFUcm^{-2}) was determined as approximately 1500, but deviations from this value can be high (Fig. 6.2). Consequently, the number of epiphytic CFU worldwide would reach 6.4×10^{21}. There are many more microbial propagules unaccounted for because: (i) many fungi cannot be cultivated on artificial media, as they are obligately biotrophic, that is, they can only thrive on living plant tissue, (ii) some fungal mycelia and propagules tightly stick to the plant surface and cannot be removed by simple washing with water, and (iii) there are various other epiphytic microorganisms, especially bacteria. The number of bacteria has been given as 10^6–10^7 per cm^2 of leaf surface (Lindow and Brandl, 2003), that is, up to 10^{26} bacterial cells worldwide on leaves of woody plants (Vorholt, 2012) or more than 1600 times the number of all stars (6×10^{22}) in the universe (Manojlovic, 2015).

Leaves and needles thus represent a huge habitat that is colonized by trillions of microorganisms. Several comprehensive reviews are available on phyllosphere microbiology (Allen, 1991; Bulgarelli et al., 2013; Burkhardt, 2010; Fokkema, 1991; Gilani and Innes, 2020; Leveau, 2019; Lindow and Brandl, 2003; Osono, 2006; Petrini, 1991; Schlaeppi and Bulgarelli, 2015; Stone et al., 2004; Turner et al., 2013; Vorholt, 2012). It is therefore difficult to be innovative without being repetitive. Here, I will concentrate on fungi on and in the leaves and needles of woody plants in temperate zones. Regarding both the ecology of endophytic fungi and their role in plant defense, some reports from subtropical and tropical regions will also be included. Epiphytic fungi and selected pathogens are only treated briefly. The focus is on endophytic

Forest Microbiology. https://doi.org/10.1016/B978-0-12-822542-4.00003-6

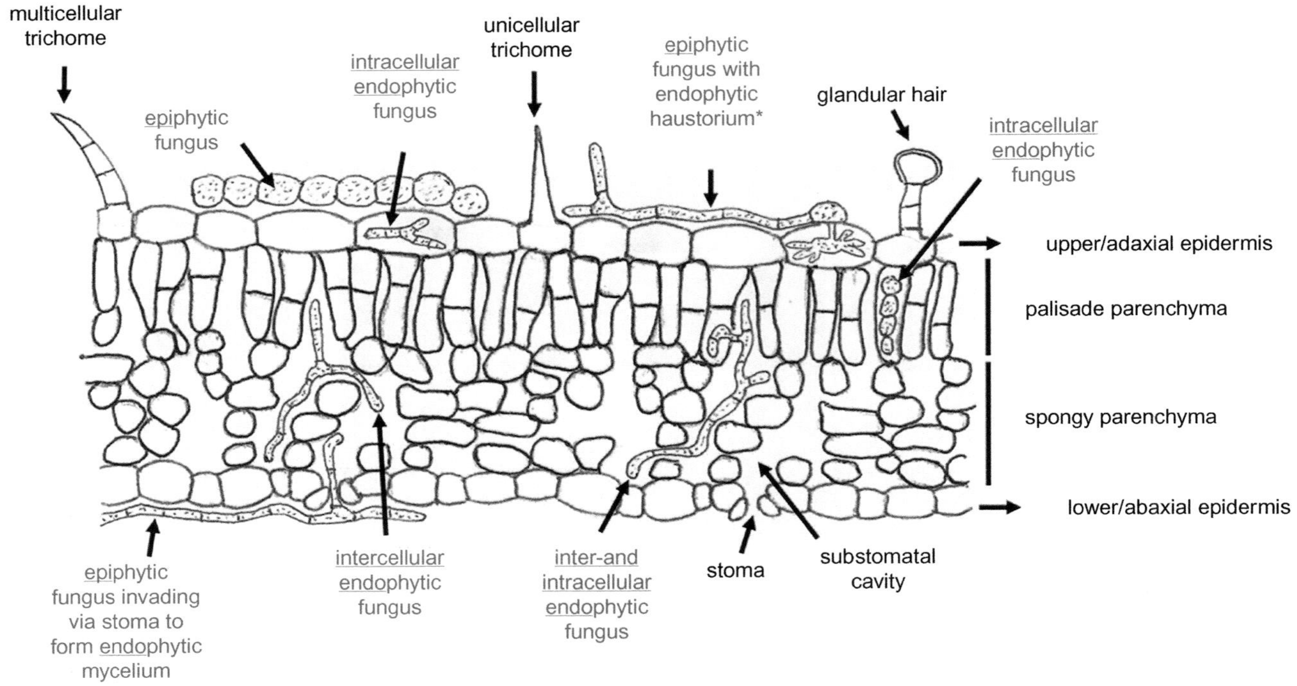

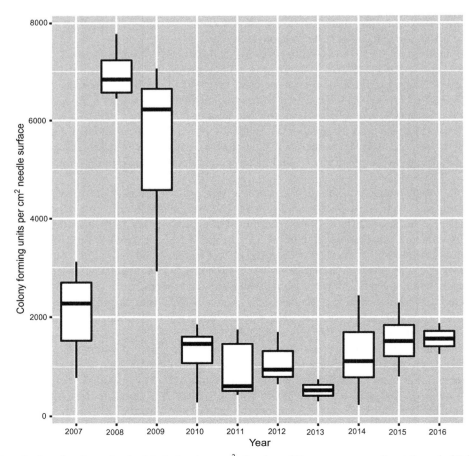

FIG. 6.1   Schematic cross-section of a leaf showing different leaf tissues and surface structures as well as the possible locations of leaf-colonizing fungi. Plant structures are in *black*, fungal structures in *blue*. * Typical structure of a powdery mildew.

FIG. 6.2   The number of colony-forming units of epiphytic fungi per cm^2 of surface of Norway spruce needles at the end of February each year from 2007 to 2016. Five hundred ($\geq$ 5-year-old) needles were randomly collected each year from 40 to 50 trees in a forest near Zurich, Switzerland.

fungi, in particular on their diversity and ecology. I will conclude with some thoughts and hypotheses about how plants "domesticate" endophytic fungi. "Leaf (leaves)" and "needle (needles)" will be used synonymously except when the difference is of importance.

2. Epiphytic fungi

In temperate zones, about 120 fungal species (*Alternaria* and *Cladosporium* species, *Epicoccum purpurascens*, *Aureobasidium pullulans*, yeast-like asco- and basidiomycetes, etc.) are known to colonize plant surfaces (Cabral, 1985; Cobban et al., 2016; Menkis et al., 2019). Few species are very common and many occur only sporadically. The number of fungal species and individuals increases during the vegetation period (Fokkema and Schippers, 1986) and on perennial leaves/needles over the years. Seasonal variation was detected for epiphytic *Aureobasidium pullulans* and a species each of *Pestalotiopsis*, *Phoma*, and *Ramichloridium* on the leaves of *Camellia japonica* (Osono, 2008). Yeasts and bacteria are more common on young leaves than old ones (Kinkel, 1991). The colonization of plant surfaces by fungi depends on many interacting factors (Allen et al., 1991; Burkhardt, 2010; Cooke and Rayner, 1984; Dik, 1991; Juniper, 1991). The most important are climate (especially extreme events), microclimate, radiation, the availability of nutrients, and the presence of competitors. On one hand, the microclimate depends on the large-scale climate. But on the other hand, it depends on the nature of the plant surface such as the surface texture, the density and type of trichomes, and the composition and structure of the cuticle (wax layer). The temperature on leaf surfaces is usually higher during the day (often 5–10° higher), colder than the ambient air at night, and colder with a good water supply (transpiration cold!) than when there is a lack of water. The air humidity on leaf surfaces is higher than in the surroundings and is subject to a diurnal rhythm (Andrews and Harris, 2000; Lindow and Brandl, 2003). The hairiness, the presence of glandular hairs, and the nature of the wax layer influence the attachment and wettability, which play a role in the germination of spores and the growth of hyphae (Monier and Lindow, 2004; Yadav et al., 2005). Leaf surfaces per se are an oligotrophic habitat, and the nutrients are irregularly distributed on them (Lindow and Brandl, 2003). Nutrients either originate directly from the host plant (exudates, guttation fluids, cuticular substances, leaching) or from the environment. Epicuticular waxes can affect the growth of both bacteria (Marcell and Beattie, 2002) and fungi (Aragon et al., 2017; Zabka et al., 2008). Pollen, spores, organic and inorganic dusts (aerosols) introduced from the environment, and sugars (honeydew) released by sucking insects (aphids, cicadas) can serve as food sources for epiphytic fungi. There are even epiphytic fungi that are entirely specialized on pollen, using pollen as the only source of food (Olivier, 1978).

3. Pathogenic and endophytic fungi

Many species of fungi have developed strategies to penetrate leaves. Some use natural openings, others penetrate the tissue by means of infection hyphae through the cuticle and epidermis. Pathogenic invaders usually have a short latent period and cause symptoms after only a few days or weeks. These include in particular rusts and powdery mildew pathogens. It can take several months or years for symptoms to develop in endophytic colonizers, and many endophytes never develop symptoms at all. It is debatable whether endophytic fungi are not simply pathogens with long latency periods (Sinclair and Cerkauskas, 1996). In many cases, the formation of symptoms must/can be triggered by a stress factor (e.g., herbivory, hail, other pathogens, drought, lack of light or nutrients). Without stress, the plant remains symptom-free. For these cases, it makes sense to use the term "endophyte" instead of "pathogen."

3.1 Pathogenic fungi

There is a great deal of literature and numerous textbooks on groups of fungi classified as pathogens (Agrios, 2005; Butin, 1995; Tainter and Baker, 1996). In near-natural, nonmanaged forests, native pathogens do not harm the host population. On the contrary, a healthy forest needs a certain amount of disease (Holdenrieder and Pautasso, 2014). In managed and non-managed forests, only introduced fungi can develop into malignant pathogens such as *Cronartium ribicola*, which is native to Eurasia but was introduced into North America, where it caused an ecological catastrophe on five-needled pine species (*Pinus* subgenus *Haploxylon*) (Loo, 2009). *Hymenoscyphus fraxineus* is another example. *H. fraxineus* is a mostly harmless endophyte in the leaves of several *Fraxinus* species in its native East Asian range, but it causes the dieback of European ash in Europe, where it has been introduced (Gross et al., 2014). Over the last 40 years, many so-called pathogens have turned out to be mostly harmless endophytic fungi that can only become a problem because of poor management.

Powdery mildews are fungi at the interface among the epiphyte, endophyte, and pathogen (Spencer-Phillips, 1997). The largest proportion of biomass of these fungi is epiphytic on the leaves. However, the haustoria that make the life of powdery mildew fungi possible are found endophytically in the epidermal cells. Therefore, powdery mildew fungi should actually be treated as epiphytes with endogenous (endophytic) nutrition or "ep-endophytes" (Fig. 6.1). Because powdery mildews can cause severe damage in crop monocultures, they are considered pathogens. However, native powdery mildews of woody plant species in natural settings can hardly be considered pathogens, especially because losses in growth increment are small, if at all. An infestation of powdery mildew in forest trees can often only be detected upon closer inspection, as the epiphytic mycelium is usually very delicate and inconspicuous and the visibility of the underlying healthy, green leaf tissue is not reduced. Many different powdery mildews attacking woody plant species have been introduced to Europe. Some of them behave or behaved as pathogens because they have not coevolved with the new hosts: *Erysiphe alphitoides*, *E. hypophylla*, and *E. quercicola* on oaks; or *E. flexuosa* on *Aesculus hippocastanum* (Ale-Agha et al., 2000; Desprez-Loustau et al., 2018). Others, though also introduced, remain inconspicuous such as *E. carpinicola* on *Carpinus betulus* (Braun et al., 2006).

When amplicon sequencing is used to study endophyte communities, powdery mildew fungi are very often detected as endophytes in completely healthy leaves, both on known and hitherto unknown hosts (Kälin, 2019; Schlegel et al., 2018) (Fig. 6.3). This raises the question of whether the epiphytic mycelium of these fungi on the hitherto unknown host has been overlooked, such as *Phyllactinia fraxini* on the Norway spruce and Douglas fir or *Sawadaea bicornis* on the European ash (Kälin, 2019; Schlegel et al., 2018). Alternatively, it cannot be excluded that some powdery mildews may have a pure endophytic lifestyle. A purely endophytic colonization without any sporulation would be a dead end for a mildew, but could be a kind of evolutionary trial and error in which the plant's immune system is challenged and tested. It is possible that the fungus will one day be successful, in the course of evolution, in completing its life cycle on the new host.

Most snow molds are considered pathogens. Some of them have an endophytic phase during their life cycle. They are seasonal parasites and take advantage of the weakened immune system or defenses of plants in winter. These fungi are

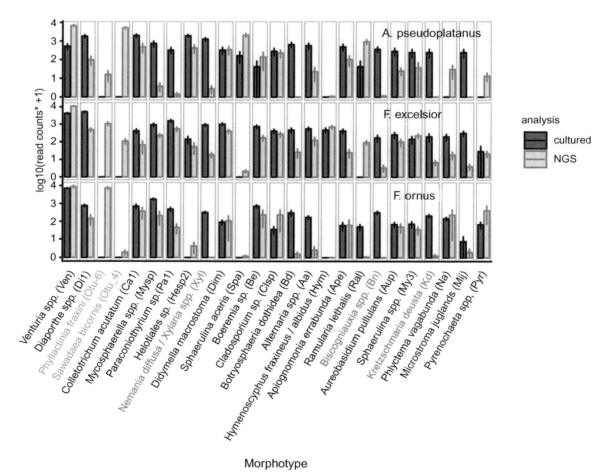

FIG. 6.3 Mean abundances of the most frequent taxa obtained by culturing or DNA barcoding (next-generation sequencing, NGS) from leaves of European ash (*Fraxinus excelsior*), manna ash (*F. ornus*), and sycamore maple (*Acer pseudoplatanus*) collected at eight sites in Switzerland and Italy (Schlegel et al., 2018). *Blue* = obligate biotrophic fungi (powdery mildews); *Red* = xylariaceous endophytes.

cold-tolerant and grow even at temperatures below 0°C. Above all, however, the development of the epiphytic mycelium requires 100% humidity, a requirement found at temperatures just above the freezing point in the boundary layer between vegetation and snow cover. The best-known snow molds on woody plants are the brown felt blight caused by *Herpotrichia pinetorum* and the white snow blight caused by *Gremmenia infestans*. *Allantophomopsis cytisporea*, previously considered a postharvest pathogen (black rot) of the cranberry (*Vaccinium macrocarpum*), has been identified in Switzerland on several occasions as a snow mold on various woody plant species in subalpine and alpine regions (Schneider et al., 2009; Sieber, 2019). The main host is common heather (*Calluna vulgaris*) in which the fungus is a common leaf endophyte (Petrini, 1985). Recently, *A. cytisporea* was also found as an endophyte in *Juniperus chinensis* var. *sargentii* in Korea (Park and Eom, 2018). On dwarf mountain pine (*Pinus mugo* subsp. *mugo*) and juniper (*Juniperus communis* subsp. *nana*) the symptoms are macroscopically indistinguishable from those of brown felt blight caused by *H. pinetorum*. *A. cytisporea* is (still) rare or has been overlooked (Schneider et al., 2009). At one location, it was responsible for all the alleged *H. pinetorum* symptoms on the dwarf mountain pine. Another location is suspected of being an evolutionary hot spot because two snow mold species that are closely related to *A. cytisporea* occurred together with *A. cytisporea* in the same sample: *Pseudophacidium ledi* and *Phacidium lacerum* (Sieber, 2019). Interestingly, *A. cytisporea* can switch from a harmless endophytic lifestyle during the vegetation period to a pathogenic lifestyle below the snow during winter, where it first kills leaves and whole shoots of common heather and then expands to adjacent vegetation, including shoots of conifers.

The Swiss needle cast of the Douglas fir (*Pseudotsuga menziesii*) is caused by *Nothophaeocryptopus gaeumannii*. The Pacific Northwest of North America is the native range of both *N. gaeumannii* and the Douglas fir, where the fungus was an inconspicuous endophyte until about 1990. Since then, it has been epidemically present in some plantations, reducing the growth increment by 25%–50% (Hansen et al., 2000; Lan et al., 2019; Ritokova et al., 2016). Because the pseudothecia of the fungus form exactly above the stomata, the gas exchange is inhibited, which leads to reduced photosynthesis and ultimately to needle cast (Ritokova et al., 2016). The Douglas fir has been cultivated as an exotic plant in Europe for more than 150 years (Schmid et al., 2014), where *N. gaeumannii* was discovered and described because it caused needle cast. In a recent study, *N. gaeumannii* was the most frequent endophyte in healthy Douglas fir needles in Switzerland and was also a very abundant colonizer of the needles of neighboring Norway spruce (*Picea abies*), on which the fungus has never been observed to sporulate so far (Kälin, 2019). Is Norway spruce a dead end for *N. gaeumannii*, similar to that of powdery mildews on "nonhosts"?

There are numerous other fungi that are considered pathogenic on the leaves of forest trees. But for many of them, it has been shown that they are primarily endophytes that only lead to symptoms after the occurrence of an inciting factor, such as *Sphaeropsis sapinea* (Smith et al., 1996), *Apiognomonia errabunda* (Sieber and Hugentobler, 1987), and *A. veneta* (Sogonov et al., 2007). Probably, *Dothistroma septospora* also belongs to this group of pathogens, but has never been identified as an endophyte so far (Drenkhan et al., 2016; Woods et al., 2016). Global warming and the resulting increasing frequency of droughts in many places is an increasingly important inciting factor.

3.2 Endophytic fungi

Many excellent reviews about endophytic fungi in woody plants have been published since my last review in 2007 (Sieber, 2007): (Hardoim et al., 2015; Kaul et al., 2012; Nisa et al., 2015; Pautasso et al., 2015; Saikkonen, 2007; Sridhar, 2019; Witzell and Martin, 2018). It is therefore difficult to provide new findings about this special group of fungi.

Most tree-leaf endophytes form disjunct inter- and/or intracellular thalli in the leaf tissues, which are the result of single infections (horizontal transmission) (see Box 6.1). It is not known whether certain tree endophytes such as clavicipitalean grass endophytes can colonize the host systemically. Systemic colonization allows certain grass endophytes to colonize the seeds directly via the flowering culm (Tintjer et al., 2008). In this case, the endophytes are transmitted vertically and sporulation is not necessary. For some tree endophytes, vertical transmission via the seeds has been proven, but it is not known whether the colonization of the seeds occurred systemically or via single infections directly from the environment (horizontally transmitted) (Decourcelle et al., 2015; Lefort et al., 2016; Sieber et al., 2007).

3.2.1 Factors influencing the abundance of endophytic fungi and the diversity of endophyte communities

The factors that control the abundance of endophytic fungi and the diversity of endophyte communities are well known: climate, host tree species, location, age of tissue (season), age of tree, presence of herbivores and pathogens, air pollution, and tree physiology. In addition, the method used to examine endophyte communities is decisive for the species diversity and the frequency of the individual species recorded, as illustrated in this chapter.

Box 6.1 Groups of endophytic fungi

Grass endophytes:
- members of the *Clavicipitaceae* (family of *Ascomycota*)
- grasses (*Poaceae*) as hosts
- systemic
- horizontal and/or vertical transmission
- intercellular, rather large thalli in aerial plant tissues
- endophyte biomass up to 1% of plant biomass
- mutualists

Other endophytes:
- members of *Ascomycota* and *Basidiomycota*
- all plant species including grasses (nonclavicipitalean grass endophytes)
- mostly nonsystemic
- horizontal transmission, vertical transmission rare
- intra- or intercellular, small, disjointed thalli
- endophyte biomass ≪ 1% of plant biomass
- mutualism often assumed but rarely proven

Detection of endophytes: Cultivation versus DNA barcoding

Ideally, the method should allow the endophyte community to be mapped as it exists in nature. However, we will never know the true composition of endophyte communities. Until a few years ago, diversity and frequency depended on the type and intensity of the surface sterilization of plant tissues, culture media, and incubation conditions (Schulz et al., 1993). Although DNA barcoding (amplicon sequencing, next-generation sequencing) does away with culture media and incubation conditions as potential problems, this method also has disadvantages.

A major advantage of DNA barcoding is that it allows the detection of significantly more organisms (Dissanayake et al., 2018; Johnston et al., 2017; Schlegel et al., 2018). It also detects nonculturable, obligately biotrophic organisms and those that grow slowly on standard media and are partially or completely inhibited by faster-growing ones. For example, the powdery mildews *Phyllactinia fraxini* and *Sawadea bicornis* were frequently detected as endophytes in apparently mildew-free ash (*Fraxinus excelsior* and *F. orni*) and sycamore maple leaves (*Acer pseudoplatanus*). Likewise, the tar spot-causing fungus *Rhytisma acerinum* occurred endophytically in healthy sycamore maple leaves (Schlegel et al., 2018).

However, there are also a series of disadvantages: (i) If endophytes are isolated and cultured, surface sterilization is sufficient when it kills the epiphytes. With DNA barcoding, the DNA of the epiphytically growing organisms must also be destroyed. In order to kill the epiphytes, hydrogen peroxide is sufficient, but to degrade their DNA as well, sodium hypochlorite (NaOCl; bleach) is considered more reliable (Small et al., 2007). Highly concentrated ethanol does not destroy DNA but on the contrary, it acts as a preservative (Bressan et al., 2014). (ii) The DNA of dead endophytic thalli is also detected. (iii) Sequences in databases can be wrongly identified. The wrong names then propagate in the literature over many years if they are uncritically adopted (Leray et al., 2019; Nilsson et al., 2006). A wrong determination can also be fatal, such as if poisonous mushrooms are identified as edible (Bridge et al., 2003). (iv) Which primers should be used to ensure that only DNA from members of the target group is amplified? Especially in the case of endophytic fungi, plant DNA is not wanted. Therefore, DNA barcoding requires a primer selection (Schlegel, 2019). (v) Depending on where the minimal degree of similarity is set to consider the target and database sequence as identical, sequence-based identification can lead to completely different names. (vi) Certain organisms escape detection with DNA barcoding or appear underrepresented in terms of quantity. In the case of endophytic fungi, this applies in particular to representatives of the *Xylariaceae* (Fig. 6.3) (Kälin, 2019; Schlegel, 2019; Schlegel et al., 2018). In this family, there are numerous soft-rot causing fungi that grow within cell walls (Fig. 6.4). During DNA extraction (Debeljak et al., 2017; Holden et al., 2003; Van Deynze and Stoffel, 2006), parts of the cell walls remain intact and therefore the DNA of these soft rot fungi probably escapes extraction. (vii) The databases where the found sequences are compared are incomplete.

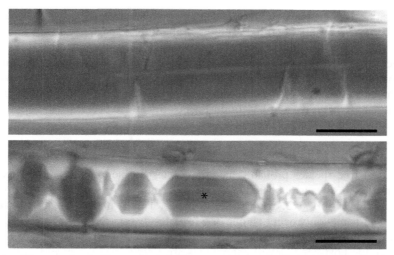

FIG. 6.4 Segments of leaf hairs of European beech (*Fagus sylvatica*). *Upper panel*: unimpaired hair with thick cell walls and a thin channel-like lumen. *Lower panel*: a fungus causing typical soft rot characterized by diamond-shaped cavities in the cell walls of a hair. * = fungal hypha. Scale bars = 10 μm. *(Courtesy of O. Holdenrieder.)*

Climate

Besides the host species, climate has the most important influence on the composition of the endophyte community in leaves. Temperature and precipitation are the main determinants of the climate and depend on the latitude and the altitude. Temperature, precipitation, latitude, and altitude are inextricably linked and cannot be studied separately in the field (The factor "temperature" can never be excluded, not even in an experiment under controlled conditions!).

Latitude Species richness and species frequencies are negatively correlated with latitude, that is, they increase from north to south in the northern hemisphere (Sokolski et al., 2007; Terhonen et al., 2011). Balint et al. (2015) reported contrary findings. However, this work was carried out with a clone of the Balsam poplar (*Populus balsamifera*) that originated from the southern limit of the distribution range and was relocated from the south to the north. The significantly higher diversity of endophytes in the north is explained by the fact that plants in the north are released from pathogens in the south. Alternatively, the southern poplars are not adapted to the northern climate and have weak defenses such that many opportunists can infect and establish as endophytes.

Altitude Just as with increasing latitude, the temperature decreases with increasing altitude. It would therefore be expected that the species richness and species frequencies of endophytic fungi are also negatively correlated with altitude above sea level. This is often true (Siddique and Unterseher, 2016; Unterseher et al., 2016; Yang et al., 2016), but there are some reports of deviating findings. For example, Bowman and Arnold (2018) found homogeneous conditions for the diversity of endophytes in needles of *Pinus ponderosa* along an altitudinal gradient of 635 m, with a tendency toward greater diversity in mid-to-high altitudes. In many studies, diversity decreases with increasing altitude, but the frequency of individual, even dominant endophytes may increase with altitude and consequently with decreasing temperature. For example, *Ascochyta fagi* was dominant at all sites in leaves of *Fagus crenata*, but its frequency increased with altitude (Hashizume et al., 2010). Similar findings were also made for leaf endophytes of *Quercus acuta* (Hashizume et al., 2008). Some species increased at higher altitude while others decreased.

Precipitation The amount of precipitation influences the composition of endophyte communities. Leaf communities of *Populus trichocarpa* differed among sites and on whether the site was located on the rainy west side or the dry east side of the Cascade Mountains in the Pacific Northwest of North America (Barge et al., 2019). The amount of rainfall explained more variation in community composition than geographic distance. The alpha diversity was higher on the more humid west side of the Cascades, indicating that rainfall promotes colonization by endophytes. In contrast, great differences in the species composition of endophyte communities of leaves of various *Betulaceae* existed among sites in Japan, although the annual amount of precipitation is approximately the same at all sites (between 2300 and 2500 mm) (Osono and Masuya, 2012). However, the sites differ according to altitude and the availability of hosts. Thus, the main factors shaping the endophyte community were temperature and host species. Annual rainfall but not temperature had a great influence on endophyte communities in needles of *Pinus radiata* in Tasmania (Prihatini et al., 2015). Similarly, among-site variation in the endophyte community composition of *Metrosideros polymorpha* in Hawaii correlated strongly with rainfall but also with temperature (Zimmerman and Vitousek, 2012). Sandstorms and the resulting deposition of sand also have an influence on the diversity of endophytic fungi. Increased amounts of dust deposition led to decreased endophytic fungal diversity of the Persian oak (*Quercus brantii*) (Doust et al., 2017).

Host specificity

Besides the climate, the host tree species probably has the greatest influence on the endophyte community. The endophyte community of most tree species is characteristic (Hoffman and Arnold, 2008; Martinez-Alvarez et al., 2012; Osono and Masuya, 2012; Sun et al., 2012; Yoo and Eom, 2012). This is understandable, as the dominant endophyte species usually are host-specific and only occur on one or a few closely related hosts. For example, all three tree species studied by Schlegel et al. (2018) were colonized by at least one strictly host-specific *Venturia* species. Other examples of host-specific (at least on the host-genus level) endophytes are *Lophodermium piceae* in the needles of the Norway spruce or *Picea glauca* (Müller-Using and Bartsch, 2003; Sieber, 1988; Stefani and Bérubé, 2006) and *L. pinastri* in pine needles. In most studies comparing different hosts and sites, the host usually has a greater influence than the site (Johnston et al., 2012; Lau et al., 2013; Matsumura and Fukuda, 2013; Persoh, 2013; Schlegel et al., 2018). In addition, there are also numerous species of fungi that are not host-specific at all (generalists) and that have been found on a wide variety of hosts such as *Alternaria alternata*, *Cladosporium cladosporioides*, and *Epicoccum purpurascens* or various representatives of the *Xylariaceae* such as *Nemania serpens*, *Biscogniauxia nummularia*, or *Hypoxylon fragiforme* (Jumpponen and Jones, 2009; Kälin, 2019; Langenfeld et al., 2013; Schlegel et al., 2018; Sieber and Hugentobler, 1987; Sieber et al., 1999).

Site specificity

Climate, soil, and management are the most important characteristics of a site. When nearby sites with almost identical climates are compared, no site effects on the endophyte community are to be expected. Endophyte communities in the leaves of *Alnus incana* and *Corylus avellana* collected at sites maximally 60 km apart in Estonia did not show any site effects (Küngas et al., 2020). Similarly, Taudière et al. (2018) could not detect any site effects for the Corsican pine (*Pinus nigra* subsp. *laricio*) between forest stands up to 70 km apart. Even with a distance of 200 km between the locations, only slight differences could be found between the fungal communities of both *Viscum album* ssp. *austriacum* and *Pinus sylvestris* (Persoh, 2013). However, the more common pattern is that the differences increase significantly with increasing geographical distance. For example, Schlegel et al. (2018) observed greater variation in the foliar mycobiomes of *Fraxinus excelsior*, *F. ornus*, and *Acer pseudoplatanus* between than within sampling sites that were up to 300 km apart. Likewise, the species compositions of the endophyte community in the needles of *Taxus chinensis* var. *mairei* collected at sites up to 1300 km apart were very different, although the diversity was approximately the same at all sites (Wu et al., 2013). However, if there are steep gradients in temperature and/or precipitation over short geographical distances, the endophyte communities can differ significantly even over short distances (Eusemann et al., 2016; Zimmerman and Vitousek, 2012). This is the case along altitudinal gradients regarding temperature, and on the luff and lee side of a mountain range regarding precipitation.

Depending on the climate and soil conditions, a site-specific mixture of tree species is formed in nonmanaged forests. This mixture significantly influences the endophyte community. For example, the diversity of endophytes increased as the diversity of tree species surrounding the host trees (*Picea mariana*) increased along a north-south gradient of about 800 km from north to south (Sokolski et al., 2007). The distance from spore sources can also play a role. The frequency of endophytic fungi in leaves of *Betula pubescens* and *B. pendula* mainly depended on the size of the island and its distance from the mainland in an archipelago in southwestern Finland (Helander et al., 2007). The birch trees on the largest islands near the mainland had the highest endophyte frequencies.

Native versus nonnative hosts

When woody species are planted outside their area of distribution, they may adopt a completely different endophyte community. While in their area of origin they host a host-specific, endemic endophyte community, they are colonized by little or no host-specific species in the area of immigration. For example, the endophyte community of the Japanese plum-yew

(*Cephalotaxus harringtonia*), endemic in Japan, has been compared to that established in France (distance 9000 km) (Langenfeld et al., 2013). As expected, the French and Japanese endophyte communities differed significantly. There was almost no overlap. In Japan, the majority of species were unknown to science. In France, the majority of species were known as cosmopolitan species. Still, the species diversity in both countries was about the same. In contrast, the species diversity of the Chinese thuja (*Platycladus orientalis*), endemic in East Asia and planted in North America, was significantly lower in both North Carolina and Arizona in the United States compared to two native *Cupressacean* species, and was characterized by cosmopolitan endophyte species (Hoffman and Arnold, 2008). If woody species are planted not far from their natural range, the endophyte community in the area of origin and outside may be similar. Manna ash (*Fraxinus ornus*) is native south of the Alps, but is often planted north of the Alps as an ornamental. For the trees planted north of the Alps, it was expected that opportunistic generalists found on neighboring plants and cosmopolitan endophytes would be the main colonizers. However, *Venturia orni*, an *F. ornus*-specific endophyte, was most frequently isolated irrespective of the side of the Alps. It must have arrived either in the form of spores carried by southerly winds over the Alps or together with *F. ornus* seedlings and/or saplings. For example, the introduction of fungi by means of hitchhiking on their host plants to other regions or continents was shown for *Colletotrichum acutatum*, *Neofabraea alba*, and *Venturia fraxini*, which were unintentionally introduced together with *Fraxinus excelsior* in New Zealand (Chen, 2012).

Leaf tissue specificity

The distinct organ specificity of endophyte communities has been demonstrated in many studies. Because this paper deals only with endophytes in leaves, the different endophyte communities in roots, stems, and branches will not be discussed further, and only differences in different leaf tissues are addressed below.

The endophyte communities of leaf laminae are expected to differ from those of leaf petioles. In fact, Schlegel et al. (2018) found significant differences between the two leaf parts in all three examined hosts (*Fraxinus excelsior*, *F. ornus*, and *Acer pseudoplatanus*). For example, members of the *Mycosphaerellaceae* (i.e., *Mycosphaerella* spp., *Ramularia lethalis*, *R. vizellae*) were much more abundant in laminae than petioles of *F. excelsior* and *A. pseudoplatanus*. In contrast, one OTU of an undescribed *Venturia* species occurred significantly more frequently in petioles than the laminae of *A. pseudoplatanus*.

A special form of specificity is exhibited by endophytes that colonize insect galls because these endophytes seem to be both gall tissue-specific and insect-specific. Galls of arthropods on the same leaf can host different endophyte communities, as demonstrated for aphid galls on *Populus deltoides* (Nashville, United States) (Lawson et al., 2014). The endophyte community within the galls of each aphid species was distinct, and differed not only from that of the galls of other aphid species but also from that of the adjacent "healthy" leaf tissue. The specific endophyte community in the gall tissue possibly interacts with the plant-specific endophytes and, thus, prevents gall abortion, which can be caused by the reactivation of plant-specific endophytes (Pehl and Butin, 1994).

Tree age, age of the leaf tissue, and season

The density of colonization of plant tissues by endophytic fungi increases with the age of the tissue. This finding is confirmed by numerous recent studies (Ladjal et al., 2013; Oono et al., 2015; Osono, 2008; Sadeghi et al., 2019; Tateno et al., 2015; Terhonen et al., 2011; Thongsandee et al., 2012; Zamora et al., 2008). The density increases continuously over the whole life span of leaves/needles, that is, in deciduous woody plants over a single growing season. In evergreen woody plants, the increase lasts for years. At the onset of senescence, many endophytes "awaken" and begin to colonize the leaf tissue more intensively until in some cases, leaf symptoms become visible even before leaf fall (Sieber, 2007). Some endophytes sporulate before leaf fall, others after. Some hibernate in the litter and sporulate when the next leaf generation flushes in the spring. The fresh leaves can then be colonized, closing the life cycle. In contrast to density, in many cases the diversity of endophytes does not increase with the age of the leaf tissue. Diversity also does not necessarily increase with the age of the trees. While Taudière et al. (2018) found a slightly higher richness in older Corsican pine (*Pinus nigra* subsp. *laricio*), the diversity in older *Pinus taeda* was lower than in young seedlings in studies by Oono et al. (2015).

Herbivores and pathogens

Endophytic fungi are known to modulate their host's resistance against pathogens and herbivores. But the reverse is equally plausible, that the frequency and diversity of endophytic fungi communities are affected by the presence of pathogens or herbivores.

The endophyte communities of the European ash south of the Alps, where the ash dieback pathogen *H. fraxineus* had not yet arrived, differed significantly from endophyte communities north of the Alps, where the ash dieback pathogen has been causing damage for several years (Schlegel et al., 2018). Certain OTUs of *Venturia fraxini* and a *Cladosporium* species were significantly more common north of the Alps, and *Preussia minima* could only be detected north of the Alps. Conversely, *Colletotrichum goedetiae*, *Paraconiothyrium* sp., and *Diaporthe eres* were significantly more common south of the Alps. One might argue that the difference is due to climatic differences. However, there was no difference between north and south of the Alps in the leaf mycobiome of the Sycamore maple (*Acer pseudoplantanus*), which does not suffer from a major disease, and which was studied at the same sites to "control" for the climate effect. Consequently, the ash mycobiome is subject to changes caused by an emerging disease of the host tree. The introduced *H. fraxineus* takes over the habitat and niche of the native *H. albidus*, which was found on both sides of the Alps before the arrival of the pathogen (before 2007), but after the arrival of the pathogen only south of the Alps (Queloz et al., 2011; Schlegel et al., 2018; Senn-Irlet et al., 2016). Similarly, before 2005, McKinney et al. (2012) frequently found *H. albidus* in Denmark, but not in 2010 when only *H. fraxineus* was detected. In the meantime, the pathogen has also arrived south of the Alps (first detection in 2014 (Sieber, 2014)). In one or two decades, it would be worthwhile to study how the endophyte community of *F. excelsior* south of the Alps has changed compared to the findings of Schlegel et al. (2018).

Communities of endophytic fungi in the leaves of aspen clones (*Populus tremula*) were related to herbivore damage (Albrectsen et al., 2010). One of the endophytes, an *Aureobasidium* species, occurred significantly less frequently in leaf tissues damaged by the brassy willow beetle (*Phratora vitellinae*), indicating that the beetle avoids leaves with dense colonization by the endophyte. A similar relationship of leaf endophytes with symptoms of *Venturia tremulae* could, however, not be established.

Lee et al. (2014) exposed seedlings of *Pinus koraiensis* to water stress and artificial inoculation with the endophyte *Cenangium ferruginosum*. Before the plants were inoculated with *C. ferruginosum*, 32.6% of the endophyte species belonged to the *Rhytismatales* and 52.2% to the *Xylariales*. After inoculation of *C. ferruginosum*, the number of Rhytismatales decreased to 21.0%, and the *Capnodiales* (40.7%) became the dominant group. The application of water stress to plants inoculated with *C. ferruginosum* resulted in another strong shift of the endophyte community to 48.2% *Diaporthales* and 30.7% *Botryosphaeriales*.

Air pollution—Urbanization

Air pollution by ozone, sulfur dioxide, nitrogen oxides, or heavy metals generally leads to the impoverishment of endophyte communities and changes in species composition (Helander et al., 2011; Lappalainen et al., 1999; Likar and Regvar, 2009). Air pollution is particularly high in some cities. For example, the diversity of endophytic fungi in the leaves of *Quercus macrocarpa* was significantly lower inside compared to outside the urban center of a small town in Kansas (United States) (Jumpponen and Jones, 2009). Matsumura and Fukuda (2013) report similar results for *Quercus myrsinaefolia*, *Quercus serrata*, *Chamaecyparis obtusa*, and *Eurya japonica* from Japan. A significant decrease in host-specific endophyte species in urban areas compared to rural areas was found. This fact was attributed to urbanization and the associated fragmentation of forests. In contrast, Wolfe et al. (2018) found only marginally significant differences in fungal community diversity in red alder leaves (*Alnus rubra*) between urban areas and a forest area in Oregon. Ozone is likely to affect epiphytes and propagules of potential endophytes prior and during infection (Barengo et al., 2000; Magan et al., 1995). However, ozone is most likely not responsible for the impoverishment of the endophyte community in urban areas because ozone accumulates in rural areas, whereas in the city it is neutralized by nitrogen oxides from car exhaust and heating systems (Cailleret et al., 2018).

Leaf physiology

The leaf physiology influences the endophyte community and vice versa. The frequency of endophytic fungi in the leaves of *Platanus orientalis* increased at higher concentrations of iron (Fe) and potassium (K) (Khorsandy et al., 2016). Iron and zinc also had an influence on the endophyte community in the leaves of *Populus* × *euramericana* (Martin-Garcia et al., 2011). Similarly, the amount of chlorophyll and flavonoids in the leaves of the European beech (*Fagus sylvatica*) had a significant influence on mycobiome biodiversity (Unterseher et al., 2016), whereas the influence of leaf age was insignificant (Siddique and Unterseher, 2016). In *Betula ermanii*, increased carbon availability led to greater alpha diversity (Yang et al., 2016). However, carbon availability was correlated with elevation and, therefore, it remains unclear whether carbon, elevation, or both combined influenced the diversity.

3.2.2 The role of leaf endophytes in plant defense—Endophyte-mediated resistance

While the frequency and diversity of endophytic fungi are well known, relatively little is known about the "function" of these endophytes. This is mainly due to the fact that Koch's postulates must be fulfilled to prove the effect of an endophytic fungus. However, it is impossible to fulfill the third postulate, i.e., "the endophyte must cause the postulated effect when reintroduced into an endophyte-free plant" (modified), for endophytes in adult trees, as it is impossible to produce endophyte-free adult trees. Endophyte-free grown seedlings and cuttings can serve as surrogates, but cannot really replace a fully grown tree.

There are only a few studies fulfilling Koch's postulates about tree endophytes controlling pathogens and herbivores. Some important work from the time before 2007 is summarized in Sieber et al. (2007): Arnold et al., 2003; Calhoun et al., 1992; Carroll, 1995; Faeth and Hammon, 1996, 1997a,b; Lappalainen and Helander, 1997; Pehl and Butin, 1994; Preszler et al., 1996; Schulz et al., 1995; Schwarz et al., 2004; Webber, 1981; Wilson, 1995.

In several recent studies, a strong toxic effect of rugulosin on various herbivores could be shown. The eastern spruce budworm (*Choristoneura fumiferana*) is a serious defoliator of North American conifers, especially black spruce (*Picea mariana*) and balsam fir (*Abies balsamea*), and can be controlled with rugulosin (Frasz et al., 2014; Quiring et al., 2019; Sumarah et al., 2008). Rugulosin is produced by the endophyte *Phialocephala scopiformis*, which occurs in the needles and branches of various conifers, including black spruce (Kowalski and Kehr, 1995; Tanney et al., 2016).

Herbivores try to avoid leaves or leaf areas densely packed with endophytic fungi. *Atta colombica* leaf-cutting ants cut more than 2.5 times the leaf area of Ecuador laurel seedlings (*Cordia alliodora*) with low endophyte density compared to seedlings with high endophyte density (Bittleston et al., 2011). Moreover, more ants were recruited to leaves low in endophyte density.

Endophytic fungi are also active against pathogenic fungi. The endophytes *Alternaria* sp., *Cladosporium* sp., *Fusarium* sp., and *Penicillium* sp. from the Chilean firetree (*Embothrium coccineum*) and Chilean wineberry (*Aristotelia chilensis*) significantly inhibited the growth of the common fungal pathogen *Botrytis cinerea* (Gonzalez-Teuber et al., 2020). An endophytic isolate of *Trichoderma koningiopsis* from the rubber tree (*Hevea guianensis*) inhibited the causal agent of *Corynespora* leaf fall disease (*Corynespora cassiicola*) both in culture and in plants (Pujade-Renaud et al., 2019).

Sanzio Pimenta et al. (2012) made two interesting discoveries. First, they found that *Phaeosphaeria nodorum* was the most frequent endophyte in the leaves of *Prunus domestica* in an orchard of the Appalachian Fruit Research Station (Kearneysville, West Virginia, United States). *P. nodorum* is known as the causal agent of glume blotch of wheat and the most frequent endophyte in wheat (Sieber et al., 1988; Weber, 1922). Thus, this was a rather surprising host jump. Alternatively, *P. nodorum* originated from *P. domestica* or another *Prunus* species and jumped to wheat. Second, some *P. nodorum* isolates produced inhibitory volatiles to *Monilinia fructicola*, the cause of the brown rot postharvest disease of plums, cherries, and peaches. The fungal volatiles (ethyl acetate, 3-methyl-1-butanol, acetic acid, 2-propyn-1-ol, and 2-propenenitrile) inhibited the growth and reduced the width of the hyphae, causing the disintegration of the hyphal content in culture. *P. nodorum* volatiles were, however, not inhibitory against *Colletotrichum gloeosporioides*, the causal agent of leaf anthracnoses.

Exudates of leaf endophytes (*Venturia fraxini*, *Paraconiothyrium* sp., *Boeremia exigua*, *Kretzschmaria deusta*, *Pezicula* sp., *Ampelomyces quisqualis*, and *Neofabraea alba*) of the European ash (*Fraxinus excelsior*) inhibited the germination of the ascospores of the ash dieback pathogen *Hymenoscyphus fraxineus* (Schlegel et al., 2016). The *Paraconiothyrium* species is related but not conspecific to *Paraconiothyrium variabile* and *P. brasiliense*, which are known for the production of the anticancer drug taxol (Garyali et al., 2014; Somjaipeng et al., 2015) and for their antibiotic effect against a multitude of pathogens (Combès et al., 2012; Nicoletti et al., 2013). *Kretzschmaria deusta* is an aggressive soft rot-causing fungus capable of attacking several deciduous tree species, and it is regularly isolated, though at low frequencies, as a leaf/needle endophyte from various tree species (Ibrahim et al., 2017; Kälin, 2019; Schlegel et al., 2018). The strong fungitoxic property of this species may be the evolutionary result of competition against other wood-decay fungi, but the endophytic colonization of leaves seems to be a dead end for the fungus. *Hypoxylon rubiginosum*, another representative of *Xylariaceae*, produces the antifungal metabolites phomopsidin and hydroxyphomopsidin that inhibit beta-tubulin, which is responsible for the formation of the eukaryotic cytoskeleton and has an antagonistic effect on *Hymenoscyphus fraxineus* (Halecker et al., 2020). *H. rubiginosum* would therefore be well suited as a biocontrol for the ash-dieback pathogen. However, *H. rubiginosum* is a branch and stem colonizer and was detected by DNA metabarcoding in only four of 1500 *F. excelsior* leaflets (0.27%) from different locations in Switzerland and Italy (Schlegel et al., 2018). Like *K. deusta*, *H. rubiginosum* causes soft rot. On beech, such fungi occasionally infect leaf hairs and cause soft rot in the cell walls (Fig. 6.4). Another endophytic *Xylaria* species from *Pinus strobus* and *Vaccinium angustifolium* produces the antimycoticum griseofulvin (Richardson et al., 2014). The endophytic *Pezicula cinnamomea* from *Castanea sativa* strongly impeded the mycelial growth of *Cryphonectria parasitica*, the causal agent of chestnut blight (Bissegger and Sieber, 1994). Similarly, endophytic *Pezicula sporulosa* from the needles of *Picea rubens* produces the antifungal metabolites cryptosporiopsin, hydroxycryptosporiopsin, cryptosporiopsinol, and mellein (McMullin et al., 2017).

The endophytes of *P. monticola* were shown to increase the survival of their host against pine blister rust (*Cronartium ribicola*) (Ganley et al., 2008). Similarly, *P. ponderosa* needles were found to be less affected by *Dothistroma* needle blight if pretreated with endophytic fungi (Ridout and Newcombe, 2015). In contrast, endophyte-colonized leaves and endophyte-free leaves of both *F. excelsior* and *F. ornus* did not differ in susceptibility to the ash-dieback pathogen (*H. fraxineus*), either in the field or under controlled conditions in a climate chamber (Schlegel et al., 2016).

One might argue that the inhibitive strength of endophytes correlates with their growth rates. However, the majority of inhibitory endophytes grow slowly. The production of inhibitory metabolites by slow-growing organisms may have evolved to compete with organisms specialized in rapid resource capture (Mille-Lindblom et al., 2006).

3.2.3 Production of secondary metabolites by leaf endophytes

Endophytic fungi produce secondary metabolites that serve to colonize their hosts, attract vectors, and inhibit competitors. Many of these metabolites are also pharmaceutically promising. The most famous are probably the various metabolites from the endophytes of the yew (*Taxus*) species: taxane, taxol, and baccatin (Dai et al., 2017; Ishino et al., 2012; Strobel, 2003). These metabolites have antimicrobial properties and are effective against certain forms of cancer. They are produced by various *Taxus* endophytes: *Taxomyces andreanae*, *Pestalotiopsis microspora*, or *Paraconiothyrium brasiliense* (Garyali et al., 2014; Somjaipeng et al., 2015; Stierle et al., 1993; Strobel et al., 1996).

A strain of a hitherto undescribed endophytic fungus from the leaves of *Psychotria horizontalis* (Rubiaceae) produced Cercosporin, which was shown to be active against leishmaniasis (*Leishmania donovani*), chagas disease (*Trypanosoma cruzi*), malaria (*Plasmodium falciparum*), and some cancer cell lines (Moreno et al., 2011). Crude extracts of endophytic fungi of the leaves of rudraksha (*Elaeocarpus sphaericus*) and fragrant nutmeg (*Myristica fragrans*) showed antimicrobial activity against *Escherichia coli*, *Staphylococcus aureus*, and *Klebsiella pneumoniae*. These bacteria belong to the usual microbiota of the body and can lead to complications for sick and immunocompromised persons (Deepthi et al., 2018). Similarly, crude extracts from an endophytic *Colletotrichum* isolate from the Pili nut tree (*Canarium ovatum*) were antagonistic against several bacteria: *Staphylococcus aureus*, *E. coli*, *Serratia marcescens*, *Micrococcus luteus*, *Bacillus subtilis*, and *B. megaterium* (Torres and dela Cruz, 2015).

3.2.4 Endophytism as "pole position"

Many endophytes resume growth when the leaves or needles start to senesce. Their biomass increases and they sporulate while the leaves are still attached to the tree or after the leaves have fallen to the ground. Their only goal seems to be in "pole position" when the leaf tissues die and are available for degradation. A typical example is *Cytospora pinastri* (teleomorphic form *Valsa friesii*) on silver fir needles (*Abies alba*) (Sieber-Canavesi and Sieber, 1993). *C. pinastri* colonizes silver fir needles endophytically. Their frequency increases strongly with the age of the needles and reaches a maximum when the needles start to senesce. *C. pinastri* sporulates abundantly on dead needles still attached to the branches. After the needles have fallen, the fungus is replaced almost completely within a year by purely saprotrophic litter fungi. The often dominant *Lophodermium* endophytes in conifer needles have similar life cycles. Even if endophytes in fresh needle litter are replaced within months, they apparently contribute significantly to needle degradation. The biomass of the litter of *Pinus massoniana* decreased by 60% within only 7 months due to the degradation activity of endophytes; the endophytic *Lophodermium* species contributed 14%–22% to this loss of biomass within only 2 months (Yuan and Chen, 2014).

Often, endophyte diversity decreases after leaf fall, but some endophytes remain active in the litter throughout the entire degradation process (Szink et al., 2016; Unterseher et al., 2013). Endophytic fungi were even more active than saprotrophic fungi in 1-year-old beech litter (Guerreiro et al., 2018). Consequently, certain endophyte species are not only able to switch from an endophytic lifestyle to a saprotrophic one in pure cultures, but they also remain competitive in litter. Thus, a lack of competitiveness with pure litter degraders as a universally valid explanation for the endophytism of certain fungal species is questioned.

4. "Domestication" of endophytic fungi

The defense of plants against pathogens and herbivores is based on preinfection defence by the epi- and endophytic microbiome, preinfection structures, defense substances, and inducible postinfection mechanisms.

Epiphytic microorganisms can act antagonistically against each other and against new arrivals. The antagonistic effects are often based on competition for nutrients (Dik, 1991), but antibiosis and parasitism are also important, such as *Ampelomyces quisqualis* parasitizing other fungi or *Beauveria bassiana* parasitizing arthropods (Cao et al., 2009; McKinnon et al., 2017). For example, *Pseudomonas fluorescens* antagonizes *Erwinia amylovora*, the causal agent of the fire blight of Maloideae (Pusey et al., 2009), or epiphytic fungi such as *Aureobasidium* sp., *Cladosporium* spp., and *Epicoccum nigrum* control the apple scab *Venturia inaequalis* (Fiss et al., 2000).

Constitutive defense includes the surface structure of the leaves/needles (hairs, texture), the cuticle (waxes), and the cell wall (toughness, thickness). Antimicrobial plant metabolites may also be constitutively present (Dadakova et al., 2015; Gonzalez-Teuber et al., 2020). For example, the number of reads of fungal endophytes decreased with the increasing content of cell wall polysaccharides (i.e., increasing cell wall thickness) while endophyte diversity decreased with increasing anthocyanin levels. In contrast, endophyte frequency was positively associated with terpenoids. European beech leaves with the highest chlorophyll and flavonoid content revealed the lowest OTU richness of the fungal community (Unterseher et al., 2016).

Although we know that endophytic fungi live inside plant tissues, we often do not know in which of the numerous microhabitats they are living. There are lots of microhabitats in a leaf: trichomes, cuticles, substomatal spaces, epidermal cells, in and/or between the cells of the palisade or spongy parenchyma, in or between vessels, and in cell walls (Fig. 6.1). The microhabitat is only known for a relatively few endophytic fungi. Depending on the microhabitat colonized, the plant reacts according to a different pattern.

The host may not react to endophytes located between the cuticle and epidermis (Viret et al., 1993), or only when the fungus tries to penetrate further into the leaf tissue. The same might be true for fungi that colonize trichomes. *Asteroma padi* makes extensive, richly branched mycelial cords between the cuticle and the epidermis of *Prunus padus* leaves (Fig. 6.5). The host does not seem to react to the intruder because the underlying leaf tissue remains completely green and healthy for a long time, although photosynthesis is restricted.

If fungi penetrate the epidermis and colonize the mesophyll, or if they penetrate through the stomata into the substomatal spaces, a reaction of the host plant can be expected, even if no symptoms are visible. *Rhabdocline parkeri* develops multicellular thalli within the lumen of single epidermal cells of Douglas fir needles (Stone, 1987). *Lophdermium piceae*, the most common endophytic fungus in spruce needles, grows intercellularly in the palisade parenchyma and substomatal cavities of spruce needles (Suske and Acker, 1989) (Fig. 6.1), similar to a leaf endophyte in the substomatal cavity of *Kunzea ericoides* (Johnston et al., 2006). *Cytospora pinastri* behaves similarly in fir needles (Fig. 6.6).

Obviously, endophytes are able to trick the plant's various defense systems and infect the plant. Why does the host tolerate these intruders? There is still no conclusive answer to this question.

Pathogens and endophytes of the same host are often very closely related (Romao et al., 2011; Sieber, 2007). This suggests that endophytes have evolved and are still evolving from pathogens and vice versa (Carroll, 1988; Saikkonen et al., 1998; Sieber, 2007). Consequently, it can be assumed that host-endophyte interactions are very similar to host-pathogen interactions, but that the endophyte at least partially escapes or weakens the plant's defense (Hardoim et al., 2015).

There are several ways to escape the immune response of the host. For instance, the ectomycorrhiza fungus *Laccaria* bicolor lacks entire gene families that encode enzymes necessary for the degradation of plant cell walls (Martin et al., 2008). Thus, the degradation products of cell walls, which would activate the plant's immune defense, cannot form. Similarly, some endophytic and obligate biotrophic fungi (e.g., powdery mildews) do not possess class II peroxidases (CIIPrx), which are useful in lignin degradation (Mathe et al., 2019). In contrast, many other endophytic fungi readily produce cell wall-degrading enzymes (Carroll and Petrini, 1983; Peng and Chen, 2007). The degradation products activate the immune

FIG. 6.5 Hyphal cords of *Asteroma padi* formed between the cuticle and epidermis of a leaf of hackberry (*Prunus padus*) in autumn. *(Courtesy of O. Holdenrieder.)*

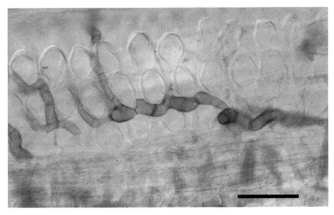

FIG. 6.6 Endophytic thallus in the mesophyll of a European white fir needle (*Abies alba*). Scale bar = 50 µm.

defense in the plant (De Wit, 2007; Dodds and Rathjen, 2010; Jones and Dangl, 2006), which induces the production of enzymes (effectors) in the endophyte that inhibit the plant's defense proteins (Rafiqi et al., 2012). In contrast to the effectors of a pathogen, these endophyte enzymes probably do not activate the resistance genes of the plant or only to the extent that the endophyte is put into a "dormant state." This state requires a minimal nutrient supply by the host. Unknown in most endophyte-plant symbioses is what the host receives as compensation for keeping the endophyte alive. However, even if the thallus' biomass hardly increases, this does not mean that nothing is happening. On the contrary, possibly stimulated by the plant, the endophyte may produce (secondary) metabolites that promote the growth of the host or that are useful for the defense against herbivores or pathogens, that is, leading to endophyte-mediated resistance.

The "game" between attack and defense is subject to constant coevolution. The fungus eliminates effector genes or develops new ones while the host reacts by developing new resistance genes (De Wit, 2007; Jones and Dangl, 2006). Powdery mildews and endophytic fungi are probably the most similar in nutritional and ecological behavior. Endophytes that form disjointed, intracellular thalli (Stone, 1987) could be regarded as powdery mildews reduced to the haustoria. In natural or seminatural forests, native powdery mildews do not pose a threat. On the contrary, because they are obligately biotrophic, they need their hosts to survive. They are beneficial by increasing the resistance of the host to pathogens (Gruner et al., 2020) and even esthetically attractive by their delicate appendices surrounding the ascomata and the formation of "green islands" (Figs 6.7 and 6.8). The mildew fungus prevents aging by producing auxins and gibberellins in the colonized areas (Coghlan and Walters, 1992; Scott, 1972). *Mlo* genes apparently play a major role in resistance against powdery mildew (Acevedo-Garcia et al., 2014; Büschges et al., 1997; Feechan et al., 2008; Jiwan et al., 2013; Jorgensen, 1992). Plants

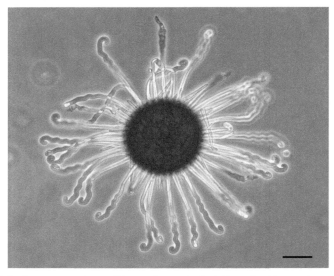

FIG. 6.7 Chasmothecium of *Erysiphe flexuosa*, powdery mildew on horse chestnut (*Aesculus hippocastanum*). Scale bar = 50 µm. *(Courtesy of O. Holdenrieder.)*

FIG. 6.8 "Green islands" resulting from colonization by a powdery mildew on a leaf of a sycamore maple (*Acer pseudoplatanus*) in autumn. *(Courtesy of O. Holdenrieder).*

carrying *Mlo* mutations exhibit durable resistance against powdery mildews. In nature, it can be assumed that the *Mlo* genes are present in the unmutated form, or if several such genes are present, only some are mutated while the others are unmutated. Thus, the plant accepts getting colonized to some extent.

The results of future research on endophytes associated with woody plants are eagerly expected. Of particular interest will be more information on the role of these fungi in biological control and the molecular mechanisms involved in the domestication of these fungi by the plant.

Acknowledgments

Many thanks to friend and colleague Ottmar Holdenrieder, who read my text critically and provided valuable input and excellent photos.

References

Acevedo-Garcia, J., Kusch, S., Panstruga, R., 2014. Magical mystery tour: MLO proteins in plant immunity and beyond. New Phytol. 204, 273–281.

Agrios, G., 2005. Plant Pathology. Academic Press, London.

Albrectsen, B.R., Bjorken, L., Varad, A., Hagner, A., Wedin, M., Karlsson, J., Jansson, S., 2010. Endophytic fungi in European aspen (*Populus tremula*) leaves-diversity, detection, and a suggested correlation with herbivory resistance. Fungal Divers. 41, 17–28.

Ale-Agha, N., Braun, U., Feige, B., Jage, H., 2000. A new powdery mildew disease on *Aesculus* spp. introduced in Europe. Cryptogam. Mycol. 21, 89–92.

Allen, M.F., 1991. The Ecology of Mycorrhizae. Cambridge University Press, Cambridge, UK.

Allen, E.A., Hoch, H.C., Steadman, J.R., Stavely, R.J., 1991. Influence of leaf surface features on spore deposition and the epiphytic growth of phytopathogenic fungi. In: Andrews, J.H., Hirano, S.S. (Eds.), Microbial Ecology of Leaves. Springer, New York, Berlin, pp. 87–110.

Andrews, J.H., Harris, R.F., 2000. The ecology and biogeography of microorganisms of plant surfaces. Annu. Rev. Phytopathol. 38, 145–180.

Aragon, W., Juan Reina-Pinto, J., Serrano, M., 2017. The intimate talk between plants and microorganisms at the leaf surface. J. Exp. Bot. 68, 5339–5350.

Arnold, A.E., Mejia, L.C., Kyllo, D., Rojas, E.I., Maynard, Z., Robbins, N., Herre, E.A., 2003. Fungal endophytes limit pathogen damage in a tropical tree. Proc. Natl. Acad. Sci. U. S. A. 100, 15649–15654.

Balint, M., Bartha, L., O'Hara, R.B., Olson, M.S., Otte, J., Pfenninger, M., Robertson, A.L., Tiffin, P., Schmitt, I., 2015. Relocation, high-latitude warming and host genetic identity shape the foliar fungal microbiome of poplars. Mol. Ecol. 24, 235–248.

Barengo, N., Sieber, T.N., Holdenrieder, O., 2000. Diversity of endophytic mycobiota in leaves and twigs of pubescent birch (*Betula pubescens*). Sydowia 52, 305–320.

Barge, E.G., Leopold, D.R., Peay, K.G., Newcombe, G., Busby, P.E., 2019. Differentiating spatial from environmental effects on foliar fungal communities of *Populus trichocarpa*. J. Biogeogr. 46, 2001–2011.

Bissegger, M., Sieber, T.N., 1994. Assemblages of endophytic fungi in coppice shoots of *Castanea sativa*. Mycologia 86, 648–655.

Bittleston, L.S., Brockmann, F., Wcislo, W., Van Bael, S.A., 2011. Endophytic fungi reduce leaf-cutting ant damage to seedlings. Biol. Lett. 7, 30–32.

Bowman, E.A., Arnold, A.E., 2018. Distributions of ectomycorrhizal and foliar endophytic fungal communities associated with *Pinus ponderosa* along a spatially constrained elevation gradient. Am. J. Bot. 105, 687–699.

Braun, U., Takamatsu, S., Heluta, V., Limkaisang, S., Divarangkoon, R., Cook, R., Boyle, H., 2006. Phylogeny and taxonomy of powdery mildew fungi of Erysiphe sect. Uncinula on Carpinus species. Mycol. Progr. 5, 139–153.

Bressan, E.A., Rossi, M.L., Gerald, L.T.S., Figueira, A., 2014. Extraction of high-quality DNA from ethanol-preserved tropical plant tissues. BMC. Res. Notes 7, 268.

Bridge, P.D., Roberts, P.J., Spooner, B.M., Panchal, G., 2003. On the unreliability of published DNA sequences. New Phytol. 160, 43–48.

Bulgarelli, D., Schlaeppi, K., Spaepen, S., van Themaat, E.V.L., Schulze-Lefert, P., 2013. Structure and functions of the bacterial microbiota of plants. In: Merchant, S.S. (Ed.), Annual Review of Plant Biology. vol. 64, pp. 807–838.

Burkhardt, J., 2010. Hygroscopic particles on leaves: nutrients or desiccants? Ecol. Monogr. 80, 369–399.

Büschges, R., Hollricher, K., Panstruga, R., Simons, G., Wolter, M., Frijters, A., van Daelen, R., van der Lee, T., Diergaarde, P., Groenendijk, J., Töpsch, S., Vos, P., Salamini, F., Schulze-Lefert, P., 1997. The barley *Mlo* gene: a novel control element of plant pathogen resistance. Cell 88, 695–705.

Butin, H., 1995. Tree diseases and disorders. In: Causes, Biology and Control in Forest and Amenity Trees. Oxford University Press, Oxford.

Cabral, D., 1985. Phyllosphere of *Eucalyptus viminalis*—dynamics of fungal populations. Trans. Br. Mycol. Soc. 85, 501–511.

Cailleret, M., Ferretti, M., Gessler, A., Rigling, A., Schaub, M., 2018. Ozone effects on European forest growth—towards an integrative approach. J. Ecol. 106, 1377–1389.

Calhoun, L.A., Findlay, J.A., Miller, J.D., Whitney, N.J., 1992. Metabolites toxic to spruce budworm from balsam fir needle endophytes. Mycol. Res. 96, 281–286.

Cao, R., Liu, X., Gao, K., Mendgen, K., Kang, Z., Gao, J., Dai, Y., Wang, X., 2009. Mycoparasitism of endophytic fungi isolated from reed on soilborne phytopathogenic fungi and production of cell wall-degrading enzymes in vitro. Curr. Microbiol. 59, 584–592.

Carroll, G.C., 1988. Fungal endophytes in stems and leaves—from latent pathogen to mutalistic symbiont. Ecology 69, 2–9.

Carroll, G.C., 1995. Forest endophytes: pattern and process. Can. J. Bot. 73, S1316–S1324.

Carroll, G., Petrini, O., 1983. Patterns of substrate utilization by some fungal endophytes from coniferous foliage. Mycologia 75, 53–63.

Chen, J., 2012. Fungal community survey of *Fraxinus excelsior* in New Zealand. http://stud.epsilon.slu.se/4172/. (Accessed 10 June 2020).

Cobban, A., Edgcomb, V.P., Burgaud, G., Repeta, D., Leadbetter, E.R., 2016. Revisiting the pink-red pigmented basidiomycete mirror yeast of the phyllosphere. Microbiology 5, 846–855.

Coghlan, S.E., Walters, D.R., 1992. Photosynthesis in green-islands on powdery mildew-infected barley leaves. Physiol. Mol. Plant Pathol. 40, 31–38.

Combès, A., Ndoye, I., Bance, C., Bruzaud, J., Djediat, C., Dupont, J., Nay, B., Prado, S., 2012. Chemical communication between the endophytic fungus *Paraconiothyrium variabile* and the phytopathogen *Fusarium oxysporum*. PLoS ONE 7, e47313.

Cooke, R.C., Rayner, A.D.M., 1984. Ecology of Saprotrophic Fungi. Longman, London and New York.

Crowther, T.W., Glick, H.B., Covey, K.R., Bettigole, C., Maynard, D.S., Thomas, S.M., Smith, J.R., Hintler, G., Duguid, M.C., Amatulli, G., Tuanmu, M.N., Jetz, W., Salas, C., Stam, C., Piotto, D., Tavani, R., Green, S., Bruce, G., Williams, S.J., Wiser, S.K., Huber, M.O., Hengeveld, G.M., Nabuurs, G.J., Tikhonova, E., Borchardt, P., Li, C.F., Powrie, L.W., Fischer, M., Hemp, A., Homeier, J., Cho, P., Vibrans, A.C., Umunay, P.M., Piao, S.L., Rowe, C.W., Ashton, M.S., Crane, P.R., Bradford, M.A., 2015. Mapping tree density at a global scale. Nature 525, 201–207.

Dadakova, K., Havelkova, M., Kurkova, B., Tlolkova, I., Kasparovsky, T., Zdrahal, Z., Lochman, J., 2015. Proteome and transcript analysis of *Vitis vinifera* cell cultures subjected to *Botrytis cinerea* infection. J. Proteome 119, 143–153.

Dai, H.-y., Liu, M.-z., Duan, Z.-g., Ma, X.-l., Lu, Z.-c., 2017. Screening and identification twenty-five strains of taxane-producing endophytic fungi from *Taxus chinensis* var. *mairei*. J. Trop. Subtrop. Bot. 25, 271–278.

De Wit, P.J.G.M., 2007. Visions & reflections (minireview)—how plants recognize pathogens and defend themselves. Cell. Mol. Life Sci. 64, 2726–2732.

Debeljak, P., Pinto, M., Proietti, M., Reisser, J., Ferrari, F.F., Abbas, B., van Loosdrecht, M.C.M., Slat, B., Herndl, G.J., 2017. Extracting DNA from ocean microplastics: a method comparison study. Anal. Methods 9, 1521–1526.

Decourcelle, T., Piou, D., Desprez-Loustau, M.L., 2015. Detection of *Diplodia sapinea* in Corsican pine seeds. Plant Pathol. 64, 442–449.

Deepthi, V.C., Sumathi, S., Faisal, M., Elyas, K.K., 2018. Isolation and identification of endophytic fungi with antimicrobial activities from the leaves of *Elaeocarpus sphaericus* (Gaertn.) K. Schum. and *Myristica fragrans* Houtt. Int. J. Pharm. Sci. Res. 9, 2783–2791.

Desprez-Loustau, M.-L., Massot, M., Toigo, M., Fort, T., Kaya, A.G.A., Boberg, J., Braun, U., Capdevielle, X., Cech, T., Chandelier, A., Christova, P., Corcobado, T., Dogmus, T., Dutech, C., Fabreguettes, O., d'Arcier, J.F., Gross, A., Jung, M.H., Iturritxa, E., Jung, T., Junker, C., Kiss, L., Kostov, K., Lehtijarvi, A., Lyubenova, A., Marcais, B., Oliva, J., Oskay, F., Pastircak, M., Pastircakova, K., Piou, D., Saint-Jean, G., Sallafranque, A., Slavov, S., Stenlid, J., Talgo, V., Takamatsu, S., Tack, A.J.M., 2018. From leaf to continent: the multi-scale distribution of an invasive cryptic pathogen complex on oak. Fungal Ecol. 36, 39–50.

Dik, A.J., 1991. Interactions among fungicides, pathogens, yeasts, and nutrients in the phyllosphere. In: Andrews, J.H., Hirano, S.S. (Eds.), Microbial Ecology of Leaves. Springer-Verlag, New York, Berlin, pp. 412–429.

Dissanayake, A.J., Purahong, W., Wubet, T., Hyde, K.D., Zhang, W., Xu, H.Y., Zhang, G.J., Fu, C.Y., Liu, M., Xing, Q.K., Li, X.H., Yan, J.Y., 2018. Direct comparison of culture-dependent and culture-independent molecular approaches reveal the diversity of fungal endophytic communities in stems of grapevine (*Vitis vinifera*). Fungal Divers. 90, 85–107.

Dodds, P.N., Rathjen, J.P., 2010. Plant immunity: towards an integrated view of plant-pathogen interactions. Nat. Rev. Genet. 11, 539–548.

Doust, N.H., Akbarinia, M., Safaie, N., Yousefzadeh, H., Balint, M., 2017. Community analysis of Persian oak fungal microbiome under dust storm conditions. Fungal Ecol. 29, 1–9.

Drenkhan, R., Tomesova-Haataja, V., Fraser, S., Bradshaw, R.E., Vahalik, P., Mullett, M.S., Martin-Garcia, J., Bulman, L.S., Wingfield, M.J., Kirisits, T., Cech, T.L., Schmitz, S., Baden, R., Tubby, K., Brown, A., Georgieva, M., Woods, A., Ahumada, R., Jankovsky, L., Thomsen, I.M., Adamson, K., Marcais, B., Vuorinen, M., Tsopelas, P., Koltay, A., Halasz, A., La Porta, N., Anselmi, N., Kiesnere, R., Markovskaja, S., Kacergius, A., Papazova-Anakieva, I., Risteski, M., Sotirovski, K., Lazarevic, J., Solheim, H., Boron, P., Braganca, H., Chira, D., Musolin, D.L., Selikhovkin, A.V., Bulgakov,

T.S., Keca, N., Karadzic, D., Galovic, V., Pap, P., Markovic, M., Pajnik, L.P., Vasic, V., Ondruskova, E., Piskur, B., Sadikovic, D., Diez, J.J., Solla, A., Millberg, H., Stenlid, J., Angst, A., Queloz, V., Lehtijarvi, A., Dogmus-Lehtijarvi, H.T., Oskay, F., Davydenko, K., Meshkova, V., Craig, D., Woodward, S., Barnes, I., 2016. Global geographic distribution and host range of *Dothistroma* species: a comprehensive review. For. Pathol. 46, 408–442.

Eusemann, P., Schnittler, M., Nilsson, R.H., Jumpponen, A., Dahl, M.B., Wuerth, D.G., Buras, A., Wilmking, M., Unterseher, M., 2016. Habitat conditions and phenological tree traits overrule the influence of tree genotype in the needle mycobiome-*Picea glauca* system at an arctic treeline ecotone. New Phytol. 211, 1221–1231.

Faeth, S.H., Hammon, K.E., 1996. Fungal endophytes and phytochemistry of oak foliage: determinants of oviposition preference of leafminers? Oecologia 108, 728–736.

Faeth, S.H., Hammon, K.E., 1997a. Fungal endophytes in oak trees: experimental analyses of interactions with leafminers. Ecology 78, 820–827.

Faeth, S.H., Hammon, K.E., 1997b. Fungal endophytes in oak trees: long-term patterns of abundance and associations with leafminers. Ecology 78, 810–819.

Feechan, A., Jermakow, A.M., Torregrosa, L., Panstruga, R., Dry, I.B., 2008. Identification of grapevine MLO gene candidates involved in susceptibility to powdery mildew. Funct. Plant Biol. 35, 1255–1266.

Fiss, M., Kucheryava, N., Schonherr, J., Kollar, A., Arnold, G., Auling, G., 2000. Isolation and characterization of epiphytic fungi from the phyllosphere of apple as potential biocontrol agents against apple scab (*Venturia inaequalis*). Z. Pflanzenkrankh. Pflanzenschutz 107, 1–11.

Fokkema, N.J., 1991. The phyllosphere as an ecologically neglected milieu: a plant pathologist's point of view. In: Andrews, J.H., Hirano, S.S. (Eds.), Microbial Ecology of Leaves. Springer, New York, Berlin, pp. 3–18.

Fokkema, N.J., Schippers, B., 1986. Phyllosphere versus rhizosphere as environments for saprophytic colonization. In: Fokkema, N.J., Van den Heuvel, J. (Eds.), Microbiology of the Phyllosphere. Cambridge University Press, Cambridge, UK, pp. 137–159.

Frasz, S.L., Walker, A.K., Nsiama, T.K., Adams, G.W., Miller, J.D., 2014. Distribution of the foliar fungal endophyte *Phialocephala scopiformis* and its toxin in the crown of a mature white spruce tree as revealed by chemical and qPCR analyses. Can. J. For. Res. 44, 1138–1143.

Ganley, R.J., Sniezko, R.A., Newcombe, G., 2008. Endophyte-mediated resistance against white pine blister rust in *Pinus monticola*. For. Ecol. Manag. 255, 2751–2760.

Garyali, S., Kumar, A., Reddy, M.S., 2014. Diversity and antimitotic activity of taxol-producing endophytic fungi isolated from Himalayan yew. Ann. Microbiol. 64, 1413–1422.

Gilani, H.R., Innes, J.L., 2020. The state of British Columbia's forests: a global comparison. Forests 11.

Gonzalez-Teuber, M., Vilo, C., Jose Guevara-Araya, M., Salgado-Luarte, C., Gianoli, E., 2020. Leaf resistance traits influence endophytic fungi colonization and community composition in a South American temperate rainforest. J. Ecol. 108, 1019–1029.

Gross, A., Holdenrieder, O., Pautasso, M., Queloz, V., Sieber, T.N., 2014. *Hymenoscyphus pseudoalbidus*, the causal agent of European ash dieback. Mol. Plant Pathol. 15, 5–21.

Gruner, K., Esser, T., Acevedo-Garcia, J., Freh, M., Habig, M., Strugala, R., Stukenbrock, E., Schaffrath, U., Panstruga, R., 2020. Evidence for allele-specific levels of enhanced susceptibility of wheat mlo mutants to the hemibiotrophic fungal pathogen *Magnaporthe oryzae* pv. *triticum*. Genes 11, 517.

Guerreiro, M.A., Brachmann, A., Begerow, D., Persoh, D., 2018. Transient leaf endophytes are the most active fungi in 1-year-old beech leaf litter. Fungal Divers. 89, 237–251.

Habermann, B., 2004. Behms Fichte hat 2800130 Nadeln. Braunschweiger Zeitung 59, e150290283. https://www.braunschweiger-zeitung.de/braunschweig/article150290283/Behms-Fichte-hat-150290282-150290800-150290130-Nadeln.html. (Accessed June 15, 2020).

Halecker, S., Wennrich, J.-P., Rodrigo, S., Andree, N., Rabsch, L., Baschien, C., Steinert, M., Stadler, M., Surup, F., Schulz, B., 2020. Fungal endophytes for biocontrol of ash dieback: the antagonistic potential of *Hypoxylon rubiginosum*. Fungal Ecol. 45.

Hansen, E.M., Stone, J.K., Capitano, B.R., Rosso, P., Sutton, W., Winton, L., Kanaskie, A., McWilliams, M.G., 2000. Incidence and impact of Swiss needle cast in forest plantations of Douglas-fir in coastal Oregon. Plant Dis. 84, 773–778.

Hardoim, P.R., van Overbeek, L.S., Berg, G., Pirttila, A.M., Compant, S., Campisano, A., Doering, M., Sessitsch, A., 2015. The hidden world within plants: ecological and evolutionary considerations for defining functioning of microbial endophytes. Microbiol. Mol. Biol. Rev. 79, 293–320.

Hashizume, Y., Sahashi, N., Fukuda, K., 2008. The influence of altitude on endophytic mycobiota in *Quercus acuta* leaves collected in two areas 1000 km apart. For. Pathol. 38, 218–226.

Hashizume, Y., Fukuda, K., Sahashi, N., 2010. Effects of summer temperature on fungal endophyte assemblages in Japanese beech (*Fagus crenata*) leaves in pure beech stands. Botany 88, 266–274.

Helander, M., Ahlholm, J., Sieber, T.N., Hinneri, S., Saikkonen, K., 2007. Fragmented environment affects birch leaf endophytes. New Phytol. 175, 547–553.

Helander, M., Vesterlund, S.-R., Saikkonen, K., 2011. Responses of foliar endophytes to pollution. In: Pirttila, A.M., Frank, A.C. (Eds.), Endophytes of Forest Tree: Biology and Applications, pp. 175–188.

Hoffman, M.T., Arnold, A.E., 2008. Geographic locality and host identity shape fungal endophyte communities in cupressaceous trees. Mycol. Res. 112, 331–344.

Holden, M.J., Blasic, J.R., Bussjaeger, L., Kao, C., Shokere, L.A., Kendall, D.C., Freese, L., Jenkins, G.R., 2003. Evaluation of extraction methodologies for corn kernel (*Zea mays*) DNA for detection of trace amounts of biotechnology-derived DNA. J. Agric. Food Chem. 51, 2468–2474.

Holdenrieder, O., Pautasso, M., 2014. Wie viel Krankheit braucht der Wald? Bündner Wald 67, 5–10.

Ibrahim, M., Sieber, T.N., Schlegel, M., 2017. Communities of fungal endophytes in leaves of *Fraxinus ornus* are highly diverse. Fungal Ecol. 29, 10–19.

Ishino, T., Terada, T., Samejima, M., Kamoda, S., 2012. Ability to synthesize taxoids of endophytic fungi in Japanese Taxaceae plants. Bull. Tokyo Univ. For., 45–58.

Jiwan, D., Roalson, E.H., Main, D., Dhingra, A., 2013. Antisense expression of peach mildew resistance locus O (PpMlo1) gene confers cross-species resistance to powdery mildew in *Fragaria x ananassa*. Transgenic Res. 22, 1119–1131.

Johnston, P.R., Sutherland, P.W., Joshee, S., 2006. Visualising endophytic fungi within leaves by detection of (1 -> 3)-beta-D-glucans in fungal cell walls. Mycologist 20, 159–162.

Johnston, P.R., Johansen, R.B., Williams, A.F.R., Paula Wikie, J., Park, D., 2012. Patterns of fungal diversity in New Zealand *Nothofagus* forests. Fung. Biol. 116, 401–412.

Johnston, P.R., Park, D., Smissen, R.D., 2017. Comparing diversity of fungi from living leaves using culturing and high-throughput environmental sequencing. Mycologia 109, 643–654.

Jones, J.D.G., Dangl, J.L., 2006. The plant immune system. Nature 444, 323–329.

Jorgensen, J.H., 1992. Discovery, characterization and exploitation of Mlo powdery mildew resistance in barley. Euphytica 63, 141–152.

Jumpponen, A., Jones, K.L., 2009. Massively parallel 454 sequencing indicates hyperdiverse fungal communities in temperate *Quercus macrocarpa* phyllosphere. New Phytol. 184, 438–448.

Juniper, B.E., 1991. The leaf from the inside and the outside. In: Andrews, J.H., Hirano, S.S. (Eds.), Microbial Ecology of Leaves. Springer-Verlag, New York, Berlin, pp. 21–42.

Kälin, P., 2019. Endophytic Communities in Douglas fir and Norway Spruce Needles—Differences in Abundance and Occurrence Between Hosts and the Methods Used. ETH Zurich, Zurich, Switzerland (MSc thesis).

Kaul, S., Gupta, S., Ahmed, M., Dhar, M.K., 2012. Endophytic fungi from medicinal plants: a treasure hunt for bioactive metabolites. Phytochem. Rev. 11, 487–505.

Khorsandy, S., Nikbakht, A., Sabzalian, M.R., Pessarakli, M., 2016. Effect of fungal endophytes on morphological characteristics, nutrients content and longevity of plane trees (*Platanus orientalis* L.). J. Plant Nutr. 39, 1156–1166.

Kinkel, L., 1991. Fungal community dynamics. In: Andrews, J.H., Hirano, S.S. (Eds.), Microbial Ecology of Leaves. Springer, New York, NY, pp. 253–270.

Kowalski, T., Kehr, R.D., 1995. Two new species of *Phialocephala* occurring on *Picea* and *Alnus*. Can. J. Bot. 73, 26–32.

Küngas, K., Bahram, M., Poldmaa, K., 2020. Host tree organ is the primary driver of endophytic fungal community structure in a hemiboreal forest. FEMS Microbiol. Ecol. 96, fiz199.

Ladjal, S., Harzallah, D., Dahamna, S., Bouamra, D., Bouharati, S., Khennouf, S., 2013. Endophytic fungi isolated from *Pinus halepensis* needles in M'sila (Algeria) region and their bioactivities. Commun. Agric. Appl. Biol. Sci. 78, 625–631.

Lan, Y.-H., Shaw, D.C., Beedlow, P.A., Lee, E.H., Waschmann, R.S., 2019. Severity of Swiss needle cast in young and mature Douglas-fir forests in western Oregon, USA. For. Ecol. Manag. 442, 79–95.

Langenfeld, A., Prado, S., Nay, B., Cruaud, C., Lacoste, S., Bury, E., Hachette, F., Hosoya, T., Dupont, J., 2013. Geographic locality greatly influences fungal endophyte communities in *Cephalotaxus harringtonia*. Fung. Biol. 117, 124–136.

Lappalainen, J.H., Helander, M.L., 1997. The role of foliar microfungi in mountain birch—insect herbivore relationships. Ecography 20, 116–122.

Lappalainen, J.H., Koricheva, J., Helander, M.L., Haukioja, E., 1999. Densities of endophytic fungi and performance of leafminers (Lepidoptera: Eriocraniidae) on birch along a pollution gradient. Environ. Pollut. 104, 99–105.

Lau, M.K., Arnold, A.E., Johnson, N.C., 2013. Factors influencing communities of foliar fungal endophytes in riparian woody plants. Fungal Ecol. 6, 365–378.

Lawson, S.P., Christian, N., Abbot, P., 2014. Comparative analysis of the biodiversity of fungal endophytes in insect-induced galls and surrounding foliar tissue. Fungal Divers. 66, 89–97.

Lee, S.K., Lee, S.K., Bae, H., Seo, S.-T., Lee, J.K., 2014. Effects of water stress on the endophytic fungal communities of *Pinus koraiensis* needles infected by *Cenangium ferruginosum*. Mycobiology 42, 331–338.

Lefort, M.C., McKinnon, A.C., Nelson, T.L., Glare, T.R., 2016. Natural occurrence of the entomopathogenic fungi *Beauveria bassiana* as a vertically transmitted endophyte of *Pinus radiata* and its effect on above- and below-ground insect pests. N. Z. Plant Protect. 69, 68–77.

Leray, M., Knowlton, N., Ho, S.-L., Nguyen, B.N., Machida, R.J., 2019. GenBank is a reliable resource for 21st century biodiversity research. Proc. Natl. Acad. Sci. U. S. A. 116, 22651–22656.

Leveau, J.H.J., 2019. A brief from the leaf: latest research to inform our understanding of the phyllosphere microbiome. Curr. Opin. Microbiol. 49, 41–49.

Likar, M., Regvar, M., 2009. Application of temporal temperature gradient gel electrophoresis for characterisation of fungal endophyte communities of *Salix caprea* L. in a heavy metal polluted soil. Sci. Total Environ. 407, 6179–6187.

Lindow, S.E., Brandl, M.T., 2003. Microbiology of the phyllosphere. Appl. Environ. Microbiol. 69, 1875–1883.

Loo, J.A., 2009. Ecological impacts of non-indigenous invasive fungi as forest pathogens. Biol. Invasions 11, 81–96.

Magan, N., Kirkwood, I.A., McLeod, A.R., Smith, M.K., 1995. Effect of open-air fumigation with sulfur-dioxide and ozone on phyllosphere and endophytic fungi of conifer needles. Plant Cell Environ. 18, 291–302.

Manojlovic, L.M., 2015. Photometry-based estimation of the total number of stars in the Universe. Appl. Opt. 54, 6589–6591.

Marcell, L.M., Beattie, G.A., 2002. Effect of leaf surface waxes on leaf colonization by *Pantoea agglomerans* and *Clavibacter michiganensis*. Mol. Plant-Microbe Interact. 15, 1236–1244.

Martin, F., Aerts, A., Ahren, D., Brun, A., Danchin, E.G.J., Duchaussoy, F., Gibon, J., Kohler, A., Lindquist, E., Pereda, V., Salamov, A., Shapiro, H.J., Wuyts, J., Blaudez, D., Buee, M., Brokstein, P., Canback, B., Cohen, D., Courty, P.E., Coutinho, P.M., Delaruelle, C., Detter, J.C., Deveau, A., DiFazio, S., Duplessis, S., Fraissinet-Tachet, L., Lucic, E., Frey-Klett, P., Fourrey, C., Feussner, I., Gay, G., Grimwood, J., Hoegger, P.J., Jain, P., Kilaru, S., Labbe, J., Lin, Y.C., Legue, V., Le Tacon, F., Marmeisse, R., Melayah, D., Montanini, B., Muratet, M., Nehls, U., Niculita-Hirzel, H., Oudot-Le Secq, M.P., Peter, M., Quesneville, H., Rajashekar, B., Reich, M., Rouhier, N., Schmutz, J., Yin, T., Chalot, M., Henrissat, B.,

Kuees, U., Lucas, S., Van de Peer, Y., Podila, G.K., Polle, A., Pukkila, P.J., Richardson, P.M., Rouze, P., Sanders, I.R., Stajich, J.E., Tunlid, A., Tuskan, G., Grigoriev, I.V., 2008. The genome of *Laccaria bicolor* provides insights into mycorrhizal symbiosis. Nature 452, 88–93.

Martinez-Alvarez, P., Martin-Garcia, J., Rodriguez-Ceinos, S., Diez, J.J., 2012. Monitoring endophyte populations in pine plantations and native oak forests in Northern Spain. For. Syst. 21, 373–382.

Martin-Garcia, J., Espiga, E., Pando, V., Javier Diez, J., 2011. Factors influencing endophytic communities in poplar plantations. Silva Fennica 45, 169–180.

Mathe, C., Fawal, N., Roux, C., Dunand, C., 2019. In silico definition of new ligninolytic peroxidase sub-classes in fungi and putative relation to fungal life style. Sci. Rep. 9, 20373.

Matsumura, E., Fukuda, K., 2013. A comparison of fungal endophytic community diversity in tree leaves of rural and urban temperate forests of Kanto district, eastern Japan. Fung. Biol. 117, 191–201.

McKinney, L.V., Thomsen, I.M., Kjaer, E.D., Bengtsson, S.B.K., Nielsen, L.R., 2012. Rapid invasion by an aggressive pathogenic fungus (*Hymenoscyphus pseudoalbidus*) replaces a native decomposer (*Hymenoscyphus albidus*): a case of local cryptic extinction? Fungal Ecol. 5 (6), 663–669.

McKinnon, A.C., Saari, S., Moran-Diez, M.E., Meyling, N.V., Raad, M., Glare, T.R., 2017. *Beauveria bassiana* as an endophyte: a critical review on associated methodology and biocontrol potential. BioControl 62, 1–17.

McMullin, D.R., Green, B.D., Prince, N.C., Tanney, J.B., Miller, J.D., 2017. Natural products of *Picea* endophytes from the Acadian forest. J. Nat. Prod. 80, 1475–1483.

Menkis, A., Povilaitiene, A., Marciulynas, A., Lynikiene, J., Gedminas, A., Marciulyniene, D., 2019. Occurrence of common phyllosphere fungi of horse-chestnut (*Aesculus hippocastanum*) is unrelated to degree of damage by leafminer (*Cameraria ohridella*). Scand. J. For. Res. 34, 26–32.

Mille-Lindblom, C., Fischer, H., Tranvik, L.J., 2006. Antagonism between bacteria and fungi: substrate competition and a possible tradeoff between fungal growth and tolerance towards bacteria. Oikos 113, 233–242.

Monier, J.M., Lindow, S.E., 2004. Frequency, size, and localization of bacterial aggregates on bean leaf surfaces. Appl. Environ. Microbiol. 70, 346–355.

Moreno, C., Varughese, T., Spadafora, C., Arnold, A.E., Coley, P.D., Kursar, T.A., Gerwick, W.H., Cubilla-Rios, L., 2011. Chemical constituents of the new endophytic fungus *Mycosphaerella* sp nov and their anti-parasitic activity. Nat. Prod. Commun. 6, 835–840.

Müller-Using, S., Bartsch, N., 2003. Totholzdynamik eines Buchenbestandes (*Fagus sylvatica*) im Solling. Nachlieferung, Ursache und Zersetzung von Totholz. Allg. For. Jagdztg. 174, 122–130.

Nicoletti, R., Filippis, A.d., Buommino, E., 2013. Antagonistic aptitude and antiproliferative properties on tumor cells of fungal endophytes from the Astroni Nature Reserve, Italy. Afr. J. Microbiol. Res. 7, 4073–4083.

Nilsson, R.H., Ryberg, M., Kristiansson, E., Abarenkov, K., Larsson, K.-H., Koljalg, U., 2006. Taxonomic reliability of DNA sequences in public sequence databases: a fungal perspective. PLoS ONE 1.

Nisa, H., Kamili, A.N., Nawchoo, I.A., Shafi, S., Shameem, N., Bandh, S.A., 2015. Fungal endophytes as prolific source of phytochemicals and other bioactive natural products: a review. Microb. Pathog. 82, 50–59.

Olivier, D.L., 1978. *Retiarius* gen-nov—phyllosphere fungi which capture wind-borne pollen grains. Trans. Br. Mycol. Soc. 71, 193–201.

Oono, R., Lefevre, E., Simha, A., Lutzoni, F., 2015. A comparison of the community diversity of foliar fungal endophytes between seedling and adult loblolly pines (*Pinus taeda*). Fung. Biol. 119, 917–928.

Osono, T., 2006. Role of phyllosphere fungi of forest trees in the development of decomposer fungal communities and decomposition processes of leaf litter. Can. J. Microbiol. 52, 701–716.

Osono, T., 2008. Endophytic and epiphytic phyllosphere fungi of *Camellia japonica*: seasonal and leaf age-dependent variations. Mycologia 100, 387–391.

Osono, T., Masuya, H., 2012. Endophytic fungi associated with leaves of Betulaceae in Japan. Can. J. Microbiol. 58, 507–515.

Park, H., Eom, A., 2018. First report on four novel endophytic fungal species isolated from leaves of *Juniperus chinensis* var. *sargentii* in Korea. Korean J. Mycol. 46, 114–121.

Pautasso, M., Schlegel, M., Holdenrieder, O., 2015. Forest health in a changing world. Microb. Ecol. 69, 826–842.

Pehl, L., Butin, H., 1994. Endophytische Pilze in Blättern von Laubbäumen und ihre Beziehungen zu Blattgallen (Zoocecidien). Mitteilun. Biol. Bundesanst. Land Forstwirt. Berlin-Dahlem 297, 1–56.

Peng, X.-W., Chen, H.-Z., 2007. Microbial oil accumulation and cellulase secretion of the endophytic fungi from oleaginous plants. Ann. Microbiol. 57, 239–242.

Persoh, D., 2013. Factors shaping community structure of endophytic fungi-evidence from the *Pinus-Viscum*-system. Fungal Divers. 60, 55–69.

Petrini, O., 1985. Host specificity of endophytic fungi of some European Ericaceae. Bot. Helv. 95, 213–238.

Petrini, O., 1991. Fungal endophytes of tree leaves. In: Andrews, J.H., Hirano, S.S. (Eds.), Microbial Ecology of Leaves. Springer, New York, NY, pp. 179–197.

Preszler, R.W., Gaylord, E.S., Boecklen, W.J., 1996. Reduced parasitism of a leaf-mining moth on trees with high infection frequencies of an endophytic fungus. Oecologia 108, 159–166.

Prihatini, I., Glen, M., Wardlaw, T.J., Ratkowsky, D.A., Mohammed, C.L., 2015. Needle fungi in young Tasmanian *Pinus radiata* plantations in relation to elevation and rainfall. N. Z. J. For. Sci. 45.

Pujade-Renaud, V., Deon, M., Gazis, R., Ribeiro, S., Dessailly, F., Granet, F., Chaverri, P., 2019. Endophytes from wild rubber trees as antagonists of the pathogen *Corynespora cassiicola*. Phytopathology 109, 1888–1899.

Pusey, P.L., Stockwell, V.O., Mazzola, M., 2009. Epiphytic bacteria and yeasts on apple blossoms and their potential as antagonists of *Erwinia amylovora*. Phytopathology 99, 571–581.

Queloz, V., Grünig, C.R., Berndt, R., Kowalski, T., Sieber, T.N., Holdenrieder, O., 2011. Cryptic speciation in *Hymenoscyphus albidus*. For. Pathol. 41, 133–142.

Quiring, D., Adams, G., Flaherty, L., McCartney, A., Miller, J.D., Edwards, S., 2019. Influence of a foliar endophyte and budburst phenology on survival of wild and laboratory-reared eastern spruce budworm, *Choristoneura fumiferana* on white spruce (*Picea glauca*). Forests 10, 503.

Rafiqi, M., Ellis, J.G., Ludowici, V.A., Hardham, A.R., Dodds, P.N., 2012. Challenges and progress towards understanding the role of effectors in plant-fungal interactions. Curr. Opin. Plant Biol. 15, 477–482.

Rajala, T., Velmala, S.M., Tuomivirta, T., Haapanen, M., Muller, M., Pennanen, T., 2013. Endophyte communities vary in the needles of Norway spruce clones. Fung. Biol. 117, 182–190.

Richardson, S.N., Walker, A.K., Nsiama, T.K., McFarlane, J., Sumarah, M.W., Ibrahim, A., Miller, J.D., 2014. Griseofulvin-producing *Xylaria* endophytes of *Pinus strobus* and *Vaccinium angustifolium*: evidence for a conifer-understory species endophyte ecology. Fungal Ecol. 11, 107–113.

Ridout, M., Newcombe, G., 2015. The frequency of modification of *Dothistroma* pine needle blight severity by fungi within the native range. For. Ecol. Manag. 337, 153–160.

Ritokova, G., Shaw, D.C., Filip, G., Kanaskie, A., Browning, J., Norlander, D., 2016. Swiss needle cast in western Oregon Douglas-fir plantations: 20-year monitoring results. Forests 7, 155.

Romao, A.S., Sposito, M.B., Andreote, F.D., Azevedo, J.L., Araujo, W.L., 2011. Enzymatic differences between the endophyte *Guignardia mangiferae* (Botryosphaeriaceae) and the citrus pathogen *G. citricarpa*. Genet. Mol. Res. 10, 243–252.

Sadeghi, F., Samsampour, D., Seyahooei, M.A., Bagheri, A., Soltani, J., 2019. Diversity and spatiotemporal distribution of fungal endophytes associated with *Citrus reticulata* cv. *siyahoo*. Curr. Microbiol. 76, 279–289.

Saikkonen, K., 2007. Forest structure and fungal endophytes. Fung. Biol. Rev. 21, 67–74.

Saikkonen, K., Faeth, S.H., Helander, M., Sullivan, T.J., 1998. Fungal endophytes: a continuum of interactions with host plants. Annu. Rev. Ecol. Syst. 29, 319–343.

Sanzio Pimenta, R., Moreira da Silva, J.F., Buyer, J.S., Janisiewicz, W.J., 2012. Endophytic fungi from plums (*Prunus domestica*) and their antifungal activity against *Monilinia fructicola*. J. Food Prot. 75, 1883–1889.

Schlaeppi, K., Bulgarelli, D., 2015. The plant microbiome at work. Mol. Plant-Microbe Interact. 28, 212–217.

Schlegel, M., 2019. Diversity and Ecology of Fungal Endophytes in Face of an Emerging Tree Disease (PhD thesis). ETH Zurich, Zurich, Switzerland, https://doi.org/10.3929/ethz-b-000331332.

Schlegel, M., Dubach, V., von Buol, L., Sieber, T.N., 2016. Effects of endophytic fungi on the ash dieback pathogen. FEMS Microbiol. Ecol. 92, fiw142.

Schlegel, M., Queloz, V., Sieber, T.N., 2018. The endophytic mycobiome of European ash and sycamore maple leaves—geographic patterns, host specificity and influence of ash dieback. Front. Microbiol. 9, 2345.

Schmid, M., Pautasso, M., Holdenrieder, O., 2014. Ecological consequences of Douglas fir (*Pseudotsuga menziesii*) cultivation in Europe. Eur. J. For. Res. 133, 13–29.

Schneider, M., Grünig, C.R., Holdenrieder, O., Sieber, T.N., 2009. Cryptic speciation and community structure of *Herpotrichia juniperi*, the causal agent of brown felt blight of conifers. Mycol. Res. 113, 887–896.

Schulz, B., Wanke, U., Draeger, S., Aust, H.-J., 1993. Endophytes from herbaceous plants and shrubs: effectiveness of surface sterilization methods. Mycol. Res. 97, 1447–1450.

Schulz, B., Sucker, J., Aust, H.J., Krohn, K., Ludewig, K., Jones, P.G., Döring, D., 1995. Biologically active secondary metabolites of endophytic *Pezicula* species. Mycol. Res. 99, 1007–1015.

Schwarz, M., Kopcke, B., Weber, R.W.S., Sterner, O., Anke, H., 2004. 3-Hydroxypropionic acid as a nematicidal principle in endophytic fungi. Phytochemistry 65, 2239–2245.

Scott, K.J., 1972. Obligate parasitism by phytopathogenic fungi. Biol. Rev. 47, 537–572.

Senn-Irlet, B.J., Gross, A., Blaser, S., 2016. SwissFungi: National Data and Information Center for the Fungi of Switzerland (Database). Version 2. Swiss Federal Institute WSL, Birmensdorf, Switzerland.

Siddique, A.B., Unterseher, M., 2016. A cost-effective and efficient strategy for Illumina sequencing of fungal communities: a case study of beech endophytes identified elevation as main explanatory factor for diversity and community composition. Fungal Ecol. 20, 175–185.

Sieber, T., 1988. Endophytic fungi in needles of healthy-looking and diseased Norway spruce (*Picea abies* L. Karsten). Eur. J. For. Pathol. 18, 321–342.

Sieber, T.N., 2007. Endophytic fungi in forest trees: are they mutualists? Fung. Biol. Rev. 21, 75–89.

Sieber, T.N., 2014. Neomyzeten—eine anhaltende Bedrohung für den Schweizer Wald. Schweiz. Z. Forstwes. 165, 173–182.

Sieber, T.N., 2019. The endophyte *Allantophomopsis cytisporea* is associated with snow blight on *Calluna vulgaris* in the Alps—an effect of climate change? Arct. Antarct. Alp. Res. 51, 460–470.

Sieber, T., Hugentobler, C., 1987. Endophytic fungi in leaves and twigs of healthy and diseased beech trees (*Fagus sylvatica* L.). Eur. J. For. Pathol. 17, 411–425.

Sieber, T.N., Riesen, T.K., Müller, E., Fried, P.M., 1988. Endophytic fungi in four winter wheat cultivars (*Triticum aestivum* L.) differing in resistance against *Stagonospora nodorum* (Berk.) Cast. & Germ. = *Septoria nodorum* (Berk.) Berk. J. Phytopathol. 122, 289–306.

Sieber, T.N., Rys, J., Holdenrieder, O., 1999. Mycobiota in symptomless needles of *Pinus mugo* ssp. *uncinata*. Mycol. Res. 103, 306–310.

Sieber, T.N., Jermini, M., Conedera, M., 2007. Effects of the harvest method on the infestation of chestnuts (*Castanea sativa*) by insects and moulds. J. Phytopathol. 155, 497–504.

Sieber-Canavesi, F., Sieber, T.N., 1993. Successional patterns of fungal communities in needles of European silver fir (*Abies alba* Mill.). New Phytol. 125, 149–161.

Sinclair, J.B., Cerkauskas, R.F., 1996. Latent infection vs. endophytic colonization by fungi. In: Redlin, S.C., Carris, L.M. (Eds.), Endophytic Fungi in Grasses and Woody Plants. APS Press, St. Paul, MN, pp. 3–29.

Small, D.A., Chang, W., Toghrol, F., Bentley, W.E., 2007. Comparative global transcription analysis of sodium hypochlorite, peracetic acid, and hydrogen peroxide on *Pseudomonas aeruginosa*. Appl. Microbiol. Biotechnol. 76, 1093–1105.

Smith, H., Wingfield, M.J., Crous, P.W., Coutinho, T.A., 1996. *Sphaeropsis sapinea* and *Botryosphaeria dothidea* endophytic in *Pinus* spp and *Eucalyptus* spp in South Africa. S. Afr. J. Bot. 62, 86–88.

Sogonov, M.V., Castlebury, L.A., Rossman, A.Y., White, J.F., 2007. The type species of *Apiognomonia, A. veneta*, with its *Discula* anamorph is distinct from *A. errabunda*. Mycol. Res. 111, 693–709.

Sokolski, S., Bernier-Cardou, M., Piche, Y., Berube, J.A., 2007. Black spruce (*Picea mariana*) foliage hosts numerous and potentially endemic fungal endophytes. Can. J. For. Res. 37, 1737–1747.

Somjaipeng, S., Medina, A., Kwasna, H., Ordaz Ortiz, J., Magan, N., 2015. Isolation, identification, and ecology of growth and taxol production by an endophytic strain of *Paraconiothyrium variabile* from English yew trees (*Taxus baccata*). Fung. Biol. 119, 1022–1031.

Spencer-Phillips, P.T.N., 1997. Function of fungal haustoria in epiphytic and endophytic infections. In: Andrews, J.H., Tommerup, I.C. (Eds.), Advances in Botanical Research Incorporating Advances in Plant Pathology. vol. 24, pp. 309–333.

Sridhar, K.R., 2019. Diversity, ecology, and significance of fungal endophytes. In: Jha, S. (Ed.), Endophytes and Secondary Metabolites, pp. 61–100.

Stefani, F.O.P., Bérubé, J., 2006. Biodiversity of foliar fungal endophytes in white spruce (*Picea glauca*) from southern Quebec. Can. J. Bot. 84, 777–790.

Stierle, A., Strobel, G., Stierle, D., 1993. Taxol and taxane production by *Taxomyces andreanae*, an endophytic fungus of Pacific yew. Science 260, 214–216.

Stone, J.K., 1987. Initiation and development of latent infections by *Rhabdocline parkeri* on Douglas fir. Can. J. Bot. 65, 2614–2621.

Stone, J.K., Polishook, J.D., White, J.F., 2004. Endophytic fungi. In: Mueller, G.M., Bills, G.F., Foster, M.S. (Eds.), Biodiversity of Fungi. Elsevier, Amsterdam, pp. 241–270.

Strobel, G.A., 2003. Endophytes as sources of bioactive products. Microbes Infect. 5, 535–544.

Strobel, G., Yang, X.S., Sears, J., Kramer, R., Sidhu, R.S., Hess, W.M., 1996. Taxol from P*estalotiopsis microspora*, an endophytic fungus of *Taxus wallachiana*. Microbiology-Uk 142, 435–440.

Sumarah, M.W., Adams, G.W., Berghout, J., Slack, G.J., Wilson, A.M., Miller, J.D., 2008. Spread and persistence of a rugulosin-producing endophyte in *Picea glauca* seedlings. Mycol. Res. 112, 731–736.

Sun, X., Ding, Q., Hyde, K.D., Guo, L.D., 2012. Community structure and preference of endophytic fungi of three woody plants in a mixed forest. Fungal Ecol. 5, 624–632.

Suske, J., Acker, G., 1989. Identification of endophytic hyphae of *Lophodermium piceae* in tissues of green, symptomless Norway spruce needles by immunoelectron microscopy. Can. J. Bot. 67, 1768–1774.

Szink, I., Davis, E.L., Ricks, K.D., Koide, R.T., 2016. New evidence for broad trophic status of leaf endophytic fungi of *Quercus gambelii*. Fungal Ecol. 22, 2–9.

Tainter, F.H., Baker, F.A., 1996. Principles of Forest Pathology. John Wiley & Sons, Inc., New York.

Tanney, J.B., Douglas, B., Seifert, K.A., 2016. Sexual and asexual states of some endophytic Phialocephala species of Picea. Mycologia 108, 255–280.

Tateno, O., Hirose, D., Osono, T., Takeda, H., 2015. Beech cupules share endophytic fungi with leaves and twigs. Mycoscience 56, 252–256.

Taudière, A., Bellanger, J.-M., Carcaillet, C., Hugot, L., Kjellberg, F., Lecanda, A., Lesne, A., Moreau, P.-A., Scharmann, K., Leidel, S., Richard, F., 2018. Diversity of foliar endophytic ascomycetes in the endemic Corsican pine forests. Fungal Ecol. 36, 128–140.

Terhonen, E., Marco, T., Sun, H., Jalkanen, R., Kasanen, R., Vuorinen, M., Asiegbu, F., 2011. The effect of latitude, season and needle-age on the mycota of Scots pine (*Pinus sylvestris*) in Finland. Silva Fennica 45, 301–317.

Thongsandee, W., Matsuda, Y., Ito, S., 2012. Temporal variations in endophytic fungal assemblages of *Ginkgo biloba* L. J. For. Res. 17, 213–218.

Tintjer, T., Leuchtmann, A., Clay, K., 2008. Variation in horizontal and vertical transmission of the endophyte *Epichloe elymi* infecting the grass *Elymus hystrix*. New Phytol. 179, 236–246.

Torres, J.M.O., dela Cruz, T.E.E., 2015. Antibacterial activities of fungal endophytes associated with the Philippine endemic tree, *Canarium ovatum*. Mycosphere 6, 266–273.

Turner, T.R., James, E.K., Poole, P.S., 2013. The plant microbiome. Genome Biol. 14.

Unterseher, M., Persoh, D., Schnittler, M., 2013. Leaf-inhabiting endophytic fungi of European Beech (*Fagus sylvatica* L.) co-occur in leaf litter but are rare on decaying wood of the same host. Fungal Divers. 60, 43–54.

Unterseher, M., Siddique, A.B., Brachmann, A., Persoh, D., 2016. Diversity and composition of the leaf mycobiome of beech (*Fagus sylvatica*) are affected by local habitat conditions and leaf biochemistry. PLoS ONE 11, e0152878.

Van Deynze, A., Stoffel, K., 2006. High-throughput DNA extraction from seeds. Seed Sci. Technol. 34, 741–745.

Viret, O., Scheidegger, C., Petrini, O., 1993. Infection of beech leaves (*Fagus sylvatica*) by the endophyte *Discula umbrinella* (teleomorph: *Agiognomonia errabunda*): low-temperature scanning electron microscopy studies. Can. J. Bot. 71, 1520–1527.

Vorholt, J.A., 2012. Microbial life in the phyllosphere. Nat. Rev. Microbiol. 10, 828–840.

Webber, J., 1981. A natural biological-control of Dutch elm disease. Nature 292, 449–451.

Weber, G.F., 1922. *Septoria* diseases of wheat. Phytopathology 12, 537–585.

Wilson, D., 1995. Fungal endophytes which invade insect galls—insect pathogens, benign saprophytes, or fungal inquilines. Oecologia 103, 255–260.

Witzell, J., Martin, J.A., 2018. Endophytes and forest health. In: Pirttila, A.M., Frank, A.C. (Eds.), Endophytes of Forest Trees: Biology and Applications, second ed, pp. 261–282.

Wolfe, E.R., Kautz, S., Singleton, S.L., Ballhorn, D.J., 2018. Differences in foliar endophyte communities of red alder (*Alnus rubra*) exposed to varying air pollutant levels. Botany 96, 825–835.

Woods, A.J., Martin-Garcia, J., Bulman, L., Vasconcelos, M.W., Boberg, J., La Porta, N., Peredo, H., Vergara, G., Ahumada, R., Brown, A., Diez, J.J., 2016. *Dothistroma* needle blight, weather and possible climatic triggers for the disease's recent emergence. For. Pathol. 46, 443–452.

Wu, L., Han, T., Li, W., Jia, M., Xue, L., Rahman, K., Qin, L., 2013. Geographic and tissue influences on endophytic fungal communities of *Taxus chinensis* var. *mairei* in China. Curr. Microbiol. 66, 40–48.

Yadav, R.K.P., Karamanoli, K., Vokou, D., 2005. Bacterial colonization of the phyllosphere of Mediterranean perennial species as influenced by leaf structural and chemical features. Microb. Ecol. 50, 185–196.

Yang, T., Weisenhorn, P., Gilbert, J.A., Ni, Y., Sun, R., Shi, Y., Chu, H., 2016. Carbon constrains fungal endophyte assemblages along the timberline. Environ. Microbiol. 18, 2455–2469.

Yoo, J.-J., Eom, A.-H., 2012. Molecular identification of endophytic fungi isolated from needle leaves of conifers in Bohyeon Mountain, Korea. Mycobiology 40, 231–235.

Yuan, Z., Chen, L., 2014. The role of endophytic fungal individuals and communities in the decomposition of *Pinus massoniana* needle litter. PLoS ONE 9, e105911.

Zabka, V., Stangl, M., Bringmann, G., Vogg, G., Riederer, M., Hildebrandt, U., 2008. Host surface properties affect prepenetration processes in the barley powdery mildew fungus. New Phytol. 177, 251–263.

Zamora, P., Martinez-Ruiz, C., Diez, J.J., 2008. Fungi in needles and twigs of pine plantations from northern Spain. Fungal Divers. 30, 171–184.

Zimmerman, N.B., Vitousek, P.M., 2012. Fungal endophyte communities reflect environmental structuring across a Hawaiian landscape. Proc. Natl. Acad. Sci. U. S. A. 109, 13022–13027.

Chapter 7

Tree leaves as a habitat for phyllobacteria

Teresa A. Coutinho and Khumbuzile N. Bophela

Department of Biochemistry, Genetics and Microbiology, Centre for Microbial Ecology and Genomics/Forestry and Agricultural Biotechnology Institute, University of Pretoria, Pretoria, South Africa

1. Introduction

The phyllosphere was initially described as the external leaf surface of a plant as a habitat for microorganisms (Ruinen, 1961). This definition was later redefined to not only include the leaf surface, the phylloplane, but also the leaf surface "waterscape" termed the phyllotelma (Doan and Leveau, 2015). Morris et al. (2002) included the endosphere in her definition. The rationale was that natural openings allow the ingress and egress of microorganisms between the internal and external domains, suggesting that epiphytes and endophytes are really a "continuum" in the phyllosphere. In this review, the focus will be on phyllobacteria, those occupying the leaf surface of forest trees. However, studies where researchers have not distinguished between the two domains will still be included.

The phyllosphere includes mostly leaf, bark, stem, flower, and fruit surfaces and has been suggested to occupy an area of $4 \times 10^8 \, km^2$ which is almost equivalent to the total earth surface (Morris and Kinkel, 2002). Bacteria are the dominant microorganisms occurring in this habitat, but archaea, filamentous fungi, yeasts, algae, nematodes, and viruses may also be present (Lindow and Brandl, 2003). The phylloplane, both the abaxial and adaxial leaf surfaces (Fig. 7.1), has been said to support as many as 10^{26} bacterial cells (Vorholt, 2012).

Microbial communities occurring on plant surfaces are commonly referred to as epiphytes. In the case of bacterial epiphytes, the majority are commensals not affecting their host, while others form a mutualistic association. The host supplies the bacteria with nutrients and shelter, while the bacteria can promote plant growth, suppress or stimulate phytopathogens, contribute to the carbon and nitrogen cycles (Lindow and Brandl, 2003), and host stress tolerance (Stone et al., 2018). In some cases they can be pathogenic. On the leaf surface they are faced with an extreme and unstable environment which has a direct effect on their diversity and abundance. Both abiotic factors, time of the year, water availability, leaf age and position, ultraviolet light, relative humidity, and biotic factors, the presence of pathogens and pests, are the major drivers of this diversity. Host genotype and the presence of other plants in the surrounding area are also contributing factors (Balint-Kurti et al., 2010; Hunter et al., 2010; Humphrey et al., 2014; Agler et al., 2016).

In recent years the development of environmental genomics, metagenomics, and metaproteogenomics has increased our knowledge on the composition and the role played by phyllosphere bacterial communities (Delmotte et al., 2009; Rastogi et al., 2013; Yu et al., 2013). There has also been a move to study the fate of single bacterial cells on a spatially heterogeneous leaf surface (Remus-Emsermann and Schlechter, 2018). Much of this research has focused on plants of agricultural importance or trees growing in urban and temperate environments, although some research has been conducted on tropical forest tree species (Griffin and Carlson, 2015). Recently, it was calculated that 46% of all trees are grown in tropical and subtropical regions of the world (Crowther et al., 2015). One would thus expect that in the tropics where it is hot and humid or very dry, the phyllosphere bacterial community would be unique (Table 7.1). In this review, we focus on the leaf surface as a habitat for phyllobacteria, the bacterial community composition of tree leaves, and the role they play in this ecosystem.

Forest Microbiology. https://doi.org/10.1016/B978-0-12-822542-4.00001-2

FIG. 7.1 Adaxial *(left)* and abaxial *(right)* surface of a leaf.

TABLE 7.1 Comparison between the relative abundance of phyllobacteria on the foliage of tropical and temperate tree species.

	Tropical rainforest in an arboretum in Malaysia[a] Trees sampled: *Arytera littoralis, Schizostachyum brachycladum, Dillenia excelsa, Gnetum* sp., and *Shorea maxima*	Neotropical forest in Panama[b] 57 tree species sampled	Quebec temperate forest[c] Trees sampled: *Acer saccharum, A. rubrum, Betula papyrifera, Abies balsamea,* and *Picea glauca*	Boulder, Colorado[d] 56 tree species sampled
	Relative abundance (% of sequences)	Relative abundance (% of sequences)	Relative abundance (% of sequences)	Relative abundance (% of sequences)
Acidobacteria	17%		6%	
Actinobacteria	9%	5.5%	5%	9%
Bacteroidetes	8%	10.9%	6%	22.5%
Firmicutes				5.3%
Saccharibacteria (TM7)				9%
Alphaproteobacteria	27%	11.9%	68%	24.5%
Betaproteobacteria	Between 1% and 4%	8.1%	6%	16.4%
Deltaproteobacteria	5%		3%	
Gammaproteobacteria	13%	12%	5%	7.9%
Members of the core microbiome	Acidobacteria occurred between 11.6% and 33.5% across all host species	*Beijerinckia* (6.5%), *Leptothrix* (3.9%), *Stenotrophomonas* (3.4%), *Niastella* (3.3%) and *Spirosoma* (2.9%)	72% Alphaproteobacteria, 9% Cytophagia, 7.8% Betaproteobacteria, 5% Acidobacteria, 2% Gammaproteobacteria, and 2% Actinobacteria	Sphingobacteriales (21.3%)

[a]Kim et al. (2012).
[b]Kembel et al. (2014).
[c]Laforest-Lapointe et al. (2016).
[d]Redford et al. (2010).
Yellow—tropical trees.
Green—temperate trees.

2. Leaf surface as an extreme environment

Leaves have a heterogeneous topography, with elevated areas, the epidermal cells, and grooves, the regions between these cells. These areas are interspersed with stomata, trichomes, hydathodes, and glandular trichomes (Fig. 7.2). These leaf traits differ in presence, distribution, and density between plant species, within a plant species, and even within leaves on the

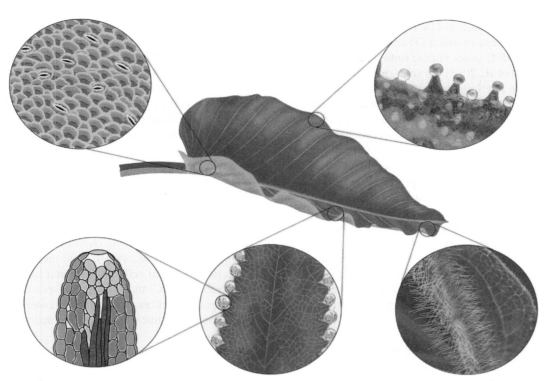

FIG. 7.2 Position of stomata *(top left)*, glandular trichomes *(top right)*, and hydathodes *(bottom and middle)* and trichomes *(bottom right)* on a leaf.

same host (Beattie, 2002). Bacteria present on the leaf surface are more commonly found in areas where they are slightly protected from abiotic stress and where they can form closer interactions with their host. They are thus less commonly found on the elevated epidermal cells and usually occur in the indentations between epidermal cells, at the trichome bases, close to leaf veins and stomatal openings or in stomata (Esser et al., 2015). In the case of tropical trees, longer leaf span, large leaf surface area, low degrees of deciduousness, and increased hydathode density have been shown to favor bacterial survival, colonization, and abundance (Griffin and Carlson, 2015).

Trichomes have several functions. These include controlling the leaf temperature, protecting the leaf against ultraviolet (UV) radiation, and releasing secondary metabolites that discourage herbivore browsing and the development of pathogens (Vacher et al., 2016). The plant cuticle, a waxy layer, is also a determining factor which may limit bacterial colonization of the leaf surface. It can change as the leaf ages which is thought to be due to microbial modifications (Knoll and Schreiber, 1998, 2000). Cuticle thickness varies among plant species and may be as thin as 0.05 μm or as thick as 5 μm in the case of black pine (*Pinus nigra*) (Morris, 2002). If this layer is thick it may limit the diffusion of nutrients and water vapor from the leaf interior on to the surface. The leaf structure also differs between the adaxial and abaxial surfaces. The adaxial surface has a thicker cuticle while the abaxial surface is involved in gas exchange and regulation of transpiration (Vacher et al., 2016).

The critical limitation of bacterial growth on the leaf surface is moisture availability (Beattie, 2002). The cuticle both prevents moisture from exiting the leaf interior and restricts the amount of water remaining on the leaf surface (Neinhuis and Barthlott, 1997). Bacteria produce extracellular polysaccharides, form aggregates and biofilms, which can resist desiccation (Wilson and Lindow, 1994; Morris et al., 2002). Others may produce surfactants to increase the leaf surface moisture (Knoll and Schreiber, 2000; Schreiber et al., 2005). The leaf surface also has a boundary layer with its own microclimate, usually with higher humidity than the rest of the phylloplane, allowing bacteria to survive there (Burrage, 1971). This layer may range from less than 1 mm to as much as 10 mm (Morris, 2002).

Epiphytes need to adapt to high fluxes of UV radiation. A large proportion of the epiphytes are pigmented to confer protection and increase their survival in the phyllosphere. The most tolerant strains produce either pink or orange pigments (Sundin and Jacobs, 1999; Jacobs et al., 2005). They can also express enzymes that control reactive oxygen species produced by solar radiation and produce DNA protection proteins (Delmotte et al., 2009). The rapid expression of DNA repair mechanisms after UV exposure may be an important determinant of bacterial survival in the phyllosphere (Sundin, 2002).

Interestingly, exposure to UV radiation has been shown to both increase bacterial diversity (Kadivar and Stapleton, 2003) or does not have an affect (Truchado et al., 2017). In some locations, temperature on the phylloplane can reach between 40°C and 55°C under intense sunlight (Yang et al., 2001) and this directly impacts on microorganism (fungal) development (Bernard et al., 2013). As solar radiation occurs in a diurnal cycle, day length and season may affect the growth and abundance of the phyllosphere microbial community. Temperature has been shown to differ within a canopy of different tree species (Leuzinger and Korner, 2007) and even across a single leaf (Stokes et al., 2006). Variation in temperature across a single leaf can be as much as 4°C with the leaf edges generally being cooler than the center of the leaf (Morris, 2002). Fluttering of leaves is another factor that regulates the leaf temperature. In the case of poplar leaves, fluttering leaves were 2–4°C cooler than leaves that were prevented from fluttering (Roden and Pearcy, 1993).

Leaf surfaces are nutrient-poor habitats. Thus, the growth of the phyllobacterial community is limited by the available nutrients, notably carbon and nitrogen (Wilson and Lindow, 1994; Mercier and Lindow, 2000). They survive by searching for minor substrates, including volatile organic compounds (VOCs) or methylated compounds, which are frequently used in cometabolism (Redford et al., 2010; Vorholt, 2012). Usually, these bacteria are oligotrophs that can tolerate low nutrient availability or they can interact with the host to obtain nutrients (Beattie and Lindow, 1999). They have been shown to modify the leaf surface to gain access to nutrients, for example, thinning the cuticle (Knoll and Schreiber, 2000), secreting plant hormones (Brandl et al., 2001), or producing surfactants (Bunster et al., 1989). Plant metabolites, organic and inorganic nutrients, can be translocated to the leaf surface by excretion from leaf cells (Mercier and Lindow, 2000) or due to osmotic pressure when the leaf is wet (Tukey, 1970). Plants also release VOCs that can support specific populations of bacteria. These compounds are either released constitutively or in response to either biotic or abiotic stresses, and have been shown to mediate plant-plant interactions, plant-insect interactions, and plant-microorganism interactions (Junker and Tholl, 2013).

Phyllosphere bacteria can obtain nitrogen from leached plant-produced amino acids or inorganic forms of nitrogen leached from the apoplast (Tejera et al., 2006). Nitrogen can also arrive on the leaf surface through atmospheric deposition. Nitrate and nitrite are usually available during rain (Papen et al., 2002; Guerrieri et al., 2015). The number of nitrogen-fixing bacteria increases during dry conditions, suggesting that their recruitment may extend the ability of the plant to adapt to the environment (Rico et al., 2014). Measured rates of nitrogen fixation in the phyllosphere have been shown to vary widely, but in the case of some tropical trees rates of over 26 ng N/cm leaf area/h have been reported (Freiberg, 1998). However, in the phyllosphere of temperate trees this amount is significantly lower (Freiberg, 1998). The phyllobacteria that fix nitrogen include both Proteobacteria, for example, *Beijerinckia, Azotobacter,* and *Klebsiella* and Cyanobacteria, for example, *Nostoc, Scytonema,* and *Stigonema* (Vacher et al., 2016).

Another necessary microbial process that occurs on the phyllosphere is methanol degradation (Corpe and Rheem, 1989; Van Aken et al., 2004). Methanol is an important carbon source for phyllosphere microorganisms, notably methanotrophic epiphytes, that is formed as a by-product of cell wall metabolism by pectin methyl esterases (Galbally and Kirstine, 2002). All methylotrophic phyllobacteria belong to a single genus *Methylobacterium* and are facultative methylotrophs with a restricted substrate range (Knief et al., 2010; Wellner et al., 2011).

Macro- and microelements are needed by phyllobacteria for growth. The expression of phosphate, sulfate, and iron transport systems has been observed *in planta* (Delmotte et al., 2009; Gourion et al., 2006; Marco et al., 2005). Siderophore production is involved in the growth of epiphytes. Studies have also shown that there may be moderate or no iron limitations on the leaf surface (Wensing et al., 2010; Joyner and Lindow, 2000).

3. Phyllobacterial community composition

Epiphytic phyllobacteria can be both transient and resident. Transient epiphytes are those that do not, or to a limited extent, multiply on the leaf surface while resident epiphytes are those that can multiple on this surface in the absence of wounds (Suslow, 2002). The origin of these phyllobacteria can be from a number of sources and have been shown to occupy this niche at different stages of plant growth. They can originate from seed tissue (Barret et al., 2015), bioaerosols (Bulgarelli et al., 2013), rainfall and irrigation water (Morris, 2002), and animals, in particular, insects (Shapiro et al., 2012). Leaf composition, chemical composition, and/or VOC emissions may be related to interspecies differences in phyllobacteria (Redford et al., 2010). Climatic factors such as temperature, seasons, exposure to sand storms (Rastogi et al., 2012) or anthropogenic factors such as use of pesticides (Shade et al., 2013) play a role in structuring the community. Warm, moist, and humid tropical habitats allow phyllobacteria to reach concentrations that are much higher and more persistent than in temperate regions where below-freezing temperatures decrease their numbers each year (Griffin and Carlson, 2015). Shade and sun leaves represent distinct microbial environments even though they exist in close proximity. Individual plant species have also been shown to select for distinct phyllobacteria on the leaf surface (Redford et al., 2010; Kim et al., 2012).

Phyllobacteria numbers vary in size among and between plant species (Hirano and Upper, 1989) and geographical location (Rastogi et al., 2013; Dong et al., 2019). Generally, the greatest numbers of bacteria are found on the abaxial surfaces (Surico, 1993), possibly because that surface has the greatest density of stomata and trichomes and/or a thinner cuticle (Beattie and Lindow, 1999). Over the growing season, bacteria dominate in the early stages followed by yeasts and finally filamentous fungi (Kinkel, 1997). Vokou et al. (2019) found that there was no difference in the level of colonization of phyllobacteria between summer and winter, although seasonal differences for individual taxonomic groups were recorded. In the case of spruce (*Picea* spp.), geographical location was found to affect microbial taxonomic composition, but it had no effect on the community functional structure, i.e., the phyllosphere microbial community composed of different taxa might be functionally similar (Li et al., 2019).

Phyllobacterial communities are usually dominated by α- and γ-proteobacteria, and bacteroidetes while the β-proteobacteria and firmicutes can also form part of the community (Bodenhausen et al., 2013; Delmotte et al., 2009; Fierer et al., 2011; Kembel et al., 2014; Redford and Fierer, 2009; Redford et al., 2010) (Fig. 7.3). A few bacterial genera, notably *Pseudomonas*, *Sphingomonas*, *Methylobacterium*, *Bacillus*, *Massilia*, *Arthrobacter*, and *Pantoea*, appear to compose the core of phyllosphere communities (Bulgarelli et al., 2013). In adult tropical trees, only 1.4% of the bacterial diversity was present on over 90% of all individuals and made up 73% of the total sequences (Griffin and Carlson, 2015). This implies that a small group of bacteria are best at surviving in the phyllosphere or at colonizing these tree species, or both. Interestingly, cyanobacteria and diazotrophic γ-proteobacteria provide significant nitrogen input in rainforest ecosystems (Frünkranz et al., 2008). Kembel et al. (2014) studied the bacterial communities on tropical tree leaves and found approximately 400 bacterial taxa dominated by Actinobacteria; α-, β-, and γ-proteobacteria; and bacteroidetes. Bacteroidetes and β-proteobacteria are more common in gymnosperms, while Actinobacteria and γ-proteobacteria are more common on angiosperms (Redford et al., 2010).

In a study by Lambais et al. (2006), the phyllobacterial community in a tropical Brazilian rainforest was dominated by undescribed species and it was estimated that between 2 and 13 million of these species inhabit this habitat. They also showed that between 0% and 5% of the bacterial species in tropical tree canopies were common to all tree species studied. Thus, phyllobacteria on different tree species are phylogenetically diverse. However, their metaproteomes are functionally convergent concerning traits for survival on the leaf surface, i.e., they share a common set of core functional proteins that are required for survival and fitness (Lambais et al., 2017).

The similarity of phyllobacteria on different trees species showed a significant tendency to follow host-tree phylogeny, with similar communities on more closely related species (Ruppel et al., 2008; Kim et al., 2012; Izhaki et al., 2013; Yao

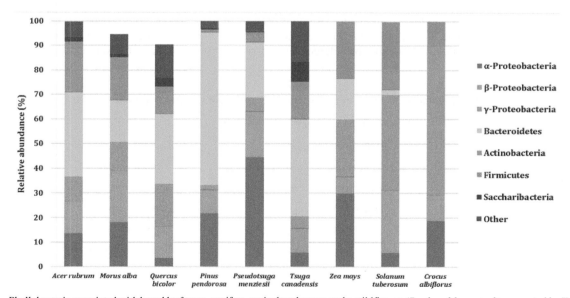

FIG. 7.3 Phyllobacteria associated with broad leaf trees, conifers, agricultural crops, and a wildflower. *(Produced from results presented by Kadivar, H., Stapleton, A., 2003. Ultraviolet radiation alters maize phyllosphere bacterial diversity. Microb. Ecol. 45, 353–361; Rasche, F., Marco-Noales, E., Velvis, H., van Overbeek, L.S., Lopez, M.M., van Elsas, J.D., Sessitsch, A., 2006. Structural characteristics and plant-beneficial effects of bacteria colonizing the shoots of field grown conventional and genetically modified T4-lysozyme producing potatoes. Plant Soil 289, 123–140; Redford, A.J., Bowers, R.M., Knight, R., Linhart, Y., Fierer, N., 2010. The ecology of the phyllosphere: geographic and phylogenetic variability in the distribution of bacteria on tree leaves. Environ. Microbiol. 12, 2885–2893; Reiter, B., Sessitsch, A., 2006. Bacterial endophytes of the wildflower Crocus albiflorus analyzed by characterization of isolates and by a cultivation-independent approach. Can. J. Microbiol. 52, 140–149.)*

et al., 2020). However, each tree species had its own unique bacterial identity—some bacteria were abundant and present on every leaf in the forest, while others were rare and only found on the leaves of a single host species (Kembel et al., 2014). Community composition does, however, differ between plant tissue types and exhibits strong spatial patterns within individual trees that relate to their anatomical structure (Leff et al., 2015). Rather than geographical location, environmental heterogeneity and varying abiotic factors across sites appear to be the most influential drivers of variation in the foliar microbiome (Carper, 2018). However, soil type, plant genotype and species, age, climate, and the geographic region are the factors determining the bacterial community assembly (Copeland et al., 2015; Leff et al., 2015). Redford et al. (2010) showed that the phyllobacteria on *Pinus ponderosa* do not vary with geographical location whereas in *Tamarix* the reverse was true (Qvit-Raz et al., 2012; Finkel et al., 2012).

Stone and Jackson (2019) found that canopy position in tropical trees after rain is a more reliable determinant of phyllobacteria composition and diversity than environmental disturbance on the phyllosphere. Richer bacterial composition was observed in the lower canopy and a more even distribution occurred in the upper canopy. In a study by Hermann et al. (2020), the phyllosphere microbiome of tropical trees consisted of high numbers of Actinobacteria (up to 46%). They also showed that the position in the canopy has strong effects on phyllobacteria in a floodplain hardwood forest. There was consistently lower bacterial diversity at the top of the canopy compared to the canopy middle and bottom positions. This difference in bacterial diversity is in contrast to the situation in temperate and tropical forests. In the canopy of large trees, some environmental factors may restrict microbial diversity and are likely along vertical gradients (Laforest-Lapointe et al., 2016) with severe stress due to abiotic factors acting at the top of the canopy.

Several studies on the host-associated microbiome have shown that biodiversity is a trait that is part of the host phenotype affecting both host plant fitness and function (Vandenkoornhuyse et al., 2015; Vorholt, 2012; Bringel and Couée, 2015). The drivers of the host-associated microbiome assembly have now become a topic of interest. According to Vacher et al. (2016) four eco-evolutionary processes are shaping the phyllosphere community, viz. dispersal, evolutionary diversification, selection, and drift.

4. Role of the phyllobacterial community

The phyllosphere bacterial community can have a positive, neutral, or negative effect on their host (Table 7.2). They have been shown to modify their microenvironment and to positively influence plant health and development (Abanda-Nkpwatt et al., 2006; Innerebner et al., 2011). Saleem et al. (2017) found that bacterial species richness and antagonistic interactions regulate host development and fitness. They achieve this in many ways (Fig. 7.4). One method phyllobacteria use is to produce plant hormones, i.e., indole acetic acid, cytokinins, etc. (Ryu et al., 2006). The production of indole-3-acetic acid, for example, is widespread among phyllobacteria (Brandl et al., 2001; Glickmann et al., 1998; Lindow et al., 1998). The biosynthesis of plant hormones may either be a defensive response to pathogens or shift promoted by them in order to

TABLE 7.2 Examples of phyllobacteria having a negative, neutral, and positive effect on their tree host.

Negative effect	Neutral (no effect)	Positive effect
Pantoea agglomerans[a]	*Pantoea agglomerans*	*Pantoea agglomerans*[b]
Pseudomonas syringae[a]	*Pseudomonas syringae*	*Pantoea ananatis*[b]
Xanthomonas spp.[a]	*Stenotrophomonas* spp.	*Sphingomonas* spp.[c]
Pectobacterium spp.[a]	*Bacillus* spp.	*Klebsiella* spp.[d]
Pantoea ananatis[a]		*Beijerinckia* spp.[d]
		Methylobacterium spp.[b]
		Streptomyces spp.[e]
		Bacillus spp.[e]

[a] As potential plant pathogens, some strains can have no effect while others can be nonpathogenic and have a positive effect.
[b] As growth promoters (produce plant growth promoters).
[c] Competitor with Ps. syringae for nutrients.
[d] Free-living nitrogen-fixing bacteria.
[e] Produce antimicrobial compounds.

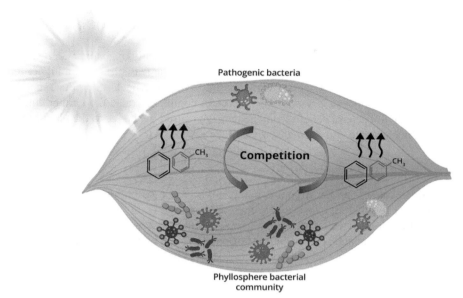

FIG. 7.4 A graphical representation of interactive events occurring on the leaf surface between the phyllosphere bacteria, the environment and with pathogenic bacteria.

increase their virulence and growth (Beattie, 2011). Abscisic acid has, for example, been shown to control stomatal closure thereby preventing the entry of plant pathogenic bacteria and fungal pathogens (Munemasa et al., 2015) and confer drought tolerance to plants (Sussmilch and McAdam, 2017). Phyllobacteria can also modify plant hormone production by the host (Bodenhausen et al., 2014). Indole acetic acid-producing bacteria can increase plant productivity (Romero et al., 2016), while those producing cytokinins can stimulate the transportation of nitrogen to above-ground plant tissues (Holland, 2011).

The second method used by phyllobacteria is to secrete biosurfactants and this is highly prevalent among epiphytic bacteria (Oso et al., 2019). For example, the plant pathogen *Pseudomonas syringae* releases syringafactin which increases the cuticle permeability (Burch et al., 2014; Schreiber et al., 2005) and establishes its association with its host. This surfactant facilitates water availability and alters sugar availability that can improve conditions for phyllobacterial growth (Lindow and Brandl, 2003; Van der Wal and Leveau, 2011). Alternatively, biosurfactants enable bacterial movement on the phylloplane (Hutchison and Johnstone, 1993). The water film created by the surfactant can spread the bacteria over the leaf surface to areas where the nutrients are more abundant.

The third method that phyllobacteria use to improve plant growth is by increasing plant resistance to abiotic and biotic stresses (Zamiousdis and Pieterse, 2012; Laforest-Lapointe et al., 2017). Phyllobacteria may alter the plants' ability to tolerate abiotic or environmental stressors such as drought, temperature, and salinity extremes (Stone et al., 2018). They may also remediate chemical pollutants such as phenol (Sandhu et al., 2007), but the extent to which they do this is currently unknown (Scheublin et al., 2014). With regards to biotic stresses, they can inhibit microbial pathogens through competition between invading pathogens and resident microbes for resources or by producing antibiotics (Pusey et al., 2011; Berg, 2009; Rastogi et al., 2013). *Streptomyces*, a known antibiotic producer, is the fifth most common genus in the phyllosphere of 57 tropical tree species in Panama (Kembel et al., 2014) and may function to ward off pathogens in this habitat. Alternatively, they can prime the host defense system (Vogel et al., 2016)—known as plant-mediated systemic-induced resistance—that triggers the plant's defenses thereby causing a systemic resistance against pathogens. The plant immune response is either systemic or locally confined and this affects the colonization pattern of the microbes (Schlechter et al., 2019). One example, where priming was identified as a mode of action, is when a *Pseudomonas fluorescens* strain was applied to *Arabidopsis thaliana* plants and found to stimulate host defense against *Pseudomonas syringae* (Van Wees et al., 2000).

Plants produce a wide range of antimicrobial active and structurally diverse secondary metabolites (Wink, 2003). These secondary metabolites include a great variety of volatile organic compounds or their precursors that may promote or inhibit specific bacterial species (Bringel and Couée, 2015). This process contributes to numerous biotic interactions and will shape the bacterial community. Phyllobacteria have also been shown to breakdown plant defense chemicals thus reducing its defense against insect defoliators (Mason et al., 2014). However, it remains unknown how and to what extent volatile organic compounds emitted by plants affect phyllobacteria. According to Farre-Armengol et al. (2016), epiphytes can affect volatile organic compound emission in several ways, viz. by producing or emitting their own VOCs, by altering their host's physiology and modifying the production and emission of VOCs, or by matching the VOCs emitted by the plant.

Epiphytic bacteria may negatively affect plant growth. Some species, notably in genera *Pseudomonas* and *Pantoea*, may play a role in frost damage of leaves. They induce freezing of water on the leaf surface at a higher temperature ($-2°C$ to $4°C$) than what would usually occur at $-7°C$ or $-8°C$ (Morris, 2002). These bacteria produce membrane-bound proteins with a crystalline structure resembling ice crystals. These protein complexes raise the temperature at which ice crystals form (Lindow et al., 1982). Phyllobacteria benefit from the damage caused by the ice to the plant as this allows nutrients to become more accessible (Zachariassen and Kristiansen, 2000; Pearce, 2001).

5. Conclusions and future perspectives

There is limited information available on the structure, composition, and distribution of the phyllobacterial community on the leaves of their tree hosts. The functional consequences of the community on the fitness of individual hosts are also not known. By combining multiple 'omics technologies, i.e., metagenomics, proteomics, and metabolomics, an understanding of the composition, the physiological and ecological roles of phyllobacteria on trees could eventually be elucidated. There is also a move to consider the plant holobiont as opposed to specific sites or individuals as a habitat for shaping the phyllosphere microbial communities.

References

Abanda-Nkpwatt, D., Musch, M., Tschiersch, J., Boettner, M., Schwab, W., 2006. Molecular interaction between *Methylobacterium extorquens* and seedlings: growth promotion, methanol consumption and localization of the methanol emission site. J. Exp. Bot. 57, 4025–4032.

Agler, M.T., Rije, K., Kroll, S., Morhenn, C., Kim, S.T., Weigel, D., Kemen, E., M., 2016. Microbial hub taxa link host and abiotic factors in plant microbiome variation. PLoS Biol. 14, e1002352.

Balint-Kurti, P., Simmons, S.J., Blum, J.E., Ballare, C.L., Stapleton, A.E., 2010. Phylloepiphytic interaction between bacteria and different plant species in a tropical agricultural system. Can. J. Microbiol. 54, 918–931.

Barret, M., Briand, M., Bonneau, S., Preveaux, A., Valiere, S., Bouchez, O., Hunault, G., Simoneau, P., Jacques, M.-A., 2015. Emergence shapes the structure of the seed-microbiota. Appl. Environ. Microbiol. 81, 1257–1266.

Beattie, G.A., 2002. Leaf surface waxes and the process of leaf colonization by microorganisms. In: Lindow, S.E., Hect-Poinar, E.I., Elliott, V.J. (Eds.), Phyllosphere Microbiology. APS Press, St Paul, MN, USA, pp. 3–26.

Beattie, G.A., 2011. Water relations in the interaction between foliar bacterial pathogens with plants. Annu. Rev. Phytopathol. 49, 533–555.

Beattie, G.A., Lindow, S.E., 1999. Bacterial colonization of leaves: a spectrum of strategies. Phytopathology 89, 353–359.

Berg, G., 2009. Plant-microbe interactions promoting plant growth and health: perspectives for controlled use of microorganisms in agriculture. Appl. Microbiol. Biotechnol. 84, 11–18.

Bernard, F., Sache, I., Suffert, F., Chelle, M., 2013. The development of a foliar pathogen does react to leaf temperature. New Phytol. 198, 232–240.

Bodenhausen, N., Horton, M.W., Bergelson, J., 2013. Bacterial communities associated with the leaves and the roots of *Arabidopsis thaliana*. PLoS ONE 8, 356329.

Bodenhausen, N., Bortfeld-Miller, N., Ackermann, M., Vorholt, J.A., 2014. A synthetic community approach levels plant genotypes affecting the phyllosphere microbiota. PLoS Genet. 10, e1004283.

Brandl, M.T., Quinones, B., Lindow, S.E., 2001. Heterogeneous transcription of an indole acetic acid biosynthetic gene in *Erwinia herbicola* on plant surfaces. Proc. Natl. Acad. Sci. U. S. A. 98, 3454–3459.

Bringel, F., Couée, I., 2015. Pivotal roles of phyllosphere microorganisms at the interface between plant functioning and atmospheric trace gas dynamics. Front. Microbiol. 6, 486.

Bulgarelli, D., Schlaeppi, K., Spaepen, S., Ver Loren van Themaat, E., Schulze-Lefert, P., 2013. Structure and functions of the bacterial microbiota of plants. Annu. Rev. Plant Biol. 64, 807–838.

Bunster, L., Fokkema, N.J., Schippers, B., 1989. Effect of surface-active *Pseudomonas* spp. on leaf wettability. Appl. Environ. Microbiol. 55, 1340–1345.

Burch, A.Y., Zeisler, V., Yokota, K., Schreiber, L., Lindow, S.E., 2014. The hygroscopic biosurfactant syringafactin produced by *Pseudomonas syringae* enhances fitness on leaf surfaces during fluctuating humidity. Environ. Microbiol. 16, 2086–2098.

Burrage, S.W., 1971. The micro-climate at the leaf surface. In: Preece, T.J., Dickinson, C.H. (Eds.), Ecology of Leaf Surface Micro-Organisms. Academic Press, London, pp. 91–101.

Carper, D.L., 2018. Abiotic and biotic factors structuring the microbiomes of conifers in the family Pinaceae. PhD thesis, In: Quantitative and Systems Biology. University of California, USA.

Copeland, J.K., Yuan, L., Layeghilard, M., Wang, P.W., Guttman, D.S., 2015. Seasonal community succession of the phyllosphere microbiome. Mol. Plant-Microbe Interact. 28, 274–285.

Corpe, W.A., Rheem, S., 1989. Ecology of the methyltrophic bacteria on living leaf surfaces. FEMS Microbiol. Ecol. 62, 243–250.

Crowther, T.W., Glick, H.B., Bradford, M.A., 2015. Mapping tree diversity on a global scale. Nature 525, 201–205.

Delmotte, N., Knief, C., Chaffron, S., Innerebner, G., Roschitzki, B., Schlapbach, R., von Mering, C., Vorholt, J.A., 2009. Community proteogenomics reveals insights into the physiology of the phyllosphere bacteria. Proc. Natl. Acad. Sci. U. S. A. 106, 16428–16433.

Doan, H.K., Leveau, J.H.J., 2015. Artificial surfaces in phyllosphere microbiology. Phytopathology 105, 1036–1042.

Dong, C.-J., Wang, L.-L., Li, Q., Shang, Q.-M., 2019. Bacterial communities in the rhizosphere, phyllosphere and endosphere of tomato plants. PLoS ONE 14, e0223847.

Esser, D.S., Leveau, J.H.J., Meyer, K.M., Weigand, K., 2015. Spatial scales of interactions among bacteria and between bacteria on the leaf surface. FEMS Microbiol. Ecol. 91, 1–13.

Farre-Armengol, G., Filella, I., Llusia, J., Penuelas, J., 2016. Bidirectional interaction between phyllospheric microbiotas and plant volatile emissions. Trends Plant Sci. 21, 854–860.

Fierer, N., McCain, C.M., Meir, P., Zimmermann, M., Rapp, J.M., Silman, M.R., Knight, R., 2011. Microbes do not follow the elevational diversity patterns of plants and animals. Ecology 92, 797–804.

Finkel, O.M., Burch, A.Y., Elad, T., Huse, S.M., Lindow, S.E., Post, A.F., Belkin, S., 2012. Distance-decay relationships partially determine diversity patterns of phyllosphere bacteria on *Tamarix* trees across the Sonoran Desert. Appl. Environ. Microbiol. 78, 6187–6193.

Freiberg, E., 1998. Microclimatic parameters influencing nitrogen fixation in the phyllosphere in a Costa Rican premontane rain forest. Oecologia 117, 9–18.

Frünkranz, M., Wanek, W., Richter, A., Abell, G., Rasche, F., Sessitsch, A., 2008. Nitrogen fixation by phyllosphere bacteria associated with higher plants and their colonizing epiphytes of a tropical lowland rainforest of Costa Rica. ISME J. 2, 561–570.

Galbally, I.E., Kirstine, W., 2002. The production of methanol by flowering plants and the global cycle of methanol. J. Atmos. Chem. 43, 195–229.

Glickmann, E., Gardan, L., Jacquet, S., Hussain, S., Elasri, M., Petit, A., Dessaux, Y., 1998. Auxin production is a common feature of most pathovars of *Pseudomonas syringae*. Mol. Plant-Microbe Interact. 11, 156–162.

Gourion, B., Rossignol, M., Vorholt, J.A., 2006. A proteomic study of *Methylobacterium extorquens* reveals a response regulator essential for epiphytic growth. Proc. Natl. Acad. Sci. U. S. A. 103, 13186–13191.

Griffin, E.A., Carlson, W.P., 2015. The ecology and natural history of foliar bacteria with a focus on tropical forests and agroecosystems. Bot. Rev. 81, 105–149.

Guerrieri, R., Vanguelova, E., Michalski, G., 2015. Isotopic evidence for the occurrence of biological nitrification and nitrogen deposition processing in forest canopies. Glob. Chang. Biol. 21, 4612–4626.

Hermann, M., Geesink, P., Richter, R., Kusel, K., 2020. Canopy position has a stronger effect that tree identity on phyllosphere bacteria diversity in a floodplain hardwood forest. bioRxiv. https://doi.org/10.1101/2020.02.07.939058.

Hirano, S.S., Upper, C.D., 1989. Diel variation in population size and ice nucleation activity of *Pseudomonas syringae* on snap bean leaflets. Appl. Environ. Microbiol. 55, 623–630.

Holland, M., 2011. Nitrogen: give and take from phyllosphere microbes. In: Ploacco, J.C., Todd, C.D. (Eds.), Ecological Aspects of Nitrogen Metabolism in Plants, first ed. Wiley, pp. 217–230.

Humphrey, P.T., Nguyen, T.T., Villalobos, M.M., Whiteman, N.K., 2014. Diversity and abundance of phyllosphere bacteria are linked to insect herbivory. Mol. Ecol. 23, 1497–1515.

Hunter, P.J., Hand, P., Pink, D., Whipps, J.M., Bending, G.D., 2010. Both leaf properties and microbe-microbe interactions influence within-species variation in bacterial population diversity and structure in the lettuce (*Lactuca* species) phyllosphere. Appl. Environ. Microbiol. 76, 8117–8125.

Hutchison, M.L., Johnstone, K., 1993. Evidence for the involvement the surface active properties of the extracellular toxin tolaasin in the manifestation of brown blotch disease symptoms by *Pseudomonas tolaasii* on *Agaricus bisporus*. Physiol. Mol. Plant Pathol. 42, 373–384.

Innerebner, G., Knief, C., Vorholt, J.A., 2011. Protection of *Arabidopsis thaliana* against leaf-pathogenic *Pseudomonas syringae* by *Sphingomonas* strains in a controlled model system. Appl. Environ. Microbiol. 77, 3202–3210.

Izhaki, I., Svetlana, F., Yoram, G., Malka, H., 2013. Variability of bacterial community composition on leaves between and within plant species. Curr. Microbiol. 66, 227–235.

Jacobs, J.L., Carroll, T.L., Sundin, G.W., 2005. The role of pigmentation, ultraviolet radiation tolerance, and leaf colonization strategies in the epiphytic survival of phyllosphere bacteria. Microb. Ecol. 49, 104–113.

Joyner, D.C., Lindow, S.E., 2000. Heterogeneity of iron bioavailability on plants assessed with a whole-cell GFP-based bacterial sensor. Microbiology 146, 2435–2445.

Junker, R.R., Tholl, D., 2013. Volatile organic compound mediated interactions at the plant-microbe interface. J. Chem. Ecol. 39, 810–825.

Kadivar, H., Stapleton, A., 2003. Ultraviolet radiation alters maize phyllosphere bacterial diversity. Microb. Ecol. 45, 353–361.

Kembel, S.W., O'Connor, T.K., Arnold, H.K., Hubbell, S.P., Wright, S.J., Green, J.L., 2014. Relationships between phyllosphere bacterial communities and plant functional traits in a neotropical forest. PNAS 111, 13715–13720.

Kim, M., Singh, D., Lai-Hoc, A., Go, R., Rahim, R.A., Aninuddin, A.N., Chun, J., Adams, J.M., 2012. Distinctive phyllosphere bacterial communities in tropical trees. Microb. Ecol. 63, 674–681.

Kinkel, L., 1997. Microbial population dynamics on leaves. Annu. Rev. Phytopathol. 35, 327–347.

Knief, C., Ramette, A., Frances, L., Alonso-Blanco, C., Vorholt, J.A., 2010. Site and plant species are important determinants of the *Methylbacterium* community composition in the plant phyllosphere. ISME J. 4, 719–728.

Knoll, D., Schreiber, L., 1998. Influence of epiphytic microorganisms on leaf wettability: wetting of the upper leaf surface of *Juglans regia* and of model surfaces in relation to colonization by microorganisms. New Phytol. 140, 271–282.

Knoll, D., Schreiber, L., 2000. Plant-microbe interactions: wetting of ivy (*Hedera helix* L.) leaf surfaces in relation to colonization by epiphytic microorganisms. Microbial. Ecol. 41, 33–42.

Laforest-Lapointe, I., Messier, C., Kembel, S.S.W., 2016. Host species identity, site and time drive temperate tree phyllosphere bacterial community structure. Microbiome 4, 27.

Laforest-Lapointe, I., Paquette, A., Messier, C., Kembel, S.W., 2017. Leaf bacterial diversity mediates plant diversity and ecosystem function relationships. Nature 546, 145.

Lambais, M.R., Crowley, D.E., Cury, J.C., Bull, R.C., Rodrigues, R.R., 2006. Bacterial diversity in tree canopies of the Atlantic forest. Science 312, 1917.

Lambais, M.R., Barrera, S.E., Santos, E.C., Crowley, D.E., Jumpponen, A., 2017. Phyllosphere metaproteomes of trees from the Brazialian Atlantic forest show high levels of functional redundancy. Microb. Ecol. 73, 123–134.

Leff, J.W., Del Tedici, P.D., Friedman, W.E., Fierer, N., 2015. Spatial structuring of bacterial communities with individual Ginkgo biloba trees. Environ. Microbiol. 17, 2352–2361.

Leuzinger, S., Korner, C., 2007. Tree species diversity affects canopy leaf temperatures in a mature temperate forest. Agric. For. Meteorol. 146, 29–37.

Li, Y., Wu, X., Wang, W., Wang, M., Zhao, C., Chen, T., Liu, G., Zhang, W., Li, S., Zhou, H., Wu, M., Yang, R., Zhang, G., 2019. Microbial taxonomical composition in spruce phyllosphere, but not community functional structure, varies by geographical location. PeerJ 7, e7376.

Lindow, S.E., Brandl, M.T., 2003. Microbiology of the phyllosphere. Appl. Environ. Microbiol. 69, 1875–1883.

Lindow, S.E., Amy, D.C., Upper, C.D., 1982. Bacterial ice nucleation: a factor in frost injury in plants. Plant Physiol. 70, 1084–1089.

Lindow, S.E., Desurmont, C., Elkins, R., McGourty, G., Clark, E., Brandl, M.T., 1998. Occurrence of indole-3-acetic acid–producing bacteria on pear trees and their association with fruit russet. Phytopathology 88, 1149–1157.

Marco, M.L., Legac, J., Lindow, S.E., 2005. *Pseudomonas syringae* genes induced during colonization of leaf surfaces. Environ. Microbiol. 7, 1379–1391.

Mason, C.J., Couture, J.J., Raffa, K.F., 2014. Plant-associated bacteria degrade defense chemicals and reduce their adverse effects on an insect defoliator. Oecologia 175, 901–910.

Mercier, J., Lindow, S.E., 2000. Role of leaf surface sugars in colonization of plants by bacterial epiphytes. Appl. Environ. Microbiol. 66, 369–374.

Morris, C.E., 2002. Phyllosphere. eLS. https://doi.org/10.1038/npg.els.0000400.

Morris, C.E., Kinkel, L.L., 2002. Fifty years of phyllosphere microbiology: significant contributions to research in related fields. In: Lindow, S.E., Hect-Poinar, E.I., Elliott, V.J. (Eds.), Phyllosphere Microbiology. APS Press, St Paul, MN, USA, pp. 365–375.

Morris, C.E., Barnes, M.B., McLean, R.C.J., 2002. Biofilsm on leaf surfaces: implications for the biology, ecology and management of populations of epiphytic bacteria. In: Lindow, S.E., Hect-Poinar, E.I., Elliott, V.J. (Eds.), Phyllosphere Microbiology. APS Press, St Paul, MN, USA, pp. 139–155.

Munemasa, S., Hauser, F., Park, J., Waadt, R., Brandt, B., Schroeder, J.I., 2015. Mechanisms of abscisic acid-mediated control of stomatal aperture. Curr. Opin. Plant Biol. 28, 134–162.

Neinhuis, C., Barthlott, W., 1997. Characterization and distribution of water-repellent, self-cleaning plant surfaces. Ann. Bot. 79, 667–677.

Oso, S., Walters, M., Schlechter, R.O., Remus-Emsermann, M.N.P., 2019. Utilization of hydrocarbons and production of surfactants by bacterial isolates from plant leaf surfaces. FEMS Microbiol. Lett. 366, fnz061.

Papen, H., Gessler, A., Zumbusch, E., Rennenberg, H., 2002. Chemolithoautotrophic nitrifers in the phyllosphere of a spruce ecosystem receiving high atmospheric nitrogen input. Curr. Microbiol. 44, 56–60.

Pearce, R.S., 2001. Plant freezing and damage. Ann. Bot. 87, 417–424.

Pusey, P.L., Stockwell, V.O., Reardon, C.L., Smits, T.H.M., Duffy, B., 2011. Antibiosis activity of *Pantoea agglomerans* biocontrol strain E325 against *Erwinia amylovora* on apple flower stigmas. Phytopathology 101, 1234–1241.

Qvit-Raz, N., Finkel, O.M., Al-Deeb, T.M., Malkawi, H.I., Hindiyeh, M.Y., Jurgevitch, E., Belkin, S., 2012. Biogeographical diversity of leaf-associated microbial communities from salt-secreting *Tamarix* trees of the Dead Sea region. Res. Microbiol. 163, 142–150.

Rastogi, G., Sbodio, A., Tech, J.J., Suslow, T.V., Coaker, G.L., Leveau, J.H., 2012. Leaf microbiota in an agroecosystem: spatiotemporal variation in bacterial community composition on field-grown lettuce. ISME J. 6, 1812–1822.

Rastogi, G., Coaker, G.L., Leveau, J.H.J., 2013. New insights into the structure and function of phyllosphere microbiota through high-throughput molecular approaches. FEMS Microbiol. Lett. 348, 1–10.

Redford, A.J., Fierer, N., 2009. Bacterial succession on the leaf surface: a novel system for studying successional dynamics. Microb. Ecol. 58, 189–198.

Redford, A.J., Bowers, R.M., Knight, R., Linhart, Y., Fierer, N., 2010. The ecology of the phyllosphere: geographic and phylogenetic variability in the distribution of bacteria on tree leaves. Environ. Microbiol. 12, 2885–2893.

Remus-Emsermann, M.N.P., Schlechter, R.O., 2018. Phyllosphere microbiology: at the interface between microbial individuals and the plant host. New Phytol. 218, 1327–1333.

Rico, U., Ogaya, R., Terradas, J., Penuelas, J., 2014. Community structures of N2-fixing bacteria associated with the phyllosphere of a holm oak forest and their response to drought. Plant Biol. 16, 586–593.

Roden, J.S., Pearcy, R.W., 1993. The effect of flutter on the temperature of poplar leaves and its implication for carbon gain. Plant Cell Environ. 16, 571–577.

Romero, F.M., Marina, M., Piecksenstain, F.L., 2016. Novel components of leaf bacterial communities of field-grown tomato plants and their potential for plant growth promotion and biocontrol of tomato diseases. Res. Microbiol. 167, 222–233.

Ruinen, J., 1961. The phyllosphere. I. An ecological neglected millieu. Plant Soil 15, 81–109.

Ruppel, S., Krumbein, A., Schreiner, M., 2008. Composition of the phyllospheric microbial populations on vegetable plants with different glucosinolate and caroteoid compositions. Microb. Ecol. 56, 364–372.

Ryu, J., Madhaiyan, M., Poonguzhali, S., Yim, W., Indirgandhi, P., Kim, K., Anandham, R., Yun, J., Sa, T., 2006. Plant growth substances produced by *Methylbacterium* spp. and their effect on tomato (*Lycopersicon esculentum* L.) and red pepper (*Capsicum annuum* L.) growth. J. Microbiol. Biotechnol. 16, 1622–1628.

Saleem, M., Meckes, N., Pervaiz, Z.H., Traw, M., 2017. Microbial interactions in the phyllosphere increase plant performance under herbivore biotic stress. Front. Microbiol. 8, 41.

Sandhu, A., Halverson, L.J., Beattie, G.A., 2007. Bacterial degradation of airborne phenol in the phyllosphere. Environ. Microbiol. 9, 383–392.

Scheublin, T.R., Deusch, S., Moreno-Forero, S.K., Müller, J.A., van der Meer, J.R., Leveau, J.H., 2014. Transcriptional profiling of gram-positive Arthrobacter in the phyllosphere: induction of pollutant degradation genes by natural plant phenolic compounds. Environ. Microbiol. 16, 2212–2225.

Schlechter, R.O., Miebach, M., Remus-Emsermann, M.N.P., 2019. Driving factors of epiphytic bacterial communities: a review. J. Adv. Res. 19, 57–65.

Schreiber, L., Krimm, U., Knoll, D., Sayed, M., Auling, G., Kroppenstredt, R.M., 2005. Plant-microbe interactions: identification of epiphytic bacteria and their ability to alter leaf surface permeability. New Phytol. 166, 589–594.

Shade, A., McManus, P.S., Handelsman, J., 2013. Unexpected diversity during community succession in the apple flower microbiome. MBio 4, 602–612.

Shapiro, L., De Moraes, C.M., Stephenson, A.G., Mescher, M.C., van der Putten, W., 2012. Pathogen effects on vegetative and floral odours mediate vector attraction and host exposure in a complex pathosystem. Ecol. Lett. 15, 1430–1438.

Stokes, V.J., Morecroft, M.D., Morison, J.I.L., 2006. Boundary layer conductance for contrasting leaf shapes in a deciduous broadleaved forest canopy. Agric. For. Meteorol. 139, 40–54.

Stone, W.G., Jackson, C.R., 2019. Canopy position is a stronger determinant of bacterial community composition and diversity than environmental disturbance in the phyllosphere. FEMS Microbiol. Ecol. 95, 1–11.

Stone, B.W.G., Weingarten, E.A., Jackson, C.R., 2018. The role of the phyllosphere microbiome in plant health and function. Annu. Plant Rev. 1, 1–24.

Sundin, G.W., 2002. Ultraviolet radiation on leaves: its influence on microbial communities and their adaptation. In: Lindow, S.E., Hect-Poinar, E.I., Elliott, V.J. (Eds.), Phyllosphere Microbiology. APS Press, St Paul, MN, USA, pp. 27–41.

Sundin, G.W., Jacobs, J.L., 1999. Ultraviolet radiation (ULR) sensitivity analysis and UVR survival strategies of a bacterial community from the phyllosphere of field-grown peanut (Arachis hypogeae L.). Microb. Ecol. 38, 27–38.

Surico, G., 1993. Scanning electron microscopy of olive and oleander leaves colonized by Pseudomonas syringae subsp. savastanoi. J. Phytopathol. 138, 31–40.

Suslow, T.V., 2002. Production practices affecting the potential for production practices affecting the potential for persistent contamination of plants by microbial foodborne pathogens. In: Lindow, S.E., Hect-Poinar, E.I., Elliott, V.J. (Eds.), Phyllosphere Microbiology. APS Press, St Paul, MN, USA, pp. 241–246.

Sussmilch, F.C., McAdam, S.A.M., 2017. Surviving a dry future: abscisic acid (ABA)-mediated plant mechanisms for conserving water under low humidity. Plan. Theory 6, 54–76.

Tejera, N., Ortega, E., Rodes, R., Lluch, C., 2006. Nitrogen compounds in the apoplast sap of sugarcane stem: some implications in the association with endophytes. J. Plant Physiol. 163, 80–85.

Truchado, P., Gil, M.I., Reboleira, P., Rodelas, B., Allende, A., 2017. Impact of solar radiation exposure on phyllosphere bacterial community of red-pigmented baby leaf lettuce. Food Microbiol. 66, 77–85.

Tukey, H.B., 1970. The leaching of substrates from plants. Annu. Rev. Plant Physiol. 21, 305–324.

Vacher, C., Hampe, A., Porte, A., Sauer, U., Compant, S., Morris, C., 2016. The phyllosphere: microbial jungle at the plant-climate interface. Annu. Rev. Ecol. Evol. Syst. 47, 1–24.

Van Aken, B., Peres, C.M., Doty, S.L., Yoon, Y.M., Schnoor, J.L., 2004. Methylobacterium populi sp. nov., a novel aerobic, pink pigmented, facultatively methylotrophic, methane utilizing bacterium isolated from poplar trees (Populus deltoides x nigra DN34). Int. J. Syst. Evol. Microbiol. 54, 1191–1196.

Van der Wal, A., Leveau, J.H.J., 2011. Modelling sugar diffusion across plant leaf cuticles: the effect of free water on substrate availability to phyllosphere bacteria. Environ. Microbiol. 13, 792–797.

Van Wees, S.C., De Swardt, E.A., Van Pelt, J.A., Van Loon, L.C., Pieterse, C.M., 2000. Enhancement of induced disease resistance by simultaneous activation of salicylate- and jasmonate-dependent defense pathways in Arabidopsis thaliana. Proc. Natl. Acad. Sci. U. S. A. 97, 8711–8716.

Vandenkoornhuyse, P., Quaiser, A., Duhamel, M., Le Van, A., Dufresne, A., 2015. The importance of the microbiome of the plant holobiont. New Phytol. 206, 1196–1206.

Vogel, C., Bodenhausen, N., Gruissem, W., Vorholt, J.A., 2016. The Arabidopsis leaf transcriptome reveals distinct but also overlapping responses to colonization by phyllosphere commensals and pathogen infection with impact on plant health. New Phytol. 202, 192–207.

Vokou, D., Genitsaris, S., Karamanoli, K., Vareli, K., Zachari, M., Voggoli, D., Monokrousos, N., Halley, J.M., Sainis, I., 2019. Metagenomic characterization reveals pronounced seasonality in the diversity and structure of the phyllosphere bacterial community in a Mediterranean ecosystem. Microorganisms 7, 518.

Vorholt, J.A., 2012. Microbial life in the phyllosphere. Nat. Rev. Microbiol. 10, 828–840.

Wellner, S., Lodders, N., Kampfer, P., 2011. Diversity and biogeography of selected phyllosphere bacteria with special emphasis on Methylobacterium spp. Syst. Appl. Microbiol. 34, 621–630.

Wensing, A., Braun, S.D., Büttner, P., Expert, D., Völksch, B., Ullrich, M.S., Weingart, H., 2010. Impact of siderophore production by Pseudomonas syringae pv. syringae 22d/93 on epiphytic fitness and biocontrol activity against Pseudomonas syringae pv. glycinea 1a/96. Appl. Environ. Microbiol. 76, 2704–2711.

Wilson, M., Lindow, S.E., 1994. Coexistance among epiphytic bacterial populations mediated through nutritional resource partitioning. Appl. Environ. Microbiol. 60, 4468–4477.

Wink, M., 2003. Evolution of secondary metabolites from an ecological and molecular phylogenetic perspective. Phytochemistry 64, 3–19.

Yang, C., Crowley, D.E., Borneman, J., Keen, N.T., 2001. Microbial phyllosphere populations are more complex than previously thought. Proc. Natl. Acad. Sci. U. S. A. 98, 3889–3894.

Yao, H., Sun, X., He, C., Li, X.-C., Guo, L.-D., 2020. Host identity is more important in structuring bacterial epiphytes than endophytes in a tropical mangrove forest. FEMS Microbiol. Ecol. 96, fiaa038.

Yu, X., Lund, S.P., Scott, R.A., Greenwald, J.W., Records, A.H., Nettleton, D., Lindow, S.E., Gross, D.C., Beattie, G.A., 2013. Transcriptional responses of *Pseudomonas syringae* to growth in epiphytic versus apoplastic leaf sites. Proc. Natl. Acad. Sci. U. S. A. 29, E425–E434.

Zachariassen, K.E., Kristiansen, E., 2000. Ice nucleation and antinucleation in nature. Cryobiology 41, 257–279.

Zamiousdis, C., Pieterse, C.M.J., 2012. Modulation of host immunity by beneficial microbes. Mol. Plant-Microbe Interact. 25, 139–150.

Chapter 8

Microbiome of reproductive organs of trees

Fei Ren[a,*], Dong-Hui Yan[b,†], and Wei Dong[c,‡]

[a]*Forestry Experiment Center in North China, Chinese Academy of Forestry, Beijing, China,* [b]*Research Institute of Forest Ecology, Environment and Protection, Key Laboratory of Biodiversity Conservation of National Forestry and Grassland Administration, Chinese Academy of Forestry, Beijing, China,* [c]*China Electric Power Research Institute, Beijing, China*

Chapter Outline

In this chapter, the term "reproductive organs" in the case of forest trees mainly refers to flowers, fruits, seeds, and cones. These organs are associated with a variety of microorganisms, mostly bacteria and fungi, that can be transmitted by regeneration. The microbes can come from different sources from parents through vertical transmission, or from the environment and other hosts through horizontal transfer (Perlmutter and Bordenstein, 2020). These microbes of tree reproductive organs could be vital for the fitness and health of the host trees and offspring, owing to their benefits for tree adaption to environmental stresses. The benefits can also be vertically passed. Therefore, it is crucial to understand their composition, transmission routes, and functional effects on host evolution and ecology.

The microbiome of reproductive organs could be significantly affected by many factors, including microbiome structure, their transmission routes, tree maternal effects, relationships with root and soil, etc. The microbiome of flowers could be regulated by the insect vector and the ecological function of pollination as well as the impacts of the symbiotic microbiome on the evolution and ecology of the hosts and microbes. At the end of the chapter, we will put forward questions and trends for future studies in the field. There is an urgent and increasing need for the continued microbiome sequencing of reproductive organs. More attention should be paid to the ecological functions of other members of the microbiome besides bacteria and fungi. The study of interactions among the microbiome in reproductive organs is expected to be intensified. We believe that studies in this area will provide a comprehensive view of the influences of the reproductive organ microorganisms on hosts, and also make a fundamental scientific basis for future applications.

---

[*] Dr. Fei Ren, majors in Microbiology, is interested in plant microbial diversity, endophytes, new taxa of microbes and microbial applications.

[†] Dr. Dong-Hui Yan, majors in Forest Pathology, is interested in microbial-host interaction, fungal pathogens and endophytes.

[‡] Mr. Wei Dong, majors in electricity engineering, is interested in computer and data analysis.

Box 8.1 The microbiome of reproductive organs.

The definition of the reproductive microbiome could broadly include fungi, bacteria, viruses, protozoans, etc., inhabiting *endo-* or epi- any reproductive organ of the host (Muller et al., 2016; Perlmutter and Bordenstein, 2020). Because the existence of the microbiome in reproductive organs suggests that fertilization is taking place in a nonsterile environment, the view of embryonic development in a sterile environment has been challenged for animals and plants (Schoenmakers et al., 2019).

Most researches pay attention to a few specific host species. As the reproductive microbiome diversity develops, many definitions focused and specialized more to different biological systems might occur. These concepts might be more specific with the microbiome of the host or organ parts such as the arthropod reproductive microbiome and the apple flower microbiome (Perlmutter and Bordenstein, 2020; Rowe et al., 2020; Shade et al., 2013). For animals, the concept comprises all the microbes associated with the host reproductive system, including the whole reproductive process as well as all the microbes of both the male and female reproductive structures (Rowe et al., 2020). However, for plants, especially in this chapter, the microorganisms of the reproductive organs are those residing in or being transmitted through the reproductive organs of trees, mainly including bacteria and fungi associated with flowers, fruits, seeds, and cones.

Presently, many questions regarding reproductive microbiomes are still uncertain, such as how these microbiomes of reproductive organs are formed and preserved and what effect their functions might have on the mating and reproductive processes of hosts. Moreover, it is not obvious to what extent the reproductive microbiomes could represent real native communities because of the exchanges with external environments and other parts of the host (Muller et al., 2016; Perlmutter and Bordenstein, 2020; Rowe et al., 2020). Studies on the ecological processes might result in more views and knowledge regarding reproductive microbiomes in the future.

1. Composition of the microbiome of reproductive organs

The microbiome of plant reproductive organs refers to a large diversity of microorganisms and includes various protective and beneficial endophytes, a group of microbial symbionts inhabiting plants, phytopathogens parasitizing plants, and epiphytic microbes (Azevedo et al., 2000; Gao et al., 2010; Muller et al., 2016; Perlmutter and Bordenstein, 2020). Endophytes are essential components of the tree microbiome (Muller et al., 2016; Ren et al., 2019a,b). They are widespread, living within the plant hosts without causing obvious harm (Azevedo et al., 2000). Endophytes have received extraordinary attention owing to their ability to generate bioactive metabolites, helping hosts resist environmental stresses and accelerating host growth (Chen et al., 2016; Muller et al., 2016; Ren et al., 2019b; Xu et al., 2010). With the advances of high-throughput sequencing technology (HTS), cultivation-independent analyses have given profound insight into the plant microbial community structure compared to traditional cultivation-dependent methods (Muller et al., 2016). In the past decade, one critical finding from the analyses is that plant-associated microorganisms have been gathered with phylogenetic structures but were not in random assemblages (Muller et al., 2016). Bacteria are highly abundant groups in the microbial communities, and fungi are also essential contributors (Hacquard et al., 2015; Kemen, 2014; Muller et al., 2016).

Seeds play a significant part in the spermatophyte life cycle. They are able to persist in a dormant state for a very long time, and not grow into a new plant until the growth conditions become appropriate (Shahzad et al., 2018). Microbes are diverse and have been identified in each seed part: the embryonic tissues, endosperm, and coat (Glassner et al., 2017). A total of 131 bacterial genera from four phyla were found in seeds from 25 different plants (Truyens et al., 2014). Seventeen species of plant seeds illustrated various bacterial phyla while the remaining eight species had only one phylum based on cultivation and high-throughput sequencing methods (Truyens et al., 2014). Different seeds might also share some same taxa (Cankar et al., 2005). For example, members of the *Rahnella* genus were cultivated from the seed embryo and endosperm as bacterial endophytes from different cones of the Norway spruce trees (Cankar et al., 2005). The quantity and cultivability of bacteria in spruce seeds declined with time (Cankar et al., 2005). The most common bacterial phyla are Proteobacteria, Actinobacteria, and Firmicutes (Muller et al., 2016; Truyens et al., 2014). Bacteria commonly identified in seeds include certain common genera, specifically *Bacillus* and *Pseudomonas*, and also *Acinetobacter*, *Micrococcus*, *Paenibacillus*, *Pantoea*, and *Staphylococcus* (Compant et al., 2011; Truyens et al., 2014). But cultivation-dependent isolation procedures resulted in the isolation of only a few bacterial species in numerous plant species (Mundt and Hinkle, 1976). This may be the result of either the impossible cultivation of these seeds' microbes or the existence of too few bacterial cells. For the seeds of sessile oak populations, fungi exist in all sampled seeds as well as embryos (Fort et al., 2019). In the Scots pine, many more endophytes were found in seed embryos, but fewer were colonized in the cones (Pirttila and Frank, 2011). Fungal communities of acorns from different oak populations differed significantly, and even differed among trees of the same population (Fort et al., 2019). Tree seeds have microbes of pathogens and the antagonists, for microbial members of acorns, some are plant pathogens such as *Gnomoniopsis paraclavulata*, others are antagonists like *Cladosporium delicatulum* (Fort et al., 2019). Factors affecting seed microbiome diversity include the tree taxa and genetics, the tree population, the seed location, and the seed maturation degree (Fort et al., 2019). The composition of the microbiome of seeds is still not totally understood, especially for forest tree species (Pirttila and Frank, 2011; Verma and White, 2019). Continued studies on microbiome composition and diversity as well as their functional roles in seed germination and development need to be further explored (Verma and White, 2019).

Owing to their ephemerality and exquisite anatomy, flowers and fruits provide unique habitats to microbes (Aleklett et al., 2014). The surfaces of floral organs host diverse communities of microbes (Hardoim et al., 2015). Studies have described microorganisms by either culture methods or sequencing of nectars, petals, pollens, and fruits (Aizenberg-Gershtein et al., 2013; Álvarez-Pérez et al., 2012; Baruzzi et al., 2012; Fridman et al., 2012; Fürnkranz et al., 2012; Glassner et al. 2017; Glassner et al., 2015; Heydenreich et al., 2012; Jacquemyn et al., 2013; Madmony et al., 2005). Research on the leaf and flower bacterial community diversity of the grapevine illustrated that the flower community was less diverse, and mostly made up of Proteobacteria, mainly of *Pseudomonas* and *Erwinia* (Bowers et al., 2011). *Enterobacter cloacae* was cultivated from the pollen of *Pinus halepensis* and *P. pinea*, as well as the ovules of *P. brutia* (Madmony et al., 2005). However, an analysis of the apple flower microbiome showed higher bacterial diversity, including the phyla Deinococcus-Thermus, TM7, Bacteroidetes, Firmicutes, and Proteobacteria (Shade et al., 2013). Interestingly, Cyanobacteria were most abundant in Jingbai pear trees, followed by Proteobacteria and Actinobacteria, which also dominated in the flowers and fruits of trees (Ren et al., 2019a). For fungi, Ren et al. (2019b) found that the fungal richness and diversity in flowers and fruits was lowest in the Jingbai pear tree organs. Ascomycota was the most abundant group, followed by Basidiomycota and Zygomycota. However, for the fungi of peach trees, except for the roots, the fungal richness and diversity of flowers was the highest while the leaves were the lowest (Ren et al., 2019c), Ascomycota and Basidiomycota showed the highest abundance among all sampled organs. Some reproductive organs show site-specific microbial communities such as flowers (Shade et al., 2013), fruits (Compant et al., 2011), and pollen (Madmony et al., 2005). Various pollens have lots of shared bacterial genera such as *Rosenbergiella*, *Pseudomonas*, *Methylobacterium*, *Friedmanniella*, and *Bacillus*, representing some common and unique symbionts with those commonly existing in seeds (Frank et al., 2017). The microbial community composition in these organs can be affected by biotic and abiotic factors such as environmental circumstances, the host genetic background, the interactions and associations among plant microbiota, etc. (Muller et al., 2016; Ren et al., 2019b). For example, the three peach cultivars held different fungal community structures (Ren et al., 2019c). Sites affect the microbial communities of the Jingbai pear (Ren et al., 2019b).

Generally, at the phylum level, the bacterial composition of reproductive organs is inconsistent with the phyla in other organs—Proteobacteria were dominant, followed by Actinobacteria, Firmicutes, and Bacteroidetes, indicating the conserved taxa plant organs and species (Muller et al., 2016; Truyens et al., 2014). For fungi, similar to other plant organs, Ascomycota were the most abundant, followed by Basidiomycota in the reproductive organs. Due to different structures and ecological niches, different tree organs and parts might host different microbial communities (Muller et al., 2016; Ren et al., 2019b). The individual microbiome of compartments consists of a selective gradient among the soil, exterior root, rhizoplane, interior roots, and other endocompartments such as leaves, stems, etc. (Muller et al., 2016). The belowground microbial communities were significantly different from those aboveground. Bacterial communities of the leaf, flower, and grape fruits shared a higher similar taxa proportion with those of soils than with each other, indicating a soil borne for below- and above-ground communities (Zarraonaindia et al., 2015). Conditions changed during the seed maturation process and could affect the microbial groups that are capable of inhabiting the seed, so some characteristics found in seed microbes were not in microbes from other organs such as roots or twigs (Truyens et al., 2014). Seed microbes may produce amylase to enable the seed to use starch and induce seed germination after a long dormant time (Mano et al., 2006), and they may also use phytate as a phosphate source for the seed (López-López et al., 2010). A predominance of Gram-negative microbes such as *Methylobacterium* and *Sphingomonas* was found in the early stages of seed development, but more Gram-positive ones came up with seed maturation such as *Bacillus* and *Curtobacterium* (Mano et al., 2006).

Seeds host microbes that are prone to pathogen defense, plant development, and nutrient uptake for seedlings (Verma and White, 2019). The plant microbiomes of reproductive organs are likely to be those vital for the host metabolism, stimulating host development and reproduction. Some main microbes of plant organs are shown in Table 8.1. (See Box 8.1.)

Nowadays, the plant microbiome has attracted considerable interest. However, a lot of studies have focused on a plant's belowground parts (Muller et al., 2016). The microbiomes of reproductive organs, especially those of flowers and fruits, are still poorly understood. Thus, a comprehensive view of their microbiome composition requires further research. It is important to highlight how diverse the microbiomes are within tree reproductive organs and their dynamics, how much the community compositions differ among individual species. Additionally, it is also important to know for plant reproductive organs why Archaea are obviously not frequent on many terrestrial plants and whether they have important metabolic functions relevant for carbon cycling (Vorholt, 2012). Similarly, studies on viruses and protozoa were also scarce in plant reproductive organs. These questions need to be further addressed. So, it is vital to further explore the identifications and

TABLE 8.1 Main microbes of plant reproductive organs and other organs.

Reproductive organs/ other organs	Microbes	References
Seed	*Rahnella, Bacillus, Pseudomonas* *Acinetobacter, Micrococcus, Paenibacillus, Pantoea, Staphylococcus,* *Enterococcus, Paracoccus, Frankiaceae* *Gnomoniopsis, Cladosporium, Stromatoseptoria, Taphrina, Epicoccum, Myc osphacrella,Mycosphaerella* *Cylindrium, Fusarium*	Compant et al., (2011) Truyens et al. (2014) Cankar et al. (2005) Fort et al. (2019)
Flower	*Rosenbergiella, Pseudomonas, Methylobacterium, Friedmanniella, Bacillus* *Lactobacillus, Acetobacter*	Frank et al. (2017) Shade et al. (2013)
Fruit	*Alternaria, Penicillium, Talaromyces, Gibberella, Pseudallescheria, Fusarium,* *Guehomyces* *Xanthmonas, Bacillus* Gammaproteobacteria, Moraxellaece *Alternaria, Penicillium, Talaromyces, Gibberella, Fusarium*	Ren et al. (2019b) Ren et al. (2019c) Verma and White (2019) Ren et al. (2019a) Ren et al. (2019b)
Root	*Bacillus, Sphingomonas, Pseudomonas, Staphylococcus,* *Stenothrophomonas, Enterobacter* *Curtobacterium, Methylobacterium, Streptomyces, Burkholderia,* *Acinetobacteria* *Alternaria, Penicillium, Talaromyces, Fusarium* *Leptosphaeria, Aureobasidum, Guehomyces, Trichoderma*	Muller et al. (2016) Ren et al. (2019a) Verma and White (2019) Ren et al. (2019b) Ren et al. (2019c)
Leaf	*Bacillus, Duganella, Pseudomonas* *Sphingomonas, Xanthomonas, Xylophilus* *Acinetobacteria* *Alternaria, Penicillium, Talaromyces, Fusarium, Aureobasidum, Chaetomium*	Muller et al. (2016) Ren et al. (2019a) Verma and White (2019) Ren et al. (2019b) Ren et al. (2019c)
Stem	*Bacillus, Erwinia, Pseudomonas* *Sphingomonas, Xanthmonas, Acinetobacter* *Curtobacterium, Stenothrophomonas, Burkholderia* *Alternaria, Penicillium, Talaromyces, Fusarium* *Leptosphaeria, Aureobasidum, Paraphaeosphaeria*	Muller et al. (2016) Ren et al. (2019a) Verma and White (2019) Ren et al. (2019b) Ren et al. (2019c)

dynamics of all microbial groups of the reproductive organs of more and more plants, and also to pay more attention to studies on host-microbiome interactions (Perlmutter and Bordenstein, 2020). The identification of core taxa and strain collections will also be very vital for further research and applied use.

2. Vertical transmission of the microbiome of reproductive organs and microbiome maternal effects on trees

Microorganisms can be transmitted from parents to the next generation vertically, and also horizontally acquired from the environments. A better comprehension of microbial transmission routes and modes will be beneficial to the study of plant-microorganism interactions. This part discusses vertical transmission via seeds and pollen, and microbiome maternal effects on trees. Horizontal transmission will be discussed in the following part.

2.1 Vertical transfer through seeds and pollen

As mentioned in the above part, the seed microbiome is attracting more interest and attention (Nelson, 2017; Shade et al., 2017; Truyens et al., 2014). Microbes were found in the surface-disinfected seeds of a variety of hosts such as crops (Hardoim et al., 2012; Liu et al., 2013; Zawoznik et al., 2014); they were also found in several Eucalyptus (Ferreira et al., 2008), Norway spruce (Cankar et al., 2005), and Curupaú trees (Alibrandi et al., 2018). Some seed endophytes showed beneficial impacts on the host. For example, seed endophytes might produce cytokinins and interact with microbial and plant hormones, contributing to seed dormancy release (Goggin et al., 2015). Seed endophytes could also help plants germinate and grow under harsh environments (Rout et al., 2013).

TABLE 8.2 Bacteria taxa that are transferred by vertical and horizontal transmission.

Microbes	Vertical transmission /horizontal transmission	References
Burkholderia	Vertical transmission	Miller (1990)
Enterobacter	Vertical transmission	Madmony et al. (2005)
Bacillus	Vertical transmission	Compant et al. (2011)
Pantoea, Sphingomonas	Vertical transmission	Liu et al. (2012)
Morganelli, Enterobacter	Vertical transmission	Mukhopadhyay et al. (1996)
Metschnikowia reukaufii	Horizontal transmission	Schaeffer et al. (2019)
Erwinia amylovora	Horizontal transmission	Spinelli et al. (2005)
Phytoplasma, Xylella	Horizontal transmission	Weintraub and Beanland (2006)
Bacillus pumilus	Horizontal transmission	Adams et al. (2008)

With these benefits, it is very reasonable that trees may create mutualisms with microbes vertically transmitted via seeds, guaranteeing continuous transmission, similar to the plants and endophytes as defensive mutualisms (Hodgson et al., 2014; Saikkonen et al., 2010). One example of the vertical transfer of plant-microorganism symbiosis is the leaf nitrogen-fixing *Burkholderia* in the angiosperm genera inhabiting all the vegetative shoot tips and every new leaf (Miller, 1990). These bacteria were observed to be transferred to the floral shoot tips, the developing ovule's embryo sac, and on the embryo epicotyl, where they are enclosed in the seedling shoot tips (Miller, 1990). Even in this vertical transfer symbiosis confirmation, it is not yet easy to discover the symbionts in seeds because of the low amount of microbial DNA in seeds (Lemaire et al., 2012). The existence of microbes in seeds does not imply that they come from the parents because not all seed-living microbes naturally settle in seedlings (Frank et al., 2017). If the seed-associated microbial community is determined by both the plant and bacterium, long stable correlations and high similarities should be kept among seed microorganisms in a host species and also in its related hosts, without considering environmental factors such as soil property and geographies (Frank et al., 2017). Many more differences will be seen among plant species and locations if the microbes are determined by a neutral process (Frank et al., 2017). The best evidence supporting the vertical transfer of the microbiome through seeds demonstrates overlap and consistency in microbial taxa between seeds and seedlings (Ferreira et al., 2008; Gagne-Bourgue et al., 2013; Johnston-Monje and Raizada, 2011). Other reports also supported vertical transfer between generations, finding continuity in the existence of special taxa among generations (Liu et al., 2012; Mukhopadhyay et al., 1996) (Table 8.2). A study on the long-time preservation in seed microorganisms pointed out that seeds from genetic relationship hosts shared similar microbial taxa (Liu et al., 2012). Another study with 16S rDNA sequencing demonstrated the existence of the same genera among many genotypes as well as their ancestors (Johnston-Monje and Raizada, 2011). One piece of indirect evidence on vertical transmission comes from seedborne microbes on plants growing from aseptic seeds—the vertically transferred diazotrophs through the seeds made the plants acquire nitrogen without an external nitrogen source (Rout et al., 2013). Fungal communities of seeds in oak trees could be shaped vertically by mother trees (Fort et al., 2019). Microbes could take possession of developing seeds vertically from parent trees, then transmit or move from other parts as well as through pollen (Agarwal and Sinclair, 1997). Endophytes were observed in seedlings from sterile seeds and in surface-sterile fruits (Puente et al., 2009b). An overlap of microbial taxa among seeds and fruits was also observed, mainly with *Bacillus* members, between pulp cells and xylem, and along cell walls (Compant et al., 2011).

Microbes could also come into seeds through the male gametophytes (Frank et al., 2017). Microorganisms have been found in and on the surface of pollen from various plants (Ambika Manirajan et al., 2016; Fürnkranz et al., 2012; Jojima et al., 2004; Madmony et al., 2005; Obersteiner et al., 2016). If the microbes in or on pollen come from inside the tree, the transfer to seeds and seedlings should also belong to vertical transfer (Frank et al., 2017). *Enterobacter cloacae* cultivated from surface-disinfected pine pollens such as *Pinus halepensis*, *P. pinea*, and *P. brutia* indicates its origin from the parent tree, and the same microbial species isolates were from *P. brutia* ovules (Madmony et al., 2005). This might indicate a vertical transmission of *Enterobacter* members in pines through pollen.

Vertical transmission could evolve from parents to offspring to make sure the beneficial transmission between mutualistic plant—microbes (Bright and Bulgheresi, 2010). The vertical transmission of microbes from the parent to the offspring via seed and pollen is common while not all these plant-microbes are obligate (Frank et al., 2017).

2.2   Microbiome between parent and offspring trees—Maternal effects on trees

Vertical transmission makes us think of microbiome acquisition between parent and offspring trees. Genetics and environmental circumstances may be the most important factors of the individual tree's phenotype (Vivas et al., 2015). Accumulating evidence shows that the biotic and abiotic factors the parent tree experiences could regulate the development and resistance ability of offspring (Burgess and Marshall, 2014; Germain and Gilbert, 2014). Especially, maternal plants have a remarkably stronger influence on progeny phenotype and health status because a vast number of substances could be directly transferred to seedlings (Rix et al., 2012; Roach and Wulff, 1987).

The maternal biotic environment may have significant effects on the plant phenotype, but the impact of associated microbial communities has hardly been explored (Vivas et al., 2015). It is increasingly realized that the microbiome is an indivisible part of the extended plant genotype and phenotype (Zilber-Rosenberg and Rosenberg, 2008). The microbiome can influence many elements of trees, including physiology, metabolism, and ecological interactions (Vivas et al., 2015).

Environmental maternal effects could affect the offspring without changing the genome DNA sequences (Donohue, 2009). These have impacts on seed traits, germination, seedling performance, tree-pathogen interactions, etc. (Donohue, 2009; Elwell et al., 2011; Holeski et al., 2012; Lopez et al., 2003). Maternal effects could also weaken the negative effects of climate change (Burgess and Marshall, 2014).

The reciprocal effects of the microbiome and host plants would also affect both the plant and the microbes in the next generation by maternal effects. As mentioned above, microbes can be vertically transferred from the mother tree to the offspring through seeds. The transferred maternal endophytes could promote seedling quality and stress resistance, presenting abilities to the offspring (Clay, 1987; Davitt et al., 2011; Novas et al., 2003; Peng et al., 2013). Endophyte vertical transmission in trees was infrequently observed (Ganley and Newcombe, 2006). Fort et al. (2019) reported that fungi varied remarkably among oak populations, and even among trees of the same population. The maternal effects were still vital after seed falling, and both maternal effects and environments can act on the seed microbiome of the sessile oak (Fort et al., 2019).

The mother could distribute different nutrients to the offspring, so the maternal microbes may influence seed provisioning (Vivas et al., 2015). By giving increased nutrients for the growth of flowers, fruits, and seeds, mycorrhizal symbionts could put an important effect on the subsequent generation performance (Nuortila et al., 2004; Stanley et al., 1993; Varga et al., 2013). These influences could also impact the resource investment of the mother seeds.

Epigenetic modifications in trees interacting with microorganisms can change the offspring phenotype (Vivas et al., 2015). For example, for some fungal pathogens, their toxins can induce histone acetylation, affecting plant regulation genes for jasmonic acid and ethylene signaling pathways (Jeon et al., 2014; Zhou et al., 2005). The pathways play vital roles in seed development and the regulation of nutrient transfer, with a subsequent effect on the offspring growth and resistance (Creelman and Mullet, 1995; Matilla, 2000; Nonogaki, 2014).

In short, microbiomes are vital components of the extended phenotypes for trees. The maternal effect influences the structure and composition of plant microbiomes by affecting vertical transmission and nutrient distribution.

3.   Microorganisms of seed from soils

Microbes can also be settled in seeds from horizontal external environments through soils (Frank et al., 2017). Root colonization from soils is the most-studied transfer route. Soil is also the most significant source for plant endophytes (Hardoim et al., 2008; Turner et al., 2013), and a reservoir for the tree microbiome.

When seeds are settling through soil, their development for rooting and growing starts, and a microbial active area encircles the germinating seeds around the spermosphere, where microbes could benefit from germination (Frank et al., 2017; Nelson, 2004; Schiltz et al., 2015). Seeds exude carbon such as sugars, proteins, and fatty acids (Nelson, 2004). These compounds could be energy for microbes, showing the potential to select and form the microbial community of the soil encircling the seeds (Kageyama and Nelson, 2003; Roberts et al., 2009; Simon et al., 2001). These relations and potential selection by the host might be necessary to establish beneficial correlations (Schiltz et al., 2015). The microbiome of soils and roots will be discussed in detail in other chapters of the book.

4.   The relationship between the flower microbiome of trees, insect vectors, pollinators, and other factors

Many microorganisms could spread over limited distances and need help for dispersal such as wind, currents, or relying on other organisms for long-distance distribution (Barton et al., 2010; Belisle et al., 2012; Lussenhop, 1992; Muller et al., 2014;

Whitaker et al., 2003). Pathogens need vectors to spread among plants (Mauck et al., 2012). But it is still not obvious as to what extent the distribution of pathogens depends on specific vectors or their distribution patterns and effects on the microbiome (Burns et al., 2016; Lindström and Langenheder, 2012; Morris et al., 2020).

Microbes are very rich in flowers (Fridman et al., 2012; Vorholt, 2012), and the microbial metabolism could also affect floral properties (de Vega and Herrera, 2012; Herrera et al., 2008) and pollinators (Vannette and Fukami, 2016). Floral nectar microorganisms have limited distribution distance (Belisle et al., 2012). Animals are vital vectors of microbes (Brysch-Herzberg, 2004; Canto and Herrera, 2012; Gilbert, 1980; Herrera et al., 2010; Vannette and Fukami, 2016). Visiting from insects such as bees and birds usually increases the microbial community in nectar, and may alter the microbial structure (Belisle et al., 2012). Environments and microbial interactions may also help shape the nonrandom microorganism community (Álvarez-Pérez et al., 2012; Herrera et al., 2010).

Microorganisms that finally colonized nectar could change the nectar properties such as amino acids, sugars, pH, and volatile organic compounds (Canto and Herrera, 2012; Rering et al., 2019; Vannette and Fukami, 2018) due to pollinator visiting (Schaeffer et al., 2014; Vannette et al., 2013). A nectar yeast, *Metschnikowia reukaufii*, could make it attractive for bumblebees to visit flowers (Schaeffer et al., 2019, 2015; Yang et al., 2019). However, the *Neokomagataea* colonization might decrease pollinator visiting owing to the different impacts on floral attractiveness (Good et al., 2014; Vannette et al., 2013). These studies connected microbes in nectar with impacts on the nectar environments and visiting frequency to variance in microbiome structure (Vannette and Fukami, 2016). It was still not easy to make a prediction on what conditions and variations in distribution will result in specific microbial community composition and function (Germain et al., 2019).

4.1 Floral microbiome

The microbes colonizing flower surfaces have the same origins as those on other plant organs. They can be from the air, dust, wind, soil, and nearby plants or insects (Aleklett et al., 2014; Frank et al., 2017). Wind acts as a distribution agent to microbial communities associated with blossoms on apple flowers (Shade et al., 2013). The fire blight pathogen of apples and pears, *Erwinia amylovora*, primarily infects flowers and could enter floral tissue (Spinelli et al., 2005; Vanneste, 2000; Wilson and Lindow, 1993). *E. amylovora* infections are on the stigmas or hypanthium-secreting nectar, and the microbes go into the plants through the nectary hones (Buban and Orosz-Kovacs, 2003; Farkas et al., 2012; Spinelli et al., 2005). Moreover, flower petals have stomata where microbes might gain entrance.

Microbial horizontal acquirement through flowers can colonize developing seeds and inhabit the offspring (Frank et al., 2017). Mitter et al. (2017) introduced endophytes into seeds by splashing bacterial inoculations directly on the flower. The used bacterial strains can be found in the embryos/seedlings, and the microbes could reproduce and be transmitted to the progeny.

Flowers might provide trees with two transfer ways for a microbiome of aerial origin: insect vectors and wind-pollinated species (Frank et al., 2017). Pollen could also be a vector for horizontal transfer for the host microbiome in wind-pollinated plants (Frank et al., 2017). Pollinators, predators, insects, and herbivores feeding on flowers can all visit flowers (Louda, 1982; McCall and Irwin, 2006; Pellmyr and Thien, 1986). Actually, flowers could be the centers of invertebrate biodiversity, many times (10–10,000) higher than on the leaves (Wardhaugh et al., 2012). Pollinator visiting affects flower microbes, and flowers could act as transfer hubs of pollinator microbes (Aizenberg-Gershtein et al., 2013; McFrederick et al., 2017; Ushio et al., 2015). Evidence indicates that fungal pathogens adopt the pollen transfer route (Marques et al., 2013).

A study has reported microbial abundance, community diversity, and colonization with wind-pollinated and insect-pollinated plant species (Ambika Manirajan et al., 2016). The communities varied remarkably between plants, possibly because of the differences of pollen anatomy, nutrients, or antimicrobial substances of the pollen coat (Ambika Manirajan et al., 2016; Zasloff, 2017). Microbial communities of insect-pollinated plants were more similar with each other than those of wind-pollinated plants, indicating potential pollinator effects on the pollen microbial community (Frank et al., 2017).

4.2 Microbiome and plant-feeding insects

Sap-feeding insects such as Phytoplasma and Xylella can also be plant disease vectors (Nault, 1997; Nault and Ammar, 1989; Weintraub and Beanland, 2006). The insects have piercing-sucking mouth structures enabling them to penetrate tree cells, absorb the components, and pass on pathogens during the process (Frank et al., 2017). The intracellular symbionts of the sap-feeding insects *Cardinium* and *Wolbachia* related with the phloem-feeding whitefly could be passed on between different phloem sap-feeding insects horizontally (Frank et al., 2017; Li et al., 2017; Montillet et al., 2013). Some insects could get phytoplasmas passively during feeding on the phloem of infected plants (Weintraub and Beanland, 2006). Microbes can inhabit the host as endophytes or just for a limited time, and nonpathogens can be transferred among trees through sap feeders (Frank et al., 2017). Sap-feeding insects could also be vectors of beneficial microbes among plants. A beneficial endophyte, *Bacillus pumilus* isolated from the lodgepole pine, showed antagonism to a fungal species of the mountain pine beetle (Adams et al., 2008), and could be transmitted via this way. Moreover, a lot of sap feeders may host some depauperate microbes (Weintraub and Beanland, 2006). The tree microbiome might also affect the microbes and traits of the insect vectors. The research suggests phytoplasmas might bestow some increased fitness to their insect vectors (Weintraub and Beanland, 2006). Beanland et al. (2000) determined that exposure to one microbial strain adds to both the lifespan and fecundity of its female vector *Matsumuratettix quadrilineatus*. More investigations are needed to confirm the significance of sap-feeding vectors of tree microbiomes and their functions.

5. Microbial evolutionary and ecological functional impacts

Microorganisms inhabiting tree reproductive organs represent an astonishingly broad group, and are present in the reproductive organs of all plants around the world (Perlmutter and Bordenstein, 2020). Microorganism symbiotic relationships span the entire range with hosts and differ from transient pathogens to obligate mutualists, performing a variety of functions. Fungal reproductive parasites can manipulate host reproduction; harmful bacteria can damage the reproductive organs; microbial nutritional symbionts give vital nutrients to trees, including reproductive organs; commensal beneficial microbes protect trees and their reproductive organs such as seeds from predation; and many other functions (Frank et al., 2017). Although many specific microorganisms and symbiotic relationships are present in the trees, the whole microbial diversity or microbial community interaction within the host reproductive organs is still scarcely known. Microbe-host symbionts, especially hereditary ones, represent a major research field (Perlmutter and Bordenstein, 2020).

Plant-microorganism symbioses may share many evolutionary principles in other host-microorganism symbioses, such as the host gene expression of reproductive organs by symbiotic modulation (Perlmutter and Bordenstein, 2020). One case is phytoplasma (Sugio et al., 2011a, b). Phytoplasma causes symptoms such as leaf yellowing (Sugio et al., 2011a, b) and generates SAP54, an effector protein, altering the flowers to vegetative structures similar to leaves (MacLean et al., 2014). SAP54 and SAP11 control the plant transcription factors regulating the flower to grow normally (Sugio et al., 2011a, b). Leafhoppers could select infected hosts for oviposition according to the physical changes of plants (MacLean et al., 2014). What's more, bacterial seed endophytes could decide upon the plant environmental niche. A study showed transfers of seed-borne rock-degrading microbes of *cardon actus* (Puente et al., 2009a, b), and the microbes help the host settle on rock surfaces. If treated with antibiotics, the seeds failed to develop; when reinoculated with the microbes, the seed growth then recovered (Puente et al., 2009a). Endophytes can accomplish this by fixing nitrogen, generating many organic acids, and releasing the minerals for growth in harsh conditions (Puente et al., 2009b). Vertically inherited seed bacterial endophytes may include microbes contributing to growth and fitness, promoting the host's ability to live in metal-contaminated mining locations (Sánchez-López et al., 2018). Moreover, some beneficial relationships might lead to microbial genome erosion and a stable association development (Perlmutter and Bordenstein, 2020; Toh et al., 2006). One case is the *Burkholderia* endosymbionts of *Psychotria* plants, endosymbionts transferred vertically and make functions in protection from pathogens (Carlier and Eberl, 2012). In addition, different microorganism taxa might make cooperative interactions in host reproductive organs. The fungal pathogen *Rhizopus* contains endosymbiotic bacteria producing a phytotoxin. With the toxin, the fungal and bacterial symbionts could benefit from plant nutrients, lest microbial dependence and function in host reproductive organs parasitization (Partida-Martinez and Hertweck, 2005).

5.1 Microbial evolution and ecology of the reproductive microbiome

The structure of the microbiome can be formed by inner interactions, transmigration from other microbes, and external environments (Miller et al., 2018; Rowe et al., 2020). Transmigration from other microbes may play a significant role in forming reproductive microbiomes at a large scale by transmission (Rowe et al., 2020). The inclination for rapid microbe evolution could also be a key to shaping the ecological dynamics, and will drive evolutionary changes in turn (Ellner et al., 2011).

The holobiont or hologenome is one great view of a structurally dynamic microbiome and host, assuming the host and all the microbes are a selection unit (Moran and Sloan, 2015). Microbiome mixing also impacts how the structure and functions of the reproductive microbiome covary with tree genetics. With an ecological view, increased microbiome mixing will reduce this covariance. Mixing has fewer impacts on the covariance between the tree and microbial genotypes. Microbes are prone to adapt to tree conditions quickly, increasing host-microbe gene covariances (Gandon et al., 1997; Rowe et al., 2020). The migration influences of local adaption could be enhanced by adjustment to cooccurring microbial members, limiting foreign microbes from invading (Tikhonov, 2016).

Chance events could also largely influence microbial community structure and functions, alleviating the microbiome specificity to host genotypes (Sprockett et al., 2018). The phenotypes of the reproductive microbiome on hosts might not match the tree genotype more than the community structure (Rowe et al., 2020). Functional sameness among taxa could alleviate structure-function connections (Martiny et al., 2015). Microbiomes are complicated ecosystems, usually predominated by strong competitive relations (Coyte et al., 2015). The interactions might be vital in forming microbiome phenotypes on hosts (Rowe et al., 2020). Disturbance and exploitation among microbes could be temporally different traits with coevolution, and can affect pathogenesis through reducing the population and community sizes (Inglis et al., 2012). Several traits chosen for competitions of the microbiome could also directly influence their hosts (O'Brien et al., 2017).

6.   Conclusions and future study

Microbes of reproductive organs have unique associations with their hosts. Microorganism heritability in reproductive organs has the great potential to show multigenerational effects from individual physiology to speciation impacts.

In this chapter, the composition of the microbiome, the routes of transmission, and the evolutionary and ecological effects for microorganisms in the reproductive organs of trees were integrated in Fig. 8.1. Although a lot of progress and achievement has been made, there are still many aspects to investigate in the future.

It is essential to identify and find dynamic composition traits in all microbial groups (besides bacteria and fungi) to the reproductive organ microbiome of many more host trees to intensify additional investigations on symbiotic interactions. For communities in reproductive organs, it will also be significant to characterize the microbial identities as well as the core microbiome taxa vital for evolution, vector control, and conservation efforts.

Regarding microbiome transmission and the evolutionary and ecological effects in reproductive organs of trees, much more work will be needed. Significant research for continued study will include: (a) to what extent does the microbiome really originate from the mother tree? How much is the microbiome transferred horizontally? (b) Does coevolution strongly affect tree-microbiome interactions? What are the key factors for the existence or loss of microbes? (c) What and how does the maternal microbiome affect the formation and microbial interactions of the offspring? (d) How much does the microbiome community contribute to the offspring phenotypes with comparison to the genotype and environment? (e) How do these symbioses vary, considering the correlations with tree host and other microbes, in the reproductive organs? (f) What is the influence extent of the paternal effect?

As shown in this chapter, plant maternal conditions, including microbial communities, affect progeny such as germination, seedlings, and resistance. The promotion of tree fitness is vital for living under hard stressful environments and for tree plantation productivity. Choosing trees resistant to pests and pathogens with a view to the reproductive organ microbiome

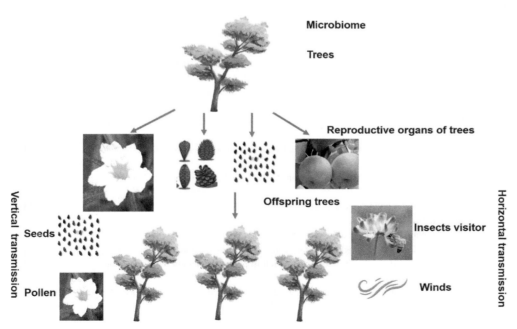

FIG. 8.1 Microbiome in reproductive organs of trees. Reproductive organs of trees include flowers, seeds, fruits, cones, etc. Microbiome from parents to offspring; Vertical transmission via seed and pollen; horizontal transmission by insect visitors and winds.

or microbiome may be a useful measure to enhance plant abilities in harsh environments (Vivas et al., 2015). Understanding the mechanisms of the composition and possible influences of the microbiome interacting with the mother tree's environments and offspring by exposing mother trees to proper environmental conditions could also show great potential as a tool to promote breeding efficiency.

References

Adams, A.S., Six, D.L., Adams, S.M., Holben, W.E., 2008. In vitro interactions between yeasts and bacteria and the fungal symbionts of the mountain pine beetle (*Dendroctonus ponderosae*). Microb. Ecol. 56, 460–466.

Agarwal, V.K., Sinclair, J.B., 1997. Principles of Seed Pathology, second ed. Lewis Publishers, Boca Raton, FL.

Aizenberg-Gershtein, Y., Izhaki, I., Halpern, M., 2013. Do honeybees shape the bacterial community composition in floral nectar? PLoS ONE 8. https://doi.org/10.1371/journal.pone.0067556, e67556.

Aleklett, K., Hart, M., Shade, A., 2014. The microbial ecology of flowers: an emerging frontier in phyllosphere research. Botany 92, 253–266.

Alibrandi, P., Cardinale, M., Rahman, M.M., Strati, F., CinÃ¡, P., de Viana, M.L., et al., 2018. The seed endosphere of *Anadenanthera colubrina* is inhabited by a complex microbiota, including *Methylobacterium* spp. and *Staphylococcus* spp. with potential plant-growth promoting activities. Plant and Soil 422 (1–2), 81–99.

Álvarez-Pérez, S., Herrera, C.M., de Vega, C., 2012. Zooming-in on floral nectar: a first exploration of nectar-associated bacteria in wild plant communities. FEMS Microbiol. Ecol. 80, 591–602.

Ambika Manirajan, B., Ratering, S., Rusch, V., Schwiertz, A., Geissler-Plaum, R., Cardinale, M., Schnell, S., 2016. Bacterial microbiota associated with flower pollen is influenced by pollination type, and shows a high degree of diversity and species-specificity. Environ. Microbiol. 18, 5161–5174.

Azevedo, J.L., Maccheroni, W., Pereira, J.A., Araujo, W.L., 2000. Endophytic microorganisms: a review on insect control and recent advances on tropical plants. Electron. J. Biotechnol. 3, 15–16.

Barton, A.D., Dutkiewicz, S., Flierl, G., Bragg, J., Follows, M.J., 2010. Patterns of diversity in marine phytoplankton. Science 327, 1509–1511.

Baruzzi, F., Cefola, M., Carito, A., Vanadia, S., Calabrese, N., 2012. Changes in bacterial composition of zucchini flowers exposed to refrigeration temperatures. Sci. World J. 2012, 127805. https://doi.org/10.1100/2012/127805.

Beanland, L., Hoy, C.W., Miller, S.A., Nault, L.R., 2000. Influence of aster yellows phytoplasma on the fitness of the aster leafhopper (Homoptera: Cicadellidae). Ann. Entomol. Soc. Am. 93, 271–276.

Belisle, M., Peay, K.G., Fukami, T., 2012. Flowers as islands: spatial distribution of nectarinhabiting microfungi among plants of *Mimulus aurantiacus*, a hummingbird-pollinated shrub. Microb. Ecol. 63, 711–718.

Bowers, R.M., McLetchie, S., Knight, R., Fierer, N., 2011. Spatial variability in airborne bacterial communities across land-use types and their relationship to the bacterial communities of potential source environments. ISME J. 5, 601–612.

Bright, M., Bulgheresi, S., 2010. A complex journey: transmission of microbial symbionts. Nat. Rev. Microbiol. 8, 218–230.

Brysch-Herzberg, M., 2004. Ecology of yeasts in plant–bumblebee mutualism in Central Europe. FEMS Microbiol. Ecol. 50, 87–100.

Buban, T., Orosz-Kovacs, Z., 2003. The nectary as the primary site of infection by *Erwinia amylovora* (Burr.) Winslow et al.: a mini review. Plant Syst. Evol. 238, 183–194.

Burgess, S.C., Marshall, D.J., 2014. Adaptive parental effects: the importance of estimating environmental predictability and offspring fitness appropriately. Oikos 123, 769–776.

Burns, A.R., Stephens, W.Z., Stagaman, K., Wong, S., Rawls, J.F., Guillemin, K., Bohannan, B.J., 2016. Contribution of neutral processes to the assembly of gut microbial communities in the zebrafish over host development. ISME J. 10, 655–664.

Cankar, K., Kraigher, H., Ravnikar, M., Rupnik, M., 2005. Bacterial endophytes from seed of Norway spruce (*Picea abies* L. Karst). FEMS Microbiol. Lett. 244, 341–345.

Canto, A., Herrera, C.M., 2012. Micro-organisms behind the pollination scenes: microbial imprint on floral nectar sugar variation in a tropical plant community. Ann. Bot. 110, 1173–1183.

Carlier, A.L., Eberl, L., 2012. The eroded genome of a *Psychotria* leaf symbiont: hypotheses about lifestyle and interactions with its plant host. Environ. Microbiol. 14, 2757–2769.

Chen, L., Zhang, Q.Y., Jia, M., Ming, Q.L., Yue, W., Rahman, K., Han, T., 2016. Endophytic fungi with antitumor activities: their occurrence and anticancer compounds. Crit. Rev. Microbiol. 42, 454–473.

Clay, K., 1987. Effects of fungal endophytes on the seed and seedling biology of *Lolium perenne* and *Festuca arundinacea*. Oecologia 73, 358–362.

Compant, S., Mitter, B., Colli-Mull, J.G., Gangl, H., Sessitsch, A., 2011. Endophytes of grapevine flowers, berries, and seeds: identification of cultivable bacteria, comparison with other plant parts, and visualization of niches of colonization. Microb. Ecol. 62, 188–197.

Coyte, K.Z., Schluter, J., Foster, K.R., 2015. The ecology of the microbiome: networks, competition, and stability. Science 350, 663–666.

Creelman, R.A., Mullet, J.E., 1995. Jasmonic acid distribution and action in plants: regulation during development and response to biotic and abiotic stress. Proc. Natl. Acad. Sci. U. S. A. 92, 4114–4119.

Davitt, A.J., Chen, C., Rudgers, J.A., 2011. Understanding context-dependency in plant–microbe symbiosis: the influence of abiotic and biotic contexts on host fitness and the rate of symbiont transmission. Environ. Exp. Bot. 71, 137–145.

de Vega, C., Herrera, C.M., 2012. Relationships among nectar-dwelling yeasts, flowers and ants: patterns and incidence on nectar traits. Oikos 121, 1878–1888.

Donohue, K., 2009. Completing the cycle: maternal effects as the missing link in plant life histories. Philos. Trans. R. Soc. Lond. B Biol. Sci. 364, 1059–1074.

Ellner, S.P., Geber, M.A., Hairston Jr., N.G., 2011. Does rapid evolution matter? Measuring the rate of contemporary evolution and its impacts on ecological dynamics. Ecol. Lett. 14, 603–614.

Elwell, A.L., Gronwall, D.S., Miller, N.D., Spalding, E.P., Brooks, T.L.D., 2011. Separating parental environment from seed size effects on next generation growth and development in *Arabidopsis*. Plant Cell Environ. 34, 291–301.

Farkas, Á., Mihalik, E., Dorgai, L., Bubán, T., 2012. Floral traits affecting fire blight infection and management. Trees 26, 47–66.

Ferreira, A., Quecine, M.C., Lacava, P.T., Oda, S., Azevedo, J.L., Araújo, W.L., 2008. Diversity of endophytic bacteria from *Eucalyptus* species seeds and colonization of seedlings by *Pantoea agglomerans*. FEMS Microbiol. Lett. 287, 8–14.

Fort, T., Pauvert, C., Zanne, A.E., Ovaskainen, O., Caignard, T., Barret, M., et al., 2019. Maternal effects and environmental filtering shape seed fungal communities in oak trees. bioRxiv, 691121. https://doi.org/10.1101/691121.

Frank, A.C., Guzman, J.P.S., Shay, J.E., 2017. Transmission of bacterial endophytes. Microorganisms 5 (4), 70. https://doi.org/10.3390/microorganisms5040070.

Fridman, S., Izhaki, I., Gerchman, Y., Halpern, M., 2012. Bacterial communities in floral nectar. Environ. Microbiol. Rep. 4, 97–104.

Fürnkranz, M., Lukesch, B., Müller, H., Huss, H., Grube, M., Berg, G., 2012. Microbial diversity inside pumpkins: microhabitat-specific communities display a high antagonistic potential against phytopathogens. Microb. Ecol. 63, 418–428.

Gagne-Bourgue, F., Aliferis, K.A., Seguin, P., Rani, M., Samson, R., Jabaji, S., 2013. Isolation and characterization of indigenous endophytic bacteria associated with leaves of switchgrass (*Panicum virgatum* L.) cultivars. J. Appl. Microbiol. 114, 836–853.

Gandon, S., Capowiez, Y., Dubois, Y., Michalakis, Y., Olivieri, I., 1997. Local adaptation and gene for-gene coevolution in a metapopulation model. Proc. R. Soc. B 263, 1003–1009.

Ganley, R.J., Newcombe, G., 2006. Fungal endophytes in seeds and needles of *Pinus monticola*. Mycol. Res. 110, 318–327.

Gao, F.K., Dai, C.C., Liu, X.Z., 2010. Mechanisms of fungal endophytes in plant protection against pathogens. Afr. J. Microbiol. Res. 4, 1346–1351.

Germain, R.M., Gilbert, B., 2014. Hidden responses to environmental variation: maternal effects reveal species niche dimensions. Ecol. Lett. 17, 662–669.

Germain, R.M., Jones, N.T., Grainger, T.N., 2019. Cryptic dispersal networks shape biodiversity in an invaded landscape. Ecology. https://doi.org/10.1002/ecy.2738, e02738.

Gilbert, D.G., 1980. Dispersal of yeasts and bacteria by *Drosophila* in a temperate forest. Oecologia 46, 135–137.

Glassner, H., Zchori-Fein, E., Compant, S., Sessitsch, A., Katzir, N., Portnoy, V., Yaron, S., 2015. Characterization of endophytic bacteria from cucurbit fruits with potential benefits to agriculture in melons (*Cucumis melo* L.). FEMS Microbiol. Ecol. 91. https://doi.org/10.1093/femsec/fiv074.

Glassner, H., Zchori-Fein, E., Yaron, S., Sessitsch, A., Sauer, U., Compant, S., 2017. Bacterial niches inside seeds of *Cucumis melo* L. Plant and Soil 422 (1–2), 101–113.

Goggin, D.E., Emery, R.N., Kurepin, L.V., Powles, S.B., 2015. A potential role for endogenous microflora in dormancy release, cytokinin metabolism and the response to fluridone in *Lolium rigidum* seeds. Ann. Bot. 115, 293–301.

Good, A.P., Gauthier, M.P.L., Vannette, R.L., Fukami, T., 2014. Honey bees avoid nectar colonized by three bacterial species, but not by a yeast species, isolated from the bee gut. PLoS One 9. https://doi.org/10.1371/journal.pone.0086494, e86494.

Hacquard, S., Garrido-Oter, R., González, A., Spaepen, S., Ackermann, G., Lebeis, S., et al., 2015. Microbiota and host nutrition across plant and animal kingdoms. Cell Host Microbe 17, 603–616.

Hardoim, P.R., Hardoim, C.C., Van Overbeek, L.S., Van Elsas, J.D., 2012. Dynamics of seed-borne rice endophytes on early plant growth stages. PLoS ONE 7, e30438.

Hardoim, P.R., Van Overbeek, L.S., Berg, G., Pirttilä, A.M., Compant, S., Campisano, A., et al., 2015. The hidden world within plants: ecological and evolutionary considerations for defining functioning of microbial endophytes. Microbiol. Mol. Biol. Rev. 79 (3), 293–320.

Herrera, C.M., García, I.M., Pérez, R., 2008. Invisible floral larcenies: microbial communities degrade floral nectar of bumble bee-pollinated plants. Ecology 89, 2369–2376.

Hardoim, P.R., van Overbeek, L.S., Elsas, J.D., 2008. Properties of bacterial endophytes and their proposed role inplant growth. Trends Microbiol. 16, 463–471.

Herrera, C.M., Canto, A., Pozo, M.I., Bazaga, P., 2010. Inhospitable sweetness: nectar filtering of pollinator-borne inocula leads to impoverished, phylogenetically clustered yeast communities. Proc. Biol. Sci. 277, 747–754.

Heydenreich, B., Bellinghausen, I., König, B., Becker, W.M., Grabbe, S., Petersen, A., Saloga, J., 2012. Gram-positive bacteria on grass pollen exhibit adjuvant activity inducing inflammatory T cell responses. Clin. Exp. Allergy 42, 76–84.

Hodgson, S., de Cates, C., Hodgson, J., Morley, N.J., Sutton, B.C., Gange, A.C., 2014. Vertical transmission of fungal endophytes is widespread in forbs. Ecol. Evol. 4, 1199–1208.

Holeski, L.M., Jander, G., Agrawal, A.A., 2012. Transgenerational defense induction and epigenetic inheritance in plants. Trends Ecol. Evol. 27, 618–626.

Inglis, R.F., Brown, S.P., Buckling, A., 2012. Spite versus cheats: competition among social strategies shapes virulence in *Pseudomonas aeruginosa*. Evolution 66 (11), 3472–3484. https://doi.org/10.1111/j.1558-5646.2012.01706.x.

Jacquemyn, H., Lenaerts, M., Brys, R., Willems, K., Honnay, O., Lievens, B., 2013. Among-population variation in microbial community structure in the floral nectar of the bee-pollinated forest herb *Pulmonaria officinalis* L. PLoS One 8, e56917.

Jeon, J., Kwon, S., Lee, Y.H., 2014. Histone acetylation in fungal pathogens of plants. Plant Pathol. J. 30, 1–9.

Johnston-Monje, D., Raizada, M.N., 2011. Conservation and diversity of seed associated endophytes in Zea across boundaries of evolution, ethnography and ecology. PLoS ONE 6, e20396.

Jojima, Y., Mihara, Y., Suzuki, S., Yokozeki, K., Yamanaka, S., Fudou, R., 2004. *Saccharibacter floricola* gen. nov., sp. nov., a novel osmophilic acetic acid bacterium isolated from pollen. Int. J. Syst. Evol. Microbiol. 54, 2263–2267.

Kageyama, K., Nelson, E.B., 2003. Differential inactivation of seed exudate stimulation of *Pythium ultimum* sporangium germination by *Enterobacter cloacae* influences biological control efficacy on different plant species. Appl. Environ. Microbiol. 69, 1114–1120.

Kemen, E., 2014. Microbe-microbe interactions determine oomycete and fungal host colonization. Curr. Opin. Plant Biol. 20, 75–81.

Lemaire, B., Janssens, S., Smets, E., Dessein, S., 2012. Endosymbiont transmission mode in bacterial leaf nodulation as revealed by a population genetic study of *Psychotria leptophylla*. Appl. Environ. Microbiol. 78, 284–287.

Li, S.J., Ahmed, M.Z., Lv, N., Shi, P.Q., Wang, X.M., Huang, J.L., Qiu, B.L., 2017. Plant mediated horizontal transmission of Wolbachia between whiteflies. ISME J. 11, 1019–1028.

Lindström, E.S., Langenheder, S., 2012. Local and regional factors influencing bacterial community assembly. Environ. Microbiol. Rep. 4, 1–9.

Liu, Y., Zuo, S., Xu, L., Zou, Y., Song, W., 2012. Study on diversity of endophytic bacterial communities in seeds of hybrid maize and their parental lines. Arch. Microbiol. 194, 1001–1012.

Liu, Y., Zuo, S., Zou, Y., Wang, J., Song, W., 2013. Investigation on diversity and population succession dynamics of endophytic bacteria from seeds of maize (*Zea mays* L., Nongda 108) at different growth stages. Ann. Microbiol. 63, 71–79.

Lopez, G.A., Potts, B.M., Vaillancourt, R.E., Apiolaza, L.A., 2003. Maternal and carryover effects on early growth of *Eucalyptus globulus*. Can. J. For. Res. 33, 2108–2115.

López-López, A., Rogel, M.A., Ormeño-Orillo, E., Martínez-Romero, J., Martínez-Romero, E., 2010. *Phaseolus vulgaris* seed-borne endophytic community with novel bacterial species such as *Rhizobium endophyticum* sp. nov. Syst. Appl. Microbiol. 33, 322–327.

Louda, S.M., 1982. Inflorescence spiders: a cost/benefit analysis for the host plant, *Haplopappus venetus* Blake (Asteraceae). Oecologia 55, 185–191.

Lussenhop, J., 1992. Mechanisms of microarthropod-microbial interactions in soil. Adv. Ecol. Res. 23, 1–33.

MacLean, A.M., Orlovskis, Z., Kowitwanich, K., Zdziarska, A.M., Angenent, G.C., Immink, R.G., Hogenhout, S.A., 2014. Phytoplasma effector SAP54 hijacks plant reproduction by degrading MADS-box proteins and promotes insect colonization in a RAD23-dependent manner. PLoS Biol. 12. https://doi.org/10.1371/journal.pbio.1001835, e1001835.

Madmony, A., Chernin, L., Pleban, S., Peleg, E., Riov, J., 2005. *Enterobacter cloacae*, an obligatory endophyte of pollen grains of Mediterranean pines. Folia Microbiol. 50, 209–216.

Mano, H., Tanaka, F., Watanabe, A., Kaga, H., Okunishi, S., Morisaki, H., 2006. Culturable surface and endophytic bacterial flora of the maturing seeds of rice plants (*Oryza sativa*) cultivated in a paddy field. Microbes Environ. 2, 86–100.

Marques, J.P.R., Amorim, L., Spósito, M.B., Marin, D., Appezzato-da-Glória, B., 2013. Infection of citrus pollen grains by *Colletotrichum acutatum*. Eur. J. Plant Pathol. 136, 35–40.

Martiny, J.B.H., Jones, S.E., Martiny, A.C., 2015. Microbiomes in light of traits: a phylogenetic perspective. Science 350. https://doi.org/10.1126/science.aac9323, aac9323.

Matilla, A.J., 2000. Ethylene in seed formation and germination. Seed Sci. Res. 10, 111–126.

Mauck, K., Bosque-Pérez, N.A., Eigenbrode, S.D., De Moraes, C.M., Mescher, M.C., 2012. Transmission mechanisms shape pathogen effects on host-vector interactions: evidence from plant viruses. Funct. Ecol. 26, 1162–1175.

McCall, A.C., Irwin, R.E., 2006. Florivory: the intersection of pollination and herbivory. Ecol. Lett. 9, 1351–1365.

McFrederick, Q.S., Thomas, J.M., Neff, J.L., Vuong, H.Q., Russell, K.A., Hale, A.R., Mueller, U.G., 2017. Flowers and wild megachilid bees share microbes. Microb. Ecol. 73, 188–200.

Miller, I.M., 1990. Bacterial leaf nodule symbiosis. Adv. Bot. Res. 17 (8), 163–234.

Miller, E.T., Svanbäck, R., Bohannan, B.J.M., 2018. Microbiomes as metacommunities: understanding host-associated microbes through metacommunity ecology. Trends Ecol. Evol. 33, 926–935.

Mitter, B., Pfaffenbichler, N., Flavell, R., Compant, S., Antonielli, L., Petric, A., Berninger, T., Naveed, M., Sheibani-Tezerji, R., von Maltzahn, G., 2017. A new approach to modify plant microbiomes and traits by introducing beneficial bacteria at flowering into progeny seeds. Front. Microbiol. 8. https://doi.org/10.3389/fmicb.2017.00011.

Montillet, J.L., Leonhardt, N., Mondy, S., Tranchimand, S., Rumeau, D., Boudsocq, M., et al., 2013. An abscisic acid-independent oxylipin pathway controls stomatal closure and immune defense in Arabidopsis. PLoS Biol. 11, e1001513.

Moran, N.A., Sloan, D.B., 2015. The hologenome concept: helpful or hollow? PLoS Biol. 13, e1002311.

Morris, M.M., Frixione, N.J., Burkert, A.C., Dinsdale, E.A., Vannette, R.L., 2020. Microbial abundance, composition, and function in nectar are shaped by flower visitor identity. FEMS Microbiol. Ecol. 3, 3. https://doi.org/10.1093/femsec/fiaa003.

Mukhopadhyay, K., Garrison, N.K., Hinton, D.M., Bacon, C.W., Khush, G.S., Peck, H.D., Datta, N., 1996. Identification and characterization of bacterial endophytes of rice. Mycopathologia 134, 151–159.

Muller, A.L., de Rezende, J.R., Hubert, C.R.J., Kjeldsen, K.U., Lagkouvardos, I., Berry, D., Jørgensen, B.B., Loy, A., 2014. Endospores of thermophilic bacteria as tracers of microbial dispersal by ocean currents. ISME J. 8, 1153–1165.

Muller, D.B., Vogel, C., Bai, Y., Vorholt, J.A., 2016. The plant microbiota: systems-level insights and perspectives. Annu. Rev. Genet. 50, 211–234.

Mundt, J.O., Hinkle, N.F., 1976. Bacteria within ovules and seeds. Appl. Environ. Microbiol. 32, 694–698.

Nault, L.R., 1997. Arthropod transmission of plant viruses: a new synthesis. Ann. Entomol. Soc. Am. 90, 521–541.

Nault, L.R., Ammar, E.D., 1989. Leafhopper and planthopper transmission of plant viruses. Annu. Rev. Entomol. 34, 503–529.

Nelson, E.B., 2004. Microbial dynamics and interactions in the spermosphere. Annu. Rev. Phytopathol. 42, 271–309.

Nelson, E.B., 2017. The seed microbiome: origins, interactions, and impacts. Plant and Soil 422, 7–34. https://doi.org/10.1007/s11104-017-3289-7.

Nonogaki, H., 2014. Seed dormancy and germination—emerging mechanisms and new hypotheses. Front. Plant Sci. 5, 233.

Novas, M.V., Gentile, A., Cabral, D., 2003. Comparative study of growth parameters on diaspores and seedlings between populations of *Bromus setifolius* from Patagonia, differing in Neotyphodium endophyte infection. Flora 198, 421–426.

Nuortila, C., Kytöviita, M.M., Tuomi, J., 2004. Mycorrhizal symbiosis has contrasting effects on fitness components in *Campanula rotundifolia*. New Phytol. 164, 543–553.

O'Brien, S., Luján, A.M., Paterson, S., 2017. Adaptation to public goods cheats in *Pseudomonas aeruginosa*. Proc. Biol. Sci. 284, 20171089. https://doi.org/10.1098/rspb.2017.1089.

Obersteiner, A., Gilles, S., Frank, U., Beck, I., Häring, F., Ernst, D., et al., 2016. Pollen-associated microbiome correlates with pollution parameters and the allergenicity of pollen. PLoS One 11, e0149545.

Partida-Martinez, L.P., Hertweck, C., 2005. Pathogenic fungus harbours endosymbiotic bacteria for toxin production. Nature 437, 884.

Pellmyr, O., Thien, L.B., 1986. Insect reproduction and floral fragrances: keys to the evolution of the angiosperms? Taxon 35, 76.

Peng, Q.Q., Li, C.J., Song, M.L., 2013. Effects of seed hydropriming on growth of *Festuca sinensis* infected with Neotyphodium endophyte. Fungal Ecol. 6, 83–91.

Perlmutter, J.I., Bordenstein, S.R., 2020. Microorganisms in the reproductive tissues of arthropods. Nat. Rev. Microbiol. 18, 97–111.

Pirttila, A.M., Frank, A.C., 2011. Endophytes of Forest Trees. Springer, https://doi.org/10.1007/978-94-007-1599-8 (ebook).

Puente, M.E., Li, C.Y., Bashan, Y., 2009a. Endophytic bacteria in cacti seeds can improve the development of cactus seedlings. Environ. Exp. Bot. 66, 402–408.

Puente, M.E., Li, C.Y., Bashan, Y., 2009b. Rock-degrading endophytic bacteria in cacti. Environ. Exp. Bot. 66, 389–401.

Ren, F., Dong, W., Yan, D.H., 2019a. Endophytic bacterial communities of Jingbai pear trees in North China analyzed with Illumina sequencing of 16S rDNA. Arch. Microbiol. 201 (2), 199–208.

Ren, F., Dong, W., Sun, H., Yan, D.H., 2019b. Endophytic mycobiota of Jingbai Pear trees in North China. Forests 10 (3), 260.

Ren, F., Dong, W., Yan, D.H., 2019c. Organs, cultivars, soil, and fruit properties affect structure of endophytic mycobiota of Pinggu peach trees. Microorganisms 7 (9), 322.

Rering, C.C., Beck, J.J., Hall, G.W., McCartney, M.M., Vannette, R.L., 2019. Nectar-inhabiting microorganisms influence nectar volatile composition and attractiveness to a generalist pollinator. New Phytol. 220 (3), 750–759.

Rix, K.D., Gracie, A.J., Potts, B.M., 2012. Paternal and maternal effects on the response of seed germination to high temperatures in *Eucalyptus globulus*. Ann. For. Sci. 69, 673–679.

Roach, D.A., Wulff, R.D., 1987. Maternal effects in plants. Annu. Rev. Ecol. Syst. 18, 209–235.

Roberts, D.P., Baker, C.J., McKenna, L., Liu, S., Buyer, J.S., Kobayashi, D.Y., 2009. Influence of host seed on metabolic activity of Enterobacter cloacae in the spermosphere. Soil Biol. Biochem. 41, 754–761.

Rout, M.E., Chrzanowski, T.H., Westlie, T.K., Deluca, T.H., Callaway, R.M., Holben, W.E., 2013. Bacterial endophytes enhance competition by invasive plants. Am. J. Bot. 100, 1726–1737.

Rowe, M., Veerus, L., Trosvik, P., Buckling, A., Pizzari, T., 2020. The reproductive microbiome: an emerging driver of sexual selection, sexual conflict, mating systems, and reproductive isolation. Trends Ecol. Evol. 35, 220–234.

Saikkonen, K., Saari, S., Helander, M., 2010. Defensive mutualism between plants and endophytic fungi? Fungal Divers. 41, 101–113.

Sánchez-López, A.S., Sofie, T., Bram, B., 2018. Community structure and diversity of endophytic bacteria in seeds of three consecutive generations of *Crotalaria pumila* growing on metal mine residues. Plant and Soil 422, 51–66.

Schaeffer, R.N., Phillips, C.R., Duryea, M.C., Andicoechea, J., Irwin, R.E., 2014. Nectar yeasts in the tall larkspur *Delphinium barbeyi* (Ranunculaceae) and effects on components of pollinator foraging behavior. PLoS ONE 9, e108214.

Schaeffer, R.N., Vannette, R.L., Irwin, R.E., 2015. Nectar yeasts in *Delphinium nuttallianum* (Ranunculaceae) and their effects on nectar quality. Fungal Ecol. 18, 100–106.

Schaeffer, R.N., Rering, C.C., Maalouf, I., Beck, J.J., Vannette, R.L., 2019. Microbial metabolites mediate bumble bee attraction and feeding. bioRxiv, 549279.

Schiltz, S., Gaillard, I., Pawlicki-Jullian, N., Thiombiano, B., Mesnard, F., Gontier, E., 2015. A review: what is the spermosphere and how can it be studied? J. Appl. Microbiol. 119, 1467–1481.

Schoenmakers, S., Steegers-Theunissen, R., Faas, M., 2019. The matter of the reproductive microbiome. Obstet. Med. 12 (3), 107–115.

Shade, A., McManus, P.S., Handelsman, J., 2013. Unexpected diversity during community succession in the apple flower microbiome. MBio 4, e00602–e00612.

Shade, A., Jacques, M.A., Barret, M., 2017. Ecological patterns of seed microbiome diversity, transmission, and assembly. Curr. Opin. Microbiol. 37, 15–22.

Shahzad, R., Khan, A.L., Bilal, S., Asaf, S., Lee, I.J., 2018. What is there in seeds? Vertically transmitted endophytic resources for sustainable improvement in plant growth. Front. Plant Sci. 9, 24. https://doi.org/10.3389/fpls.2018.00024.

Simon, H.M., Smith, K.P., Dodsworth, J.A., Guenthner, B., Handelsman, J., Goodman, R.M., 2001. Influence of tomato genotype on growth of inoculated and indigenous bacteria in the spermosphere. Appl. Environ. Microbiol. 67, 514–520.

Spinelli, F., Ciampolini, F., Cresti, M., Geider, K., Costa, G., 2005. Influence of stigmatic morphology on flower colonization by *Erwinia amylovora* and *Pantoea agglomerans*. Eur. J. Plant Pathol. 113, 395–405.

Sprockett, D., Fukami, T., Relman, D.A., 2018. Role of priority effects in the early-life assembly of the gut microbiota. Nat. Rev. Gastroenterol. Hepatol. 15, 197–205.

Stanley, M.R., Koide, R.T., Shumway, D.L., 1993. Mycorrhizal symbiosis increases growth, reproduction and recruitment of *Abutilon theophrasti* medic. in the field. Oecologia 94, 30–35.

Sugio, A., Kingdom, H.N., MacLean, A.M., Grieve, V.M., Hogenhout, S.A., 2011a. Phytoplasma protein effector SAP11 enhances insect vector reproduction by manipulating plant development and defense hormone biosynthesis. Proc. Natl. Acad. Sci. U. S. A. 108 (48), 1254–1263.

Sugio, A., MacLean, A.M., Kingdom, H.N., Grieve, V.M., Manimekalai, R., Hogenhout, S.A., 2011b. Diverse targets of phytoplasma effectors: from plant development to defense against insects. Annu. Rev. Phytopathol. 49, 175–195.

Tikhonov, M., 2016. Community-level cohesion without cooperation. Elife 5, e15747.

Toh, H., Weiss, B.L., Perkin, S.A.H., Yamashita, A., Oshima, K., Hattori, M., Aksoy, S., 2006. Massive genome erosion and functional adaptations provide insights into the symbiotic lifestyle of *Sodalis glossinidius* in the tsetse host. Genome Res. 16, 149–156.

Truyens, S., Weyens, N., Cuypers, A., Vangronsveld, J., 2014. Bacterial seed endophytes: genera, vertical transmission and interaction with plants. Environ. Microbiol. 7, 40–50.

Turner, T.R., James, E.K., Poole, P.S., 2013. The plant microbiome. Genome Biol. 14, 209.

Ushio, M., Yamasaki, E., Takasu, H., Nagano, A.J., Fujinaga, S., Honjo, M.N., et al., 2015. Microbial communities on flower surfaces act as signatures of pollinator visitation. Sci. Rep. 5 (1), 1–7.

Vanneste, J.L., 2000. Fire Blight: The Disease and its Causative Agent, *Erwinia amylovora*. CABI, New York.

Vannette, R.L., Fukami, T., 2016. Nectar microbes can reduce secondary metabolites in nectar and alter effects on nectar consumption by pollinators. Ecology 97, 1410–1419.

Vannette, R.L., Fukami, T., 2018. Contrasting effects of yeasts and bacteria on floral nectar traits. Ann. Bot. 121, 1343–1349.

Vannette, R.L., Gauthier, M.P., Fukami, T., 2013. Nectar bacteria, but not yeast, weaken a plant-pollinator mutualism. Proc. Natl. Acad. Sci. 280, 20122601.

Varga, S., Vega-Frutis, R., Kytöviita, M.M., 2013. Transgenerational effects of plant sex and arbuscular mycorrhizal symbiosis. New Phytol. 199, 812–821.

Verma, S.K., White, J.F., 2019. Seed Endophytes. Springer, Cham, Switzerland.

Vivas, M., Kemler, M., Slippers, B., 2015. Maternal effects on tree phenotypes: considering the microbiome. Trends Plant Sci. 20 (9), 541–544.

Vorholt, J.A., 2012. Microbial life in the phyllosphere. Nat. Rev. Microbiol. 10, 828.

Wardhaugh, C.W., Stork, N.E., Edwards, W., Grimbacher, P.S., 2012. The overlooked biodiversity of flower-visiting invertebrates. PLoS ONE 7, e45796.

Weintraub, P.G., Beanland, L., 2006. Insect vectors of phytoplasmas. Annu. Rev. Entomol. 51, 91–111.

Whitaker, R.J., Grogan, D.W., Taylor, J.W., 2003. Geographic barriers isolate endemic populations of hyperthermophilic archaea. Science 301, 976–978.

Wilson, M., Lindow, S.E., 1993. Interactions between the biological control agent *Pseudomonas fluorescens* A506 and *Erwinia amylovora* in pear blossoms. Phytopathology 83, 117.

Xu, J., Ebada, S.S., Proksch, P., 2010. *Pestalotiopsis* a highly creative genus: chemistry and bioactivity of secondary metabolites. Fungal Divers. 44, 15–31.

Yang, M., Deng, G.C., Gong, Y.B., Huang, S.Q., 2019. Nectar yeasts enhance the interaction between *Clematis akebioides* and its bumblebee pollinator. Plant Biol. 21, 732–737.

Zarraonaindia, I., Owens, S.M., Weisenhorn, P., West, K., Hampton-Marcell, J., Lax, S., et al., 2015. The soil microbiome influences grapevine-associated microbiota. MBio 6, e02527-14.

Zasloff, M., 2017. Pollen has a microbiome: implications for plant reproduction, insect pollination and human allergies: pollen has a microbiome. Environ. Microbiol. 19, 1–2.

Zawoznik, M.S., Vázquez, S.C., Díaz Herrera, S.M., Groppa, M.D., 2014. Search for endophytic diazotrophs in barley seeds. Braz. J. Microbiol. 45, 621–625.

Zhou, C., Zhang, L., Duan, J., Miki, B., Wu, K., 2005. Histone deacetylase19 is involved in jasmonic acid and ethylene signaling of pathogen response in Arabidopsis. Plant Cell 17, 1196–1204.

Zilber-Rosenberg, I., Rosenberg, E., 2008. Role of microorganisms in the evolution of animals and plants: the hologenome theory of evolution. FEMS Microbiol. Rev. 32, 723–735.

Section C

Endosphere microbiome

Chapter 9

Bacterial biota of forest trees

Bethany J. Pettifor and James E. McDonald

School of Natural Sciences, Bangor University, Bangor, Gwynedd, United Kingdom

1. Introduction

Bacteria are ubiquitous throughout above- and below-ground compartments of forest biomes, inhabiting the foliage, outer bark and woody tissues of living trees, and the roots, rhizosphere, and forest soils (Baldrian, 2017a; Griffin and Carson, 2015). In addition to wider roles in ecosystem processes such as nutrient cycling and mineralization in forest environments (Kaiser et al., 2016; Lladó et al., 2017; Ren et al., 2015), bacteria in combination with other Archaea, fungi, and protists are also key members of the tree microbiota (the collection of living organisms in an ecosystem), with supporting functions in promoting tree health and productivity through nutrient acquisition, stress resistance, immune regulation, and disease suppression (Turner et al., 2013). The tree microbiome can therefore be defined as the collection of living microorganisms (the microbiota) and viruses that occupy reasonably well-defined habitats within the different tree compartments and have distinct physicochemical properties (Berg et al., 2020). Although definitions of the microbiome are evolving in line with methodological and conceptual advances in the field, the original definition "a characteristic microbial community occupying a reasonably well-defined habitat which has distinct physio-chemical properties. The term thus not only refers to the microorganisms involved but also encompasses their theatre of activity" described by Whipps et al. (1988) is arguably the most comprehensive and widely accepted, as it takes into account the "theatre of activity" of the microbiota and also integrates host, evolutionary, and ecological factors into the concept of the microbiome (Berg et al., 2020).

All tree-associated tissues harbor microbiota, and tree microbiomes occupy the phyllosphere (the aerial plant surfaces), the endosphere (the internal tree tissues), and the rhizosphere (below-ground surfaces and soil regions associated with tree roots, and influenced by nutrient exchange) (Turner et al., 2013). The phyllosphere includes all parts of the tree that are present above the ground level, including stem, foliage, flowers, and fruit, and experience extremes of temperature, rainfall, and moisture, in addition to relatively low nutrient availability (Vorholt, 2012). The rhizosphere refers to all parts of the tree system that are below the ground level, and can be comprised of the root endosphere, the rhizoplane (interface between the root and the rhizosphere soil), the rhizosphere soil, and the bulk soil (Hacquard et al., 2015), and is associated with nutrient exchange through tree root exudation (Turner et al., 2013). The endosphere comprises the above- and below-ground internal tissues of the tree, where some bacteria and other microorganisms such as fungi are able to colonize and survive (Fernández-González et al., 2019). These heterogeneous compartments each have distinctive roles in the functioning of the tree system and are subject to a wide range of physical conditions, influenced by factors such as tree species and geographic region, which can result in differences in bacterial community composition (Beckers et al., 2016; Haas et al., 2018).

The tree microbiome is a major determinant of plant health and productivity, with roles in the competitive exclusion of pathogens, immune priming, nitrogen fixation, nutrient acquisition, modification of root architecture, and modulating systemic immune responses and metabolism to increase pathogen resistance or avoid herbivory (reviewed by Hacquard and Schadt (2015)). Consequently, the microbiota are considered as an extended phenotype of the host, and eukaryotes are

Forest Microbiology. https://doi.org/10.1016/B978-0-12-822542-4.00019-X

increasingly considered as meta-organisms that have coevolved with their associated microbiota for thousands of years and represent an inseparable functional unit, called the "holobiont" (Berg et al., 2020; Rosenberg et al., 2016). The composition and function of the tree microbiome is therefore influenced by the host, its environment, and interactions with other microorganisms (both pathogens and mutualistic symbionts). Tree species are distributed across tropical, temperate, and boreal biomes, each with varying ecological properties that ultimately influence the assembly and composition of the tree microbiota due to the prevailing above- and below-ground physical and chemical conditions, and host genetic influences. In this chapter, we consider the bacterial biota associated with the phyllosphere, endosphere, and rhizosphere of forest tree species across Mediterranean, temperate, and boreal forest biomes.

1.1 Aims

The overarching aim of this work was to present a synthesis of the available literature regarding (1) the composition of bacterial biota in tree microbiomes, and (2) the functional roles of bacterial biota in tree microbiomes. Additional aims were to (3) compare the composition of bacterial biota across different tree species and forest biomes (Mediterranean, temperate, and boreal) and (4) to identify key knowledge gaps in our understanding of the bacterial biota of forest trees and make recommendations for future research. To address these aims we compiled published data on the composition and function of bacterial biota in Mediterranean, temperate, and boreal tree species. Out of 243 publications that discussed the bacterial biota of forest trees, 43 peer-reviewed publications were selected on the basis that they described the association of specific named bacterial taxa with tree species found in the biomes studied. The remaining publications did not name the association of specific taxa with tree species, for example they discussed the differences in alpha and beta diversity across different conditions or after climatic events but didn't describe taxonomic composition and could therefore not be included in the analysis.

2. Composition and function of bacterial biota of the above- and below-ground compartments of forest tree species

In order to understand the composition and function of bacterial biota associated with forest trees, the tree system must be broken down into its individual ecosystems that harbor microbiomes. These different compartments of the tree host include the foliage, stem and the root and rhizosphere systems, which are physiologically diverse habitats, subject to a range of conditions, and the bacterial communities that inhabit these compartments are just as diverse. A total of 46 phyla (included in this count were the classes Alphaproteobacteria, Betaproteobacteria, Gammaproteobacteria, and Deltaproteobacteria) were identified in the bacterial communities of forest trees, with a number of phyla being unique to particular compartments (Fig. 9.1) with diverse functions (Table 9.1).

2.1 Foliage

Analysis of the foliar bacterial community across a number of tree species and three geographical regions (Fig. 9.2) highlighted that bacteria associated with tree foliage are diverse, both with regards to taxa and function, including species with roles in pathogen defense, growth promotion, and nutrient acquisition (Peñuelas and Terradas, 2014). At this interface between the plant and the aerosphere, there is a high availability of gasses such as oxygen and carbon dioxide, as well as low

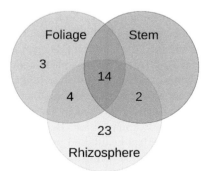

FIG. 9.1 Venn diagram illustrating overlap in bacterial Phylum composition across tree compartments in Mediterranean, boreal, and temperate biomes (combined).

TABLE 9.1 Functional roles of key bacterial biota across tree compartments.

Key bacterial biota	Functional roles	Tree compartment
Actinobacteria	Plant pathogens Mutualistic symbionts Antimicrobial production (biological control agents) Plant growth promotion (e.g., nutrient acquisition/ammonia production)	Foliage, stem, rhizosphere
Acidobacteria	Carbon cycling Carbohydrate metabolism Nitrogen fixation Plant growth promotion (e.g., nutrient acquisition/ammonia production)	Stem, rhizosphere
Bacteroidetes	Organic matter degradation Mutual symbionts Plant growth promotion (e.g., nutrient acquisition/ammonia production)	Foliage, stem, rhizosphere
Proteobacteria	Isoprene degradation Nutrient cycling Indicator species for soil condition Plant growth promotion (e.g., nutrient acquisition/ammonia production)	Foliage, stem, rhizosphere
Deinococcus-Thermus	Radiation resistance	Foliage
Firmicutes	Diverse survival strategies Plant growth promotion (e.g., nutrient acquisition/ammonia production)	Foliage, rhizosphere
Chloroflexi	Carbon dioxide fixation Nitrification	Rhizosphere
Nitrospirae	Carbon dioxide fixation Nitrification	Rhizosphere

levels of available water and nutrients, which along with factors such as leaf age, positioning, and chemical characteristics, can heavily influence bacterial community composition both on the surface and within the tissues (Hacquard et al., 2015; Yadav et al., 2005, 2011).

A total of 21 phyla (including Proteobacteria presented as separate classes) were found to be present in the foliar bacterial community. Table 9.2 highlights the percentage range of total 16S rRNA gene sequences assigned to each taxon in studies where this data was presented. Most commonly observed was the phylum Actinobacteria, members of which hold a number of roles in the foliar microbiome, including plant pathogens and mutualistic symbionts. Members of this phylum have also been identified as abundant producers of secondary metabolites and antimicrobial compounds, making them ideal competitors for the nutrient-limited environment of the leaf surface (Ait Barka et al., 2015; Sharma and Thakur, 2020). Another phylum common to the leaf microbiome is the diverse Bacteroidetes, members of which have been isolated from a number of forest environments, and are known to perform a number of key forest ecosystem functions including organic matter degradation and symbioses with various organisms (Hahnke et al., 2016; Shaffer et al., 2017). The phylum Proteobacteria (classes Alphaproteobacteria, Betaproteobacteria, Gammaproteobacteria, and Deltaproteobacteria) is common to the leaf microbiome, independent of tree species or geographical region; however, they have been studied in close association with tree species known to produce isoprene, a climate-active volatile compound. Some members of the phylum Proteobacteria have been found to have the ability to degrade isoprene, making them a high priority for climate research (Crombie et al., 2018; Fini et al., 2017; Larke-Mejía et al., 2019). Members of the phylum Deinococcus-Thermus, although known primarily to be extremophiles, resistant to various types of radiation (Theodorakopoulos et al., 2013), have also been identified as core members of the leaf microbiome across a number of plant species (Espenshade et al., 2019). The Firmicutes is also a phylum common to the leaf microbiome across a number of plant species and geographical regions; this diverse phylum has members capable of a number of survival strategies and functions, making them ideal for surviving

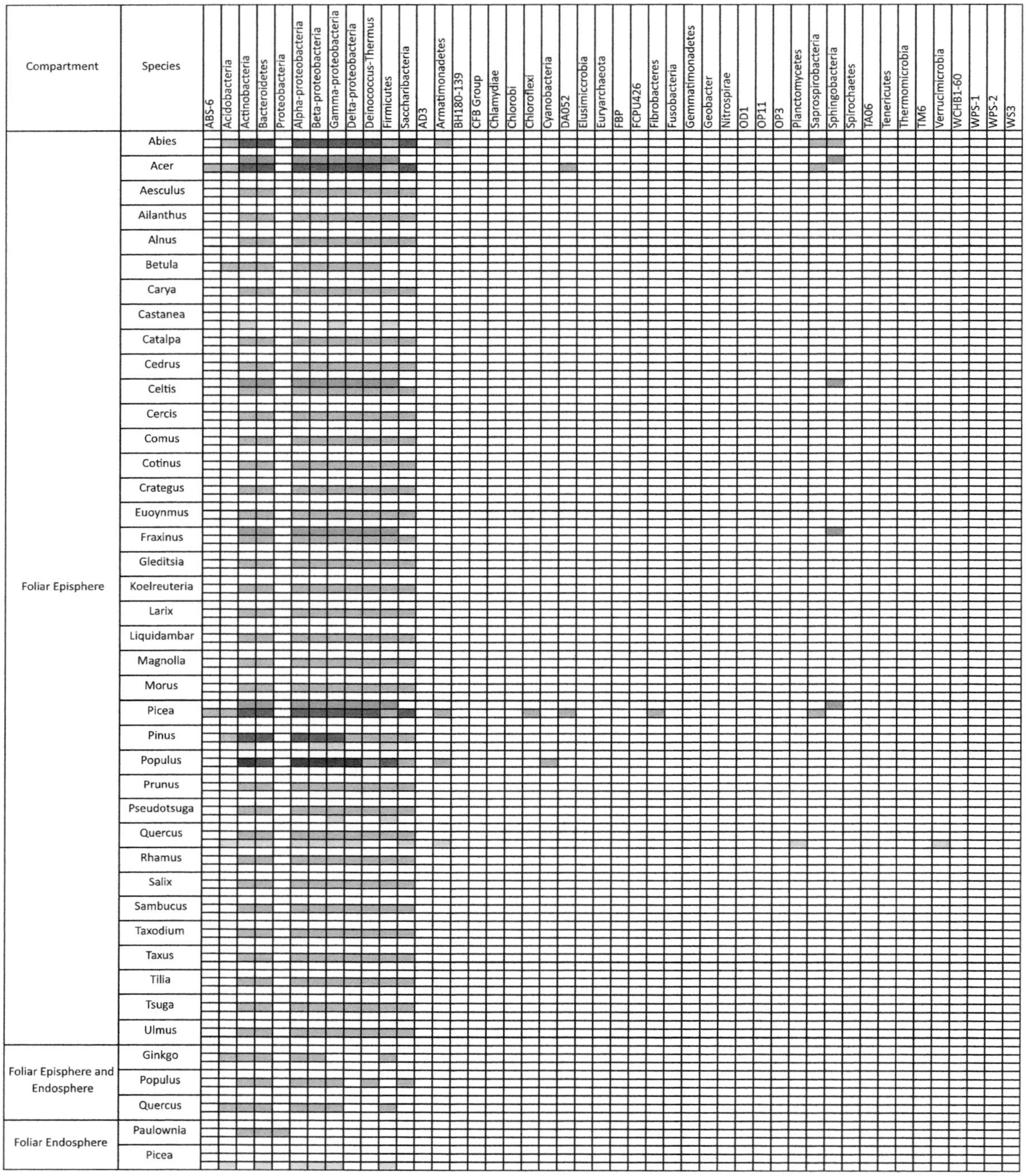

FIG. 9.2 Heatmap displaying the presence of bacterial phyla in the foliar microbiome of different tree species. *Blue squares* indicate data from a Boreal forest, *green squares* indicate data from a Temperate forest, and *yellow squares* indicate data from a Mediterranean forest. The intensity of the color is illustrative of the number of studies represented (light=1 study, mid=2 studies, dark=3 or more studies).

TABLE 9.2 Approximate relative abundance (%) of bacterial phyla across tree compartments.

Bacterial phylum	Tree compartment		
	Leaves	Stems	Rhizosphere
Acidobacteria	<1–2	<1–13	2–38
Actinobacteria	4.5–59	2–42	2–57
Bacteroidetes	2.5–70	<1–38	<1–42
Proteobacteria	85–95	36–85	22–80
Alphaproteobacteria	40–68	20–58	4–27
Betaproteobacteria	3–45	<1–30	4–27
Gammaproteobacteria	1–65	3–59	3–52
Deltaproteobacteria	<1–10	<1	1–6
Deinococcus-Thermus	9–32	12	
Firmicutes	<1–15	1–10	<1–60
Saccharibacteria (TM7)	<1–50	<1–4	<1–5
AD3	<1	<1	<1–3
Armatimonadetes	<1–17	<1–6	<1
CFB Group			3
Chlamydiae			<1
Chlorobi			<1–16
Chloroflexi	<1	<1	<1–25
Cyanobacteria	2		<1–6.5
Euryarchaeota			<1
FBP	<1	<1	<1
Fusobacteria	<1	<1	<1–2.5
Gemmatimonadetes	<1	<1	<1–15
Nitrospirae	<1	<1	<1–5
OP10			<1–1
Planctomycetes	<1–12	<1–2	<1–5
Saprospirobacteria	<1–5		4–6
Sphingobacteria	<1–27		13
Spirochaetes			<1
TM6	<1	<1	<1–1
Verrucomicrobia	<1–10	<1	<1–8.5
WPS-1			<1
WPS-2			<1
WS3	<1	<1	<1–1.5

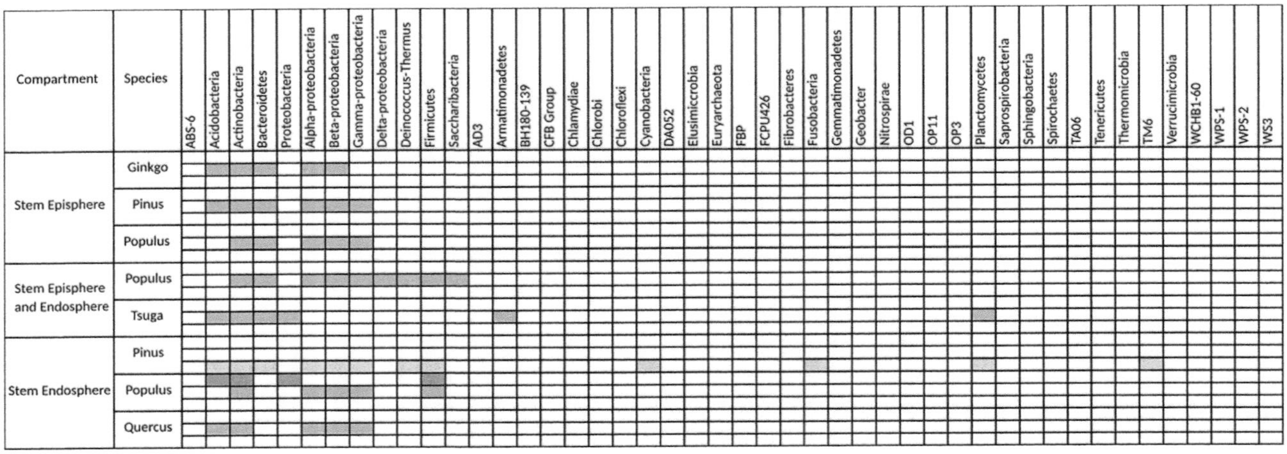

FIG. 9.3 Heatmap displaying the presence of bacterial phyla in the stem microbiome of different tree species. *Blue squares* indicate data from a Boreal forest, *green squares* indicate data from a Temperate forest, and *yellow squares* indicate data from a Mediterranean forest. The intensity of the color is illustrative of the number of studies represented (light = 1 study, mid = 2 studies, dark = 3 or more studies).

in the ever-changing foliar environment of a forest tree (Filippidou et al., 2016). The phylum Saccharibacteria, previously known as candidate phylum TM7, although common across a number of ecosystems (Starr et al., 2018) and the majority of foliar microbiomes, is relatively understudied with regards to functionality in the foliar environment.

2.2 Stem

Analysis of stem bacterial community composition in tree species (Fig. 9.3) revealed that, although classed as part of the phyllosphere, the stem microbiome is considerably different to the leaf microbiome. A lower species richness indicates that stem bacteria are generally more specialized to inhabiting such a harsh, nutrient-poor environment (Beckers et al., 2017; Kobayashi and Aoyagi, 2019).

Of the total 46 phyla (including Proteobacteria presented as separate classes) identified as members of the tree microbiome, there was a relatively low number of phyla associated with the stem tissues and an absence of phyla unique to the stem microbiome. Phyla common to the stem microbiome include Acidobacteria, Actinobacteria, Bacteroidetes, and three classes of Proteobacteria (Alphaproteobacteria, Betaproteobacteria, and Gammaproteobacteria); however, none were exclusive to the stem microbiome. Members of the phylum Acidobacteria have generally been isolated from soil and sediment, however are known to be key members of a number of microbiomes across the forest ecosystem, performing functions such as carbohydrate metabolism and nitrogen fixation (Eichorst et al., 2018). The phylum Actinobacteria is known for its production of secondary metabolites and antimicrobial compounds (Ait Barka et al., 2015), although this has not been explored with regards to functionality in the stem.

2.3 Root, rhizosphere, soil, and litter

The analysis of the bacterial community composition of the below-ground rhizosphere system, including roots, rhizosphere soil, bulk soil, and forest litter (Fig. 9.4) emphasized how the rhizosphere microbiome of forest trees is high in species richness and diversity (Cregger et al., 2018). Bacteria in this environment are responsible for activities such as nutrient acquisition, nutrient cycling, plant growth promotion, and abiotic stress tolerance (Manpoong et al., 2020).

Of the total 46 phyla (including Proteobacteria presented as separate classes) identified as members of the forest microbiome, there was a large number of phyla associated solely with this tree compartment. Twenty-three of these phyla were found to be unique to the below-ground rhizosphere system. The phyla common to the rhizosphere, independent of associated tree species and geographical region, included the phylum Acidobacteria, members of which have been isolated from numerous ecosystems on a global scale, and along with being involved in carbon cycling in soils. Acidobacteria are also known to act as plant growth-promoting bacteria (Kielak et al., 2016b; Rawat et al., 2012). Other phyla including the Proteobacteria, Actinobacteria, Firmicutes, and Bacteroidetes have also been studied for their plant growth-promoting abilities such as nutrient acquisition and ammonia production (Flores-Núñez et al., 2018; Radhapriya et al., 2018). The phylum Proteobacteria makes up a large proportion of the rhizosphere microbiome, with key roles in functions such as nutrient

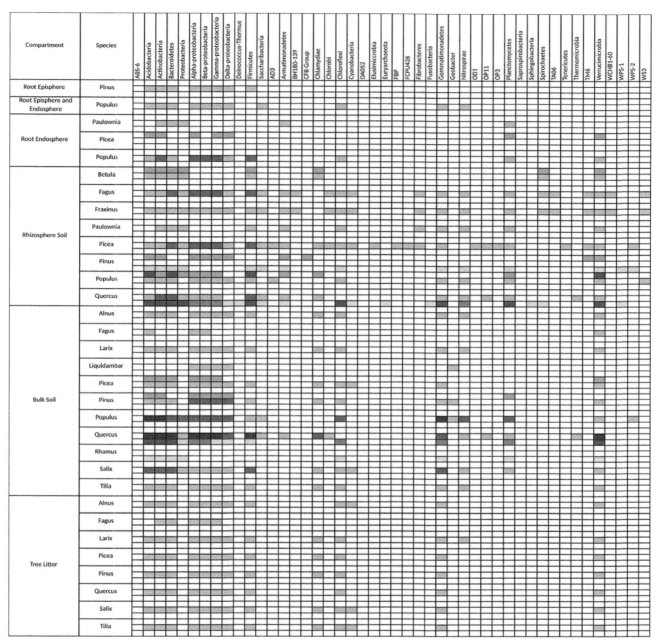

FIG. 9.4 Heatmap displaying the presence of bacterial phyla in the below-ground rhizosphere microbiome of different tree species. *Blue squares* indicate data from a Boreal forest, *green squares* indicate data from a Temperate forest, and *yellow squares* indicate data from a Mediterranean forest. The intensity of the color is illustrative of the number of studies represented (light = 1 study, mid = 2 studies, dark = 3 or more studies).

cycling (Pérez-Izquierdo et al., 2019). A link has been identified between the ratio of Acidobacteria and Proteobacteria and the nutrient content of a soil, meaning that the abundance of these phyla can be used as an indicator for identifying soil nutrient content, with a higher ratio of Acidobacteria being linked to nutrient-poor soil and a higher ratio of Proteobacteria being linked to nutrient-rich soils (Kielak et al., 2016a). The phylum Actinobacteria is also common to rhizosphere systems independent of tree species and geographical range. Members of this phylum are generally responsible for nutrient cycling and plant growth promotion; however, there has also been research into using members of this phylum as biological control agents against pathogens such as *Armillaria* due to their ability to produce antimicrobial compounds (de Vasconcellos and Cardoso, 2009; Sharma and Thakur, 2020). The phyla Chloroflexi and Nitrospirae are known to be key members of the rhizosphere microbiome of a number of species, with functions such as carbon dioxide fixation and nitrification, key to plant growth (Zhang et al., 2013). Other phyla common to the rhizosphere microbiome included the Armatimonadetes, Gemmatimonadetes, Planctomycetes, and Verrucomicrobia.

3. Comparing bacterial biota composition across tree species and forest biomes

Forest biomes cover a significant proportion of Earth's terrestrial surface, with important functions in carbon storage, geochemical cycles, and climate regulation (Baldrian, 2017b). Forest microbiomes are therefore dynamic, both at temporal and spatial scales, and are influenced by ecological factors that comprise short-term seasonal dynamics and long-term events such as disease outbreaks, insect attack, and fire (Baldrian, 2017a; Jackson and Denney, 2011). In addition, tree species composition and host genetics will also play a role in forest microbiome composition, and while there is less variation in microbial community composition between tree species, seasonal effects play a significant role in the composition of the phyllosphere microbiome (Jackson and Denney, 2011). Consequently, a complex combination of host, environmental, and ecological factors drive tree microbiome structure and function. In this section, we consider the distribution and co-occurrence of major taxonomic groups of bacterial biota in above- and below-ground compartments of tree species across Mediterranean, temperate, and boreal forest biomes.

A total of 46 phyla (including Proteobacteria presented as separate classes) and candidate bacterial phyla were detected in the microbiomes of tree species from Mediterranean, temperate, and boreal forest biomes. It is clear that temperate tree genera have received much greater research attention in terms of bacterial biota composition ($n=39$ tree genera studied) when compared with Mediterranean and boreal biomes ($n=6$ and $n=7$ tree genera studied, respectively) (Table 9.3). Across those biomes, foliar bacterial microbiota were the most frequently studied (in total, 52 genera across the three biomes), followed by roots/rhizosphere/soil ($n=21$ genera studied) and the stem bacterial biota ($n=7$ studies). Only seven tree genera had comparative data from two different biomes for the foliage and roots/rhizosphere/soil compartments, and only a single genus (*Populus*) had comparative data for stem bacterial microbiota across two biomes (boreal and temperate). This likely reflects greater research focus on foliar and rhizosphere microbiomes, and the relative ease of sampling these compartments in comparison to stem microbiomes, which require destructive sampling, have low associated bacterial biomass, and are arguably more challenging materials for molecular microbiome analyses (Broberg and McDonald, 2019).

Phylum-level comparisons of microbiota composition are often used to provide broad overviews of taxonomic composition of microbiomes, but lack taxonomic and functional resolution of the community. In addition, due to differences in the methodological approaches, sampling depth, and replication levels across the studies collated here, it is difficult to infer rigorous conclusions on similarities and differences between tree microbiota across biomes. However, we discuss the composition of bacterial microbiota across tree compartments here at the phylum level to provide an overview of key taxonomic patterns in these environments, with the caveat that further fine-scale taxonomic and functional assessments of tree microbiome composition and biomes using standardized methodological approaches, such as those used in the Earth Microbiome Project (Thompson et al., 2017), would be advantageous and should be a future research priority.

Three phyla (Actinobacteria, Bacteroidetes, Firmicutes) and four classes of the Proteobacteria (Alphaproteobacteria, Betaproteobacteria, Gammaproteobacteria, Deltaproteobacteria) were detected in the foliage of tree genera across all three biomes, and three phyla (Acidobacteria, Actinobacteria, and Firmicutes) were detected in stem tissue of trees across all three biomes (Fig. 9.5). In the root/rhizosphere/soil compartment, eight phyla (Acidobacteria, Actinobacteria, Bacteroidetes, Firmicutes, Armatimonadetes, Planctomycetes, Spirochaetes, Verrucomicrobia) and four classes of

TABLE 9.3 Overview of the number of tree genera studied across Mediterranean, temperate, and boreal forest biomes, and number of genera studied for each tree compartment.

Number of tree genera studied	Biome		
	Mediterranean	Temperate	Boreal
Total[a]	6	39	7
Foliage (endosphere and episphere)	5	38	3
Stem (endosphere and episphere)	1	5	1
Root, rhizosphere, and soil	3	14	4

[a] The cumulative number of tree genera for all tree compartments (foliage, stem, or root/rhizosphere/soil) does not match the "total" number of tree genera studied, as some publications presented data on more than one compartment for the same genus.

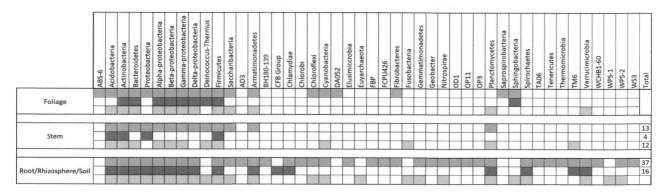

FIG. 9.5 Diagram illustrating the occurrence of bacterial Phyla (or classes for Proteobacteria) shared between tree different compartments in Mediterranean, boreal, and temperate biomes. *Green squares*, phylum detected in temperate tree species; *blue squares*, phylum detected in boreal tree species; *yellow squares*, phylum detected in Mediterranean tree species.

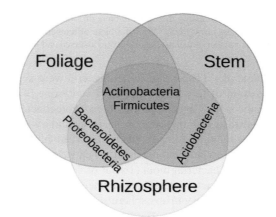

FIG. 9.6 Venn diagram illustrating overlap in bacterial Phyla shared between tree compartments across species in Mediterranean, boreal, and temperate biomes.

Proteobacteria (Alphaproteobacteria, Betaproteobacteria, Gammaproteobacteria, and Deltaproteobacteria) were detected across all three biomes (Fig. 9.5). In addition, two phyla (Actinobacteria and Firmicutes) were shared between all three compartments across all biomes, one phylum (Acidobacteria) was shared between the stem and root/rhizosphere/soil compartments across all biomes, and one phylum (Bacteroidetes) and four classes of the Proteobacteria (Alphaproteobacteria, Betaproteobacteria, Gammaproteobacteria, and Deltaproteobacteria) were found in both the foliar and root/rhizosphere/soil compartments across all three biomes (Fig. 9.5). These data highlight the connectivity of core microbiome members across tree species and biomes across global scales (Fig. 9.6).

4.   Conclusions and future research priorities

The microbiota of forest trees is a major determinant of health and productivity in the plant holobiont, with critical roles in nutrient acquisition and mineralization, disease suppression and pathogen exclusion, immune and metabolic regulation, and development of root architecture and driving microbiome interactions (Hacquard and Schadt, 2015; Turner et al., 2013). The plant holobiont comprises assemblages of bacteria, fungi, Archaea, protists, and viruses, with complex ecological interactions that support the host. Here, we assessed the composition and function of bacterial biota in Mediterranean, temperate, and boreal tree species from 43 relevant peer reviewed publications to review the composition and function of bacterial biota of tree microbiomes. Additional aims were to compare the composition of bacterial biota across different tree species and forest biomes (Mediterranean, temperate, and boreal) and identify key knowledge gaps in our understanding of the bacterial biota of forest trees and make recommendations for future research.

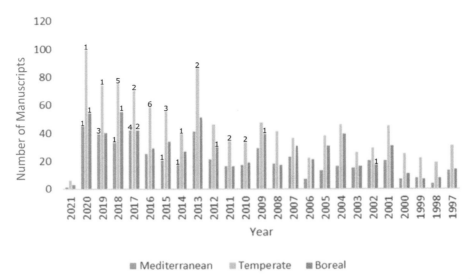

FIG. 9.7 Number of manuscripts published since 1997 discussing the bacterial biota of forest trees according to a search on Science Direct using the following search terms: "bacterial biota of Mediterranean forest trees," "bacterial biota of temperate forest trees," and "bacterial biota of boreal forest trees." The number at the top of the bars indicates the number of manuscripts that were assessed and utilized in the analysis of this study.

In total, 46 bacterial phyla or candidate phyla (note that Proteobacteria were separated into four classes) were identified in available datasets on the composition of forest tree species. Twenty-one of these phyla were detected in foliage, with Actinobacteria, Bacteroidetes, Deinococcus-Thermus, Firmicutes, Proteobacteria, and Saccharibacteria representing the most commonly observed phyla. These taxa play important roles in secondary metabolite and antimicrobial compound production, promoting symbiotic interactions and isoprene production, and are highly adapted to the relatively harsh conditions (e.g., UV irradiation, high competition, low nutrient availability) experienced by foliar microbiomes. In contrast, the stem microbiome exhibits a lower diversity of bacterial biota (16 out of 46 phyla), and no taxa were exclusive to the stem compartment. Acidobacteria, Actinobacteria, Bacteroidetes, and Proteobacteria were the main phyla detected. Twenty-three of the 46 phyla detected were unique to the root/rhizosphere/soil compartment and play important roles in nutrient acquisition, plant growth promotion, and stress tolerance. For example, Acidobacteria were strongly associated with this compartment and have roles in nutrient cycling and promoting plant growth (Kielak et al., 2016a).

Comparing the composition of bacterial biota across different tree species and forest biomes, it is clear that most research effort has focused on temperate tree species, with 39 genera of trees studied, compared to six and seven for Mediterranean and boreal biomes, respectively (Fig. 9.7). Future research efforts should therefore aim to increase the breadth of knowledge surrounding the microbiota associated with boreal and Mediterranean trees. A further interesting observation was that foliar microbiomes have received greatest attention, followed by the root/rhizosphere/soil compartment, and the stem, which likely reflects ease of sampling, the destructive nature of stem tissue sampling, and the low microbial biomass in stem tissues (Broberg and McDonald, 2019; Denman et al., 2018). Given the contrasting composition and function of tree microbiomes across the above- and below-ground compartments, future studies should aim, where possible, to study microbiome composition and function across different compartments of individual tree species and across geographic scales, in order to gain further information about the assembly and dynamics of tree microbiome composition and function.

Despite significant advances in our knowledge and understanding of the role of microorganisms in forest health and ecosystem function, due to the complex ecological and host-associated factors that drive holobiont function across spatial and temporal scales, integrated systems-based approaches will be required to address these multiscale interactions (Broberg et al., 2018). Although this chapter focuses on the bacterial biota of forest trees, interactions between the bacterial, archaeal, fungal, protozoan, and viral components of the tree microbiome should be studied in tandem, and integrated with relevant abiotic datasets using ecological approaches. Advances in methodological and technical approaches for culture-based, molecular and computational analysis of the microbiome in tandem with ecological research now make this goal possible. Where possible, future studies should therefore focus on integration of selected complementary approaches for isolation and cultivation of microbiota, combined with single gene community profiling (e.g., 16S or 18S rRNA gene and ITS sequencing), genomics, transcriptomics, proteomics, and metabolomics to address these questions. In addition, fine-scale taxonomic and functional assessments of tree microbiome composition using standardized methodological approaches, such as those used in the Earth Microbiome Project (Thompson et al., 2017), would be advantageous for comparative meta-analysis of datasets, and should be a future research priority. Such data would facilitate the development of a global picture of forest microbiome composition and function similar to that recently presented to generate an atlas of global soil bacterial diversity (Delgado-Baquerizo et al., 2018). Systems approaches that use "big data" to characterize the complex ecological processes that underpin the dynamics of global forest biomes will therefore be important to address the existential threat that global forest biomes face in light of changing climate and pest and pathogen attack.

References

Ait Barka, E., Vatsa, P., Sanchez, L., Gaveau-Vaillant, N., Jacquard, C., Klenk, H.-P., Clément, C., Ouhdouch, Y., Van Wezel, G.P., 2015. Taxonomy, physiology, and natural products of actinobacteria. Microbiol. Mol. Biol. Rev. 80 (1), 1–43. https://doi.org/10.1128/MMBR.00019-15.

Baldrian, P., 2017a. Forest microbiome: diversity, complexity and dynamics. FEMS Microbiol. Rev. 41, 109–130. https://doi.org/10.1093/femsre/fuw040.

Baldrian, P., 2017b. Microbial activity and the dynamics of ecosystem processes in forest soils. Curr. Opin. Microbiol. 37, 128–134. https://doi.org/10.1016/j.mib.2017.06.008.

Beckers, B., De Beeck, M.O., Weyens, N., Van Acker, R., Van Montagu, M., Boerjan, W., Vangronsveld, J., 2016. Lignin engineering in field-grown poplar trees affects the endosphere bacterial microbiome. Proc. Natl. Acad. Sci. U. S. A. 113, 2312–2317. https://doi.org/10.1073/pnas.1523264113.

Beckers, B., De Beeck, M.O., Weyens, N., Boerjan, W., Vangronsveld, J., 2017. Structural variability and niche differentiation in the rhizosphere and endosphere bacterial microbiome of field-grown poplar trees. Microbiome 5, 1–17. https://doi.org/10.1186/s40168-017-0241-2.

Berg, G., Rybakova, D., Fischer, D., Cernava, T., Vergès, M.C.C., Charles, T., Chen, X., Cocolin, L., Eversole, K., Corral, G.H., Kazou, M., Kinkel, L., Lange, L., Lima, N., Loy, A., Macklin, J.A., Maguin, E., Mauchline, T., McClure, R., Mitter, B., Ryan, M., Sarand, I., Smidt, H., Schelkle, B., Roume, H., Kiran, G.S., Selvin, J., de Souza, R.S.C., van Overbeek, L., Singh, B.K., Wagner, M., Walsh, A., Sessitsch, A., Schloter, M., 2020. Microbiome definition re-visited: old concepts and new challenges. Microbiome 8, 103. https://doi.org/10.1186/s40168-020-00875-0.

Broberg, M., McDonald, J.E., 2019. Extraction of microbial and host DNA, RNA, and proteins from oak bark tissue. Methods Protoc. https://doi.org/10.3390/mps2010015.

Broberg, M., Doonan, J., Mundt, F., Denman, S., McDonald, J.E., 2018. Integrated multi-omic analysis of host-microbiota interactions in acute oak decline. Microbiome 6, 21. https://doi.org/10.1186/s40168-018-0408-5.

Cregger, M.A., Veach, A.M., Yang, Z.K., Crouch, M.J., Vilgalys, R., Tuskan, G.A., Schadt, C.W., 2018. The Populus holobiont: dissecting the effects of plant niches and genotype on the microbiome. Microbiome 6, 1–14. https://doi.org/10.1186/s40168-018-0413-8.

Crombie, A.T., Larke-Mejia, N.L., Emery, H., Dawson, R., Pratscher, J., Murphy, G.P., McGenity, T.J., Murrell, J.C., 2018. Poplar phyllosphere harbors disparate isoprene-degrading bacteria. Proc. Natl. Acad. Sci. U. S. A. 115, 13081–13086. https://doi.org/10.1073/pnas.1812668115.

de Vasconcellos, R.L.F., Cardoso, E.J.B.N., 2009. Rhizospheric streptomycetes as potential biocontrol agents of *Fusarium* and *Armillaria* pine rot and as PGPR for Pinus taeda. BioControl 54, 807–816. https://doi.org/10.1007/s10526-009-9226-9.

Delgado-Baquerizo, M., Oliverio, A.M., Brewer, T.E., Benavent-González, A., Eldridge, D.J., Bardgett, R.D., Maestre, F.T., Singh, B.K., Fierer, N., 2018. A global atlas of the dominant bacteria found in soil. Science (80-). https://doi.org/10.1126/science.aap9516.

Denman, S., Doonan, J., Ransom-Jones, E., Broberg, M., Plummer, S., Kirk, S., Scarlett, K., Griffiths, A.R., Kaczmarek, M., Forster, J., Peace, A., Golyshin, P.N., Hassard, F., Brown, N., Kenny, J.G., McDonald, J.E., 2018. Microbiome and infectivity studies reveal complex polyspecies tree disease in Acute Oak Decline. ISME J. 12. https://doi.org/10.1038/ismej.2017.170.

Eichorst, S.A., Trojan, D., Roux, S., Herbold, C., Rattei, T., Woebken, D., 2018. Genomic insights into the *Acidobacteria* reveal strategies for their success in terrestrial environments. Environ. Microbiol. 20, 1041–1063. https://doi.org/10.1111/1462-2920.14043.

Espenshade, J., Thijs, S., Gawronski, S., Bové, H., Weyens, N., Vangronsveld, J., 2019. Influence of urbanization on epiphytic bacterial communities of the platanus × hispanica tree leaves in a biennial study. Front. Microbiol. 10. https://doi.org/10.3389/fmicb.2019.00675.

Fernández-González, A.J., Villadas, P.J., Gómez-Lama Cabanás, C., Valverde-Corredor, A., Belaj, A., Mercado-Blanco, J., Fernández-López, M., 2019. Defining the root endosphere and rhizosphere microbiomes from the World Olive Germplasm Collection. Sci. Rep. 9, 1–13. https://doi.org/10.1038/s41598-019-56977-9.

Filippidou, S., Wunderlin, T., Junier, T., Jeanneret, N., Dorador, C., Molina, V., Johnson, D.R., Junier, P., 2016. A combination of extreme environmental conditions favor the prevalence of endospore-forming *Firmicutes*. Front. Microbiol. 7, 1707. https://doi.org/10.3389/fmicb.2016.01707.

Fini, A., Brunetti, C., Loreto, F., Centritto, M., Ferrini, F., Tattini, M., 2017. Isoprene responses and functions in plants challenged by environmental pressures associated to climate change. Front. Plant Sci. https://doi.org/10.3389/fpls.2017.01281.

Flores-Núñez, V.M., Amora-Lazcano, E., Rodríguez-Dorantes, A., Cruz-Maya, J.A., Jan-Roblero, J., 2018. Comparison of plant growth-promoting rhizobacteria in a pine forest soil and an agricultural soil. Soil Res. 56, 346. https://doi.org/10.1071/sr17227.

Griffin, E.A., Carson, W.P., 2015. The ecology and natural history of foliar bacteria with a focus on tropical forests and agroecosystems. Bot. Rev. 81, 105–149. https://doi.org/10.1007/s12229-015-9151-9.

Haas, J.C., Street, N.R., Sjödin, A., Lee, N.M., Högberg, M.N., Näsholm, T., Hurry, V., 2018. Microbial community response to growing season and plant nutrient optimisation in a boreal Norway spruce forest. Soil Biol. Biochem. 125, 197–209. https://doi.org/10.1016/j.soilbio.2018.07.005.

Hacquard, S., Schadt, C.W., 2015. Towards a holistic understanding of the beneficial interactions across the Populus microbiome. New Phytol. 205, 1424–1430. https://doi.org/10.1111/nph.13133.

Hacquard, S., Garrido-Oter, R., González, A., Spaepen, S., Ackermann, G., Lebeis, S., McHardy, A.C., Dangl, J.L., Knight, R., Ley, R., Schulze-Lefert, P., 2015. Microbiota and host nutrition across plant and animal kingdoms. Cell Host Microbe 17, 603–616. https://doi.org/10.1016/j.chom.2015.04.009.

Hahnke, R.L., Meier-Kolthoff, J.P., García-López, M., Mukherjee, S., Huntemann, M., Ivanova, N.N., Woyke, T., Kyrpides, N.C., Klenk, H.P., Göker, M., 2016. Genome-based taxonomic classification of *Bacteroidetes*. Front. Microbiol. 7, 2003. https://doi.org/10.3389/fmicb.2016.02003.

Jackson, C.R., Denney, W.C., 2011. Annual and seasonal variation in the phyllosphere bacterial community associated with leaves of the southern Magnolia (Magnolia grandiflora). Microb. Ecol. 61, 113–122. https://doi.org/10.1007/s00248-010-9742-2.

Kaiser, K., Wemheuer, B., Korolkow, V., Wemheuer, F., Nacke, H., Schöning, I., Schrumpf, M., Daniel, R., 2016. Driving forces of soil bacterial community structure, diversity, and function in temperate grasslands and forests. Sci. Rep. 6, 1–12. https://doi.org/10.1038/srep33696.

Kielak, A.M., Barreto, C.C., Kowalchuk, G.A., van Veen, J.A., Kuramae, E.E., 2016a. The ecology of Acidobacteria: moving beyond genes and genomes. Front. Microbiol. https://doi.org/10.3389/fmicb.2016.00744.

Kielak, A.M., Cipriano, M.A.P., Kuramae, E.E., 2016b. *Acidobacteria* strains from subdivision 1 act as plant growth-promoting bacteria. Arch. Microbiol. 198, 987–993. https://doi.org/10.1007/s00203-016-1260-2.

Kobayashi, K., Aoyagi, H., 2019. Microbial community structure analysis in Acer palmatum bark and isolation of novel bacteria IAD-21 of the candidate division FBP. PeerJ 2019, e7876. https://doi.org/10.7717/peerj.7876.

Larke-Mejía, N.L., Crombie, A.T., Pratscher, J., McGenity, T.J., Murrell, J.C., 2019. Novel isoprene-degrading proteobacteria from soil and leaves identified by cultivation and metagenomics analysis of stable isotope probing experiments. Front. Microbiol. 10, 2700. https://doi.org/10.3389/fmicb.2019.02700.

Lladó, S., López-Mondéjar, R., Baldrian, P., 2017. Forest soil bacteria: diversity, involvement in ecosystem processes, and response to global change. Microbiol. Mol. Biol. Rev. 81, 1–27. https://doi.org/10.1128/mmbr.00063-16.

Manpoong, C., De Mandal, S., Bangaruswamy, D.K., Perumal, R.C., Benny, J., Beena, P.S., Ghosh, A., Kumar, N.S., Tripathi, S.K., 2020. Linking rhizosphere soil biochemical and microbial community characteristics across different land use systems in mountainous region in Northeast India. Meta Gene 23, 100625. https://doi.org/10.1016/j.mgene.2019.100625.

Peñuelas, J., Terradas, J., 2014. The foliar microbiome. Trends Plant Sci. https://doi.org/10.1016/j.tplants.2013.12.007.

Pérez-Izquierdo, L., Zabal-Aguirre, M., González-Martínez, S.C., Buée, M., Verdú, M., Rincón, A., Goberna, M., 2019. Plant intraspecific variation modulates nutrient cycling through its below ground rhizospheric microbiome. J. Ecol. 107, 1594–1605. https://doi.org/10.1111/1365-2745.13202.

Radhapriya, P., Ramachandran, A., Palani, P., 2018. Indigenous plant growth-promoting bacteria enhance plant growth, biomass, and nutrient uptake in degraded forest plants. 3 Biotech 8, 154. https://doi.org/10.1007/s13205-018-1179-1.

Rawat, S.R., Männistö, M.K., Bromberg, Y., Häggblom, M.M., 2012. Comparative genomic and physiological analysis provides insights into the role of *Acidobacteria* in organic carbon utilization in Arctic tundra soils. FEMS Microbiol. Ecol. 82, 341–355. https://doi.org/10.1111/j.1574-6941.2012.01381.x.

Ren, G., Zhang, H., Lin, X., Zhu, J., Jia, Z., 2015. Response of leaf endophytic bacterial community to elevated CO_2 at different growth stages of rice plant. Front. Microbiol. 6, 1–13. https://doi.org/10.3389/fmicb.2015.00855.

Rosenberg, E., Zilber-Rosenberg, I., Manwani, D., Mortha, A., Xu, C., Faith, J., Burk, R., Kunisaki, Y., Jang, J., Scheiermann, C., 2016. Microbes drive evolution of animals and plants: the hologenome concept. MBio 7. https://doi.org/10.1128/mBio.01395-15. e01395-15.

Shaffer, J.P., U'Ren, J.M., Gallery, R.E., Baltrus, D.A., Arnold, A.E., 2017. An endohyphal bacterium (*Chitinophaga, Bacteroidetes*) alters carbon source use by *Fusarium keratoplasticum* (*F. solani* species complex, *Nectriaceae*). Front. Microbiol. 8, 350. https://doi.org/10.3389/fmicb.2017.00350.

Sharma, P., Thakur, D., 2020. Antimicrobial biosynthetic potential and diversity of culturable soil *Actinobacteria* from forest ecosystems of Northeast India. Sci. Rep. 10, 1–18. https://doi.org/10.1038/s41598-020-60968-6.

Starr, E.P., Shi, S., Blazewicz, S.J., Probst, A.J., Herman, D.J., Firestone, M.K., Banfield, J.F., 2018. Stable isotope informed genome-resolved metagenomics reveals that *Saccharibacteria* utilize microbially-processed plant-derived carbon. Microbiome 6, 122. https://doi.org/10.1186/s40168-018-0499-z.

Theodorakopoulos, N., Bachar, D., Christen, R., Alain, K., Chapon, V., 2013. Exploration of *Deinococcus-Thermus* molecular diversity by novel group-specific PCR primers. Microbiologyopen 2. https://doi.org/10.1002/mbo3.119.

Thompson, L.R., Sanders, J.G., McDonald, D., Amir, A., Ladau, J., Locey, K.J., Prill, R.J., Tripathi, A., Gibbons, S.M., Ackermann, G., Navas-Molina, J.A., Janssen, S., Kopylova, E., Vázquez-Baeza, Y., González, A., Morton, J.T., Mirarab, S., Xu, Z.Z., Jiang, L., Haroon, M.F., Kanbar, J., Zhu, Q., Song, S.J., Kosciolek, T., Bokulich, N.A., Lefler, J., Brislawn, C.J., Humphrey, G., Owens, S.M., Hampton-Marcell, J., Berg-Lyons, D., McKenzie, V., Fierer, N., Fuhrman, J.A., Clauset, A., Stevens, R.L., Shade, A., Pollard, K.S., Goodwin, K.D., Jansson, J.K., Gilbert, J.A., Knight, R., Agosto Rivera, J.L., Al-Moosawi, L., Alverdy, J., Amato, K.R., Andras, J., Angenent, L.T., Antonopoulos, D.A., Apprill, A., Armitage, D., Ballantine, K., Bárta, J., Baum, J.K., Berry, A., Bhatnagar, A., Bhatnagar, M., Biddle, J.F., Bittner, L., Boldgiv, B., Bottos, E., Boyer, D.M., Braun, J., Brazelton, W., Brearley, F.Q., Campbell, A.H., Caporaso, J.G., Cardona, C., Carroll, J.L., Cary, S.C., Casper, B.B., Charles, T.C., Chu, H., Claar, D.C., Clark, R.G., Clayton, J.B., Clemente, J.C., Cochran, A., Coleman, M.L., Collins, G., Colwell, R.R., Contreras, M., Crary, B.B., Creer, S., Cristol, D.A., Crump, B.C., Cui, D., Daly, S.E., Davalos, L., Dawson, R.D., Defazio, J., Delsuc, F., Dionisi, H.M., Dominguez-Bello, M.G., Dowell, R., Dubinsky, E.A., Dunn, P.O., Ercolini, D., Espinoza, R.E., Ezenwa, V., Fenner, N., Findlay, H.S., Fleming, I.D., Fogliano, V., Forsman, A., Freeman, C., Friedman, E.S., Galindo, G., Garcia, L., Garcia-Amado, M.A., Garshelis, D., Gasser, R.B., Gerdts, G., Gibson, M.K., Gifford, I., Gill, R.T., Giray, T., Gittel, A., Golyshin, P., Gong, D., Grossart, H.P., Guyton, K., Haig, S.J., Hale, V., Hall, R.S., Hallam, S.J., Handley, K.M., Hasan, N.A., Haydon, S.R., Hickman, J.E., Hidalgo, G., Hofmockel, K.S., Hooker, J., Hulth, S., Hultman, J., Hyde, E., Ibáñez-Álamo, J.D., Jastrow, J.D., Jex, A.R., Johnson, L.S., Johnston, E.R., Joseph, S., Jurburg, S.D., Jurelevicius, D., Karlsson, A., Karlsson, R., Kauppinen, S., Kellogg, C.T.E., Kennedy, S.J., Kerkhof, L.J., King, G.M., Kling, G.W., Koehler, A.V., Krezalek, M., Kueneman, J., Lamendella, R., Landon, E.M., Lanede Graaf, K., LaRoche, J., Larsen, P., Laverock, B., Lax, S., Lentino, M., Levin, I.I., Liancourt, P., Liang, W., Linz, A.M., Lipson, D.A., Liu, Y., Lladser, M.E., Lozada, M., Spirito, C.M., MacCormack, W.P., MacRae-Crerar, A., Magris, M., Martín-Platero, A.M., Martín-Vivaldi, M., Martínez, L.M., Martínez-Bueno, M., Marzinelli, E.M., Mason, O.U., Mayer, G.D., McDevitt-Irwin, J.M., McDonald, J.E., McGuire, K.L., McMahon, K.D., McMinds, R., Medina, M., Mendelson, J.R., Metcalf, J.L., Meyer, F., Michelangeli, F., Miller, K., Mills, D.A., Minich, J., Mocali, S., Moitinho-Silva, L., Moore, A., Morgan-Kiss, R.M., Munroe, P., Myrold, D., Neufeld, J.D., Ni, Y., Nicol, G.W., Nielsen, S., Nissimov, J.I., Niu, K., Nolan, M.J., Noyce, K., O'Brien, S.L., Okamoto, N.,

Orlando, L., Castellano, Y.O., Osuolale, O., Oswald, W., Parnell, J., Peralta-Sánchez, J.M., Petraitis, P., Pfister, C., Pilon-Smits, E., Piombino, P., Pointing, S.B., Pollock, F.J., Potter, C., Prithiviraj, B., Quince, C., Rani, A., Ranjan, R., Rao, S., Rees, A.P., Richardson, M., Riebesell, U., Robinson, C., Rockne, K.J., Rodriguezl, S.M., Rohwer, F., Roundstone, W., Safran, R.J., Sangwan, N., Sanz, V., Schrenk, M., Schrenzel, M.D., Scott, N.M., Seger, R.L., Seguinorlando, A., Seldin, L., Seyler, L.M., Shakhsheer, B., Sheets, G.M., Shen, C., Shi, Y., Shin, H., Shogan, B.D., Shutler, D., Siegel, J., Simmons, S., Sjöling, S., Smith, D.P., Soler, J.J., Sperling, M., Steinberg, P.D., Stephens, B., Stevens, M.A., Taghavi, S., Tai, V., Tait, K., Tan, C.L., Taş, N., Taylor, D.L., Thomas, T., Timling, I., Turner, B.L., Urich, T., Ursell, L.K., Van Der Lelie, D., Van Treuren, W., Van Zwieten, L., Vargas-Robles, D., Thurber, R.V., Vitaglione, P., Walker, D.A., Walters, W.A., Wang, S., Wang, T., Weaver, T., Webster, N.S., Wehrle, B., Weisenhorn, P., Weiss, S., Werner, J.J., West, K., Whitehead, A., Whitehead, S.R., Whittingham, L.A., Willerslev, E., Williams, A.E., Wood, S.A., Woodhams, D.C., Yang, Y., Zaneveld, J., Zarraonaindia, I., Zhang, Q., Zhao, H., 2017. A communal catalogue reveals Earth's multiscale microbial diversity. Nature 551, 457–463. https://doi.org/10.1038/nature24621.

Turner, T.R., James, E.K., Poole, P.S., 2013. The plant microbiome. Genome Biol. 14, 209. https://doi.org/10.1186/gb-2013-14-6-209.

Vorholt, J.A., 2012. Microbial life in the phyllosphere. Nat. Rev. Microbiol. https://doi.org/10.1038/nrmicro2910.

Whipps, J.M., Karen, L., Cooke, R.C., 1988. Mycoparasitism and plant disease control. In: Fungi in Biological Control Systems. Manchester University Press, pp. 161–187.

Yadav, R.K.P., Karamanoli, K., Vokou, D., 2005. Bacterial colonization of the phyllosphere of mediterranean perennial species as influenced by leaf structural and chemical features. Microb. Ecol. 50, 185–196. https://doi.org/10.1007/s00248-004-0171-y.

Yadav, R.K.P., Karamanoli, K., Vokou, D., 2011. Bacterial populations on the phyllosphere of Mediterranean plants: influence of leaf age and leaf surface. Front. Agric. China 5, 60–63. https://doi.org/10.1007/s11703-011-1068-4.

Zhang, X., Chen, Q., Han, X., 2013. Soil bacterial communities respond to mowing and nutrient addition in a steppe ecosystem. PLoS One 8. https://doi.org/10.1371/journal.pone.0084210, e84210.

Chapter 10

Fungi inhabiting woody tree tissues

Gitta Jutta Langer[a], Johanna Bußkamp[a], Eeva Terhonen[b], and Kathrin Blumenstein[b]

[a]Section Mycology and Complex Diseases, Department of Forest Protection, Northwest German Forest Research Institute (NW-FVA), Göttingen, Germany, [b]Forest Pathology Research Group, Department of Forest Botany and Tree Physiology, Faculty of Forest Sciences and Forest Ecology, University of Göttingen, Göttingen, Germany

Chapter Outline

1. Introduction

The species-rich assemblages of microorganisms associated with plants are dominated by fungi (Stone and Petrini, 1997). Besides the mycobiome of the rhizosphere, fungal species colonize foliar and twig surfaces (epiphytes), internal foliage space (foliar endophytes), young and old bark (bark endophytes), and woody tissues (nonfoliar or xylotrophic endophytes and wood decomposers). The interplay between trees and fungi inhabiting living woody tissues is a key process in forest ecosystems. Wood provides habitat, shelter, and nutrition to diverse organisms, especially fungi and saproxylic insects. Therefore, it is an important habitat and structural component (Floren et al., 2015; Harmon et al., 1986). Wood-inhabiting fungi play a major role in forest ecosystems because they are able to decompose wood and recycle nutrients, and they may initiate successional dynamics for saproxylic arthropods and soil formation (Lonsdale et al., 2008; Šamonil et al., 2020). A study on wood-inhabiting fungi in stems of the European ash (*Fraxinus excelsior*) (Lygis et al., 2005) demonstrated that the species richness was similar in trees of different health stages, but community assemblies were different. These results suggest that the species composition of wood-inhabiting species may shift in correlation with tree health and vitality. The wood-inhabiting fungi are primarily defined as wood-decaying fungi that decompose wood, causing it to rot. Fungi inhabiting woody tree tissues are to be distinguished from wood-inhabiting fungi. Several wood-decay fungi (wood-inhabiting fungi) are saprotrophic and attack/grow only dead wood or coarse woody debris (CWD). However, fungi inhabiting woody tree tissues do not necessarily have to be wood-decaying fungi. Often, they are endophytic and latently parasitic or their function is unknown. But they also are typical wood-decay fungi of their hosts, colonizing dying or dead branches (Boddy and Rayner, 1983), or are members of the fungal community of the natural pruning of branches (Butin and Kowalski, 1983). The wood-inhabiting fungi usually include wood-decay fungi, but fungi inhabiting wood also include true endophytes, latent saprophytes, parasites, and wood-decay fungi with an endophytic stage.

Parasitic wood-decay species colonize living trees, but might have initially invaded the woody tree tissues in an endophytic stage. The wood-decay fungi can have an endophytic stage in their life cycle to invade their niches in the host before decomposition is initiated.

The ecological and functional impacts of endophytes on wood decay have been studied and discussed for birch (Cline et al., 2018). The pathogenic stages of xylem endophytes, such as *Neonectria coccinea* and the white rot fungus *Fomes fomentarius*, seem to be controlled by the limitation of oxygen and nutrients (Rodríguez et al., 2011). Often, wood decay starts with stain-associated, ascomycetous fungi, which are derived from latent propagules in the wood and bark (Tables 10.1–10.3). They are followed and replaced by basidiomycetous white rot and/or brown rot fungi or by soft rot fungi, particularly Ascomycota (Chapela and Boddy, 1988a; see Section 8 and Table 10.4).

Healthy woody tissues have symbiotic associations with endophytic fungi forming quiescent microthalli (Sieber, 2007). It is assumed that latent fungal propagules (such as hyphal fragments or chlamydospores) are distributed widely but scantly in the sapwood (Oses et al., 2008; Schwarze et al., 2000). The community assembly that occurs in woody roots, stems, twigs, and foliar tissues can be variable (Table 10.1), but the species diversity of endophyte communities can be high, with more than 100 species in a special tissue type (Sieber, 2007). Especially fungal communities of the inner bark (phloem) can be rich with a mean of 117–171 operational taxonomic units (OTUs) per tree (Pellitier et al., 2019). For example, from *Pinus sylvestris*, 16 endophytic fungi were isolated from stems and only 10 from the xylem (Petrini and Fisher, 1988) whereas 103 fungal endophytes were isolated from the woody twig tissues of *P. sylvestris* (Bußkamp et al., 2020) and an average of 171 OTUs were found from the inner bark of *Pinus strobus* (Pellitier et al., 2019).

In this article, the used fungal names follow the Index Fungorum (www.indexfungorum.org). The authors and orders of taxa mentioned are provided in the appended index to this chapter.

2. Endophytes

Endophytes are considered to be mutualistic, neutral (commensals), opportunistic (able to cause disease after the host has been weakened by another factor), or deleterious (parasitism or pathogenicity) to their hosts (Sieber, 2007; Terhonen et al., 2019). Tree endophytes have evolved together with their hosts for millions of years, excluding the high virulence (Sieber, 2007). However, the lifestyles (mutualism, commensalism, and parasitism) of most endophytic fungi inhabiting woody tree tissues are still unknown and there are species with lifestyle transitions. Depending on the senescence, vitality, environmental, and health conditions of their hosts, tree endophytes may exhibit pathogenic, putatively mutualistic (Sieber, 2007), or saprotrophic behavior. For this reason, endophytic assemblages may bear latent pathogens or dormant saprotrophs of their hosts (Osono, 2006). Opportunistic pathogens within the endophytic assemblages may benefit from altered physiological or biochemical parameters of their hosts, for example due to senescence or water deficiency (Boyer, 1995; Houston, 1984; Sieber, 2007). Endophytic, latent pathogens are in close and continuous relation with their hosts. They are sensitive to minor changes in the metabolic pathways and vigor of host trees and may take advantage when plant defenses weaken (Gonthier et al., 2005). This behavior is hypothesized for endophytic, wood-decaying *Xylaria* species that are latent before

TABLE 10.1 Endophytes in different types of tissues.

Endophyte species or genus	Inner bark (phloem)	Sapwood (xylem)	Heartwood	References
Ascocoryne cylichnium, A. sarcoides			*Picea abies*	Leonhardt et al. (2019), Roll-Hansen and Roll-Hansen (1979), Wal et al. (2016)
Bjerkandera adusta		Broadleaf trees		Oses et al. (2008)
Chaetomium globosum	*Terminalia arjuna*			Tejesvi et al. (2005)
Fomes fomentarius		*Fagus sylvatica*		Rodríguez et al. (2011)
Liberomyces	Broadleaf trees, such as, *Quercus, Alnus, Salix*			Pažoutová et al. (2012)
Myrothecium	*Terminalia arjuna*			Tejesvi et al. (2005)
Neonectria coccinea		*Fagus sylvatica*		Rodríguez et al. (2011)
Pestalotiopsis spp.	*Terminalia arjuna*			Tejesvi et al. (2005)
Phialophora spp.			*Eucalyptus globulus*	Simeto et al. (2005)

TABLE 10.2 Comparison of common endophytic fungi of selected angiosperms and gymnosperm tree species.

	Tree species	Frequent endophyte	Division, order	Tissue type	References
Angiosperm	*Fagus sylvatica*	*Coniophora puteana*	B, Boletales	Woody tissues stem	Baum et al. (2003)
		Vuilleminia comedens	B, Corticiales	Woody tissues	Parfitt et al. (2010)
		Apiognomonia errabunda	A, Diaporthales	Leaves, buds, twigs, stems	Toti et al. (1993), Sieber and Hugentobler (1987), ibid
		Diaporthe eres		Leaves, twigs, stems	Sieber (2007), ibid
		Dicarpella dryina		Leaves	Sieber (2007)
		Phomopsis spp.		Woody tissues	Hendry et al. (2002)
		Xylodon radula, Schizopora paradoxa	B, Hymenochaetales	Woody tissues	Hendry et al. (2002)
		Clonostachys rosea	A, Hypocreales	Woody tissues	Hendry et al. (2002)
		Sphaerostilbella penicillioides		Woody tissues	Hendry et al. (2002)
		Asterosporium asterospermum	A, Incertae sedis	Twigs, branches and stems	Butin (2011), Chapela and Boddy (1988a), Senanayake et al. (2018)
		Virgariella sp.		Woody tissues	Hendry et al. (2002)
		Neonectria coccinea[a]	A, Nectriaceae	Sapwood (Xylem), woody tissues	Chapela (1989), Chapela and Boddy (1988a), Hendry et al. (2002)
		Fomes fomentarius	B, Polyporales	Woody tissues, stem	Baum et al. (2003)
		Phlebia radiata		Woody tissues	Hendry et al. (2002)
		Hericium cirrhatum, Stereum gausapatum, Stereum rugosum	B, Russulales	Woody tissues	Parfitt et al. (2010)
		Biscogniauxia nummularia[a] *Daldinia concentrica, Eutypa spinosa, Hypoxylon fragiforme*[a] *Hypoxylon fuscum, Nemania serpens*	A, Xylariales	Woody tissues	Chapela (1989), Chapela and Boddy (1988a), Hendry et al. (2002), Parfitt et al. (2010)
Gymnosperm	*Pinus sylvestris*	*Sydowia polyspora*	A, Dothideales	Woody tissues	Bußkamp et al. (2020), Kowalski and Kehr (1992), Sanz-Ros et al. (2015)
		Aureobasidium pullulans	A, Dothideales	Woody tissues	Martínez-Álvarez et al. (2012)
		Microsphaeropsis olivacea	A, Pleosporales	Woody tissues	Bußkamp et al. (2020), Kowalski and Kehr (1992), Petrini and Fisher (1988)
		Pezicula spp.	A, Helotiales	Woody tissues	Kowalski and Kehr (1992), Sanz-Ros et al. (2015)
		Pestalotiopsis spp.	A, Amphisphaeriales	Woody tissues	Bußkamp et al. (2020)
		Diaporthe spp.	A, Diaporthales		Kowalski and Kehr (1992), Petrini and Fisher (1988)
		Sphaeropsis sapinea	A, Botryosphaeriales	Woody tissues	Kowalski and Kehr (1992), Martínez-Álvarez et al. (2012), Petrini and Fisher (1988)
		Peniophora pini	B, Russulales	Woody tissues	Bußkamp et al. (2020)
		Bjerkandera adusta	B, Polyporales	Woody tissues	Giordano et al. (2009), Schlechte (1986)
		Heterobasidion annosum	B, Russulales	Woody tissues	Giordano et al. (2009)
		Sistotrema coroniferum	B, Cantharellales	Woody tissues	Giordano et al. (2009)
		Rhizoctonia solani	B, Cantharellales	Woody tissues	Giordano et al. (2009)

A, Ascomycota; *B*, Basidiomycota.
[a] *Most common endophytes of woody tissues of the host species.*

TABLE 10.3 Endophytes of woody tree tissues except vascular wilt pathogens.

Endophyte	Division, order	Host trees	Lifestyle	Disease	References
Alternaria	A, Pleosporales	*Pinus sylvestris* twigs	Endophyte		Suryanarayanan (2011), Bußkamp et al., (2020)
Alternaria alternata	A, Pleosporales	*Q. brantii, Q. macranthera*	Generalist, pathogen	Leaf spot and other diseases	Ghasemi et al. (2019), Samson et al. (2010)
Apiognomonia errabunda = *Discula umbrinella* = *Apiognomonia quercina*	A, Diaporthales	*Fagus sylvatica* (leaves, buds, twigs, stems) *Quercus*	Endophyte, opportunistic pathogen	Anthracnose	Jaklitsch and Voglmayr (2019), Linaldeddu et al. (2011), Moricca and Ragazzi (2008, 2011), Senanayake et al. (2018), Butin (1980, 2011), Toti et al. (1993), Sieber and Hugentobler (1987)
Aposphaeria spp.	A, Pleosporales	*Betula pendula*	Endophyte		Kowalski (1998)
Ascocoryne cylichnium, A. sarcoides	A, Helotiales	*Picea abies*	Endophyte and subsequent early stage decomposer		Leonhardt et al. (2019), Roll-Hansen and Roll-Hansen (1979), Wal et al. (2016)
Asterosporium spp.	A, Incertae sedis	Betulaceae, Fagaceae, Juglandaceae, and Sapindaceae, such as, *Betula* and *Alnus*	Endophytes and saprobes		Wijayawardene et al. (2016), Tanaka et al. (2010)
Asterosporium asterospermum	A, Incertae sedis	*Fagus sylvatica* (on twigs, branches and stems)	Host-specific endophyte and subsequent decomposer		Butin (2011), Chapela and Boddy (1988a), Senanayake et al. (2018)
Aureobasidium pullulans	A, Dothideales	*Pinus sylvestris*	Endophyte		Martínez-Álvarez et al. (2012)
Biscogniauxia mediterranea	A, Xylariales	*Fagus sylvatica, Quercus* ssp., *Q. suber, Q. brantii, Q. macranthera, Pinus sylvestris*	Endophyte, latent pathogen, wood decay fungus	Cause of charcoal canker on several *Quercus* and other hardwood species	Jurc and Ogris (2006), Mirabolfathy (2013), Linaldeddu et al. (2011), Ghasemi et al. (2019), Henriques et al., (2014a, b, 2015), Ragazzi et al., (2011), Bußkamp et al. (2020)
Biscogniauxia nummularia	A, Xylariales	*Fagus sylvatica, Pinus sylvestris* and various other plant species	Endophyte, latent pathogen, wood-decay fungus originating from xylem	Wood decay, cause of charcoal canker and strip cankering on European beech	Petrini-Klieber (1985), Bußkamp et al. (2020), Chapela (1989), Chapela and Boddy (1988a)
Bjerkandera adusta	B, Polyporales	Broadleaf trees, *Fagus sylvatica, Pinus sylvestris*	Endophyte, opportunistic pathogen, mainly saprotrophicwood decay fungus	WR	Giordano et al. (2009), Oses et al., (2008), Schlechte (1986), ibid
Botryosphaeria corticola	A, Botryosphaeriales	Oak	Endophyte, latent, weak pathogen	Bot (ryosphaeria) canker	Linaldeddu et al. (2011)
Chaetomium globosum	A, Sordariales	*Terminalia arjuna*	Endophyte		Tejesvi et al. (2005)
Cladosporium	A, Capnodiales	*Pinus sylvestris* twigs	Endophyte		Suryanarayanan (2011), Bußkamp et al., (2020)
Clonostachys rosea	A, Hypocreales	*Fagus sylvatica*	Endophyte		Hendry et al. (2002)
Colletotrichum	A, Glomerellales	Various tree species	Endophyte	Anthracnose diseases	Suryanarayanan (2011)
Colpoma quercinum	A, Rhytismatales	*Quercus robur, Q. petraea,* twigs	Endophyte, subsequent decompmoser		Kowalski and Kehr (1992), Moricca et al. (2012), Halmschlager et al. (1993)

Coniella sp.	A, Diaporthales	Eucalyptus globolus	Endophyte		Bettucci and Saravay (1993)
Coniophora puteana	B, Boletales	Fagus (frequent in the lower canopy)	Endophyte, wood decay fungus, originating from the bark, BRF	BR	Baum et al. (2003), Hendry et al. (2002)
Coprinellus sp.	B, Agaricales	Pinus sylvetris	Endophyte WRF	WR	Bußkamp et al. (2020)
Cryptococcus sp.	B, Tremellales	Quercus brantii	Endophyte, yeast-like		Ghobad-Nejhad et al. (2018)
Cryptosporella betulae	A, Diaporthales	Betula pendula, Betula pubescens	Endophyte, pathogen, saprobiont		Barengo et al. (2000), Kowalski and Kehr (1992), Hanso and Drenkhan (2010)
Cryptosporella suffusa = Disculina vulgaris	A, Diaporthales	Alnus glutinosa, Alnus incana, Alnus viridis	Endophyte, pathogen, saprobiont	Alder canker	Fisher and Petrini (1990), Sieber et al. (1991a,b), Mejía et al. (2008)
Cytospora chrysosperma	A, Diaporthales	Populus tremuloides, various host tree species	Endophyte, pathogen	Stem and branch canker widely spread in Asia, North and South America, Africa, and Oceania	Bagherabadi et al. (2017), Chapela (1989)
Cytospora populina ≡ Cryptosphaeria populina	A, Diaporthales	Populus tremuloides	Endophyte, pathogen	Stem and branch canker	Chapela (1989)
Cytospora spp.	A, Diaporthales	Quercus. brantii twigs	Endophyte		Ghasemi et al. (2019)
Daldinia concentrica	A, Xylariales	Angiosperm trees, such as, Fagus sylvatica, Quercus, Betula	Endophyte, latent pathogen, saprobiont	Wood decay	Parfitt et al. (2010)
Dendrostoma leiphaemia = Discula quercina = Fusicoccum quercinum	A, Diaporthales	Quercus	Endophytes and opportunistic pathogens	Anthracnose	Jaklitsch and Voglmayr (2019), Linaldeddu et al. (2011), Moricca and Ragazzi (2008, 2011), Senanayake et al. (2018), Butin (1980, 2011)
Desmazierella acicola	A, Pezizales	Pinus spp.	Endophyte, Sapropyhte		Kowalski and Kehr (1992), Kowalski and Zych (2002), Petrini and Fisher (1988), Ponge (1991)
Diaporthe	A, Diaporthales	Pinus sylvestris	Endophyte		Suryanarayanan (2011), Bußkamp et al., (2020)
Diaporthe eres = Phomopsis velata	A, Diaporthales	Fagus sylvatica	Endophyte, latent secondary pathogen		Sieber (2007), ibid
Diatrype spp.	A, Xylariales	Quercus. Brantii twigs	Endophyte and subsequent decomposer	Wood decay	Ghasemi et al. (2019)
Dicarpella dryina (Anamorph: Tubakia dryina)	A, Diaporthales	Quercus	Endophyte, weak pathogen, saprobiont	Leaf spots	Gennaro et al. (2003), Bressem et al. (2013)
Durandiella gallica	A, Helotiales	Abies alba	Endophyte, weak pathogen, subsequent decomposer		Kowalski and Butin (1989)
Entoleuca mammata ≡ Hypoxylon mammatum	A, Xylariales	Populus tremuloides, Populus spp., Salix sp.	Endophyte, latent pathogen, subsequent decomposer	Native to North America and causing hypoxylon canker in hardwood trees, particularly in aspen, poplars, and willows	Chapela (1989), Jeger et al. (2017)

Continued

TABLE 10.3 Endophytes of woody tree tissues except vascular wilt pathogens.—cont'd

Endophyte	Division, order	Host trees	Lifestyle	Disease	References
Eutypa spinosa	A, Xylariales	Betula, Fagus sylvatica, Quercus	Endophyte, latent pathogen, saprobiont	Wood decay fungus	Parfitt et al. (2010)
Fomes fomentarius	B, Polyporales	Betula, Fagus sylvatica, Quercus brantii	Endophytic latent wood decay fungus	WR	Baum et al. (2003), Parfitt et al. (2010), Rodriguez et al. (2011)
Fusicoccum betulae	A, Botryosphaeriales	Betula pendula	Endophyte		Kowalski (1998)
Grovesiella abieticola	A, Helotiales	Abies alba	Endophyte and subsequent decomposer		Kowalski and Butin (1989)
Hericium cirrhatum	B, Russulales	Fagus sylvatica	Endophytic latent wood decay fungus	WR	Parfitt et al. (2010)
Hericium sp.	B, Russulales	Fagus sylvatica	Endophytic latent wood decay fungus	WR	Parfitt et al. (2010)
Heterobasidion annosum	B, Russulales	Pinus sylvestris	Endophyte, parasite and wood decay fungus	WR, root and stem rot	Giordano et al. (2009)
Hypoxylon	A, Xylariales	Pinus sylvestris twigs	Endophyte	Wood decay	Suryanarayanan (2011), Bußkamp et al., (2020)
Hypoxylon fragiforme	A, Xylariales	Fagus sylvatica, Quercus (twigs, branches, and stems)	Endophyte, latent pathogen, saprobiont; originating from xylem, and fructificate with its anamorph or teleomorph (red cushion hypoxylon) in clusters on the bark of dead beech	Wood decay	Hendry et al. (2002), Parfitt et al. (2010), Chapela (1989), Chapela and Boddy (1988a)
Hypoxylon fuscum	A, Xylariales	Fagus sylvatica	Endophytic latent wood decay fungus	Wood decay	Parfitt et al. (2010)
Lasiodiplodia theobromae	A, Botryosphaeriales	Various tropical and subtropical trees	Endophyte and latent pathogene	Bot (ryosphaeria) canker and other disease symptoms on several trees	Mohali et al. (2005), Suryanarayanan (2011), Úrbez-Torres (2011)
Lentinus tigrinus	B, Polyporales	Quercus brantii	Endophytic latent wood decay fungus	WR	Ghobad-Nejhad et al. (2018)
Liberomyces	A, Incertae sedis	Broadleaf trees such as Quercus, Alnus, Salix	Endophyte		Pažoutová et al. (2012)
Melanconis alni = Melanconium apiocarpum	A, Diaporthales	Alnus glutinosa	Endophyte and subsequent decomposer		Fisher and Petrini (1990), Sieber et al. (1991a,b)
Microsphaeropsis olivacea ≡ Coniothyrium olivaceum Bonord.	A, Pleosporales	Quercus. brantii twigs, Pinus sylvestris twigs, twigs and branches of Acer Cytisus, Hedera, Laurus, Lycium, Quercus Sambucus	Endophyte, pathogen, and saprophyte		Ghasemi et al. (2019), Kowalski (1993), Kowalski and Kehr (1992), Bußkamp et al., (2020), Petrini and Fisher (1988), Ellis and Ellis (1985), Sun et al. (2011), Alidadi et al. (2019)

Species	A/B, Order	Host	Role	Wood decay	References
Nemania serpens	A, Xylariales	Fagus sylvatica, Quercus, Pinus sylvetris	Endophyte, latent pathogen, saprobiont	Wood decay	Parfitt et al. (2010), Bußkamp et al., (2020)
Neocucurbitaria cava	A, Xylariales	Oak	Endophyte, weak pathogen		Linaldeddu et al. (2011)
Neonectria coccinea	A, Hypocreales	Fagus sylvatica	Endophyte, latent pathogen originating from xylem	Key causal agent in BBD, bark necrosis	Chapela (1989), Chapela and Boddy (1988a), Hendry et al. (2002), Langer et al. (2020)
Oedocephalum sp.	A, Pezizales	Quercus robur (sapwood of trunks)	Endophyte		Gonthier et al. (2005)
Ophiognomonia intermedia	A, Diaporthales	Betula pendula	Endophyte		Kowalski (1998)
Paecilomyces maximus	A, Eurotiales	Q. brantii, Q. macranthera	Endophyte		Ghasemi et al. (2019)
Peniophora pini	B, Russulales	Pinus sylvestris	Endophyte WRF	WR	Bußkamp et al. (2020)
Peniophora spp.	B, Russulales	Abies beshanzuensis, Pinus sylvestris	Endophytic latent wood decay fungus	WR	Yuan et al. (2011), Giordano et al. (2009)
Pestalotiopsis besseyi	A, Amphisphaeriales	Pinus halepensis, P. tabuliformis	Endophyte		Guo et al. (2008), Wang and Guo (2007), Botella and Diez (2011)
Pestalotiopsis citrina	A, Amphisphaeriales	Pinus tabuliformis	Endophyte		Wang and Guo (2007)
Pestalotiopsis funerea	A, Amphisphaeriales	Pinus pinaster	Endophyte		Martínez-Álvarez et al. (2012)
Pestalotiopsis guepinii	A, Amphisphaeriales	Eucalyptus globolus, Pinus elliotti, and P. taeda	Endophyte		Bettucci and Saravay (1993), Alonso et al. (2011)
Pestalotiopsis spp.	A, Amphisphaeriales	Various tree species, Terminalia arjuna, Pinus sylvestris	Endophytes, pathogens, and saprotrophs		Tejesvi et al. (2005), Maharachchikumbura et al. (2011), Suryanarayanan (2011), Bußkamp et al. (2020)
Pezicula alni	A, Helotiales	Alnus incana, Alnus viridis	Endophyte		Fisher and Petrini (1990)
Pezicula aurantiaca	A, Helotiales	Alnus incana, Alnus viridis, Pinus tabuliformis	Endophyte		Fisher and Petrini (1990), Guo et al. (2003)
Pezicula cinnamomea	A, Helotiales	Abies alba, Betula pendula, Quercus, Picea abies, and Pinus sylvestris	Endophyte, latent and weak pathogen	Pezicula-canker on oak	Kowalski (1998), Kehr (1991), Bußkamp et al., (2020), Kowalski and Kehr (1992)
Pezicula eucrita	A, Helotiales	Pinus nigra and P. sylvestris	Endophyte, latent and weak pathogen	Pezicula-canker on Pinaceae	Verkley (1999), Kowalski and Zych (2002), Bußkamp et al., (2020), Sanz-Ros et al. (2015)
Pezicula livida	A, Helotiales	Abies alba, Picea abies, and Pinus sylvestris	Endophyte		Bußkamp (2018), Kowalski and Kehr (1992), Sieber-Canavesi and Sieber (1987)
Pezicula spp.	A, Helotiales	Pinus sylvestris	Endophyte		Kowalski and Kehr (1992), Sanz-Ros et al. (2015)
Pezicula. sporulosa	A, Helotiales	Abies beshanzuensis	Endophyte		Yuan et al. (2011)
Phanerochaete sp.	B, Polyporales	Eucalyptus globolus	Endophyte, WRF	WR	Bettucci and Saravay (1993)

Continued

TABLE 10.3 Endophytes of woody tree tissues except vascular wilt pathogens.—cont'd

Endophyte	Division, order	Host trees	Lifestyle	Disease	References
Phialocephala sp.	A, Helotiales	*Betula pendula*	Endophyte		Kowalski (1998)
Phialophora spp.	A, Chaetothyriales	*Eucalyptus globulus*	Endophyte		Simeto et al. (2005)
Phlebia radiata	B, Polyporales	*Fagus sylvatica, Quercus brantii*	Endophytic latent wood-decay fungus	WR	Hendry et al. (2002), Ghasemi et al. (2019)
Phomopsis archeri	A, Diaporthales	*Pinus elliotti, P. tabuliformis*	Endophyte		Wang and Guo (2007), Guo et al. (2008), Alonso et al. (2011)
Phomopsis occulta	A, Diaporthales	*Abies alba Picea abies Pinus sylvestris, Pinus nigra*	Endophyte		Kowalski and Kehr (1992), Kowalski and Zych (2002), Sieber (1989), Sieber-Canavesi and Sieber (1987)
Phomopsis spp.	A, Diaporthales	*Abies alba, Fagus sylvatica, Pinus halepensis, and P. wallichiana*	Endophyte		Kowalski and Kehr (1992), Botella and Diez (2011), Qadri et al. (2014), Hendry et al. (2002)
Phyllachora eucalypti ≡ *Plectosphaera eucalypti*	G, Xylariales	*Eucalyptus globolus*	Endophyte		Bettucci and Saravay (1993)
Pleurophomopsis lignicola	A, Pleosporales	*Alnus glutinosa*	Endophyte		Fisher and Petrini (1990)
Prosthemium stellare	A, Pleosporales	*Alnus incana, Alnus viridis*	Endophyte		Fisher and Petrini (1990)
Pseudovalsa umbonata	A, Diaporthales	oak	Endophyte, weak pathogen		Linaldeddu et al. (2011)
Pycnoporus sanguineus	B, Polyporales	*Eucalyptus globulus*	Endophyte, WRF	WR	Bettucci and Saravay (1993)
Rhizoctonia solani	B, Cantharellales	*Pinus sylvestris*	Endophyte, parasite	Various plant diseases, collar and root rots, wire stems, and damping off	Giordano et al. (2009)
Rhizosphaera kalkhoffii	A, Venturiales	*Abies beshanzuensis, Picea abies*	Endophyte, weak pathogen	Needle cast	Yuan et al. (2011), Sieber (1989), Diamandis and Minter (1980), Scattolin and Montecchio (2009)
Sarocladium strictum	A, Hypocreales	*Quercus robur* (sapwood of trunks)	Endophyte		Gonthier et al. (2005)
Schizophyllum commune	B, Agaricales	*Pinus sylvestris*	Endophyte, weak parasite and wood decay fungus; often opportunistic after sunburn of the host tree	WR	Giordano et al. (2009)
Schizopora paradoxa	B, Hymenochaetales	*Fagus sylvatica*	Endophyte and latent wood decay fungus	WR	Hendry et al. (2002)
Scytalidium lignicola	A, Helotiales	*Quercus robur* twig and stem	Endophyte		Gonthier et al. (2005)
Sistotrema coroniferum	B, Cantharellales	*Pinus sylvestris*	Endophyte, latent wood decay fungus	WR	Giordano et al. (2009)

Species	Phylum, Order	Host	Function	Type	References
Sordaria	A, Sordariales	*Pinus sylvestris* twigs	Endophyte		Suryanarayanan (2011), Bußkamp et al., (2020)
Sphaeropsis sapinea	A, Botryosphaeriales	*Pinus sylvestris*, pines and several other conifers, *Fagus sylvatica*; *Quercus suber*	Endophyte, latent pathogen, subsequent decomposer	Diplodia tip blight	Zlatković et al. (2017), Bußkamp et al., (2020), Smahi et al. (2017)
Sphaerostilbella penicillioides	A, Hypocreales	*Fagus sylvatica*	Endophyte		Hendry et al. (2002)
Stereum gausapatum	B, Russulales	*Fagus sylvatica, Quercus brantii*	Endophytic latent wood decay fungus	WR	Linaldeddu et al. (2011), Parfitt et al. (2010)
Stereum rugosum	B, Russulales	*Betula, Fagus sylvatica, Quercus brantii*	Endophytic latent wood decay fungus	WR	Linaldeddu et al. (2011), Parfitt et al. (2010), Linaldeddu et al. (2011)
Sydowia polyspora	A, Dothideales	Conifers, *Pinus sylvestris, Abies alba, Picea abies*	Endophyte, latent *pathogen and saprophyte*	Current season needle necrosis (CSNN)	Müller and Hallaksela (2000), Bußkamp et al., (2020), Sieber (1989), Lygis et al. (2014), Menkis et al. (2006), Müller et al. (2001), Pan et al. (2018), Talgo et al. (2010)
Trametes versicolor	B, Polporales	*Abies beshanzuensis; Eucalyptus globolus*	Endophytic latent wood decay fungus	WR	Yuan et al. (2011), Bettucci and Saravay (1993)
Trichoderma spp.	A, Hypocreales	*Fagus, Populus, Quercus*	Epiphytes of many plant surfaces, endophytes and as saprotrophic decomposers of woody tissues originating from the bark, soil fungi		Baum et al. (2003), Chapela (1989), Cotter and Blanchard (1982), Hendry et al. (2002), Ragazzi et al. (1999), Samuels (1996)
Trichothecium roseum	A, Hypocreales	*Q. brantii, Q. macranthera*	Endophyte		Ghasemi et al. (2019)
Truncatella conorum-piceae ≡ Pestalotia conorum-piceae	A, Amphisphaeriales	*Pinus, Pinus sylvestris* twigs	Endophyte, weak pathogen, mainly saprotrophic and subsequent decomposer of predamaged needles		Landeskompetenzzentrum Forst Eberswalde (2016), Bußkamp et al., (2020)
Tryblidiopsis pinastri	A, Rhytismatales	*Picea abies*	Endophyte, weak pathogen, and subsequent decomposer		Livsey and Minter (1994), Tanney and Seifert (2019)
Virgariella sp.	A, Incertae sedis	*Fagus sylvatica*	Endophyte		Hendry et al. (2002)
Vuilleminia comedens	B, Corticiales	*Fagus sylvatica, Quercus brantii*	Endophytic latent wood decay fungus	WRF	Parfitt et al. (2010)
Xylaria	A, Xylariales	*Pinus sylvestris* twigs	Endophyte	Wood decay	Suryanarayanan (2011), Bußkamp et al., (2020)
Xylaria spp.	A, Xylariales	Various tree species, *Fagus sylvatica*	Endophytic latent wood-decay fungus	Wood decay	Hendry et al. (2002), Suryanaravanan (2011)
Xylodon radula	B, Hymenochaetales	*Fagus sylvatica*	Endophytic latent wood decay fungus	WR	Hendry et al. (2002)

A, Ascomycota; B, Basidiomycota.

TABLE 10.4 Types of wood decay and examples of wood-decaying fungi.

	White rot (WR)	Brown rot (BR)	Soft rot (SR)	
Type of wood decay	Complete degradation of all wood compounds with the most effectively ability to degrade lignin	Degradation of carbohydrates (cellulose and hemicellulose) without lignin-degradation; only able to demethylate lignin	Enzymatically breaks down cellulose. Some ascomycetous species are able to degrade lignin	
Leaving wood residues	Bleached to whitish and fibrous	Brownish, not fibrillary often breaking cubical or there mains can be ground into powder	Bleached or discolorated, sometimes similar to brown-rotted wood, often with minute cavities inside	
Causal agents	**Basidiomycota**	**Basidiomycota**	**Mainly Ascomycota**	**References**
Armillaria spp.	Root, butt, and stem rot			Lygis et al. (2005)
Botryobasidium botryosum	Intermediate type of wood decay in the final stages of wood degradation; able to degrade lignin and polysaccharides, but lacks the typical high-oxidation potential peroxidases			Alfaro et al. (2016), Riley et al. (2014)
Cadophora (C. malorum, C. luteo-olivacea, C. fastigiata)			x	Blanchette et al. (2004), Savory (2008)
Ceratocystis			x	Blanchette et al. (2004), Savory (2008)
Chaetomium			x	Blanchette et al. (2004), Savory (2008)
Coniophora puteana		x		Várnai et al. (2014)
Dacrymycetales such as, *Dacrymyces stillatus, Dacryopinax* sp.		x		Várnai et al. (2014)
Dichomitus squalens	White pocket rot, with the initial decay stage giving a red coloration to the wood followed by the full discoloration of the wood tissue with extensive damage of the structure due to lignin degradation			Rytioja et al. (2015), Casado López et al. (2019)
Fomes fomentarius	Trunk and stem rot, simultaneous			Blanchette (1984)
Fomitopsis pinicola		Trunk and stem rot		Várnai et al. (2014)
Ganoderma applanatum	Root and butt rot, simultaneous	Trunk and stem rot		Rodríguez et al. (2011), Blanchette (1984)
Heterobasidion annosum s.l.	Root rot, stem rot. Rot step by step or simultaneous			Baldrian (2008), Blanchette (1991)
Inonotus hispidus	White and Soft rot		x	Schwarze et al. (1995)
Kretzschmaria deusta			Root and butt rot with significant lignin degradation	Schwarze (2018)
Laetiporus sulphureus		Trunk and stem rot		Lee et al. (2009), ibid
Mucidula mucida	Stem rot with cavity forming		Transient stem rot	Daniel et al. (1992)

TABLE 10.4 Types of wood decay and examples of wood-decaying fungi.—ccont'd

Causal agents	Basidiomycota	Basidiomycota	Mainly Ascomycota	References
Phellinus, such as, Phellinus pini ≡ Porodaedalea pini	Pocket rot			Rodríguez et al. (2011)
Phlebia radiata	+			Mattila et al. (2020)
Pleurotus ostreatus	Trunk and stemrot			
Rigidoporus microporus	White root rot disease, decay type intermediate between WR and BR			Oghenekaro et al. (2020)
Schizophyllum commune	Trunk and stem rot without high-oxidation potential peroxidases			Riley et al. (2014), Schmidt and Liese (1980)
Trametes versicolor	Simultaneous			Blanchette (1984)
Xylaria polymorpha			Root and butt rot with significant lignin degratation	Schwarze (2018)
Xylariales family			Wood decay with significant lignin rot	Schwarze (2018)
Xylobolus frustulatus	Pocket rot			Rodríguez et al. (2011)

decomposing the host cell wall compounds (Petrini and Petrini, 1985; Whalley, 1996). The lifestyle changes when their hosts are aged or have less vigor (Petrini and Petrini, 1985; Whalley, 1996).

The ecological role of endophytes is still poorly understood. Many endophytes are considered to be endured parasites of their host or opportunistic or latent pathogens (Sanz-Ros et al., 2015). Another role of endophytes is as a primary colonizer, that is, they are initially in a physiological resting phase in the tree. During the senescence of the host, they can form fruiting bodies and spores (Chapela and Boddy, 1988a; Griffith and Boddy, 1988; Kehr, 1998; Oses et al., 2008; Osono, 2006; Sanz-Ros et al., 2015). The question is, how do endophytes overcome a plant's defense and colonize their host's tissues? This could be done either through metabolites secreted by the endophyte (Peters et al., 1998; Schulz et al., 2002, 2015) or by changing the phytohormone balance of the tree (Navarro-Meléndez and Heil, 2014). Because fungal endophytes share the same ecological niche with other living organisms, they compete with one another and can produce substances with antagonistic effects. These can act against fungi (Romeralo et al., 2015), bacteria (Schulz et al., 2002) or insects (Azevedo et al., 2000). Therefore, endophytes are potential agents to improve the host tree health or possible biological tools to control tree pathogens and pests (Rabiey et al., 2019). Additionally, tree endophytes could represent sources of pharmacologically effective metabolites with antitumor or antibiotic impacts, among others (Fatima et al., 2016).

3. Fungi inhabiting living woody tissues: Are Basidiomycetes underrepresented or not yet discovered?

The majority of Basidiomycota in woody tissues appears to be wood-decay fungi, especially white rotters such as *Stereum rugosum* (Parfitt et al., 2010). Studies based on classical isolation methods indicate the low species richness of endophytic basidiomycetous fungi in woody tissues (Bußkamp et al., 2020). This could be due to methodological limitations of the isolation or the low abundance of Basidiomycota in these ecological niches. Moreover, Basidiomycota, especially wood-decay fungi, need longer incubation periods than Ascomycota for outgrowing from sampled woody tissues (Oses et al., 2008). High-throughput sequencing (HTS) methods might overcome such technical limitations.

The diversity of the endophytic mycobiome of temperate trees was discussed by Unterseher (2011), and for tropical trees by Suryanarayanan (2011) and Arnold et al. (2001). Usually, Ascomycota are significantly predominant endophyte communities in woody tissues, followed by Basidiomycota (Bußkamp, 2018; Ghobad-Nejhad et al., 2018; Singh et al., 2017). A divergence of the fungal communities in angiosperms and gymnosperms was found. Angiosperm endophyte communities are often dominated by species of Diaporthales (Fig. 10.4D–F), whereas in gymnosperms, Helotiales prevail (Sieber, 2007).

However, these differences are not necessarily obvious in woody tissues in comparison to leaves and needles (Table 10.2). Woody tissues of branches of deciduous trees seem to have higher species richness than in coniferous trees (Kowalski and Kehr, 1992). The frequency of the most common branch endophytes is influenced host-specifically by the branch diameter (Kowalski and Kehr, 1992). Indeed, in culture-based studies on endophytes in branches of Scots pines, only a few Basidiomycota at all were detected, such as *Peniophora pini* and *Coprinellus* sp. among 103 pine endophytes (Bußkamp et al., 2020). Similarly, previous studies also noted that very few basidiomycetous endophytes were isolated (Kowalski and Kehr, 1992; Martínez-Álvarez et al., 2012; Peršoh et al., 2010; Petrini and Fisher, 1988; Sanz-Ros et al., 2015). Kowalski and Kehr (1992) found very few basidiomycetous fungi from *Larix decidua*. Three Basidiomycota, *Trametes versicolor* and two *Peniophora* species, were isolated from *Abies beshanzuensis* (Yuan et al., 2011). Giordano et al. (2009) found that among 143 species isolated from the sapwood of *P. sylvestris* in Spain, 17 species belonged to Basidiomycota (12% of the species) such as *Bjerkandera adusta, Heterobasidion annosum, Peniophora* sp., *Schizophyllum commune, Sistotrema coroniferum,* and *Rhizoctonia solani*. Because a large part of the research on fungi inhabiting woody tissues was performed on diseased trees, (weak) parasites and saprobionts were identified. Typical pine species were *B. adusta, H. annosum, R. solani,* and *Tr. versicolor* (Butin, 2011; Schlechte, 1986; Woodward et al., 1998). Based on these results, living branches and twigs of conifers do not appear to have a large variety of basidiomycetous endophytes.

Overall, endophyte communities of woody tree tissues are dominated by a few host-specific species (Sieber, 2007), generalists, or typical species for the tissue type of the host (Bußkamp, 2018). Generalists are multihost endophytes that occur in taxonomically unrelated tree species and have lost their host specificity (Suryanarayanan, 2011). However, there are endophytes with a wide host range that sporulate only on a single or a few host tree species (Baayen et al., 2002; Petrini and Petrini, 1985). *Biscogniauxia nummularia* is a widespread endophyte in coniferous twigs (e.g., pine, Douglas fir, and fir) and other hosts (Petrini-Klieber, 1985). It is one of the most common endophytes of the woody tissues of the European beech (Chapela, 1989; Chapela and Boddy, 1988a). The multihost cosmopolitic endophyte *Lasiodiplodia theobromae* has a worldwide distribution in tropical and subtropical regions and it causes "Bot(ryosphaeria) canker" as well as other disease symptoms on several trees (Mohali et al., 2005; Suryanarayanan, 2011; Úrbez-Torres, 2011). Typical tree endophyte generalists are species such as *Colletotrichum, Pestalotiopsis,* and *Xylaria* (Suryanarayanan, 2011). The typical endophyte of the twigs of *Pinus sylvestris* is *Sphaeropsis sapinea* (Fig. 10.1), which was found in other conifers and recently in the beech *Fagus sylvatica* (Zlatković et al., 2017). Two other commonly observed endophytes in Scots pine twigs are generalists, *Microsphaeropsis olivacea* (Fig. 10.3) in angio- and gymnosperms and *Sydowia polyspora* (Fig. 10.2) in conifers (Bußkamp, 2018). Indeed, the observed endophyte assemblages of temperate trees contain a great many species, usually colonizing ubiquitously or opportunistically the phyllosphere and other substrates (Unterseher, 2011) such as species of *Alternaria* (Fig. 10.4I), *Cladosporium, Hypoxylon* (Fig. 10.4C), *Diaporthe, Sordaria,* and *Xylaria* in Scots pine twigs (Bußkamp, 2018; Suryanarayanan, 2011).

Most of the endophytic species in trees are found in the outer bark (Barklund and Kowalski, 1996). However, the inner bark also hosts a great diversity of fungal endophytes, for example 32 species on *Prosopis cineraria* (Gehlot, 2008). In a study by Bußkamp (2018), 26 different species were detected in the bark of *P. sylvestris*; 20 were found in the area of the cambium and only nine species could be detected in the woody tissue, including typical endophytes such as *S. polyspora* (Fig. 10.2) and *M. olivacea* (Fig. 10.3). Similarly, Petrini and Fisher (1988) isolated 16 fungi in the stems of *P. sylvestris* and only 10 in the xylem (excluding rare isolates). Wang and Guo (2007) examined different tissues of *Pinus tabuliformis* and found that the bark harbors more endophytic fungi than the xylem. The species density of endophytic fungi was higher in the stem bark (colonization density 64%–68%) than in the stem xylem (15%–35%) of *Alnus* (Fisher and Petrini, 1990).

In addition, the endophyte assemblages are tissue-specific and mainly composed of a small number of dominant species accompanied by a multitude of rare isolates and singletons. Within a host species, tissue type had the strongest effect on the species evenness of the endophyte community, followed by the geographical location of the sampling site and the season (Juybari et al., 2019; Singh et al., 2017). Fisher and Petrini (1990) discovered interspecific differences in the fungal communities of different host species within the genus *Alnus*. However, congeneric host species are often colonized by the same fungal endophyte or by closely related species, so-called "sister-species" (Sieber, 2007). For example, *Apiognomonia errabunda* s. l. (Fig. 10.4D) is an endophyte of the *Quercus* and *Fagus* hosts. Regardless of the influences of season and site locality, the endophyte diversity in stems, branches, and twigs seems to be less than in leaves (Singh et al., 2017).

4. Fungi causing tree diseases

Besides abiotic factors such as wind, drought, and fire, biotic factors such as insects and microbes are agents causing tree mortality (Harmon et al., 1986). Fungal diseases are among the most important factors affecting tree health; often, fungi inhabiting woody tree tissues initiate the disease process. Manion's (1981) "Decline Spiral" describes a model for the multifactorial complex diseases of forest trees. According to Manion, there are three categories of damaging factors: predisposing, inciting, and contributing factors. Predisposed, already weakened trees are vulnerable to secondary pathogens that give the trees the "mortal blow." Manion endophytic fungi inhabiting woody tree tissues may be activated by triggering environmental factors such as drought or heat. A switch from an endophytic to a parasitic lifestyle of latent pathogens may be induced and finally lead to tree disease or mortality. If enough photosynthetically active tissue is killed due to a pathogen, the tree dies if no resources to repair the damage are available or the tree can no longer preserve living tissue. Especially, shoot blights in coniferous trees are serious because they store a lot of carbon and nutrients in their needles (Krause and Raffa, 1996). Similarly, the most destructive tree diseases are vascular wilts that are generally caused by soil-borne bacteria, oomycetes, and fungi (Yadeta and Thomma, 2013).

4.1   Lifestyle switch from endophytic to pathogenic

Many fungal species can occur as symptomless endophytes in one host and cause disease in another (Sanz-Ros et al., 2015). Under certain circumstances, endophytes can become pathogens while residing in the same host (Hyde and Soytong, 2008; Ragazzi et al., 2003). Potential pathogens and typical endophytes are fungi of the genus *Diaporthe/Phomopsis* (Fig. 10.4F). *Diaporthe* spp. have a wide range of host plants and disease symptoms and they act as plant pathogens, endophytes, and saprophytes (Gomes et al., 2013). The species of this genus are often isolated from trees such as *Phomopsis occulta* from *P. sylvestris* (Kowalski and Kehr, 1992) as well as from shoots of *Pinus nigra* (Kowalski and Zych, 2002), *Picea abies* (Sieber, 1989), and *Abies alba* (Sieber, 1989; Sieber-Canavesi and Sieber, 1987). Other *Diaporthe* species, mainly *Ph. archeri*, were isolated from *P. tabuliformis* (Wang and Guo, 2007; Guo et al., 2008) and from twigs of *P. elliottii* (Alonso et al., 2011). *Phomopsis* spp. was isolated also from *A. alba* (Kowalski and Kehr, 1992), *P. halepensis* (Botella and Diez, 2011), and *P. wallichiana* (Qadri et al., 2014).

Fungi of the genus *Pezicula*, such as *Pe. cinnamomea* (Fig. 10.4H) are weak pathogens on oaks (Kehr, 1991). However, they are typical endophytes such as *Pe. livida* and *Pe. cinnamomea* on *A. alba*, *P. abies*, and *P. sylvestris* (Bußkamp, 2018; Kowalski and Kehr, 1992). *Pe. eucrita* can cause cankers on Pinaceae (Verkley, 1999), but was isolated as an endophyte from *P. nigra* (Kowalski and Zych, 2002) and *P. sylvestris* (Bußkamp, 2018; Sanz-Ros et al., 2015). Sieber-Canavesi and Sieber (1987) most frequently isolated *Pe. livida* from the branches of *A. alba* in Switzerland. *Pe. livida* was isolated from spruce, pine, and fir in branch bases according to Kowalski and Kehr (1992). *Pe. livida* is the most important fungus for natural pruning dynamics (Kowalski and Butin, 1989). Several other *Pezicula* species were isolated, including *Pe. aurantiaca* on *P. tabuliformis* (Guo et al., 2003) and *Pe. sporulosa* from twigs of *A. beshanzuensis* (Yuan et al., 2011).

Fungi of the genus *Pestalotiopsis* occur worldwide as pathogens, endophytes, and saprotrophs (Maharachchikumbura et al., 2011). Numerous *Pestalotiopsis* species have been detected frequently in conifers, for example *Pestalotiopsis besseyi* in the bark of *P. tabuliformis* (Guo et al., 2008; Wang and Guo, 2007) and in the twigs and needles of *P. halepensis* (Botella and Diez, 2011); *Pes. citrina* isolated from the bark of *P. tabuliformis* (Wang and Guo, 2007); *Pes. funerea* in the twigs of *Pinus pinaster* (Martínez-Álvarez et al., 2012); and *Pes. guepinii* from the bark of *P. elliottii* and *P. taeda* (Alonso et al., 2011).

4.1.1   Sphaeropsis sapinea–emerging endophyte switching to pathogenic lifestyle

The anamorphic species *S. sapinea* (= *Diplodia pinea*) occurs endophytically in conifers worldwide (Bihon et al., 2012; Brodde et al., 2019; Flowers et al., 2001; Langer et al., 2011; Luchi et al., 2014; Stanosz et al., 2001) and lives saprophytically on dead host tissues such as stems, twigs, bark, cones, or needles where pycnidia are developed (Fig. 10.1A).

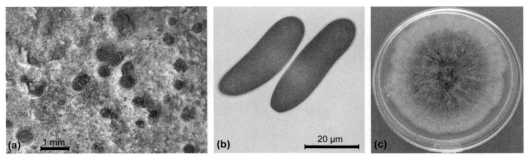

FIG. 10.1   *Sphaeropsis sapinea*, (A) pycnidia on the bark of *Abies grandis*; (B) Conidia; (C) pure culture on MYP after 7 days in ambient daylight at room temperature.

However, *S. sapinea* is a warmth-loving opportunistic pathogen and the causal agent of *Diplodia* tip blight (Brodde et al., 2019). To switch to a parasitic lifestyle, an injury of the host tissue or predisposing or triggering stress factors such as a lack of water of the host tree are needed (Swart and Wingfield, 1991; Blodgett et al., 1997; Stanosz et al., 2001; Luchi et al., 2014). Worldwide, 75 host plants and other plants are known to be affected (CABI, 2019) (such as, *Pinus* spp. with 45 species listed, *Abies* spp., *Larix* spp., *Picea* spp., and *Pseudotsuga menziesii*; Kaya et al., 2014). It was also detected in diseased tissues of *F. sylvatica* (Zlatković et al., 2017) and *Quercus suber* (Smahi et al., 2017).

Typical disease symptoms are shoot dieback and resin flow in shoots, which can lead to crown death as well as bark damage, blue stain, bark necrosis, root rot, and even tree death (such as, Swart and Wingfield, 1991). *Diplodia* tip blight first occurred on nonnative pine plantations in South Africa and New Zealand at the beginning of the 20th century, and later spread to South and Central Europe (Swart and Wingfield, 1991). It was detected in Sweden in 2013 and in 2015. *S. sapinea* was discovered in Finland for the first time (Müller et al., 2018; Oliva et al., 2013). It was most likely distributed worldwide by infected plant material such as seedlings or insects carrying *S. sapinea* (Drenkhan et al., 2017; Feci et al., 2002; Luchi et al., 2012; Whitehill et al., 2007). Predisposing and triggering events for *Diplodia* tip blight could be drought and other sources of stress (e.g., infestation with *Viscum album* subsp. *austriacum*, edaphic factors/nutrient supply (Dijk et al., 1992; Diminić et al., 2012)). It is presumed that damage caused by *S. sapinea* in climate change will increase at high air temperatures and drought stress. Second, elevated temperature has a positive effect on the growth of *S. sapinea* (Bosso et al., 2017). The optimal temperature for the growth of *S. sapinea in vitro* is 25–30°C (Bußkamp, 2018; Keen and Smits, 1989; Milijašević, 2006; Palmer et al., 1987). In several studies, the association between temperature and disease incidence could be observed. Indeed, warm May and June temperatures were associated with higher damage in Sweden (Brodde et al., 2019) and low winter temperatures correlated negatively with its occurrence in France (Fabre et al., 2011).

For inciting factors, hail (Langer et al., 2011; Schumacher, 2012; Zwolinski et al., 1995) and insects feeding on shoots are known. *Diplodia* tip blight damage has been noted to be connected with insects and especially shoot-injuring insects such as cicadas or bark beetles (Feci et al., 2003; Haddow and Newman, 1942; Nicholls and Ostry, 1990; Swart et al., 1987; Wingfield and Knox-Davies, 1980). These insects cause wounds on twigs and create entry points for pathogens and/or devitalize the pines, thus making them more susceptible to *S. sapinea*.

Infections are mainly airborne and conidia are transported by wind and water droplets or carried by vectors such as bark beetles (Coleoptera: Scolytinae; e.g., Whitehill et al., 2007) or *Hylobius abietis* (Drenkhan et al., 2017). Conidia (Fig. 10.1B) are oblong to clavate, straight to slightly curved, first hyaline and aseptate, mature brownish, sometimes becoming 1-septate, (21-) 28-38 (-45)×(9-) 11-15 (-17) µm (own measurements).

5. Fungi in woody tissues of conifers

Investigations on the fungal diversity of conifers have mainly focused on needles (Carroll et al., 1977; Hata and Futai, 1996; Lee et al., 2014; Millberg et al., 2015; Romeralo et al., 2012; Sieber et al., 1999; Terhonen et al., 2011). Studies on stems, twigs, and branches are limited and have only focused on spruce (*Picea*), pine (*Pinus*), and fir (*Abies*) species.

According to Sieber (2007), endophytic fungi isolated from the foliage of Gymnosperms mostly belong to Helotiales, such as *Cenangium* and *Cryptocline* (Carroll and Carroll, 1978; Jurc et al., 2000; Sieber et al., 1999). From conifer stems, branches, and twigs (including bark and xylem), the isolated genera often belong to the ubiquitous or generalist endophytes, including *Alternaria, Aspergillus, Cladosporium, Epicoccum, Nigrospora, Penicillium, Pestalotiopsis, Phoma, Phomopsis, Sordaria*, and *Xylaria* (Kowalski and Kehr, 1992; Petrini and Fisher, 1988; Sanz-Ros et al., 2015).

Sydowia polyspora (Fig. 10.2) is a common endophyte of conifers (Bußkamp, 2018; Lygis et al., 2014; Menkis et al., 2006). The fungus mainly lives as a saprotroph (Müller et al., 2001), but Boberg et al. (2011) found that *S. polyspora*

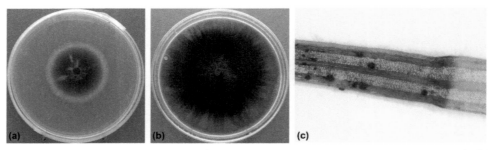

FIG. 10.2 *Sydowia polyspora*; (A and B) pure culture on MYP after 7 days, reverse (A) and 28 days, obverse (B) in ambient daylight at room temperature; (C) Pycnidia on the needle of *Abies grandis*.

only consumes soluble compounds in needles. It lives as an endophyte in the twigs of *A. alba, P. abies,* and *P. sylvestris* (Bußkamp, 2018; Müller and Hallaksela, 2000; Sieber, 1989). There are also reports on the pathogenic lifestyle of *S. polyspora* (Pan et al., 2018; Talgø et al., 2010). *S. polyspora* is airborne and also carried by insects (*Tomicus* spp. and *Hylurgus ligniperda*) that can transmit the spores from tree to tree (Davydenko et al., 2014). *S. polyspora* is distributed worldwide in North America, Europe, China, Australia, and Africa. Common hosts are the species of Cupressaceae and Pinaceae.

Microsphaeropsis olivacea (≡ *Coniothyrium olivaceum*) is a common endophyte in *P. sylvestris* (Bußkamp, 2018; Kowalski, 1993; Kowalski and Kehr, 1992; Petrini and Fisher, 1988). But *M. olivacea* is not restricted to conifers, as it is a ubiquitous, plurivorous saprotroph that occurs on, for example, the twigs and branches of *Cytisus, Hedera, Laurus, Lycium,* and *Sambucus* (Ellis and Ellis, 1985), *Acer* (Sun et al., 2011) or *Quercus* (Alidadi et al., 2019). As a member of the Pleosporales anamorphic species, *M. olivacea* has worldwide distribution and usually forms dark conidia in more or less globular pycnidia occurring on living or dead plants. In pure culture, the fungus is able to form different morphotypes on artificial media (Fig. 10.3).

As described earlier, a widespread endophyte of the genus *Pinus* is *S. sapinea* (Bußkamp, 2018). *Desmazierella acicola* is also a typical twig endophyte on *Pinus* spp. in Europe (Bußkamp, 2018; Kowalski and Kehr, 1992; Kowalski and Zych, 2002; Petrini and Fisher, 1988). *D. acicola* is a saprotroph that colonizes needles after fall (Ponge, 1991).

Truncatella conorum-piceae (≡ *Pestalotia conorum-piceae*) is mainly saprotrophic and is known as a subsequent decomposer of predamaged pine needles (Landeskompetenzzentrum Forst Eberswalde 2016). *Truncatella* spp. are isolated from Scots pine rarely (Menkis et al., 2006; Terhonen et al., 2011). However, numerous Amphisphaeriales species of the genus *Pestalotiopsis* have been detected endophytically in conifers. Fungi of the genus *Pestalotiopsis* occur worldwide as pathogens, endophytes, and saprophytes (Maharachchikumbura et al., 2011).

Rhizosphaera kalkhoffii has been detected in the twigs of *A. beshanzuensis* (Yuan et al., 2011) and *P. abies* (Sieber, 1989). Remarkably, *R. kalkhoffii* is usually isolated in needles and seeds (Deckert et al., 2019; Magan and Smith, 1996; Sieber, 1988). *R. kalkhoffii* can also be pathogenic for the hosts when it occurs as a needle cast pathogen (Diamandis and Minter, 1980; Scattolin and Montecchio, 2009).

Very few host- and organ-specific endophytes apart from leaves of conifers are described in the literature. *Tryblidiopsis pinastri* is a common Ascomycete on the coniferous twigs and branches of *P. abies*, mainly in northern Europe (Livsey and Minter, 1994). Different lifestyles from saprophytic to pathogenic were assumed for *T. pinastri*, but it is probably a weak pathogen (Livsey and Minter, 1994; Tanney and Seifert, 2019). It is presumed that *T. pinastri* is host- and organ-specific for *P. abies* branches (Barklund and Kowalski, 1996; Kowalski and Kehr, 1992). It was detected in wood as well as in the outer and inner bark of branches of Norway spruce collected in Germany, Sweden, and Finland (Barklund and Kowalski, 1996; Müller and Hallaksela, 2000). In contrast, *T. pinastri* was not detected by Sieber (1989) in a study on spruce branches in Switzerland.

Kowalski and Kehr (1992) analyzed branch bases from *A. alba* in Poland and Germany. They discovered two host-specific endophytes: *Durandiella gallica* and *Grovesiella abieticola*. *D. gallica* is a typical primary colonizer of dead branches on fir (Kowalski and Butin, 1989). However, it was also described as a weak pathogen (Kowalski and Butin, 1989). *G. abieticola* is also a typical primary colonizer on dead fir branches (Kowalski and Butin, 1989).

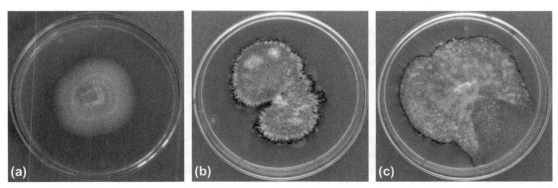

FIG. 10.3 *Microsphaeropsis olivacea*; pure culture exhibiting different morphotypes (A–C) growing 7 days on MYP in ambient daylight at room temperature.

6. Fungi inhabiting woody tissues of deciduous trees

According to Sieber (2007), endophyte assemblages of woody tissues of deciduous trees are often dominated by species of Diaporthales (Ascomycota). For example, *Dicarpella dryina* (Anamorph: *Tubakia dryina*, Fig. 10.4E) is one of the most common leaf endophytes in oak but is less frequent in twig woody tissues (Gennaro et al., 2003). The *Asterosporium* species are endophytes and saprobes of woody tissues of Betulaceae, Fagaceae, Juglandaceae, and Sapindaceae (Wijayawardene et al., 2016). These anamorphic coelomycetous species with stellate conidia were often found as endophytes in the twigs of *Betula* and *Alnus* (Tanaka et al., 2010). Common endophytic, wood-decaying Xylariaceae in angiosperm woody tissues are *Eutypa spinosa, Daldinia concentrica, Hypoxylon fragiforme, Hypoxylon fuscum,* and *Nemania serpens* (Parfitt et al., 2010). The transient endophytic lifestyle of wood-decay fungi may be preparation for the subsequent saprotrophic wood-decay activity (Parfitt et al., 2010; Boddy and Rayner, 1983). The fungal assemblage of woody tissues differs from leaves (Gennaro et al., 2003) but also between healthy and diseased trees (Ghobad-Nejhad et al., 2018).

Baum et al. (2003) studied the endophytes of healthy European beech stems and the most frequent (88%) of all isolated strains was *Trichoderma* spp. The probability that the *Trichoderma* species were contaminants during the isolation process is high because they are known as epiphytes of many plant surfaces and as saprotrophic decomposers of woody tissues and soils (Baum et al., 2003; Samuels, 1996). Nevertheless, *Trichoderma* spp. were found as endophytes of aspen, beech, and oak (Chapela, 1989; Cotter and Blanchard, 1982; Ragazzi et al., 1999).

6.1 Fungal species in *Quercus* tissues

The Diaporthales species *Apiognomonia errabunda* (anamorph: *Discula umbrinella,* syn. *Apiognomonia quercina,* Fig. 10.4D) and *Dendrostoma leiphaemia* (syn. e.g., *Discula quercina* and *Fusicoccum quercinum*) are members of the endophytic community of the woody tissues of European oaks, especially in the Mediterranean area (Jaklitsch and Voglmayr, 2019; Linaldeddu et al., 2011; Moricca and Ragazzi, 2011; Senanayake et al., 2018). The total number of culturable twig endophytes varies from 2 to 30 between different oak species (Ghasemi et al., 2019; Gonthier et al., 2005). A typical twig endophyte of *Q. robur* and *Q. petraea* is *Colpoma quercinum* (Kowalski and Kehr, 1992; Halmschlager et al., 1993; Moricca et al., 2012). A commonly observed endophyte in *Q. robur* twigs and stem woody tissues is *Scytalidium lignicola* (Gonthier et al., 2005). The most frequently isolated taxon in the twigs of *Quercus brantii* was *Cytospora* spp. (50% of all isolates, Diaporthales), whereas the most frequent endophyte in *Q. brantii, Q. macranthera,* and *Q. suber* was *Biscogniauxia mediterranea* (Fig. 10.4A) (Ghasemi et al., 2019; Linaldeddu et al., 2011). *B. mediterranea* is a latent pathogen and wood decay species that causes major losses in oak (Jurc and Ogris, 2006; Mirabolfathy, 2013). Endophytic, weak pathogens of oak can be isolated sporadically, such as *Botryosphaeria corticola, Pseudovalsa umbonata,* and *Neocucurbitaria cava* (Linaldeddu et al., 2011). Endophytic Basidiomycota isolated from *Quercus brantii* were a yeast-like species of *Cryptococcus* (Tremellales) (Ghobad-Nejhad et al., 2018) and several white rot fungi such as *F. fomentarius* (Parfitt et al., 2010) or *Vuilleminia comedens.* The common endophytic latent wood-decaying Ascomycota of oak are *Da. concentrica, Eu. spinosa, Ne. serpens,* and *H. fragiforme* (Parfitt et al., 2010).

6.2 Fungal species in *Fagus* tissues

The composition of beech endophytes is different in the stem as well as the lower and upper canopies (Baum et al., 2003), branches, and twigs (Boddy and Griffith, 1989; Griffith and Boddy, 1990; Toti et al., 1993). *Asterosporium asterospermum* seems to be a host-specific endophyte that fructificates saprobically on the twigs, branches, and stems of *F. sylvatica* (Butin, 2011; Chapela and Boddy, 1988b; Senanayake et al., 2018). *H. fragiforme* (Fig. 10.4C), *B. nummularia* (Fig. 10.4B), and *N. coccinea* (Fig. 10.4G) (all Ascomycota) were the most common latent-developing fungi in the woody tissues of beech (Chapela, 1989; Chapela and Boddy, 1988a; Hendry et al., 2002). Frequent endophytic basidiomycetous wood decay fungi of the beech are *F. fomentarius* (white-rot fungus = WRF) and *Coniophora puteana* (brown-rot fungus = BRF). They seem to be mutually exclusive, in which *C. puteana* was most frequent in the lower canopy where *F. fomentarius* was not to be found. In contrast, *C. puteana* was not isolated from stems where *F. fomentarius* dominated (Baum et al., 2003). In contrast to *B. nummularia, N. coccinea,* and *H. fragiforme,* which originate from the xylem, *C. puteana* and *Trichoderma* spp. originate from the bark (Hendry et al., 2002). Additional ascomycetous and basidiomycetous endophytes of the beech are presented in Table 10.2.

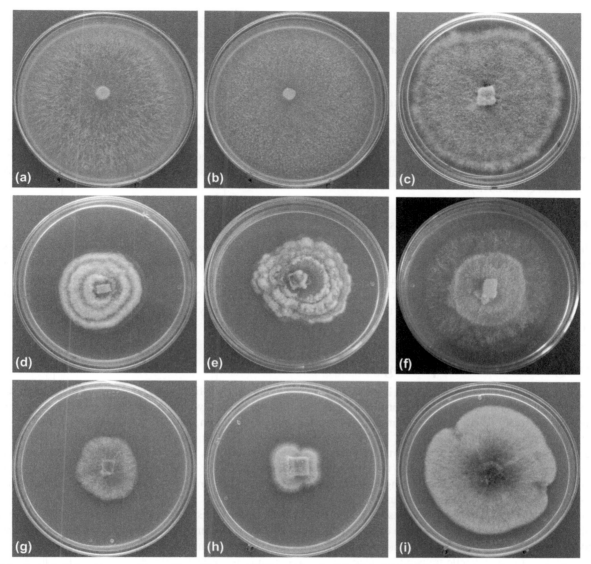

FIG. 10.4 Endophytic fungi in pure culture cultivated 7 days on MYP in ambient daylight; Xylariales (A) *Biscogniauxia mediterranea*, (B) *Biscogniauxia nummularia*, (C) *Hypoxylon fragiforme*, Diaporthales: (D) *Apiognomonia errabunda*, (E) *Dicarpella dryina*, (F) *Diaporthe/Phomopsis* sp., (G) *Neonectria coccinea* (Hypocreales), (H) *Pezicula cinnamomea* (Helotiales), (I) *Alternaria alternata* (Pleosporales).

6.3   Fungal species in *Alnus* tissues

The screening of the woody and bark tissues of three different alder species resulted in 85 different endophytic fungal taxa, whereas only 30 species exhibited an appreciable isolation frequency (Fisher and Petrini, 1990). The endophyte community of the studied alders was dominated by Ascomycota, but only two basidiomycetous strains were isolated. *Cryptosporella suffusa* (= *Disculina vulgaris*) is the most common endophyte in the bark of *Alnus glutinosa* in Europe and *Al. rubra* in North America (Fisher and Petrini, 1990; Sieber et al., 1991a,b). *Cr. suffusa* initially has an endophytic lifestyle; it can also cause canker in alder trees and becomes saprobic as plant tissues die (Mejía et al., 2008).

6.4   Fungal species in *Betula* tissues

From the stems, branches, and twigs of *Betula pendula*, 69 fungal species have been isolated (Kowalski, 1998). The most-frequent endophytes in the xylem of the stems were *Aposphaeria* spp., *Ophiognomonia intermedia*, *Fusicoccum betulae*, *Pe. cinnamomea* (Fig. 10.4H), and *Phialocephala* sp. (Kowalski, 1998). Common wood decay endophytes found in birch are *F. fomentarius* (Baum et al., 2003), *St. rugosum, Da. concentrica,* and *Eu. spinosa* (Parfitt et al., 2010). The dominant endophytes of twigs and branches are *Cryptosporella species* with their *Disculina* anamorphs (Mejía et al., 2011). For example, *Cry. betulae* is the most dominant branch endophyte of *B. pendula* and *B. pubescens* in Europe (Barengo et al., 2000; Kowalski and Kehr, 1992).

6.5 Fungal species in *Populus* tissues

The most abundant endophytic species in the woody tissues of the stems and branches of *Populus tremuloides* were *Cytospora populina* (≡ *Cryptosphaeria populina*), *Cy. chrysosperma,* and *Entoleuca mammata* (≡ *Hypoxylon mammatum*) (Chapela, 1989). All three ascomycetous species are also pathogenic to several host species. The *Cytospora* species are anamorphs of *Valsa* spp. and are common inhabitants of woody plants, including stem and branch canker pathogens (Adams et al., 2006; Fan et al., 2015).

6.6 Fungal species in *Eucalyptus* tissues

According to Bettucci and Saravay (1993), the most common endophyte of *Eucalyptus globulus,* including seedling stems, shoot stems, and the xylem, is *Phyllachora eucalypti* (≡ *Plectosphaera eucalypti*). Altogether, 41 endophytic fungal taxa from *Eucalyptus* seedling stems were isolated, ranging from 16 to 19 species per studied tissue. The fungal endophytic assemblages of the studied tissue seedlings, shoot xylem, and complete shoot stems were tissue-specific and dominated by Ascomycota. Several Basidiomycota were isolated too, but mainly from the xylem such as *Tr. versicolor, Phanerochaete* sp., and *Pycnoporus sanguineus.*

7. Fungal vascular wilt pathogens

Vascular wilt fungi are predominantly restricted to grow in the xylem in their life cycle, but some of these fungi degrade the xylem vessel walls to colonize adjacent parenchyma cells. Most vascular wilt fungi produce resting and overwintering structures such as microsclerotia, chlamydospores, thick-walled mycelium, and spore-bearing coremia in the soil or on dead host tissues (Yadeta and Thomma, 2013). Once other parts of the host tissues die, these are also occupied by the vascular wilt pathogens (VWPs), which starts to form resting structures (Agrios, 2005). VWPs may acquire nourishment by parasitizing parenchyma cells in addition to the nutritionally poor xylem cells or by inducing nutrient leakage from ambient tissues (Yadeta and Thomma, 2013). During host colonization, some VWPs produce various phytotoxins that have often been associated with host wilting (Yadeta and Thomma, 2013).

Most infections happen through the roots via wounds or cracks, and the parasite enters the water-conducting xylem vessels. But fungal xylem pathogens may also enter their host tree tissues via natural openings of leaves such as stomata and hydathodes or be transferred by vectors (Yadeta and Thomma, 2013), or they may be incorporated directly into the xylem by insects such as the bark beetle transmitting *Ophiostoma* spp. in the case of Dutch elm disease (Moser et al., 2009). The transmission of wilting fungi is also possible via root contacts from diseased to nonaffected roots. After host penetration, VWPs grow inside the cortical root cells. Their hyphae spread out intercellularly to the vascular parenchyma cells and colonize the xylem (Yadeta and Thomma, 2013). Conidiospores are produced in the xylem and disseminated acropetally with the xylem sap flow. Growing inside the xylem, VWPs obstruct the transportation of water and minerals and lead to wilting symptoms. Depending on the virulence of the pathogen species and the vitality and resistance of the host, trees may become withered partially or completely, finally dying (Yadeta and Thomma, 2013). The majority of fungal VWPs belong to one of the following four ascomycetous genera: *Ceratocystis, Fusarium, Ophiostoma,* or *Verticillium* (Yadeta and Thomma, 2013).

7.1 *Ceratocystis* vascular wilt

The *Ceratocystis* species may cause vascular wilt of the oak, cocoa, and eucalyptus (Yadeta and Thomma, 2013). A serious forest pathogen is the oak wilt fungus *Ceratocystis fagacearum,* first noticed in 1942; it caused dramatic losses of red oak in North America (Juzwik et al., 2008). The fungus produces sporulation mats emanating fruity odors and forming asexual barrel-shaped spores in chains called endoconidia. Especially, certain species of sap beetles, birds, and other animals are attracted to the fruity smell. On these mats, perithecia are formed producing sexual ascospores. Conidia and ascospores are mainly dispersed by air, rain, or insects such as the sap-feeding vector *Colopterus truncatus* (Juzwik et al., 2008). The disease is spread, for example, by the sap beetle picking up spores from diseased red oaks and later transferring them while feeding on healthy oaks.

7.2 *Fusarium* vascular wilt

Fusarium VWPs have a wide host range; most of them belong to the species *Fusarium oxysporum* (Yadeta and Thomma, 2013). This species contains morphologically indistinguishable pathogenic and nonpathogenic strains, the so-called *formae speciales* (Lievens et al., 2008). Pathogenic strains of *F. oxysporum* are causal agents of vascular wilt—foot rot, root rot, or bulb rot—in numerous different host species (Di Pietro et al., 2003). The sexual reproductive stage (teleomorph) of *F. oxysporum* is unknown but it is closely related to the teleomorphic genus *Gibberella*.

7.3 *Ophiostoma* vascular wilt

Three pleomorphic *Ophiostoma* species cause the mainly scolytid bark beetle-borne vascular wilt of elm trees, called Dutch elm disease (DED). Besides *Ulmus*, DED affects species in the genus *Zelkova* (Ahmadi, 2014). The first known DED epidemic in the *Ulmus* species was caused by the fungus *Ophiostoma ulmi* from the 1920s onward (Harwood et al., 2011). The disease was named after pioneering Dutch phytopathologists (Holmes et al., 1990). A related highly aggressive species, *O. novo-ulmi*, was first recognized in the 1970s; it caused a second and ongoing pandemic (Brasier, 1991; Gibbs and Brasier, 1973). Both species have spread across North America, Europe, and central Asia and have led to the deaths of billions of mature elms (Brasier and Kirk, 2010). The less aggressive *O. ulmi* is replaced by the *O. novo-ulmi*, although transient hybrids are formed between the two species. *O. novo-ulmi* exists also as two subspecies, which are currently hybridizing in Europe: *americana* and *novo-ulmi* (Konrad et al., 2002). The third species, *Ophiostoma himal-ulmi*, is a species endemic to the western Himalaya (Brasier and Mehrotra, 1996; Greig and Gibbs, 1983). Vector insects, bark-boring beetles of the genus *Scolytus* spp. or *Hylurgopinus rufipes*, transfer *Ophiostoma* spores to trees (Webber and Brasier, 1984). After infection, the pathogen grows in the xylem vessels with a yeast-like propagation phase (Webber and Brasier, 1984), causing a cavitation in the vessel. Finally, the VWP grows saprotrophically and sporulates in the inner bark and phloem of dying elms (Blumenstein, 2015).

7.4 *Verticillium* vascular wilt

Verticillium vascular wilt species have a broad host range worldwide. Seven species of *Verticillium* s. str. (Plectosphaerellaceae, Glomerellales) are known that cause severe wilting in eudicotyledon trees, herbaceous plants, and plantation crops (Inderbitzin et al., 2011; Johansson, 2006; Pegg and Brady, 2002; Zare et al., 2007): *Verticillium albo-atrum*, *V. alfalfae*, *V. dahliae* (Fig. 10.5), *V. longisporum*, *V. nonalfalfae*, *V. nubilum*, and *V. tricorpus* (Maschek and Halmschlager, 2017; Yadeta and Thomma, 2013). The basic life cycle of the soil-borne *Verticillium* wilt fungi is similar across species, but they are very diverse in their survival structures. For example, *V. albo-atrum* only forms mycelium while *V. dahliae* and *V. longisporum* produce microsclerotia, *V. nubilum* forms chlamydospores, and *V. tricorpus* is able to form mycelium, microsclerotia, and chlamydospores. The germination of resting structures in the soil was induced by exudates of the adjacent host roots (Mol and van Riessen, 1995). Outgrowing hyphae enter the root, mainly via natural wounds due to soil abrasion. In early disease stages, the fungus grows endophytically in the vascular tissue. Later, the fungus switches to a seminecrotrophic lifestyle, especially when the senescence of the host plant has increased (Johansson, 2006). *Verticillium* can quickly kill its hosts, especially small plants and seedlings. In large and more developed plants, the disease severity may vary, for example only

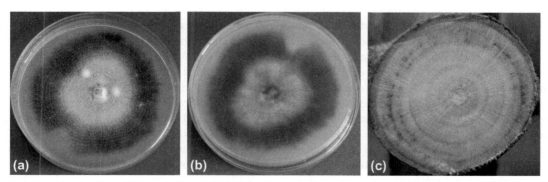

FIG. 10.5 *Verticillium dahliae*; (A and B) pure culture isolated from *Acer pseudoplatanus* collected in Germany (strain NW-FVA 1193) cultivated 7 days on MYP, (A) obverse, (B) reverse; (C) stem of *Acer pseudoplatanus* in a cross-section with visible internal vascular circular tissue discoloration caused by the strain NW-FVA 1193.

parts of a tree stem may be infected. Besides wilt symptoms, a *Verticillium* infection includes stunting, chlorosis or yellowing of the leaves, vein clearing, necrosis or tissue death, and defoliation. Often, an internal vascular circular tissue discoloration is visible in the cross-section of the stems or shoots (Fig. 10.5C).

As inhabitants of woody tree tissues, *V. albo-atrum, V. nonalfalfae,* and *V. dahliae* are known (Butin, 2011; Inderbitzin et al., 2011; Johansson, 2006; Maschek and Halmschlager, 2015, 2017; Nair et al., 2015; Sinclair and Lyon, 2005; Taylor, 1968). These species infect about 270 different plant species worldwide with various host-specific races and varieties. Woody hosts are species of *Acer, Fraxinus, Tilia, Aesculus hippocastanum, Ailanthus altissima, Catalpa bignonioides, Cercis siliquastrum, Cotinus coggygria, Prunus dulcis, Rhus typhina*, and *Sorbus aucuparia* (Butin, 2011). Besides natural infection, the *Verticillium* species can be transferred during pruning and root cuts via infected instruments (Butin, 2011).

The life cycle of *V. dahliae* comprises three major phases besides the endophytic stage: the dormant, the parasitic, and the saprotrophic phases. During the dormant phase, the fungus exists as elongated or globose, thick-walled, melanized microsclerotia in soils. The microsclerotia remain viable and may germinate when stimulated by host root exudates and the conditions are favorable. After germinating, the fungus infects the root tips of its host plant. The parasitic phase affects the plant by growing inside the xylem, enhancing its deterioration. Fungal pectinolytic enzymes damage parts of the xylem, and a glue-like substance affects the discoloration and blockage of vessels and parenchymal cells. Additionally, the vessels are continuously filled with fungal hyphae. Ovoid to ellipsoid, single-celled conidia are formed in the vascular system and transported to other parts of the plant. Fungal toxins affect the photosynthesis and respiration of the host plant and ultimately result in their wilting. When the host deteriorates and dies, the saprophytic stage of VWP starts. The fungus colonizes various parts of the plant and produces microsclerotia that are left in the soil (Soesanto, 2000). In culture, *V. dahlia* grows with short-celled hyaline to dark hyphae (Fig. 10.5A and B).

8. Wood-decay fungi

Fungi are the only organisms able to degrade lignin effectively (Mester et al., 2004), and they mainly decompose wood. An overview of lignicolous fungi and their decay skills is given in Worrall et al. (1997). The ability to decompose lignin is generally restricted to the species of Agaricomycotina in Basidiomycota (Floudas et al., 2012) while ascomycetous fungi are rarely able to degrade or modify lignin, such as certain Xylariales (Worrall et al., 1997). Endophytic fungi trigger the early stages of wood decay, and usually ascomycetous endophytes initiate decomposition by wood rot fungi or wood-decaying fungi. Certain primary species cause priority effects and nonrandom cooccurrence patterns in assembly following wood-decay species. These successional patterns are influenced by substrate modifications and species interactions (Ottosson et al., 2014). Wood-decaying fungi colonize and degrade woody tissues (mainly lignin, cellulose, and hemicellulose) with enzymatic and nonenzymatic systems (Rodríguez et al., 2011). There are variations in their enzymatic systems to degrade the different chemical wood compounds. Depending on the type of wood decay (Fig. 10.6) and the ability to degrade lignin, three major groups of wood-decaying fungi can be classified: White rot fungi (WRF, Fig. 10.7), brown rot fungi (BRF, Fig. 10.8), and soft rot fungi (SRF, Fig. 10.9) (Blanchette, 1991).

8.1 White rot fungi

White rot fungi (WRF) are able to degrade lignin most effectively due to the production of ligninolytic extracellular oxidative enzymes; they belong to Basidiomycota. They decompose wood, leaving wood residues that are usually whitish in color and fibrous in texture. Because these fungi can produce a wide variety of polysaccharide- (cellulose/hemicellulose/pectins) and lignin-degrading enzymes (e.g., laccases, manganese peroxidases, and lignin peroxidases), they are capable of the complete degradation of wood compounds. The so-called high-oxidation potential class II peroxidases such as lignin peroxidases are key enzymes usually characterizing WRF (Riley et al., 2014). WRF are typically associated with hardwood decay (Goodell and Jellison, 2008), but they can also degrade coniferous wood.

There are two subtypes of WRF. Wood residues can vary in color and texture depending on the decay species. Most WRF (subtype 1) exhibit a simultaneous, corrosive white rot. These fungi degrade all wood compounds simultaneously. A gradual increase of cell wall degradation toward the middle lamella is assumed (Rodríguez et al., 2011). The minor part of WRF can be assigned to subtype 2 (selective or successive white rot), which exhibits selective delignification. These fungi initially attack lignin and later hemicellulose and cellulose, often leaving residues enriched with cellulose. Typical species of WRF subtype 2 are *Ganoderma* spp. and *Armillaria* spp. as well as the causes of pocket and spongy white rots, for example, *Xylobolus frustulatus* or *Dichomitus squalens* (Rytioja et al., 2015). However, there are species such as *Heterobasidion* sp. that combine the step-by-step use of different wood compounds and simultaneous decomposition (Baldrian, 2008; Blanchette, 1991). Others such as *Mucidula mucida* (Fig. 10.7) exhibit a transient soft rot.

FIG. 10.6 (A) White-rotted wood due to *Fomes fomentarius*, (B) brown-rotted wood, (C) soft-rotted wood.

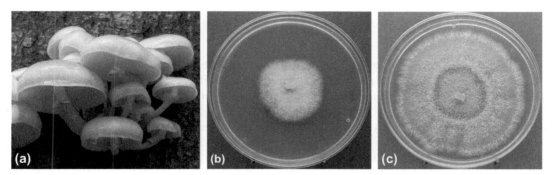

FIG. 10.7 WRF and transient SRF *Mucidula mucida*; (A) Basidiocarps, (B) and (C) pure culture of strain NW-FVA 2772 cultivated 7 days on MYP (B) and several weeks on MYP (C).

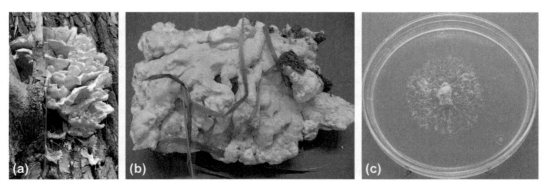

FIG. 10.8 *Laetiporus sulphureus*, BRF; (A and B) Basidiocarps, (C) pure culture (strain NW-FVA4106) cultivated 7 days on MYP in ambient daylight.

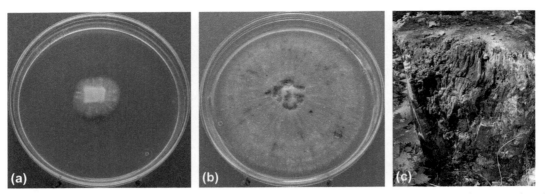

FIG. 10.9 (A and B) Pure culture of *Kretzschmaria deusta* (strain NW-FVA 4062) cultivated (A) 7 days on MYP and (B) several weeks on MYP; (C) stump rotted by several fungi, including *Kretzschmaria deusta*.

8.2 Brown rot fungi

Brown rot fungi (BRF) are assigned to Basidiomycota, but they don't produce lignin-degrading enzymes. Approximately 7% of all known wood-decaying fungal species are BRF, mainly growing in conifers (Renvall, 1995). A common BRF that also grows on broad-leaved trees is *Laetiporus sulphureus* (Fig. 10.8). In contrast to WRF, their decay is characterized by the degradation of carbohydrates (cellulose and hemicellulose) and an extensive demethylation of lignin, leaving a brownish-appearing wood residue. This brown-rotten wood is stained by oxidized lignin and has no fibrous texture due to the loss of cellulose. It shrinks upon drying and often breaks cubically. When brown rot fungi have finally decomposed their substrate, most often the remains of the wood can be manually ground into powder. The initial stage of brown rot is nonenzymatic (Goodell et al., 2017) because these fungi (Imami, 2015) produce low-molecular mass iron-reducing substances, oxalic acid, and hydroxyl radicals through the Fenton reaction. The latter attacks wood cell walls and snips cellulose and hemicellulose into smaller, partly soluble compounds. These are enzymatically degraded and absorbed by the fungus (Rodríguez et al., 2011).

8.3 Soft rot fungi

Soft rot fungi (SRF) are assigned to Ascomycota. In order to enzymatically break down cellulose in woody tissues, their hyphae secrete cellulases. In certain life stages, several Basidiomycota exhibit soft rot wood decay (Martínez et al., 2005). Ascomycetous SRF grow inside cell walls (Schwarze et al., 2000) and are able to colonize wood in conditions that are too wet, too hot, or too cold for brown or white rot fungi (Stokland et al., 2012). But usually SRF are less effective decomposers than WRF (Vane et al., 2005). The enzymatic activity of SRF leads to minute cavities inside the wood (Savory, 2008). It also sometimes leads to a bleaching, brownish discoloration (Fig. 10.6C) and a cracking pattern similar to brown rot, for example in the case of *Kretzschmaria deusta* (Fig. 10.9).

The above historical classification of the three main types of wood decay has been recently discussed because it does not circumscribe the whole fungal diversity of wood-degrading pathways (Mattila et al., 2020; Riley et al., 2014). The latter authors conducted phylogenetically informed principal component analysis (PCA) of fungal genes encoding biomass-degrading enzymes. In these analyses, the corticicoid, wood-inhabiting species *Botryobasidium botryosum* (Cantharellales) was grouped close to the model WRF *Phanerochaete chrysosporium* (Polyporales). *B. botryosum*, which usually occurs in the final stages of wood degradation, was able to degrade lignin and polysaccharides, but lacks the typical high-oxidation potential peroxidases. Therefore, the type of wood decay performed by *B. botryosum* seems to be intermediate between WFR and BRF (Alfaro et al., 2016; Riley et al., 2014). A similar pattern of wood decay was determined for *Rigidoporus microporus* (Hymenochaetales), a serious plant pathogen and the cause of white root rot disease on *Hevea brasiliensis* (Oghenekaro et al., 2020). Oghenekaro et al. (2020) based their assessment on a complete genome sequence and a principle coordinate analysis of wood-degrading enzymes. Also, the opportunistic white rot species *Schizophyllum commune* (Agaricales) did not produce high-oxidation potential peroxidases (Riley et al., 2014). On the other hand, the basidiomycetous *Mucidula mucida* (Agaricales, Fig. 10.7) produces high-oxidation potential class II peroxidases. But as mentioned above, the rot from this fungus resembles the soft rot type and has cavities in the residues (Daniel et al., 1992).

Moreover, some of the xylarialean soft rot species degrade lignin more efficiently than Basidiomycota, for example *K. deusta* (Schwarze, 2018, Fig. 10.9).

The previous doctrine was that the fungal wood decay process is strictly aerobic. But recent results on the WRF *Phlebia radiata* concerning a so-called "hypoxia response mechanism" calls this into question. However, recent findings on wood-decaying fungi to produce ethanol from various lignocelluloses under oxygen-depleted conditions have led us to question this (Mattila et al., 2020).

9. Conclusion

In summary, fungi inhabiting woody tree tissues, especially stems, branches, and twigs, play an important role in forest ecosystems. On the one hand, they may be beneficial to the health of their hosts as mutualistic endophytes. On the other hand, as latent pathogens, they can severely influence tree health and lead to diseases and mortality. In addition, fungi inhabiting living or dead woody tissues are key factors in the nutrition cycle of forest ecosystems. They also play an important role in soil formation and within the biotic interactions of forests. Wood-decaying fungi form habitat structures, especially for saproxylic insects. But the knowledge of fungi inhabiting living or dead wood of forest trees is still insufficient. Therefore, detailed studies on endophytes and latent fungal pathogens of forest trees, especially in light of climate change and globalization, are needed. Knowledge about predisposing, inciting, and trigging factors that facilitate the lifestyle switch from the endophytic to the parasitic stage of pathogens will be increasingly relevant.

References

Adams, G.C., Roux, J., Wingfield, M.J., 2006. Cytospora species (Ascomycota, Diaporthales, Valsaceae): introduced and native pathogens of trees in South Africa. Australas. Plant Pathol. 35, 521–548. https://doi.org/10.1071/AP06058.

Agrios, G., 2005. Plant Pathology, fifth ed. Academic Press.

Ahmadi, A., 2014. Zelkova carpinifolia reservoir from Hyrcanian Forests, Northern Iran, a new sacrifice of *Ophiostoma novo-ulmi*. Biodiversitas 15, 48–52. https://doi.org/10.13057/biodiv/d150107.

Alfaro, M., Castanera, R., Lavín, J.L., Grigoriev, I.V., Oguiza, J.A., Ramírez, L., Pisabarro, A.G., 2016. Comparative and transcriptional analysis of the predicted secretome in the lignocellulose-degrading basidiomycete fungus *Pleurotus ostreatus*: functional study of *P. ostreatus* bioinfosecretome. Environ. Microbiol. 18, 4710–4726. https://doi.org/10.1111/1462-2920.13360.

Alidadi, A., Kowsari, M., Javan-Nikkhah, M., Jouzani, G.R.S., Rastaghi, M.E., 2019. New pathogenic and endophytic fungal species associated with Persian oak in Iran. Eur. J. Plant Pathol. https://doi.org/10.1007/s10658-019-01830-y.

Alonso, R., Tiscornia, S., Bettucci, L., 2011. Fungal endophytes of needles and twigs from *Pinus taeda* and *Pinus elliottii* in Uruguay. Sydowia 63, 141–153.

Arnold, A., Maynard, Z., Gilbert, G., 2001. Fungal endophytes in dicotyledonous neotropical trees: patterns of abundance and diversity. Mycol. Res. 105, 1502–1507. https://doi.org/10.1017/S0953756201004956.

Azevedo, J.L., Maccheroni Jr., W., Pereira, J.O., de Araújo, W.L., 2000. Endophytic microorganisms: a review on insect control and recent advances on tropical plants. Electron. J. Biotechnol. 3, 15–16.

Baayen, R., Bonants, P., Verkley, G., Carroll, G., Aa, H., Weerdt, M., Brouwershaven, I., Schutte, G., Maccheroni, W., Glienke, C., Azevedo, J., 2002. Nonpathogenic isolates of the Citrus Black Spot Fungus, *Guignardia citricarpa*, identified as a cosmopolitan endophyte of woody plants, *G. mangiferae* (*Phyllosticta capitalensis*). Phytopathology 92, 464–477. https://doi.org/10.1094/PHYTO.2002.92.5.464.

Bagherabadi, S., Zafari, D., Soleimani, M.J., 2017. Morphological and molecular identification of *Cytospora chrysosperma* causing canker disease on *Prunus persica*. Australas. Plant Dis. Notes 12, 26. https://doi.org/10.1007/s13314-017-0250-9.

Baldrian, P., van West, P., 2008. Chapter 2: Enzymes of saprotrophic basidiomycetes. In: Boddy, L., Frankland, J.C. (Eds.), British Mycological Society Symposia Series, Ecology of Saprotrophic Basidiomycetes. Academic Press, pp. 19–41, https://doi.org/10.1016/S0275-0287(08)80004-5.

Barengo, N., Sieber, T.N., Holdenrieder, O., 2000. Diversity of endophytic mycobiota in leaves and twigs of pubescent birch (*Betula pubescens*). Sydowia 52, 305–320.

Barklund, P., Kowalski, T., 1996. Endophytic fungi in branches of Norway spruce with particular reference to *Tryblidiopsis pinastri*. Can. J. Bot. 74, 673–678.

Baum, S., Sieber, T., Schwarze, F., Fink, S., 2003. Latent infections of Fomes fomentarius in the xylem of European beech (*Fagus sylvatica*). Mycol. Prog. 2, 141–148. https://doi.org/10.1007/s11557-006-0052-5.

Bettucci, L., Saravay, M., 1993. Endophytic fungi of *Eucalyptus globulus*: a preliminary study. Mycol. Res. 97, 679–682. https://doi.org/10.1016/S0953-7562(09)80147-0.

Bihon, W., Slippers, B., Burgess, T.I., Wingfield, M.J., Wingfield, B.D., 2012. Diverse sources of infection and cryptic recombination revealed in South African *Diplodia pinea* populations. Fungal Biol. 116, 112–120.

Blanchette, R.A., 1984. Screening wood decayed by white rot fungi for preferential lignin degradation. Appl. Environ. Microbiol. 48, 647–653.

Blanchette, R., 1991. Delignification by wood-decay fungi. Annu. Rev. Phytopathol. 29, 381–403. https://doi.org/10.1146/annurev.py.29.090191.002121.

Blanchette, R.A., Held, B.W., Jurgens, J.A., McNew, D.L., Harrington, T.C., Duncan, S.M., Farrell, R.L., 2004. Wood-destroying soft rot fungi in the historic expedition huts of Antarctica. Appl. Environ. Microbiol. 70, 1328–1335. https://doi.org/10.1128/AEM.70.3.1328-1335.2004.

Blodgett, J.T., Kruger, E.L., Stanosz, G.R., 1997. Effects of moderate water stress on disease development by *Sphaeropsis sapinea* on red pine. Phytopathology 87, 422–428.

Blumenstein, K., 2015. Endophytic Fungi in Elms: Implications for the Integrated Management of Dutch Elm Disease. Swedish University of Agricultural Sciences, Alnarp, Sweden, pp. 1–84.

Boberg, J.B., Ihrmark, K., Lindahl, B.D., 2011. Decomposing capacity of fungi commonly detected in *Pinus sylvestris* needle litter. Fungal Ecol. 4, 110–114. https://doi.org/10.1016/j.funeco.2010.09.002.

Boddy, L.M., Griffith, G.S., 1989. Role of endophytes and latent invasion in the development of decay communities in sapwood of angiospermous trees. Sydowia 41, 41–73.

Boddy, L., Rayner, A.D.M., 1983. Origins of decay in living deciduous trees: the role of moisture content and a re-appraisal of the expanded concept of tree decay. New Phytol. 94, 623–641. https://doi.org/10.1111/j.1469-8137.1983.tb04871.x.

Bosso, L., Luchi, N., Maresi, G., Cristinzio, G., Smeraldo, S., Russo, D., 2017. Predicting current and future disease outbreaks of *Diplodia sapinea* shoot blight in Italy: species distribution models as a tool for forest management planning. For. Ecol. Manage. 400, 655–664. https://doi.org/10.1016/j.foreco.2017.06.044.

Botella, L., Diez, J.J., 2011. Phylogenic diversity of fungal endophytes in Spanish stands of *Pinus halepensis*. Fungal Divers. 47, 9–18.

Boyer, J.S., 1995. Biochemical and biophysical aspects of water deficits and the predisposition to disease. Annu. Rev. Phytopathol. 33, 251–274.

Brasier, C.M., 1991. *Ophiostoma novo-ulmi* sp. nov., causative agent of current Dutch elm disease pandemics. Mycopathologia 115, 151–161. https://doi.org/10.1007/BF00462219.

Brasier, C.M., Kirk, S.A., 2010. Rapid emergence of hybrids between the two subspecies of *Ophiostoma novo-ulmi* with a high level of pathogenic fitness. Plant Pathol. 59, 186–199. https://doi.org/10.1111/j.1365-3059.2009.02157.x.

Brasier, C.M., Mehrotra, M.D., 1996. *Ophiostoma himal-ulmi* sp. nov., a new species of Dutch elm disease fungus endemic to the Himalayas. Mycol. Res. 99 (2), 205–215. https://doi.org/10.1016/S0953-7562(09)80887-3.

Bressem, U., Langer, G., Habermann, M., 2013. Anhaltende Belastungen und Schäden bei älteren Eichen. AFZ-Der Wald 19 (2013), 38–40.

Brodde, L., Adamson, K., Julio Camarero, J., Castaño, C., Drenkhan, R., Lehtijärvi, A., Luchi, N., Migliorini, D., Sánchez-Miranda, Á., Stenlid, J., Özdağ, Ş., Oliva, J., 2019. Diplodia tip blight on its way to the North: drivers of disease emergence in Northern Europe. Front. Plant Sci. 9. https://doi.org/10.3389/fpls.2018.01818.

Bußkamp, J., 2018. Schadenserhebung, Kartierung und Charakterisierung des "Diplodia-Triebsterbens" der Kiefer, insbesondere des endophytischen Vorkommens in den klimasensiblen Räumen und Identifikation von den in Kiefer (*Pinus sylvestris*) vorkommenden Endophyten. Universität Kassel, Kassel.

Bußkamp, J., Langer, G.J., Langer, E.J., 2020. *Sphaeropsis sapinea* and fungal endophyte diversity in twigs of Scots pine (*Pinus sylvestris*) in Germany. Mycol. Progr. 19, 985–999. https://doi.org/10.1007/s11557-020-01617-0.

Butin, H., 1980. Über einige Phomopsis-Arten der Eiche einschliesslich *Fusicoccum quercus* Oudemans. Sydowia 33, 18–28.

Butin, H., 2011. Krankheiten der Wald- und Parkbäume – Diagnose, Biologie, Bekämpfung. Eugen Ulmer, Stuttgart (Hohenheim), pp. 1–318.

Butin, H., Kowalski, T., 1983. Die natürliche Astreinigung und ihre biologischen Voraussetzungen. Eur. J. For. Pathol. 13, 428–439. https://doi.org/10.1111/j.1439-0329.1983.tb00145.x.

CABI, 2019. *Sphaeropsis sapinea* (*Sphaeropsis* blight). Invasive Species Compendium.

Carroll, G.C., Carroll, F.E., 1978. Studies on the incidence of coniferous needle endophytes in the Pacific Northwest. Can. J. Bot. 56, 3034–3043.

Carroll, F.E., Muller, E., Sutton, B.C., 1977. Preliminary studies on the incidence of needle endophytes in some European conifers. Sydowia 29, 87–103.

Casado López, S., Peng, M., Daly, P., Andreopoulos, B., Pangilinan, J., Lipzen, A., Riley, R., Ahrendt, S., Ng, V., Barry, K., Daum, C., Grigoriev, I.V., Hildén, K.S., Mäkelä, M.R., de Vries, R.P., 2019. Draft genome sequences of three monokaryotic isolates of the white-rot basidiomycete fungus *Dichomitus squalens*. Microbiol. Resour. Announc. 8. https://doi.org/10.1128/MRA.00264-19.

Chapela, I.H., 1989. Fungi in healthy stems and branches of American beech and aspen: a comparative study. New Phytol. 113, 65–75. https://doi.org/10.1111/j.1469-8137.1989.tb02396.x.

Chapela, I.H., Boddy, L., 1988a. Fungal colonization of attached beech branches. I. Early stages of development of fungal communities. New Phytol. 110, 39–45. https://doi.org/10.1111/j.1469-8137.1988.tb00235.x.

Chapela, I.H., Boddy, L., 1988b. Fungal colonization of attached beech branches. New Phytol. 110, 47–57.

Cline, L.C., Schilling, J.S., Menke, J., Groenhof, E., Kennedy, P.G., 2018. Ecological and functional effects of fungal endophytes on wood decomposition. Funct. Ecol. 32, 181–191. https://doi.org/10.1111/1365-2435.12949.

Cotter, H.V.T., Blanchard, R.O., 1982. The fungal flora of bark of *Fagus grandifolia*. Mycologia 74, 836–843. https://doi.org/10.2307/3792872.

Daniel, G., Volc, J., Nilsson, T., 1992. Soft rot and multiple T-branching by the basidiomycete *Oudemansiella mucida*. Mycol. Res. 96, 49–54. https://doi.org/10.1016/S0953-7562(09)80995-7.

Davydenko, K., Vasaitis, R., Meshkova, V., Menkis, A., 2014. Fungi associated with the red-haired bark beetle, *Hylurgus ligniperda* (Coleoptera: Curculionidae) in the forest-steppe zone in eastern Ukraine. EJE 111, 561–565. https://doi.org/10.14411/eje.2014.070.

Deckert, R.J., Gehring, C.A., Patterson, A., 2019. Pine seeds carry symbionts: endophyte transmission re-examined. In: Verma, S.K., White, J., Francis, J. (Eds.), Seed Endophytes: Biology and Biotechnology. Springer International Publishing, Cham, pp. 335–361, https://doi.org/10.1007/978-3-030-10504-4_16.

Di Pietro, A., Madrid, M., Caracuel, Z., Delgado-Jarana, J., Roncero, M.I., 2003. *Fusarium oxysporum*: exploring the molecular arsenal of a vascular wilt fungus. Mol. Plant Pathol. 4, 315–325. https://doi.org/10.1046/j.1364-3703.2003.00180.x.

Diamandis, S., Minter, D.W., 1980. Rhizosphaera kalkhoffii. [Descriptions of Fungi and Bacteria]. IMI Descriptions of Fungi and Bacteria.

Dijk, H.F.G.V., der Gaag, M.V., Perik, P.J.M., Roelofs, J.G.M., 1992. Nutrient availability in Corsican pine stands in The Netherlands and the occurrence of *Sphaeropsis sapinea*: a field study. Can. J. Bot. 70, 870–875.

Diminić, D., Potočić, N., Seletković, I., 2012. The role of site in predisposition of Austrian Pine (*Pinus nigra* Arnold) to pathogenic fungus *Sphaeropsis sapinea* (Fr.) Dyko et Sutton in Istria (Croatia). Šumarski List 136, 19–35.

Drenkhan, T., Voolma, K., Adamson, K., Sibul, I., Drenkhan, R., 2017. The large pine weevil *Hylobius abietis* (L.) as a potential vector of the pathogenic fungus *Diplodia sapinea* (Fr.) Fuckel. Agric. For. Entomol. 19, 4–9. https://doi.org/10.1111/afe.12173.

Ellis, M.B., Ellis, J.P., 1985. Microfungi on Land Plants. An Identification Handbook, first ed. Croom Helm Ltd, London, UK; Sydney, pp. 1–818.

Fabre, B., Piou, D., Desprez-Loustau, M.-L., Marçais, B., 2011. Can the emergence of pine Diplodia shoot blight in France be explained by changes in pathogen pressure linked to climate change? Glob. Change Biol. 17, 3218–3227.

Fan, X., Hyde, K., Liu, M., Liang, Y.-M., Tian, C., 2015. *Cytospora* species associated with walnut canker disease in China, with description of a new species *C. gigalocus*. Fungal Biol. 119. https://doi.org/10.1016/j.funbio.2014.12.011.

Fatima, N., Muhammad, S.A., Khan, I., Qazi, M.A., Shahzadi, I., Mumtaz, A., Hashmi, M.A., Khan, A.K., Ismail, T., 2016. *Chaetomium* endophytes: a repository of pharmacologically active metabolites. Acta Physiol. Plant. 38, 136. https://doi.org/10.1007/s11738-016-2138-2.

Feci, E., Battisti, A., Capretti, P., Tegli, S., 2002. An association between the fungus *Sphaeropsis sapinea* and the cone bug *Gastrodes grossipes* in cones of *Pinus nigra* in Italy. For. Pathol. 32, 241–247. https://doi.org/10.1046/j.1439-0329.2002.00286.x.

Feci, E., Smith, D., Stanosz, G.R., 2003. Association of *Sphaeropsis sapinea* with insect-damaged red pine shoots and cones. For. Pathol. 33, 7–13.

Fisher, P.J., Petrini, O., 1990. A comparative study of fungal endophytes in xylem and bark of *Alnus* species in England and Switzerland. Mycol. Res. 94, 313–319. https://doi.org/10.1016/S0953-7562(09)80356-0.

Floren, A., Krüger, D., Müller, T., Dittrich, M., Rudloff, R., Hoppe, B., Linsenmair, K.E., 2015. Diversity and interactions of wood-inhabiting fungi and beetles after deadwood enrichment. PLoS One 10. https://doi.org/10.1371/journal.pone.0143566.

Floudas, D., Binder, M., Riley, R., Barry, K., Blanchette, R., Henrissat, B., Martinez, A.T., Otillar, R., Spatafora, J., Yadav, J., Aerts, A., Benoit, I., Boyd, A., Carlson, A., Copeland, A., Coutinho, P., Vries, R.P., Ferreira Neila, P., Findley, K., Hibbett, D., 2012. The Paleozoic origin of enzymatic lignin decomposition reconstructed from 31 fungal genomes. Science 336, 1715–1719.

Flowers, J., Nuckles, E., Hartman, J., Vaillancourt, L.J., 2001. Latent infection of Austrian and Scots pine tissues by *Sphaeropsis sapinea*. Plant Dis. 85, 1107–1112.

Gehlot, P., 2008. Endophytic mycoflora of inner bark of *Prosopis cineraria* – a key stone tree species of Indian desert. Am. Eur. J. Bot. 1, 1–4.

Gennaro, M., Gonthier, P., Nicolotti, G., 2003. Fungal endophytic communities in healthy and declining *Quercus robur* L. and *Q. cerris* L. trees in Northern Italy. J. Phytopathol. 151, 529–534. https://doi.org/10.1046/j.1439-0434.2003.00763.x.

Ghasemi, S., Khodaei, S., Karimi, K., Tavakoli, M., Pertot, I., Arzanlou, M., 2019. Biodiversity study of endophytic fungi associated with two *Quercus* species in Iran. For. Syst. 28. https://doi.org/10.5424/fs/2019281-14528.

Ghobad-Nejhad, M., Meyn, R., Langer, E., 2018. Endophytic fungi isolated from healthy and declining Persian oak (*Quercus brantii*) in western Iran. Nova Hedwigia 107. https://doi.org/10.1127/nova_hedwigia/2018/0470.

Gibbs, J., Brasier, C., 1973. Correlation between cultural characters and pathogenicity in *Ceratocystis ulmi* from Britain, Europe and America. Nature 241, 381–383. https://doi.org/10.1038/241381a0.

Giordano, L., Gonthier, P., Varese, G.C., Miserere, L., Nicolotti, G., 2009. Mycobiota inhabiting sapwood of healthy and declining Scots pine (*Pinus sylvestris* L.) trees in the Alps. Fungal Divers. 38, 69–83.

Gomes, R.R., Glienke, C., Videira, S.I.R., Lombard, L., Groenewald, J.Z., Crous, P.W., 2013. *Diaporthe* a genus of endophytic, saprobic and plant pathogenic fungi. Persoonia 31, 1–41.

Gonthier, P., Gennaro, M., Nicolotti, G., 2005. Effects of water stress on the endophytic mycota of *Quercus robur*. Fungal Divers. 21, 69–80.

Goodell, B., Jellison, J., Militz, H., Freeman, M.H., 2008. Fungal decay of wood: soft rot—brown rot—white rot. In: Schultz, T.P., Goodell, B., Nicholas, D.D. (Eds.), Development of Commercial Wood Preservatives: Efficacy, Environmental, and Health Issues. vol. 982. American Chemical Society, Washington, DC, pp. 9–31, https://doi.org/10.1021/bk-2008-0982.ch002.

Goodell, B., Zhu, Y., Kim, S., Kafle, K., Eastwood, D., Daniel, G., Jellison, J., Yoshida, M., Groom, L., Pingali, S.V., O'Neill, H., 2017. Modification of the nanostructure of lignocellulose cell walls via a non-enzymatic lignocellulose deconstruction system in brown rot wood-decay fungi. Biotechnol. Biofuels 10, 179. https://doi.org/10.1186/s13068-017-0865-2.

Greig, B., Gibbs, J., 1983. 3. Control of Dutch elm disease in Britain. In: Burdekin, B.A. (Ed.), Research on Dutch Elm Disease in Europe, Forestry Commision Bulletin. HSMO, London, pp. 10–16.

Griffith, G.S., Boddy, L., 1988. Fungal communities in attached ash (*Fraxinus excelsior*) twigs. Trans. Br. Mycol. Soc. 91, 599–606.

Griffith, G.S., Boddy, L., 1990. Fungal decomposition of attached angiosperm twigs I. Decay community development in ash, beech and oak. New Phytol. 116, 407–415. https://doi.org/10.1111/j.1469-8137.1990.tb00526.x.

Guo, L.-D., Huang, G.R., Wang, Y., He, W.H., Zheng, W.H., Hyde, K.D., 2003. Molecular identification of white morphotype strains of endophytic fungi from *Pinus tabulaeformis*. Mycol. Res. 107, 680–688.

Guo, L.-D., Huang, G.-R., Wang, Y., 2008. Seasonal and tissue age influences on endophytic fungi of *Pinus tabulaeformis* (*Pinaceae*) in the Dongling Mountains, Beijing. J. Integr. Plant Biol. 50, 997–1003.

Haddow, W.R., Newman, F.S., 1942. A disease of the Scots Pine (*Pinus sylvestris* L.) caused fey the fungus *Diplodia pinea* Kickx, associated with the Pine Spittle-bug (*Aphrophora parallela* Say.). I. Symptoms and Etiology. Trans. R. Canad. Inst., Toronto 24, 1–18.

Halmschlager, E.V., Butin, H., Donaubauer, E., 1993. Endophytische Pilze in Blättern und Zweigen von *Quercus petraea*. Eur. J. For. Pathol. 23, 51–63. https://doi.org/10.1111/j.1439-0329.1993.tb00805.x.

Hanso, M., Drenkhan, R., 2010. Two new ascomycetes on twigs and leaves of silver birches (*Betula pendula*) in Estonia. Folia Cryptogam. Estonica 47, 21–26.

Harmon, M., Franklin, J., Swanson, F., Sollins, P., Gregory, S., Lattin, J., Anderson, N., Cline, S., Aumen, N., Sedell, J.R., Lienkaempeer, G., Cromack, K., 1986. Ecology of coarse woody debris in temperate ecosystems. Adv. Ecol. Res. 15. https://doi.org/10.1016/S0065-2504(03)34002-4.

Harwood, T.D., Tomlinson, I., Potter, C.A., Knight, J.D., 2011. Dutch elm disease revisited: past, present and future management in Great Britain. Plant Pathol. 60, 545–555. https://doi.org/10.1111/j.1365-3059.2010.02391.x.

Hata, K., Futai, K., 1996. Variation in fungal endophyte populations in needles of the genus *Pinus*. Can. J. Bot. 74, 103–114.

Hendry, S.J., Boddy, L., Lonsdale, D., 2002. Abiotic variables effect differential expression of latent infections in beech (*Fagus sylvatica*). New Phytol. 155, 449–460. https://doi.org/10.1046/j.1469-8137.2002.00473.x.

Henriques, J., Barrento, M., Bonifacio, L., Gomes, A., Lima, A., Sousa, E., 2014a. Factors affecting the dispersion of *Biscogniauxia mediterranea* in Portuguese cork oak stands. Silva Lusit. 22, 83–97.

Henriques, J., Nóbrega, F., Sousa, E., Lima, A., 2014b. Diversity of *Biscogniauxia mediterranea* within single stromata on cork oak. J. Mycol., 1–5. https://doi.org/10.1155/2014/324349.

Henriques, J., Nóbrega, F., Sousa, E., Lima, A., 2015. Morphological and genetic diversity of *Biscogniauxia mediterranea* associated to *Quercus suber* in the Mediterranean Basin. Rev. Cienc. Agrar. 38, 166–175.

Holmes, F.W., Heybroek, H.M., Society, A.P., 1990. Dutch Elm Disease: The Early Papers: Selected Works of Seven Dutch Women Phytopathologists. APS Press.

Houston, D.R., 1984. Stress related to diseases. Arboric. J. 8, 137–149. https://doi.org/10.1080/03071375.1984.9746670.

Hyde, K.D., Soytong, K., 2008. The fungal endophyte dilemma. Fungal Divers. 33, 163–173.

Imami, A., 2015. Biotransformationen von Lignosulfonaten und Herbiziden durch Basidiomyceten. Justus-Liebig-Universität Gießen, Gießen.

Inderbitzin, P., Bostock, R.M., Davis, R.M., Usami, T., Platt, H.W., Subbarao, K.V., 2011. Phylogenetics and taxonomy of the fungal vascular wilt pathogen *Verticillium*, with the descriptions of five new species. PLoS One 6. https://doi.org/10.1371/journal.pone.0028341, e28341.

Jaklitsch, W.M., Voglmayr, H., 2019. European species of *Dendrostoma* (Diaporthales). MycoKeys 59, 1–26. https://doi.org/10.3897/mycokeys.59.37966.

Jeger, M., Bragard, C., Caffier, D., Candresse, T., Chatzivassiliou, E., Dehnen-Schmutz, K., Gilioli, G., Grégoire, J.-C., Miret, J., Macleod, A., Navajas, M., Niere, B., Parnell, S., Potting, R., Rafoss, T., Rossi, V., Urek, G., van Bruggen, A., Werf, W., Pautasso, M., 2017. Pest categorisation of *Entoleuca mammata*. EFSA J. 15. https://doi.org/10.2903/j.efsa.2017.4925.

Johansson, A., 2006. *Verticillium longisporum*, Infection, Host Range, Prevalence and Plant Defence Responses [WWW Document]. https://pub.epsilon.slu.se/1119/. (Accessed 31 March 2020).

Jurc, D., Ogris, N., 2006. First reported outbreak of charcoal disease caused by *Biscogniauxia mediterranea* on Turkey oak in Slovenia. Plant Pathol. 55, 299. https://doi.org/10.1111/j.1365-3059.2005.01297.x.

Jurc, D., Jurc, M., Sieber, T.N., Bojovic, S., 2000. Endophytic *Cenangium ferruginosum* (Ascomycota) as a reservoir for an epidemic of cenangium dieback in Austrian pine. Phyton (Horn) 40, 103–108.

Juybari, H.Z., Ghanbary, M.A.T., Rahimian, H., Karimi, K., Arzanlou, M., 2019. Seasonal, tissue and age influences on frequency and biodiversity of endophytic fungi of *Citrus sinensis* in Iran. For. Pathol. 49. https://doi.org/10.1111/efp.12559, e12559.

Juzwik, J., Harrington, T.C., MacDonald, W.L., Appel, D.N., 2008. The origin of *Ceratocystis fagacearum*, the oak wilt fungus. Annu. Rev. Phytopathol. 46, 13–26. https://doi.org/10.1146/annurev.phyto.45.062806.094406.

Kaya, A.G.A., Lehtijärvi, A., Kaya, Ö., Doğmuş-Lehtijärvi, T., 2014. First report of *Diplodia pinea* on *Pseudotsuga menziesii* in Turkey. Plant Dis. 98, 689.

Keen, A., Smits, T.F.C., 1989. Application of a mathematical function for a temperature optimum curve to establish differences in growth between isolates of a fungus. Neth. J. Plant Pathol. 95, 37–49.

Kehr, R.D., 1991. Pezicula canker of *Quercus rubra* L., caused by *Pezicula cinnamomea* (DC.) Sacc. I. Symptoms and pathogenesis. Eur. J. For. Pathol. 21, 218–233. https://doi.org/10.1111/j.1439-0329.1991.tb00973.x.

Kehr, R., 1998. Zur Bedeutung pilzlicher Endophyten bei Waldbäumen. Biologische Bundesanstalt für Land- und Forstwirtschaft Braunschweig, Institut für Pflanzenschutz im Forst 100 Jahre Pflanzenschutzforschung Aktuelle Forschungsschwerpunkte im Forst-und Rebschutz, pp. 8–30.

Konrad, H., Kirisits, T., Riegler, M., Halmschlager, E., Stauffer, C., 2002. Genetic evidence for natural hybridization between the Dutch elm disease pathogens *Ophiostoma novo-ulmi* ssp. *novo-ulmi* and *O. novo-ulmi* ssp. *americana*. Plant Pathol. 51, 78–84. https://doi.org/10.1046/j.0032-0862.2001.00653.x.

Kowalski, T., 1993. Fungi in living symptomless needles of *Pinus sylvestris* with respect to some observed disease processes. J. Phytopathol. Phytopathol. Z. 139, 129–145.

Kowalski, T., 1998. Endophytic mycobiota in stems and branches of *Betula pendula* to a different degree affected by air pollution. Österr. Z. Pilzk. 7, 13–24.

Kowalski, T., Butin, H., 1989. Die natürliche Astreinigung und ihre biologischen Voraussetzungen. IV. Die Pilzflora der Tanne (*Abies alba* Mill.). Z. Mykol. 55, 189–193.

Kowalski, T., Kehr, R., 1992. Endophytic fungal colonization of branch bases in several forest tree species. Sydowia 44, 137–168.

Kowalski, T., Zych, P., 2002. Fungi isolated from living symptomless shoots of *Pinus nigra* growing in different site conditions. Österr. Z. Pilzk. 11, 107–116.

Krause, S.C., Raffa, K.F., 1996. Differential growth and recovery rates following defoliation in related deciduous and evergreen trees. Trees 10, 308–316.

Landeskompetenzzentrum Forst Eberswalde (Ed.), 2016. Diagnose Report 2015, Diagnostische Arbeiten unter besonderer Berücksichtigung pilzlicher Organismen.

Langer, G., Bressem, U., Habermann, M., 2011. *Diplodia*-Triebsterben der Kiefer und endophytischer Nachweis des Erregers *Sphaeropsis sapinea*. AFZ-Der Wald, 28–31.

Langer, G.J., Bußkamp, J., Langer, E.J., 2020. Absterbeerscheinungen bei Rotbuche durch Trockenheit und Wärme. AFZ-Der Wald 4, 24–27.

Lee, J.-W., Park, J.-Y., Kwon, M., Choi, I.-G., 2009. Purification and characterization of a thermostable xylanase from the brown-rot fungus *Laetiporus sulphureus*. J. Biosci. Bioeng. 107 (1), 33–37.

Lee, S.K., Lee, S.K., Bae, H., Seo, S.-T., Lee, J.K., 2014. Effects of water stress on the endophytic fungal communities of *Pinus koraiensis* needles infected by *Cenangium ferruginosum*. Mycobiology 42, 331–338.

Leonhardt, S., Hoppe, B., Stengel, E., Noll, L., Moll, J., Bässler, C., Dahl, A., Buscot, F., Hofrichter, M., Kellner, H., 2019. Molecular fungal community and its decomposition activity in sapwood and heartwood of 13 temperate European tree species. PLoS One 14. https://doi.org/10.1371/journal.pone.0212120, e0212120.

Lievens, B., Rep, M., Thomma, B.P.H.J., 2008. Recent developments in the molecular discrimination of formae speciales of *Fusarium oxysporum*. Pest Manage. Sci. 64, 781–788.

Linaldeddu, B., Costantino, S., Spano, D., Franceschini, A., 2011. Variation of endophytic cork oak-associated fungal communities in relation to plant health and water stress. For. Pathol. 41, 193–201. https://doi.org/10.1111/j.1439-0329.2010.00652.x.

Livsey, S., Minter, D.W., 1994. The taxonomy and biology of *Tryblidiopsis pinastri*. Can. J. Bot. 72, 549–557.

Lonsdale, D., Pautasso, M., Holdenrieder, O., 2008. Wood-decaying fungi in the forest: conservation needs and management options. Eur. J. For. Res. 127. https://doi.org/10.1007/s10342-007-0182-6.

Luchi, N., Mancini, V., Feducci, M., Santini, A., Capretti, P., 2012. *Leptoglossus occidentalis* and *Diplodia pinea*: a new insect-fungus association in Mediterranean forests. For. Pathol. 42, 246–251. https://doi.org/10.1111/j.1439-0329.2011.00750.x.

Luchi, N., Oliveira Longa, C.M., Danti, R., Capretti, P., Maresi, G., 2014. *Diplodia sapinea*: the main fungal species involved in the colonization of pine shoots in Italy. For. Path. 44, 372–381.

Lygis, V., Vasiliauskas, R., Larsson, K.-H., Stenlid, J., 2005. Wood-inhabiting fungi in stems of *Fraxinus excelsior* in declining ash stands of northern Lithuania, with particular reference to *Armillaria cepistipes*. Scand. J. For. Res. 20, 337–346. https://doi.org/10.1080/02827580510036238.

Lygis, V., Vasiliauskaite, I., Matelis, A., Pliūra, A., Vasaitis, R., 2014. Fungi in living and dead stems and stumps of *Pinus mugo* on coastal dunes of the Baltic Sea. Plant Prot. Sci. 50, 221–226.

Magan, N., Smith, M.K., 1996. Isolation of the endophytes *Lophodermium piceae* and *Rhizosphaera kalkhoffii* from Sitka spruce needles in poor and good growth sites and in vitro effects of environmental factors. Phyton, Austria, pp. 103–110.

Maharachchikumbura, S.S.N., Guo, L.-D., Chukeatirote, E., Bahkali, A.H., Hyde, K.D., 2011. *Pestalotiopsis*—morphology, phylogeny, biochemistry and diversity. Fungal Divers. 50, 167–187.

Manion, P.D., 1981. Tree Disease Concepts. Prentice-Hall.

Martínez, Á.T., Speranza, M., Ruiz-Dueñas, F.J., Ferreira, P., Camarero, S., Guillén, F.S., Martínez, M.J., Gutiérrez, A., del Río, J.C., 2005. Biodegradation of lignocellulosics: microbial, chemical, and enzymatic aspects of the fungal attack of lignin. Int. Microbiol. 8, 195–204. https://doi.org/10.13039/501100003339.

Martínez-Álvarez, P., Rodríguez-Ceinós, S., Martín-García, J., Diez, J.J., 2012. Monitoring endophyte populations in pine plantations and native oak forests in Northern Spain. For. Syst. 21, 373.

Maschek, O., Halmschlager, E., 2015. First report of Verticillium wilt on Ailanthus altissima in Europe caused by Verticillium nonalfalfae. Plant Dis. 100. https://doi.org/10.1094/PDIS-07-15-0733-PDN.

Maschek, O., Halmschlager, E., 2017. Natural distribution of Verticillium wilt on invasive *Ailanthus altissima* in eastern Austria and its potential for biocontrol. For. Pathol. https://doi.org/10.1111/efp.12356, e12356.

Mattila, H.K., Mäkinen, M., Lundell, T., 2020. Hypoxia is regulating enzymatic wood decomposition and intracellular carbohydrate metabolism in filamentous white rot fungus. Biotechnol. Biofuels 13, 26. https://doi.org/10.1186/s13068-020-01677-0.

Mejía, L.C., Castlebury, L.A., Rossman, A.Y., Sogonov, M.V., White, J.F., 2008. Phylogenetic placement and taxonomic review of the genus *Cryptosporella* and its synonyms *Ophiovalsa* and *Winterella* (Gnomoniaceae, Diaporthales). Mycol. Res. 112, 23–35. https://doi.org/10.1016/j.mycres.2007.03.021.

Mejía, L.C., Rossman, A.Y., Castlebury, L.A., Wight Jr., J.F., 2011. New species, phylogeny, host-associations and geographic distribution of genus *Cryptosporella* (Gnomoniaceae, Diaporthales). Mycologia 103, 379–399. https://doi.org/10.3852/10-134.

Menkis, A., Vasiliauskas, R., Taylor, A.F.S., Stenström, E., Stenlid, J., Finlay, R., 2006. Fungi in decayed roots of conifer seedlings in forest nurseries, afforested clear-cuts and abandoned farmland. Plant Pathol. 55, 117–129.

Mester, T., Varela, E., Tien, M., 2004. Wood degradation by brown-rot and white-rot fungi. Genet. Biotechnol., 355–368. https://doi.org/10.1007/978-3-662-07426-8_17.

Milijašević, T., 2006. Effect of temperature on the mycelial growth of the fungus *Sphaeropsis sapinea*. Bulletin. Faculty of Forestry, Glasnik Šumarskog Fakulteta 94, 211–222.

Millberg, H., Boberg, J., Stenlid, J., 2015. Changes in fungal community of Scots pine (*Pinus sylvestris*) needles along a latitudinal gradient in Sweden. Fungal Ecol. 17, 126–139.

Mirabolfathy, M., 2013. Outbreak of charcoal disease on *Quercus spp.* and *Zelkova carpinifolia* trees in forests of Zagros and Alborz mountains in Iran. Iranian J. Plant Pathol. 492, 77–79.

Mohali, S., Burgess, T.I., Wingfield, M.J., 2005. Diversity and host association of the tropical tree endophyte *Lasiodiplodia theobromae* revealed using simple sequence repeat markers. For. Pathol. 35, 385–396. https://doi.org/10.1111/j.1439-0329.2005.00418.x.

Mol, L., van Riessen, H., 1995. Effect of plant roots on the germination of microsclerotia of *Verticillum dahliae*. Eur. J. Plant Pathol. 101, 673–678.

Moricca, S., Ragazzi, A., 2008. Fungal endophytes in Mediterranean oak forests: a lesson from *Discula quercina*. Phytopathology 98, 380–386. https://doi.org/10.1094/PHYTO-98-4-0380.

Moricca, S., Ragazzi, A., 2011. The holomorph *Apiognomonia quercina/Discula quercina* as a Pathogen/Endophyte in Oak. In: Pirttilä, A.M., Frank, A.C. (Eds.), Endophytes of Forest Trees, Forestry Sciences. Springer, Netherlands, pp. 47–66.

Moricca, S., Beatrice, G., Ragazzi, A., 2012. Species- and organ-specificity in endophytes colonizing healthy and declining Mediterranean oaks. Phytopathol. Mediterr. 51, 587–598. https://doi.org/10.14601/Phytopathol_Mediterr-11705.

Moser, J., Konrad, H., Blomquist, S., Kirisits, T., 2009. Do mites phoretic on elm bark beetles contribute to the transmission of Dutch elm disease? Naturwissenschaften 97, 219–227. https://doi.org/10.1007/s00114-009-0630-x.

Müller, M.M., Hallaksela, A.-M., 2000. Fungal diversity in Norway spruce: a case study. Mycol. Res. 104, 1139–1145. https://doi.org/10.1017/S0953756200003105.

Müller, M.M., Valjakka, R., Suokko, A., Hantula, J., 2001. Diversity of endophytic fungi of single Norway spruce needles and their role as pioneer decomposers. Mol. Ecol. 10, 1801–1810.

Müller, M.M., Hantula, J., Wingfield, M., Drenkhan, R., 2018. *Diplodia sapinea* found on Scots pine in Finland. For. Pathol. https://doi.org/10.1111/efp.12483, e12483.

Nair, P.V.R., Wiechel, T.J., Crump, N.S., Taylor, P.W.J., 2015. First report of *Verticillium tricorpus* causing *Verticillium* wilt in potatoes in Australia. Plant Dis. 99, 731. https://doi.org/10.1094/PDIS-10-14-1014-PDN.

Navarro-Meléndez, A.L., Heil, M., 2014. Symptomless endophytic fungi suppress endogenous levels of salicylic acid and interact with the jasmonate-dependent indirect defense traits of their host, lima bean (*Phaseolus lunatus*). J. Chem. Ecol. 40, 816–825. https://doi.org/10.1007/s10886-014-0477-2.

Nicholls, T., Ostry, M., 1990. *Sphaeropsis sapinea* cankers on stressed red and jack pines in Minnesota and Wisconsin. Plant Dis. 74, 54–56.

Oghenekaro, A.O., Kovalchuk, A., Raffaello, T., Camarero, S., Gressler, M., Henrissat, B., Lee, J., Liu, M., Martínez, A.T., Miettinen, O., Mihaltcheva, S., Pangilinan, J., Ren, F., Riley, R., Ruiz-Dueñas, F.J., Serrano, A., Thon, M.R., Wen, Z., Zeng, Z., Barry, K., Grigoriev, I.V., Martin, F., Asiegbu, F.O., 2020. Genome sequencing of *Rigidoporus microporus* provides insights on genes important for wood decay, latex tolerance and interspecific fungal interactions. Sci. Rep. 10, 5250. https://doi.org/10.1038/s41598-020-62150-4.

Oliva, J., Boberg, J., Stenlid, J., 2013. First report of *Sphaeropsis sapinea* on Scots pine (*Pinus sylvestris*) and Austrian pine (*P. nigra*) in Sweden. New Dis. Rep. 27, 23.

Oses, R., Valenzuela, S., Freer, J., Sanfuentes, E., Rodríguez, J., 2008. Fungal endophytes of healthy Chilean trees and their possible role in early wood decay. Fungal Divers. 33, 77–86.

Osono, T., 2006. Role of phyllosphere fungi of forest trees in the development of decomposer fungal communities and decomposition processes of leaf litter. Can. J. Microbiol. 52, 701–716.

Ottosson, E., Nordén, J., Dahlberg, A., Edman, M., Jönsson, M., Larsson, K.-H., Olsson, J., Penttilä, R., Stenlid, J., Ovaskainen, O., 2014. Species associations during the succession of wood-inhabiting fungal communities. Fungal Ecol. 11, 17–28. https://doi.org/10.1016/j.funeco.2014.03.003.

Palmer, M.A., Stewart, E.L., Wingfield, M.J., 1987. Variation among isolates of *Sphaeropsis sapinea* in the North Central United States. Phytopathology 77, 944–948.

Pan, Y., Ye, H., Lu, J., Chen, P., Zhou, X.-D., Qiao, M., Yu, Z.-F., 2018. Isolation and identification of *Sydowia polyspora* and its pathogenicity on *Pinus yunnanensis* in Southwestern China. J. Phytopathol. 166, 386–395. https://doi.org/10.1111/jph.12696.

Parfitt, D., Hunt, J., Dockrell, D., Rogers, H.J., Boddy, L., 2010. Do all trees carry the seeds of their own destruction? PCR reveals numerous wood decay fungi latently present in sapwood of a wide range of angiosperm trees. Fungal Ecol. 3, 338–346. https://doi.org/10.1016/j.funeco.2010.02.001.

Pažoutová, S., Šrůtka, P., Holuša, J., Chudíčková, M., Kubátová, A., Kolařík, M., 2012. *Liberomyces* gen. nov. with two new species of endophytic coelomycetes from broadleaf trees. Mycologia 104, 198–210. https://doi.org/10.3852/11-081.

Pegg, G.F., Brady, B.F., 2002. Verticillium wilts. CAB INTL, Wallingford, Oxon, UK; New York.

Pellitier, P.T., Zak, D.R., Salley, S.O., 2019. Environmental filtering structures fungal endophyte communities in tree bark. Mol. Ecol. 28, 5188–5198. https://doi.org/10.1111/mec.15237.

Peršoh, D., Melcher, M., Flessa, F., Rambold, G., 2010. First fungal community analyses of endophytic ascomycetes associated with *Viscum album* ssp. *austriacum* and its host *Pinus sylvestris*. Fungal Biol. 114, 585–596.

Peters, S., Draeger, S., Aust, H.-J., Schulz, B., 1998. Interactions in dual cultures of endophytic fungi with host and nonhost plant calli. Mycologia 90, 360–367.

Petrini, O., Fisher, P., 1988. A comparative study of fungal endophytes in xylem and whole stem of *Pinus sylvestris* and *Fagus sylvatica*. Trans. Br. Mycol. Soc. 91, 233–238.

Petrini, L., Petrini, O., 1985. Xylariaceous fungi as endophytes. Sydowia 28, 216–234.

Petrini-Klieber, L.E., 1985. Untersuchungen über die Gattung *Hypoxylon* (Ascomycetes) und Verwandte Arten. ETH Zürich.

Ponge, J.F., 1991. Succession of fungi and fauna during decomposition of needles in a small area of Scots pine litter. Plant Soil 138, 99–113. https://doi.org/10.1007/BF00011812.

Qadri, M., Rajput, R., Abdin, M.Z., Vishwakarma, R.A., Riyaz-Ul-Hassan, S., 2014. Diversity, molecular phylogeny, and bioactive potential of fungal endophytes associated with the Himalayan Blue Pine (*Pinus wallichiana*). Microb. Ecol. 67, 877–887.

Rabiey, M., Hailey, L.E., Roy, S.R., Grenz, K., Al-Zadjali, M.A.S., Barrett, G.A., Jackson, R.W., 2019. Endophytes vs tree pathogens and pests: can they be used as biological control agents to improve tree health? Eur. J. Plant Pathol. 155, 711–729. https://doi.org/10.1007/s10658-019-01814-y.

Ragazzi, A., Ginetti, B., Moricca, S., 2011. First report of *Biscogniauxia mediterranea* on English ash in Italy. Plant Dis. 96. https://doi.org/10.1094/PDIS-05-12-0442-PDN.

Ragazzi, A., Moricca, S., Capretti, P., Dellavalle, I., 1999. Endophytic presence of *Discula quercina* on Declining *Quercus cerris*. J. Phytopathol. 147, 437–440. https://doi.org/10.1111/j.1439-0434.1999.tb03847.x.

Ragazzi, A., Moricca, S., Capretti, P., Dellavalle, I., Turco, E., 2003. Differences in composition of endophytic mycobiota in twigs and leaves of healthy and declining *Quercus* species in Italy. For. Pathol. 33, 31–38.

Renvall, P., 1995. Community structure and dynamics of wood-rotting Basidiomycetes on decomposing conifer trunks in northern Finland. Karstenia 32, 1–51. https://doi.org/10.29203/ka.1995.309.

Riley, R., Salamov, A.A., Brown, D.W., Nagy, L.G., Floudas, D., Held, B.W., Levasseur, A., Lombard, V., Morin, E., Otillar, R., Lindquist, E.A., Sun, H., LaButti, K.M., Schmutz, J., Jabbour, D., Luo, H., Baker, S.E., Pisabarro, A.G., Walton, J.D., Blanchette, R.A., Henrissat, B., Martin, F., Cullen, D., Hibbett, D.S., Grigoriev, I.V., 2014. Extensive sampling of basidiomycete genomes demonstrates inadequacy of the white-rot/brown-rot paradigm for wood decay fungi. Proc. Natl. Acad. Sci. U.S.A. 111, 9923–9928. https://doi.org/10.1073/pnas.1400592111.

Rodríguez, J., Elissetche, J., Valenzuela, S., 2011. Tree endophytes and wood. Biodegradation, 81–93. https://doi.org/10.1007/978-94-007-1599-8_5.

Roll-Hansen, F., Roll-Hansen, H., 1979. Microflora of sound-looking wood in *Picea abies* stems. Eur. J. For. Pathol. 9, 308–316. https://doi.org/10.1111/j.1439-0329.1979.tb00693.x.

Romeralo, C., Diez, J.J., Santiago, N.F., 2012. Presence of fungi in Scots pine needles found to correlate with air quality as measured by bioindicators in northern Spain. For. Pathol. 42, 443–453.

Romeralo, C., Santamaría, O., Pando, V., Diez, J.J., 2015. Fungal endophytes reduce necrosis length produced by *Gremmeniella abietina* in *Pinus halepensis* seedlings. Biol. Control 80, 30–39. https://doi.org/10.1016/j.biocontrol.2014.09.010.

Rytioja, J., Hildén, K., Mäkinen, S., Vehmaanperä, J., Hatakka, A., Mäkelä, M.R., 2015. Saccharification of lignocelluloses by carbohydrate active enzymes of the white rot fungus *Dichomitus squalens*. PLoS One 10. https://doi.org/10.1371/journal.pone.0145166, e0145166.

Šamonil, P., Daněk, P., Baldrian, P., Tláskal, V., Tejnecký, V., Drábek, O., 2020. Convergence, divergence or chaos? Consequences of tree trunk decay for pedogenesis and the soil microbiome in a temperate natural forest. Geoderma 376. https://doi.org/10.1016/j.geoderma.2020.114499, 114499.

Samson, R.A., Houbraken, J., Thrane, U., Frisvad, J.C., Andersen, B., 2010. Food and Indoor Fungi, CBS Laboratory Manual Series. CBS-KNAW Fungal Biodiversity Centre, Utrecht.

Samuels, G.J., 1996. Trichoderma: a review of biology and systematics of the genus. Mycol. Res. 100, 923–935.

Sanz-Ros, A.V., Müller, M.M., San Martín, R., Diez, J.J., 2015. Fungal endophytic communities on twigs of fast and slow growing Scots pine (*Pinus sylvestris* L.) in northern Spain. Fungal Biol. 119, 870–883.

Savory, J., 2008. Breakdown of timber by ascomycetes and fungi imperfecti. Ann. Appl. Biol. 41, 336–347. https://doi.org/10.1111/j.1744-7348.1954.tb01126.x.

Scattolin, L., Montecchio, L., 2009. *Lophodermium piceae* and *Rhizosphaera kalkhoffii* in Norway spruce: correlations with host age and climatic features. Phytopathol. Mediterr. 48, 226–239.

Schlechte, G., 1986. Holzbewohnende Pilze: 240 Arten in Farbe. Jahn u. Ernst.

Schmidt, O., Liese, W., 1980. Variability of wood degrading enzymes of *Schizophyllum commune*. Holzforschung 34, 67–72. https://doi.org/10.1515/hfsg.1980.34.2.67.

Schulz, B., Boyle, C., Draeger, S., Römmert, A.-K., Krohn, K., 2002. Endophytic fungi: a source of novel biologically active secondary metabolites. Mycol. Res. 106, 996–1004.

Schulz, B., Haas, S., Junker, C., Andree, N., Schobert, M., 2015. Fungal endophytes are involved in multiple balanced antagonisms. Curr. Sci. 109, 39–45.

Schumacher, J., 2012. Auftreten und Ausbreitung neuartiger Baumkrankheiten in Mitteleuropa unter Berücksichtigung klimatischer Aspekte. Forstwiss. Beiträge Tharandt, Ulmer, Stuttgart.

Schwarze, F.W.M.R., 2018. Diagnose und Prognose der Fäuledynamik in Stadtbäumen. MycoSolutions AG, ST. Gallen.

Schwarze, F.W.M.R., Lonsdale, D., Fink, S., 1995. Soft rot and multiple T-branching by the basidiomycete *Inonotus hispidus* in ash and London plane. Mycol. Res. 99, 813–820. https://doi.org/10.1016/S0953-7562(09)80732-6.

Schwarze, F., Engels, J., Mattheck, C., 2000. Fungal Strategies of Wood Decay in Trees. Springer-Verlag, Berlin, Heidelberg, Freiburg, pp. 1–185.

Senanayake, I.C., Jeewon, R., Chomnunti, P., Wanasinghe, D.N., Norphanphoun, C., Karunarathna, A., Pem, D., Perera, R.H., Camporesi, E., McKenzie, E.H.C., Hyde, K.D., Karunarathna, S.C., 2018. Taxonomic circumscription of Diaporthales based on multigene phylogeny and morphology. Fungal Divers. 93, 241–443. https://doi.org/10.1007/s13225-018-0410-z.

Sieber, T., 1988. Endophytische Pilze in Nadeln von gesunden und geschädigten Fichten (*Picea abies* [L.] Karsten). Eur. J. For. Pathol. 18, 321–342.

Sieber, T.N., 1989. Endophytic fungi in twigs of healthy and diseased Norway spruce and white fir. Mycol. Res. 92, 322–326. https://doi.org/10.1016/S0953-7562(89)80073-5.

Sieber, T.N., 2007. Endophytic fungi in forest trees: are they mutualists? Fungal Biol. Rev. 21, 75–89. https://doi.org/10.1016/j.fbr.2007.05.004.

Sieber, V.T., Hugentobler, C., 1987. Endophytische Pilze in Blättern und Ästen gesunder und geschädigter Buchen (*Fagus sylvatica* L.). Eur. J. For. Pathol. 17, 411–425. https://doi.org/10.1111/j.1439-0329.1987.tb01119.x.

Sieber, T.N., Sieber-Canavesi, F., Dorworth, C., 1991a. Endophytic fungi of red alder (*Alnus rubra*) leaves and twigs in British Columbia. Can. J. Bot. 69, 407–411. https://doi.org/10.1139/b91-056.

Sieber, T.N., Sieber-Canavesi, F., Petrini, O., Ekramoddoullah, A.K.M., Dorworth, C.E., 1991b. Characterization of Canadian and European Melanconium from some Alnus species by morphological, cultural and biochemical studies. Can. J. Bot. 69, 2170–2176.

Sieber, T., Rys, J., Holdenrieder, O., 1999. Mycobiota in symptomless needles of *Pinus mugo* ssp. *uncinata*. Mycol. Res. 103, 306–310.

Sieber-Canavesi, F., Sieber, T.N., 1987. Endophytische Pilze in Tanne (*Abies alba* Mill.). Vergleich zweier Standorte im Schweizer Mittelland (Naturwald-Aufforstung). Sydowia 40, 250–273.

Simeto, S., Alonso, R., Tiscornia, S., Bettucci, L., 2005. Fungal community of *Eucalyptus globulus* and *Eucalyptus maidenii* stems in Uruguay. Sydowia 57, 13.

Sinclair, W.A., Lyon, H.H., 2005. Diseases of Trees and Shrubs, second ed. Comstock Publishing.

Singh, D.K., Sharma, V.K., Kumar, J., Mishra, A., Verma, S.K., Sieber, T.N., Kharwar, R.N., 2017. Diversity of endophytic mycobiota of tropical tree *Tectona grandis* Linn.f.: spatiotemporal and tissue type effects. Sci. Rep. 7, 1–14. https://doi.org/10.1038/s41598-017-03933-0.

Smahi, H., Belhoucine-Guezouli, L., Berraf-Tebbal, A., Chouih, S., Arkam, M., Franceschini, A., Linaldeddu, B.T., Phillips, A.J.L., 2017. Molecular characterization and pathogenicity of *Diplodia corticola* and other Botryosphaeriaceae species associated with canker and dieback of Quercus suber in Algeria. Mycosphere 8, 1261–1272.

Soesanto, L., 2000. Ecology and Biological Control of *Verticillium dahliae*. University Wagenigen, Wagenigen, pp. 1–120.

Stanosz, G.R., Blodgett, J.T., Smith, D.R., Kruger, E.L., 2001. Water stress and *Sphaeropsis sapinea* as a latent pathogen of red pine seedlings. New Phytol. 149, 531–538.

Stokland, J., Siitonen, J., Jonsson, B., 2012. Biodiversity in Dead Wood (Ecology, Biodiversity and Conservation). Cambridge University Press, Cambridge, pp. 1–509, https://doi.org/10.1017/CBO9781139025843.

Stone, J., Petrini, O., 1997. Endophytes of forest trees: a model for fungus-plant interactions. In: Carroll, G.C., Tudzynski, P. (Eds.), Plant Relationships Part B: Part B, The Mycota. Springer, Berlin, Heidelberg, pp. 129–140, https://doi.org/10.1007/978-3-642-60647-2_8.

Sun, X., Guo, L.-D., Hyde, K.D., 2011. Community composition of endophytic fungi in *Acer truncatum* and their role in decomposition. Fungal Divers. 47, 85–95.

Suryanarayanan, T.S., 2011. Diversity of fungal endophytes in tropical trees. In: Pirttilä, A.M., Frank, A.C. (Eds.), Endophytes of Forest Trees: Biology and Applications, Forestry Sciences. Springer Netherlands, Dordrecht, pp. 67–80, https://doi.org/10.1007/978-94-007-1599-8_4.

Swart, W.J., Wingfield, M.J., 1991. Biology and control of *Sphaeropsis sapinea* on *Pinus* species in South Africa. Plant Dis. 75, 761–766.

Swart, W.J., Wingfield, M.J., Knox-Davies, P.S., 1987. Factors associated with *Sphaeropsis sapinea* infection of pine trees in South Africa. Phytophylactica 19, 505–510.

Talgø, V., Chastagner, G., Thomsen, I.M., Cech, T., Riley, K., Lange, K., Klemsdal, S.S., Stensvand, A., 2010. *Sydowia polyspora* associated with current season needle necrosis (CSNN) on true fir (*Abies* spp.). Fungal Biol. 114, 545–554.

Tanaka, K., Mel'nik, V., Kamiyama, M., Hirayama, K., Shirouzu, T., 2010. Molecular phylogeny of two coelomycetous fungal genera with stellate conidia, *Prosthemium* and *Asterosporium*, on Fagales trees. Botany 88, 1057–1071. https://doi.org/10.1139/B10-078.

Tanney, J.B., Seifert, K.A., 2019. *Tryblidiopsis magnesii* sp. nov. from *Picea glauca* in Eastern Canada. Fungal Syst. Evol. 4, 13–20.

Taylor, J.B., 1968. Host range of *Verticillium tricorpus* (Isaac). N. Z. J. Agric. Res. 11, 521–523. https://doi.org/10.1080/00288233.1968.10431449.

Tejesvi, M.V., Mahesh, B., Nalini, M.S., Prakash, H.S., Kini, K.R., Subbiah, V., Shetty, H.S., 2005. Endophytic fungal assemblages from inner bark and twig of *Terminalia arjuna* W. & A. (Combretaceae). World J. Microbiol. Biotechnol. 21, 1535–1540. https://doi.org/10.1007/s11274-005-7579-5.

Terhonen, E., Marco, T., Sun, H., Jalkanen, R., Kasanen, R., Vuorinen, M., Asiegbu, F., 2011. The effect of latitude, season and needle-age on the mycota of Scots Pine (*Pinus sylvestris*) in Finland. Silva Fenn. 45, 301–317.

Terhonen, E., Kovalchuk, A., Zarsav, A., Asiegbu, F.O., 2019. Forest tree microbioms and associated fungal Endophytes: functional roles and impact on forest health. Forests, 10. https://doi.org/10.1007/978-3-319-89833-9_13.

Toti, L., Viret, O., Horat, G., Petrini, O., 1993. Detection of the endophyte *Discula umbrinella* in buds and twigs of *Fagus sylvatica*. Eur. J. For. Pathol. 23, 147–152. https://doi.org/10.1111/j.1439-0329.1993.tb00954.x.

Unterseher, M., 2011. Diversity of fungal endophytes in temperate forest. Trees, 31–46. https://doi.org/10.1007/978-94-007-1599-8_2.

Úrbez-Torres, J.R., 2011. The status of Botryosphaeriaceae species infecting grapevines. Phytopathol. Mediterr. 50, S5–S45. https://doi.org/10.14601/Phytopathol_Mediterr-9316.

van der Wal, A., Gunnewiek, P.J.A.K., Cornelissen, J.H.C., Crowther, T.W., de Boer, W., 2016. Patterns of natural fungal community assembly during initial decay of coniferous and broadleaf tree logs. Ecosphere 7. https://doi.org/10.1002/ecs2.1393, e01393.

Vane, C.H., Drage, T.C., Snape, C.E., Stephenson, M.H., Foster, C., 2005. Decay of cultivated apricot wood (*Prunus armeniaca*) by the ascomycete *Hypocrea sulphurea*, using solid state 13C NMR and off-line TMAH thermochemolysis with GC–MS. Int. Biodeterior. Biodegradation 55, 175–185.

Várnai, A., Mäkelä, M.R., Djajadi, D.T., Rahikainen, J., Hatakka, A., Viikari, L., 2014. Binding modules of fungal cellulases: occurrence in nature, function, and relevance in industrial biomass conversion. Adv. Appl. Microbiol. 88, 103–165.

Verkley, G.J.M., 1999. A monograph of the genus *Pezicula* and its anamorphs. Stud. Mycol. 44, 1–180.

Wang, Y., Guo, L., 2007. A comparative study of endophytic fungi in needles, bark, and xylem of *Pinus tabulaeformis*. Can. J. Bot. 85, 911–917.

Webber, J.F., Brasier, C.M., 1984. The transmission of Dutch elm disease: a study of the process involved. In: Anderson, J.M., Rayner, A.D.M., Walton, D.W.H. (Eds.), Invertebrate-Microbial Interactions 1984. Cambridge University Press, Cambridge UK, pp. 271–306.

Whalley, A.J.S., 1996. The xylariaceous way of life. Mycol. Res. 100, 897–922.

Whitehill, J.G.A., Lehman, J.S., Bonello, P., 2007. *Ips pini* (Curculionidae: Scolytinae) is a vector of the fungal pathogen, *Sphaeropsis sapinea* (Coelomycetes), to Austrian Pines, *Pinus nigra* (Pinaceae). Environ. Entomol. 36, 114–120. https://doi.org/10.1603/0046-225X(2007)36[114:IPCSIA]2.0.CO;2.

Wijayawardene, N.N., Hyde, K.D., Wanasinghe, D.N., Papizadeh, M., Goonasekara, I.D., Camporesi, E., Bhat, D.J., McKenzie, E.H.C., Phillips, A.J.L., Diederich, P., Tanaka, K., Li, W.J., Tangthirasunun, N., Phookamsak, R., Dai, D.-Q., Dissanayake, A.J., Weerakoon, G., Maharachchikumbura, S.S.N., Hashimoto, A., Matsumura, M., Bahkali, A.H., Wang, Y., 2016. Taxonomy and phylogeny of dematiaceous coelomycetes. Fungal Divers. 77, 1–316.

Wingfield, M.J., Knox-Davies, P.S., 1980. Observations on diseases in pine and eucalyptus plantations in South Africa. Phytophylactica 12, 57–63.

Woodward, S., Stenlid, J., Karjalainen, R., Hüttermann, A. (Eds.), 1998. *Heterobasidion annosum*: Biology, Ecology, Impact and Control. CAB International, Wallingford.

Worrall, J.J., Anagnost, S.E., Zabel, R.A., 1997. Comparison of wood decay among diverse lignicolous fungi. Mycologia 89, 199–219.

Yadeta, K.A., Thomma, B.P.H.J., 2013. The xylem as battleground for plant hosts and vascular wilt pathogens. Front. Plant Sci. 4. https://doi.org/10.3389/fpls.2013.00097.

Yuan, Z.-L., Rao, L.-B., Chen, Y.-C., Zhang, C.-L., Wu, Y.-G., 2011. From pattern to process: species and functional diversity in fungal endophytes of *Abies beshanzuensis*. Fungal Biol. 115, 197–213. https://doi.org/10.1016/j.funbio.2010.11.002.

Zare, R., Gams, W., Starink-Willemse, M., Summerbell, R.C., 2007. Gibellulopsis, a Suitable Genus for *Verticillium nigrescens*, and Musicillium, a New Genus for *V. theobromae* [WWW Document]. https://www.ingentaconnect.com/content/schweiz/novh/2007/00000085/F0020003/art00015. (Accessed 4 January 2020).

Zlatković, M., Keča, N., Wingfield, M.J., Jami, F., Slippers, B., 2017. New and unexpected host associations for *Diplodia sapinea* in the Western Balkans. For. Pathol. 47. https://doi.org/10.1111/efp.12328.

Zwolinski, J.B., Swart, W.J., Wingfield, M.J., 1995. Association of *Sphaeropsis sapinea* with insect infestation following hail damage of *Pinus radiata*. For. Ecol. Manage. 72, 293–298.

Chapter 11

Dark septate endophytes of forest trees

Eeva Terhonen

Forest Pathology Research Group, Department of Forest Botany and Tree Physiology, Faculty of Forest Sciences and Forest Ecology, University of Göttingen, Göttingen, Germany

Chapter Outline

1. Historical perspectives on dark septate endophytes

Septate endophytes have been observed in plant roots already over 100 years ago (Gallaud, 1905) and later they were described as *Mycelium radicis atrovirens* (MRA) (Melin, 1922, 1923). These MRA were reported to intracellularly colonize (no Hartig net or mantle was present) boreal tree roots (*Pinus sylvestris* and *Picea abies*) (Melin, 1922, 1923). Taxonomic identification of these dark sterile fungi was challenging and Gams (1963) was able to suggest first taxonomic identities in the root MRA complex. However, later it was pointed that most likely the first identifications were wrong (Wang and Wilcox, 1985). Currently the sequencing analyses allow more sophisticated and trustworthy methods to identify dark septate endophytes (DSE) and the new applications (genetics) have replaced the time-consuming morphological identification.

2. Endophytes and forest trees

Forest trees have a balanced relationship with diverse mycobiome that share overlapping habitats. These relationships usually vary from latent to mutualism and/or pathogenic interactions. One important, but yet unknown, group are the endophytic microbes, which play important roles in the growth and health of forest trees, as well as a role in nutrient cycling (Sieber, 2007; Porras-Alfaro and Bayman, 2011; Terhonen et al., 2019). Endophytes are microorganisms that do not cause visible damage when colonizing their host for the whole or in part of their life cycle (Petrini, 1991; Saikkonen et al., 1998). In that sense, definition of "endophyte" excludes beneficial microorganisms and pathogens. However, pathogens can switch from symptomless endophyte to pathogenic. This phenomenon is known as cryptic life cycle (Wilson, 1995; Porras-Alfaro and Bayman, 2011; Bußkamp, 2018). Indeed, endophytes do not inevitably form symptomless associations with their hosts. Rather endophytes have cryptic life cycles and depending on environment and the host genotype, they can switch from commensalistic/mutualistic to pathogenic and/or saprotrophic lifestyles (Saikkonen et al., 2010; Rajala et al., 2013; Reininger et al., 2012; Bußkamp, 2018).

Fungal endophytes have been found from all studied plant species (Jumpponen and Trappe, 1998a). Although huge range of plants hosts fungal endophytes, many concepts of endophytic interactions with their host are still not well understood (Sieber and Grünig, 2013). It has demonstrated that endophytic colonization can benefit the host, and controversially the negative effects are documented constantly. In this chapter, I will focus on root fungal tree endophytes and more specifically on DSE and their functions in forest trees.

3. Ecology of DSE

Fungal endophytes are classified in two major groups (clavicipitaceous and nonclavicipitaceous) and four different classes (1–4) by Rodriguez et al. (2009). Class 1 includes only the clavicipitaceous endophytic fungi and the nonclavicipitaceous endophytic fungi are divided to Classes 2, 3, and 4 depending on the colonized plant tissue, biodiversity, transmission, and fitness benefits (Rodriguez et al., 2009). DSE are listed in Class 4 as nonclavicipitaceous endophytes that are observed only in the roots (Rodriguez et al., 2009). DSE are primarily ascomycetous fungi and even if extensively studied, their ecological roles are not well understood (Rodriguez et al., 2009; Sieber and Grünig, 2013). As mentioned before, most likely DSE were observed one hundred years ago (Melin, 1922, 1923), but the judgment of a fungi to be actually a DSE has still deficiencies, especially in the absence of unity in taxonomy and evolutionary origins of these fungi (Mandyam and Jumpponen, 2015). DSE have been observed from the tropics to arctic and from all plant species studied (Jumpponen and Trappe, 1998a; Sieber, 2007; Rodriguez et al., 2009; Rodriguez and Redman, 2008; Sieber and Grünig, 2013). It is very likely that DSEs can be isolated from all plant species (terrene) (see review by Jumpponen and Trappe, 1998a) and they do present dominant core component of fungal communities within plant roots worldwide (Knapp et al., 2019). Notable is that endophytes have been studied only from ~1% to 2% of the known plant species (Strobel, 2018).

The typical character of the DSE is the production of melanin (black polymers) found in septate hyphae (McGonigle et al., 1990; Jumpponen and Trappe, 1998a; Barrow and Aaltonen, 2001). DSE form specialized structures in the host roots, round microsclerotias, that grow both inter- and intracellularly within the cortex (Jumpponen and Trappe, 1998a; Barrow and Osuna, 2002; Barrow, 2003; Mandyam and Jumpponen, 2005; Heinonsalo et al., 2017, Fig. 11.1). Indeed, to classify fungi as DSE hyphae, it should be found in plant roots as asexual, melanized, and septate, as well as the microsclerotia should be observed (Jumpponen and Trappe, 1998a; Jumpponen, 2001; Mandyam and Jumpponen, 2005). DSE are conidial or sterile helotialean ascomycete fungi (Stoyke and Currah, 1991; Jumpponen and Trappe, 1998a).

4. Beneficial effects of DSE on their hosts

DSE are able to provide benefits to their hosts in various ways. They can increase the fitness of the host in harsh environments as they tolerate a range of environmental stresses such as higher temperatures and drought (Li et al., 2018; He et al., 2019; Li et al., 2019; Rayment et al., 2020), pollution (Ruotsalainen et al., 2007; Likar and Regvar, 2013; Hou et al., 2020), and/or increasing elevation (Newsham et al., 2009; Kotilínek et al., 2017). DSEs have also been noted to increase their host plant growth (Newsham, 2011; Terhonen et al., 2016; Li et al., 2019; Hou et al., 2020) and performance against pathogens (Schulz et al., 1998, 2002; Mandyam and Jumpponen, 2005; Tellenbach and Sieber, 2012; Schulz et al., 2015; Terhonen et al., 2016, Fig. 11.1). Several mechanisms through which DSE may increase the host resistance against pathogens have been suggested by Mandyam and Jumpponen (2005). These include: (1) niche competition (physical competition between fungi or better nutrient utilization); (2) pathogen inhibition due to metabolites produced by DSE (Fig. 11.1); (3) DSE may

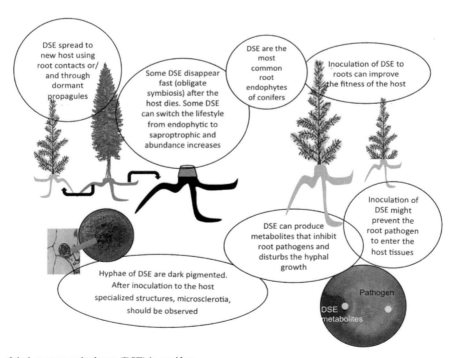

FIG. 11.1 Life cycle of dark septate endophytes (DSE) in conifers.

induce plant defense responses (Mandyam and Jumpponen, 2005). Even with considerable dominance of DSEs within individual host roots (Mandyam and Jumpponen, 2005; Green et al., 2008; Stroheker et al., 2018a,b; Landolt et al., 2020), the functional interplay between DSEs and their hosts in nature has not been resolved (Compant et al., 2016; Sieber and Grünig, 2013). More hypothesis-driven research is needed that will help to answer questions related to DSE functions, and how these intimate associations between plants and DSE do not result in disease (Compant et al., 2016).

5. *Phialocephala fortinii* s.l.-*Acephala applanata* species complex

The most commonly isolated and observed group of fungi in DSE is species belonging in the *Phialocephala fortinii* s.l.-*Acephala applanata* species complex (PAC) (Sieber and Grünig, 2006; Grünig et al., 2008b). The PAC members are cryptic species (CSP) as the species separation cannot be performed based on the morphology of each species. Until now, 22 genetically unique species (CSP) have been observed (Queloz et al., 2008; Queloz et al., 2011; Landolt et al., 2020), of which eight species have been formally described (Grünig et al., 2001, 2002a, 2003, 2004, 2008a; Queloz et al., 2010). Described species include *Phialocephala fortinii* s.s., *Acephala applanata*, *Phialocephala turiciensis*, *Phialocephala letzii*, *Phialocephala europaea*, *Phialocephala helvetica*, *Phialocephala uotilensis*, and *Phialocephala subalpina* (Grünig et al., 2001, 2002a, 2003, 2004; Grünig and Sieber, 2005; Queloz et al., 2010). The members of the PAC can be separated to species level based on amplification of 13 loci applying the multiplex method commenced by Queloz et al. (2008, 2010) (Landolt et al., 2020).

DSE and PAC distribution has been studied in temperate and boreal forests in Europe and North America (Stoyke and Currah, 1991; Ahlich and Sieber, 1996; Ahlich-Schlegel, 1997; Hambleton and Currah, 1997; Ahlich et al., 1998; Addy et al., 2000; Grünig et al., 2004; Piercey et al., 2004; Menkis et al., 2004; Grünig et al., 2006; Terhonen et al., 2014; Sietiö et al., 2018). PAC species have only been detected a few times in the southern hemisphere (Grünig et al., 2008b). Whereas DSE overall have not shown any host specificity, some PAC species seem to prefer conifers (Ahlich and Sieber, 1996; Ahlich-Schlegel, 1997; Addy et al., 2000; Grünig et al., 2002b; Addy et al., 2005; Grünig et al., 2006; Sieber, 2007; Saikkonen, 2007; Grünig et al., 2008a; Terhonen et al., 2014; Sietiö et al., 2018; Landolt et al., 2020, Table 11.1). However, PAC are also found commonly in the roots of deciduous trees (Wilson et al., 2004; Halmschlager and Kowalski, 2004; Kwasna et al., 2008; Lygis et al., 2004; Reininger et al., 2012; Landolt et al., 2020) and ericaceous hosts (Ahlich and Sieber, 1996; Sietiö et al., 2018). From members of PAC, *P. fortinii* s.s. don't seem to favor no plant host (Ahlich and Sieber, 1996; Grünig et al., 2006, 2008b; Tejesvi et al., 2010, 2013; Sietiö et al., 2018, Table 11.1). However, *A. applanata* is more commonly observed in *Picea abies* (Grünig and Sieber, 2005; Grünig et al., 2006, 2008b; Stroheker et al., 2016, 2018a,b, Table 11.1). Similarly, host preferences have been found for PAC members *P. europaea* and *P. helvetica* as their abundance was significantly higher in *Pinus sylvestris* compared to broadleaved host species (*Quercus pubescens*) (Landolt et al., 2020).

6. Changes in PAC/DSE communities

PAC are well established in the northern hemisphere (Queloz et al., 2011) and they are defined as the most widespread root inhabitants (Sieber and Grünig, 2013). The communities of PAC are greatly varied (Queloz et al., 2005; Queloz et al., 2008; Grünig et al., 2006; Queloz et al., 2010; Stroheker et al., 2018a,b). The resident PAC community in roots of living trees in natural forests has demonstrated to be stable for years (Queloz et al., 2005; Stroheker et al., 2016). Indeed, it can take decades that changes can be seen in undisturbed PAC communities (Queloz et al., 2005; Stroheker et al., 2016). Several PAC species occur sympatrically in same tree roots (Queloz et al., 2005; Grünig et al., 2008b; Queloz et al., 2008), and most likely different PAC strains compete against each other for space and nutrients (Stroheker et al., 2016). There are observations for changes in PAC communities in living tree roots that take place between PAC members, which compete for the one and the same niche (Reininger et al., 2012; Stroheker et al., 2018b). However, the environmental changes and severe disturbances (e.g., human influence) can have effect on the PAC communities. Indeed, Stroheker et al. (2018b) showed that clear-cutting changed significantly the major PAC community. Members of the PAC are dependent on living root material, since PAC presence is decreasing after clear-cut (dependent on living host) (Stroheker et al., 2018b, Fig. 11.1). Similarly, Kluting et al. (2019) found out that DSE are present more frequently/abundantly in organic and mineral soils that have root material compared to mineral soil without root material. These observations indicate that DSE and PAC are dependent on root material and highlight their commensalism/mutualistic nature. However, notable is that *A. applanata* presence is increasing in dead root material and in that sense it is benefitting from clear-cutting (Stroheker et al., 2018b). Stroheker et al. (2018b) suspect that *A. applanata* is associated with dying or dead plant material (roots), especially when compared to further species of the PAC, indicating the switch from neutral symbiont to saprotroph. As stated before, it is the only

TABLE 11.1 List of *Phialocephala fortinii* s.l.-*Acephala applanata* species complex (PAC) observed in different hosts.

DSE	Tree host	References
Phialocephala fortinii[a]	*Abies alba*	Ahlich and Sieber (1996)
	Alnus rubra	Ahlich and Sieber (1996)
	Betula pendula	Menkis et al. (2004)
	Fagus sylvatica	Ahlich and Sieber (1996)
	Picea abies	Ahlich and Sieber (1996)
		Grünig et al. (2002b, 2006, 2008a)
		Menkis et al. (2004)
		Queloz et al. (2005)
		Stenström et al. (2014)
		Terhonen et al. (2014)
		Stroheker et al. (2018b)
	Pinus contorta	Jumpponen et al. (1998)
		O'Dell et al. (1993)
	Pinus sylvestris	Wang and Wilcox (1985)
		Menkis et al. (2004)
		Grünig et al. (2008a)
		Tienaho et al. (2019)
Acephala applanata	*Picea abies*	Grünig and Sieber (2005)
		Grünig et al. (2006, 2008a)
		Terhonen et al. (2014)
		Stroheker et al. (2016)
		Stroheker et al. (2018b)
Phialocephala subalpina[b]	*Pinus sylvestris*	Landolt et al. (2020)
	Quercus pubescens	Landolt et al. (2020)
	Picea abies	Stroheker et al. (2018b)
Phialocephala europaea	*Pinus sylvestris*	Landolt et al. (2020)
	Quercus pubescens	Landolt et al. (2020)
	Picea abies	Stroheker et al. (2016)
		Stroheker et al. (2018b)
		Santschi (2015)
	Fraxinus excelsior	Santschi (2015)
	Acer pseudoplatanus	Santschi (2015)

[a] Defined as sensu stricto but some cases can be sensu lato.
[b] Genome sequenced (Schlegel et al., 2016).

host-specific (Norway spruce) PAC species (Grünig et al., 2008b). The genome of PAC member, *P. subalpina*, includes functional gene pools typical for saprotrophs, mutualistic symbionts, or pathogens, which could explain the PAC/DSE community's adaptation to different environments and lifestyle switch after disturbance (Schlegel et al., 2016; Stroheker et al., 2018b).

It has been suggested that changes in environment (higher temperature) can alter a faster change of the PAC communities and reform the competitive power of individual PAC strains (Reininger et al., 2012; Stroheker et al., 2018b). Studies of colonization intensity of PAC members indicate that lower temperature tends to increase the infection intensity (18°C vs 14°C) (Stroheker et al., 2018a). The emergence of (axenic) primary roots is strongest in spring time (Hendrick and Pregitzer, 1996; Sword et al., 2000; Pregitzer et al., 2008; Stroheker et al., 2018a,b) and it can be assumed that the uncolonized primary roots are easily colonized by PAC species (Stroheker et al., 2018a,b) leading to higher infection rate (Stroheker et al., 2018a). However, forest soil pH can have effect on DSE composition. Wilson et al. (2004) isolated DSE, *Phialocephala sphaeroides*, as an associate of roots of plants growing in highly acidic sites. Similarly, *P. sphaeroides* has been isolated from Norway spruce and Scots pine from highly acid sites (Terhonen et al., 2014; Sietiö et al., 2018) but it has never been observed from more alkaline sites (Wilson et al., 2004; Grünig et al., 2008a,b; Queloz et al., 2011).

Notable is that despite DSE and PAC ubiquity, their dispersal and behavioral biology remain uncharacterized to date. So far, researchers have not come across with teleomorph (sexual state) of DSE. However, Grünig et al. (2004, 2006) stated that mating-type loci and evidences for reproduction exist. *A. applanate* is only PAC/DSE whose mating-type genes are organized in a homothallic fashion in the same individual (both MAT1-1 and MAT1-2) (Zaffarano et al., 2010). Furthermore, PAC members have operating of mating-type loci (Zaffarano et al., 2010, 2011). Asexual spores (conidia) are formed rarely after long incubation time (~ 1 year) of cultures in the darkness at 4°C (Ahlich and Sieber, 1996; Grünig et al., 2008b; Terhonen et al., 2014). The spores are not known to be airborne (Kauserud et al., 2005) and conidia do not germinate in vitro (Grünig et al., 2004; Grünig et al., 2008b). Locally, PAC were shown to be transferred via root contacts (Stroheker et al., 2016, 2018a,b). Stroheker et al. (2018b) stated that PAC are spread via root contacts between living roots, as in *Heterobasidion* species (Piri and Hamberg, 2015), and/or by means of dormant propagules within root debris (Day and Currah, 2011; Stroheker et al., 2016). Day and Currah (2011) suggested that DSE fungi can persist and produce propagules, i.e., microsclerotia and conidia, in the absence of living host roots. In that sense, Day and Currah (2011) hypothesized that DSE fungi are already present in forest sites (dormant propagules) before plant hosts during primary succession events, e.g., after clear-cuts (Day and Currah, 2011). Relocation to new sites may occur through planting of previously colonized seedlings (Brenn et al., 2008; Stenström et al., 2014).

7. DSE and host interactions in harsh environments

The DSE have been noted to be strikingly present in plant roots in harsh environments (Likar and Regvar, 2009; Regvar et al., 2010; Likar and Regvar, 2013). In metal-polluted sites, it has been noted that DSE are progressively increasing in these habitats, indicating they have important function for maintaining host fitness (Cevnik et al., 2000; Deram et al., 2008; Likar and Regvar, 2009; Regvar et al., 2010; Likar and Regvar, 2013). Indeed, high abundance of DSE has been noted to improve their plant host performance in heavy metals polluted soils (Ruotsalainen et al., 2007; Zhang et al., 2008; Deram et al., 2008; Likar and Regvar, 2009; Regvar et al., 2010; Likar and Regvar, 2013; Yamaji et al., 2016; Wang et al., 2016). This can be partly explained by the melanin-enriched cell walls that decrease heavy metal toxicity due to the ability of melanin to bind heavy metal ions (Gadd, 1993; Fogarty and Tobin, 1996; Ban et al., 2012). Similarly, the higher number of DSE sclerotia can indicate the survival strategy in heavy metal-polluted sites (Ruotsalainen et al., 2007). Likar and Regvar (2013) observed that inoculation of DSE to *Salix caprea* cuttings reduced the metal uptake in metal-enriched soils. Similar observation was made by Zhu et al. (2018), as DSE-inoculated tomatoes had a lower Cd and Zn accumulation. Under metal stress conditions, DSE inoculation, and due to the melanin produced, can enhance the activities of antioxidant enzymes increasing the fitness of host (Zhan et al., 2011; Ban et al., 2012; Zhu et al., 2018).

It has been shown that DSE have positive effects on several plant species under water-limited conditions (Zhang et al., 2010; Kivlin et al., 2013; dos Santos et al., 2017; Valli and Muthukumar, 2018; Landolt et al., 2020). Mandyam and Jumpponen (2005) assume that the main role of DSE is to improve water balance and drought tolerance. In a meta-analysis of the effects of higher temperatures and drought, DSE consistently exhibited a positive effect on plant biomass production (Kivlin et al., 2013). The high melanin concentrations and microsclerotia in DSE hyphae can protect them against extreme temperatures (heat/cold) (Jumpponen and Trappe, 1998a; Ahlich-Schlegel, 1997). DSEs are more frequently observed in polar regions than mycorrhizal fungi (Newsham et al., 2009; Day and Currah, 2011). Similarly, DSE colonization is higher than mycorrhizal at higher elevations (Schmidt et al., 2008; Kotilínek et al., 2017) and the mycorrhizal symbiosis was observed to be absent at the highest elevations in Himalayas and only DSE were present (Kotilínek et al., 2017). Systemic

DSE colonization seen in native plants in an arid ecosystem has been suggested to help plants to overcome the severe drought conditions (Barrow, 2003). In living forest tree roots, the colonization of DSE (here as PAC) was significantly higher in density in pubescent oak (*Quercus pubescens*) in drier, nonirrigated plots than irrigated ones (Landolt et al., 2020). A similar observation was made by Stroheker et al. (2018b) for Norway spruce (*P. abies*) seedlings as the roots were significantly more frequently colonized by an artificially introduced PAC strain in dry rather than in wet plots. The reasons for higher PAC abundance in tree roots at drier sites can be due to presence of microsclerotia in the root cortex (pseudo-sclerotia) (Grünig et al., 2008a; Sieber and Grünig, 2013; Landolt et al., 2020). Indeed, the harsh conditions (extreme temperatures/high elevations) could be too extreme for mycorrhiza, when DSE can be found rather evenly throughout all habitats (Barrow, 2003; Jumpponen and Trappe, 1998a; Kotilínek et al., 2017). This might be due to melanin-containing microsclerotia that are waterproof, resistant to drought, as well as they can repeatedly be frozen and thawed (Ahlich-Schlegel, 1997). Thus the microsclerotia are able to survive adverse conditions and in that sense also protect roots against desiccation (Mandyam and Jumpponen, 2005; Grünig et al., 2008b). However, notable is that is not clear how adaptation of DSE to harsh environment can truly protect also the hosts in these harsh environments. Similarly, Landolt et al. (2020) highlight that further studies are needed to confirm that the higher frequency of PAC in oak roots in nonirrigated plots is actually due to the higher formation of waterproof microsclerotia. This would reveal if PAC could protect tree roots against desiccation and improve drought resistance (Landolt et al., 2020). In future, extreme climate events, e.g., drought and higher temperatures, are expected to increase in frequency, duration, and intensity (Seidl et al., 2017). In that sense, future experiments exploring the function of DSE and PAC in tree roots are needed to reveal if they can improve the fitness of host.

8.   DSE-tree host interactions

Outcomes of artificial infections of DSE to tree roots have varied from beneficial (Jumpponen and Trappe, 1998b; Terhonen et al., 2016), to neutral (Reininger et al., 2012) and sometimes even to detrimental (Tellenbach et al., 2011; Reininger et al., 2012; Terhonen et al., 2016) effects to the host growth. Similarly, the meta-analysis of host reactions to artificial inoculations with root endophytes (including DSE) has shown detrimental (lower biomass) effects to the host (Mayerhofer et al., 2013). Positive impacts of DSE to the host have shown on total, shoot and root biomass, and on shoot nitrogen (N) and phosphorus contents (Newsham, 2011). The observed positive responses of the host could be explained by the nutrient mineralization (Jumpponen and Trappe, 1998b; Jumpponen et al., 1998; Jumpponen, 2001; Mandyam and Jumpponen, 2005; Newsham, 2011; Reininger and Sieber, 2013) and/or by the production of phytohormones (Schulz et al., 1998, 2002; Schulz and Boyle, 2005) that improve the plant growth. Reininger et al. (2012) observed that inoculation of Norway spruce (*P. abies*) and silver birch (*Betula pendula*) with different combinations of four PAC species had plant growth reduction that was dependent on strain/host combination used. PAC was detrimental on Norway spruce, but interestingly no pathogenic effect was observed on silver birch (Reininger et al., 2012). Other two PAC strains synergistically had less unfavorable effects on Norway spruce but again these strains together were able to reduce silver birch biomass (Reininger et al., 2012). Similar alternating observations were made by Tellenbach et al. (2011) as the virulence of *P. subalpina* genotypes varied. *P. subalpina* strains collected from outside the native range of Norway spruce were noted to be less severe than the strains collected from Norway spruce natural habitat (Tellenbach et al., 2011). The other PAC member had neutral to highly virulent interaction with Norway spruce and was primarily isolate dependent (Tellenbach et al., 2011; Tellenbach and Sieber, 2012). Tellenbach et al. (2011) observed that the virulence is dependent on strains within PAC species. However, Stroheker et al. (2018a) could not observe any negative affects to the health status of 3-year-old Norway spruce saplings, even if in previous experiments same strain was detrimental to 5–7-month-old seedlings (Tellenbach et al., 2011; Reininger et al., 2012). Dual inoculation of PAC and ectomycorrhizal fungi (ECM), *Hebeloma crustuliniforme*, in Norway spruce could neutralize plant growth suppression that was caused by using only one fungus during inoculation (Reininger and Sieber, 2013). Inoculation of DSE, *P. sphaeroides*, in the Norway spruce seedlings increased the biomass but, in this experiment, only one strain was used (Terhonen et al., 2016). Wilcox and Wang (1987a) showed that inoculation with PAC *P. fortinii* s.s. had slight pathogenic influence for *Pinus resinosa* and *Picea rubens* seedlings whereas inoculation with *Phialophora finlandia* improved the growth (Wilcox and Wang, 1987b). However, *P. fortinii* s.s. could improve the growth of *Pinus contorta* seedlings when glucose was provided (Jumpponen and Trappe, 1998b). Based on these results, the observed positive/negative plant growth effects depend on host (age and species), fungal species, and strains within the species as well as the environment. However, artificial infections do not reflect the situation in the field and further studies should concentrate on more natural conditions.

The competition of PAC strains between each other's and other fungi can reduce their negative impacts to the host (Reininger et al., 2012; Reininger and Sieber, 2013) as the DSE in the roots compete for their niche with other fungi (Reininger et al., 2012; Reininger and Sieber, 2013; Stroheker et al., 2018a,b). In theory, DSE could also compete with root pathogens and they could be utilized in integrated pest management (IPM). Tellenbach and Sieber (2012) showed that inoculation of *P. subalpina* into Norway spruce roots could reduce mortality and intensity of diseases caused by *Phytophthora plurivora* and *Elongisporangium undulatum*. However, suppression of pathogens was strongly dependent on the *P. subalpina* strain (Tellenbach and Sieber, 2012). Similarly, DSE *P. sphaeroides* could prevent *Heterobasidion parviporum* infection in Norway spruce seedling roots (Terhonen et al., 2016). The possible protection provided by DSE against more harmful necrotrophic pathogens might explain why trees host and tolerate wide range of DSE/PAC. Similarly, as stated before these hypotheses should be tested under field conditions.

9. DSE-mycorrhiza-host interactions

The presence of endophytic fungi is assumed to be mutualistic/commensal as they are found in healthy roots and together with mycorrhizas (Kernaghan et al., 2003; Kernaghan and Patriquin, 2011; Terhonen et al., 2014). Artificial inoculations of DSE in plants have revealed increased performance on total biomass as well as on phosphorus (P) and nitrogen (N) contents in shoots (Newsham, 2011). Mycorrhizal fungi colonize unlignified roots, while DSE is present everywhere in the root system (Grünig et al., 2011; Landolt et al., 2020) and they promptly infect the newly emerging roots (Terhonen et al., 2016; Stroheker et al., 2018a,b). Indeed, members of PAC can occur in the root cortex anywhere in a tree's root system in contrast to mycorrhizal fungi that are limited to absorptive roots (Landolt et al., 2020). PAC colonize roots undergoing secondary growth and they can therefore occur in the root cortex anywhere in a plant root system (Grünig et al., 2011; Landolt et al., 2020). As PAC and DSE colonize roots anywhere and of all age classes, they can also be found in stem bases (Grünig et al., 2008a; Menkis et al., 2004), e.g., *Cadophora* sp. are observed in stem bases of *Fraxinus excelsior* (Langer, 2017). DSE and PAC have to compete with ectomycorrhizal fungi during colonization of host roots, especially in the tips (Kernaghan et al., 2003; Menkis et al., 2004; Grünig et al., 2008b; Wagg et al., 2008; Stroheker et al., 2018a,b; Landolt et al. 2020). Ahlich et al. (1998) noted that freshly planted, sterile conifer seedlings are colonized more quickly by PAC than ectomycorrhizal fungi. Consequently, this indicates that the function of DSE is as genuine as the role of mycorrhizal fungi even if the host-endophyte interaction in the plant roots is different from mycorrhizal symbioses. The DSEs colonize intracellular spaces of ectomycorrhizal hosts and form microsclerotial structures (O'Dell et al., 1993). Similarly, as the DSE are thought to lack structures specialized for nutrient transfer (compared to mycorrhiza) with their hosts, the reported positive influence of inoculation with DSE remains unclear.

Wilcox and Wang (1987a,b) described *P. finlandia*-colonized roots to be ectendomycorrhizal in ectomycorrhizal host. Sclerotia-like structures (*P. finlandia, P. fortinii*) could be observed in the inner cortical cells (Wilcox and Wang, 1987a,b) and the Hartig net formed by *P. finlandia* extended to the endodermis in the spruce (*P. rubens*), but it surrounded only the epidermis in birch (*Betula alleghaensis*) (Wilcox and Wang, 1987b). DSE, *Acephala macrosclerotiorum*, has been observed to form structures that resemble ectomycorrhizae with its host tree (conifers) (Lukešová et al., 2015). This fungus was observed as root endophyte in nonmycorrhizal roots in Norway spruce (Terhonen et al., 2014). Notably, plant hosts which are not ecto- or ectendomycorrhizal can be colonized by these DSE (Jumpponen and Trappe, 1998a; Jumpponen, 2001) and they are found in nonmycorrhizal roots (Terhonen et al., 2014). PAC *P. fortinii* formed thin layer of hyphae surrounding the root surface, and the hyphae (intercellular and intracellular) grew in cortical and cortex cells in Scots pine roots (Heinonsalo et al., 2017). This *P. fortinii* intracellular colonization did not cause any detrimental damages to the host tissue and mainly healthy root tissue could be found (Heinonsalo et al., 2017). *P. fortinii* being there caused cell wall thickening and formation of structures within or between cortical cells (Heinonsalo et al., 2017).

As stated before, the morphology of DSE after inoculation in tree roots can resemble mycorrhizal associations. However, these structures formed by DSE do differ from the conventional types of mycorrhizas (Jumpponen, 2001) as specific structures for nutrient/carbon exchange have not been found (O'Dell et al., 1993; Heinonsalo et al., 2017, Fig. 11.2). This kind of interaction (morphology) in the roots has been described to be as ectendomycorrhizal (Wilcox and Wang, 1987a,b). The DSE hyphae in cortical and epidermal cells can form intracellularly microsclerotial structures (O'Dell et al., 1993; Jumpponen, 2001). Compared to ectomycorrhizal fungi, DSE cannot form a complete mantle or Hartig net when inoculated to a susceptible ectomycorrhizal host (O'Dell et al., 1993; Jumpponen, 2001; Heinonsalo et al., 2017). DSE fungi can grow intracellularly in roots without causing any visible damage to the host (Heinonsalo et al., 2017).

To increase P concentration in plants, they have evolved to form symbiosis with mycorrhizas that help plants to access phosphorus (P) present in the soil (Smith and Read, 1997). Similarly, it has been shown that inoculating DSE in *P. contorta* seedling roots increases the P concentration in the host (Jumpponen et al., 1998). It was recorded that *Phialocephala glacialis* and *P. turiciensis* are able to mineralize organic P and increasing the pool of P in the soil (Della Monica et al., 2015); this result suggests that indeed DSE fungi could release P from organic sources that can be used by their host plants (Della Monica et al., 2015). Similarly, it was demonstrated that DSE fungi are clearly capable of producing the extracellular enzymes necessary to process major detrital C, N, and P polymers into usable subunits (Caldwell et al., 2000). This would also

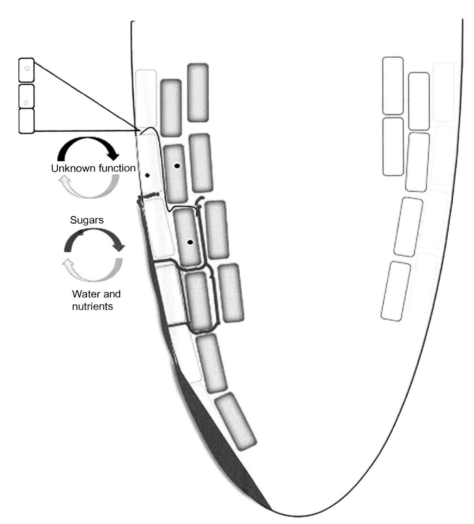

FIG. 11.2 DSE fungi form melanized microsclerotios *(black dots)* inside epidermis and cortex cells of their host plant. DSE also grow as melanized hyphae inter- and intracellularly *(black)*. The hyphae are septate and hyaline structures can be observed (*). Ectomycorrhiza produces mantle and Hartig net *(purple)*. The function between plant and DSE is currently unknown. Ectomycorrhiza and plant host exchange sugars (from plant) and water and mineral nutrients (fungi). DSE = *black*, ECM = *purple*.

allow the host plant to have an access to N and P due to the activities of DSE in environments where nutrients accumulate in organic pool (Caldwell et al., 2000). These observations suggest that DSE could supply, similar to mycorrhiza, nutrients to its host. Heinonsalo et al. (2017) stated that their results do not support this hypothesis (DSE and mycorrhiza would have related positive effects on the host plant). Moreover, they indicated that fungal endophytes (DSE) are ecologically distinct group of fungi that have typical functional characteristics, different from ectomycorrhizal fungi (Heinonsalo et al., 2017). Della Monica et al. (2015) suggested that a close relationship between DSE and arbuscular mycorrhizal fungi (AMF) is possible in relation to P availability and uptake in their host. Whereas DSE increased the pool of P in the rhizosphere, AMF are responsible for the transfer of P to the host (Della Monica et al., 2015), resulting in cocolonization of plants by DSE and AMF to have a synergistic outcome.

It has been suggested that mycorrhizas and DSE colonization are equal (Mandyam and Jumpponen, 2005, 2008; Dolinar and Gaberščik, 2010; Uma et al., 2012; Zhang et al., 2010), and/or the presence of DSE could be more abundant (Mandyam and Jumpponen, 2008; Sieber and Grünig, 2013). PAC and DSE seem to be the faster and first colonizers of nonsuberized roots of Norway spruce (Terhonen et al., 2014; Stroheker et al., 2018a,b), which might suggest an essential role against pathogens early in the life of a host (Terhonen et al., 2016), but overall the root might be controlled by delayed colonization by ectomycorrhiza (Stroheker et al., 2018a,b). It is still mystery what can be the roles of DSE in the DSE-mycorrhiza-host continuum. At the moment, despite DSE high plenitude in plants, their functions and role in ecosystems have not been fully solved (Sieber and Grünig, 2013).

10. DSE and metabolites

DSE are able to produce novel secondary metabolites that provide possible benefits to the host through stimulating plant growth or restricting pathogens (Schulz et al., 1999, 2002; Mandyam and Jumpponen, 2005; Tellenbach and Sieber, 2012; Schulz et al., 2015; Terhonen et al., 2016). In larch (*Larix* sp.), an increase in phenolic defense metabolites (proanthocyanidins) followed by DSE infection has been observed (Schulz et al., 1999; Schulz et al., 2002). From *P. europaea* (PAC) four compounds were isolated, out of which sclerin and sclerotinin A were showed to significantly lower the growing of *Phytophthora citricola* sensu *lato* (Tellenbach et al., 2013). Similarly, mixture of metabolites (extracted from liquid cultures of DSE), including sclerin that was detected only in the metabolite profiles of *P. sphaeroides*, could inhibit the growth of several plant pathogens (Terhonen et al., 2016).

Tienaho et al. (2019) screened with UPLC-Orbitrap-MS method the metabolic profiles of DSE isolated from Scots pine (*P. sylvestris*). This revealed several novel compounds that possess interesting bioactive properties (Tienaho et al., 2019). Altogether 184 metabolites were observed from *P. fortinii* (Tienaho et al., 2019). These included fucose, guanine, acetylcitrulline, disaccharides, dinucleotides, acetylleucine or acetylisoleucine, the endophytic fungi metabolite orsellinic acid ester, and the plant metabolites blumeoside A and asperulosidic acid as well as dipeptides (Tienaho et al., 2019). Terhonen et al. (2016) using the UPLC-QTOF/MS method observed a total of 214 unique metabolites from *P. sphaeroides* and 79 from PAC member. *P. sphaeroides* seemed to produce dichlorodiaportin (Larsen and Breinholt, 1999) and cryptosporiopsin (Strunz et al., 1969; Terhonen et al., 2016). Large number of these metabolites may contribute to host fitness, increasing biomass (Terhonen et al., 2016), helping with nutrient uptake (Tienaho et al., 2019), or increase resistance of the host plant (Tellenbach et al., 2013; Terhonen et al., 2016; Tienaho et al., 2019).

Heinonsalo et al. (2017) observed that *P. fortinii* infection in Scots pine roots had the highest enzymatic activities compared to mycorrhizal or saprotrophic fungal infections. *P. fortinii* formed very high-density hyphae within and around pine roots, which could be the reason for these high levels of activities (Heinonsalo et al., 2017). The production of enzymes build up significantly in root tips of Scots pine due to presence of *P. fortinii*, but the importance and outcome of higher enzymatic activity for the host plant is not resolved (Heinonsalo et al., 2017).

11. DSE-tree interaction under changing environment

The growth conditions for trees can become unfavorable due to the changes in temperature, increased solar radiation, discrepancy in soil ground water level (drought/flood), and variability in nutrition levels (poor). Forest trees are one of the most economically important sources for bioeconomy in general. DSE in tree roots are ubiquitous, exist together with mycorrhizal fungi, produce bioactive metabolites, and can have profound effects on their host. Despite the positive responses of hosts and ubiquitous nature of DSE, only few number of studies has been done over DSE-tree interactions. This endophytic stage of DSE represents a balanced interaction between the fungus and its tree host. However, there are evidences that endophytic fungal species can become pathogens or saprotrophs when this balance is disturbed, e.g., due to environmental change (Rodriguez et al., 2009; Bußkamp, 2018; Terhonen et al., 2019). Understanding how the lifestyle of fungi switches due to environmental stress is critical for deciphering the evolution of tree-DSE interactions (Kuo et al., 2014). Fungal lifestyles are not in that sense stable but dynamic and are likely influenced by the genetics of the fungal species, host factors, and changing environments (Kuo et al., 2014). Therefore endophytes have likely evolved to switch their lifestyles to adapt to different environmental conditions (Kuo et al., 2014). DSE are known to contribute to the well-being of plants, acting as growth promoters that synthesize phytohormones and they can potentially protect plants from pathogenic fungi by their antifungal activity (Terhonen et al., 2018, 2019). DSE are found virtually everywhere, hence, understanding the functioning of DSE-host interaction is highly needed. In future, extreme climate events harmful for forest ecosystems (drought and higher temperatures) are increasing (Seidl et al., 2017). Climate change also affects not only trees but also communities of fungi, as well as fungal activity and distribution (Talbot, 2017). The changes of local environment due to temperature/drought have impact on DSE-host interaction and can shift the DSE communities (Spagnoletti and Giacometti, 2020). The root endophytes, containing also DSE fungi, e.g., *Phialocephala* sp., diversity increased with warming temperature (positive correlation) (Geml et al., 2015). Moreover, severe disturbances can change the PAC species composition significantly (Stroheker et al., 2018b). Local environmental change can also relate to nutrient balance change. Kiheri et al. (2020) showed that long-term addition of N and P decreased the DSE on roots of two ericaceous shrubs, *Calluna vulgaris* and *Erica tetralix*. They highlighted that reduction in melanized DSE biomass may have implications for peatland C sequestration under nutrient addition (Kiheri et al., 2020).

12. Future studies needed

Endophytic fungi are found virtually everywhere, and as stated before their ecological roles are not yet fully understood. The high interest in plant fungal endophytic communities is obvious as there is apparent potential of the endophytes to shape and modulate the stress tolerance in host plants (Rodriguez et al., 2008; Eyles et al., 2010; Albrectsen and Witzell, 2012; Witzell et al., 2014; Blumenstein et al., 2015). Moreover, fungal endophytes can serve as a major source of biologically active compounds (Schulz et al., 2002; Terhonen et al., 2016; Tienaho et al., 2019) providing new sources to be exploited for industrial, pharmaceutical, or agricultural purposes (Schulz et al., 2002; Tienaho et al., 2019). Future studies should concentrate to reveal the functions of DSE in natural forest ecosystem. In future as the local environments can become more stressful for the forest trees (Seidl et al., 2017), DSE and their possible beneficial effect to the host should be revealed (Landolt et al., 2020). Perhaps DSE fungi could be used tackling the environmental stress that seedlings face after reforestation. This would secure trees to overcome the stress caused by changes and disturbances in their environment.

References

Addy, H.D., Hambleton, S., Currah, R.S., 2000. Distribution and molecular characterization of the root endophyte *Phialocephala fortinii* along an environmental gradient in the boreal forest of Alberta. Mycol. Res. 104, 1213–1221.

Addy, H.D., Piercey, M.M., Currah, R.S., 2005. Microfungal endophytes in roots. Can. J. Bot. 83, 1–13.

Ahlich, K., Sieber, T.N., 1996. The profusion of dark septate endophytic fungi in non-ectomycorrhizal fine roots of forest trees and shrubs. New Phytol. 132, 259–270.

Ahlich, K., Rigling, D., Holdenrieder, O., Sieber, T.N., 1998. Dark septate hyphomycetes in swiss conifer forest soils surveyed using Norway-spruce seedlings as bait. Soil Biol. Biochem. 30, 1069–1075.

Ahlich-Schlegel, K., 1997. Vorkommen und Charakterisierung von dunklen, septierten Hyphomyceten (DSH) in Gehölzwurzeln. Ph.D. thesis, Department of Forest Sciences, Forest Pathology and Dendrology, Swiss Federal Institute of Technology, Zürich, Switzerland.

Albrectsen, B., Witzell, J., 2012. Disentangling functions of fungal endophytes in forest trees. In: Paz Silva, A., Sol, M. (Eds.), Fungi: Types, Environmental Impact and Role in Disease. Nova Science Publishers, Inc, New York, pp. 235–246. Chapter 12.

Ban, Y., Tang, M., Chen, H., Xu, Z., Zhang, H., Yang, Y., 2012. The response of dark septate endophytes (DSE) to heavy metals in pure culture. PLoS One 7 (10), e47968.

Barrow, J.R., 2003. Atypical morphology of dark septate fungal root endophytes of *Bouteloua* in arid southwestern USA rangelands. Mycorrhiza 13, 239–247.

Barrow, J.R., Aaltonen, R.E., 2001. Evaluation of the internal colonization of *Atriplex canascens* (Pursh) Nutt. roots by dark septate fungi and the influence of host physiological activity. Mycorrhiza 11, 199–205.

Barrow, J.R., Osuna, P., 2002. Phosphorus solubilization and uptake by dark septate fungi in fourwing saltbush, *Atriplex canescens* (Pursh) Nutt. J. Arid Environ. 51, 449–459.

Blumenstein, K., Albrectsen, B.R., Martín, J.A., Hultberg, M., Sieber, T.N., Helander, M., Witzell, J., 2015. Nutritional niche overlap potentiates the use of endophytes in biocontrol of a tree disease. Biocontrol 60, 655–667.

Brenn, N., Menkis, A., Grünig, C.R., Sieber, T.N., Holdenrieder, O., 2008. Community structure of *Phialocephala fortinii* s. lat. in European tree nurseries, and assessment of the potential of the seedlings as dissemination vehicles. Mycol. Res. 112, 650–662.

Bußkamp, J., 2018. Schadenserhebung, Kartierung und Charakterisierung des, "Diplodia-Triebsterbens" der Kiefer, *insbesondere* des endophytischen Vorkommens in den klimasensiblen Räumen und Identifikation von den in Kiefer (*Pinus sylvestris*) vorkommenden Endophyten. Dissertation, Universität Kassel, Deutschland.

Caldwell, B., Jumpponen, A., Trappe, J., 2000. Utilization of major detrital substrates by dark-septate root endophytes. Mycologia 92. https://doi.org/10.1080/00275514.2000.12061149.

Cevnik, M., Jurc, M., Vodnik, D., 2000. Filamentous fungi associated with the fine roots of *Erica herbacea* L. from the area influenced by the *Zerjav* lead smelter (Slovenia). Phyton. Ann. Rei. Bot. 40, 61–64.

Compant, S., Saikkonen, K., Mitter, B., Campisano, A., Mercado-Blanco, J., 2016. Editorial special issue: soil, plants and endophytes. Plant Soil 405, 1–11.

Day, M., Currah, R., 2011. Role of selected dark septate endophyte species and other hyphomycetes as saprobes on moss gametophytes. Botany 89, 349–359.

Della Monica, I.F., Saparrat, M.C.N., Godeas, A.M., Scervino, J.M., 2015. The co-existence between DSE and AMF symbionts affects plant P pools through P mineralization and solubilization processes. Fungal Ecol. 17, 10–17.

Deram, A., Languereau-Leman, F., Howsam, M., Petit, D., Van Haluwyn, C., 2008. Seasonal patterns of cadmium accumulation in *Arrhenatherum elatius* (Poaceae): influence of mycorrhizal and endophytic fungal colonisation. Soil Biol. Biochem. 40, 845–848.

Dolinar, N., Gaberščik, A., 2010. Mycorrhizal colonization and growth of Phragmites australis in an intermittent wetland. Aquat. Bot. 93, 93–98. https://doi.org/10.1016/j.aquabot.2010.03.012.

dos Santos, S.G., Alves da Silva, P.R., Garcia, A.C., Zilli, J.E., Louro Berbara, R.L., 2017. Dark septate endophyte decreases stress on rice plants. Braz. J. Microbiol. 48, 333–341.

Eyles, A., Bonello, P., Ganley, R., Mohammed, C., 2010. Induced resistance to pests and pathogens in trees. New Phytol. 185, 893–908.

Fogarty, R.V., Tobin, J.M., 1996. Fungal melanins and their interactions with metals. Enzyme Microb. Tech. 19, 311–317.

Gadd, G.M., 1993. Interactions of fungi with toxic metals. New Phytol. 124, 25–60.

Gallaud, T., 1905. Etudes sur *les* mycorhizes endotrophes. Rev. Gen. Bot. 17, 4–500.

Gams, W., 1963. *Mycelium radicis atrovirens* in forest soils, isolation from soil microhabitats and identification. In: Doeksen, J., van der Drift, J. (Eds.), Soil Organisms. North-Holland Publications, Amsterdam, The Netherlands, pp. 176–182.

Geml, J., Morgado, L.N., Semenova, T.A., Welker, J.M., Walker, M.D., Smets, E., 2015. Long-term warming alters richness and composition of taxonomic and functional groups of arctic fungi. FEMS Microbiol. Ecol. 91, fiv095.

Green, L.E., Porras-Alfaro, A., Sinsabaugh, R.L., 2008. Translocation of nitrogen and carbon integrates biotic crust and grass production in desert grassland. J. Ecol. 96, 1076–1085.

Grünig, C.R., Sieber, T.N., 2005. Molecular and phenotypic description of the widespread root symbiont *Acephala applanata* gen. et sp. nov., formerly known as dark-septate endophyte Type 1. Mycologia 97, 628–640.

Grünig, C.R., Sieber, T.N., Holdenrieder, O., 2001. Characterisation of dark septate endophytic fungi (DSE) using inter-simple-sequence-repeat-anchored polymerase chain reaction (ISSR-PCR) amplification. Mycol. Res. 105, 24–32.

Grünig, C.R., Sieber, T.N., Rogers, S.O., Holdenrieder, O., 2002a. Genetic variability among strains of *Phialocephala fortinii* and phylogenetic analysis of the genus *Phialocephala* based on rDNA ITS sequence comparisons. Can. J. Bot. 80, 1239–1249.

Grünig, C.R., Sieber, T.N., Rogers, S.O., Holdenrieder, O., 2002b. Spatial distribution of dark septate endophytes in a confined forest plot. Mycol. Res. 106, 832–840.

Grünig, C.R., Linde, C.C., Sieber, T.N., Rogers, S.O., 2003. Development of single-copy RFLP markers for population genetic studies of *Phialocephala fortinii* and closely related taxa. Mycol. Res. 107, 1332–1341.

Grünig, C.R., McDonald, B.A., Sieber, T.N., Rogers, S.O., Holdenrieder, O., 2004. Evidence for subdivision of the root-endophyte *Phialocephala fortinii* into cryptic species and recombination within species. Fungal Genet. Biol. 41, 676–687.

Grünig, C.R., Duò, A., Sieber, T.N., 2006. Population genetic analysis of *Phialocephala fortinii s.l.* and *Acephala applanata* in two undisturbed forests in Switzerland and evidence for new cryptic species. Fungal Genet. Biol. 43, 410–421.

Grünig, C.R., Duo, A., Sieber, T.N., Holdenrieder, O., 2008a. Assignment of species rank to six reproductively isolated cryptic species of the *Phialocephala fortinii* s.l – *Acephala applanata* species complex. Mycologia 100, 47–67.

Grünig, C.R., Queloz, V., Sieber, T.N., Holdenrieder, O., 2008b. Dark septate endophytes (DSE) of the *Phialocephala fortinii* s.l.–*Acephala applanata* species complex in tree roots: classification, population biology and ecology. Botany 86, 1355–1369.

Grünig, C.R., Queloz, V., Sieber, T.N., 2011. Structure of diversity in dark septate endophytes: from species to genes. In: Pirttilä, A.M., Frank, C. (Eds.), Endophytes of Forest Trees: Biology and Applications. Springer Forestry Series, Berlin, pp. 3–30.

Halmschlager, E., Kowalski, T., 2004. The mycobiota in nonmycorrhizal roots of healthy and declining oaks. Can. J. Bot. 82, 1446–1458.

Hambleton, S., Currah, R.S., 1997. Fungal endophytes from the roots of alpine and boreal Ericaceae. Can. J. Bot. 75, 1570–1581.

He, C., Wang, W., Hou, J., 2019. Plant growth and soil microbial impacts of enhancing licorice with inoculating dark septate endophytes under drought stress. Front. Microbiol. 10, 2277.

Heinonsalo, J., Buee, M., Vaario, L.-M., 2017. Root-endophytic fungi cause morphological and functional differences in Scots pine roots in contrast to ectomycorrhizal fungi. Botany 95, 203–210.

Hendrick, R.L., Pregitzer, K.S., 1996. Temporal and depth-related patterns of fine root dynamics in northern hardwood forests. J. Ecol. 84, 167–176.

Hou, L., Yu, J., Zhao, L., He, X., 2020. Dark septate endophytes improve the growth and the tolerance of *Medicago sativa* and *Ammopiptanthus mongolicus* under cadmium stress. Front. Microbiol. 10, 3061.

Jumpponen, A.M., 2001. Dark septate endophytes—are they mycorrhizal? Mycorrhiza 11, 207–211.

Jumpponen, A., Trappe, J.M., 1998a. Dark septate endophytes: a review of facultative biotrophic root-colonizing fungi. New Phytol. 140, 295–310.

Jumpponen, A., Trappe, J.M., 1998b. Performance of *Pinus contorta* inoculated with two strains of root endophytic fungus, *Phialocephala fortinii*: effects of synthesis system and glucose concentration. Can. J. Bot. 76, 1205–1213.

Jumpponen, A., Mattson, K.G., Trappe, J.M., 1998. Mycorrhizal functioning of *Phialocephala fortinii* with *Pinus contorta* on glacier forefront soil: interactions with soil nitrogen and organic matter. Mycorrhiza 7, 261–265.

Kauserud, H., Lie, M., Stensrud, O., Ohlson, M., 2005. Molecular characterization of airborne fungal spores in boreal forests of contrasting human disturbance. Mycologia 97, 1215–1224.

Kernaghan, G., Patriquin, G., 2011. Host associations between fungal root endophytes and boreal trees. Microb. Ecol. 62, 460–473.

Kernaghan, G., Sigler, L., Khasa, D., 2003. Mycorrhizal and root endophytic fungi of containerized *Picea glauca* seedlings assessed by rDNA sequence analysis. Microb. Ecol. 45, 128–136.

Kiheri, H., Velmala, S., Pennanen, T., Timonen, S., Sietiö, O.-M., Fritze, H., Heinonsalo, J., van Dijk, N., Dise, N., Larmola, T., 2020. Fungal colonization patterns and enzymatic activities of peatland ericaceous plants following long-term nutrient addition. Soil Biol. Biochem. 147, 107833.

Kivlin, S., Emery, S., Rudgers, J., 2013. Fungal symbionts alter plant responses to global change. Am. J. Bot. 100, 1445–1457.

Kluting, K., Clemmensen, K., Jonaitis, S., Vasaitis, R., Holmström, S., Finlay, R., Rosling, A., 2019. Distribution patterns of fungal taxa and inferred functional traits reflect the non-uniform vertical stratification of soil microhabitats in a coastal pine forest. FEMS Microbiol. Ecol. 95, fiz149.

Knapp, D.G., Imrefi, I., Boldpurev, E., Csíkos, S., Akhmetova, G., Berek-Nagy, P.J., Otgonsuren, B., Kovács, G.M., 2019. Root-colonizing endophytic fungi of the dominant grass Stipa krylovii from a Mongolian steppe grassland. Front. Microbiol. 10, 1–13.

Kotilínek, M., Hiiesalu, I., Košnar, J., Šmilauerová, M., Šmilauer, P., Altman, J., Dvorský, M., Kopecký, M., Doležal, J., 2017. Fungal root symbionts of high-altitude vascular plants in the Himalayas. Sci. Rep. 7, 6562.

Kuo, H.C., Hui, S., Choi, J., Asiegbu, F.O., Valkonen, J., Lee, Y.-H., 2014. Secret lifestyles of Neurospora crassa. Sci. Rep. 4, 5135.

Kwasna, H., Bateman, G.L., Ward, E., 2008. Determining species diversity of microfungal communities in forest tree roots by pure-culture isolation and DNA sequencing. Appl. Soil Ecol. 40, 44–56.

Landolt, M., Stroheker, S., Queloz, V., Gall, A., Sieber, T.N., 2020. Does water availability influence the abundance of species of the Phialocephala fortinii s.l. – Acephala applanata complex (PAC) in roots of pubescent oak (Quercus pubescens) and Scots pine (Pinus sylvestris)? Fungal Ecol. 44, 100904.

Langer, G.J., 2017. Collar rots in forests of Northwest Germany affected by ash dieback. Balt. For. 23, 4–19.

Larsen, T.O., Breinholt, J., 1999. Dichlorodiaportin, diaportinol, and diaportinic acid: three novel isocoumarins from Penicillium nalgiovense. J. Nat. Prod. 62, 1182–1184.

Li, X., He, X., Hou, L., Ren, Y., Wang, S., Su, F., 2018. Dark septate endophytes isolated from a xerophyte plant promote the growth of Ammopiptanthus mongolicus under drought condition. Sci. Rep. 8, 7896.

Li, X., He, X.-L., Zhou, Y., Hou, Y.-T., Zuo, Y.-L., 2019. Effects of dark septate endophytes on the performance of Hedysarum scoparium under water deficit stress. Front. Plant Sci. 10, 903.

Likar, M., Regvar, M., 2009. Application of temporal temperature gradient gel electrophoresis for characterisation of fungal endophyte communities of Salix caprea L. in a heavy metal polluted soil. Sci. Total Environ. 407, 6179–6187.

Likar, M., Regvar, M., 2013. Isolates of dark septate endophytes reduce metal uptake and improve physiology of Salix caprea L. Plant Soil 370, 593–604.

Lukešová, T., Kohout, P., Větrovský, T., Vohník, M., 2015. The potential of dark septate endophytes to form root symbioses with ectomycorrhizal and ericoid mycorrhizal middle European forest plants. PLoS One 10, e0124752.

Lygis, V., Vasiliauskas, R., Stenlid, J., 2004. Planting Betula pendula on pine sites infested by Heterobasidion annosum: disease transfer, silvicultural evaluation, and community of wood-inhabiting fungi. Can. J. For. Res. 34, 120–130.

Mandyam, K., Jumpponen, A., 2005. Seeking the elusive function of the root-colonizing dark septate endophytic fungi. Stud. Mycol. 53, 173–189.

Mandyam, K., Jumpponen, A., 2008. Seasonal and temporal dynamics of arbuscular mycorrhizal and dark septate endophytic fungi in a tallgrassprairie ecosystem are minimally affected by nitrogen enrichment. Mycorrhiza 18, 145–155.

Mandyam, K., Jumpponen, A., 2015. Mutualism–parasitism paradigm synthesized from results of root-endophyte models. Front. Microbiol. 5, 1–13.

Mayerhofer, M.S., Kernaghan, G., Harper, K.A., 2013. The effects of fungal root endophytes on plant growth: a meta-analysis. Mycorrhiza 23, 119–128.

McGonigle, T.P., Miller, M.H., Evans, D.G., Fairchild, G.L., Swan, J.A., 1990. A new method which gives an objective measure of colonization of roots by vesicular—arbuscular mycorrhizal fungi. New Phytol. 115, 495–501.

Melin, E., 1922. On the mycorrhizas of Pinus sylvestris L. and Picea abies Karst. A preliminary note. J. Ecol. 9, 254–257.

Melin, E., 1923. Experimentelle Untersuchungen über die Konstitution und Ökologie der Mykorrhizen von Pinus silvestris und Picea abies. In: Falck, R. (Ed.), Mykologische Untersuchungen und Berichte. vol. 2. Druck und Verlag G. Gottheilt, Kassel, pp. 73–331.

Menkis, A., Allmer, J., Vasiliauskas, R., Lygis, V., Stenlid, J., Finlay, R., 2004. Ecology and molecular characterization of dark septate fungi from roots, living stems, coarse and fine woody debris. Mycol. Res. 108, 965–973.

Newsham, K.K., 2011. A meta-analysis of plant responses to dark septate root endophytes. New Phytol. 190, 783–793.

Newsham, K.K., Upson, R., Read, D.J., 2009. Mycorrhizas and dark septate root endophytes in polar regions. Fungal Ecol. 2, 10–20.

O'Dell, T.E., Massicotte, H.B., Trappe, J.M., 1993. Root colonization of Lupinus latifolius Agardh. and Pinus contorta Dougl. by Phialocephala fortinii Wang & Wilcox. New Phytol. 124, 93–100.

Petrini, O., 1991. Fungal endophytes of tree leaves. In: Microbial Ecology of Leaves. Springer, pp. 179–197.

Piercey, M.M., Graham, S.W., Currah, R.S., 2004. Patterns of genetic variation in Phialocephala fortinii across a broad latitudinal transect in Canada. Mycol. Res. 108, 955–964.

Piri, T., Hamberg, L., 2015. Persistence and infectivity of Heterobasidion parviporum in Norway spruce root residuals following stump harvesting. For. Ecol. Manage. 353, 49–58.

Porras-Alfaro, A., Bayman, P., 2011. Hidden fungi, emergent properties: endophytes and microbiomes. Annu. Rev. Phytopathol. 49, 291–315.

Pregitzer, K.S., King, J.S., Burton, A.J., Brown, S.E., 2008. Responses of tree fine roots to temperature. New Phytol. 147, 105–115.

Queloz, V., Grünig, C.R., Sieber, T.N., Holdenrieder, O., 2005. Monitoring the spatial and temporal dynamics of a community of the tree-root endophyte Phialocephala fortinii sl. New Phytol. 168, 651–660.

Queloz, V., Duo, A., Grünig, C.R., 2008. Isolation and characterization of microsatellite markers for the tree-root endophytes Phialocephala subalpina and Phialocephala fortinii s.s. Mol. Ecol. Resour. 8, 1322–1325.

Queloz, V., Duo, A., Sieber, T.N., Grünig, C.R., 2010. Microsatellite size homoplasies and null alleles do not affect species diagnosis and population genetic analysis in a fungal species complex. Mol. Ecol. Resour. 10, 348–367.

Queloz, V., Sieber, T.N., Holdenrieder, O., McDonald, B.A., Grünig, C.R., 2011. No biogeographical pattern for a root-associated fungal species complex. Glob. Ecol. Biogeogr. 20, 160–169.

Rajala, T., Velmala, S.M., Tuomivirta, T., Haapanen, M., Muller, M., Pennanen, T., 2013. Endophyte communities vary in the needles of Norway spruce clones. Fungal Biol. 117, 182–190.

Rayment, J., Jones, S., French, K., 2020. Seasonal patterns of fungal colonisation in Australian native plants of different ages. Symbiosis 80, 169–182.

Regvar, M., Likar, M., Piltaver, A., Kogonič, N., Smith, J.E., 2010. Fungal community structure under goat willows (*Salix caprea* L.) growing at a metal-polluted site: the potential of screening in a model phytostabilisation study. Plant Soil 330, 345–356.

Reininger, V., Sieber, T.N., 2013. Mitigation of antagonistic effects on plant growth due to root co-colonization by dark septate endophytes (DSE) and ectomycorrhiza (ECM). Environ. Microbiol. Rep. 5, 892–898.

Reininger, V., Gruenig, C.R., Sieber, T.N., 2012. Host species and strain combination determine growth reduction of spruce and birch seedlings colonized by root associated dark septate endophytes. Environ. Microbiol. 14, 1064–1076.

Rodriguez, R.J., Henson, J., Van Volkenburgh, E., Hoy, M., Wright, L., Beckwith, F., Kim, Y., Redman, R.S., 2008. Stress tolerance in plants via habitat-adapted symbiosis. ISME J. 2, 404–416.

Rodriguez, R., Redman, R., 2008. More than 400 million years of evolution and some plants still can't make it on their own: plant stress tolerance via fungal symbiosis. J. Exp. Bot. 59 (5), 1109–1114. https://doi.org/10.1093/jxb/erm342.

Rodriguez, R.J., White Jr., J.F., Arnold, A.E., Redman, R.S., 2009. Fungal endophytes: diversity and functional roles. New Phytol. 182, 314–330.

Ruotsalainen, A.L., Markkola, A., Kozlov, M.V., 2007. Root fungal colonisation in *Deschampsia flexuosa*: effects of pollution and neighbouring trees. Environ. Pollut. 147, 723–728.

Saikkonen, K., 2007. Forest structure and fungal endophytes. Fungal Biol. Rev. 21, 67–74.

Saikkonen, K., Faeth, S.H., Helander, M., Sullivan, T.J., 1998. Fungal endophytes: a continuum of interactions with host plants. Annu. Rev. Ecol. Syst. 29, 319–343.

Saikkonen, K., Saari, S., Helander, M., 2010. Defensive mutualism between plants and endophytic fungi? Fungal Divers. 41, 101–113.

Santschi, F., 2015. Endophytic Microbiota of *Picea abies*, *Fraxinus excelsior* and *Acer pseudoplatanus* and the Host- and Site-Specificity of Species of the *Phialocephala fortinii* s.l. – *Acephala applanata* Species Complex. Semester thesis, ETH Zürich, Zürich, Switzerland.

Schlegel, M., Münsterkötter, M., Güldener, U., Bruggmann, R., Duò, A., Hainaut, M., Henrissat, B., Sieber, C.M.K., Hoffmeister, D., Grünig, C.R., 2016. Globally distributed root endophyte Phialocephala subalpina links pathogenic and saprophytic lifestyles. BMC Genomics 17, 1015.

Schmidt, S.K., Sobieniak-Wiseman, L.C., Kageyama, S., Halloy, S.R.P., Schadt, C.W., 2008. Mycorrhizal and dark-septate fungi in plant roots above 4270 meters elevation in the Andes and Rocky Mountains. Arct. Antarct. Alp. Res. 40, 576–583.

Schulz, B., Boyle, C., 2005. The endophytic continuum. Mycol. Res. 109, 661–686.

Schulz, B.J.E., Guske, S., Dammann, U., Boyle, C., 1998. Endophyte-host interactions II. Defining symbiosis of the endophyte-host interaction. Symbiosis 25, 213–227.

Schulz, B., Boyle, C., Draeger, S., Rommert, A.K., Krohn, K., 2002. Endophytic fungi: a source of novel biologically active secondary metabolites. Mycol. Res. 106, 996–1004.

Schulz, B., Haas, S., Junker, C., Andree, N., Schobert, M., 2015. Fungal endophytes are involved in multiple balanced antagonisms. Curr. Sci. 109, 39–45.

Schulz, B., Rommert, A.K., Dammann, U., Aust, H.J., Strack, D., 1999. The endophyte–host interaction: a balanced antagonism? Mycol. Res. 10, 1275–1283.

Seidl, R., Thom, D., Kautz, M., Martin-Benito, D., Peltoniemi, M., Vacchiano, G., Wild, J., Ascoli, D., Petr, M., Honkaniemi, J., Lexer, M.J., Trotsiuk, V., Mairota, P., Svoboda, M., Fabrika, M., Nagel, T.A., Reyer, C.P.O., 2017. Forest disturbances under climate change. Nat. Clim. Change 7, 395–402.

Sieber, T.N., 2007. Endophytic fungi in forest trees: are they mutualists? Fungal Biol. Rev. 21, 75–89.

Sieber, T.N., Grünig, C.R., 2006. Biodiversity of fungal root-endophyte communities and populations, in particular of the dark septate endophyte *Phialocephala fortinii* s.l. In: Schulz, B.J.E., Boyle, C.J.C., Sieber, T.N. (Eds.), Microbial Root Endophytes, Soil Biology. Springer Berlin Heidelberg, Berlin, Heidelberg, pp. 107–132.

Sieber, T.N., Grünig, C.R., 2013. Fungal root endophytes. In: Eshel, A., Beeckman, T. (Eds.), Plant Roots – The Hidden Half, fourth ed. CRC Press, Taylor & Francis Group, Boca Raton, FL, USA, pp. 38,1–38,49.

Sietiö, O.M., Tuomivirta, T., Santalahti, M., Kiheri, H., Timonen, S., Sun, H., Fritze, H., Heinonsalo, J., 2018. Ericoid plant species and *Pinus sylvestris* shape fungal communities in their roots and surrounding soil. New Phytol. 218, 738–751.

Smith, S.E., Read, D.J., 1997. Mycorrhizal Symbiosis, second ed. Academic Press, London.

Spagnoletti, F.N., Giacometti, R., 2020. Dark septate endophytic fungi (DSE) response to global change and soil contamination. In: Hasanuzzaman, M. (Ed.), Plant Ecophysiology and Adaptation Under Climate Change: Mechanisms and Perspectives II. Springer, Singapore, pp. 629–642.

Stenström, E., Ndobe, N.E., Jonsson, M., Stenlid, J., Menkis, A., 2014. Root-associated fungi of healthy-looking *Pinus sylvestris* and *Picea abies* seedlings in Swedish forest nurseries. Scand. J. For. Res. 29, 12–21.

Stoyke, G., Currah, R.S., 1991. Endophytic fungi from the mycorrhizae of alpine ericoid plants. Can. J. Bot. 69, 347–352.

Strobel, G., 2018. The emergence of endophytic microbes and their biological promise. J. Fungi 4, 57.

Stroheker, S., Queloz, V., Sieber, T.N., 2016. Spatial and temporal dynamics in the *Phialocephala fortinii* s.l. – *Acephala applanata* species complex (PAC). Plant Soil 407, 231–241.

Stroheker, S., Dubach, V., Sieber, T.N., 2018a. Competitiveness of endophytic *Phialocephala fortinii* s.l. – *Acephala applanata* strains in Norway spruce roots. Fungal Biol. 122, 345–352.

Stroheker, S., Dubach, V., Queloz, V., Sieber, T.N., 2018b. Resilience of *Phialocephala fortinii* s.l.—*Acephala applanata* communities—effects of disturbance and strain introduction. Fungal Ecol. 31, 19–28.

Strunz, G.M., Court, A.S., Komlossy, J., Stillwell, A., 1969. Structure of cryptosporiopsin: a new antibiotic produced by a species of *Cryptosporiopsis*. Can. J. Chem. 47, 2087–2094.

Sword, M.A., Kuehler, E.A., Tang, Z., 2000. Seasonal fine root carbohydrate relations of plantation loblolly pine after thinning. J. Sustain. For. 10 (3/4), 295–305.

Talbot, J.M., 2017. Fungal communities and climate change. In: The Fungal Community, Its Organization and Role in the Ecosystem, fourth ed. CRC Press, CRC Press Taylor & Francis Group, pp. 471–488. Chapter 32.

Tejesvi, M.V., Ruotsalainen, A.L., Markkola, A.M., Pirttilä, A.M., 2010. Root endophytes along a primary succession gradient in northern Finland. Fungal Divers. 41, 125–134.

Tejesvi, M.V., Sauvola, T., Pirttila, A.M., Ruotsalainen, A.L., 2013. Neighboring *Deschampsia flexuosa* and *Trientalis europaea* harbor contrasting root fungal endophytic communities. Mycorrhiza 23, 1–10.

Tellenbach, C., Sieber, T.N., 2012. Do colonization by dark septate endophytes and elevated temperature affect pathogenicity of oomycetes? FEMS Microbiol. Ecol. 82, 157–168.

Tellenbach, C., Grünig, C.R., Sieber, T.N., 2011. Negative effects on survival and performance of Norway spruce seedlings colonized by dark septate root endophytes are primarily isolate-dependent. Environ. Microbiol. 13, 2508–2517.

Tellenbach, C., Sumarah, M.W., Grünig, C.R., Miller, D.J., 2013. Inhibition of *Phytophthora* species by secondary metabolites produced by the dark septate endophyte *Phialocephala europaea*. Fungal Ecol. 6, 12–18.

Terhonen, E., Keriö, S., Sun, H., Asiegbu, F.O., 2014. Endophytic fungi of Norway spruce roots in boreal pristine mire, drained peatland and mineral soil and their inhibitory effect on *Heterobasidion parviporum in vitro*. Fungal Ecol. 9, 17–24.

Terhonen, E., Sipari, S., Asiegbu, F.O., 2016. Inhibition of phytopathogens by fungal root endophytes of Norway spruce. Biol. Control 99, 53–63.

Terhonen, E., Kovalchuk, A., Zarsav, A., Asiegbu, F.O., 2018. Biocontrol potential of forest tree endophytes. In: Endophytes of Forest Trees. Springer, pp. 283–318.

Terhonen, E., Blumenstein, K., Kovalchuk, A., Asiegbu, F.O., 2019. Forest tree microbiomes and associated fungal endophytes: functional roles and impact on forest health. Forests 10, 42.

Tienaho, J., Karonen, M., Muilu-Mäkelä, R., Wähälä, K., Denegri, E.L., Franzen, R., Karp, M., Santala, V., Sarjala, T., 2019. Metabolic profiling of water-soluble compounds from the extracts of dark septate endophytic fungi (DSE) isolated from Scots Pine (*Pinus sylvestris* L.) seedlings using UPLC-Orbitrap-MS. Molecules 24, 2330.

Uma, E., Sathiyadash, K., Loganathan, J., Muthukumar, T., 2012. Tree species as hosts for arbuscular mycorrhizal and dark septate endophyte fungi. J. For. Res. 23, 641–649. https://doi.org/10.1007/s11676-012-0267-z.

Valli, P.P.S., Muthukumar, T., 2018. Dark septate root endophytic fungus *Nectria haematococca* improves tomato growth under water limiting conditions. Indian J. Microbiol. 58, 489–495.

Wagg, C., Pautler, M., Massicotte, H.B., Peterson, R.L., 2008. The co-occurrence of ectomycorrhizal, arbuscular mycorrhizal, and dark septate fungi in seedlings of four members of the Pinaceae. Mycorrhiza 18, 103–110.

Wang, C., Wilcox, H., 1985. New species of ectendomycorrhizal and pseudomycorrhizal fungi: *Phialophora finlandia*, *Chloridium paucisporum*, and *Phialocephala fortinii*. Mycologia 77, 951–958.

Wang, J.L., Li, T., Liu, G.Y., Smith, J.M., Zhao, Z.W., 2016. Unraveling the role of dark septate endophyte (DSE) colonizing maize (*Zea mays*) under cadmium stress: physiological, cytological and genic aspects. Sci. Rep. 6, 22028.

Wilcox, H.E., Wang, C.J.K., 1987a. Mycorrhizal and pathological associations of dematiaceous fungi in roots of 7-month-old tree seedlings. Can. J. For. Res. 17, 884–889.

Wilcox, H.E., Wang, C.J.K., 1987b. Ectomycorrhizal and ectendomycorrhizal associations of *Phialophora finlandia* with *Pinus resinosa*, *Picea rubens*, and *Betula alleghaensis*. Can. J. For. Res. 17, 976–990.

Wilson, D., 1995. Endophyte – the evolution of a term, and clarification of its use and definition. Oikos 73, 274–276.

Wilson, B.J., Addy, H.D., Tsuneda, A., Hambleton, S., Currah, R.S., 2004. *Phialocephala sphaeroides* sp. nov., a new species among the dark septate endophytes from a boreal wetland in Canada. Can. J. Bot. 82, 607–617.

Witzell, J., Martín, J.A., Blumenstein, K., 2014. Ecological aspects of endophyte-based biocontrol of forest diseases. In: Verma, V.C., Gange, A.C. (Eds.), Advances in Endophytic Research. Springer-Verlag, Berlin, pp. 321–333.

Yamaji, K., Watanabe, Y., Masuya, H., Shigeto, A., Yui, H., Haruma, T., 2016. Root fungal endophytes enhance heavy-metal stress tolerance of *Clethra barbinervis* growing naturally at mining sites via growth enhancement, promotion of nutrient uptake and decrease of heavy-metal concentration. PLoS One 11, e0169089.

Zaffarano, P.L., Duò, A., Grünig, C.R., 2010. Characterization of the mating type (MAT) locus in the *Phialocephala fortinii* s.l. – *Acephala applanata* species complex. Fungal Genet. Biol. 47, 761–772.

Zaffarano, P.L., Queloz, V., Duò, A., Grünig, C.R., 2011. Sex in the PAC: a hidden affair in dark septate endophytes? BMC Evol. Biol. 11, 282.

Zhan, F.D., He, Y.M., Zu, Y.Q., Li, T., Zhao, Z.W., 2011. Characterization of melanin isolated from a dark septate endophyte (DSE), *Exophiala pisciphila*. World J. Microb. Biot. 27, 2483–2489.

Zhang, Y.J., Zhang, Y., Liu, M.J., Shi, X.D., Zhao, Z.W., 2008. Dark septate endophyte (DSE) fungi isolated from metal polluted soils: their taxonomic position, tolerance, and accumulation of heavy metals in vitro. J. Microbiol. 46, 624–632.

Zhang, H., Tang, M., Chen, H., Wang, Y., Ban, Y., 2010. Arbuscular mycorrhizas and dark septate endophytes colonization status in medicinal plant *Lycium barbarum* L. in arid Northwestern China. African J. Microbiol. Res. 4, 1914–1920.

Zhu, L., Li, T., Wang, C., Zhang, X., Xu, L., Xu, R., Zhao, Z., 2018. The effects of dark septate endophyte (DSE) inoculation on tomato seedlings under Zn and Cd stress. Environ. Sci. Pollut. Res. 25, 35232–35241.

Section D

Rhizosphere microbiome

Chapter 12

Nature and characteristics of forest soils and peat soils as niches for microorganisms

Mike Starr and Harri Vasander

Department of Forest Sciences, University of Helsinki, Helsinki, Finland

Chapter Outline

1. Forest soils and peat soils defined

Soil is a medium in which plant roots grow and is formed from unconsolidated mineral or organic material containing varying amounts of water and air and is a habitat for soil fauna and a vast diversity of microorganisms. Depending on the organic matter content, soils can broadly be classified as being either mineral or organic (FAO, 2001). Trees grow on both mineral and organic soils. Mineral soils that support forest are broadly referred to as *forest soils*, while peat soils, which are formed from organic matter, may or may not support trees. Most forest soils are formed in unconsolidated, weathered mineral deposits and contain varying amounts of organic matter (Binkley and Fisher, 2019; Comerford and Fox, 2016). Although trees can grow on lowland mineral soils that are influenced by groundwater, forest soils are generally taken to be freely draining and not influenced by groundwater and are thus sometimes referred to as *upland* forest soils. As most forests grow on mineral soils, forest cover indicates the extent of forest soils. Currently, forests cover 4.06 billion ha (FAO and UNEP, 2020) but covered as much as half the Earth's land surface in prehistoric times. As a result of the global distribution of forests, forest soils occur in most major soil groups (FAO, 2001; Van Rees, 2016).

Peat soils contain very little mineral matter and are formed from the residues of vegetation, specifically mire vegetation, in varying degrees of decomposition. In terms of soil classification, peat is an organogenic hydromorphic soil, having formed under the influence of water saturation and belong to the major soil group of histosols. Globally, histosols cover some 325–375 million ha but are predominantly found in the boreal zone (FAO, 2001). Microbial decomposition of the peat is retarded by anaerobic conditions resulting from high water tables. As a result, not all peat soils support trees or can only do so after drainage. While peat is formed in situ in mire ecosystems, peat soils are associated with the broader concept of a *peatland*, an area where peat has accumulated but which may have lost its status as a pristine mire through disturbance and land use (Joosten and Clarke, 2002). At a higher level of classification, peatlands (including mires) are *wetlands*. Wetlands include other hydromorphic soils having high mineral matter contents and not supporting mire vegetation. The focus in this chapter is on soils in the boreal zone that have given rise to a cover of trees, that is, upland forest soils (podzols, or at least podzolized, developed in glacial till or sorted glaciofluvial deposits) and peat soils (treed mires or ditched peatlands that were drained for forestry purposes).

As a result of the fundamental difference in parent material, the physical and chemical properties of forest mineral and peat soils are very different (Table 12.1), resulting in a large difference in the levels of edaphic factors (temperature, moisture, aeration, nutrients, and acidity) that influence microorganisms (Fig. 12.1).

2. Characteristics of forest soils

Primary forest soils (i.e., those with native tree cover and where there has been little or no disturbance) have soil profiles, horizonation, and properties that reflect the integrated effects of soil formation factors and processes. This differentiates

TABLE 12.1 Ranges in bulk density, pH, organic carbon (OC), and elemental concentrations reported for forest soils (humus layer and mineral soil) and surface peat soils in Finland.

Soil	BD (g/cm³)	pH (water)	OC (%)	N (%)	P (mg/kg)	K (mg/kg)	Ca (mg/kg)	References
Humus layer	0.09–0.17	3.5–4.1	35.8–53.7	0.87–1.31	608–1075	–	–	Bergström et al. (1995)
	0.06–0.64	3.1–6.2	8.0–47.5	0.34–3.00	370–2600	440–2700	700–16,000	Tamminen (1991)
	–	3.6–5.8	16.0–47.5	–	–	–	–	Tamminen and Starr (1990)
	–	–	–	–	436–2330	346–2590	841–14,600	Tamminen et al. (2004)
	0.07–0.13	3.1–3.9	17.4–47.8	0.64–1.30	–	–	–	Westman et al. (1985)
Mineral soil	0.72–1.68	4.1–5.8	0.2–5.0	0.01–0.17	87–1604	–	–	Bergström et al. (1995)
	0.80–1.80	3.5–5.5	0.5–12.7	0.01–0.53	4.1–340	26–2200	25–18,400	Tamminen (1991)
	–	3.6–7.4	0.1–11.0	–	–	–	–	Tamminen and Starr (1990)
	0.48–1.84	–	0.1–10.3	–	–	–	–	Tamminen and Starr (1994)
	1.06–1.81	3.5–5.3	0.1–4.5	0.01–0.11	–	–	–	Westman et al. (1985)
Peat	0.07–0.15	3.3–4.1	53.1–53.2	0.97–1.60	466–681	–	–	Bergström et al. (1995)
	0.05–0.09	4.1–4.6	46.0–48.5	1.00–2.17	450–1000	42–1690	2090–5280	Westman (1981)
	0.06–0.11	3.3–4.5	41.9–51.3	0.89–3.11	–	–	–	Westman et al. (1985)

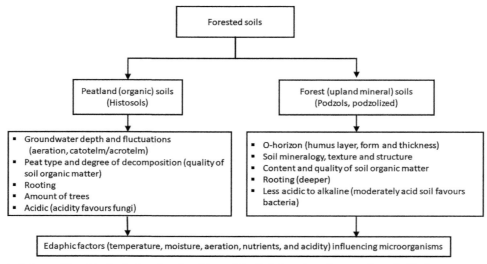

FIG. 12.1 The main features of forest (upland mineral) and peat (organic) soils influencing microorganism edaphic factors.

forest soils from cultivated and cropped soils where the natural soil profile has been considerably disturbed. Consequently, soil properties and conditions for microorganisms differ among the soil horizons, especially between the surface organic layer and the underlying mineral soil horizons. Although pedogenically not part of the soil proper, forest soils typically have a surface organic layer (O horizon). This layer is often the "soil" that is sampled and studied by microbiologists, especially in boreal forests where the O horizon can be several centimeters thick and rich in roots and microorganisms. The surface organic layer may also be formed of a thin layer of peat when it is referred to as a histic organic horizon (H horizon) and represents the transition from mineral soil groups to histosols (peat soil) typically lower down in the landscape. The boundary between the surface O horizon and the mineral soil is often difficult to distinguish, especially in the more fertile and less acidic soils where there is more soil fauna activity and mixing of the organic matter with the upper mineral soil (bioturbation). Forest soils also tend to be marginal in that they have low fertility or are difficult to cultivate for crop production because of texture and stoniness. For example, in the boreal zone, forest soils higher up in the landscape are typically developed in coarse-textured glaciofluvial deposits with a low clay content or in stony till, the properties of which do not favor cultivation or crop production. Even the soils of luxuriant tropical rain forests are often poor in nutrients per se, as the parent material is often old, highly weathered, and lacking in primary (relatively nutrient-rich) minerals; the forests being maintained by the rapid microbial decomposition of litterfall and tight nutrient cycling.

3. Physicochemical properties of forest soils

The physicochemical properties of forest soils vary hugely at all spatial scales, reflecting differences and interactions between climate, parent material, and topography (Binkley and Fisher, 2019; Comerford and Fox, 2016). The fertility of forest soils is to a great extent determined by the nature of the O horizon and content of organic matter in the underlying mineral soil, soil organic matter (SOM) content. Dead organic matter from plants and soil organisms is the source of energy and nutrients for plants and microorganisms, and plays an important role in determining the water holding and cation exchange capacity of the soil. The nature of the O horizon and SOM contents are therefore directly related to moisture retention and contents, aeration, and the availability of nutrients, especially in soils with low clay and silt contents. Through the dissociation of organic acids, the O horizon and SOM content of the mineral soil also determine soil acidity. All these soil properties depend on the quality and quantity of the organic matter forming the O horizon and SOM, which in turn depends on the species composition and characteristics of the forest cover. In addition to the incorporation of aboveground litterfall into the O horizon and mixing with the underlying mineral soil by soil fauna, organic matter is transported into the soil with percolating water in the form of dissolved organic matter (DOM) derived from the O horizon and deposited on the surface of mineral particles, and produced in situ in the form of fine root exudates and mortality. The amount of organic matter in and on the soil is determined by the balance between the above-mentioned inputs and decomposition losses, a microbial process that is strongly dependent on temperature. By shading the soil, forest soils tend to be cooler than adjacent nonforest soils.

As finer textured and nonstony soils are preferred for cultivation, remaining forest soils—at least in the boreal and temperate zone—are coarse textured and often stony. Soil texture—the proportions of sand, silt, and clay particle size fractions—largely determines the pore size distribution within the soil and therefore water retention and moisture conditions and, its complement, aeration, vital for aerobic microorganisms. Soil structure—how the soil particles are aggregated into secondary structural soil units—tends to be limited in forest soils, at least in coarser textured soils, and is therefore of less importance in determining soil moisture and aeration levels compared to the situation in agricultural soils. Stones formed from igneous and associated metamorphic rocks are nonporous and so the pore space, moisture content, and aeration of the soil are reduced by their presence. Stones formed from sedimentary rocks tend to be porous, and so may contain water that is available to plants and microorganisms.

While fungi, bacteria, and archaea thrive best in well-aerated soils, fungi are the major agents of decay in the O horizon because of its acidic nature and high lignin content, and specialized communities of bacteria and archaea are the major agents of decomposition in anoxic environments (Bates et al., 2011; Brockett et al., 2012). However, in terms of ecological niches for microorganisms, it is necessary to divide the soil into two parts: the rhizosphere and the bulk soil (Fig. 12.2). The rhizosphere—or more accurately, the ectorhizosphere—is the volume of soil surrounding the fine roots that is strongly influenced by root activity in comparison to the outer bulk soil (McNear, 2013). It extends from millimeters to a few centimeters out from the root surface into the soil and occupies only a small fraction of the bulk soil volume. Because of the greater supply and range of carbon-containing compounds in the form of fine root exudates and turnover (Philippot et al., 2013), soil microorganism numbers are much greater in the rhizosphere than in the bulk soil (Reinhold-Hurek et al., 2015; Kuzyakov and Blagodatskaya, 2015). Even though occupying only a small fraction of the bulk soil volume, rhizosphere soil has been shown to account for as much as 20% of forest soil total respiration and that microbial respiration in the rhizosphere is a significant sink for photosynthetically fixed C in forests (Kelting et al., 1998).

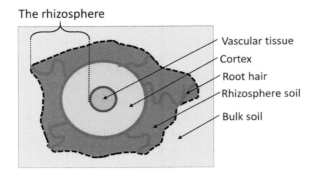

FIG. 12.2 Schematic cross-section of a root showing the division between the rhizosphere (ecto- and endorhizosphere) and bulk soil.

Nutrient availability is also greater in rhizosphere soil, influencing the whole biogeochemistry of forest ecosystems (Gobran et al., 1998). Even forest soils contaminated with high levels of heavy metals have been shown to have significantly higher contents of SOM, available nitrogen and phosphorus, and greater enzyme activity and bacterial diversity in their rhizosphere soil than in the bulk soil (Yang et al., 2017). Rhizosphere soil also shows marked differences in physical characteristics compared to the bulk soil, including the realignment of mineral soil particles by the pressure exerted by growing roots and increased macro-aggregation and development of soil structure, mineralogy, and weathering (April and Keller, 1990; Baumert et al., 2018).

4. Mire and peat formation

Peat is an autochthonic deposit, i.e., formed in situ, having formed as a result of the infilling of water bodies or through the paludification of wet mineral soils (Damman, 1986). During mire succession, a shift in the composition of the *Sphagnum*-associated active methanotroph bacterial (SAM) community appears to take place (Putkinen et al., 2014). Instead of the presence of *Sphagnum* species, environmental conditions became more important in determining the composition of the SAM community.

Two environmental gradients determine the characteristics of mires—wetness and nutrient levels—resulting in a spectrum of mire site types (Eurola et al., 1984; Moen et al., 2017). Drier mires are usually forested and wet mires are usually open, i.e., without trees. Between open and forested mires there exist composite site types where higher surfaces (mounds or hummocks and strings) have vegetation typical for forested sites and lower surfaces (depressions or lawns, hollows, and flarks) have vegetation typical for open mires (Eurola et al., 1984; Moen et al., 2017). The formation of macro- and microtopography in mires results in the formation of microhabitats having different water levels and anaerobicity for methanotrophs and methanogens (Vasander and Kettunen, 2006).

The term *trophy* or *trophic status* denotes nutrient status. *Minerotrophy* refers to mineral nutrients brought into a mire or peatland with water inflowing from the surrounding mineral soils. *Ombrotrophy* refers to the area of a mire or peatland that receives mineral nutrients only via precipitation. Thus the wetness and nutrient status of mires and peatlands are determined by their ecohydrology (Eggelsmann et al., 1993). At the beginning of their formation, mires are usually minerotrophic but gradually the peat-forming plant communities lose their connection to the groundwater and become ombrotrophic. Minerotrophic mires are often referred to as *fens* and ombrotrophic mires as *bogs*.

Minerotrophy is usually divided into three classes: *oligotrophy* (poor), *mesotrophy* (medium), and *eutrophy* (rich in nutrients). When moving from oligotrophy toward eutrophy the concentrations and amounts of nitrogen and mineral nutrients (especially Ca and Mg) in the peat increase. *Oligotrophic* sites are often dominated by sedges (e.g., *Carex rostrata*, *C. lasiocarpa*, *C. chordorrhiza*) and by some *Sphagnum* species from the subgenus Cuspidata (e.g., *S. fallax*) (Laine et al., 2018). The pH of the interstitial water is usually ≤4.5. *Mesotrophic* sites are usually dominated by herbs and grasses (e.g., *Comarum palustre*, *Trichophorum alpinum*, *Lysimachia* spp.) and by some *Sphagnum* species from the subgenus Subsecunda (e.g., *S. subsecundum*). Interstitial water pH is usually 4.5–5.5. *Eutrophic* sites are usually dominated by herbs and grasses (e.g., *Saussurea alpina*, *Rhynchospora fusca*, *Eriophorum latifolium*) and typically by brown mosses (e.g., *Scorpidium* spp., *Campylium stellatum*). The interstitial water pH is usually 5.5–7.5 but can be higher. Mineral-rich waters inflowing from the mineral soils bordering the mire often give rise to a minerotrophic zone around the edge of the mire (lagg), but groundwater from the underlying mineral soil can also bring mineral nutrients into the mire. The mineral nutrients entering the edge of the mire are taken up by the vegetation and therefore interstitial water concentrations rapidly decline with distance and the lagg zone is usually narrow.

Ombrotrophic sites usually have thick peat layers, and interstitial water has low pH (usually $\leq$ pH 4) and concentrations of calcium and other solutes (Tolonen and Hosiaisluoma, 1978; Eurola et al., 1984). However, the water chemistry of mires and peatlands is also influenced by the distance from the sea and marine inputs (Moen et al., 2017) and by air pollution and deposition. Typical species are *Sphagnum fuscum* on the hummocks, *S. balticum* in the hollows, and *S. rubellum* on the lawns (the surfaces between hummocks and hollows). Typical species in the field layer are *Eriophorum vaginatum*, *Trichophorum cespitosum*, and *Scheuchzeria palustris*.

5. Peat types and decomposition

The characteristics of peat are mainly determined by the plant species of the mire vegetation and its degree of decomposition. The type of peat is thus determined by the plant communities growing on the mire, and therefore the peat type varies both across the mire surface and with peat depth. Peat type can be considered analogous to parent material mineralogy in mineral soils.

Generally, three types of peat constituents are identified: mosses, sedges, and wood. On sites that are dominated by mosses, the peat is overgrown by new growth, and older stems and leaves (mosses do not have roots) are gradually buried deeper and deeper. Both *Sphagnum* peat and brown moss peat are formed principally in the same way. On sites dominated by sedges, peat is mainly formed from the roots of sedges (*Carex* peat). The main part of sedge roots is confined to the more oxic top 30 cm peat but a small amount grows deeper to the anoxic peat, even down to a depth of 2 m (Saarinen, 1996). The penetration of sedge roots into the deeper peat means that the younger peat can be formed within a matrix of older peat. Furthermore, oxygen is transported down aerenchyma cells in the stems, rhizomes, and roots of sedges so that the rhizosphere can be oxic while the peat matrix mostly remains anoxic. These different niches may be expected to be reflected in the populations of fungi, methanogens, and methanotrophs. In forested areas, peat is formed mainly by the stems and roots of trees but also other remains of trees and dwarf shrubs. Moss remains are typically also found in woody peat. However, as tree roots cannot grow very deep in the peat due to anoxic conditions, oxic rhizospheres are confined to the surface peat layer.

As the chemical composition of plant species and their parts differ widely (Straková et al., 2010) the microbial decomposition of their remains also varies (Straková et al., 2011, 2012). The decomposition of organic matter can be considered to be analogous to weathering of mineral and rock particles in mineral soils—there is a loss of organic matter, loss of physical structure, and changes in chemical state (Clymo, 1983). The degree of peat decomposition is usually assessed in the field using von Post's humification classes (1–10) where 1 is completely undecomposed and 10 is completely decomposed (von Post, 1922; Päivänen and Hånell, 2012).

The degree of decomposition is affected by the length of time the plant remains are in the oxic acrotelm layer before entering the anoxic catotelm where microbial decomposition is much slower. This transition time depends on the wetness of the mire, whether the residues are deposited on the surface or at depth, as with sedges, and the height growth of the mire (Clymo, 1983). The best predictors of peat carbon: nutrient ratios are the degree of decomposition and depth, variables which are strongly correlated to each other (Wang et al., 2015). Significant changes in C:nutrient ratios occur around a von Post value of 4 (Wang et al., 2015), which generally corresponds to the acrotelm/catotelm boundary (Fig. 12.3), and thus indicates a significant change in the structure, distribution, and composition of microbial communities. For example, nitrogen fixation by microbes thriving in the hyaline cells of *Sphagnum* leaves and partial oxidation of methane rising from the catotelm are linked (Kox et al., 2020; Larmola et al., 2014).

6. Drained peatlands and peat

The ground vegetation of mires drained to harvest the peat for fuel or horticulture is removed and kept without vegetation while that of mires drained for agricultural use is replaced by monocultures of crop species. Except for mires in the tropics drained for oil palm production, the ground vegetation of mires drained for forestry purposes is usually not changed so dramatically. Ditches dug around the mire to prevent the inflow of surface waters from the surrounding mineral soil changes the minerotrophic lagg to being more ombrotrophic and the vegetation correspondingly changes. Drainage also initiates a secondary succession of ground vegetation towards that of upland forest ecosystems. The first species to vanish from drained mires are those specialized for wet conditions, mire species are gradually substituted by upland forest species, and the amount and share of above- and belowground litter from trees increases (Laiho et al., 2003). But regardless of changes in the ground vegetation, the lowering of the water table as a result of drainage has a huge impact on edaphic factors and hence on conditions for microorganisms. The fundamental changes in aeration, pH, and the amounts and composition of organic matter create new niches for microorganisms. For example, denitrifying bacteria are especially active in drained and cultivated peat soils, releasing significant amounts of N_2O to the atmosphere (Leppelt et al., 2014).

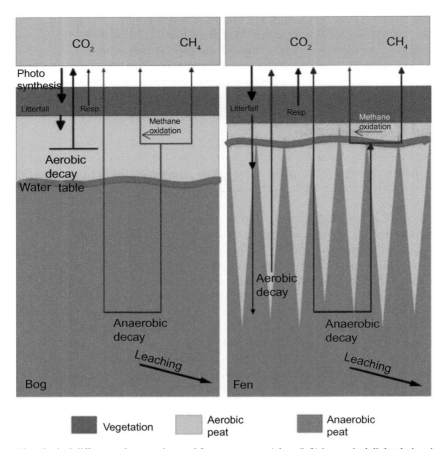

FIG. 12.3 The fundamental ecological differences between bog and fen ecosystems. A bog (left) is a typical diplotelmic mire ecosystem where the upper aerobic peat (acrotelm) is separated from the lower anaerobic peat layer (catotelm). A fen (right) has more seepage where minerotrophic water flows along the surface of the ecosystem. Beneath the water level, the peat layer also has areas of aerobic decay due to oxygen transport by sedge roots. The decrease in the availability of oxygen with depth implies that the peat above the water level is decomposed by aerobic microbes, whereas below the water level decomposition is dominated by anaerobic microorganisms. The anaerobic decay of organic matter in this anoxic region leads to the release of both methane and carbon dioxide (Laine and Vasander, 1996). *(Reproduced with the permission of The Finnish Peatland Society.)*

References

April, R., Keller, D., 1990. Mineralogy of the rhizosphere in forest soils of the eastern United States. Biogeochemistry 9, 1–18.

Bates, S., Berg-Lyons, D., Caporaso, J., Walters, W.A., Knight, R., Fierer, N., 2011. Examining the global distribution of dominant archaeal populations in soil. ISME J. 5, 908–917. https://doi.org/10.1038/ismej.2010.171.

Baumert, V.L., Vasilyeva, N.A., Vladimirov, A.A., Meier, I.C., Kögel-Knabner, I., Mueller, C.W., 2018. Root exudates induce soil macroaggregation facilitated by fungi in subsoil. Front. Environ. Sci. 6 (140), 1–17.

Bergström, I., Mäkelä, K., Starr, M. (Eds.), 1995. Integrated Monitoring Programme in Finland. Ministry of the Environment, Environmental Policy Department. First National Report. Report 1. Ministry of the Environment, pp. 138 (+7 appendices). ISSN 1236-5610, ISBN 951-731-042-0.

Binkley, D., Fisher, R.F., 2019. Ecology and Management of Forest Soils, fifth ed. Wiley-Blackwell.

Brockett, B.F.T., Prescott, C.E., Grayston, S.J., 2012. Soil moisture is the major factor influencing microbial community structure and enzyme activities across seven biogeoclimatic zones in western Canada. Soil Biol. Biochem. 44 (1), 9–20. https://doi.org/10.1016/j.soilbio.2011.09.003.

Clymo, R.S., 1983. Peat. In: Gore, A.J.P. (Ed.), Mires: Swamp, Bog, Fen and Moor. Ecosystems of the World 4A. Elsevier, Amsterdam, pp. 159–224.

Comerford, N., Fox, T., 2016. Forest soils. In: Lal, R. (Ed.), Encyclopedia of Soil Science. CRC Press, Boca Raton, https://doi.org/10.1081/e-ess3. third ed. 3068 pp.

Damman, A.W.H., 1986. Hydrology, development, and biogeochemistry of ombrogenous peat bogs with special reference to nutrient relocation in western Newfoundland bog. Can. J. Bot. 64, 384–394.

Eggelsmann, R., Heathwaite, A.I., Grosse-Brauckmann, G., Küster, F., Naucke, W., Schuch, M., Schweikle, V., 1993. Physical processes and properties of mires. In: Heathwaite, A.L. (Ed.), Mires: Process, Exploitation and Conservation. John Wiley & Sons Ltd, pp. 171–262.

Eurola, S., Hicks, S., Kaakinen, E., 1984. Key to Finnish mire types. In: Moore, P.D. (Ed.), European Mires. Academic Press, London, pp. 11–117.

FAO, 2001. Lecture notes on the major soils of the world. World Soil Resources reports, 94, Food and Agriculture Organization of the United Nations, Rome. 334 pp.

FAO and UNEP, 2020. The State of the World's Forests 2020. Forests, biodiversity and people. Rome, https://doi.org/10.4060/ca8642en.

Gobran, G.R., Clegg, S., Courchesne, F., 1998. Rhizospheric processes influencing the biogeochemistry of forest ecosystems. Biogeochemistry 42, 107–120.

Joosten, H., Clarke, D., 2002. Wise Use of Mires and Peatlands – Background and Principles Including a Framework for Decision-Making. International Mire Conservation Group, International Peat Society, Finland. 304 pp. ISBN 9519774483 9789519774480.

Kelting, D.L., Burger, J.A., Edwards, G.S., 1998. Estimating root respiration, microbial respiration in the rhizosphere, and root-free soil respiration in forest soils. Soil Biol. Biochem. 30 (7), 961–968.

Kox, M.A.R., van den Elzen, E., Lamers, L.P.M., Jetten, M.S.M., van Kessel, M.A.H.J., 2020. Microbial nitrogen fixation and methane oxidation are strongly enhanced by light in *Sphagnum* mosses. AMB Express 10 (61), 1–11.

Kuzyakov, Y., Blagodatskaya, E., 2015. Microbial hotspots and hot moments in soil: concept & review. Soil Biol. Biochem. 83, 184–199. https://doi.org/10.1016/j.soilbio.2015.01.025.

Laiho, R., Vasander, H., Penttila, T., Laine, J., 2003. Dynamics of plant-mediated organic matter and nutrient cycling following water-level drawdown in boreal peatlands. Glob. Biogeochem. Cycles 17 (2), 1–11.

Laine, J., Vasander, H., 1996. Ecology and vegetation gradients of peatlands. In: Vasander, H. (Ed.), Peatlands in Finland. Finnish Peatland Society, Helsinki, Finland, pp. 10–19.

Laine, J., Flatberg, K.I., Harju, P., Timonen, T., Minkkinen, K.J., Laine, A., Tuittila, E.-S., Vasander, H., 2018. Sphagnum Mosses: The Stars of European Mires. Sphagna Ky, Helsinki.

Larmola, T., Leppänen, S.M., Tuittila, E.-S., Aarva, M., Merilä, P., Fritze, H., Tiirola, M., 2014. Methanotrophy induces nitrogen fixation during peatland development. Proc. Natl. Acad. Sci. 111 (2), 734–739. https://doi.org/10.1073/pnas.1314284111.

Leppelt, T., Dechow, R., Gebbert, S., Freibauer, A., Lohila, A., Augustin, J., Drösler, M., Fiedler, S., Glatzel, S., Höper, H., Järveoja, J., Lærke, P.E., Maljanen, M., Mander, Ü., Mäkiranta, P., Minkkinen, K., Ojanen, P., Regina, K., Strömgren, M., 2014. Nitrous oxide emission budgets and land-use-driven hotspots for organic soils in Europe. Biogeosciences 11, 6595–6612.

McNear Jr., D.H., 2013. The rhizosphere – roots, soil and everything in between. Nat. Educ. Knowl. 4 (3), 1. https://www.nature.com/scitable/knowledge/library/the-rhizosphere-roots-soil-and-67500617/.

Moen, A., Joosten, H., Tanneberger, F., 2017. Mire diversity in Europe: mire regionality. In: Joosten, H., Tanneberger, F., Moen, A. (Eds.), Mires and Peatlands of Europe: Status, Distribution and Conservation. Schweizerbart Science Publishers, Stuttgart, pp. 97–149.

Päivänen, J., Hånell, B., 2012. Peatland Ecology and Forestry – A Sound Approach. vol. 3 University of Helsinki Department of Forest Sciences Publications, pp. 1–267.

Philippot, L., Raaijmakers, J.M., Lemanceau, P., van der Putten, W.H., 2013. Going back to the roots: the microbial ecology of the rhizosphere. Nat. Rev. Microbiol. 11, 789–799. https://doi.org/10.1038/nrmicro3109.

Putkinen, A., Larmola, T., Tuomivirta, T., Siljanen, H.M.P., Bodrossy, L., Tuittila, E.-S., Fritze, H., 2014. Peatland succession induces a shift in the community composition of Sphagnum-associated active methanotrophs. FEMS Microbiol. Ecol. 88, 596–611.

Reinhold-Hurek, B., Bünger, W., Burbano, C.S., Sabale, M., Hurek, T., 2015. Roots shaping their microbiome: global hotspots for microbial activity. Annu. Rev. Phytopathol. 53, 403–424. https://doi.org/10.1146/annurev-phyto-082712-102342.

Saarinen, T., 1996. Biomass and production of two vascular plants in boreal mesotrophic fen. Can. J. Bot. 74, 934–938.

Straková, P., Anttila, J., Spetz, P., Kitunen, V., Tapanila, T., Laiho, R., 2010. Litter quality and its response to water level drawdown in boreal peatlands at plant species and community level. Plant Soil 335 (1–2), 501–520.

Straková, P., Niemi, M., Freeman, C., Peltoniemi, K., Toberman, H., Heiskanen, I., Fritze, H., Laiho, R., 2011. Litter type affects the activity of aerobic decomposers in a boreal peatland more than site nutrient and water level regimes. Biogeosciences 8, 2741–2755.

Straková, P., Penttilä, T., Laine, J., Laiho, R., 2012. Disentangling direct and indirect effects of water table drawdown on above and belowground plant litter decomposition: consequences for accumulation of organic matter in boreal peatlands. Glob. Chang. Biol. 18 (1), 322–335.

Tamminen, P., 1991. Kangasmaan ravinnetunnusten ilmaiseminen ja viljavuuden alueellinen vaihtelu. Summary: Expression of soil nutrient status and regional variation in soil fertility of forested sites in southern Finland. Folia For. 777, 1–40.

Tamminen, P., Starr, M.R., 1990. A survey of forest soil properties related to soil acidification in southern Finland. In: Kauppi, P., Kenttämies, K., Anttila, P. (Eds.), Acidification in Finland. Springer-Verlag, Berlin, Heidelberg, pp. 231–247.

Tamminen, P., Starr, M., 1994. Bulk density of forested mineral soils. Silva Fenn. 28 (1), 53–60.

Tamminen, P., Starr, M., Kubin, E., 2004. Element concentrations in boreal, coniferous forest humus layers in relation to concentrations in ectohydric mosses and soil factors. Plant Soil 259 (1–2), 51–58.

Tolonen, K., Hosiaisluoma, V., 1978. Chemical properties of surface-water in Finnish ombrotrophic mire complexes with special reference to algal growth. Ann. Bot. Fenn. 15, 55–72.

Van Rees, K., 2016. Forest soils: major. In: Lal, R. (Ed.), Encyclopedia of Soil Science. CRC Press, Boca Raton, p. 3068, https://doi.org/10.1081/e-ess3. third ed.

Vasander, H., Kettunen, A., 2006. Carbon in boreal peatlands. In: Wieder, R.K., Vitt, D.H. (Eds.), Boreal Peatland Ecosystems. Ecological Studies, vol. 188. Springer, Berlin, pp. 165–194.

von Post, L., 1922. Sveriges geologiska undersöknings torvinventering och några av dess hittills vunna resultat. Sven. Mosskulturfören. Tidskr. 1, 1–27.

Wang, M., Moore, T.R., Talbot, J., Riley, J.L., 2015. The stoichiometry of carbon and nutrients in peat formation. Glob. Biogeochem. Cycles 29 (2), 113–121.

Westman, C.J., 1981. Fertility of surface peat in relation to the site type and potential stand growth. Acta For. Fenn. 172, 1–77.

Westman, C.J., Starr, M., Laine, J., 1985. A comparison of gravimetric and volumetric soil properties in peatland and upland sites. Silva Fenn. 19 (1), 73–80.

Yang, Y., Dong, M., Cao, Y., Wang, J., Tang, M., Ban, Y., 2017. Comparisons of soil properties, enzyme activities and microbial communities in heavy metal contaminated bulk and rhizosphere soils of *Robinia pseudoacacia* L. in the northern foot of Qinling Mountain. Forests 8, 430. https://doi.org/10.3390/f8110430.

Chapter 13

Fungal community of forest soil: Diversity, functions, and services

Leticia Pérez-Izquierdo[a], Ana Rincón[b], Björn D. Lindahl[a], and Marc Buée[c]

[a]Department of Soil and Environment, SLU, Uppsala, Sweden, [b]Institute of Agricultural Sciences (ICA), Spanish National Research Council (CSIC), Madrid, Spain, [c]French National Research Institute for Agriculture, Food and the Environment (INRAE), Lorraine University, Department of "Tree-Microbe Interactions", Champenoux, France

Chapter Outline

1. Introduction

Funi constitute a significant fraction of the soil microbiome and account for most of the microbial biomass in forest soils. One gram of soil can contain several hundreds of fungal species. This high ecological and taxonomical fungal diversity and biomass of forest soils (Bardgett and van der Putten, 2014; Bahram et al., 2018) has important implications for the functioning and dynamics of forest ecosystems (Clemmensen et al., 2013). Fungi mediate key ecosystem processes such as carbon and nutrient cycling through organic matter decomposition, and form mutualistic relationships with roots, affecting the nutrition, productivity, regeneration, and health of trees (Treseder and Lennon, 2015; van der Heijden et al., 2015; Baldrian, 2017). In forest, multiple trophic fungal guilds—saprotrophs, mycorrhizal, endophytes, pathogens—exert wide-range effects on soil biogeochemistry, health of trees, and nutrient dynamics (Baldrian, 2017; Frac et al., 2018). These four ecological lifestyles are characterized by the strategies and functions developed during their evolution to acquire carbon (sugars) from a living or dead host, i.e., organic matter (Box 13.1). As such, saprotrophic fungi are free-living decomposers that obtain carbon and nutrients by breaking down nonliving organic matter, while mycorrhizal fungi form symbiosis with the plant roots facilitating the nutrient uptake to their host plant in exchange for photosynthetic carbon. Different types of mycorrhizas have been described based on their structural and functional characteristics together with the identity of the plants and fungal species involved (Smith and Read, 2008). Fungal endophytes inhabit living plants and do not harm them, in contrast to fungal pathogens that usually cause disease to their host (Zanne et al., 2020). Although the taxonomic attributes facilitate the classification of fungal species, these ecological categories are based on functional data, and comparative genomics has considerably improved our knowledge in this research field (Kohler et al., 2015; Stajich, 2017; Martino et al., 2018; Miyauchi et al., 2020). However, the boundaries between these categories remain porous and some species can shift from one trophic status to another or belong to two different types of guilds depending on their development, on the host they interact with, or the stage of their life cycle (Delaye et al., 2013; Lindahl and Tunlid, 2015; Thines, 2019; Thoen et al., 2020).

As in forest ecosystems, mycorrhizal fungi have a central position in this chapter. Indeed, majority of the nitrogen and phosphorus in plants is supplied by mycorrhizal fungi, which receive in exchange the photosynthesis products from

Forest Microbiology. https://doi.org/10.1016/B978-0-12-822542-4.00022-X

BOX 13.1 Square identifying ecological groups and interaction types that soil forest fungi can establish with trees.

Biotrophic fungi colonize living plants as a source of nutrients, while **Necrotrophic** fungi kill their hosts and live off the dead tissue. Biotrophic fungi can be mutualistic organisms (**endophytic** and **mycorrhizal** fungi) or **pathogens**.

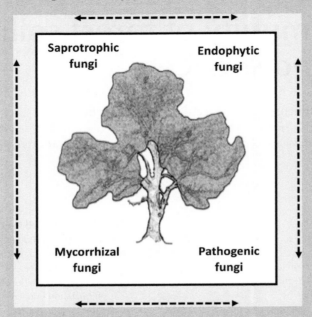

Endophytic fungi colonize living plant tissues without causing any visible morphological changes or manifestation of disease. These fungi live in mutualistic association with plants for at least a part of their life cycle. During this endophytic phase, the carbon nutrition of these fungi remains unclear for many species.

Mycorrhizal fungi must establish mutual symbiotic association with plant roots for their nutrition. The plants supply sugars (carbon) to these soil fungi from photosynthesis, in return the fungal partners provide to the plant water and mineral nutrients, such as nitrogen and phosphorus. Mycorrhizal fungi are divided into ectomycorrhizas and endomycorrhizas, including arbuscular, ericoid, and orchid mycorrhiza.

Phytopathogens, or **plant pathogenic fungi**, colonize host plant to mobilize part of photosynthetic carbon for their own nutrition or to complete their life cycle. Through these interactions, fungi cause plant diseases, killing their hosts and feed on dead material (**necrotrophs**) or colonizing the living tissue (**biotrophs**).

Saprotrophic fungi, also called **saprobes** or **saprotrophs**, are organisms that decay nonliving organic matter for their carbon nutrition and other nutrient mobilization. Using a wide range of extracellular enzymes digesting the carbohydrate polymers, saprotrophic fungi are critical to cell wall decomposition process and nutrient cycling.

the plant. Almost all terrestrial plants form symbiotic associations with mycorrhizal fungi, and studies estimate that approximately 50,000 fungal species could form mycorrhizal associations with approximately 250,000 plant species (van der Heijden et al., 2015). Although this nutritional exchange is universal, these symbiotic interactions can be divided into four main types that have been described based on their structure and function: arbuscular mycorrhiza, ectomycorrhiza, ericoid mycorrhiza, and orchid mycorrhiza (Smith and Reads, 2008). The majority of plant species form arbuscular mycorrhizas with fungi of the Glomeromycota clade (Smith and Read, 2008; Brundrett, 2009). These fungi are able to form numerous fungal hyphae inside the root cortex that feed arbuscules, tree-like structures, that are formed by the fungus inside cortical root cells. These arbuscules are the seats of the exchange of nutrients between the two partners. Ectomycorrhizal symbiosis groups ectomycorrhizal fungi belonging to Basidiomycota and Ascomycota (Tedersoo et al., 2010). Although only 2% of plants form ectomycorrhizal associations, this symbiosis concerns the majority of trees which represent Boreal, Temperate, and Mediterranean forests (Brundrett, 2009). Ectomycorrhizal fungi colonize lateral roots forming a mantle around the root and a Hartig net around the epidermal root cells. It is always this chimeric organ that allows the fungus and the tree to exchange signals and nutrients. Some fungi form ericoid mycorrhizas also with species of the *Ericaceae*, plants which are mostly common under acid and infertile soils, such as acid heathland, peatlands, and part of the Boreal forests. Well-known ericoid fungi belong to the Helotiales (Ascomycetes) that are soil saprotrophs also. Some Basidiomycetes, such as

Sebacinales, are also able to form ericoid mycorrhizal associations (Selosse et al., 2007). In these harsh habitats, fungi and plant survival relies on nutrient mobilization from soil organic matter (SOM) by their fungal partners. More and more studies are paying attention to this mycorrhizal type, and recent genomics data revealed the ericoid mycorrhizal fungi possess gene repertoire with a capacity for a dual saprotrophic and biotrophic lifestyle (Martino et al., 2018). In this chapter, less attention will be paid to fungi forming mycorrhizas with orchids, but it should be noted that the *Orchidaceae* is one of the largest plant families on Earth, including almost 10% of all flowering plant species (McCormick et al., 2004). Orchids are fully dependent on mycorrhizal fungi because their seeds are extremely small, containing very few reserves, and orchid protocorms lack chlorophyll. Different morphological steps of a protocorm have been reported and its development to the seedling stage can be long (Yeung, 2017). Several fungi forming mycorrhizas with orchids can live also as saprotrophs or establish endophytic or ectomycorrhizal interactions with trees (Selosse and Martos, 2014). Consecutively, it has been shown that some achlorophyllous orchids, and even certain green orchids and *Ericaceae (mycoheterotrophic and/or mixotrophic plants)*, obtain their carbon thanks to these fungi from dead wood (Suetsugu et al., 2020) or through mycelial connections with neighboring trees (Martos et al., 2009; Selosse and Roy, 2009).

Here, we describe the principal abiotic and biotic factors driving the assemblage of fungal communities in forest soils, as well as potential underlying mechanisms. The biogeography and distribution of forest soil fungi across different scales is further addressed followed by a detailed overview of the major roles of fungal communities in forest soils. We also describe the ecological roles and functions that characterize these different fungal guilds and the potential interactions among them. Finally, we show how anthropic factors, particularly forest management, may shape soil fungal communities, modify their functions, and affect their associated ecosystem services.

2. Fungal community structure

Fungal communities in forest soils are hyperdiverse (Buée et al., 2009; Wu et al., 2019), typically consisting of a few dominant and many less abundant species that encompass some of the major fungal phyla, i.e., Basidiomycota, Ascomycota, Glomeromycota, Chytridiomycota, and Zygomycota. Both richness and species composition, and their shifts in response to environmental factors, define the structure of soil fungal communities. Determining how functional redundancy and hyperdiversity may occur simultaneously remains challenging (Smith et al., 2018). The great increase in technology advances is revolutionizing our knowledge of the true dimension of fungal biodiversity and its potential functional outcomes (Kohler et al., 2015; Pérez-Izquierdo et al., 2019; Hawksworth and Lücking, 2017; Wu et al., 2019), although there is still much to be understood (Baldrian, 2017; Smith and Peay, 2020; Tedersoo et al., 2020a). Biodiversity metrics based on species counts are commonly used to describe the fungal community structure. However, other community parameters that take into account similarities in the ecological niche of species, such as fungal trait patterns (López-García et al., 2018; Treseder and Lennon, 2015; Courty et al., 2016; Zanne et al., 2020) or phylogenetic relatedness of taxa (Pérez-Izquierdo et al., 2020a,b; Pérez-Valera et al., 2015; Rincón et al., 2014; Tedersoo et al., 2012a) are gaining in importance.

Multiple factors directly or indirectly contribute to the high spatial and temporal heterogeneity of soil fungal communities in forests both in terms of structural and functional variations (Fig. 13.1). The type of forest and climatic region are the main determinants of fungal community structure (Tedersoo et al., 2014; Bahram et al., 2018; van der Linde et al., 2018). At finer scales, soil chemistry and local climate (Tedersoo et al., 2020b; Castaño et al., 2018; Rincón et al., 2015; Coince et al., 2013), together with tree species that determine root exudates, leaf litter, and wood (Nguyen et al., 2020; Pérez-Izquierdo et al., 2017, 2019; Kennedy and Bruns, 2005), are common habitat drivers of the different soil fungal trophic guilds. In addition, disturbances, such as forestry, fires, or pathogen attacks, are key determinants of the distribution and abundance of soil fungi in forests (Holden and Treseder, 2013). Studies about the architecture of networks, which represents the linkage between plant and fungal communities, have shown that arbuscular mycorrhizal, ectomycorrhizal, and saprotrophic/endophytic plant-fungal associations have moderate or low levels of host-fungus specificity, but they are structured to avoid overlap of host plant ranges (Toju et al., 2018).

The assembly of forest fungal communities is governed by different ecological mechanisms and processes, which can be deterministic, i.e., biotic and abiotic factors that filter the regional fungal species pool to shape the local community, and stochastic, i.e., based on probabilistic events as consequence of disturbance or dispersal (Peay et al., 2008; Peay, 2016). Abiotic gradients act as environmental filters that promote niche differentiation across species leading to structural clustering patterns in fungal communities, i.e., selection of functionally similar fungal taxa (Tedersoo et al., 2020b; Pérez-Izquierdo et al., 2019; Rincón et al., 2014). On the other hand, niche limitation may result in structural overdispersion of fungal communities due to the increase of biotic interactions and competition among species, i.e., selection of functionally different fungal guilds or taxa (van Nuland and Peay, 2020; Tedersoo et al., 2020a; Kennedy, 2010). Sequential assembly rules seem to regulate the diversity of microbial communities, with abiotic filtering as the pervasive structuring force, and biotic interactions subsequently acting (Pérez-Valera et al., 2018).

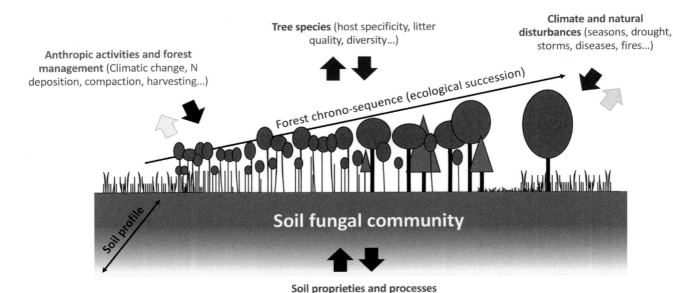

FIG. 13.1 Drawing of a forest consisting of trees with different ages (chronosequence) that associate with soil fungal community. The *arrows* represent the main abiotic and biotic factors, including anthropogenic and natural disturbances, driving the assemblage of fungal communities in forest soil and, in return, the main effects of fungi on these parameters.

2.1 Biotic and abiotic drivers of soil fungal communities

Multiple direct and indirect interactions between the above- and belowground compartments drive a range of key processes and functions in forest ecosystems (Baldrian, 2017). Plant and soil fungal communities have an extraordinary influence on each other, and changes in any of them lead to reciprocal shifts in community composition and functioning (Fig. 13.1). Habitat characteristics such as land use and disturbances, together with vegetation type, plant species, forest age, climate, and edaphic properties, modulate the structure of forest fungal communities and determine their functional responses (Gehring et al., 2014; Rincón et al., 2015; Pérez-Izquierdo et al., 2017; Kennedy et al., 2018; Frac et al., 2018; Nguyen et al., 2020). These effects and processes are dynamic and may occur at different spatial—vertical soil profile, local, regional, landscape—and temporal—from days to many years—scales (Baldrian, 2017; Tedersoo et al., 2020b). In parallel, the evolutionary differences among different fungal groups also determine their responses to abiotic and biotic filters, which usually results in a considerable variety of responses—at least at the fungal family and order levels—to a given environmental factor (Treseder and Lennon, 2015; Rincón et al., 2015). Thus the drivers of any multitaxa fungal community or trophic guild are multiple and mostly context dependent (Tedersoo et al., 2016; Nguyen et al., 2020). This is particularly relevant in the case of ectomycorrhizal fungal communities, because shifts in their composition may have important consequences for host plants and ultimately for forest dynamics (Prieto et al., 2016; Pérez-Izquierdo et al., 2017, 2019). Indeed, recent studies identified links between plant-soil feedback, plant growth, and the diversity of fungal guilds in soil (Semchenko et al., 2018). Some of them indicate that the presence and absence of pathogenic and mycorrhizal fungi (Klironomos, 2002; Bennett et al., 2017) have been considered to be key regulators of plant-soil feedbacks. Comparing the relative effect of conspecific (versus heterospecific) trees on seedling survival between ectomycorrhizal or arbuscular mycorrhizal tree species, Bennett et al. (2017) demonstrated that trees associated with arbuscular mycorrhizal fungi exhibited strong local conspecific inhibition, whereas ectomycorrhizal trees exhibited conspecific facilitation locally and less severe conspecific inhibition for a long distance (regionally). Like some pathogenic agents, mycorrhizal fungi could be important contributors to plant community structure in temperate forests through effects on plant-soil feedbacks. These interactions, and potential feedbacks, plant-fungi-soil, will be extensively developed in the third paragraph of this chapter.

Concerning biotic factors, traits of the dominant tree species shape fungal communities in forest soil through microclimatic variations and the organic inputs provided, with potential cascading effects on the ecosystem functioning (Flores-Rentería et al., 2016). Plant community composition and plant traits outline the quantity and chemistry of their organic matter inputs (i.e., litter and root exudates) and are often good predictors of soil biogeochemistry and fungal community structure (van der Heijden et al., 2015; Aponte et al., 2010). This is particularly evident in the case of mycorrhizal fungal communities, tightly interconnected to the tree host by specific mycelial structures (Smith and Read, 2008). For example,

Nguyen et al. (2020) demonstrated that root chemistry carbohydrates structured the root-associated fungal assemblages. Likewise, significant differences in the diversity and species composition of ectomycorrhizal fungal communities are commonly observed across tree host species and tree genotypes. For example, Pérez-Izquierdo et al. (2017, 2019) showed that *Pinus pinaster* Ait. genotypes differing in their productivity had dissimilar ectomycorrhizal fungal communities, i.e., Basidiomycetes were overrepresented under the most productive genotypes, and this was causally related to a different carbon, nitrogen, or phosphorus mobilization ability under the trees. Reciprocally, positive plant-fungal relations such as mycorrhizal associations may drive plant population and community biology by influencing plant traits and regulating plant-plant interactions through signaling and nutrient transfer by common mycelial networks (Tedersoo et al., 2020a; Pickles and Simard, 2017).

There is ample evidence for competitive biotic interactions in soil fungal communities, but direct indication for competition as a determinant of fungal diversity (i.e., through competitive exclusion) is scarce (Wardle, 2006). A nice example is given by Peay et al. (2012) who demonstrated that competition through inhibitory priority effects increased the phylogenetic diversity of nectar yeast communities because the negative effect of early arriving on late-arriving species was stronger between closer relatives. Other studies reveal that the presence of a given saprotrophic fungal species on a substrate (litter, dead wood) may exclude the colonization of new saprotrophic fungal species on that substrate, but also facilitate the arrival of specialized fungi, e.g., those able to breakdown recalcitrant carbohydrates enabling subsequent colonization by fungi using simple carbon forms (Cline and Zak, 2015). Competition-colonization trade-offs have been demonstrated to be also crucial for ectomycorrhizal fungal assembly (Smith et al., 2018).

Although fungal communities are predominantly regulated by bottom-up (e.g., resource availability) rather than top-down (e.g., predation) factors, invertebrate grazers have been demonstrated to exert significant selective pressures on soil fungal decomposer communities able to reverse the outcomes of competitive interactions, for example by stimulating growth of less competitive fungi (Crowther et al., 2019). The importance of deciphering the role that grazing intensity exerts on mycorrhizal communities in a range of ecosystems worldwide has been recently recognized (van der Heyde et al., 2019). Besides plant-fungal and animal-fungal relations, other interkingdom biotic interactions, such as strong bacterial-fungal antagonism in different habitats, seem to play an important role in determining the distributions of functional diversity and biomass of soil fungi (Bahram et al., 2018).

Regarding abiotic factors, a range of edaphic and climatic gradients has been reported to influence soil fungal communities. For example, climate variations along elevation gradients can promote significant changes in the soil fungal communities (Bahram et al., 2012; Coince et al., 2014; Rincón et al., 2015; Truong et al., 2019). By analyzing fungi of root tips and soils of *Pinus sylvestris* L. forests along elevation gradients at regional scale, Rincón et al. (2015) found that overall soil fungal richness increased with pH but did not vary with climate, whereas that of representative fungal groups, i.e., Ascomycetes, Basidiomycetes, differentially responded to these variables indicating selective impacts of environmental variations on these communities. The structure and functioning of forest fungal communities also greatly varies over time depending on climate variation and the availability of resources (Baldrian, 2017; Pérez-Izquierdo et al., 2017; Castaño et al., 2018). Local microclimate variations may promote fungi better adapted to those conditions, e.g., in terms of enzymatic capacity, which may, in turn, alter decomposition rates, nutrient mobilization, and tree nutrition. In fact, soil moisture is a main factor regulating forest fungal communities (Buée et al., 2005; Flores-Rentería et al., 2016), which can be crucial for forest dynamics affecting important processes such as decomposition, natural regeneration, and facilitation (Richard et al., 2011; Montesinos-Navarro et al., 2019). The context dependency of soil fungal communities has been often highlighted (Cox et al., 2010; Tedersoo et al., 2016; Glassman et al., 2017; Pérez-Izquierdo et al., 2017), with pH, organic matter, and nitrogen availability among the most influential edaphic factors. In a multiyear regional-scale survey, Tedersoo et al. (2020b) have showed that, together with plant species, soil pH is the strongest factor affecting the diversity of fungi and the multiple taxonomic and fungal functional groups.

2.2 Impact of disturbances on soil fungal communities

Forests are subjected to a range of disturbances—natural such as insect outbreaks, wind throw, drought, and wildfire, and anthropogenic such as climate change, nitrogen deposition, air pollution, invasive forest pests, and harvesting regimes—over different temporal and spatial scales, which have main consequences for living organisms and ecosystem functioning and dynamics (Anderegg et al., 2020). Forest ecosystems are usually adapted to their natural disturbance regimes. As such, increases in the frequency and intensity of disturbances forecasted in future climatic scenarios can impact above- and belowground communities in unprecedented ways (Rodriguez-Ramos et al., 2020). For example, in relation to climate change, forest ecotones move up in elevation and shifts in plant species composition due to temperature rise may drive clear changes in fungal community assembly and soil biogeochemistry (van Nuland et al., 2020). Many forest disturbances usually cause the loss of vegetation and the disruption of the forest floor directly impacting the structure and functioning of soil microbial communities. A large portion of the carbon fixed by plants is transferred to roots and immobilized in fungal mycelium (Clemmensen et al., 2013), and postdisturbance shifts in the dominance of soil fungi may have important consequences in the long term for carbon storage in soils. However, while abiotic disturbances consistently reduce soil fungal biomass, biotic disturbances often have less drastic effects (Holden and Treseder, 2013), likely related to the different carbon resources outstanding in soils. Yet, dramatic decreases of decomposition and fungal biomass due to bark beetle attack were reported in *Picea abies* (L.) Karst., together with a shift in the dominance from ectomycorrhizal to saprotrophic fungi (Štursová et al., 2014). Similar fungal shifts and important losses of carbon stored in soil fungal biomass were also reported in lodgepole pine forests of Canada after different abiotic disturbances (Rodriguez-Ramos et al., 2020), as well as an increased frequency of arbuscular mycorrhizal fungi in the case of wildfire.

Soil properties are particularly affected by disturbance causing an indirect general reduction of fungal diversity, biomass and activity, and deep changes on the cycling of nutrients and the vegetation recovery (Pérez-Valera et al., 2020; Pérez-Izquierdo et al., 2020a,b). The degradation of soils after disturbance, and in particular of organic horizons, significantly alters fungal community structure, the dominance of certain fungal groups, and trade-offs and functions of the different fungal guilds. For example, in maritime pine forests in Spain, high phylogenetic diversity (overdispersion) of the ectomycorrhizal community was found in burned sites, while low phylogenetic diversity (clustering) was observed in unburned control sites (Rincón et al., 2014), likely indicating a strong habitat filtering exerted by fire. Fire causes drastic shifts in the species composition of fungal communities, particularly affecting ectomycorrhizal fungi and deeply alters the cycling of nutrients (Rincón et al., 2014; Pérez-Izquierdo et al., 2020a,b). Severe crown fires usually lead to replacement of ectomycorrhizal- and litter-associated fungi by stress-tolerant ascomycetes, both in Mediterranean (Pérez-Izquierdo et al., 2020a; Rincón and Pueyo, 2010; Rincón et al., 2014) and Boreal (Pérez-Izquierdo et al., 2020b) forest ecosystems. Likewise, a prevalence of ectomycorrhizal ascomycetes, e.g., *Cenococcum geophilum* Fr., is usually found in drought-prone forests (Querejeta et al., 2009; Taniguchi et al., 2018). Some fungal traits such as hydrophobicity, melanin, and mycelium exploration types have been associated with the protection of roots against stress and desiccation (Treseder and Lennon, 2015). In maritime pine forests, ectomycorrhizal fungi of short-distance exploration type were more abundant under dry summer conditions than those of long-distance exploration type that predominated during wet periods (Castaño et al., 2018). Interestingly, the ectomycorrhizal fungal species are distributed in different mycelial exploration types (Agerer, 2001) and we will illustrate in Section 3 of this chapter how these different fungal structures could be adapted to soil exploration for the mobilization of nutrients and water.

Shifts in the abundance and biomass of specific fungal groups after disturbance may have multiple consequences on the ecosystem functioning, including nutrient cycling and carbon dynamics (Holden and Treseder, 2013; Treseder and Lennon, 2015). A dominant imbalance of saprotrophs over ectomycorrhizal fungi may lead to accelerated decomposition, altered nitrogen pools, and reduced possibilities for seedling establishment, which can be critical for forest regeneration and dynamics and should be taken into account in postdisturbance forest recovery programs.

2.3 Biogeography and spatial distribution across different scales

In forest soils, saprotrophic free-living fungi typically dominate litter layers, where freshly cellulose-rich aboveground litter is deposited on the surface, while mycorrhizal fungi typically dominate at lower depths, in the well-decomposed humic layer where tree roots abound, and in the mineral soil. This spatial structure is thought to be the result of niche partitioning through the substrate use specialization and/or competitive exclusion (Bödeker et al., 2016). Saprotrophic fungi are more efficient in colonizing and degrading energy-rich litter (Voříšková and Baldrian, 2012) and they outcompete mycorrhizal fungi from the upper part of the forest floor. As decomposition continues, the substrate becomes depleted in available energy, carbon to nitrogen ratio decreases, and saprotrophs become less competitive, opening a niche that is occupied by mycorrhizal fungi which are not dependent on litter-derived energy (Lindahl et al., 2007). Although this pattern is common in boreal (Lindahl et al., 2007; Kyaschenko et al., 2017a,b), temperate (Voříšková et al., 2014; Carteron et al., 2020), and tropical (McGuire et al., 2013) forests, it is more or less evident depending on plant-soil interactions determined by the type of symbiosis (arbuscular or ectomycorrhizal), climate, belowground carbon allocation, and nutrient availability. Ectomycorrhizal fungi, with access to host-derived carbon, are able to perform energy-demanding oxidation of recalcitrant organic matter to retrieve nitrogen and phosphorus enhancing the depletion of nutrients (Lindahl and Tunlid, 2015), while arbuscular mycorrhizal fungi lack the enzymatic capability and normally scavenge for nutrients released by saprotrophic microorganisms (Read and Perez-Moreno, 2003). Then, in nitrogen-limited boreal and temperate forests where the recirculation of nutrients is based on ectomycorrhizal fungi, the vertical stratification seems to be more evident. On the contrary, in forest systems where the mineralization rates and the availability of nutrients are higher, such as arbuscular mycorrhizal-dominated forest, this transition is less abrupt. Accordingly, in a hemi-boreal carbon to nitrogen gradient, Sterkenburg et al. (2018) found that litter-associated saprotrophic fungi inhabited lower humus layers on the fertile end of the gradient, suggesting a relaxation of competition for nitrogen between fungal guilds. In the same way, Carteron et al. (2020) observed that saprotrophic fungi tended to be more abundant in organic horizons of mixed and arbuscular mycorrhizal temperate forests compared to ectomycorrhizal temperate forests. Likewise, some studies have shown ectomycorrhizal fungal growth on litter, mainly in temperate forests (Buée et al., 2007; Anderson et al., 2014).

Vertical variation of ectomycorrhizal and arbuscular mycorrhizal fungal communities between soil horizons has been also observed (Dickie et al., 2002; Rosling et al., 2003; Courty et al., 2008; Bahram et al., 2015). In a litterbag transplant experiment, Bödeker et al. (2016) observed that different ectomycorrhizal fungi showed different colonization patterns depending on both substrate quality and the depth at which the substrates were incubated. These results support previous works that indicate the existence of vertical niche separation within ectomycorrhizal fungal communities (Dickie et al., 2002; Tedersoo et al., 2003), probably due to changes in organic matter quality, nutrient availability, moisture, and pH (Baier et al., 2006), and also indicate substrate-dependent competitive behavior between ectomycorrhizal fungi (Mujic et al., 2016). The niche differentiation by spatial partitioning of the forest floor appears to be one of the most important factors in structuring communities of mycorrhizal fungi in forests (Bahram et al., 2015) and to contribute to the high diversity of ectomycorrhizal fungal communities observed in forests (Dickie, 2007). Spatial segregation between mycelium and tips of ectomycorrhizal species has been also observed (Genney et al., 2006; Anderson et al., 2014) that seems to indicate competition for root tips (Kennedy and Bruns, 2005). Different relative abundances of root tips with depth and different root distribution among plant species may additionally amplify the depth effect (Dickie, 2007; Ishida et al., 2007). Vertical segregation between arbuscular and ectomycorrhizal roots has been also found down in a tropical forest (Moyersoen et al., 1998) and vertical distribution of soil niches is one of the mechanisms proposed to allow coexistence of arbuscular and ectomycorrhizal fungi (Neville et al., 2002).

Fungal communities in forest soils also exhibit horizontal spatial distribution. Mycelial individuals can vary from a few micrometers after spores germination to several meters (Douhan et al., 2011), or even hundreds of meters as it is the case of genets from the saprotrophic fungi *Armillaria* (Ferguson et al., 2003). Compared to Ascomycetes which use local resources, filamentous Basidiomycetes, irrespectively of their ecological strategy, are better adapted to search and allocate new resources foraging at a larger spatial scale (Boddy, 1999; Lindahl and Olsson, 2004; Genney et al., 2006). They form larger genetic individuals than Ascomycota, to the point to be classified as macroorganisms (Bahram et al., 2015). Communities of ectomycorrhizal fungi usually exhibit an autocorrelation range of 2–3 m (Lilleskov et al., 2004; Bahram et al., 2011, 2013, 2016; Pickles et al., 2012), although individuals can reach tens of meters as it has been shown for species of the genera *Cortinarius*, *Pisolithus*, *Rhizopogon*, *Russula*, *Suillus*, or *Xerocomus* (Douhan et al., 2011). The structure and size of extraradical mycelium in soil varies greatly among ectomycorrhizal taxa given their different exploration types (Agerer, 2001). In a temperate forest, Bahram et al. (2016) observed a stronger spatial structure of biotrophic ectomycorrhizal and pathogen fungi (<2 m) compared with saprotrophs, and the pattern corresponded to vegetation patchiness. Spatial patterns of soil fungi are well known to depend on habitat type (Bahram et al., 2013, 2015). As such, these authors showed that the spatial autocorrelation of ectomycorrhizal fungi is greater in tropical than nontropical forests (exceeding 10 m), potentially due to the greater isolation of hosts and dispersal limitation in these ecosystems, where they form monodominant stands. They also showed that the spatial autocorrelation increased with the tree age, likely related to the dominance of late successional colonizers in older forests where more stable conditions for expansion exist. In a metaanalysis, Bahram et al. (2015) reported the spatial autocorrelation range of arbuscular fungi varying from 6 to 9 m and Maherali and Klironomos (2012) showed that one fungal arbuscular genetic individual can extend over 10 m. On the contrary, Kohout et al. (2017) did not observe spatial autocorrelation among ericoid mycorrhizal fungal communities. In the same way, Oja et al. (2017) did not observe spatial autocorrelation in orchid mycorrhizal fungal distribution, but contrarily, Voyron et al. (2017) reported significant horizontal spatial autocorrelation of these fungi in soil, at distances up to 10 m. They also found a significantly higher frequency of spatial autocorrelation in orchid than in nonorchid ceratobasidioid fungi, probably related to different dispersal patterns or different trophic strategies. Discrepancies among studies might be related to habitat type. Bahram et al. (2015), in their metaanalysis, also calculated horizontal variation of mycorrhizal fungal communities separately for different soil horizons. They observed stronger community variation in topsoil compared with lower horizons. Similarly, Štursová et al. (2016) found higher dissimilarity in fungal communities in litter compared with soil. The higher variation in topsoil seems to be related to greater environmental heterogeneity, litter input from localized vegetation, climatic seasonal changes and disturbances, including grazing by soil fauna (Duan et al., 2009; Bahram et al., 2015; Štursová et al., 2016).

One of the main suggested theories to explain the high diversity or co-occurrence of fungi in forest is niche differentiation (Dickie, 2007). As mentioned before, by occupying different soil vertical and horizontal depth profiles multiple fungal species are able to coexist. Other niche factors include host specificity (Ishida et al., 2007; Tedersoo et al., 2008), proximity to other host trees (Bahram et al., 2011), seasonality (Buée et al., 2005; Koide et al., 2007), or soil nutrient concentration and pH (Bahram et al., 2016). At the local scale, Bahram et al. (2016) showed that saprotrophic fungi were mainly related to soil variables while biotrophic fungi, in particular ectomycorrhizal and pathogens, were highly influenced by the neighborhood effect of trees and understory vegetation. Moreover, other factors such as trophic interactions and disturbances as well as stochastic processes, including priority effects and dispersal limitation influence fungal diversity at the local scale (Kennedy et al., 2009; Wardle, 2006; Bahram et al., 2013). Traditionally, fungi were assumed to be cosmopolitan; however, studies from last decade indicate that fungal communities exhibit strong biogeographic patterns. Talbot et al. (2014) showed that 85% of fungal taxa were only found within a single North American bioregion. In a global study, Tedersoo et al. (2014) showed that biogeographic ranges seem to be driven by climate and dispersal limitation. They concluded that climatic factors, followed by edaphic factors such as pH, are the best predictors of soil fungal richness and community composition at the global scale. Their results also pointed to a higher endemism of fungi in tropical regions compared with higher latitudes. Moreover, similarities in fungal species among distant continents reflect relatively efficient long-distance dispersal capabilities. According to these previous studies, Egidi et al. (2019) found that only 83 phylotypes, mostly belonging to generalist Ascomycota (<0.1% of the fungi), dominate soils globally.

Species diversity usually decreases with latitude and elevation, but it seems not to be a clear pattern in the case of fungi (Coince et al., 2014; Rincón et al., 2015; Tedersoo et al., 2014). At the global scale, ectomycorrhizal fungal richness displays an unimodal relationship with latitude, which peaks in temperate forest (Kennedy et al., 2012; Tedersoo et al., 2012b) and a stronger effect of latitude than longitude (Bahram et al., 2013). Ectomycorrhizal symbiosis dominates forests at high latitudes and elevation resulting from cold and dry climates, while arbuscular mycorrhizal trees dominate in warm tropical forests and in temperate forests where they occur with ectomycorrhizal trees (Steidinger et al., 2019). The 80% of all terrestrial plant species are involved in arbuscular mycorrhizal symbiosis while only 2% form ectomycorrhizal symbiosis. As such, like fungal pathogens, the distribution of ectomycorrhizal fungi is directly influenced by the distribution and density of their host plants. In an environmental gradient in Europe, van der Linde et al. (2018) concluded that environmental and host factors explain most of the variation in ectomycorrhizal fungal diversity. In the case of arbuscular mycorrhizal fungi, these fungal species produce big spores that limit their short-distance dispersal and they tend to be phylogenetically clustered within sites (Kivlin et al., 2011). However, despite this limited dispersal ability, Davison et al. (2016) found a cosmopolitan distribution pattern of arbuscular fungal taxa at global scale, suggesting that the biogeography of arbuscular fungi is driven by efficient dispersal, probably by both abiotic and biotic vectors. As for arbuscular mycorrhizal fungi, the major fungal families that associate with orchids (Tulasnellaceae, Ceratobasidiaceae, and Serendipitaceae) seem to be ubiquitous at global scale (Tedersoo, 2017) and limited information about distribution of ericoid mycorrhizal fungi is available (Kohout, 2017).

3. Roles of soil fungi in forest ecosystems

3.1 Soil fungi and drought

The filamentous growth form of fungi enables them to handle water stress in a different way than unicellular soil organisms. Whereas the water availability of prokaryotes, protozoa, and nonfilamentous yeast fungi depends strictly on the water potential of their local environment, i.e., the soil pore in which they reside, mycelial development enables fungi to transport water across considerable distances and through air-filled pore spaces. On a microscopic scale, fine fungal hyphae may grow into narrow soil pores, where water is retained under drought, and withdraw water that may be transported through hyphae to sustain survival and growth of mycelium in larger pores with low water potential. On a larger spatial scale, soil fungi may reallocate water throughout their mycelia, to remain active in dry environments (Guhr et al., 2015). Water transport may occur via symplastic pathways, by reallocation of cytoplasm, but apoplastic transport by "wicking" in cell walls is likely to be more efficient (Allen, 2009). Basidiomycetes, including wood decomposers, litter fungi, and ectomycorrhizal species that are particularly common in forest soils, have often morphologically differentiated mycelia with cell walls impregnated in hydrophobic proteins. Frequently, hyphae assemble in parallel arrays, forming larger apoplastic conduits, in which water may flow rapidly across large distances. Such alignments of microscopic hyphae into bundles that often are visible by the naked eye are called mycelial cords or strands (Cairney, 1992). Thus, in theory, filamentous soil fungi should occupy a broader niche in terms of drought tolerance than unicellular organisms, with cord-forming basidiomycetes being the most resilient. Yet, soil fungi in a subtropical forest were more severely affected by experimental reductions in dry-season precipitation than bacteria, in terms of community diversity (Zhao et al., 2017; He et al., 2017), and arbuscular mycorrhizal fungi were sensitive to low soil water content (Maitra et al., 2019).

In forest soils, trees and fungi may respond to drought by reallocating biomass to deeper, moister soil horizons (Børja et al., 2017), but the taproots of trees should be better adapted than fungal mycelia to access deep water reservoirs. While ectomycorrhizal fungi may transport water from wet soil patches to cover the water demands of their host plant in laboratory systems (Duddridge et al., 1980), deeper root penetration should be more efficient than ectomycorrhizal associations in increasing tree access to water in the field. Thus a direct importance of fungi in alleviating drought stress in forest trees seems questionable (Lehto and Zwiazek, 2011). However, the drier surface layers are enriched in organic matter and nutrients, and reallocation of root growth to deeper soil may have negative consequences for tree nutrition. Trees malnutrition under drought stress affects photosynthesis negatively, which decreases resources available to mycorrhizal symbiosis and drought adaptations, leading to a positive feedback with fatal consequences (Leon-Sanchez et al., 2018). Mycorrhizal fungi are essential facilitators of tree nutrient uptake, and it is likely to be of critical importance for drought-exposed trees to maintain mycorrhizal activity in the dry surface layers. In the soil of a Mediterranean pine forest in Spain, fungal communities were relatively enriched in ectomycorrhizal species in the driest habitats and during the driest seasons (Castaño et al., 2018). The same pattern was observed in subtropical forest in China after experimental intensification of seasonal drought (Zhao et al., 2017; He et al., 2017). In both studies, the relative abundance of other fungi i.e., molds, yeasts, and saprotrophic ascomycetes, decreased under intense drought, suggesting that mycorrhizal fungi are less sensitive than free-living fungi. Access to host-derived sugars may be particularly advantageous under drought if poor functioning of hydrolytic enzymes hampers mobilization of metabolic resources from soil organic matter. Drought resistance of ectomycorrhizal fungi may also be related to the well-developed capacity of basidiomycetes to redistribute scarce water resources in their mycelia. From a fungal perspective, living roots may be viewed as moist islands in dry soils, potentially constituting a critical source of water for ectomycorrhizal fungi. With access to water in deeper soil horizons, trees may redistribute water by hydraulic lift to maintain a high water potential in surface roots. Particularly during night, when canopy transpiration and soil evaporation is lower, ectomycorrhizal fungi may exploit the roots as a water source to sustain proliferation of extraradical mycelium in the dry surface soils (Querejeta et al., 2003, 2007). Fungi may even release water droplets at their hyphal tips, supplemented with carbohydrates and amino acids to stimulate biological activity and nutrient mobilization in the soil, essentially using root-derived water to "wet and activate" the soil (Sun et al., 1999). Thus, in ectomycorrhizal forests, fungal redistribution of water *away from roots*, maintaining nutrient mobilization in dry surface soils and indirectly alleviating drought stress, may be more important than direct water transport *toward root* (Allen, 2007, 2009).

Ectomycorrhizal fungal species are frequently assigned to different mycelial exploration strategies (Agerer, 2001). Long- and medium-range exploration types have mycelium that differentiates into hydrophobic mycelial cords behind a hyphal front that explore the soil for resources, potentially far from the roots. Short-range types, in contrast, have hydrophilic, nonaggregated hyphae that colonize soil in the closer vicinity of the root tips. Intuitively, one would think that ectomycorrhizal fungi with hydrophobic mycelium and efficient apoplastic transport would be better adapted to drought. However, in Californian (Gehring et al., 2017) and Spanish (Castaño et al., 2018) Mediterranean pine forests, drought-adapted pine genotypes as well as pines in dry conditions associated preferably with short-range exploration types (*Geopora*, *Inocybe*), whereas long-range types (Suillaceae) were more common in drought-sensitive genotypes and in moister conditions. If root tips constitute the principal water resource for the fungi in an otherwise dry soil environment, it is plausible that mycelial production in close vicinity to the root is more advantageous than far-ranging explorative growth, which requires water transport across large distances. Furthermore, long- and medium-range types have been proposed to be more expensive in terms of host carbohydrates (Lilleskov et al., 2011), which are scarce when photosynthesis is constrained by drought.

3.2 Soil fungi and decomposition

Decomposition of organic inputs to soils has been described as a gradual transformation of initially intact plant tissues, via particulate organic matter to low molecular size compounds of microbial origin that associate more or less tightly with mineral particles (Lehmann and Kleber, 2015). Soil bacteria may be superior to fungi in depolymerization and utilization of pure cellulose (Štursová et al., 2012) and often proliferate in freshly fallen litter (Moore-Kucera and Dick, 2008). Still, litter decomposition, i.e., the conversion of plant tissues to particulate organic matter, seems primarily to be a fungal affair. In *Picea* litter, 70% of microbial gene transcripts were of fungal origin and 80% of fungal RNA was ascribed to basidiomycetes (Žifčáková et al., 2015). Fungal assimilation of ^{13}C in labeled *Pinus* litter increased in relation to bacterial C uptake during the 9 initial months of decomposition, corresponding to 60% mass loss, but decreased gradually during later decomposition stages (Moore-Kucera and Dick, 2008).

Fungal hyphae are well adapted to penetrate fresh plant tissues and proliferate within new litter substrates. Some fungi may colonize living plant tissues, providing a priority advantage when the plant dies and the fungi switch from endophytic strategies to become saprotrophs (Kohout et al., 2018). Whereas unicellular microorganisms have to adapt to the conditions of their surroundings, mycelial transport of resources provides fungi unique opportunities to condition their environment according to their preferences (Lindahl and Olsson, 2004). Mycelial transport of water may enable colonization of dry substrates, and fungi may also transport other resources, such as nitrogen and phosphorus, which are often scarce in plant litter at early stages of decomposition. Wells et al. (1998) showed that soil basidiomycetes imported phosphorus into recently colonized wood units. Similarly, Boberg et al. (2014) found that when mycelia interconnected recently deposited needles and the underlying, decomposing litter, fungi reallocated nitrogen from the old litter to accelerate proliferation and decomposition in the fresh, nitrogen poor litter at the surface.

The crystalline structure of cellulose and its interaction with nonhydrolyzable lignin implies that a major fraction of the macromolecules in forest litter may be protected from hydrolytic depolymerization (Baskaran et al., 2019). Fungal decomposition of crystalline cellulose in litter is initiated by lytic polysaccharide monooxygenases (Barbi et al., 2019), which are oxidative enzymes that can cleave chains on the surface of crystalline cellulose, providing access for subsequent depolymerization by hydrolytic cellulases (Johansen, 2016). These enzymes are widely spread among fungi and prokaryotes. However, decomposition of cellulose is hindered by its interaction with lignin and other phenolic plant protective compounds. Fungi in the basidiomycete class Agaricomycetes have unique capacities to oxidize phenolic macromolecules, such as lignin, tannins, and melanins, using potent peroxidases that often depend on manganese to oxidize organic target molecules in an unspecific manner (Sinsabaugh, 2010; Floudas et al., 2012; Keiluweit et al., 2015). Without such enzymes, decomposition of needle litter is suppressed at an early stage (Barbi et al., 2019), so extracellular peroxidases provide agaricomycete decomposers exclusive access to the vast energy resource that recalcitrant plant litter constitutes in forests. In boreal forest, organic matter accumulation in the topsoil is regulated by the activity of Mn peroxidases (Kyaschenko et al., 2017a; Stendahl et al., 2017). The combination of ligninolytic enzymes and a well-developed capacity to reallocate nutrients through mycelial cords to enable efficient exploitation of litter components should imply that agaricomycetes increase their importance in the soil system as the plant economy is shifted from acquisitive traits (low lignin) to conservative traits (high lignin) (Freschet et al., 2012). The generally more conservative traits of forest trees are also related to higher fungal to bacterial ratios in forest soils compared to for example grasslands (Fierer et al., 2009).

The fungal advantage provided by dynamic hyphal growth and efficient oxidative enzymes is likely to decline as decomposition progresses. The integrity of plant tissues breaks up, lignin and other plant phenols are gradually oxidized and disrupted, and respiratory losses of carbon decrease the energy content of the substrate, but nutrients accumulate in dead hyphae, making well-decomposed substrates more suitable for unicellular microorganisms than the original plant litter. Disruptive activities of soil fauna also play a central role in regulating the development of microbial decomposer communities. Crowther et al. (2013) demonstrated how grazing of basidiomycete mycelium by isopods decreased fungal decomposition but stimulated opportunistic molds and yeasts. Thus filamentous soil fungi seem to dominate decomposition of particulate organic matter of plant origin, but prokaryotes and yeasts are likely to be more important for the turnover of older, mineral-associated soil organic matter of microbial origin, with soil fauna accelerating the transition between the two decomposer systems.

3.3 Soil fungi and nutrient cycling

Given that fungi generally have a lower nitrogen content in their tissues than bacteria, one would expect them to be less prone than bacteria to immobilize nitrogen in their biomass and more likely to release nitrogen by mineralization. However, priority access to fresh litter feeds large quantities of carbon from cellulose into fungal communities, and mycorrhizal associations with living roots provide a direct input of recently photosynthesized sugars. Such local carbon inputs may then be redistributed throughout the entire mycelia. Thereby, nutrient hotspots can be met by reallocation of carbon to balance stoichiometric availability against local demand. Redistribution of carbon enables filamentous fungi to minimize mineralization losses, even when utilizing substrates with low carbon-to-nitrogen ratios (Boberg et al., 2010). Reciprocally, a local nitrogen surplus may be redistributed to mycelium that experiences shortage, e.g., in high carbon to nitrogen litter components, or to sinks created by mycorrhizal host plants. In this way, the connectedness of mycelia combined with efficient exploitation of carbon sources in the form of live or dead plants enables soil fungi to be conservative with nutrients (Boberg et al., 2014) and less prone to mineralize nitrogen than unicellular microorganisms, in spite of their lower nutrient content (Lindahl et al., 2002).

Mycorrhizal symbioses are generally viewed as an adaptation of plants to increase nutrient extraction from soils (Hodge, 2017). Mycorrhizal fungi may access chemical forms of nutrients that are inaccessible to plants and share them with their hosts. They may stimulate weathering of minerals to release soluble nutrients, essentially phosphate. Weathering rates are increased by exudation of organic acids and removal of weathering products, shifting chemical equilibrium toward soluble forms (Smits and Wallander, 2017; Finlay et al., 2020). Although arbuscular mycorrhizal fungi have little capacity to directly mobilize organic forms of nutrients, they may facilitate release of organic phosphorus indirectly by transporting phosphorus-solubilizing bacteria to organic substrates and stimulate their proliferation in the patch (Jiang et al., 2020).

Ectomycorrhizal fungi may produce a variety of extracellular enzymes that engage in depolymerization of organic macromolecules, which are inaccessible for plant roots without fungal aid (Abuzinadah and Read, 1989; Read and Perez-Moreno, 2003). However, in forest soils, organic nitrogen-containing substrates, such as proteins or chitin, often occur in forms that are unavailable for enzymatic hydrolysis. Free-living saprotrophic fungi have priority access to fresh plant litter (Lindahl et al., 2007), and their activity leads to the incorporation of nitrogen into recalcitrant complexes (Baskaran et al., 2019), hampering hydrolytic nutrient mobilization by mycorrhizal successors. Many soil fungi with pigmented mycelium contain melanin and after death, they form necromass (i.e., dead fungal tissues) in which nitrogen is protected from hydrolytic decomposition (Fernandez et al., 2016). These fungal residues are composed of polymeric substances, such as chitin, glucans, lipids, and a wide range of proteins (Akroume et al., 2019). Interestingly, soil chitin is derived mainly from fungal and arthropod necromass, and this biopolymer is the most abundant aminopolysaccharide in nature (Kumar, 2000). Further, tannins from plant roots may interact with proteins and chitin in fungal mycelium, protecting them from hydrolytic decomposition (Adamczyk et al., 2019). Nitrogen-containing molecules may also bind tightly to iron oxides on mineral surfaces (Wang et al., 2020). Mobilization of these recalcitrant and locked-up nitrogen stocks requires transformation by oxidation, which disrupts complexes, decreases molecular sizes, introduces polar groups that increase solubility and sometimes even oxidize carbon all the way to CO_2 (Šnajdr et al., 2010). Certain groups of ectomycorrhizal agaricomycetes have retained the potent capacities of their saprotrophic ancestors to oxidize organic matter. Some groups, notably *Cortinarius* species, oxidize organic matter in the purely organic top soils of boreal and subarctic forests, producing the same type of manganese-dependent peroxidases that facilitate lignin decomposition in wood and litter decomposers (Bödeker et al., 2009, 2014; Kyaschenko et al., 2017b). Other ectomycorrhizal fungi, particularly species in the Boletales, depend on iron-mediated Fenton reactions to generate hydroxyl radicals, which oxidize a wide range of organic targets in an unspecific manner (Op De Beeck et al., 2015; Rineau et al., 2012; Shah et al., 2016). Both mechanisms depend on external energy in the form of (host-derived) sugars (Rineau et al., 2013; Sterkenburg et al., 2018) and are induced by low levels of inorganic nitrogen (Bödeker et al., 2014; Op De Beeck et al., 2018; Nicolas et al., 2019). Evidence is, thus, accumulating that forest trees may respond to nitrogen shortage by allocating sugars to their associated ectomycorrhizal fungi and stimulate them to decompose organic matter, using a variety of oxidative mechanisms, to release nitrogen locked up in recalcitrant organic stocks (Lindahl and Tunlid, 2015). This process may be interpreted as a "priming effect," whereby input of labile carbon primes decomposition of old, recalcitrant stocks (Carney et al., 2007).

Soils are chemically complex and spatially heterogeneous (Schmidt et al., 2011), and phosphate and organic compounds bind tightly to mineral particles, in particular to iron oxides (Regelink et al., 2015). Adsorption hampers interactions between extracellular enzymes and their substrates, not only because the substrates are tightly bound, but also by incapacitation of enzymes that bind to mineral surfaces. Some ectomycorrhizal fungi have even been shown to enhance enzymatic depolymerization by excreting organic compounds that adhere to mineral surfaces, reducing their reactivity and mitigating enzyme adhesion (Wang et al., 2020). This is another example of how fungi benefit from their capacity to reallocate resources and turn mycelium into a local carbon sink to exploit nutrient hotspots (Lindahl and Olsson, 2004). Soil fungi are, thereby, fundamentally different from unicellular, saprotrophic soil microorganisms, which cannot endure a negative balance between carbon expenditure and acquisition, even locally. Hyphal transport enables acquired resources to be transported intracellularly, protected from adsorptive mineral surfaces, with diffusion rates speeded up by active, energy demanding cytoplasmic streaming of motile vacuoles (Cairney, 1992). Hyphae that extend through air-filled pore spaces further enhance nutrient mobility in dry soils (Allen, 2009; Guhr et al., 2015). Apoplastic nutrient transport in hyphal cords may be exceptionally fast and directional if fungi use osmotic regulation to create mass flow from sources to sinks (Cairney, 1992).

Mycorrhizal associations may enlarge the surface area of the root system that is active in nutrient uptake and increase the affinity of the root system for dissolved nutrients (Sa et al., 2019). In environments where fast decomposition is paralleled by rapid mineralization of nutrients, this benefit of mycorrhiza may be more important than potential access to recalcitrant nutrient pools. More efficient nutrient capture may increase the competitive strength of individual trees, and reduced leaching can increase overall ecosystem productivity. However, fungal mycelium also constitutes a major sink for nitrogen in the soil (Aber et al., 1998; Kyaschenko et al., 2019). The mycorrhizal symbiosis faces the dilemma that the more carbon the plant provides to the fungi, the more mycelium the fungi produce, increasing the proportion of the acquired nutrients retained in fungal biomass (Corrêa et al., 2008; Näsholm et al., 2013). These studies suggested that investment in mycorrhizal symbiosis may actually reduce or suppress plant growth, given a finite supply of available nutrients. Nutrient cycling in ectomycorrhizal systems may be further reduced by direct antagonism (Bödeker et al., 2016) and competition for nutrients between mycorrhizal fungi and free-living saprotrophic decomposers—the Gadgil effect (Gadgil and Gadgil, 1971;

Fernandez and Kennedy, 2016). The idea that ectomycorrhizal fungi might suppress tree growth by restricting saprotrophic activity and immobilizing nutrients depends on the assumption that trees and fungi compete for inorganic nutrients, and that nutrient release from more stable pools is regulated entirely by saprotrophic decomposition and mineralization without any direct influence of the mycorrhizal partnership. As outlined before, however, there are increasing evidence that nutrient cycling may be under direct mycorrhizal control, as energy and carbon from the plant host may be used by fungal partners to access to relatively stable nutrient pools, e.g., through biological weathering or organic matter oxidation. Thus ectomycorrhizal fungi may suppress saprotrophic nutrient release and restrict tree access to inorganic nutrients, but at the same time take control over organic matter turnover and shortcut nutrient cycling by bypassing mineralization (Lindahl et al., 2002).

4. Changing roles of soil fungi in different biomes

The interplay between trees, soil fungi, and ecosystem processes in forests is complex and has to be understood within the context of different biomes, climates, and soil types (Read and Perez-Moreno, 2003; Steidinger et al., 2019). In subarctic birch forests at the Scandinavian tundra transition, Parker et al. (2015) observed that presence of trees more than halved organic matter storage relative to tree-free tundra, and proposed that the ectomycorrhizal symbiosis of the birches increased decomposition rates to such a great extent that the higher litter input under the trees was outbalanced by stimulated turn-over. Similarly, in a northern boreal forest chronosequence, Clemmensen et al. (2013) found evidence that organic matter dynamics was under the control of roots and associated mycorrhizal fungi and linked rapid decomposition to the presence of ectomycorrhizal *Cortinarius* species (Clemmensen et al., 2015), presumably related to their capacity to produce manganese peroxidases that enable oxidation of organic matter (Bödeker et al., 2014). In old ecosystem that had not burned for centuries, abundance of ectomycorrhizal fungi declined and ericoid mycorrhizal fungi become strongly dominant in the soil fungal community. This transition was linked to retention of organic nitrogen belowground in large organic stocks, which decreased tree productivity and led to gradual ecosystem retrogression toward tundra-like stages. Thus regular disturbance, in the form of low-severity wildfires, seems necessary to maintain ectomycorrhizal-driven fertility in boreal forests. However, severe wildfire that killed all trees and more or less extinguished ectomycorrhizal fungi led to a complete loss of manganese peroxidase activity, potentially delaying ecosystem recovery (Pérez-Izquierdo et al., 2020b). Manipulating fungal communities by root severing, Sterkenburg et al. (2018) concluded that positive effects of ectomycorrhizal fungi on decomposition, associated with production of oxidative enzymes, probably override the negative Gadgil effect of competition with saprotrophs (although the latter was still evident). Thus, in boreal forest, where harsh climate and acidic soils constrain decomposition rates, well-developed ectomycorrhizal symbiosis seems like a prerequisite to maintain decomposition, avoiding accumulation of large organic stocks and associated nutrient retention, which lead to declined ecosystem productivity and retrogression toward heath vegetation dominated by ericoid mycorrhiza.

At the boreal-temperate transition, where ectomycorrhizal deciduous trees (*Betula*, *Populus*, etc.) coexist with evergreen conifers, Kyaschenko et al. (2017a) studied regulation of organic matter accumulation in a local fertility gradient and found that increased fertility was linked to higher abundance and activity of saprotrophic agaricomycetes decomposers (*Mycena* and *Galerina* species) in the organic topsoil. Here, in contrast to the boreal situation, dominance of ectomycorrhizal species increased accumulation of organic matter and led to retention of nitrogen in the root zone (Kyaschenko et al., 2019). In milder climates, saprotrophic decomposition and nutrient mineralization rates increase, and the effect of ectomycorrhizal symbiosis on tree growth may gradually shift to negative as the impetus for direct ectomycorrhizal decomposition decreases. But saprotrophic nutrient release is constrained by nutrient immobilization in extraradical mycelium (Corrêa et al., 2008; Näsholm et al., 2013) and competitive Gadgil interactions (Bödeker et al., 2016; Fernandez and Kennedy, 2016). In a theoretical model, Baskaran et al. (2017) showed that the effect of ectomycorrhizal on tree growth depends on the relative involvement of ectomycorrhizal fungi and free-living saprotrophs in decomposition of recalcitrant organic matter, and that the ectomycorrhizal fungi generally hamper tree growth at higher levels of nutrient availability. However, even if the symbiosis hampers tree growth, it may still promote tree fitness and be ecologically stable as the negative influence on cycling of inorganic nutrients may suppress competition from nonectomycorrhizal vegetation (Northup et al., 1995). Franklin et al. (2014) argued that stability in the symbiosis may be maintained in spite of negative effect on overall tree growth, because nutrient immobilization in mycorrhizal mycelium would have even larger negative effects on nonconnected neighbors—an example of the "tragedy of the commons." This reasoning may be elevated to the level of tree species, with depletion of easily accessible nutrients suppressing competing vegetation that do not have access to the mycorrhizal network. Such positive plant-soil feedback, i.e., that plants favor growth of conspecific neighbors, is indeed common among ectomycorrhizal trees, but rare among trees that form arbuscular mycorrhiza (Bennett et al., 2017; Segnitz et al., 2020), presumably due to the lower mycelial biomass of arbuscular mycorrhizal fungi. Results even suggest that arbuscular mycorrhizal type could negatively regulate plant community structure through negative plant-soil feedbacks (Kadowaki et al., 2018; Segnitz et al.,

2020). Differences in protection from antagonists would have contributed to the variation in plant-soil feedbacks between mycorrhizal types. Indeed, arbuscular mycorrhizal seedlings would have developed more lesions than ectomycorrhizal seedlings by antagonists accumulating near conspecifics, possibly because arbuscular mycorrhizal colonization gives less protection (Bennett et al., 2017). Positive plant-soil feedback decreases tree diversity and has been proposed to contribute to the formation of monospecific patches of ectomycorrhizal trees, such as Dipterocarpaceae, in highly diverse tropical forests dominated by arbuscular mycorrhizal species (Corrales et al., 2016; Segnitz et al., 2020). The control of plant diversity by fungi does not primarily depend on mycorrhizal types. The Janzen-Connell hypothesis could explain the maintenance of plant diversity by an accumulation of host-specific natural enemies near reproductive adults, such as insects, oomycetes, or pathogenic fungi (Mangan et al., 2010). These natural enemies could limit host plant species abundance and aggregation (Janzen, 1970; Mills and Bever, 1998; Legeay et al., 2020). Augspurger (1983) monitored seedlings of the tropical tree *Platypodium elegans* Vogel, for 1 year from germination in a Panama forest, surveying causes of mortality as a function of distance from the parent tree and density of seedlings. Both the incidence and the rate of damping-off were inversely correlated with the distance of seedlings from the parent tree. This author suggested that fungal pathogens causing damping-off may increase selection for the dispersal of seeds away from the parent tree, ultimately contributing to the species diversity of the tropical forest. In subtropical forest, Liu et al. (2012) demonstrated that a negative plant-soil feedback on a legume tree resulted from *Fusarium oxysporum*, a host-specific pathogen. The host-specific fungal pathogens, locally accumulated around parent trees, could thus play an important role in tree community structure and the maintenance of plant diversity, particularly in (sub)tropical forests.

The mechanism "tragedy of the commons," which may act to maintain ectomycorrhizal dominance in an ecosystem, collapses as soon as nutrient limitation is replaced by competition for light. Competition then shifts from belowground to aboveground mechanisms and the high carbon expenditure required to uphold ectomycorrhizal symbiosis becomes a competitive disadvantage, relative to arbuscular mycorrhiza. Phillips et al. (2013) proposed general patterns in how carbon and nutrient cycling shifts along with mycorrhizal types in temperate forests. Ectomycorrhizal forest was characterized by high fungal but low bacterial activity and suppressed saprotrophic decomposition (Cheeke et al., 2017). Low rates of nutrient mineralization were compensated by direct enzymatic release of nutrient by ectomycorrhizal fungi (Brzostek et al., 2015). In contrast, arbuscular mycorrhizal forest was characterized by higher bacterial activity with rapid decomposition and nutrient mineralization. These differences are linked to global-scale patterns with larger organic stocks but lower nitrogen concentrations in ectomycorrhizal forests relative to arbuscular mycorrhizal ecosystems (Averill et al., 2014; Steidinger et al., 2019). However, there is an obvious risk that such broad, linear correlations hide important context dependency. Notably, in boreal and subarctic forests, ectomycorrhizal associations are a prerequisite for efficient nutrient cycling and lead to lower organic matter stocks relative to heath and tundra ecosystems (Clemmensen et al., 2013; Parker et al., 2015). Tropical forests also deviate from the global pattern, with ectomycorrhizal forest storing less carbon in the mineral soil, presumably due to lower productivity and organic matter input (Lin et al., 2017).

5. Soil fungi and forest management in a changing world

Like all terrestrial and marine ecosystems, forests are globally strongly disturbed by processes of anthropic degradation (Ellis, 2011). This forest degradation is defined as the loss or the reduction of forest capacities to provide goods and services, such as biomass, carbon sequestration, water regulation, soil protection, and biodiversity conservation (Ghazoul et al., 2015). These rapid disturbances require appropriate and short-term management for the conservation and sustainable use of forests. Moreover, this appropriate policy also needs to be responsive to the future needs of society, as greenhouse gas mitigation or the switch to bioenergy (renewable biomass). In the precedent sections, we have described how forest fungi play basic functions providing their hosts with soil nutrients and water, helping plants to tolerate harsh environmental conditions, and playing a key role in carbon cycle. Here, we describe how soil fungi can improve ecosystem services depending on their incorporation and conservation into forest management or through restoration projects of these ecosystems (Fig. 13.2).

5.1 Restoration activities, plantation, and fungal inoculation in response to global change

In a previous section, we have detailed how various disturbances can impact soil fungal communities. Some of these disturbances have direct or indirect anthropogenic origins, including forest management. Paradoxically, forest practices could facilitate also the ecosystem adaptation to these constraints. Global warming is characterized by an increase in temperature, disturbed precipitation, and an elevated atmospheric CO_2 concentration. These perturbations alter tree phenology by increasing the length of the growing season and the photosynthesis activity (Cleland et al., 2007; Kellomäki and Wang, 1996;

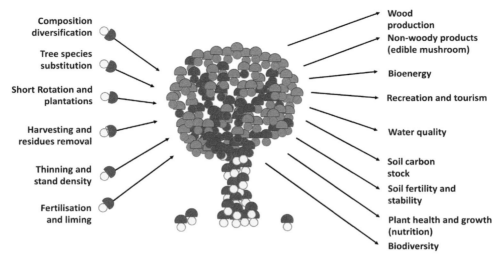

FIG. 13.2 Drawing illustrating the goods and services that tree—fungal interactions can provide under the regulation of various silvicultural practices.

Parmesan and Yohe, 2003; Walther et al., 2002). This supplemental carbohydrate flux can benefit root development and the growth of the mycelial network (Janssens et al., 2005; Norby et al., 2004; Norby and Luo, 2004; Palmroth et al., 2005). Conversely, Steidinger et al. (2019) predicted declines in ectomycorrhizal species richness in North America Pinaceae forests with associated effects on forest ecosystem. If warming of forests in these boreal ecosystems could cause ectomycorrhizal fungal species loss, the increase of greenhouse gas emissions and temperatures of eastern America temperate forests would have the opposite effect. Additionally, tree physiology changes coincide with changes in fruiting phenology of fungi also, with an increasing delay of fruitbodies production (Kauserud et al., 2008). Consecutively, climatic change can strongly change edible mushroom productions also in different areas (Yang et al., 2012; Taye et al., 2016; Parladé et al., 2017). In an experimental study, drought reduced mushroom production by 62% on average in a Spain oak forest, confirming a decrease in mushroom production in the near future, particularly in sensitive regions as for Mediterranean areas (Ogaya and Peñuelas, 2005). Mushrooms are important nontimber forest products, with relatively important economic value in many European rural areas. With a predicted drier climate in numerous areas, it will be necessary to select both tree species more adapted to drought and associated fungal strains for controlled inoculation. De Aragón et al. (2012) reported that the introduction of *Quercus ilex* L. seedlings inoculated with *Tuber melanosporum* Vittad. in perturbated sites can promote reforestation and may provide economic incentives. These authors suggest that reforestation with *T. melanosporum*-inoculated seedlings can be successful after forest fires and they confirm the competitiveness of this fungus within the ectomycorrhizal community in these soils. Restoration activities of perturbed areas have relied often on monospecific tree plantations, such as extensive plantation of *Pinus halepensis* Mill. during the 20th century in Spain (Maestre and Cortina, 2004), but these seedlings are produced in tree nurseries. Inoculating trees with selected ectomycorrhizal strains is an efficient way to improve the growth of forest seedlings and enhance drought tolerance of inoculated trees (Castellano and Trappe, 1985; Garbaye and Churin, 1997; Ortega et al., 2004). In addition, the transplantation stress could be reduced by preinoculating seedlings with selected ectomycorrhizal fungi (Garbaye and Churin, 1997; Rincón et al., 2006). Maltz and Treseder (2015) surveyed articles on the influence of mycorrhizal fungal inoculation on percent colonization of plant roots. They reported that inoculation may influence mycorrhizal colonization and provide benefits to plants in restoration projects. Across ecosystem types, they found that inoculation increased the abundance of mycorrhizal fungi in degraded ecosystems, and consecutively improved the establishment of plants.

5.2 Nitrogen deposition, liming, and other fertilization impact

Human activities are generating different sources of atmospheric reactive nitrogen (e.g., NH_3, NO, NO_2, or peroxyacetyl nitrates) that are deposited on ecosystems and could have important effects on forest ecosystems and their biodiversity. Within an anthropogenic nitrogen deposition gradient in southern California, Egerton-Warburton and Allen (2000) revealed that nitrogen enrichment induced a shift in arbuscular mycorrhizal community composition, with a significant reduction in species richness and diversity related to eutrophication. In addition, the nitrogen enrichment reduced the spore abundance. In forest regions, different authors have also demonstrated changes in composition of fungi, and particularly of ectomycorrhizal fungal community in both sporocarp and root tips (Lilleskov et al., 2001; Peter et al., 2001). Lilleskov et al. (2001,

2019) described two categories of fungi: the "nitrophobic taxa," such as *Cortinarius*, *Tricholoma*, or *Lactarius*, which declined in species richness or abundance with increasing mineral nitrogen and a second group, named "nitrophilic taxa" (e.g., *Laccaria*, *Paxillus*), that were not impacted by nitrogen availability in organic horizon. Because mycorrhizal fungi are the primary drivers of nitrogen dynamics in ecosystems, ectomycorrhizal fungi might respond more sensitively to nitrogen availability. In a large biogeographic study, van der Linde et al. (2018) described the effect of 38 host, environmental, climatic, and geographic variables on ectomycorrhizal diversity. These authors explained that with increasing nitrogen availability, metabolically costly ways of obtaining nitrogen from complex soil organic sources are less cost effective. Consequently, "nitrophobic" fungi that use these pathways (*Cortinarius*, *Piloderma*, and *Tricholoma*) have a disadvantage in comparison to fungi that use inorganic nitrogen such as *Elaphomyces* and *Laccaria* (Lilleskov et al., 2001; van der Linde et al., 2018).

Under certain conditions, fertilization of forest ecosystems is intentional. Indeed, understory removal and/or nitrogen fertilization are important forest management practices in forest plantations (Zhao et al., 2013). The major limiting factor for forest production is soil acidity and a shortage of plant-available phosphorus and potassium. Liming and/or application of specific nutrients, such as phosphorus, potassium, and nitrogen, have been proposed as countermeasures (Sikström, 2001). In addition, heavy-metal contaminated forest soils could be restored by means of liming and fertilization (Derome, 2000). Plantations of *Eucalyptus globulus* Labill. in Western Australia are commonly subjected to fertilization (especially nitrogen, phosphorus, potassium, and copper) to sustain high growth rates. Pampolina et al. (2002) studied the effect of phosphorus fertilization on ectomycorrhizal fungal dynamics in these plantations. Although fungal diversity was not modified, the addition of phosphorus reduced significantly the number of sporocarps and soil hyphal biomass. As illustrated for nitrogen deposition, this study showed the significant role of ectomycorrhizal fungi in phosphorus cycling and other nutrients. Zavišić et al. (2018) used young beech (*Fagus sylvatica* L.) trees in natural forest soil from a phosphorus-rich and phosphorus-poor site to study the impact of phosphorus amendment on mycorrhizas and beech phosphorus nutrition. They found that phosphorus fertilization shaped mycorrhizal fungal assemblages. However, phosphorus concentration did not affect microbial biomass. Because higher fungal taxa diversity in phosphorus-rich soil was correlated with lower plant phosphorus uptake efficiency, authors suggested these results might indicate a higher specialization of fungal taxa colonizing roots in a stressful than in a less stressful soil. The direct calcium and magnesium input to forest soil, which restores tree mineral nutrition, counteracts forest decline in acidic soils. Liming changed the saprotrophic and ectomycorrhizal sporocarp production in different forest stands (Agerer et al., 1998; Rineau et al., 2010). A positive effect of liming on ectomycorrhizal root tip abundance was reported in deciduous and coniferous forests (Bakker et al., 2000; Nowotny et al., 1998), but the opposite effect was observed in a mixed oak and beech forest (Blaise and Garbaye, 1983). Similarly, Rineau and Garbaye (2009a) reported higher abundance of ectomycorrhizal root tips in topsoil layers of the untreated plots compared to treated plots. Moreover, most of the ectomycorrhizal root tips in the limed plots were in the organomineral layer, whatever the tree host, and liming was the major determinant of fungal community structure in spruce and beech forests. The same authors showed that soil liming modified enzymatic activities of ectomycorrhizal community, suggesting changes in their ability to mobilize nutrients from soil organic matter (Rineau and Garbaye, 2009b).

5.3 Thinning, harvesting, and residues removal

In order to contribute to climate change mitigation, most countries aim to decrease technologies and infrastructures that cause fossil carbon emissions and they have increased the use of forest biomass for energy. Alternatively, managers must control the provisioning of the ecosystem service of carbon storage by forests, while maintaining carbon stock in the aboveground biomass. Consequently, the issue of 'carbon neutrality' has been debated with regard to the bioenergy products that are produced from forest biomass (Hanssen et al., 2020; Berndes et al., 2016). In addition, concerns have been raised on the environmental effects and biodiversity impacts of these practices. Because forest managers must find a compromise between a strategy favoring carbon storage in forests and an ever-increasing demand for wood energy, from now on they must include the belowground component of carbon stocks. As detailed before, fungi benefit from their capacity to reallocate plant carbon into their mycelium acting as a carbon sink in soil (Clemmensen et al., 2013).

One strategy proposes the use of fast-growing species, such as Short Rotation Coppice (SRC) of *Populus* or *Eucalyptus*, with an intensification of wood harvesting for the production of manufactured objects and for replacing fossil carbon for energy. Maillard et al. (2019a) evaluated the consequences of this approach, including the effects of residue management harvesting, on soil fungal functions. These authors showed that residue management in *Eucalyptus* plantations might provide disturbances on the activity of soil microorganisms. Among these microbial functions, fungal extracellular enzyme activities decreased significantly in whole-tree harvesting plots in comparison to stem only harvesting plots. This impact increased with soil depth, illustrating the durability of the disturbance. Studies that evaluate the effect of SRCs and the SRC management regime on fungal diversity focused on (ecto)mycorrhizal fungi (Vanbeveren and Ceulemans, 2019), showed large differences between mycorrhizal communities associated with SRC and agricultural land (Pellegrino et al., 2011) or between 6- and 3-year SRC, in poplar or willow plantations (Hrynkiewicz et al., 2010). Consequently, these practices correspond to land use changes and little or no study concerns the conversion of ancient forests to SRC.

An alternative approach suggests to let forest undisturbed and promote tree species efficient at storing carbon in forests. Although the availability of coarse woody debris (CWD) and distribution of dead trees can protect polypores in some unmanaged forest (Vasiliauskas et al., 2004), this nonmanagement is a risky strategy because this pool is highly vulnerable to storms, wildfires, pests, and fungal diseases (Lindroth et al., 2009; Seidl et al., 2018). In their opinion paper, Bellassen and Luyssaert (2014) advocated that this substitution-sequestration dilemma could be resolved thanks to forest practices that increase both the amount of wood produced and the carbon stock retained in forests, in harvest residues and top soil. Indeed, studies recommend that CWD and logging residues should be maintained in managed forests in order to promote the conservation of saprotrophic fungi (Lonsdale et al., 2008). Similarly, in a study conducted at several *Picea abies* forest sites in British Columbia, Hartmann et al. (2009) revealed that harvesting and removal of all organic matter (i.e., whole tree, understory vegetation, scalping of forest floor) affected both mycorrhizal and saprotrophic species of Basidiomycetes and Ascomycetes. Basidiomycete diversity was modified to a minor degree but this intensive practice had stronger effects on some Ascomycetes. Different studies reported no effect on the diversity of ectomycorrhizae one or two years after clear-cutting (Hagerman et al., 1999; Visser et al., 1998). Conversely, Kyaschenko et al. (2017b) demonstrated across a chronosequence of managed *Pinus sylvestris* stands that clear-cutting had a negative effect on ectomycorrhizal fungal abundance and diversity in the decade following harvesting. In contrast, clear-cutting favored saprotrophic fungi proliferation. In a similar study, Varenius et al. (2016) confirmed the majority of common ectomycorrhizal fungi seems to reestablish 30–50 years after tree harvest in boreal Scots pine forests. In a complementary study, Hartmann et al. (2012) demonstrated a significant and long-term impact of harvesting on soil fungi (i.e., more than a decade after tree harvest). Clear-cut harvesting increased fungal evenness and altered fungal community structure in a field experiment replicated at six forest sites in British Columbia, Canada. Authors reported that fungi were more susceptible than bacteria because fungal populations have a patchy distribution. Indeed, it has been hypothesized that organisms with high spatial variability may be more prone to disturbances of their habitat (Vucetich et al., 2000). Symbiotic and saprotrophic fungi appeared to be the most sensitive microbial groups and might be potential indicators for monitoring the recovery of forests. Moreover, these results indicated that fungal communities manifested significant long-term responses to soil compaction, which is a persisting disturbance associated to organic matter removal (Hartmann et al., 2012, 2014).

The negative impacts of CWD and biomass exports on saprophytic fungi are very intuitive. On the other hand, the consequences of this management strategy on ectomycorrhizal communities are still controversial. Walker et al. (2012) showed that retention of CWD had no detectable effect on taxon richness, evenness, or diversity of ectomycorrhizal fungi on root tips, i.e., on functional organs of ectomycorrhizal symbiosis. However, these authors confirmed a shift in ectomycorrhizal species composition, with a significantly higher abundance of Amphinema byssoides (P.Karst.) P.Karst. root tips in response to CWD removal. Mahmood et al. (1999) studied the effect of repeated forest residue harvesting on ectomycorrhizal root tip community structure. They showed no shift in ectomycorrhizal community structure as a result of harvesting and suggested that repeated removal of forest residues might have a strong effect on the quantity and development of ectomycorrhizal roots in the organic horizon, but little effect on the species composition. Conversely, another study reported long-term decline in diversity of ectomycorrhizal fungi after organic matter removal (Wilhelm et al., 2017). Finally, Maillard et al. (2019b) showed that intensive organic matter removal (leaves and residual wood) decreased the microbial functions involved in the mobilization of nutrients, including fungal activities. Nevertheless, in these same forest sites, Maillard (2018) observed an increase in the relative abundance of ectomycorrhizal fungal species 3 years after an intensive organic matter removal.

Although unthinned forests would have the higher carbon stock in these ecosystems (Garcia-Gonzalo et al., 2007), thinning regimes are essential to promote the growth of trees and give them better resistance to climatic or biotic stress. Interestingly, Buée et al. (2005) demonstrated that thinning had no impact on the richness of ectomycorrhizal root tips, nor on the composition of these communities. Nevertheless, in a thinning experiment, Egli et al. (2010) revealed that the growth reaction of trees was associated with changes in fruiting phenology of fungi, especially for the ectomycorrhizal fruitbodies production. Indeed, the fruitbody number increased significantly after thinning, leading to a significantly positive correlation between fruitbody numbers and tree-ring width. This phenology change in response to thinning has been applied to the production of edible mushrooms such as *Lactarius deliciosus* L. ex. Fr. Gray (saffron milk caps) in Spain forests (Bonet et al., 2012). The authors showed a positive response of *L. deliciosus* production to forest thinning. Production was five times greater in plots in the first year after thinning and two times greater in the second year, as compared to the nonthinned plots. All together, these studies illustrate the difficulty of finding compromises in forest management in order to produce biomass while providing various ecosystem services (Fig. 13.2).

References

Aber, J., Mcdowell, W., Nadelhoffer, K., Magill, A., Berntson, G., Kamakea, M., McNulty, S., Currie, W., Rustad, L., Fernandez, I., 1998. Nitrogen saturation intemperate forest ecosystems. Bioscience 48, 921–934.

Abuzinadah, R.A., Read, D.J., 1989. The role of proteins in the nitrogen nutrition of ectomycorrhizal plants: IV. The utilization of peptides by birch (*Betula pendula* L) infected with different mycorrhizal fungi. New Phytol. 112, 55–60.

Adamczyk, B., Sietiö, O.M., Biasi, C., Heinonsalo, J., 2019. Interaction between tannins and fungal necromass stabilizes fungal residues in boreal forest soils. New Phytol. 223, 16–21.

Agerer, R., 2001. Exploration types of ectomycorrhizae—a proposal to classify ectomycorrhizal mycelial systems according to their patterns of differentiation and putative ecological importance. Mycorrhiza 11, 107–114.

Agerer, R., Taylor, A.F.S., Treu, R., 1998. Effects of acid irrigation and liming on the production of fruit bodies by ectomycorrhizal fungi. Plant Soil 199, 83–89.

Akroume, E., Maillard, F., Bach, C., Hossann, C., Brechet, C., Angeli, N., Zeller, B., Saint-André, L., Buée, M., 2019. First evidences that the ectomycorrhizal fungus Paxillus involutus mobilizes nitrogen and carbon from saprotrophic fungus necromass. Environ. Microbiol. 21, 197–208.

Allen, M.F., 2007. Mycorrhizal fungi: highways for water and nutrients in arid soils. Vadose Zone J. 6, 291–297.

Allen, M.F., 2009. Water relations in the mycorrhizosphere. In: Lüttge, U., Beyschlag, W., Büdel, B., Francis, D. (Eds.), Progress in Botany. vol. 70. Springer, Berlin, Heidelberg.

Anderegg, W.R.L., Trugman, A.T., Badgley, G., Anderson, C.M., Bartuska, A., Ciais, P., Cullenward, D., Field, C.B., Freeman, J., Goetz, S.J., et al., 2020. Climate-driven risks to the climate mitigation potential of forests. Science 368, 6497.

Anderson, I.C., Genney, D.R., Alexander, I.J., 2014. Fine-scale diversity and distribution of ectomycorrhizal fungal mycelium in a Scots pine forest. New Phytol. 201, 1423–1430.

Aponte, C., García, L.V., Marañón, T., Gardes, M., 2010. Indirect host effect on ectomycorrhizal fungi: leaf fall and litter quality explain changes in fungal communities on the roots of co-occurring Mediterranean oaks. Soil Biol. Biochem. 42, 788–796.

Augspurger, C.K., 1983. Seed dispersal of the tropical tree, Platypodium elegans, and the escape of its seedlings from fungal pathogens. J. Ecol., 759–771.

Averill, C., Turner, B.L., Finzi, A.C., 2014. Mycorrhiza-mediated competition between plants and decomposers drives soil carbon storage. Nature 505, 543–545.

Bahram, M., Põlme, S., Kõljalg, U., Tedersoo, L., 2011. A single European aspen (*Populus tremula*) tree individual may potentially harbour dozens of Cenococcum geophilum ITS genotypes and hundreds of species of ectomycorrhizal fungi. FEMS Microbiol. Ecol. 75, 313–320.

Bahram, M., Polme, S., Koljalg, U., Zarre, S., Tedersoo, L., 2012. Regional and local patterns of ectomycorrhizal fungal diversity and community structure along an altitudinal gradient in the Hyrcanian forests of northern Iran. New Phytol. 193, 465–473.

Bahram, M., Kõljalg, U., Courty, P.E., Diédhiou, A.G., Kjøller, R., Põlme, S., Ryberg, M., Veldre, V., Tedersoo, L., 2013. The distance decay of similarity in communities of ectomycorrhizal fungi in different ecosystems and scales. J. Ecol. 101, 1335–1344.

Bahram, M., Peay, K.G., Tedersoo, L., 2015. Local-scale biogeography and spatiotemporal variability in communities of mycorrhizal fungi. New Phytol. 205, 1454–1463.

Bahram, M., Kohout, P., Anslan, S., Harend, H., Abarenkov, K., Tedersoo, L., 2016. Stochastic distribution of small soil eukaryotes resulting from high dispersal and drift in a local environment. ISME J. 10, 885–896.

Bahram, M., Hildebrand, F., Forslund, S.K., Anderson, J.L., Soudzilovskaia, N.A., Bodegom, P.M., Bengtsson-Palme, J., Anslan, S., Coelho, L.P., Harend, H., Huerta-Cepas, J., 2018. Structure and function of the global topsoil microbiome. Nature 560, 233–237.

Baier, R., Ingenhaag, J., Blaschke, H., Göttlein, A., Agerer, R., 2006. Vertical distribution of an ectomycorrhizal community in upper soil horizons of a young Norway spruce (*Picea abies* [L.] Karst.) stand of the Bavarian Limestone Alps. Mycorrhiza 16, 197–206.

Bakker, M.R., Garbaye, J., Nys, C., 2000. Effect of liming on the ectomycorrhizal status of oak. For. Ecol. Manag. 126, 121–131.

Baldrian, P., 2017. Microbial activity and the dynamics of ecosystem processes in forest soils. Curr. Opin. Microbiol. 37, 128–134.

Barbi, F., Kohler, A., Barry, K., Baskaran, P., Daum, C., Fauchery, L., Ihrmark, K., Kuo, A., LaButti, K., Lipsen, A., et al., 2019. Fungal ecological strategies reflected in gene transcription—a case study of two litter decomposers. Environ. Microbiol. 22, 1089–1103.

Bardgett, R.D., van der Putten, W.H., 2014. Belowground biodiversity and ecosystem functioning. Nature 515, 505–511.

Baskaran, P., Hyvönen, R., Berglund, S.L., Clemmensen, K.E., Ågren, G.I., Lindahl, B.D., Manzoni, S., 2017. Modelling the influence of ectomycorrhizal decomposition on plant nutrition and soil carbon sequestration in boreal forest ecosystems. New Phytol. 213, 1452–1465.

Baskaran, P., Ekblad, A., Soucémarianadin, L., Hyvönen, R., Schleucher, J., Lindahl, B.D., 2019. Nitrogen dynamics of decomposing Scots pine needle litter depends on colonizing fungal species. FEMS Microbiol. Ecol. 95, fiz059.

Bellassen, V., Luyssaert, S., 2014. Carbon sequestration: managing forests in uncertain times. Nature 506, 153–155.

Bennett, J.A., Maherali, H., Reinhart, K.O., Lekberg, Y., Hart, M.M., Klironomos, J., 2017. Plant-soil feedbacks and mycorrhizal type influence temperate forest population dynamics. Science 355, 181–184.

Berndes, G., Abt, B., Asikainen, A., Cowie, A., Dale, V., Egnell, G., Lindner, M., Marelli, L., Paré, D., Pingoud, K., Yeh, S., 2016. Forest Biomass, Carbon Neutrality and Climate Change Mitigation. From Science to Policy 3, p. 7.

Blaise, T., Garbaye, J., 1983. Effets de la fertilisation minerale sur les ectomycorhizes d'une hetraie. Acta Oecol. Série Oecologia Plantarum Montreuil 4, 165–169.

Boberg, J.B., Finlay, R.D., Stenlid, J., Lindahl, B.D., 2010. Fungal C-translocation restricts N-mineralization in heterogeneous substrates. Funct. Ecol. 24, 454–459.

Boberg, J.B., Finlay, R.D., Stenlid, J., Ekblad, A., Lindahl, B.D., 2014. Nitrogen and carbon reallocation in fungal mycelia during decomposition of boreal forest litter. PLoS One 9, e92897.

Boddy, L., 1999. Saprotrophic cord-forming fungi: meeting the challenge of heterogeneous environments. Mycologia 91, 13–32.

Bödeker, I.T.M., Nygren, C.M.R., Taylor, A.F.S., Olson, Å., Lindahl, B.D., 2009. Class II peroxidase encoding genes are present in a wide phylogenetic range of ectomycorrhizal fungi. ISME J. 3, 1387–1395.

Bödeker, I.T.M., Clemmensen, K.E., de Boer, W., Martin, F., Olson, Å., Lindahl, B.D., 2014. Ectomycorrhizal *Cortinarius* species participate in enzymatic oxidation of humus in northern forest ecosystems. New Phytol. 203, 245–256.

Bödeker, I.T.M., Lindahl, B.D., Olson, Å., Clemmensen, K.E., 2016. Mycorrhizal and saprotrophic fungal guilds compete for the same organic substrates but affect decomposition differently. Funct. Ecol. 30, 1967–1978.

Bonet, J.A., De-Miguel, S., de Aragón, J.M., Pukkala, T., Palahí, M., 2012. Immediate effect of thinning on the yield of Lactarius group deliciosus in *Pinus pinaster* forests in Northeastern Spain. For. Ecol. Manag. 265, 211–217.

Børja, I., Godbold, D.L., Světlík, J., Nagy, N.E., Gebauer, R., Urban, J., Volařík, D., Lange, H., Krokene, P., Čermák, P., Eldhuset, T.D., 2017. Norway spruce fine roots and fungal hyphae grow deeper in forest soils after extended drought. In: Lukac, M., Grenni, P., Gamboni, M. (Eds.), Soil Biological Communities and Ecosystem Resilience. Sustainability in Plant and Crop Protection. Springer, Berlin, Heidelberg.

Brundrett, M.C., 2009. Mycorrhizal associations and other means of nutrition of vascular plants: understanding the global diversity of host plants by resolving conflicting information and developing reliable means of diagnosis. Plant Soil 320, 37–77.

Brzostek, E.R., Dragoni, D., Brown, Z.A., Phillips, R.P., 2015. Mycorrhizal type determines the magnitude and direction of root-induced changes in decomposition in a temperate forest. New Phytol. 206, 1274–1282.

Buée, M., Vairelles, D., Garbaye, J., 2005. Year-round monitoring of diversity and potential metabolic activity of the ectomycorrhizal community in a beech (*Fagus silvatica*) forest subjected to two thinning regimes. Mycorrhiza 15, 235–245.

Buée, M., Courty, P.E., Mignot, D., Garbaye, J., 2007. Soil niche effect on species diversity and catabolic activities in an ectomycorrhizal fungal community. Soil Biol. Biochem. 39, 1947–1955.

Buée, M., Reich, M., Murat, C., Morin, E., Nilsson, R.H., Uroz, S., Martin, F., 2009. 454 Pyrosequencing analyses of forest soils reveal an unexpectedly high fungal diversity. New Phytol. 184, 449–456.

Cairney, J.W.G., 1992. Translocation of solutes in ectomycorrhizal and saprotrophic rhizomorphs. Mycol. Res. 96, 135–141.

Carney, K.M., Hungate, B.A., Drake, B.G., Megonigal, J.P., 2007. Altered soil microbial community at elevated CO_2 leads to loss of soil carbon. Proc. Natl. Acad. Sci. U. S. A. 104, 4990–4995.

Carteron, A., Beigas, M., Joly, S., Turner, B.L., Laliberté, E., 2020. Temperate forests dominated by arbuscular or ectomycorrhizal fungi are characterized by strong shifts from saprotrophic to mycorrhizal fungi with increasing soil depth. Microb. Ecol., 1–14.

Castaño, C., Lindahl, B.D., Alday, J.G., Hagenbo, A., Martínez de Aragón, J., Parladé, X., Pera, J., Bonet, J.A., 2018. Soil microclimate changes affect soil fungal communities in a Mediterranean pine forest. New Phytol. 220, 1211–1221.

Castellano, M.A., Trappe, J.M., 1985. Ectomycorrhizal formation and plantation performance of Douglas-fir nursery stock inoculated with *Rhizopogon* spores. Can. J. For. Res. 15, 613–617.

Cheeke, T.E., Phillips, R.P., Brzostek, E.R., Rosling, A., Bever, J.D., Fransson, P., 2017. Dominant mycorrhizal association of trees alters carbon and nutrient cycling by selecting for microbial groups with distinct enzyme function. New Phytol. 214, 432–442.

Cleland, E.E., Chuine, I., Menzel, A., Mooney, H.A., Schwartz, M.D., 2007. Shifting plant phenology in response to global change. Trends Ecol. Evol. 22, 357–365.

Clemmensen, K.E., Bahr, A., Ovaskainen, O., Dahlberg, A., Ekblad, A., Wallander, H., Stenlid, J., Finlay, R.D., Wardle, D.A., Lindahl, B.D., 2013. Roots and associated fungi drive long-term carbon sequestration in boreal forest. Science 339, 1615–1618.

Clemmensen, K.E., Finlay, R.D., Dahlberg, A., Stenlid, J., Wardle, D.A., Lindahl, B.D., 2015. Carbon sequestration is related to mycorrhizal fungal community shifts during long term succession in boreal forests. New Phytol. 205, 1525–1536.

Cline, L.C., Zak, D.R., 2015. Initial colonization, community assembly and ecosystem function: fungal colonist traits and litter biochemistry mediate decay rate. Mol. Ecol. 24, 5045–5058.

Coince, A., Cael, O., Bach, C., Lengelle, J., Cruaud, C., Gavory, F., Morin, E., Murat, C., Marcais, B., Buée, M., 2013. Below-ground fine-scale distribution and soil versus fine root detection of fungal and soil oomycete communities in a French beech forest. Fungal Ecol. 6, 223–235.

Coince, A., Cordier, T., Lengellé, J., Defossez, E., Vacher, C., Robin, C., Buée, M., Marçais, B., 2014. Below-ground and above-ground fungal assemblages do not follow similar elevational diversity pattern. PLoS One 9, e100668.

Corrales, A., Mangan, S.A., Turner, B.L., Dalling, J.W., 2016. An ectomycorrhizal nitrogen economy facilitates monodominance in a neotropical forest. Ecol. Lett. 19, 383–392.

Corrêa, A., Strasser, R., Martins-Loução, M., 2008. Response of plants to ectomycorrhizae in N-limited conditions: which factors determine its variation? Mycorrhiza 18, 413–427.

Courty, P.E., Franc, A., Pierrat, J.C., Garbaye, J., 2008. Temporal changes in the ectomycorrhizal community in two soil horizons of a temperate oak forest. Appl. Environ. Microbiol. 74, 5792–5801.

Courty, P.E., Munoz, F., Selosse, M.A., Duchemin, M., Criquet, S., Ziarelli, F., Buée, M., Plassard, C., Taudière, A., Garbaye, J., Richard, F., 2016. Into the functional ecology of ectomycorrhizal communities: environmental filtering of enzymatic activities. J. Ecol. 104, 1585–1598.

Cox, F., Barsoum, N., Lilleskov, E.A., Biartondo, M.I., 2010. Nitrogen availability is a primary determinant of conifer mycorrhizas across complex environmental gradients. Ecol. Lett. 13, 1103–1113.

Crowther, T.W., Stanton, D.W.G., Thomas, S.M., A'Bear, A.D., Hiscox, J., Jones, T.H., Voriskova, J., Baldrian, P., Boddy, L., 2013. Top-down control of soil fungal community composition by a globally distributed keystone consumer. Ecology 94, 2518–2528.

Crowther, T.W., Van den Hoogen, J., Wan, J., Mayes, M.A., Keiser, A.D., Mo, L., Averill, C., Maynard, D.S., 2019. The global soil community and its influence on biogeochemistry. Science 365 (6455), eaav0550.

Davison, J., Moora, M., Öpik, M., Adholeya, A., Ainsaar, L., Bâ, A., Burla, S., Diedhiou, A.G., Hiiesalu, I., Jairus, T., et al., 2016. Global assessment of arbuscular mycorrhizal fungus diversity reveals very low endemism. Science 351, 826.

De Aragón, J.M., Fischer, C., Bonet, J.A., Olivera, A., Oliach, D., Colinas, C., 2012. Economically profitable post fire restoration with black truffle (*Tuber melanosporum*) producing plantations. New For. 43, 615–630.

Delaye, L., García-Guzmán, G., Heil, M., 2013. Endophytes versus biotrophic and necrotrophic pathogens—are fungal lifestyles evolutionarily stable traits? Fungal Divers. 60, 125–135.

Derome, J., 2000. Detoxification and amelioration of heavy-metal contaminated forest soils by means of liming and fertilisation. Environ. Pollut. 107, 79–88.

Dickie, I.A., 2007. Host preference, niches and fungal diversity. New Phytol. 174, 230–233.

Dickie, I.A., Xu, B., Koide, R.T., 2002. Vertical niche differentiation of ectomycorrhizal hyphae in soil as shown by T-RFLP analysis. New Phytol. 156, 527–535.

Douhan, G.W., Vincenot, L., Gryta, H., Selosse, M.A., 2011. Population genetics of ectomycorrhizal fungi: from current knowledge to emerging directions. Fungal Biol. 115, 569–597.

Duan, W., Wang, J., Li, Y., 2009. Microenvironmental heterogeneity of physical soil properties in a broad-leaved *Pinus koraiensis* forest gap. Front. Forest China 4, 38–45.

Duddridge, J.A., Malibari, A., Read, D.J., 1980. Structure and function of mycorrhizal rhizomorphs with special reference to their role in water transport. Nature 287, 834–836.

Egerton-Warburton, L.M., Allen, E.B., 2000. Shifts in arbuscular mycorrhizal communities along an anthropogenic nitrogen deposition gradient. Ecol. Appl. 10, 484–496.

Egidi, E., Delgado-Baquerizo, M., Plett, J.M., Wang, J., Bardgett RD, E.D.J., Maestre, F.T., Singh, B.K., 2019. A few Ascomycota taxa dominate soil fungal communities worldwide. Nat. Commun. 10, 1–9.

Egli, S., Ayer, F., Peter, M., Eilmann, B., Rigling, A., 2010. Is forest mushroom productivity driven by tree growth? Results from a thinning experiment. Ann. For. Sci. 67 (5), 509.

Ellis, E.C., 2011. Anthropogenic transformation of the terrestrial biosphere. Philos. Trans. R. Soc. A 369, 1010–1035.

Ferguson, B.A., Dreisbach, T.A., Parks, C.G., Filip, G.M., Schmitt, C.L., 2003. Coarse-scale population structure of pathogenic Armillaria species in a mixed-conifer forest in the Blue Mountains of northeast Oregon. Can. J. For. Res. 33, 612–623.

Fernandez, C.W., Kennedy, P.G., 2016. Revisiting the 'Gadgil effect': do interguild fungal interactions control carbon cycling in forest soils? New Phytol. 209, 1382–1394.

Fernandez, C.W., Langley, J.A., Chapman, S., McCormack, M.L., Koide, R.T., 2016. The decomposition of ectomycorrhizal fungal necromass. Soil Biol. Biochem. 93, 38–49.

Fierer, N., Strickland, M.S., Liptzin, D., Bradford, M.A., Cleveland, C.C., 2009. Global patterns in belowground communities. Ecol. Lett. 12, 1238–1249.

Finlay, R.D., Mahmood, S., Rosenstock, N., Bolou-Bi, E.B., Kohler, S.J., Fahad, Z., Rosling, A., Wallander, H., Belyazid, S., Bishop, K., Lian, B., 2020. Biological weathering and its consequences at different spatial levels—from nanoscale to global scale. Biogeosciences 17, 1507–1533.

Flores-Rentería, D., Rincón, A., Valladares, F., Yuste, J.C., 2016. Agricultural matrix affects diffferently the alpha and beta structural and functional diversity of soil microbial communities in a fragmented Mediterranean holm oak forest. Soil Biol. Biochem. 92, 79–90.

Floudas, D., Binder, M., Riley, R., Barry, K., Blanchette, R.A., Henrissat, B., Martínez, A.T., Otillar, R., Spatafora, J.W., et al., 2012. The Paleozoic origin of enzymatic lignin decomposition reconstructed from 31 fungal genomes. Science 336, 1715–1719.

Frac, M., Hannula, S.E., Bełka, M., Jedryczka, M., 2018. Fungal biodiversity and their role in soil health. Front. Microbiol. 9, 707.

Franklin, O., Näsholm, T., Högberg, P., Högberg, M.N., 2014. Forests trapped in nitrogen limitation—an ecological market perspective on ectomycorrhizal symbiosis. New Phytol. 203, 657–666.

Freschet, G.T., Aerts, R., Cornelissen, J.H.C., 2012. A plant economics spectrum of litter decomposability. Funct. Ecol. 26, 56–65.

Gadgil, R.L., Gadgil, P.D., 1971. Mycorrhiza and litter decomposition. Nature 233, 133.

Garbaye, J., Churin, J.L., 1997. Growth stimulation of young oak plantations inoculated with the ectomycorrhizal fungus *Paxillus involutus* with special reference to summer drought. For. Ecol. Manag. 98, 221–228.

Garcia-Gonzalo, J., Peltola, H., Briceno-Elizondo, E., Kellomäki, S., 2007. Changed thinning regimes may increase carbon stock under climate change: a case study from a Finnish boreal forest. Clim. Chang. 81, 431–454.

Gehring, C.A., Mueller, R.C., Haskins, K.E., Rubow, T., Whitham, T.G., 2014. Convergence in mycorrhizal fungal communities due to drought, plant competition, parasitism, and susceptibility to herbivory: consequences for fungi and host plants. Front. Microbiol. 5, 1–9.

Gehring, C.A., Sthultz, C.M., Flores-Rentería, L., Whipple, A.V., Whitham, T.G., 2017. Tree genetics defines fungal partner communities that may confer drought tolerance. Proc. Natl. Acad. Sci. U. S. A. 114 (42), 11169–11174.

Genney, D.R., Anderson, I.C., Alexander, I.J., 2006. Fine-scale distribution of pine ectomycorrhizas and their extramatrical mycelium. New Phytol. 170, 381–390.

Ghazoul, J., Burivalova, Z., Garcia-Ulloa, J., King, L.A., 2015. Conceptualizing forest degradation. Trends Ecol. Evol. 30, 622–632.

Glassman, S.I., Wang, I.J., Bruns, T.D., 2017. Environmental filtering by pH and soil nutrients drives community assembly in fungi at fine spatial scales. Mol. Ecol. 26, 6960–6973.

Guhr, A., Borken, W., Spohn, M., Matzner, E., 2015. Redistribution of soil water by a saprotrophic fungus enhances carbon mineralization. Proc. Natl. Acad. Sci. U. S. A. 112, 14647–14651.

Hagerman, S.M., Jones, M.D., Bradfield, G.E., Gillespie, M., Durall, D.M., 1999. Effects of clear-cut logging on the diversity and persistence of ectomycorrhizae at a subalpine forest. Can. J. For. Res. 29, 124–134.

Hanssen, S.V., Daioglou, V., Steinmann, Z.J.N., Doelman, J.C., Van Vuuren, D.P., Huijbregts, M.A.J., 2020. The climate change mitigation potential of bioenergy with carbon capture and storage. Nat. Clim. Chang., 1–7.

Hartmann, M., Lee, S., Hallam, S.J., Mohn, W.W., 2009. Bacterial, archaeal and eukaryal community structures throughout soil horizons of harvested and naturally disturbed forest stands. Environ. Microbiol. 11, 3045–3062.

Hartmann, M., Howes, C.G., VanInsberghe, D., Yu, H., Bachar, D., Christen, R., et al., 2012. Significant and persistent impact of timber harvesting on soil microbial communities in Northern coniferous forests. ISME J. 6, 2199–2218.

Hartmann, M., Niklaus, P.A., Zimmermann, S., Schmutz, S., Kremer, J., Abarenkov, K., Lüscher, P., Widmer, F., Frey, B., 2014. Resistance and resilience of the forest soil microbiome to logging-associated compaction. ISME J. 8, 226–244.

Hawksworth, D.L., Lücking, R., 2017. Fungal diversity revisited: 2.2 to 3.8 million species. Microbial. Spectr. 5 (4), 1–17.

He, D., Shen, W., Eberwein, J., Zhao, Q., Ren, L., Wu, Q.L., 2017. Diversity and co-occurrence network of soil fungi are more responsive than those of bacteria to shifts in precipitation seasonality in a subtropical forest. Soil Biol. Biochem. 115, 499–510.

Hodge, A., 2017. Accessibility of inorganic and organic nutrients for mycorrhizas. In: Collins Johnson, N., Gehring, C., Jansa, J. (Eds.), Mycorrhizal Mediation of Soil Fertility, Structure, and Carbon Storage. Elsevier, pp. 129–148.

Holden, S.R., Treseder, K.K., 2013. A metaanalysis of soil microbial biomass responses to forest disturbances. Front. Microbiol. 4, 163.

Hrynkiewicz, K., Baum, C., Leinweber, P., Weih, M., Dimitriou, I., 2010. The significance of rotation periods for mycorrhiza formation in short rotation coppice. For. Ecol. Manag. 260, 1943–1949.

Ishida, T.A., Nara, K., Hogetsu, T., 2007. Host effects on ectomycorrhizal fungal communities: insight from eight host species in mixed conifer-broadleaf forests. New Phytol. 174, 430–440.

Janssens, I.A., Freibauer, A., Schlamadinger, B., Ceulemans, B., Ciais, P., Dolman, A.J., et al., 2005. The carbon budget of terrestrial ecosystems at country-scale – A European case study. Biogeosciences 2, 15–26.

Janzen, D.H., 1970. Herbivores and the number of tree species in tropical forests. Am. Nat. 104, 501–528.

Jiang, F., Zhang, L., Zhou, J., George, T.S., Feng, G., 2020. Arbuscular mycorrhizal fungi enhance mineralization of organic phosphorus (P) by carrying bacteria along their extraradical hyphae. New Phytol. https://doi.org/10.1111/nph.17081.

Johansen, K.S., 2016. Lytic polysaccharide monooxygenases: the microbial power tool for lignocellulose degradation. Trends Plant Sci. 21, 926–936.

Kadowaki, K., Yamamoto, S., Sato, H., Tanabe, A.S., Hidaka, A., Toju, H., 2018. Mycorrhizal fungi mediate the direction and strength of plant–soil feedbacks differently between arbuscular mycorrhizal and ectomycorrhizal communities. Commun. Biol. 1, 1–11.

Kauserud, H., Stige, L.C., Vik, J.O., Økland, R.H., Høiland, K., Stenseth, N.C., 2008. Mushroom fruiting and climate change. Proc. Natl. Acad. Sci. U. S. A. 105 (10), 3811–3814.

Keiluweit, M., Nico, P., Harmon, M.E., Mao, J., Pett-Ridge, J., Kleber, M., 2015. Long-term litter decomposition controlled by manganese redox cycling. Proc. Natl. Acad. Sci. U. S. A. 112, E5253–E5260.

Kellomäki, S., Wang, K.Y., 1996. Photosynthetic responses to needle water potentials in Scots pine after a four-year exposure to elevated CO_2 and temperature. Tree Physiol. 16, 765–772.

Kennedy, P., 2010. Ectomycorrhizal fungi and interspecific competition: species interactions, community structure, coexistence mechanisms, and future research directions. New Phytol. 187 (4), 895–910.

Kennedy, P.G., Bruns, T.D., 2005. Priority effects determine the outcome of ectomycorrhizal competition between two *Rhizopogon* species colonizing *Pinus muricata* seedlings. New Phytol. 166 (2), 631–638.

Kennedy, P.G., Peay, K.G., Bruns, T.D., 2009. Root tip competition among ectomycorrhizal fungi: are priority effects a rule or an exception? Ecology 90, 2098–2107.

Kennedy, P.G., Matheny, P.B., Ryberg, K.M., Henkel, T.W., Uehling, J.K., Smith, M.E., 2012. Scaling up: examining the macroecology of ectomycorrhizal fungi. Mol. Ecol. 21, 4151–4154.

Kennedy, P.G., Mielke, L.A., Nguyen, N.H., 2018. Ecological responses to forest age, habitat and host vary by mcyorrhizal type in boreal peatlands. Mycorrhiza 28, 315–328.

Kivlin, S.N., Hawkes, C.V., Treseder, K.K., 2011. Global diversity and distribution of arbuscular mycorrhizal fungi. Soil Biol. Biochem. 43, 2294–2303.

Klironomos, J.N., 2002. Feedback with soil biota contributes to plant rarity and invasiveness in communities. Nature 417, 67–70.

Kohler, A., Kuo, A., Nagy, L.G., Morin, E., Barry, K.W., Buscot, F., Canbäck, B., Choi, C., Cichocki, N., Clum, A., et al., 2015. Convergent losses of decay mechanisms and rapid turnover of symbiosis genes in mycorrhizal mutualists. Nat. Genet. 47, 410–415.

Kohout, P., 2017. Biogeography of ericoid mycorrhiza. In: Biogeography of Mycorrhizal Symbiosis. Springer, Cham, pp. 179–193.

Kohout, P., Bahram, M., Põlme, S., Tedersoo, L., 2017. Elevation, space and host plant species structure Ericaceae root-associated fungal communities in Papua New Guinea. Fungal Ecol. 30, 112–121.

Kohout, P., Charvátová, M., Štursová, M., Mašínová, T., Tomšovský, M., Baldrian, P., 2018. Clearcutting alters decomposition processes and initiates complex restructuring of fungal communities in soil and tree roots. ISME J. 12, 692–703.

Koide, R.T., Shumway, D.L., Xu, B., Sharda, J.N., 2007. On temporal partitioning of a community of ectomycorrhizal fungi. New Phytol. 174, 420–429.

Kumar, M.N.R., 2000. A review of chitin and chitosan applications. React. Funct. Polym. 46, 1–27.

Kyaschenko, J., Clemmensen, K.E., Karltun, E., Lindahl, B.D., 2017a. Below-ground organic matter accumulation along a boreal forest fertility gradient relates to guild interaction within fungal communities. Ecol. Lett. 20, 1546–1555.

Kyaschenko, J., Clemmensen, K.E., Hagenbo, A., Karltun, E., Lindahl, B.D., 2017b. Shift in fungal communities and associated enzyme activities along an age gradient of managed Pinus sylvestris stands. ISME J. 11, 863–874.

Kyaschenko, J., Ovaskainen, O., Ekblad, A., Hagenbo, A., Karltun, E., Clemmensen, K.E., Lindahl, B.D., 2019. Soil fertility in boreal forest relates to root-driven nitrogen retention and carbon sequestration in the mor layer. New Phytol. 221, 1492–1502.

Legeay, J., Husson, C., Boudier, B., Louisanna, E., Baraloto, C., Schimann, H., Marcais, B., Buée, M., 2020. Surprising low diversity of the plant pathogen Phytophthora in Amazonian forests. Environ. Microbiol., 1–14.

Lehmann, J., Kleber, M., 2015. The contentious nature of soil organic matter. Nature 528, 60–68.

Lehto, T., Zwiazek, J.J., 2011. Ectomycorrhizas and water relations of trees, a review. Mycorrhiza 21, 71–90.

Leon-Sanchez, L., Nicolas, E., Goberna, M., Prieto, I., Maestre, F.T., Querejeta, J.I., 2018. Poor plant performance under simulated climate change is linked to mycorrhizal responses in a semi-arid shrubland. J. Ecol. 106, 960–976.

Lilleskov, E.A., Fahey, T.J., Lovett, G.M., 2001. Ectomycorrhizal fungal aboveground community change over an atmospheric nitrogen deposition gradient. Ecol. Appl. 11, 397–410.

Lilleskov, E.A., Bruns, T.D., Horton, T.R., Taylor, D.L., Grogan, P., 2004. Detection of forest stand-level spatial structure in ectomycorrhizal fungal communities. FEMS Microbiol. Ecol. 49, 319–332.

Lilleskov, E.A., Hobbie, E.A., Horton, T.R., 2011. Conservation of ectomycorrhizal fungi: exploring the linkages between functional and taxonomic responses to anthropogenic N deposition. Fungal Ecol. 4, 174–183.

Lilleskov, E.A., Kuyper, T.W., Bidartondo, M.I., Hobbie, E.A., 2019. Atmospheric nitrogen deposition impacts on the structure and function of forest mycorrhizal communities: a review. Environ. Pollut. 246, 148–162.

Lin, G.G., McCormack, M.L., Ma, C.E., Guo, D.L., 2017. Similar below-ground carbon cycling dynamics but contrasting modes of nitrogen cycling between arbuscular mycorrhizal and ectomycorrhizal forests. New Phytol. 213, 1440–1451.

Lindahl, B.D., Olsson, S., 2004. Fungal translocation—creating and responding to environmental heterogeneity. Mycologist 18, 79–88.

Lindahl, B.D., Tunlid, A., 2015. Ectomycorrhizal fungi—potential organic matter decomposers, yet not saprotrophs. New Phytol. 205, 1443–1447.

Lindahl, B.D., Taylor, A., Finlay, R., 2002. Defining nutritional constraints on carbon cycling—towards a less "phytocentric" perspective. Plant Soil 242, 123–135.

Lindahl, B.D., Ihrmark, K., Boberg, J., Trumbore, S.E., Högberg, P., Stenlid, J., Finlay, R.D., 2007. Spatial separation of litter decomposition and mycorrhizal nitrogen uptake in a boreal forest. New Phytol. 173, 611–620.

Lindroth, A., Lagergren, F., Grelle, A., Klemedtsson, L., Langvall, O.L.A., Weslien, P.E.R., Tuulik, J., 2009. Storms can cause Europe-wide reduction in forest carbon sink. Glob. Chang. Biol. 15, 346–355.

Liu, Y., Yu, S., Xie, Z.P., Staehelin, C., 2012. Analysis of a negative plant–soil feedback in a subtropical monsoon forest. J. Ecol. 100, 1019–1028.

Lonsdale, D., Pautasso, M., Holdenrieder, O., 2008. Wood-decaying fungi in the forest: conservation needs and management options. Eur. J. For. Res. 127, 1–22.

López-García, Á., Gil-Martínez, M., Navarro-Fernández, C.M., Kjøller, R., Azcón-Aguilar, C., Domínguez, M.T., Marañón, T., 2018. Functional diversity of ectomycorrhizal fungal communities is reduced by trace element contamination. Soil Biol. Biochem. 121, 202–211.

Maestre, F.T., Cortina, J., 2004. Are *Pinus halepensis* plantations useful as a restoration tool in semiarid Mediterranean areas? For. Ecol. Manag. 198, 303–317.

Maherali, H., Klironomos, J.N., 2012. Phylogenetic and trait-based assembly of arbuscular mycorrhizal fungal communities. PLoS One, 7.

Mahmood, S., Finlay, R.D., Erland, S., 1999. Effects of repeated harvesting of forest residues on the ectomycorrhizal community in a Swedish spruce forest. New Phytol. 142, 577–585.

Maillard, F., 2018. Rôle des communautés microbiennes dans la dégradation de la matière organique en forêt dans un contexte d'exportation intense de biomasse. (PhD thesis, Nancy, France).

Maillard, F., Leduc, V., Bach, C., de Moraes Gonçalves, J.L., Androte, F.D., Saint-André, L., Laclau, J.P., Buée, M., Robin, A., 2019a. Microbial enzymatic activities and community-level physiological profiles (CLPP) in subsoil layers are altered by harvest residue management practices in a tropical *Eucalyptus grandis* plantation. Microb. Ecol. 78, 528–533.

Maillard, F., Leduc, V., Bach, C., Reichard, A., Fauchery, L., Saint-André, L., Zeller, B., Buée, M., 2019b. Soil microbial functions are affected by organic matter removal in temperate deciduous forest. Soil Biol. Biochem. 133, 28–36.

Maitra, P., Zheng, Y., Chen, L., Wang, Y.L., Ji, N.N., Lü, P.P., Gan, H.Y., Li, X.C., Sun'X, Zhou XH, Guo LD., 2019. Effect of drought and season on arbuscular mycorrhizal fungi in a subtropical secondary forest. Fungal Ecol. 41, 107–115.

Maltz, M.R., Treseder, K.K., 2015. Sources of inocula influence mycorrhizal colonization of plants in restoration projects: a meta-analysis. Restor. Ecol. 23, 625–634.

Mangan, S.A., Schnitzer, S.A., Herre, E.A., Mack, K.M., Valencia, M.C., Sanchez, E.I., Bever, J.D., 2010. Negative plant–soil feedback predicts tree-species relative abundance in a tropical forest. Nature 466, 752–755.

Martino, E., Morin, E., Grelet, G.A., Kuo, A., Kohler, A., Daghino, S., Barry, K.W., Cichocki, N., Clum, A., Dockter, R.B., et al., 2018. Comparative genomics and transcriptomics depict ericoid mycorrhizal fungi as versatile saprotrophs and plant mutualists. New Phytol. 217, 1213–1229.

Martos, F., Dulormne, M., Pailler, T., Bonfante, P., Faccio, A., Fournel, J., Dubois, M.-P., Selosse, M.-A., 2009. Independent recruitment of saprotrophic fungi as mycorrhizal partners by tropical achlorophyllous orchids. New Phytol. 184, 668–681.

McCormick, M.K., Whigham, D.F., O'neill, J., 2004. Mycorrhizal diversity in photosynthetic terrestrial orchids. New Phytol. 163, 425–438.

McGuire, K.L., Allison, S.D., Fierer, N., Treseder, K.K., 2013. Ectomycorrhizal-dominated boreal and tropical forests have distinct fungal communities, but analogous spatial patterns across soil horizons. PLoS One 8, 1–9.

Mills, K.E., Bever, J.D., 1998. Maintenance of diversity within plant communities: soil pathogens as agents of negative feedback. Ecology 79, 1595–1601.

Miyauchi, S., Kiss, E., Kuo, A., Drula, E., Kohler, A., Sanchez-García, M., Morin, E., Andreopoulos, B., Barry, K.W., Bonito, G., Buée, M., et al., 2020. Large-scale genome sequencing of mycorrhizal fungi provides insights into the early evolution of symbiotic traits. Nat. Commun. 11, 1–17.

Montesinos-Navarro, A., Valiente-Banuet, A., Verdú, M., 2019. Plant facilitation through mycorrhizal symbiosis is stronger between distantly related plant species. New Phytol. 224, 928–935.

Moore-Kucera, J., Dick, R.P., 2008. Application of 13C-labeled litter and root materials for in situ decomposition studies using phospholipid fatty acids. Soil Biol. Biochem. 40, 2485–2493.

Moyersoen, B., Fitter, A.H., Alexander, I.J., 1998. Spatial distribution of ectomycorrhizas and arbuscular mycorrhizas in Korup National Park rain forest, Cameroon, in relation to edaphic parameters. New Phytol. 139, 311–320.

Mujic, A.B., Durall, D.M., Spatafora, J.W., Kennedy, P.G., 2016. Competitive avoidance not edaphic specialization drives vertical niche partitioning among sister species of ectomycorrhizal fungi. New Phytol. 209, 1174–1183.

Näsholm, T., Högberg, P., Franklin, O., Metcalfe, D., Keel, S.G., Campbell, C., Hurry, V., Linder, S., Högberg, M.N., 2013. Are ectomycorrhizal fungi alleviating or aggravating nitrogen limitation of tree growth in boreal forests? New Phytol. 198, 214–221.

Neville, J., Tessier, J.L., Morrison, I., Scarratt, J., Canning, B., Klironomos, J.N., 2002. Soil depth distribution of ecto- and arbuscular mycorrhizal fungi associated with *Populus tremuloides* within a 3-year-old boreal forest clear-cut. Appl. Soil Ecol. 19, 209–216.

Nguyen, D.Q., Schneider, D., Brinkmann, N., Song, B., Janz, D., Schöning, I., Daniel, R., Pena, R., Polle, A., 2020. Soil and root nutriet chemistry structure root-associated fungal assemblages in temperate forests. Environ. Microbiol. 22, 3081–3095.

Nicolas, C., Martin-Bertelsen, T., Floudas, D., Bentzer, J., Smits, M., Johansson, T., Troein, C., Persson, P., Tunlid, A., 2019. The soil organic matter decomposition mechanisms in ectomycorrhizal fungi are tuned for liberating soil organic nitrogen. ISME J. 13, 977–988.

Norby, R.J., Luo, Y., 2004. Evaluating ecosystem responses to rising atmospheric CO_2 and global warming in a multi-factor world. New Phytologist. 62, 281–293.

Norby, R.J., Ledford, J., Reilly, C.D., Miller, N.E., O'Neill, E.G., 2004. Fine-root production dominates response of a deciduous forest to atmospheric CO_2 enrichment. Proc. Natl. Acad. Sci. U. S. A. 101, 9689–9693.

Northup, R.R., Yu, Z., Dahlgren, R.A., Vogt, K.A., 1995. Polyphenol control of nitrogen release from pine litter. Nature 377, 227–229.

Nowotny, I., Dähne, J., Klingelhöfer, D., Rothe, G.M., 1998. Effect of artificial soil acidification and liming on growth and nutrient status of mycorrhizal roots of Norway spruce (*Picea abies* [L.] Karst). Plant Soil 199, 29–40.

Ogaya, R., Peñuelas, J., 2005. Decreased mushroom production in a holm oak forest in response to an experimental drought. Forestry 78, 279–283.

Oja, J., Vahtra, J., Bahram, M., Kohout, P., Kull, T., Rannap, R., Kõljalg, U., Tedersoo, L., 2017. Local-scale spatial structure and community composition of orchid mycorrhizal fungi in semi-natural grasslands. Mycorrhiza 27, 355–367.

Op De Beeck, M., Lievens, B., Busschaert, P., Rineau, F., Smits, M., Vangronsveld, J., Colpaert, J.V., 2015. Impact of metal pollution on fungal diversity and community structures. Environ. Microbiol. 17, 2035–2047.

Op De Beeck, M., Troein, C., Peterson, C., Persson, P., Tunlid, A., 2018. Fenton reaction facilitates organic nitrogen acquisition by an ectomycorrhizal fungus. New Phytol. 218, 335–343.

Ortega, U., Dunabeitia, M., Menendez, S., Gonzalez-Murua, C., Majada, J., 2004. Effectiveness of mycorrhizal inoculation in the nursery on growth and water relations of *Pinus radiata* in different water regimes. Tree Physiol. 24, 65–73.

Palmroth, S., Maier, C.A., McCarthy, H.R., Oishi, A.C., Kim, H.S., Johnsen, K.H., et al., 2005. Contrasting responses to drought of forest floor CO2 efflux in a loblolly pine plantation and a nearby oak-hickory forest. Glob. Chang. Biol. 11, 421–434.

Pampolina, N.M., Dell, B., Malajczuk, N., 2002. Dynamics of ectomycorrhizal fungi in an *Eucalyptus globulus* plantation, effect of phosphorus fertilization. For. Ecol. Manag. 158, 291–304.

Parker, T., Subke, J.A., Wookey, P.A., 2015. Rapid carbon turnover beneath shrub and tree vegetation is associated with low soil carbon stocks at a sub-arctic treeline. Glob. Chang. Biol. 21, 2070–2081.

Parladé, J., Martínez-Peña, F., Pera, J., 2017. Effects of forest management and climatic variables on the mycelium dynamics and sporocarp production of the ectomycorrhizal fungus *Boletus edulis*. For. Ecol. Manag. 390, 73–79.

Parmesan, C., Yohe, G., 2003. A globally coherent fingerprint of climate change impacts across natural systems. Nature 421, 37–42.

Peay, K.G., 2016. The mutualistic niche: mycorrhizal symbiosis and community dynamics. Annu. Rev. Ecol. Evol. Syst. 47, 143–164.

Peay, K.G., Kennedy, P.G., Bruns, T.D., 2008. Fungal community ecology: a hybrid beast with a molecular master. Bioscience 58, 799–810.

Peay, K.G., Belisle, M., Fukami, T., 2012. Phylogenetic relatedness predicts priority effects in nectar yeast communities. Proc. R. Soc. B 279, 749–758.

Pellegrino, E., Di Bene, C., Tozzini, C., Bonari, E., 2011. Impact on soil quality of a 10-year-old short-rotation coppice poplar stand compared with intensive agricultural and uncultivated systems in a Mediterranean area. Agric. Ecosyst. Environ. 140, 245–254.

Pérez-Izquierdo, L., Zabal-Aguirre, M., Flores-Rentería, D., González-Martínez, S.C., Buée, M., Rincón, A., 2017. Functional outcomes of fungal community shifts driven by tree genotype and spatial-temporal factors in Mediterranean pine forests. Environ. Microbiol. 19, 1639–1652.

Pérez-Izquierdo, L., Zabal-Aguirre, M., González-Martínez, S.C., Buée, M., Verdú, M., Rincón, A., Goberna, M., 2019. Plant intraspecific variation modulates nutrient cycling through its below ground rhizospheric microbiome. J. Ecol. 107 (4), 1594–1605.

Pérez-Izquierdo, L., Zabal-Aguirre, M., Verdú, M., Buée, M., Rincón, A., 2020a. Ectomycorrhizal fungal diversity decreases in Mediterranean pine forests adapted to recurrent fires. Mol. Ecol. 19, 2463–2476.

Pérez-Izquierdo, L., Clemmensen, K.E., Strengbom, J., Granath, G., Wardle, D.A., Nilsson, M.C., Lindahl, B.D., 2020b. Crown-fire severity is more important than ground-fire severity in determining soil fungal community development in the boreal forest. J. Ecol., 1–15.

Pérez-Valera, E., Goberna, M., Verdu, M., 2015. Phylogenetic structure of soil bacterial communities predicts ecosystem functioning. FEMS Microbiol. Ecol. 91, 1–9.

Pérez-Valera, E., Verdú, M., Navarro-Cano, J.A., Goberna, M., 2018. Resilience to fire of phylogenetic diversity across biological domains. Mol. Ecol. 27, 2896–2908.

Pérez-Valera, E., Goberna, M., Verdú, M., 2020. Soil microbiome drives the recovery of ecosystem functions after fire. Soil Biol. Biochem. 129, 80–89.

Peter, M., Ayer, F., Egli, S., 2001. Nitrogen addition in a Norway spruce stand altered macromycete sporocarp production and below-ground ectomycorrhizal species composition. New Phytol. 149, 311–325.

Phillips, R.P., Brzostek, E., Midgley, M.G., 2013. The mycorrhizal-associated nutrient economy: a new framework for predicting carbon-nutrient couplings in temperate forests. New Phytol. 199, 41–51.

Pickles, B.J., Simard, S., 2017. Mycorrhizal networks and forest resilience to drought. In: Collins Johnson, N., Gehring, C., Jansa, J. (Eds.), Mycorrhizal Mediation of Soil: Fertility, Structure and Carbon Storage. Elsevier, CA.

Pickles, B.J., Genney, D.R., Anderson, I.C., Alexander, I.J., 2012. Spatial analysis of ectomycorrhizal fungi reveals that root tip communities are structured by competitive interactions. Mol. Ecol. 21, 5110–5123.

Prieto, I., Roldán, A., Huygens, D., del Mar, A.M., Navarro-Cano, J.A., Querejeta, J.I., 2016. Species-specific roles of ectomycorrhizal fungi in facilitating interplant transfer of hydraulically redistributed water between *Pinus halepensis* saplings and seedlings. Plant Soil 406, 15–27.

Querejeta, J.I., Egerton-Warburton, L.M., Allen, M.F., 2003. Direct nocturnal water transfer from oaks to their mycorrhizal symbionts during severe soil drying. Oecologia 134, 55–64.

Querejeta, J.I., Egerton-Warburton, L.M., Allen, M.F., 2007. Hydraulic lift may buffer rhizosphere hyphae against the negative effects of severe soil drying in a California Oak savannah. Soil Biol. Biochem. 39, 409–417.

Querejeta, J.I., Egerton-Warburton, L.M., Allen, M.F., 2009. Topographic position modulates the mycorrhizal response of oak trees to interannual rainfall variability. Ecology 90, 649–662.

Read, D.J., Perez-Moreno, J., 2003. Mycorrhizas and nutrient cycling in ecosystems—a journey towards relevance? New Phytol. 157, 475–492.

Regelink, I.C., Weng, L., Lair, G.J., Comans, R.N.J., 2015. Adsorption of phosphate and organic matter on metal (hydr)oxides in arable and forest soil: a mechanistic modelling study. Eur. J. Soil Sci. 66, 867–875.

Richard, F., Roy, M., Shahin, O., Sthultz, C., Duchemin, M., Joffre, R., Selosse, M.A., 2011. Ectomycorrhizal communities in a Mediterranean forest ecosystem dominated by *Quercus ilex*: seasonal dynamics and response to drought in the surface organic horizon. Ann. For. Sci. 68, 57–68.

Rincón, A., Pueyo, J.J., 2010. Effect of fire severity and site slope on diversity and structure of the ectomycorrhizal fungal community associated with post-fire regenerated *Pinus pinaster* Ait. seedlings. For. Ecol. Manag. 260 (3), 361–369.

Rincón, A., de Felipe, R., Fernández-Pascual, M., 2006. Inoculation of *Pinus halepensis* Mill. with selected ectomycorrhizal fungi improves seedling establishment 2 years after planting in a degraded gypsum soil. Mycorrhiza 18, 23–32.

Rincón, A., Santamaría, B.P., Ocaña, L., Verdú, M., 2014. Structure and phylogenetic diversity of post-fire ectomycorrhizal communities of maritime pine. Mycorrhiza 24, 131–141.

Rincón, A., Santamaría-Pérez, B., Rabasa, S.G., Coince, A., Marçais, B., Buée, M., 2015. Compartmentalized and contrasted response of ectomycorrhizal and soil fungal communities of Scots pine forests along elevation gradients in France and Spain. Environ. Microbiol. 17 (8), 3009–3024.

Rineau, F., Garbaye, J., 2009a. Effects of liming on ectomycorrhizal community structure in relation to soil horizons and tree hosts. Fungal Ecol. 2, 103–109.

Rineau, F., Garbaye, J., 2009b. Does forest liming impact the enzymatic profiles of ectomycorrhizal communities through specialized fungal symbionts? Mycorrhiza 19, 493–500.

Rineau, F., Maurice, J.P., Nys, C., Voiry, H., Garbaye, J., 2010. Forest liming durably impact the communities of ectomycorrhizas and fungal epigeous fruiting bodies. Ann. For. Sci. 67, 110.

Rineau, F., Roth, D., Shah, F., Smits, M., Johansson, T., Canbäck, B., Olsen, P.B., Persson, P., Nedergaard Grell, M., Lindquist, E., Grigoriev, I.V., Lange, L., Tunlid, A., 2012. The ectomycorrhizal fungus Paxillus involutus converts organic matter in plant litter using a trimmed brown-rot mechanism involving Fenton chemistry. Environ. Microbiol. 14, 1477–1487.

Rineau, F., Shah, F., Smits, M.M., Persson, P., Johansson, T., Carleer, R., Troein, C., Tunlid, A., 2013. Carbon availability triggers the decomposition of plant litter and assimilation of nitrogen by an ectomycorrhizal fungus. ISME J. 7, 2010–2022.

Rodriguez-Ramos, J.C., Cale, J.A., Cahill Jr., J.F., Simard, S.W., Karst, J., Erbilgin, N., 2020. Changes in soil fungal community composition depend on functional group and forest disturbance type. New Phytol. https://doi.org/10.1111/nph.16749.

Rosling, A., Landeweert, R., Lindahl, B.D., Larsson, K.H., Kuyper, T.W., Taylor, A.F.S., Finlay, R.D., 2003. Vertical distribution of ectomycorrhizal fungal taxa in a podzol soil profile. New Phytol. 159, 775–783.

Sa, G., Yao, J., Deng, C., Liu, J., Zhang, Y.N., Zhu, Z.M., Zhang, Y.H., Ma, X.J., Zhao, R., Lin, S.Z., Lu, C.F., Polle, A., Chen, S.L., 2019. Amelioration of nitrate uptake under salt stress by ectomycorrhiza with and without a Hartig net. New Phytol. 222, 1951–1964.

Schmidt, M.W., Torn, M.S., Abiven, S., Dittmar, T., Guggenberger, G., Janssens, I.A., Kleber, M., Kögel-Knabner, I., Lehmann, J., Manning, D.A., et al., 2011. Persistence of soil organic matter as an ecosystem property. Nature 478, 49–56.

Segnitz, R.M., Russo, S.E., Davies, S.J., Peay, K.G., 2020. Ectomycorrhizal fungi drive positive phylogenetic plant–soil feedbacks in a regionally dominant tropical plant family. Ecology 101, e03083.

Seidl, R., Klonner, G., Rammer, W., Essl, F., Moreno, A., Neumann, M., Dullinger, S., 2018. Invasive alien pests threaten the carbon stored in Europe's forests. Nat. Commun. 9, 1–10.

Selosse, M.A., Martos, F., 2014. Do chlorophyllous orchids heterotrophically use mycorrhizal fungal carbon? Trends Plant Sci. 19, 683–685.

Selosse, M.A., Roy, M., 2009. Green plants that feed on fungi: facts and questions about mixotrophy. Trends Plant Sci. 14, 64–70.

Selosse, M.-A., Setaro, S., Glatard, F., Richard, F., Urcelay, C., Weiss, M., 2007. Sebacinales are common mycorrhizal associates of Ericaceae. New Phytol. 174, 864–878.

Semchenko, M., Leff, J.W., Lozano, Y.M., Saar, S., Davison, J., Wilkinson, A., et al., 2018. Fungal diversity regulates plant-soil feedbacks in temperate grassland. Sci. Adv. 4, 4578.

Shah, F., Nicolás, C., Bentzer, J., Ellström, M., Smits, M., Rineau, F., et al., 2016. Ectomycorrhizal fungi decompose soil organic matter using oxidative mechanisms adapted from saprotrophic ancestors. New Phytologist. 209, 1705–1719.

Sikström, U., 2001. Growth and Nutrition of Coniferous Forests on Acidic Mineral Soils-Status and Effects of Liming and Fertilization. In: Doctoral thesis. Acta Universitatis Agriculturae Sueciae. Silvestria, pp. 1401–6230.

Sinsabaugh, R.L., 2010. Phenol oxidase, peroxidase and organic matter dynamics of soil. Soil Biol. Biochem. 42, 391–404.

Smith, G.R., Peay, K.G., 2020. Stepping forward from relevance in mycorrhizal ecology. New Phytol. 226, 292–294.

Smith, S.E., Read, D.J., 2008. Mycorrhizal Symbiosis, third ed. Academic Press, London.

Smith, G.R., Steidinger, B.S., Bruns, T.D., Peay, K.G., 2018. Competition–colonization tradeoffs structure fungal diversity. ISME J. 12, 1758–1767.

Smits, M.M., Wallander, H., 2017. Role of mycorrhizal symbiosis in mineral weathering and nutrient mining from soil parent material. In: Collins Johnson, N., Gehring, C., Jansa, J. (Eds.), Mycorrhizal Mediation of Soil Fertility, Structure, and Carbon Storage. Elsevier, pp. 35–46.

Šnajdr, J., Steffen, K.T., Hofrichter, M., Baldrian, P., 2010. Transformation of 14C-labelled lignin and humic substances in forest soil by the saprobic basidiomycetes *Gymnopus erythropus* and *Hypholoma fasciculare*. Soil Biol. Biochem. 42, 1541–1548.

Stajich, J.E., 2017. Fungal genomes and insights into the evolution of the kingdom. In: The Fungal Kingdom, pp. 619–633.

Steidinger, B.S., Crowther, T.W., Liang, J., Van Nuland, M.E., Werner, G.D.A., Reich, P.B., Nabuurs, G., de Miguel, S., Zhou, M., Picard, N., et al., 2019. Climatic controls of decomposition drive the global biogeography of forest-tree symbioses. Nature 569, 404–408.

Stendahl, J., Berg, B., Lindahl, B.D., 2017. Manganese availability is negatively associated with carbon storage in northern coniferous forest humus layers. Sci. Rep. 7, 15487.

Sterkenburg, E., Clemmensen, K.E., Ekblad, A., Finlay, R.D., Lindahl, B.D., 2018. Contrasting effects of ectomycorrhizal fungi on early and late stage decomposition in a boreal forest. ISME J. 12, 2187–2197.

Štursová, M., Žifčáková, L., Leigh, M.B., Burgess, R., Baldrian, P., 2012. Cellulose utilization in forest litter and soil: identification of bacterial and fungal decomposers. FEMS Microbiol. Ecol. 80, 735–746.

Štursová, M., Šnajdr, J., Cajthaml, T., Bárta, J., Šantrůčková, H., 2014. When the forest dies: the response of forest soil fungi to a bark beetle-induced tree dieback. ISME J. 8, 1920–1931.

Štursová, M., Bárta, J., Šantrůčková, H., Baldrian, P., 2016. Small-scale spatial heterogeneity of ecosystem properties, microbial community composition and microbial activities in a temperate mountain forest soil. FEMS Microbiol. Ecol. 92, fiw185.

Suetsugu, K., Matsubayashi, J., Tayasu, I., 2020. Some mycoheterotrophic orchids depend on carbon from dead wood: novel evidence from a radiocarbon approach. New Phytol. https://doi.org/10.1111/nph.16409.

Sun, Y.P., Unestam, T., Lucas, S.D., Johanson, J.D., Kenne, L., Finlay, R., 1999. Exudation-reabsorption in a mycorrhizal fungus, the dynamic interface for interaction with soil and soil microorganisms. Mycorrhiza 9, 137–144.

Talbot, J.M., Bruns, T.D., Taylor, J.W., Smith, D.P., Branco, S., Glassman, S.I., Erlandson, S., Vilgalys, R., Liao, H.L., Smith, M.E., et al., 2014. Endemism and functional convergence across the North American soil mycobiome. Proc. Natl. Acad. Sci. U. S. A. 111, 6341–6346.

Taniguchi, T., Kitajima, K., Douhan, G.W., Yamanaka, N., Allen, M.F., 2018. A pulse of summer precipitation after the dry season triggers changes in ectomycorrhizal formation, diversity, and community composition in a Mediterranean forest in California, USA. Mycorrhiza 28, 665–677.

Taye, Z.M., Martínez-Peña, F., Bonet, J.A., de Aragón, J.M., de-Miguel S., 2016. Meteorological conditions and site characteristics driving edible mushroom production in *Pinus pinaster* forests of Central Spain. Fungal Ecol. 23, 30–41.

Tedersoo, L. (Ed.), 2017. Biogeography of Mycorrhizal Symbiosis. vol. 828. Springer, Tartu, Estonia.

Tedersoo, L., Kõljalg, U., Hallenberg, N., Larsson, K.H., 2003. Fine scale distribution of ectomycorrhizal fungi and roots across substrate layers including coarse woody debris in a mixed forest. New Phytol. 159, 153–165.

Tedersoo, L., Jairus, T., Horton, B.M., Abarenkov, K., Suvi, T., Saar, I., Kõljalg, U., 2008. Strong host preference of ectomycorrhizal fungi in a Tasmanian wet sclerophyll forest as revealed by DNA barcoding and taxon-specific primers. New Phytol. 180, 479–490.

Tedersoo, L., May, T.W., Smith, M.E., 2010. Ectomycorrhizal lifestyle in fungi: global diversity, distribution, and evolution of phylogenetic lineages. Mycorrhiza 20, 217–263.

Tedersoo, L., Bahram, M., Toots, M., Diédhiou, A.G., Henkel, T.W., Kjoller, R., Morris, M.H., Nara, K., Nouhra, E., Peay, K.G., et al., 2012a. Towards global patterns in the diversity and community structure of ectomycorrhizal fungi. Mol. Ecol. 21, 4160–4170.

Tedersoo, L., Naadel, T., Bahram, M., Pritsch, K., Buegger, F., Leal, M., Kõljalg, U., Põldmaa, K., 2012b. Enzymatic activities and stable isotope patterns of ectomycorrhizal fungi in relation to phylogeny and exploration types in an afrotropical rain forest. New Phytol. 195, 832–843.

Tedersoo, L., Bahram, M., Põlme, S., Kõljalg, U., Yorou, N.S., Wijesundera, R., Ruiz, L.V., Vasco-Palacios, A.M., Thu, P.Q., Suija, A., Smith, M.E., et al., 2014. Global diversity and geography of soil fungi. Science 346, 6213.

Tedersoo, L., Bahram, M., Cajthaml, T., Põlme, S., Hiiesalu, I., Anslan, S., Harend, H., Buegger, F., Pritsch, K., Koricheva, J., Abarenkov, K., 2016. Tree diversity and species identity effects on soil fungi, protists and animals are context dependent. ISME J. 10, 346–362.

Tedersoo, L., Bahram, M., Zobel, M., 2020a. How mycorrhizal associations drive plant population and community biology. Science 367, 867.

Tedersoo, L., et al., 2020b. Regional-scale in depth analysis of soil fungal diversity reveals strong pH and plant species effects in northern Europe. Front. Microbiol. 11, 1953.

Thines, M., 2019. An evolutionary framework for host shifts–jumping ships for survival. New Phytol. 224 (2), 605–617.

Thoen, E., Harder, C.B., Kauserud, H., Botnen, S.S., Vik, U., Taylor, A.F., Menkis, A., Skrede, I., 2020. In vitro evidence of root colonization suggests ecological versatility in the genus *Mycena*. New Phytol. 227, 601–612.

Toju, H., Sato, H., Yamamoto, S., Tanabe, A.S., 2018. Structural diversity across arbuscular mycorrhizal, ectomycorrhizal, and endophytic plant–fungus networks. BMC Plant Biol. 18, 292.

Treseder, K.K., Lennon, J.T., 2015. Fungal traits that drive ecosystem dynamics on land. Microbiol. Mol. Biol. Rev. 79, 243–262.

Truong, C., Gabbarini, L.A., Corrales, A., Mujic, A.B., Escobar, J.M., Moretto, A., Smith, M.E., 2019. Ectomycorrhizal fungi and soil enzymes exhibit contrasting patterns along elevation gradients in southern Patagonia. New Phytol. 222, 1936–1950.

van der Heijden, M.G., Martin, F.M., Selosse, M.A., Sanders, I.R., 2015. Mycorrhizal ecology and evolution: the past, the present, and the future. New Phytol. 205, 1406–1423.

van der Heyde, M., Abbott, L.K., Gehring, C., Kokkoris, V., Hart, M.M., 2019. Reconciling disparate responses to grazing in the arbuscular mycorrhizal symbiosis. Rhizosphere 11, 100167.

van der Linde, S., Suz, L.M., Orme, C.D.L., Cox, F., Andreae, H., Asi, E., Atkinson, B., Sue, S., Carroll, C., Cools, N., et al., 2018. Environment and host as large-scale controls of ectomycorrhizal fungi. Nature 561, E42.

van Nuland, M.E., Peay, K.G., 2020. Symbiotic niche mapping reveals functional specialization by two ectomycorrhizal fungi that expands the host plant niche. Fungal Ecol. 46, 100960.

van Nuland, M.E., Smith, D.P., Bhatnagar, J.M., Stefanski, A., Hobbie, S.E., Reich, P.B., Peay, K.G., 2020. Warming and disturbance alter soil microbiome diversity and function in a northern forest ecotone. FEMS Microbiol. Ecol. https://doi.org/10.1093/femsec/fiaa108.

Vanbeveren, S.P., Ceulemans, R., 2019. Biodiversity in short-rotation coppice. Renew. Sust. Energ. Rev. 111, 34–43.

Varenius, K., Kårén, O., Lindahl, B., Dahlberg, A., 2016. Long-term effects of tree harvesting on ectomycorrhizal fungal communities in boreal Scots pine forests. For. Ecol. Manag. 380, 41–49.

Vasiliauskas, R., Vasiliauskas, A., Stenlid, J., Matelis, A., 2004. Dead trees and protected polypores in unmanaged north-temperate forest stands of Lithuania. For. Ecol. Manag. 193, 355–370.

Visser, S., Maynard, D., Danielson, R.M., 1998. Response of ecto-and arbuscular mycorrhizal fungi to clear-cutting and the application of chipped aspen wood in a mixedwood site in Alberta, Canada. Appl. Soil Ecol. 7, 257–269.

Voříšková, J., Baldrian, P., 2012. Fungal community on decomposing leaf litter undergoes rapid successional changes. ISME J. 7, 477–486.

Voříšková, J., Brabcová, V., Cajthaml, T., Baldrian, P., 2014. Seasonal dynamics of fungal communities in a temperate oak forest soil. New Phytol. 201, 269–278.

Voyron, S., Ercole, E., Ghignone, S., Perotto, S., Girlanda, M., 2017. Fine-scale spatial distribution of orchid mycorrhizal fungi in the soil of host-rich grasslands. New Phytol. 213, 1428–1439.

Vucetich, J.A., Waite, T.A., Qvarnemark, L., Ibargüen, S., 2000. Population variability and extinction risk. Conserv. Biol. 14, 1704–1714.

Walker, J.K.M., Ward, V., Paterson, C., Jones, M.D., 2012. Coarse woody debris retention in subalpine clearcuts affects ectomycorrhizal root tip community structure within fifteen years of harvest. Appl. Soil Ecol. 60, 5–15.

Walther, G.R., Post, E., Convey, P., Menzel, A., Parmesan, C., Beebee, T.J., et al., 2002. Ecological responses to recent climate change. Nature 416, 389–395.

Wang, T., Tian, Z.M., Tunlid, A., Persson, P., 2020. Nitrogen acquisition from mineral-associated proteins by an ectomycorrhizal fungus. New Phytol. 228, 697–711.

Wardle, D.A., 2006. The influence of biotic interactions on soil biodiversity. Ecol. Lett. 9, 870–886.

Wells, J.M., Harris, M.J., Boddy, L., 1998. Temporary phosphorus partitioning in mycelial systems of the cord-forming basidiomycete *Phanerochaete velutina*. New Phytol. 140, 283–293.

Wilhelm, R.C., Cardenas, E., Maas, K.R., Leung, H., McNeil, L., Berch, S., Chapman, W., Hope, G., Kranabetter, J.M., Dubé, S., et al., 2017. Biogeography and organic matter removal shape long-term effects of timber harvesting on forest soil microbial communities. ISME J. 11, 2552–2568.

Wu, B., Hussain, M., Zhang, W., Stadler, M., Liu, X., Xiang, M., 2019. Current insights into fungal species diversity and perspective on naming the environmental DNA sequences of fungi. Mycology 10, 27–140.

Yang, X., Luedeling, E., Chen, G., Hyde, K.D., Yang, Y., Zhou, D., Xu, J., Yang, Y., 2012. Climate change effects fruiting of the prize matsutake mushroom in China. Fungal Divers. 56, 189–198.

Yeung, E.C., 2017. A perspective on orchid seed and protocorm development. Bot. Stud. 58, 33.

Zanne, A.E., Abarenkov, K., Afkhami, M.E., Aguilar-Trigueros, C.A., Bates, S., Bhatnagar, J.M., Busby, P.E., Christian, N., Cornwell, W.K., Crowther, T.W., Flores-Moreno, H., et al., 2020. Fungal functional ecology: bringing a trait-based approach to plant-associated fungi. Biol. Rev. 95, 409–433.

Zavišić, A., Yang, N., Marhan, S., Kandeler, E., Polle, A., 2018. Forest soil phosphorus resources and fertilization affect ectomycorrhizal community composition, beech P uptake efficiency, and photosynthesis. Front. Plant Sci. 9, 463.

Zhao, J., Wan, S., Fu, S., Wang, X., Wang, M., Liang, C., Chen, Y., Zhu, X., 2013. Effects of understory removal and nitrogen fertilization on soil microbial communities in Eucalyptus plantations. For. Ecol. Manag. 310, 80–86.

Zhao, Q., Jian, S., Nunan, N., Maestre, F.T., Tedersoo, L., He, J., Wei, H., Tan, X., Shen, W., 2017. Altered precipitation seasonality impacts the dominant fungal but rare bacterial taxa in subtropical forest soils. Biol. Fertil. Soils 53, 231–245.

Žifčáková, L., Větrovský, T., Howe, A., Baldrian, P., 2015. Microbial activity in forest soil reflects the changes in ecosystem properties between summer and winter. Environ. Microbiol. 18, 288–301.

Chapter 14

The influence of mycorrhizal fungi on rhizosphere bacterial communities in forests

David J. Burke and Sarah R. Carrino-Kyker

The Holden Arboretum, Kirtland, OH, United States

Chapter Outline

1. Forest soil as a microbial landscape

"Everything is everywhere but the environment selects" (O'Malley, 2008). This general concept has existed since the 19th century and was meant to describe both the ubiquities of microbial species and the importance of the environment for structuring microbial community patterns and species distribution. It was thought that the smaller the organism, the more likely dispersal barriers would not exist, and the species would be universal in their distribution. But our concept of microbial species and biogeography in the early 21st century has changed, largely as a product of the new perspective that modern molecular methods have provided. Studies have shown that microbial communities can vary across environmental gradients ranging from thousands of kilometers or even within a few centimeters in soil (Berg, 2012; Fierer and Jackson, 2006). At even finer scales, microbial communities are products of strong environmental selection for characteristics and physiologies that enable microbes to coexist under a specific set of conditions (Paerl and Pinckney, 1996). Certainly, the presence of different species of soil microbes at coarse scales, such as across a landscape, could be partly the product of dispersal barriers; however, dispersal barriers would have to be especially strong to prevent microbes from moving a few centimeters in soil and this seems unlikely. Rather, environmental selection might be especially important for diversification of soil microbes at finer scales, such as within a particular soil patch. Here we define a soil patch as the area across which soil microorganisms are not dispersal limited, but where environmental filtering is a driver of community composition. As is the case for macro-organisms, contemporary environmental conditions and historical events influence the distribution and abundance of microbial species (Horner-Devine et al., 2007; Martiny et al., 2006) and local environmental conditions certainly play a major role in structuring microbial communities. But how substantially does the environment change within a few centimeters or even millimeters of forest soil?

Although microbes are small and operate at very fine spatial scales (micron level; see Berg, 2012), they nonetheless occupy a patchy environmental landscape, albeit, one that can be difficult for us to perceive (Fig. 14.1). One essential problem in understanding patterns of microbial distribution, diversity, and function is our inability to perceive and quantify the landscape features important to microbes. In fact, landscape heterogeneity at the micro-scale likely drives microbial biodiversity, functional diversity, and ecosystem function. Microbial landscape heterogeneity is affected by several factors, such as changes in soil texture, plant root activity, and the actions of other biota including soil fungi. Although fungi are also responding to the heterogeneity of the soil landscape, they, in fact, can affect that microbial landscape, especially as it

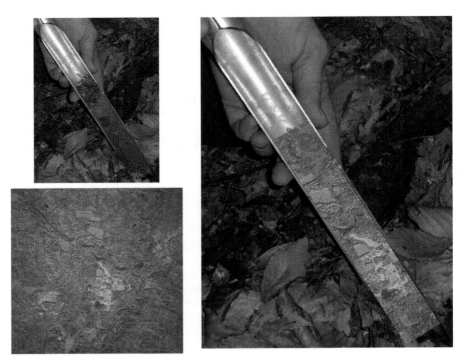

FIG. 14.1 A soil core may represent not a habitat patch, but a soil landscape, that contains many habitat patches that can be colonized by different groups of bacteria, which are altered by the inherent environment of each patch. (The landscape shown here is from an aerial photograph of the Holden Arboretum and natural areas. The soil core pictured was collected from an old growth, high quality hardwood forest within Stebbins Gulch, a 334 ha mixed-mesophytic forest located at Holden Arboretum.)

relates to soil bacteria, given that fungal hyphae intimately associate with these organisms (Artursson and Jansson, 2003; Bonafante and Anca, 2009; Sharma et al., 2008). For example, bacterial functional groups, such as denitrifying bacteria and methane-oxidizing bacteria, are significantly affected by the distribution of fungal biomass in soil and the presence of extracellular enzyme activity (Burke et al., 2012). Thus, fungi not only occupy the soil landscape and respond to it, but fungi also modify the patchiness of the landscape in such a way that resource availability and environmental conditions are changed for other soil microbes (i.e., a form of autogenic succession). There is ample evidence that forest fungi can affect the soil landscape and affect bacteria through three mechanisms: (1) the influence that fungi have on soil nutrient and carbon (C) availability through decomposition; (2) the effect of fungi on soil chemistry, especially in the rhizosphere and; (3) the effects of fungi on root growth and rhizodeposition.

It is well accepted that soil fungi are one of the most important microbial groups within temperate forest soil and are vital to many ecosystem processes. These processes include nutrient and C cycling and primary productivity due to their direct connections to plants and facilitation of plant growth. Fungi are also an important resource for other organisms both in the soil (i.e., hyphae are an important food source to soil-dwelling invertebrates) and above ground (i.e., sporocarps are an important food source to insects and other animals). The effects of fungi on bacteria and the processes they mediate tend to be less well recognized, but are nonetheless of great potential importance. Fungal communities in forests are spatially structured by the availability of suitable plant hosts and high heterogeneity of nutrient sources at fine scales (Baldrian, 2017; Burke, 2015; Burke et al., 2009). Soil pH can also strongly affect soil fungal communities and their functional activities vis-à-vis production of extracellular enzymes, and, consequently, nutrient availability (Burke et al., 2009; Carrino-Kyker et al., 2016; Kluber et al., 2012; Rousk et al., 2009, 2010). The presence of coarse woody debris can also largely shape fungal communities, including mycorrhizal fungi (Tedersoo et al., 2003) and saprotrophic fungi (Allmér et al., 2006). At small spatial scales, though, the strongest environmental drivers in soil exist within the plant root zone and through vertical stratification (Berg, 2012). Fungi not only respond to these fine-scale environmental gradients within soil patches, but also contribute to the patchy nature (Fig. 14.2).

The effects of fungi may be greatest within the region of soil immediately surrounding and influenced by the activity of plant roots, which is called the rhizosphere (Fig. 14.3). This is the case because more than 85% of all plant species form relationships with mycorrhizal fungi which colonize the plant root in a mutually beneficial relationship where nutrients and C are exchanged between the fungi and plant host (Smith and Read, 2008). Due to this intimate association with the

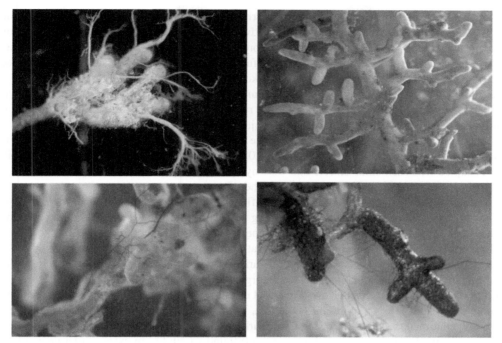

FIG. 14.2 The patchiness of the soil landscape is affected by mycorrhizal fungi and the degree of extraradical hyphae production. The species of *Tricholoma* pictured here *(top left)* forms a large mass of extraradical hyphae, including rhizomorphs, whereas the species of *Russula* pictured *(top right)* has a contact morphology with few extraradical hyphae. The species of *Cortinarius (bottom left)* and *Cenococcum (bottom right)* pictured display medium- and short-distanced exploration types, respectively, with hairy mantles and emulating hyphae and, for *Cortinarius*, filamentous rhizomorphs. These differences in mycorrhizal species and morphology can alter the extent of the soil landscape modified by them.

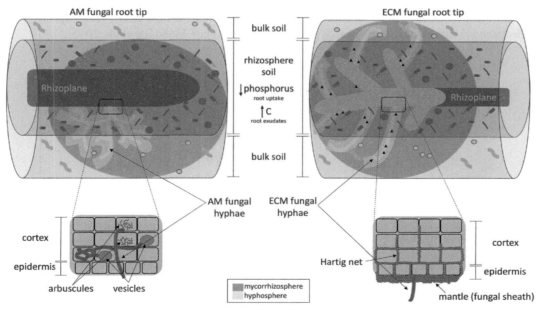

FIG. 14.3 Bacteria in forest soil are found directly on plant roots (rhizoplane), in the zone of soil immediately surrounding plant roots (rhizosphere), and in soil not penetrated by plant roots (bulk soil). Soil bacteria are most abundant and diverse in the rhizosphere. Rhizosphere soil is influenced by plant roots, such as through root exudation, which can increase carbon availability, and through nutrient uptake, which can decrease phosphorus availability. In addition, plant roots are often times colonized by mycorrhizal fungi, which also influence soil bacteria themselves. The two most common types of mycorrhizal fungi are arbuscular mycorrhizal (AM) fungi (shown on the *left*) and ectomycorrhizal (ECM) fungi (shown on the *right*). Mycorrhizal fungi extend their hyphae beyond the rhizosphere and into bulk soil. The zone of soil immediately surrounding the hyphae is known as the hyphosphere and the zone of influence of mycorrhizal fungi is known as the mycorrhizosphere; these are shown shaded in blue. The rhizosphere of roots colonized by AM fungi can differ from the rhizosphere of roots colonized by ECM fungi because ECM fungi contribute to extracellular enzyme production (shown by *black triangles*), which can be greater in the rhizosphere than surrounding bulk soil.

plant root and their presence within the rhizosphere, mycorrhizal fungi can have significant effects on chemistry and C availability and, thereby, affect rhizosphere bacteria. One advantage of mycorrhizal fungi is that their fungal filaments, called hyphae, forage and extend widely into soil and, therefore, can explore a larger volume of soil than plants roots alone (Fig. 14.3; Smith and Read, 2008). This allows mycorrhizal fungi to access nutrients outside the area immediately around the root which can be depleted of nutrients due to root uptake (i.e., the depletion zone; Karandashov and Bucher, 2005). Thus mycorrhizal fungi have access to a greater potential volume of soil and mass of soil nutrients. This means that mycorrhizal hyphae can affect a much larger volume of soil than the root can itself, often beyond the root's zone of influence. This region of soil influenced by the mycorrhizas alone or in conjunction with the root is referred to as the mycorrhizosphere (Fig. 14.3). In addition, because hyphae are very small in size (2–20 μm in diameter), they can also extend into very small soil pore spaces (Johnson and Gehring, 2007; Smith et al., 2010). The ability to affect the chemistry of soil, especially in small pore spaces, coupled with their small size, means that mycorrhizal fungi can interact intimately with soil bacteria and influence the composition of the communities and possibly their function.

The fine-scale effects of fungi on bacteria are likely also influenced by vertical soil stratification (Fig. 14.4). Fungal community changes have been consistently found with increasing soil depth (Baldrian et al., 2012; Burke, 2015; Clemmensen et al., 2013; Coince et al., 2013; Dickie et al., 2002; Jumpponen et al., 2010; Tedersoo et al., 2003; Voříšková et al., 2014). In a review by Bahram et al. (2015), fine-scale vertical stratification was found to be a stronger driver of mycorrhizal community structure than spatial (horizontal) or temporal variability. Such vertical separation is believed to be related to decomposition of organic matter where the litter layer and organic (O) horizon are supplied with C from leaf litter, while the lower A and B horizons consist of soils that are of mineral origin where high quality C substrates are depleted (Bödeker et al., 2016; Lindahl et al., 2007). It is important to note that, for many forests, the transition from organic soil to mineral soil can be over a few centimeters (i.e., 2 cm or less; see Burke, 2015) and that nutrient and C variation can exist across very small vertical gradients. For example, Šnajdr et al. (2008) found variability in extracellular enzyme activity across a depth of only 1 cm in organic soils of a hardwood forest. These changes in enzyme production corresponded with changes in bacterial taxa. Because bacteria are not dispersal limited across these small changes in soil depth, vertical changes to the bacterial community are likely due to specialized taxa that exploit the substrates available at shallow depths where soils are organic vs deeper mineral soil (Uroz et al., 2016). The influence of mycorrhizal fungi on altering the soil environment over small spatial gradients (both horizontal and vertical) is poorly understood. It is likely that certain bacterial taxa may be enriched within the O horizon by the fungi (including mycorrhizal fungi) actively decomposing leaf litter. Whereas rhizosphere bacteria in roots from deeper, more mineral soil layers are likely more influenced by rhizodeposits, root death, and turnover, and death and decay of the mycorrhizal hyphae themselves (López-Mondéjar et al., 2015).

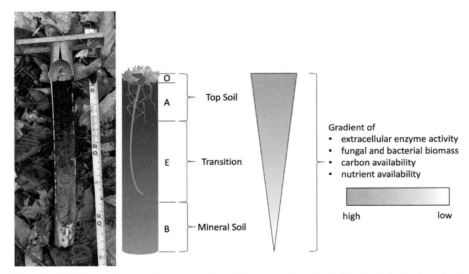

FIG. 14.4 Forest soil formation is influenced by organic matter cycling. Much organic matter in forests originates from forest plants, such as leaves or woody plant debris, and leads to nutrient-rich top soil, consisting of the O horizon, sometimes referred to as the litter layer, which is characterized as having a high percentage of organic matter, and A horizon, with soils originating from the parent material, but also containing organic matter. Below the topsoil is a transition zone (E horizon) before the mineral soil of the B horizon. Pictured here is a 20-cm soil profile (10–30 cm on the ruler) in a deciduous forest at Holden Arboretum, Ohio, United States. The O horizon has been removed from the pictured soil core. Typical soils for these forests have an A horizon that extends to a 5-cm depth. The top 5-cm of the soil are the most biologically active, as the majority of plant roots are found here. Like many other deciduous, hardwood forests, nutrient availability decreases with soil depth, due in part to decreases in decomposition.

1.1 Fungal influences on nutrient and carbon availability through decomposition

Fungi generally influence nutrient and C availability through their primary role as decomposers of organic matter. Decomposition of leaf litter is often considered the dominant pathway of organic matter cycling in forests, but root turnover can also serve as an important decomposition pathway and source of nutrients and C (Burke and Raynal, 1994; Joslin and Henderson, 1987). Therefore, decomposition is a spatially and vertically segregated process, with litter decomposition taking place close to the surface of the forest soil, while root decomposition and turnover can take place at any depth in which roots are present. Shallow, organic soils of forests have been shown to have greater decomposition, as measured with extracellular enzyme analysis (EEA), compared to mineral soils (Uroz et al., 2013). These organic soils, particularly in the litter layer, also have higher fungal and bacterial biomass, a higher ratio of fungal to bacterial biomass, higher nutrient and carbon availability, and greater metabolic diversity (Baldrian et al., 2012; Eilers et al., 2012; Uroz et al., 2013). In soil samples collected monthly for 3 years, Burke (2015) showed that the shallowest soil depth sampled (2-cm depth) consistently had the highest P (both inorganic and organic), N, and C to N ratio compared to lower soil depths likely because of the proximity to leaf litter as a nutrient source. Bacterial community composition also changes between organic and mineral soils in forests (e.g., Turlapati et al., 2013; Table 14.1). In a study by Baldrian et al. (2012), bacterial community differences were found in bulk soil between the litter layer and top soil at a lower soil depth (called the H horizon in Baldrian et al. (2012) but similar to the A horizon shown in Fig. 14.4). Baldrian et al. (2012) found higher bacterial diversity in the litter layer and some taxonomic changes between soil horizons (Table 14.1). In the bulk soil from a Norway spruce plantation, Uroz et al. (2013) found that organic soil samples were enriched with Proteobacteria (alpha-, beta-, and gamma-Proteobacteria), Bacteroidetes (Bacteroidia, Cytophagia, Flavobacteria, and Sphingobacteria), and Verrucomicrobia (Opitutae), while mineral soils mainly harbored relatively more Firmicutes (Bacilli and Clostridia) and Chloroflexi (Dehalococcoidetes) (Table 14.1). These authors suggested that these changes are due to bacterial species that are specialized for the carbon rich and complex substrates found at more shallow soil depths (Uroz et al., 2013) and could also explain the enrichment of these bacterial groups in the rhizosphere where plant exudates are abundant. However, the overall contribution of bacteria to litter decomposition in forests is considered relatively small. In a proteomic study, Schneider et al. (2012) found that the extracellular enzymes on beech leaf litter were all of fungal origin and suggested that the bacterial production of extracellular enzymes was too low to be of importance. Other studies have suggested an expanded role for bacteria during decomposition in forest soils (see Lladó et al., 2017). Regardless of their role in decomposition, bacterial abundance and diversity are known to be shaped by extracellular enzyme production and successional decay (Schneider et al., 2012; Lladó et al., 2017), which indicates that bacteria are likely to respond to changes in nutrient and C availability brought about through decomposition.

The majority of studies on how decomposition affects bacteria across fine spatial gradients have focused on bulk soil even though mineralization and extracellular enzyme production can be higher in the rhizosphere (Finzi et al., 2014). Mycorrhizal fungi contribute to decomposition within the rhizosphere because some ectomycorrhizal (ECM) fungi are known to have saprotrophic capability (Bödeker et al., 2014; Courty et al., 2010; Talbot et al., 2013). For example, taxa in the genus *Russula* can produce both phosphatase and chitinase in culture (Burke et al., 2014). ECM fungi displaying this function contribute to decomposition of organic matter inputs, such as leaf litter, but also play a role in nutrient recovery from dying roots. Mycorrhizal fungi are also known to heighten organic P degradation within the rhizosphere through phosphatase production (Cumming et al., 2015). However, this increase in phosphatase activity may not necessarily increase inorganic P availability due to much of it being transferred to the plant host (Cumming et al., 2015) and because the soil immediately surrounding plant roots tends to be limited in P because the rate of absorption by plant roots is much higher than the rate of diffusion from soil (Karandashov and Bucher, 2005). Because not all ECM fungi display saprotrophic capabilities, the species of mycorrhizal fungi colonizing a plant root is important for nutrient and C availability within the rhizosphere and mycorrhizosphere. Such nutrient and C changes likely affect bacterial abundance and diversity within the rhizosphere, as they do in bulk soil and forest litter; however, the direct effects of decomposition on bacteria within the rhizosphere are largely unexplored.

Mycorrhizal fungi themselves can be a significant source of nutrients and C within the rhizosphere. It has been estimated that anywhere from 5% to 50% of microbial biomass can come from arbuscular mycorrhizal (AM) fungi in natural systems (Johnson and Gehring, 2007) and in forests, as much as 30% of microbial biomass may originate from ECM fungi (Högberg and Högberg, 2002). Some estimates of ECM hyphae approach 200m per gram of dry soil (Read and Boyd, 1986). However, the input of C will depend on the species of mycorrhizal fungi present within the soil and on plant roots. ECM fungi, for example, vary greatly in their hyphal morphology (see Agerer, 2001, 2003). Species of ECM fungi that have contact morphotypes may input less C as hyphae compared to species with medium or long-distance exploration strategies that produce large amounts of extraradical hyphae. In our own experience with temperate deciduous forests, total fungal biomass often ranges between 3.5% and 6.0% of total microbial biomass and AM fungal biomass between 2.8% and 3.2%

TABLE 14.1 Recently published studies that found differences in bacterial taxa between soil horizons in forests.[a]

References	Habitat	Bacterial taxa in litter layer	Bacterial taxa in organic soil	Bacterial taxa in mineral soil
Baldrian et al. (2012)[b]	Spruce forest (Bohemian Forest mountain range)	DNA community: Actinobacteria (Frankineae, *Ferrithrix*, Actinomycetales), Aquificae, Bacteroidetes, Cyanobacteria, Fibrobacteres, Chlamydiae, Chloroflexi, Nitrospira, Proteobacteria (*Steroidobacter*, *Afipia*, *Burkholderia*, *Dyella*), Spirochaetes, Verrucomicrobia (other taxa: Acidobacteria Gp1, Acidobacteria Gp2, Acidobacteria Gp3) RNA community: Firmicutes, Synergistetes (other taxa: *Steroidobacter*, Acidobacteria Gp1, Acidobacteria Gp2, Acidobacteria Gp3, Frankineae, *Burkholderia*, Acetobacteraceae, *Rhodovastum*, *Phenylobacterium*, *Chondromyces*, Caulobacteraceae)	DNA community: Acidobacteria (Gp1, Gp2), Firmicutes (*Paenibacillus*, *Spromusa*), Gemmatimonadetes (other taxa: Actinomycetales, *Desulfomonile*, Rhizobiales, *Mycobacterium*) RNA community: Actinobacteria (*Actinomadura*, Actinomycetales, *Actinoallomurus*, *Mycobacterium*), Thermodesulfobacteria (other taxa: Acidobacteria Gp1, Acidobacteria Gp2, Acidobacteria Gp3, *Paenibacillus*, Rhodospirillales, Rhizobiales)	N/A (mineral soil not sampled)
Eilers et al. (2012)[c]	Upper montane forest (Gordon Gulch watershed)	N/A (litter layer not explicitly sampled)	Bacteroidetes, Alphaproteobacteria (Caulobacterales), Betaproteobacteria, Gammaproteobacteria	Verrucomicrobia
López-Mondéjar et al. (2015)[d]	Sessile oak (*Quercus petraea*) forest (Xaverovský Háj Natural Reserve)	Armatimonadetes, Bacteroidetes, Chloroflexi, Proteobacteria (Alphaproteobacteria, Betaproteobacteria, Gammaproteobacteria	Acidobacteria (Acidobacteria Gp1, Acidobacteria Gp3, Acidobacteria Gp13), Armatimonadetes, Chloroflexi, Gemmatimonadetes (other class: Gammaproteobacteria)	Acidobacteria (Acidobacteria Gp1, Acidobacteria Gp2, Acidobacteria Gp3, Acidobacteria Gp13), Firmicutes, Planctomycetes
Turlapati et al. (2013)	Mixed hardwood forest (Harvard Forest)	N/A (litter layer not sampled)	Actinobacteria	Chloroflexi, Firmicutes, Nitrospira
Uroz et al. (2013)	Norway spruce plantation (Breuil-Chenue LTO)	N/A (litter layer not sampled)	Proteobacteria (Alphaproteobacteria, Betaproteobacteria, Gammaproteobacteria), Bacteroidetes (Sphingobacteria, Cytophagia, Bacteroidia, *Flavobacterium*), Verrucomicrobia (Opitutae) (other class: Cloeobacteria)	Firmicutes (Bacilli, Clostridia, Dictyoglomia), Chloroflexi (Chloroflexi, Dehalococcoidetes), Deinococcus-Thermus (*Deinococci*), Chlorobi (*Chlorobium*, Thermomicrobia), Thermotogae, Aquificae, *Dictyoglomus*, Fusobacteria (Fusobacteria) (other classes: Solibacteres, Deltaproteobacteria, Epsilonproteobacteria)

[a]Bacterial taxa are listed under the horizon in which they were more abundant. Only taxa with statistically significant differences in nucleic acid sequence abundance are shown. Phyla are listed with lower taxonomic levels in parentheses.

[b]The statistical difference between soil horizons in Baldrian et al. (2012) was performed at the OTU level; thus some taxa are found as statistically more abundant in both horizons.

[c]Eilers et al. (2012) did not explicitly compare organic and mineral soil, but rather examined bacteria across soil depths; shown here are bacterial taxa found more often at shallow or deeper soil depths.

[d]The data reported from López-Mondéjar et al. (2015) show statistical differences at a higher taxonomic resolution; this study also compared bacteria between soil horizons at the OTU level, which can be found in their Supplemental Table 2.

based on PFLA analysis (Burke et al., 2012; Carrino-Kyker et al., 2016). The majority of microbial biomass in these forests is dominated by bacteria (bacterial fungal ratios approaching 20:1). Nonetheless, mycorrhizal biomass may be a major input of C to these soils given that hyphae excrete C, and hyphal turnover can be quite high with many ECM root tips developing and senescing within 4 months (Rygiewicz et al., 1997). An estimated 10%–20% of net primary productivity may be allocated to ECM fungi, with a range of 5%–85% depending on the system (Treseder and Allen, 2000); thus mycorrhizal fungi are an important C resource for organisms that consume fungi.

1.2 Mycorrhizal fungal influences on rhizodeposition

The rhizosphere is a biogeochemical hot spot because of rhizodeposits from plant roots. Rhizodeposits constitute C actively excreted or lost from roots within the soil matrix. C cycling is an efficient process within the rhizosphere where C is rarely a limiting nutrient, even if the bulk soil is deficient in C (Koch et al., 2001; van Overbeek et al., 1997). Bacteria that grow in environments rich in nutrients, particularly carbon (i.e., copiotrophic bacteria) are found more often in the rhizosphere than bulk soil (Semenov et al., 1999). Further, these bacteria respond to the patchy nature of nutrient availability within the rhizosphere. Even along the length of a root, the amount of C being excreted varies, with highest levels nearest the root tip where high bacterial, particularly copiotroph, abundance can be found (Semenov et al., 1999). High abundance of both copiotrophic and oligotrophic bacteria was also found in the middle part of the root, even though carbon measurements were low, and Semenov et al. (1999) attributed this to carbon consumption outweighing production. At the root tip and base, though, these authors suggest that production outweighs consumption in two different fashions. At the root tip, there may be a lag in microbial growth as bacteria exit from their dormant state once root growth is evident, while at the base of the root production of C may be driven by decay of cortical cells (Semenov et al., 1999). Based on these data, it has been further proposed that r strategists (i.e., copiotrophs) dominate areas with high nutrient availability, such as root hairs, while K strategists (i.e., oligotrophs) dominate areas with low nutrient availability, such as older roots (Buée et al., 2009).

These relationships can be altered by the type of mycorrhizal colonization in forests since mycorrhizal fungi are not only affected by plant rhizodeposition, but can alter exudation once they colonize the plant root (see review by Jones et al., 2004). For example, AM fungal colonization is often greatest on lower order roots and declines as root order increases and roots become suberized and develop secondary xylem (Eissenstat et al., 2015; Guo et al., 2008). This could lead to variation in C excretion by roots, both in terms of quantity and quality, of trees in genera such as *Acer*, *Fraxinus*, and *Liriodendron*, all of which form relationships with AM fungi with concomitant changes to bacterial population size and community structure.

Root anatomy can also affect AM colonization, with tree species in the Magnoliaceae, which have thick roots, being more heavily colonized by AM fungi than trees that form thin roots, such as species in the Sapindaceae (e.g., the genus *Acer*; Eissenstat et al., 2015). This may also be the case for most forest herbaceous plants which form relationships with AM fungi (Brundrett and Kendrick, 1990); the activities of herbaceous plants and their AM fungi likely influence the spatial patchiness of bacterial communities (Burke and Chan, 2010). On the contrary, ECM fungi colonize the root tip themselves, and once the tip is colonized, its growth is under the control of the ECM fungi and its morphology responds to the colonization as well (Smith and Read, 2008). Since ECM fungi vary greatly in terms of their hyphal morphology and extent of extraradical hyphae production (Fig. 14.2; Agerer, 2001), the identity of the ECM fungi colonizing the root could, in principle, have large effects on the bacterial communities. But, recent studies have found that plant nutrient content and microsite environment around the ECM tips may have greater effects on bacterial communities than the identity of the ECM species colonizing the root (Burke et al., 2008; Izumi et al., 2008; Kretzer et al., 2009; Uroz et al., 2012). Nonetheless, the role of ECM species and how differences in hyphal growth alter soil bacteria in the mycorrhizosphere deserve greater study.

1.3 Other chemical and physical changes in the rhizosphere influenced by mycorrhizal fungi

In addition to affecting soil nutrient and C availability, fungi can also impact the soil landscape through other chemical changes within the rhizosphere. Because these chemical changes come about through the direct actions of plants roots, fungi act more as a modifier of conditions within the rhizosphere. One prominent way for roots and fungi to modify the chemistry of the rhizosphere is through cation exchange processes that alter rhizosphere pH. It is well accepted that plant roots can acquire various forms of N from soil, such as nitrate $\left(NO_3^-\right)$ or ammonium $\left(NH_4^+\right)$, and this process results in changes to the pH of the rhizosphere. In order to maintain electro-neutrality across the root membrane, plants will release H^+ or OH^- as they absorb NH_4^+ or NO_3^- (Jaillard et al., 2003). The form of N uptake dictates the direction of rhizosphere pH changes. If the uptake of N is in the form of NH_4^+, plants release H^+ ions, which can make rhizosphere acidification more pronounced. If N uptake is in the form of NO_3^-, the release of OH^- or HCO_3^- can occur, which leads to an increase in rhizosphere pH (reviewed in Jones et al., 2004). The excretion of plant C can also result in significant rhizosphere acidification.

Organic acids can be present in root exudates and can alter rhizosphere pH. Acidic compounds are also excreted due to iron deficiency (see Hawkes et al., 2007). Mycorrhizal fungi can increase the intensity of acidification in the rhizosphere (Bago and Azcón-Aguilar, 1997) or can act as rhizosphere buffers, modifying the natural acidification that occurs due to nutrient uptake (Rygiewicz et al., 1984). It is well known that soil pH is a strong driver of soil bacterial community structure at both broad and local scales (Fierer and Jackson, 2006; Lauber et al., 2009; Rousk et al., 2010), which makes the activities of mycorrhizal fungi that influence pH in the rhizosphere important for bacteria. In temperate forests, Landesman et al. (2014) showed that pH was the strongest driver of bacterial community structure. Compositional turnover with varying soil pH was greatest for bacteria in the Acidobacteria, Verrucomicrobia, and Proteobacteria phyla (Landesman et al., 2014). This compositional turnover, however, was also influenced by the proximity to different tree species. Interestingly, this effect was related to differing soil pH associated with the tree species (i.e., soils were more acidic under *Fagus grandifolia* compared to *Acer saccharum*; Landesman et al., 2014). The authors concluded that the mechanism of pH change could be differing activities of these tree species belowground, such as greater organic acid excretion by *Fagus grandifolia*, which lowers pH, and greater calcium transfer by *Acer saccharum*, which increases pH (Landesman et al., 2014). Though this study was conducted across a latitudinal gradient, because the local effect of individual tree species was also explored, hypotheses about pH effects on bacteria at fine scales can be drawn. For example, these pH differences between tree species could be a by-product of the litter quality produced by each species, which is influenced by the fungal decomposers in the immediate vicinity of the trees. In addition, *Acer saccharum* associates with AM fungi, while *Fagus grandifolia* associates with ECM fungi; as such, pH differences between tree species may also be affected by the mycorrhizal fungi associated with the trees.

The rhizosphere also has high levels of oxygen consumption due to the high level of biological activity from plant roots, and associated fungi and bacteria. This can lead to reductions in soil oxygen levels within the rhizosphere and creation of anaerobic conditions (Hawkes et al., 2007; Sørensen, 1997). Soil fungi and mycorrhizal fungi can contribute to this depletion of soil oxygen and the creation of anaerobic to microaerobic conditions where oxygen levels are suppressed. Recent work has found a significant correlation between the presence of fungal biomass in forest soil and the distribution of microbial functional groups such as denitrifying bacteria that require microaerophilic conditions with low oxygen levels (Burke et al., 2012). This suggests that fungi can reduce soil oxygen levels and create microaerophilic conditions needed for some bacterial functional groups. These microaerophilic patches within the broader well aerated space of forest soil probably add significantly to the heterogeneity of the soil landscape.

Another way that fungi can influence soil is through their effect on soil structure brought about through particle aggregation. Mycorrhizal fungi have likely greater effects on soil structure than saprotrophic fungi as they often serve as a net input of carbon into the soil and it is the excreted carbon that effectively binds soil mineral particles, such as clay, together (see Smith and Read, 2008 for review). As saprotrophic fungi respire soil organic matter, their role in soil structure formation is probably smaller than that of mycorrhizal fungi. In addition to the influence of excreted C on soil aggregation, fungal hyphae can also serve to bind soil particles together. Particle size, and by extension pore size created by those soil particles (i.e., sand, silt, and clay), has a large influence on bacterial taxa, with Alphaproteobacteria dominating areas with larger particle sizes and Acidobacteria preferring smaller particle sizes (Sessitsch et al., 2001). How fungi alter pore size, therefore, has a critical role to play in structuring the soil landscape and facilitating different bacterial groups.

2. Interactions between mycorrhizal fungi and rhizosphere bacteria

2.1 Community structure of rhizosphere bacteria associated with ectomycorrhizal fungi

The mycorrhizosphere is the zone of soil influenced by roots colonized by mycorrhizal fungi (Fig. 14.3) and the mycorrhizosphere effect on other rhizosphere inhabitants is an area of much current attention (Priyadharsini et al., 2016). This is largely because it is nearly impossible to understand the rhizosphere effect on bacterial communities without incorporating mycorrhizal fungi. Cumming et al. (2015) made the assertion that because the majority of forest tree roots are colonized by mycorrhizal fungi, understanding the recruitment of bacteria by the plant host is nearly impossible without a consideration of recruitment by the associated mycorrhizal fungi. Our understanding of rhizosphere bacteria is that they exist predominantly in mutualistic relationship with other rhizosphere microorganisms due to root exudates that control biotic interactions within this zone of soil and promote mutualisms (Rasmann and Turlings, 2016). For example, mycorrhiza helper bacteria are commonly found in rhizosphere soils (Box 14.1). Though some examples of antagonistic interactions between mycorrhizal fungi and rhizosphere bacteria have been found (e.g., Stark and Kytöviita, 2005), much of the published literature has focused on how mycorrhizal fungi and bacteria act in synergy to promote plant health (e.g., through nitrogen uptake; Hestrin et al., 2019).

Box 14.1 Mycorrhiza helper bacteria

Mycorrhiza helper bacteria (MHB) are a subset of rhizosphere bacteria that evidence suggests can improve the function of the mycorrhizal symbiosis. This improvement may occur due to increases in nutrient availability or increases in the protection of plants from soil pathogens (Frey-Klett et al., 2007). In this last sense, MHB may be seen as a subset of plant growth promoting bacteria (PGPB) which are known to improve plant growth through their ability to increase plant disease resistance (Berendsen et al., 2012). But MHB can also affect nutrient availability and are considered to affect plants through the mycorrhizal fungi. MHB are not specific to any one type of mycorrhizal fungi and have been found associated with roots colonized by AM or ECM fungi. Major groups of bacteria that include potential MHB are the Proteobacteria and especially the genus *Pseudomonas* (Gammaproteobacteria) and gram-positive bacteria, such as the Actinobacteria (Frey-Klett et al., 2007; Franco-Correa et al., 2010). Both of these bacterial groups are often the dominant taxa found within the rhizosphere.

For forests, though, most research has focused on how mycorrhizal fungi influence the community structure of rhizosphere bacteria. For example, plant roots colonized by mycorrhizal fungi generally have different root exudates than uncolonized roots and a modified physicochemical soil environment, which can influence the structure of rhizosphere bacteria (reviewed in Duponnois et al., 2008; Hartmann et al., 2009; Uroz et al., 2016). For AM fungi, there is evidence that exudates from AM mycelia affect bacterial community composition (Toljander et al., 2007) and that the taxonomic identity of AM fungi can affect the community of associated bacteria (Toljander et al., 2006; Rillig et al., 2006). Studies that have examined the effects of AM fungi on rhizosphere and soil bacterial communities are typically conducted ex situ and with model plant species, such as leek or carrot (see Scheublin et al., 2010). However, two recent studies that examine AM fungal tree species belowground shed light on associated bacteria in forests. In a study of *Acer saccharum* using Illumina sequencing, rhizosphere communities were found to be dominated by Proteobacteria (42%), Acidobacteria (25%), Actinobacteria (10%), and Chloroflexi (4%) (Wallace et al., 2018; Table 14.2). Similar bacterial taxonomic distribution was observed in another study that compared rhizoplane communities of *Acer saccharum* and *Acer platanoides*, where communities were dominated by Actinobacteria and Proteobacteria (DeBellis et al., 2019). The majority of studies examining forest soil bacterial communities, though, have focused on trees associated with ECM fungi (Table 14.2); consequently, we focus much of our discussion on bacterial associations with ECM fungi.

Plant roots colonized by ECM fungi have been reported to harbor unique bacterial taxa compared to uncolonized roots or bulk soil (Izumi, 2019; Izumi et al., 2008; Uroz et al., 2007, 2012). Further, there is evidence that different species of ECM fungi can harbor unique communities of bacteria (Izumi and Finlay, 2010; Marupakula et al., 2016; Nguyen and

TABLE 14.2 Examples of bacterial taxa found in the rhizosphere of roots colonized by AM- and ECM-fungi.

Taxa	AM fungal rhizosphere[a]	ECM fungal rhizosphere[a]	Frequency of taxa for ECM fungal rhizosphere[b]
Acidobacteria	25	16%–19%	33%
Actinobacteria	10	14%–27%	22%
Bacteroidetes	9	NR	22%
Chloroflexi	4	NR	6%
Firmicutes	NR	NR	17%
Proteobacteria	42	29%–43%	100%
Alpha	19	28	83%
Beta	7	14	56%
Delta	5	NR	NR
Gamma	10	3	44%

NR = taxa not reported.

[a]*AM fungal rhizosphere taxa from Wallace et al. (2018) and ECM fungal rhizosphere from Burke et al. (2006).*

[b]*The frequency of taxa detected in ECM studies is based on studies reported in Table 14.3 (18 studies total).*

Bruns, 2015); however, these community differences may be driven by subtle changes in bacterial taxa (Burke et al., 2008; Izumi et al., 2008; Kretzer et al., 2009), changes to bacterial OTUs rather than groups at higher taxonomic resolution (Uroz et al., 2012), or bacterial populations (Mogge et al., 2000). The influence of ECM fungal identity on bacterial community structure may be more important for newly colonized root tips than for older ECM fungal tips with more established bacterial communities (Burke et al., 2006; Marupakula et al., 2016) and may depend on the taxa of ECM fungi studied (Nguyen and Bruns, 2015).

Most of the studies of the influence of ECM fungal species on bacterial communities conclude that structure is mostly similar between ECM fungal species (Burke et al., 2008; Izumi et al., 2008; Kretzer et al., 2009; Uroz et al., 2012). An exception is Izumi and Finlay (2010) who found that different ECM fungal taxa select for unique communities of bacteria. The mechanisms of bacterial recruitment within the mycorrhizosphere, though, are likely dependent on more than just mycorrhizal fungal identity. Marupakula et al. (2016) suggest a temporal component where the age of the ECM fungi affects their associated bacterial community, which can confound field data sets where root age is difficult to determine accurately. The genetic basis for bacterial recruitment by ECM fungi has also been explored (e.g., Ruiz-Lozano and Bonfante, 2000; Warmink and van Elsas, 2008), as has the effects of nutrients on bacterial recruitment. For example, Uroz et al. (2007) suggested that carbon metabolism by mycorrhizal fungi likely selects for a bacterial community with a high weathering potential and Nuccio et al. (2013) suggested that the ability of AM fungi to uptake inorganic N is a driver of bacterial community changes in leaf litter. Whether mycorrhizal fungi influence bacteria in the rhizosphere due to complementary actions on substrates (see Uroz et al., 2016) or competition for available substrates (e.g., Mille-Lindblom et al., 2006) is likely site specific. Only when we can perceive the fine-scale heterogeneity that bacteria respond to within a rhizosphere patch can we tease apart the effects of competition with mycorrhizal fungi from complementary functions.

We can, however, make some generalizations about the bacteria likely to be found within the rhizosphere and mycorrhizosphere of ECM fungal trees based on recent studies (Table 14.3). The Proteobacteria, Acidobacteria, and Actinobacteria are generally well represented within the rhizosphere and mycorrhizosphere, and within Proteobacteria, the Alphaproteobacteria class tends to be especially well represented. Bacteria in the Rhizobiales are also typically found in the mycorrhizosphere (Table 14.3). Obase (2019) had a similar observation that studies of bacterial communities associated with ECM fungi, primarily those after the mid-2000s, often reported bacteria in the Rhizobiales as highly represented taxa in both culture-dependent and culture-independent data sets (also discussed in Nguyen and Bruns, 2015). The Rhizobiales includes the genera *Rhizobium* and *Bradyrhizobium*, which contain a specialized group of N-fixing bacteria, collectively called rhizobia, which form root nodules on leguminous plants. Members of the genus *Burkholderia* have shown potential to fix N and this genus is also commonly found associated with ECM fungal roots (Table 14.3). Some of the most common bacterial associates of ECM fungi, then, are genera known to contain rhizobia. A number of strains within these genera, though, do not form root nodules and are, thus, referred to as nonnodulating strains (Tanaka and Nara, 2009). Evidence suggests that these nonnodulating strains are widespread in the rhizosphere (Tanaka and Nara, 2009), but their function in forests is currently unknown. Despite this broad general pattern for *Rhizobium*, *Bradyrhizobium*, and *Burkholderia*, there tends to be a great deal of variability in the bacterial groups that have been found in the mycorrhizosphere across studies (Tables 14.2 and 14.3). For example, some studies have found a high number of OTUs matching with Acidobacteria associated with mycorrhizal fungi (Uroz et al., 2012) while other research has found that this group was more poorly represented, potentially due to washing of the mycorrhizal root tips (Burke et al., 2008). The Gammaproteobacteria are frequently encountered in some studies but not others, as is the case with the genus *Pseudomonas*. This variability could be due to the methods used (e.g., sequencing platform or choice of primers for amplifying bacterial taxa). In addition, the host plant species are rarely the same between studies of species-specific mycorrhizosphere effects on bacteria, which may also contribute to variation in the bacterial taxa found. However, it is likely that underlying differences in environmental conditions or the nutrient status of the tree host have a major role in affecting bacterial communities within the rhizosphere and mycorrhizosphere (Burke et al., 2006, 2008). The underlying heterogeneity of the soil landscape, and as mentioned earlier, our difficulty in perceiving that landscape, likely contributes to our inability to completely understand the effect of mycorrhizal fungi on soil bacteria. But it also provides fertile ground for additional inquiries to expand our conceptual framework of mycorrhiza-bacteria interactions, which is necessary for a better understanding of how these interactions affect ecosystem function.

2.2 General effects of fungi on bacterial functional groups

Many studies on the relationship between bacterial functional groups and fungi in the rhizosphere have focused on N-fixing bacteria for several reasons. First, N is often considered the primary limiting nutrient in northern forests, whether those forests are dominated by gymnosperms or angiosperms (Vitousek and Howarth, 1991). The dominance of ECM trees in

TABLE 14.3 Dominant bacterial taxa associated with ectomycorrhizal fungi.

References	Method	Plant host/ecosystem	Mycorrhizal fungal taxa	Dominant bacterial taxa found
Burke et al. (2006)	Sanger sequencing	*Pinus taeda* (loblolly pine plantation)	Mixed species (frequently *Tomentella* sp. and *Cenococcum* sp.)	Alphaproteobacteria (Rhizobiales, ***Bradyrhizobium***, *Blastochloris*, *Stella*, *Rhodoplanes*, *Phenylobacterium*, *Acetobacter*, *Rhodopila*), Betaproteobacteria (***Burkholderia***), Gammaproteobacteria (*Nitrosococcus*), Bacteroidetes (*Flexibacter*), Acidobacteria (*Acidobacterium*), Actinobacteria (*Mycobacterium*), Firmicutes (*Pelotomaculum*)
Burke et al. (2008)	Sanger sequencing	*Pseudotsuga menziesii*	Primarily Russulaceae (*Russula* and *Lactarius*)	Alphaproteobacteria, Bacteroidetes, Gammaproteobacteria. ***Rhizobium***. ***Bradyrhizobium***. Sphingobacteriales
Brooks et al. (2011)	Sanger sequencing on bacterial isolates	Forest stands of *Pseudotsuga menziesii* and *Betula papyrifera*	Not determined	Proteobacteria (Alphaproteobacteria, Betaproteobacteria), Actinobacteria, Bacteroidetes (Sphingobacteria)
Deveau et al. (2016)	454 pyrosequencing	*Corylus avellana*	*Tuber melanosporum*	Alphaproteobacteria (class), **Rhizobiales** (order), Thermoleophilales (order), ***Bradyrhizobium*** (genus), *Thermoleophilum* (genus)
Izumi et al. (2006)	Sanger sequencing on bacterial isolates	*Pinus sylvestris*	*Suillus flavidus*, *Suillus variegatus*, *Russula paludosa*, *Russula* sp.	*Pseudomonas*, ***Burkholderia***, *Bacillus*
Izumi et al. (2008)	Sanger sequencing	Slash pine plantation (*Pinus elliottii*)	*Suillus cothurantus*, *Suillus subluteus*, unknown morphotype (family Atheliaceae)	***Burkholderia***, *Pantoea*, *Acinetobacter*
Izumi (2019)	Cloning and sequencing from DNA extracts of colonized root tips	*Betula pubescens*	*Leccinum scabrum*	*Lactobacillus* sp., *Methylobacterium populi*, *Paenibacillus validus*, *Bacillus* sp., *Paenibacillus* sp., *Paenibacillus hongkongensis*
Kataoka et al. (2008)	Denaturing gradient gel electrophoresis and Sanger sequencing	*Pinus thunbergii*	Mixed species; mostly *Cenococcum geophilum*	***Burkholderia*** (genus), ***Bradyrhizobium*** (genus)
Kluber et al. (2011)	Sanger sequencing	*Pseudotsuga menziesii*	*Piloderma fallax*	Alphaproteobacteria, Acidobacteria, Gammaproteobacteria
Kretzer et al. (2009)	Sanger sequencing	*Pseudotsuga menziesii*	*Rhizopogon vinicolor* and *Rhizopogon vesiculosus*	Alphaproteobacteria, Gammaproteobacteria, Acidobacteria

Continued

TABLE 14.3 Dominant bacterial taxa associated with ectomycorrhizal fungi—cont'd

References	Method	Plant host/ecosystem	Mycorrhizal fungal taxa	Dominant bacterial taxa found
Marupakula et al. (2016)	454 pyrosequencing	*Pinus sylvestris*	*Phialocephala fortinii, Meliniomyces variabilis, Russula* sp.	*Burkholderia, Dyella, Pseudomonas, Favisolibacter, Beijerinckia, Actinospica, Janthinobacterium, Aquaspirillum, Acidobacter, Gp1, Sphingomonas, Terriglobus, Enhydrobacter, Magnetospirillum,* ***Bradyrhizobium***
Nguyen and Bruns (2015)	454 pyrosequencing	*Pinus muricata* (monoculture stands of Bishop pine)	Morphotyped root tips (commonly *Rhizopogon salebrosus, Russula cerolens, Tricholoma imbricatum, Tomentella sublilacina,* and *Clavulina* sp.)	**Burkholderiales** (order), **Rhizobiales** (order), ***Burkholderia phenazinium*** (species), *Leptothirix* (genus), ***Burkholderia sordidicola*** (species), ***Bradyrhizobium elkanii*** (species)
Obase (2019)	Sanger sequencing on cultured bacterial isolates	*Castanea crenata* (Chestnut plantation)	*Laccaria laccata*	***Bradyrhizobium, Rhizobium***
Shirakawa et al. (2019)	Sanger sequencing on cultured bacterial isolates	*Pinus densiflora*	Mixed species.	***Burkholderia****, Caballeronia, Collimonas, Novosphingobium, Paraburkholderia,* ***Rhizobium***
Tanaka and Nara (2009)	Sanger sequencing on cultured bacterial isolates	*Pinus densiflora*	*Cenococcum geophilum, Pisolithus* sp., and *Suillus granulatus*	***Burkholderia, Rhizobium, Bradyrhizobium,*** *Variovorax*
Timonen and Hurek (2006)	Sanger sequencing on bacterial isolates	Boreal pine forest of *Pinus sylvestris*	*Suillus bovinus*	***Burkholderia***
Uroz et al. (2012)	454 pyrosequencing	Oak forest (*Quercus petraea*)	*Scleroderma citrinum, Xerocomus pruinatus*	*Acidobacterium,* ***Burkholderia,*** *Chitinophaga, Rhodoplanes,* ***Bradyrhizobium***
Vik et al. (2013)	Titanium sequencing	*Bistorta vivipara*	Mixed ECM species (whole roots were harvested)	*Acidobacteria, Actinobacteria, Armatimonadetes, Chloroflexi, Planctomycetes, Proteobacteria*

Shown here are studies where the effects of ectomycorrhizal fungi on bacterial community structure were examined on root tips or in rhizosphere soil within the past 15 years. We are not including studies where mycorrhizal roots were inoculated with bacterial isolates (i.e., see Frey-Klett et al., 2007). Potential nitrogen-fixing bacteria are shown in bold.

northern forests has been attributed to the enhanced ability of ECM fungi to access and transport N to the host trees, as well as their potential saprotrophic activity (Read and Perez-Moreno, 2003). Second, although as much as 90% of the N trees need on an annual basis comes from recycling of organic N within the system, 10% of the N must be imported into the systems and N fixation by bacteria in soil is the most likely route for N imports into forests (Schlesinger, 1997). In an agricultural context, N fixation is an important and highly studied process mostly because a specialized group of N-fixing bacteria (collectively called rhizobia; see earlier) live in association with the roots of legumes (family Fabaceae) in a symbiotic relationship where the rhizobia exchange fixed N for carbon (Lindström and Mousavi, 2010). This symbiosis directly benefits plant health and nutrition for a number of agriculturally important plants, including soybean and alfalfa. As such, it has garnered extensive attention by the scientific community (Cumming et al., 2015). This relationship between N-fixing bacteria and legumes is also complemented by the presence of AM fungi that supply phosphorous to the N-fixing bacteria and enhance rhizobial colonization of roots as well as N fixation (Hayman, 1986; Mikola, 1986). However, most northern forest trees do not form such symbiotic relationships with N-fixing bacteria, and the trees that do form symbiotic relationships with N-fixing bacteria (e.g., *Alnus*, *Robinia*) are relatively rare in northern forests (Menge et al., 2010). As such, most N fixation within forest soils occurs either in bulk soil, or more likely with the C-rich microsites created by roots in the rhizosphere since N fixation is energy intensive and C limited within soil environments (Lynch, 1988; Vitousek and Howarth, 1991). Genera with known N-fixating bacteria, including *Rhizobium*, *Bradyrhizobium*, and *Burkholderia* are often important components of the rhizosphere and mycorrhizosphere (Table 14.3). They have been found associated with a widespread group of nonleguminous plants (Garrido-Oter et al., 2018) and can dominate ectomycorrhizal roots (Tanaka and Nara, 2009). But, because these bacteria do not form root nodules or colonize plant roots on nonleguminous plants, they are not truly symbiotic. These bacteria can be considered associative bacteria, as they exist within the root zone and rely upon root exudates. But, their function in temperate forest ecosystems is currently unknown; thus the effects of mycorrhizal fungi in the extent and magnitude of associative N fixation in forests are an area of future study.

Other bacterial functional groups in the rhizosphere of forests have received far less attention even though they are critical for nutrient cycling in forests and, thus, indirectly benefit plant health and nutrition (Andrade, 2008). There is evidence that mycorrhizal fungi affect other bacterial functional groups. For example, Amora-Lazcano et al. (1998) found a higher abundance of autotrophic ammonium oxidizers in the roots of maize plants that were colonized by AM fungi. Kim et al. (1997) suggested that AM fungi and P-solubilizing bacteria act in synergy to enhance N and P uptake by plants in their study using tomato and Villegas and Fortin (2002) found increased P solubilization in a culture system with carrot roots when they interacted with AM fungi and P-solubilizing bacteria. However, current knowledge of the interactions between these plant growth-promoting rhizobacteria and mycorrhizal fungi is largely from agricultural systems (Backer et al., 2018). For forests, which are typically found on nutrient-poor soils (Uroz et al., 2016), fungal influences on bacterial functional groups are likely critical for overall forest health. Burke et al. (2012) found that denitrifying bacteria (DNB) and methane-oxidizing bacteria (MOB) were significantly affected by the distribution of fungal biomass in soil and the presence of extracellular enzyme activity in a northern hardwood forest. MOB were negatively correlated with fungal biomass and enzyme activity, while DNB were positively correlated. This suggests that MOB preferred sites with low levels of nutrient cycling and organic matter decomposition, while DNB were more likely to be found in sites with high levels of organic matter turnover and nutrient cycling (Burke et al., 2012). This study further elucidated that these changes in microbial distribution and function existed over fine spatial scales; DNB distribution has been reported to be highest in the top 5 cm of soil (Mergel et al., 2001) where fungal biomass is also known to be highest (Bååth and Söderström, 1982). This suggests that the fine-scale gradients that fungi create in forests are influential for bacterial functional groups. As we expand our knowledge of bacteria in forest soil and the environmental influences on them, it is imperative that we consider bacterial functional groups, which affect overall nutrient cycling in forest ecosystems.

3. Conclusions

Comparatively, there are more studies on forest fungi than forest bacteria (Baldrian, 2017; Uroz et al., 2016). Recently, there has been a push to investigate environmental drivers of soil bacterial communities at a global scale to better understand the ecology of these organisms (Delgado-Basquerizo et al., 2018). Here we have provided evidence for how mycorrhizal fungi alter the soil landscape at small spatial scales and how the activity of these fungi can, in turn, affect bacteria in forests, particularly in the rhizosphere and mycorrhizosphere. Few previous studies have examined both fungi and bacteria in the same soil samples from forests. Uroz et al. (2016) summarized studies that examined both fungi and bacteria in forest soils and concluded that, in general, bacterial communities are more diverse than fungal communities in forest soils, bacteria are more affected by plant compartment than fungi, and bacterial communities are more resilient to environmental changes, but the drivers of these community patterns remain largely unexplained. In a study where we examined both

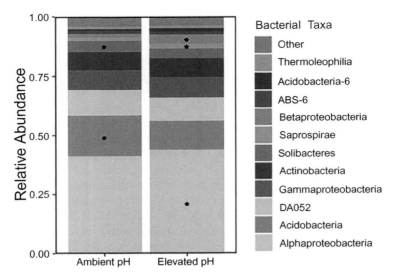

FIG. 14.5 Relative abundance of bacterial taxa found in bulk soil of temperate forests that had been manipulated to elevate pH (~6) or left at ambient pH (~4). Indicator species (following Dufrêne and Legendre, 1997) for each pH treatment are shown with asterisks. Only taxa with a relative abundance of 0.01 or greater are shown. Additional indicators of the ambient pH treatment were Deltaproteobacteria and PRR-12 from the candidate phylum WS3, but are not shown because they were less than 0.01 in relative abundance.

fungal and bacterial communities in hardwood forests that had been manipulated to elevated soil pH, we found that bacterial communities responded strongly to the changes in soil pH and environmental changes over time, while the strongest driver of fungal community structure was the forest site (Fig. 14.5; Carrino-Kyker et al., 2020). Thus our data suggest that environmental factors are a strong driver of bacterial community structure.

Here, we suggest that the environmental drivers that influence bacteria in forests may operate at very fine scales that are difficult to perceive. Additionally, this fine-scale patchiness may be created and maintained to a large degree by the activities of mycorrhizal fungi, since they affect nutrient and C availability in soil, which can be important limiting factors for bacterial growth. However, studies that examine the direct effects of decomposition on bacteria in the rhizosphere are lacking and there is a gap in our knowledge on how nutrient and C cycling impact rhizosphere bacteria in forests. The effects of decomposition on rhizosphere bacteria may very well differ between shallow organic soils that experience a high degree of leaf litter inputs and deeper mineral soils, as decomposition is a vertically stratified process (see earlier). Future research could help close this gap in our knowledge by examining rhizosphere bacteria across small-scale changes in soil depth, in conjunction with extracellular enzyme production. This would help determine if the heightened decomposition rates experienced at shallow soil depths promote or suppress bacterial taxa that are important for plant health, such as mycorrhiza helper bacteria (see Box 14.1). This seems plausible as many mycorrhiza helper bacteria are members of the Proteobacteria and Actinobacteria and bacteria in these phyla are also found more often at shallow soil depths and in litter layers (see Table 14.1), but the specific effects of decomposition on mycorrhiza helper bacteria in the rhizosphere of forests are currently unknown. Understanding how leaf litter impacts beneficial bacteria, though, can improve our restoration efforts and forest management. For example, temperate forests that have regrown on abandoned agricultural fields have an abundance of earthworms and lack sufficient leaf litter (Burke et al., 2016). If leaf litter is, in fact, important for rhizosphere bacterial communities that improve forest health, restoration efforts that focus on restoring litter layers in these secondary forests can be employed.

In addition, the effect of mycorrhizal-bacterial interactions on forest ecosystem services has been given little attention, especially for certain bacterial functions, such as denitrification, methane oxidation, or P solubilization. Studies that examine bacterial community structure in forests are important for understanding the environmental drivers that these communities respond to. But, knowledge on how these changing communities are linked to specific ecosystem functions is necessary to help preserve the ecosystem services that forests provide. This can be achieved by conducting RNA-based studies and by including specific bacterial functional groups in studies of soil bacteria. Understanding which functions are gained or lost when bacterial community structure changes can help understand the role of bacteria for ecosystem services, such as C sequestration or soil quality. It may be, though, that in order to fully understand the role of bacteria for ecosystem services, they need to be examined at fine spatial scales where the environmental drivers that influence them can be adequately perceived.

References

Agerer, R., 2001. Exploration types of ectomycorrhizae: a proposal to classify ectomycorrhizal mycelial systems according to their patterns of differentiation and putative ecological importance. Mycorrhiza 11 (2), 107–114.

Agerer, R., 2003. Color Atlas of Ectomycorrhiza. Einhorn-Verlag, Munich, Germany.

Allmér, J., Vasiliauskas, R., Ihrmark, K., Stenlid, J., Dahlberg, A., 2006. Wood-inhabiting fungal communities in woody debris of Norway spruce (*Picea abies* (L.) Karst.) as reflected by sporocarps mycelial isolations and T-RFLP identification FEMS. Microb. Ecol. 55 (1), 57–67.

Amora-Lazcano, E., Vázquez, M.M., Azcón, R., 1998. Response of nitrogen-transforming microorganisms to arbuscular mycorrhizal fungi. Biol. Fertil. Soils 27 (1), 65–70.

Andrade, G., 2008. Role of functional groups of microorganisms on the rhizosphere microcosm dynamics. In: Varma, A., Abbott, L., Werner, D., Hampp, R. (Eds.), Plant Surface Microbiology. Springer, Berlin, Heidelberg, Germany, pp. 51–69.

Artursson, V., Jansson, J.K., 2003. Use of bromodeoxyuridine immunocapture to identify active bacteria associated with arbuscular mycorrhizal hyphae. Appl. Environ. Microbiol. 69 (10), 6208–6215.

Bååth, E., Söderström, B., 1982. Seasonal and spatial variation in fungal biomass in a forest soil. Soil Biol. Biochem. 14 (4), 353–358.

Backer, R., Rokem, J.S., Ilangumaran, G., Lamont, J., Praslickova, D., Ricci, E., Subramanian, S., Smith, D.L., 2018. Plant growth-promoting rhizobacteria: context, mechanisms of action, and roadmap to commercialization of biostimulants for sustainable agriculture. Front. Plant Sci. 9, 1473.

Bago, B., Azcón-Aguilar, C., 1997. Changes in the rhizospheric pH induced by arbuscular mycorrhiza formation in onion (*Allium cepa* L.). J. Plant Nutr. Soil Sci. 160 (2), 333–339. https://doi.org/10.1002/jpln.19971600231.

Bahram, M., Peay, K.G., Tedersoo, L., 2015. Local-scale biogeography and spatiotemporal variability in communities of mycorrhizal fungi. New Phytol. 205 (4), 1454–1463.

Baldrian, P., 2017. Forest microbiome: diversity, complexity and dynamics. FEMS Microbiol. Rev. 41 (2), 109–130.

Baldrian, P., Kolařík, M., Štursová, M., Kopecký, J., Valášková, V., Větrovský, T., Žifčáková, L., Šnajdr, J., Rídl, J., Vlček, C., Voříšková, J., 2012. Active and total microbial communities in forest soil are largely different and highly stratified during decomposition. ISME J. 6 (2), 248–258.

Berendsen, R.L., Pieterse, C.M.J., Bakker, P.A.H.M., 2012. The rhizosphere microbiome and plant health. Trends Plant Sci. 17 (8), 478–486.

Berg, M.P., 2012. Patterns of biodiversity at fine and small spatial scales. In: Wall, D.H., Bardgett, R.D., Behan-Pelletier, V., Herrick, J.E., Jones, T.H., Ritz, K., Six, J., Strong, D.R., van der Putten, W.H. (Eds.), Soil Ecology and Ecosystem Services. Oxford University Press, Oxford, UK, pp. 136–152.

Bödeker, I.T.M., Clemmensen, K.E., de Boer, W., Martin, F., Olson, Å., Lindahl, B.D., 2014. Ectomycorrhizal *Cortinarius* species participate in enzymatic oxidation of humus in northern forest ecosystems. New Phytol. 203 (1), 245–256.

Bödeker, I.T.M., Lindahl, B.D., Olson, Å., Clemmensen, K.E., 2016. Mycorrhizal and saprotrophic fungal guilds compete for the same organic substrates but affect decomposition differently. Funct. Ecol. 30 (12), 1967–1978.

Bonafante, P., Anca, I.-A., 2009. Plants, mycorrhizal fungi, and bacteria: a network of interactions. Annu. Rev. Microbiol. 63 (1), 363–383.

Brooks, D.D., Chan, R., Starks, E.R., Grayston, S.J., Jones, M.D., 2011. Ectomycorrhizal hyphae structure components of the soil bacterial community for decreased phosphatase production. FEMS Microbiol. Ecol. 76 (2), 245–255.

Brundrett, M., Kendrick, B., 1990. The roots and mycorrhizas of herbaceous woodland plants. I. Quantitative aspects of morphology. New Phytol. 114 (3), 457–468.

Buée, M., De Boer, W., Martin, F., van Overbeek, L., Jurkevitch, E., 2009. The rhizosphere zoo: an overview of plant-associated communities of microorganisms, including phages, bacteria, archaea, and fungi, and of some of their structuring factors. Plant Soil 321 (1), 189–212.

Burke, D.J., 2015. Effects of annual and interannual environmental variability on soil fungi associated with an old-growth, temperate hardwood forest. FEMS Microbiol. Ecol. 91 (6), fiv053.

Burke, D.J., Chan, C.R., 2010. Effects of the invasive plant garlic mustard (Alliaria petiolata) on bacterial communities in a northern hardwood forest soil. Can. J. Microbiol. 56 (1), 81–86.

Burke, M.K., Raynal, D.J., 1994. Fine root growth phenology, production, and turnover in a northern hardwood forest ecosystem. Plant Soil 162 (1), 135–146.

Burke, D.J., Kretzer, A.M., Rygiewicz, P.T., Topa, M.A., 2006. Soil bacterial diversity in a loblolly pine plantation: influence of ectomycorrhizas and fertilization. FEMS Microbiol. Ecol. 57 (3), 409–419.

Burke, D.J., Dunham, S.M., Kretzer, A.M., 2008. Molecular analysis of bacterial communities associated with the roots of Douglas fir (*Pseudotsuga menziesii*) colonized by different ectomycorrhizal fungi. FEMS Microbiol. Ecol. 65 (2), 299–309.

Burke, D.J., López-Gutiérrez, J.C., Smemo, K.A., Chan, C.R., 2009. Vegetation and soil environmental influence the spatial distribution of root-associated fungi in a mature beech-maple forest. Appl. Environ. Microbiol. 75 (24), 7639–7648.

Burke, D.J., Smemo, K.A., López-Gutiérrez, J.C., DeForest, J.L., 2012. Soil fungi influence the distribution of microbial functional groups that mediate forest greenhouse gas emissions. Soil Biol. Biochem. 53, 112–119.

Burke, D.J., Smemo, K.A., Hewins, C.R., 2014. Ectomycorrhizl fungi isolated from old-growth northern hardwood forest display variability in extracellular enzyme activity in the presence of plant litter. Soil Biol. Biochem. 68, 219–222.

Burke, D.J., Knisely, C., Watson, M.L., Carrino-Kyker, S.R., Mauk, R.L., 2016. The effects of agricultural history on forest ecological integrity as determined by a rapid forest assessment method. For. Ecol. Manag. 378, 1–13.

Carrino-Kyker, S.R., Kluber, L.A., Petersen, S.M., Coyle, K.P., Hewins, C.R., DeForest, J.L., Smemo, K.A., Burke, D.J., 2016. Mycorrhizal fungal communities respond to experimental elevation of soil pH and P availability in temperate hardwood forests. FEMS Microbiol. Ecol. 92 (3), fiw024.

Carrino-Kyker, S.R., Coyle, K.P., Kluber, L.A., Burke, D.J., 2020. Fungal and bacterial communities exhibit consistent responses to reversal of soil acidification and phosphorus limitation over time. Microorganisms 8 (1), 1.

Clemmensen, K.E., Bahr, A., Ovaskainen, O., Dahlberg, A., Ekblad, A., Wallander, H., Stenlid, J., Finlay, R.D., Wardle, D.A., Lindahl, B.D., 2013. Roots and associated fungi drive long-term carbon sequestration in boreal forest. Science 339 (6127), 1615–1618.

Coince, A., Caël, O., Bach, C., Lengellé, J., Cruaud, C., Gavory, F., Morin, E., Murat, C., Marcais, B., Buée, M., 2013. Below-ground fine-scale distribution and soil *versus* fine root detection of fungal and soil oomycete communities in a French beech forest. Fungal Ecol. 6 (3), 223–235.

Courty, P.-E., Franc, A., Garbaye, J., 2010. Temporal and functional pattern of secreted enzyme activities in an ectomycorrhizal community. Soil Biol. Biochem. 42 (11), 2022–2025.

Cumming, J.R., Zawaski, C., Desai, S., Collart, F.R., 2015. Phosphorus disequilibrium in the tripartite plant-ectomycorrhiza-plant growth promoting rhizobacterial association. J. Soil Sci. Plant Nutr. 15 (2), 464–485.

DeBellis, T., Kembel, S.W., Lessard, J.-P., 2019. Shared mycorrhizae but distinct communities of other root-associated microbes on co-occurring native and invasive maples. PeerJ 7, e7295.

Delgado-Basquerizo, M., Oliverio, A.M., Brewer, T.E., Benavent-González, A., Eldridge, D.J., Bardgett, R.D., Maestre, F.T., Singh, B.K., Fierer, N., 2018. A global atlas of the dominant bacteria found in soil. Science 359 (6373), 320–325.

Deveau, A., Antony-Babu, S., Le Tacon, F., Robin, C., Frey-Klett, P., Uroz, S., 2016. Temporal changes of bacterial communities in the *Tuber melanosporum* ectomycorrhizosphere during ascocarp development. Mycorrhiza 26 (5), 389–399.

Dickie, I.A., Xu, B., Koide, R.T., 2002. Vertical niche differentiation of ectomycorrhizal hyphae in soil as shown by T-RFLP analysis. New Phytol. 156 (3), 527–535.

Dufrêne, M., Legendre, P., 1997. Species assemblages and indicator species: the need for a flexible asymmetrical approach. Ecol. Monogr. 67 (3), 345–366.

Duponnois, R., Galiana, A., Prin, Y., 2008. The mycorrhizosphere effect: a multitrophic interaction complex improves mycorrhizal symbiosis and plant growth. In: Siddiqui, Z.A., Akhtar, M.S., Futai, K. (Eds.), Mycorrhizae: Sustainable Agriculture and Forestry. Springer, Dordrecht, Netherlands, pp. 227–240.

Eilers, K.G., Debenport, S., Anderson, S., Fierer, N., 2012. Digging deeper to find unique microbial communities: the strong effect of depth on the structure of bacterial and archael communities in soil. Soil Biol. Biochem. 50, 58–65.

Eissenstat, D.M., Kucharski, J.M., Zadworny, M., Adams, T.S., Koide, R.T., 2015. Linking root traits to nutrient foraging in arbuscular mycorrhizal trees in a temperate forest. New Phytol. 208 (1), 114–124.

Fierer, N., Jackson, R.B., 2006. The diversity and biogeography of soil bacterial communities. Proc. Natl. Acad. Sci. USA 103 (3), 626–631.

Finzi, A.C., Abramoff, R.Z., Spiller, K.S., Brzostek, E.R., Darby, B.A., Kramer, M.A., Phillips, R.P., 2014. Rhizosphere processes are quantitatively important components of terrestrial carbon and nutrient cycles. Glob. Chang. Biol. 21 (5), 2082–2094.

Franco-Correa, M., Quintana, A., Duque, C., Suarez, C., Rodríguez, M.X., Barea, J.-M., 2010. Evaluation of actinomycete strains for key traints related with plant growth promotion and mycorrhiza helping activities. Appl. Soil Ecol. 45 (3), 209–217.

Frey-Klett, P., Garbaye, J., Tarkka, M., 2007. Tansley review: the mycorrhiza helper bacteria revisited. New Phytol. 176 (1), 22–36.

Garrido-Oter, R., Nakano, R.T., Dombrowski, N., Ma, K.-W., The AgBiome Team, McHardy, A.C., Schulze-Lefert, P., 2018. Modular traits of the Rhizobiales root microbiota and their evolutionary relationship with symbiotic rhizobia. Cell Host Microbe 24 (1), 155–167.

Guo, D., Mitchell, R.J., Withington, J.M., Fan, P.-P., Hendricks, J.J., 2008. Edogenous and exogenous controls of root life span, mortality and nitrogen flux in a longleaf pine forest: root branch order predominates. J. Ecol. 96 (4), 737–745.

Hartmann, A., Schmid, M., van Tuinen, D., Berg, G., 2009. Plant-driven selection of microbes. Plant Soil 321 (1), 235–257.

Hawkes, C.V., Deangelis, K., Firestone, M.K., 2007. Root interactions with soil microbial communities and processes. In: Cardon, Z.G., Whitbeck, J.L. (Eds.), The Rhizosphere: An Ecological Perspective. Elsevier Academic Press, San Diego, CA, USA, pp. 1–29.

Hayman, D.S., 1986. Mycorrhizae of nitrogen-fixing legumes. MIRCEN J. Appl. Microbiol. Biotechnol. 2 (1), 121–145.

Hestrin, R., Hammer, E.C., Mueller, C.W., Lehmann, J., 2019. Synergies between mycorrhizal fungi and soil microbial communities increase plant nitrogen acquisition. Commun. Biol. 2, 233.

Högberg, M.N., Högberg, P., 2002. Extramatrical ectomycorrhizal mycelium contributes one-third of microbial biomass and produces, together with associated roots, half the dissolved organic carbon in a forest soil. New Phytol. 154 (3), 791–795.

Horner-Devine, M.C., Silver, J.M., Leibold, M.A., Bohannan, B.J., Colwell, R.K., Fuhrman, J.A., Green, J.L., Kuske, C.R., Martiny, J.B.H., Muyzer, G., Øvreås, L., Reysenbach, A.-L., Smith, V.H., 2007. A comparison of taxon co-occurrence patterns for macro- and microorganisms. Ecology 88 (6), 1345–1353. https://doi.org/10.1890/06-0286.

Izumi, H., 2019. Temporal and special dynamics of metabolically active bacteria associated with ectomycorrhizal roots of *Betula pubescens*. Biol. Fertil. Soils 55 (8), 777–788.

Izumi, H., Finlay, R.D., 2010. Ectomycorrhizal roots select distinctive bacterial and ascomycete communities in Swedish subarctic forests. Environ. Microbiol. 13 (3), 819–830.

Izumi, H., Anderson, I.C., Alexander, I.J., Killham, K., Moore, E.R.B., 2006. Edobacteria in some ectomycorrhiza of Scots pine (*Pinus sylvestris*). FEMS Microbiol. Ecol. 56 (1), 34–43.

Izumi, H., Cairney, J.W.G., Killham, K., Moore, E., Alexander, I.J., Anderson, I.C., 2008. Bacteria associated with ectomycorrhizas of slash pine (*Pinus elliottii*) in South-Eastern Queensland, Australia. FEMS Microbiol. Lett. 282 (2), 196–204.

Jaillard, B., Plassard, C., Hinsinger, P., 2003. Measurements of H^+ fluxes and concentrations in the rhizosphere. In: Rengel, Z. (Ed.), Handbook of Soil Acidity. Marcel Dekker, New York, NY, USA, pp. 231–266.

Johnson, N.C., Gehring, C.A., 2007. Mycorrhizas: symbiotic mediators of rhizosphere and ecosystem processes. In: Cardon, Z.G., Whitbeck, J.L. (Eds.), The Rhizosphere: An Ecological Perspective. Elsevier Academic Press, New York, NY, USA, pp. 73–100.

Jones, D.L., Hodge, A., Kuzyakov, Y., 2004. Tansley review: plant and mycorrhizal regulation of rhizodeposition. New Phytol. 163 (3), 459–480.

Joslin, J.D., Henderson, G.S., 1987. Organic matter and nutrients associated with fine root turnover in a white oak stand. For. Sci. 33 (2), 330–346.

Jumpponen, A., Jones, K.L., Blair, J., 2010. Vertical distribution of fungal communities in tallgrass prairie soil. Mycologia 102 (5), 1027–1041.

Karandashov, V., Bucher, M., 2005. Symbiotic phosphate transport in arbuscular mycorrhizas. Trends Plant Sci. 10 (1), 22–29.

Kataoka, R., Taniguchi, T., Ooshima, H., Futai, K., 2008. Comparison of the bacterial communities established on the mycorrhizae formed on *Pinus thunbergii* roots by eight species of fungi. Plant Soil 304 (1), 267–275.

Kim, K., Jordan, D., McDonald, G., 1997. Effect of phosphate-solubilizing bacteria and vesicular-arbuscular mycorrhizae on tomato growth and soil microbial activity. Biol. Fertil. Soils 26 (2), 79–87.

Kluber, L.A., Smith, J.E., Myrold, D.D., 2011. Distinctive fungal and bacterial communities are associated with mats formed by ectomycorrhizal fungi. Soil Biol. Biochem. 43 (5), 1042–1050.

Kluber, L.A., Carrino-Kyker, S.R., Coyle, K.P., DeForest, J.L., Hewins, C.R., Shaw, A.N., Smemo, K.A., Burke, D.J., 2012. Mycorrhizal response to experimental pH and P manipulation in acidic hardwood forests. PLoS One 7 (11), e48946.

Koch, B., Worm, J., Jensen, L.E., Hojberg, O., Nybroe, O., 2001. Carbon limitation induces os-dependent gene expression in *Pseudomonas fluorescens* in soil. Appl. Environ. Microbiol. 67 (8), 3363–3370.

Kretzer, A.M., King, Z.R., Bai, S., 2009. Bacterial communities associated with tuberculate ectomycorrhizae of *Rhizopogon* spp. Mycorrhiza 19 (4), 277–282.

Landesman, W.J., Nelson, D.M., Fitzpatrick, M.C., 2014. Soil properties and tree species drive β-diversity of soil bacterial communities. Soil Biol. Biochem. 76, 201–209.

Lauber, C.L., Hamady, M., Knight, R., Fierer, N., 2009. Pyrosequencing-based assessment of soil pH as a predictor of soil bacterial community structure at the continental scale. Appl. Environ. Microbiol. 75 (15), 5111–5120.

Lindahl, B.D., Ihrmark, K., Boberg, J., Trumbore, S.E., Högberg, P., Stenlid, J., Finlay, R.D., 2007. Spatial separation of litter decomposition and mycorrhizal nitrogen uptake in a boreal forest. New Phytol. 173 (3), 611–620.

Lindström, K., Mousavi, S.A., 2010. *Rhizobium* and other N-fixing symbioses. In: Encyclopedia of Life Sciences (eLS). John Wiley & Sons, Chichester, UK, pp. 21–38.

Lladó, S., López-Mondéjar, R., Baldrian, P., 2017. Forest soil bacteria: diversity, involvement in ecosystem processes, and response to global change. Microbiol. Mol. Biol. Rev. 81 (2). e00063-16.

López-Mondéjar, R., Voříšková, J., Větrovský, T., Baldrian, P., 2015. The bacterial community inhabiting temperate deciduous forests is vertically stratified and undergoes seasonal dynamics. Soil Biol. Biochem. 87, 43–50.

Lynch, J.M., 1988. Microorganisms in their natural environments: the terrestrial environment. In: Lynch, J.M., Hobbie, J.M. (Eds.), Micro-Organisms in Action: Concepts and Applications in Microbial Ecology. Blackwell Scientific Publications, London, UK, pp. 103–131.

Martiny, J.B.H., Bohannan, B.J.M., Brown, J.H., Colwell, R.K., Fuhrman, J.A., Green, J.L., Horner-Devine, M.C., Kane, M., Adams Krumins, J., Kuske, C.R., Morin, P.J., Naeem, S., Øvreås, L., Reysenbach, A.-L., Smith, V.H., Staley, J.T., 2006. Microbial biogeography: putting microorganisms on the map. Nat. Rev. Microbiol. 4, 102–112. https://doi.org/10.1038/nrmicro1341.

Marupakula, S., Mahmood, S., Finlay, R.D., 2016. Analysis of single root tip microbiomes suggests that distinctive bacterial communities are selected by *Pinus sylvestris* roots colonized by different ectomycorrhizal fungi. Environ. Micobiol. 18 (5), 1470–1483. https://doi.org/10.1111/1462-2920.13102.

Menge, D.N.L., DeNoyer, J.L., Lichstein, J.W., 2010. Phylogenetic constraints do not explain the rarity of nitrogen-fixing trees in late-successional temperate forests. PLoS One 5 (8), e12056.

Mergel, A., Schmitz, O., Mallmann, T., Bothe, H., 2001. Relative abundance of denitrifying and dinitrogen-fixing bacteria in layers of a forest soil. FEMS Microbiol. Ecol. 36 (1), 33–42.

Mikola, P.U., 1986. Relationship between nitrogen fixation and mycorrhiza. MIRCEN J. Appl. Microbiol. Biotechnol. 2 (2), 275–282.

Mille-Lindblom, C., Fischer, H., Tranvik, L.J., 2006. Antagonism between bacteria and fungi: substrate competition and a possible tradeoff between fungal growth and tolerance towards bacteria. Oikos 113 (2), 233–242.

Mogge, B., Loferer, C., Agerer, R., Hutzler, P., Hartmann, A., 2000. Bacterial community structure and colonization patterns of *Fagus sylvatica* L. ectomycorrhizospheres as determined by fluorescence *in situ* hybridization and confocal laser scanning microscopy. Mycorrhiza 9 (5), 271–278.

Nguyen, N.H., Bruns, T.D., 2015. The microbiome of *Pinus muricata* ectomycorrhizae: community assemblages, fungal species effects, and *Burkholderia* as important bacteria in multipartnered symbioses. Microb. Ecol. 69 (4), 914–921.

Nuccio, E.E., Hodge, A., Pett-Ridge, J., Herman, D.J., Weber, P.K., Firestone, M.K., 2013. An arbuscular mycorrhizal fungus significantly modifies the soil bacterial community and nitrogen cycling during litter decomposition. Environ. Microbiol. 15 (6), 1870–1881.

O'Malley, M.A., 2008. 'Everything is everywhere: but the environment selects': ubiquitous distribution and ecological determinism in microbial biogeography. Stud. Hist. Philos. Sci. C 39 (3), 314–325.

Obase, K., 2019. Bacterial community on ectomycorrhizal roots of *Laccaria laccata* in a chestnut plantation. Mycoscience 60 (1), 40–44.

Paerl, H.W., Pinckney, J.L., 1996. A mini-review of microbial consortia: their roles in aquatic production and biogeochemical cycling. Microb. Ecol. 31 (3), 225–247.

Priyadharsini, P., Rojamala, K., Ravi, R.K., Muthuraja, R., Nagaraj, K., Muthukumar, T., 2016. Mycorrhizosphere: the extended rhizosphere and its significance. In: Choudhary, D., Varma, A., Tuteja, N. (Eds.), Plant-Microbe Interaction: An Approach to Sustainable Agriculture. Springer, Singapore, pp. 97–124.

Rasmann, S., Turlings, T.C.J., 2016. Root signals that mediate mutualistic interactions in the rhizosphere. Curr. Opin. Plant Biol. 32, 62–68.

Read, D.J., Boyd, R., 1986. Water relations of mycorrhizal fungi and their host plants. In: Ayres, P.G., Boddy, L. (Eds.), Water, Fungi and Plants. Cambridge University Press, Cambridge, UK, pp. 287–303.

Read, D.J., Perez-Moreno, J., 2003. Mycorrhizas and nutrient cycling in ecosystems—a journey towards relevance? New Phytol. 157 (3), 475–492.

Rillig, M., Mummey, D.L., Ramsey, P.W., Klironomos, J.N., Gannon, J.E., 2006. Phylogeny of arbuscular mycorrhizal fungi predicts community composition of symbiosis-associated bacteria. FEMS Microbiol. Ecol. 57 (3), 389–395.

Rousk, J., Brookes, P.C., Bååth, E., 2009. Contrasting soil pH effects on fungal and bacterial growth suggest functional redundancy in carbon mineralization. Appl. Environ. Microbiol. 75 (6), 1589–1596.

Rousk, J., Bååth, E., Brookes, P.C., Lauber, C.L., Lozupone, C., Caporaso, J.G., Knight, R., Fierer, N., 2010. Soil bacterial and fungal communities across a pH gradient in an arable soil. ISME J. 4 (10), 1340–1351.

Ruiz-Lozano, J., Bonfante, P., 2000. A *Burkholderia* strain living inside the arbuscular mycorrhizal fungus *Gigaspora margarita* possesses the *vacB* gene, which is involved in host cell colonization by bacteria. Microb. Ecol. 39 (2), 137–144.

Rygiewicz, P.T., Bledsoe, C.S., Zasoski, R.J., 1984. Effects of ectomycorrhizae and solution pH on [^{15}N] nitrate uptake by coniferous seedlings. Can. J. Bot. 14 (6), 893–899.

Rygiewicz, P.T., Johnson, M.G., Ganio, L.M., Tingey, D.T., Storm, M.J., 1997. Lifetime and temporal occurrence of ectomycorrhizae on ponderosa pine (*Pinus ponderosa* Laws.) seedlings grown under varied atmospheric CO2 and nitrogen levels. Plant Soil 189 (2), 275–287.

Scheublin, T.R., Sanders, I.R., Keel, C., van der Meer, J.R., 2010. Characterisation of microbial communities colonising the hyphal surfaces of arbuscular mycorrhizal fungi. ISME J. 4 (6), 752–763.

Schlesinger, W.H., 1997. Biogeochemistry, an Analysis of Global Change, second ed. Academic Press, San Diego, CA, USA.

Schneider, T., Keiblinger, K.M., Schmid, E., Sterflinger-Gleixner, K., Ellersdorfer, G., Roschitzki, B., Richter, A., Eberl, L., Zechmeister-Boltenstern, S., Riedel, K., 2012. Who is who in litter decomposition? Metaproteomics reveals major microbial players and their biogeochemical functions. ISME J. 6 (9), 1749–1762.

Semenov, A.M., van Bruggen, A.H.C., Zelenev, V.V., 1999. Moving waves of bacterial populations and total organic carbon along roots of wheat. Microb. Ecol. 37 (2), 116–128.

Sessitsch, A., Weilharter, A., Gerzabek, M.H., Kirchmann, H., Kandeler, E., 2001. Microbial population structures in soil particle size fractions of a long-term fertilizer field experiment. Appl. Environ. Microbiol. 67 (9), 4215–4224.

Sharma, M., Schmid, M., Rothballer, M., Hause, G., Zuccaro, A., Imani, J., Kämpfer, P., Domann, E., Schäfer, P., Hartmann, A., Kogel, K.-H., 2008. Detection and identification of bacteria intimately associated with fungi of the order Sebacinales. Cell. Microbiol. 10 (11), 2235–2246.

Shirakawa, M., Uehara, I., Tanaka, M., 2019. Mycorrhizosphere bacterial communities and their sensitivity to antibacterial activity of ectomycorrhizal fungi. Microbes Environ. 34 (2), 191–198.

Smith, S.E., Read, D.J., 2008. Mycorrhizal Symbiosis, third ed. Academic Press, San Deigo, CA, USA.

Smith, S.E., Facelli, E., Pope, S., Smith, F.A., 2010. Plant performance in stressful environments: interpreting new and established knowledge of the roles of arbuscular mycorrhizas. Plant Soil 326 (1), 3–20.

Šnajdr, J., Valášková, V., Merhautová, V., Herinková, J., Cajthaml, T., Paldrian, P., 2008. Spatial variability of enzyme activities and microbial biomass in the upper layers of *Quercus petraea* forest soil. Soil Biol. Biochem. 40 (9), 2068–2075.

Sørensen, J., 1997. The rhizosphere as a habitat for soil microorganisms. In: Elsas, J.D.V., Trevors, J.T., Wellington, E.M.H. (Eds.), Modern Microbiology. Marcel Dekker, New York, NY, USA, pp. 21–45.

Stark, S., Kytöviita, M.-M., 2005. Evidence of antagonistic interactions between rhizosphere microorganisms and mycorrhizal fungi associated with birch (*Betula pubescens*). Acta Oecol. 28 (2), 149–155.

Talbot, J.M., Bruns, T.D., Smith, D.P., Branco, S., Glassman, S.I., Erlandson, S., Vilgalys, R., Peay, K.G., 2013. Independent roles of ectomycorrhizal and saprotrophic communities in soil organic matter decomposition. Soil Biol. Biochem. 57, 282–291.

Tanaka, M., Nara, K., 2009. Phylogenetic diversity of non-nodulating *Rhizobium* associated with pine ectomycorrhizae. FEMS Microbiol. Ecol. 69 (3), 29–343.

Tedersoo, L., Hallenberg, N., Larsson, K.H., 2003. Fine scale distribution of ectomycorrhizal fungi and roots across substrate layers including coarse woody debris in a mixed forest. New Phytol. 159 (1), 153–165.

Timonen, S., Hurek, T., 2006. Characterization of culturable bacterial populations associating with *Pinus sylvestris-Suillus bovinus* mycorrhizospheres. Can. J. Microbiol. 52 (8), 769–778.

Toljander, J.F., Artursson, V., Paul, L.R., Jansson, J.K., Finlay, R.D., 2006. Attachment of different soil bacteria to arbuscular mycorrhizal fungal extraradical hyphae is determined by hyphal vitality and fungal species. FEMS Microbiol. Lett. 254 (1), 34–40.

Toljander, J.F., Lindahl, B.D., Paul, L.R., Elfstrand, M., Finlay, R.D., 2007. Influence of arbuscular mycorrhizal mycelial exudates on soil bacterial growth and community structure. FEMS Microbiol. Ecol. 61 (2), 295–304.

Treseder, K.K., Allen, M.F., 2000. Mycorrhizal fungi have a potential role in soil carbon storage under elevated CO_2 and nitrogen deposition. New Phytol. 147 (1), 189–200.

Turlapati, S.A., Minocha, R., Bhiravarasa, P.S., Tisa, L.S., Thomas, W.K., Minocha, S.C., 2013. Chronic N-amended soils exhibit an altered bacterial community structure in Harvard Forest, MA, USA. FEMS Microbiol. Ecol. 83 (2), 478–493.

Uroz, S., Calvaruso, C., Turpault, M.P., Pierrat, J.C., Mustin, C., Frey-Klett, P., 2007. Effect of the mycorrhizosphere on the genotypic and metabolic diversity of the bacterial communities involved in mineral weathering in a forest soil. Appl. Environ. Microbiol. 73 (9), 3019–3027.

Uroz, S., Oger, P., Morin, E., Frey-Klett, P., 2012. Distinct ectomycorrhizospheres share similar bacterial communities as revealed by pyrosequencing-based analysis of 16S rRNA genes. Appl. Environ. Microbiol. 78 (8), 3020–3024.

Uroz, S., Ioannidis, P., Lengelle, J., Cébron, A., Morin, E., Buée, M., Martin, F., 2013. Functional assays and metagenomic analyses reveals differences between the microbial communities inhabiting the soil horizons of a Norway spruce plantation. PLoS One 8 (2), e55929.

Uroz, S., Buée, M., Deveau, A., Mieszkin, S., Martin, F., 2016. Ecology of the forest microbiome: highlights of temperate and boreal ecosystems. Soil Biol. Biochem. 103, 471–488.

van Overbeek, L.S., van Elsas, J.D., van Veen, J.A., 1997. *Pseudomonas fluorescens* Tn5-B20 mutant RA92 responds to carbon limitation in soil. FEMS Microbiol. Ecol. 24 (1), 57–71.

Vik, U., Logares, R., Blaalid, R., Halvorsen, R., Carlsen, T., Bakke, I., Kolstø, A.-B., Økstad, O.A., Kauserud, H., 2013. Different bacterial communities in ectomycorrhizae and surrounding soil. Sci. Rep. 3, 3471.

Villegas, J., Fortin, J.A., 2002. Phosphorus solubilization and pH changes as a result of the interactions between soil bacteria and arbuscular mycorrhizal fungi on a medium containing NO_3^- as nitrogen source. Can. J. Bot. 80 (5), 571–576.

Vitousek, P.M., Howarth, R.W., 1991. Nitrogen limitation on land and in the sea: how can it occur? Biogeochemistry 13 (2), 87–115.

Voříšková, J., Brabcová, V., Cajthaml, T., Baldrian, P., 2014. Seasonal dynamics of fungal communities in a temperate oak forest soil. New Phytol. 201 (1), 269–278.

Wallace, J., Laforest-Lapointe, I., Kembel, S.W., 2018. Variation in the leaf and root microbiome of sugar maple (*Acer saccharum*) at an elevational range limit. PeerJ 6, e5293.

Warmink, J.A., van Elsas, J.D., 2008. Selection of bacterial populations in the mycosphere of *Laccaria proxima*: is type III secretion involved? ISME J. 2 (8), 887–900.

Chapter 15

Pathobiome and microbial communities associated with forest tree root diseases

Jane E. Stewart[a], Mee-Sook Kim[b], Bradley Lalande[c], and Ned B. Klopfenstein[d]

[a]Department of Agricultural Biology, Colorado State University, Fort Collins, CO, United States, [b]USDA Forest Service, Pacific Northwest Research Station, Corvallis, OR, United States, [c]USDA Forest Service, Forest Health Protection, Gunnison, CO, United States, [d]USDA Forest Service, Rocky Mountain Research Station, Moscow, ID, United States

Chapter Outline

1. Diverse drivers of microbial change in plants

Soil microbial communities occupy the most biologically diverse habitats in the world. A single gram of soil can support more than several thousand fungal taxa near the root rhizosphere (Buée et al., 2009). As mentioned in other chapters in this book, many factors can influence the microbial communities associated with tree leaves, stems, and roots. Differences in host species (Prescott and Grayston, 2013), cultivar type within a species, soil type, physiological status of host, and pathogen presence can influence variation in microbial communities (Costa et al., 2007; Aira et al., 2010; Chaparro et al., 2013; Yuan et al., 2015). Ecological balance within the associated microbial community is critical for plant health, especially in the rhizosphere, and disturbances can cause imbalances within the microbial communities. Previous studies have documented that beneficial microbial relationships can enhance seedling vigor, seed germination, plant development, and plant growth that lead to higher plant productivity, whereas attacks by plant pathogens can alter the microbiome structure, functionality, and activity (Trivedi et al., 2012). Beneficial microbial interactions can lead to improved host resistance against pathogenic bacteria and fungi. For example, beneficial microbial taxa can secrete various allelopathic chemicals and toxins that provide the plant with protective barriers that impede plant pathogens. The rhizosphere has been shown to contain diverse and complex biological communities that encompass bacteria, fungi, oomycetes, and many other microorganisms, such as archaea, nematodes, and viruses (Raaijmakers et al., 2009). Other tree organs, including leaves, branches, and stems, are also known to contain a diverse suite of microbial taxa, but overall diversities are typically lower than those found in soils (Baldrian, 2017). Although microbial diversity can vary greatly, pathogens can greatly affect microbial communities. This chapter will briefly review the concept of pathobiome, how microbial communities protect against plant disease, and various changes that can occur within microbial communities in the presence of plant pathogens. Because these research topics are recently developing in forest sciences, examples will be derived from cropping systems as diverse as wheat, apples, and forests. As expected, microbial communities can be vastly different within annual vs. perennial cropping systems; however, the influence of plant pathogens on microbial communities and their ecological roles have been documented primarily in diverse cropping systems.

Forest Microbiology. https://doi.org/10.1016/B978-0-12-822542-4.00004-8

2. Pathobiome

Historically, plant disease was thought to result from the interaction of a single host plant and a single pathogen under suitable environmental conditions; however, new sequencing technologies have helped give rise to a new paradigm that plant diseases, in some cases, are the culmination of numerous biological/microbial interactions with the plant host and pathogen that, in concert, reduce host defense mechanisms and/or increase pathogenicity of the pathogen to foster the development of disease. "Pathobiome" is a term used to describe a consortia of species associated with a host that, when present, reduce host vitality under suitable environmental conditions (Fig. 15.1). Our current capacity to identify species within plant-associated microbial communities has raised awareness on how these microbial communities can affect the fitness of their plant host in a myriad of ways, both positively (e.g., increases in survival, growth, and protection from disease) and negatively (e.g., parasitic interactions) (Garbelotto et al., 2019). Factors that drive disease or parasitic interactions include complex relationships among host-associated bacteria, eukaryotes, and viruses under suitable environmental conditions. Also, our considerations of plant disease processes are now also expanding to consider the critical roles of these microbes within these ecological communities. We are currently just beginning to understand how these interactions occur, but they are thought to be very diverse, extremely complex, and variable over time (Bass et al., 2019). Understanding the roles of individual microbes within the pathobiome could result in the development of novel disease mitigation methods. Research efforts are underway to determine how to shift the microbial communities from a "pathobiome" to a more neutral or even beneficial microbiome, which may include methods aimed at synthetically adding new members to the communities with known beneficial properties that increase microbiota-modulated immunity, or adjusting the soil environment to favor naturally occurring, beneficial microbes and/or disfavor detrimental microbes (Vannier et al., 2019). These research efforts offer potential to substantially improve disease mitigation efforts.

FIG. 15.1 The pathobiome is a term to describe a consortia of species associated with a host that, when present, reduce host vitality under suitable environmental conditions. The left side of the tree has a microbial community of nondetrimental microbes *"blue and green circles."* The pathogen *(mustard)* is present but cannot infect. On the right side of the tree, the combination of the microbial community *(red, orange, and yellow circles)* and the pathogen *(mustard)* work in concert to cause disease.

3. Soil microbiomes

We have long known that soil microbes play critical roles in forest ecosystems processes. In fact, soil microbes are essential for forest productivity, sustainability, and resilience (Bonan, 2008; Hartmann et al., 2013). Prominent examples of microbial roles in forests include (1) mycorrhizal and bacterial associations that promote plant health and improve tolerance to biotic and abiotic disturbances (Jansson and Hofmockel, 2020; Mendes et al., 2011; Panke-Buisee et al., 2015), (2) decomposition that recycles nutrients (van der Heijden et al., 2006; Yang et al., 2009), and (3) contributions to soil structure (Bahram et al., 2018; Fierer, 2017). However, determining the dynamic composition of microbes in forest soils and understanding the ecological roles of soil microbes represents one of the greatest challenges in forest biology and ecology. Furthermore, microbiota within soils are influenced by countless biotic and abiotic factors, such as temperature, moisture, pH, nutrients, other microbes, plant roots, etc. (Fierer, 2017).

Microbial communities, such as fungi and bacteria, associated with soils and plant roots have been studied for decades. Important microbial taxa associated with plant root systems have been identified, and microbial functions that contribute to plant and soil health are becoming better understood. Key discoveries include microbial functions in nitrogen fixation and ammonia oxidation (De Vries and Bardgett, 2012), and mycorrhizal fungi and/or rhizobia that provide nutrients, protective benefits, and growth stimulation (Dechassa and Schenk, 2004; Dinkelaker et al., 1995). However, with the recent advent of molecular techniques, like metagenomic and metatranscriptomic sequencing, studies examining soil microbial communities with direct sequencing of genomes and transcriptomes can identify genes and their expression in soils along with their potential functions. Studying the metabolomic profile of plant-microbial community interactions has also been established as an important approach for unraveling nonprotein-based metabolites (Chong and Xia, 2017). Mass spectrometry and high performance liquid chromatography (HPLC) are common tools for establishing metabolomic profiles in host-pathogen interactions (Pelsi et al., 2018). Studies using these technologies have markedly increased our knowledge of the chemical and molecular basis for the critical ecological functions that are provided by soil microbes (Chong and Xia, 2017; Jansson and Hofmockel, 2018). Access to the gene composition of microbial communities allows identification and detection of novel biocatalysts or enzymes, networks linking function and phylogeny of uncultured organisms, and evolutionary profiles of soil community function and structure (Jansson and Hofmockel, 2018; Ma et al., 2018).

Studies have documented a strong relationship among soil type, soil physical properties, and geographic location on characteristics and diversity of soil microbial communities, including beneficial and pathogenic fungi and bacteria (Lareen et al., 2016). Several studies have documented greater diversity of microbial communities associated with increasing levels and complexity of organic matter (Lehmann et al., 2020). Further, increasing soil pH, water retention, and nutrients, such as carbon, nitrogen, phosphorus, potassium, and others, have been shown to influence microbial diversity, which has been linked to benefits of disease suppression in some studies (Li et al., 2019; Palansooriya et al., 2019), but variable results have also been observed. Suppressive soils are those in which some members of the soil microbial community help the plant defend against fungal and bacterial pathogens (Baker and Cook, 1974). For example, when examining microbial diversity and richness in disease suppressive soils for a vanilla (*Vanilla planifolia*) cropping system, Xiong et al. (2017) found that bacterial richness and phylogenetic diversity showed positive correlations with soil pH and electrical conductivity, whereas the fungal richness and phylogenetic diversity were negatively correlated with these two soil properties. In another study, comparing microbial diversity in plots with high and low levels of the banana wilt pathogen (*Fusarium oxysporum* f. sp. *cubense*), soil organic matter varied significantly between healthy and diseased plots and was negatively correlated with disease incidence and *Fusarium* abundance (Shen et al., 2019a). Soil organic matter correlated with overall abundance of fungal microbial communities (Shen et al., 2019a). Likewise, in a study examining microbial communities associated with *Armillaria solidipes*, a root disease pathogen, higher fungal richness and phylogenetic diversity were positively correlated with available N, P, and Fe in the soils (Lalande, 2019). The work also showed differences in bacterial communities associated with the nonpathogen (*Armillaria altimontana*) compared to those associated with *A. solidipes* (Fig. 15.2). In general, diverse studies indicate the great complexity of interactions that influence the richness and overall composition of microbes in varying soil types across different geographic areas. Despite these complexities, it is known that soil microbes can help to suppress disease.

4. Plant-pathogen-microbe interactions in disease-suppressive soils

Plant-microbe interactions occur in plant layers, such as the phyllosphere, rhizoplane, and rhizosphere. The plant and plant-associated microbial communities (e.g., bacteria, archaea, viruses, and microbial eukaryotes), which occur above and below ground, are collectively known as the "holobiont"—the genomic reflection of the complex network of symbiotic interactions between the host and associated microbiome (Guerrero et al., 2013; Vandenkoornhuyse et al., 2015). The holobiont

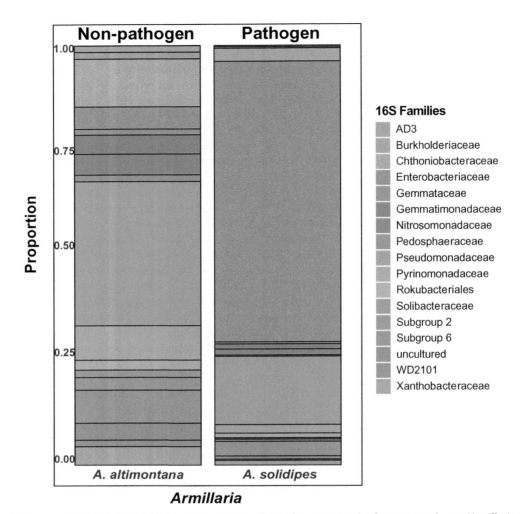

FIG. 15.2   Bacterial communities associated with a nonpathogen (*Armillaria altimontana*) and a forest root pathogen (*Armillaria solidipes*). Soils associated with *Armillaria altimontana* are dominated by beneficial Pseudomonadaceae bacteria commonly found in suppressive soils, whereas soils associated with *Armillaria solidipes* are dominated by pathogenic and saprotrophic Enterobacteriaceae bacterial species.

is formed by selective pressures likely driven by interactions among microbes and the host plant; however, little is known about the contribution of competitive and cooperative microbe-microbe interactions, as these are difficult to interpret even with recent sequencing technologies (Hassani et al., 2018). Contributing to the "rhizosphere effect," plants exude roughly 21% of their photosynthetically fixed carbon at the root-soil interface, which influences the microbial community in the rhizosphere. Typically, a shift from overall diversity of communities to high densities of selected taxa within the community occurs along the gradient from bulk soil to rhizosphere (Mendes et al., 2011). An enrichment of specialized taxa within the community and an overall reduction in the diversity of microbial communities are typically observed in the rhizosphere (Zhang et al., 2017). Plants can provide signals to conspecific, beneficial organisms by producing antifungal effectors and secreting primary and secondary metabolites, such as strigolactones, phenazines, polyketides, and siderophores into the rhizosphere (Carrión et al., 2019), which can cause shifts in the microbial communities that typically vary based on plant species and/or genotypes (Hahl et al., 2020). These plant-driven selective pressures form the basis for microbe-host interactions. Key drivers in the selection of microbes that encompass the holobiont are plant exudates, such as carbohydrates, amino acids, secondary metabolites, and plant-specific flavonoids (Moe, 2013; Weston and Mathesius, 2013). Microbial members of the holobiont can provide plant benefits, such as growth stimulation, stress resistance, nutrient mobilization and transport, and disease suppression via the production of antimicrobial products that can defend against pathogens within the rhizosphere and enhance plant resistance to pathogens (Berg et al., 2014; De Souza et al., 2016; Carrión et al., 2019). Microbes can also elicit host plant responses by symbiotic signaling molecules that can trigger cell wall thickening, programmed cell death, reactive oxygen species (ROS) generation, and production of defense phytohormones and salicylic acid (SA) (Jones and Dangl, 2006; Dodds and Rathjen, 2010; Montiel et al., 2012; De Souza et al., 2016). Additionally, bacteria and fungi produce a multitude of volatile organic compounds (VOCs), like sesquiterpenes (Kramer and Abraham, 2012). VOCs have been shown to disrupt the plants' ability to defend themselves (Bitas et al., 2013). For example, Bitas et al. (2015) showed that VOCs of *Fusarium oxysporum* affected auxin transport and signaling in *Arabidopsis thaliana*. VOCs from plant pathogens such as *Alternaria alternata* and *Pencillium charlesii* can promote starch metabolism, and it is hypothesized that this increases pathogen fitness, but the exact mechanisms are unknown (Ezquer et al., 2010; Bitas et al., 2013). However, structurally diverse VOCs are also known to play roles in defense against other microbes. *Trichoderma asperellum* was shown to significantly reduce disease symptoms of *Arabidopsis thaliana* challenged by *Botrytis cinerea* and *Alternaria brassicicola* (Kottb et al., 2015). Furthermore, Norway spruce (*Picea abies*) hosts a suite of monoterpenes and diterpenes that have antifungal properties and have been shown to decrease infections by *Heterobasidion* sp. (Kusumoto et al., 2014). In general, further studies are needed to better understand the mechanisms of microbiome-induced disease resistance/protection in herbaceous and woody plants.

5. Role of metabolites from bacteria-fungal interactions on plant disease development

Two main types of disease suppression by bacteria and fungi can occur. General suppression, which encompasses the competitive activities of the overall micro- and macroflora, is common in most soils. In contrast, specific suppression, which is typically induced by disease outbreaks on host plants grown on long rotations, results from the enrichment of specific subsets of soil microorganisms (Raaijmakers and Mazzola, 2016). Specific suppression of fungal root pathogens has been attributed to the production of antifungal metabolites by different bacterial taxa, carbon competition, and induced systemic resistance by nonpathogenic fungi. For managing disease risk, it is critical to understand the taxa within suppressive soils that drive competition against pathogens, identify indicators of suppressive soils, and determine taxa that help maintain disease-suppressive soils.

Bacterial taxa within the rhizosphere microbiome have been demonstrated to play critical roles in assimilation of nutrients, resistance to stress, and reduction of disease (Yuan et al., 2015; Zeng et al., 2019). The ability of rhizosphere taxa to reduce plant susceptibility to pathogens has been well documented. Chapelle et al. (2016) documented reduced infection of sugar beet (*Beta vulgaris*) by *Rhizoctonia solani* in suppressive soils, in which microbial communities in *Rhizoctonia solani*-inoculated soils had higher numbers of several bacterial families that have been documented to inhibit fungal growth and protect plant roots from fungal infection. Using mRNA sequencing, these same authors found that HtrA/Sec secretion systems, guanosine-3, 5-bispyrophosphate metabolism, and oxidative stress responses were upregulated by the bacterial families, indicating that bacterial responses that protected sugar beet from infection were triggered by pathogen inoculation in suppressive soils (Chapelle et al., 2016). Another study examining the microbial taxa within *Rhizoctonia*-suppressive soils identified that members of the Pseudomonadaceae protect plants via the production of putative chlorinated lipopeptides encoded by nonribosomal peptide synthetases (NRPS) genes (Mendes et al., 2011). Burkholderiaceae, Xanthomonadales, and Actinobacteria, which have documented activities against soil-borne pathogenic fungi, were also found within suppressive soils (Mendes et al., 2011). Further, Trivedi et al. (2017) determined that bacteria belonging to the phyla Actinobacteria, Firmicutes, and Acidobacteria were correlated with reduced levels of *Fusarium oxysporum* f. sp. *cucumerinum* within inoculated soils, and these microbial taxa were identified as important indicators of suppressive soils in Australia. Other studies have also implicated Actinobacteria as important in the suppression of Fusarium wilt of strawberry, caused by *Fusarium oxysporum* f. sp. *fragariae* (Cha et al., 2016). Actinobacteria are known to produce numerous antibiotics, enzymes, inhibitors, and plant growth regulators, such as auxins, cytokinins, and gibberellins, that can defend plants against pathogens and contribute to soil health (Sakure and Bhosale, 2019).

Fungi, together with their bacterial symbionts, can also play a strong role in disease suppression, and nearly 80% of land plants are associated with arbuscular mycorrhizal fungi (AMF) (Cameron et al., 2013). Several groups of bacterial symbionts, which are stimulated in the rhizosphere associated with AMF fungi, have also been shown to enhance plant growth and suppress fungal and bacterial pathogens. *Funneliformis*, a genus of AMF, has been found to protect roots against microbial pathogens by priming jasmonate (JA) signaling. *Funneliformis mosseae* has also been shown to enhance grape vine defenses against nematodes by inducing JA- and SA-inducible defense genes (Li et al., 2006), upregulating SA-independent resistance against *Gaeumannomycetes graminis* in barley (Khaosaad et al., 2007) and tomatoes (Pozo et al., 2002), and enhancing production of benzoxazinoids that defend wheat against challenges by *Rhizoctonia solani* (Song et al., 2011). Other AMF genera have been shown to prime defense-related enzymatic activity and enhance production of phenolic compounds (Abdel-Fattah et al., 2011). In a meta-analysis of 106 scientific studies, Veresoglou and Rillig (2011) demonstrated that AMF on average reduce fungal and nematode damage by 42% and 44%–57%, respectively, thus emphasizing the strong effect of AMF on disease suppression.

Similarly, other fungi, including the ectomycorrhizal (ECM) fungi have been shown to play roles in host defense against soil-borne pathogens. For example, several ECM species, including *Laccaria laccata*, *Hebeloma crustuliniforme*, *Hebeloma sinapizans*, and *Paxillus involutus*, were shown to decrease disease severity caused by root pathogens, *Phytophthora cambivora* and *Phytophthora cinnamomi* on European chestnut (*Castanea sativa*) (Branzanti et al., 1999). Another ECM fungus, *Hebeloma mesophaeum*, was also shown to reduce damage of a rust fungus, *Melampsora laricis-populina,* on the leaves of *Populus* spp. (Pfabel et al., 2012). Kope and Fortin (1989) tested seven ECM as potential inhibitors of 20 phytopathogens. They documented that *Pisolithus tinctorius* and *Tricholoma pessundatum* were antagonistic toward most phytopathogens including root pathogens (e.g., *Armillaria mellea*, *Fusarium oxysporum*, *Rhizoctonia* spp., among others). Although both exhibited inhibitory qualities, *Pisolithus tinctorius* was antagonistic to 85% of the root pathogens, whereas *Tricholoma pessundatum* only suppressed 55% (Kope and Fortin, 1989). A study assessing the differences between natural soils that suppressed *Fusarium oxysporum* and soils that were conducive to the Fusarium wilt disease, Xiong et al. (2017), identified higher levels of *Mortierella*, *Ceratobasidium*, and *Gymnopus* in association to suppressive soil compared to conducive soil. *Trichoderma harzianum*, a common biological fungicide for soil, can be used to inhibit the growth of root pathogens (Grinstein et al., 1979). In a greenhouse and field study, the application of *Trichoderma harzianum*-wheat bran inoculum to previously infested soil successfully protected crops from *Rhizoctonia solani* and *Sclerotium rolfsii* (Grinstein et al., 1979). A field study in British Columbia inoculated *Armillaria ostoyae*-infected stumps with *Hypholoma fasciculare* (an abundant fungus isolated from soil at the site) and results suggested that *Hypholoma fasciculare* can act as direct competitor of the root pathogen, while inhibiting the pathogen spread within soil (Chapman and Xiao, 2000). Two years after the study, one of the sites showed a large reduction in roots infected by *Armillaria ostoyae*, but more time was needed to determine if *Armillaria* could be controlled over the long term with *Hypholoma fasciculare* (Chapman and Xiao, 2000). The use of both bacterial and fungal antagonists, naturally occurring in the soil, may provide an essential biocontrol tool to assist in the overall management of root pathogens.

6. Plant infections by fungal pathogens result in changes in beneficial taxa

Few studies have documented microbial changes that occur after emergence of a plant pathogen, especially in forest ecosystems. Further, studying the interactive effects of plant pathogens on microbial communities is difficult because of the myriad of diverse microbial taxa that are producing a wide array of compounds, and most of the compounds produced by plant pathogens, symbionts, and other associated microbes have not yet been identified and/or characterized. However, evidence shows that some plant pathogens have the ability to shut down beneficial symbionts by deactivating genes involved in antimicrobial production and further weakening symbionts by releasing a suite of secondary metabolites, such as hydrogen cyanide, alkaloids, and bacteriocins (Benizri et al., 2005; Antunes et al., 2008; Holtsmark et al., 2008). This arsenal of compounds can modify the community structure and function, resulting in deleterious impacts for the host plant (De Souza et al., 2016). For example, infections by plant pathogens can cause the upregulation and production of broad-spectrum resistance compounds in the host that can result in the reduction of diverse groups of fungi, bacteria, and viruses residing in the host. This has been observed as pathogen-triggered oxidative bursts in hosts that can result in the production of ROS (De Souza et al., 2016). This activation also results in subsequent induction of endogenous SA. Together, ROS and SA contribute to the production of broad spectrum, multiple-resistance mechanisms. Several studies have found that ROS elevation can result in the decrease of nodulation of rhizobial bacteria in several host plant species (Cárdenas et al., 2008; Lohar et al., 2007; Muñoz et al., 2012). Further, exogenous SA application in alfalfa (*Medicago sativa*) plants resulted in the inhibition of nodule formation (Mabood and Smith, 2007; Hayat et al., 2010), and blocked nodulation by *Rhizobium leguminosarum* in common vetch (*Vicia sativa*) in Norway (Martínez-Abarca et al., 1998).

When examining effects of pathogens on microbial communities, many studies have focused on characterizing bacterial and fungal communities present within rhizospheres of healthy and diseased plants. It is difficult to define trends in changes of microbial community after a pathogen becomes established. In some cases, beneficial taxa within a community are reduced, whereas, in other cases, beneficial taxa remain in association with diseased plants. However, a common result of plant disease is an overall reduction in community diversity and beneficial bacterial/fungal taxa, including Acidobacteria, AMF, and ECM fungi. For example, Huanglongbing (HLB) of citrus, caused by *Candidatus Liberibacter asiaticus* infection, has been shown to reduce bacterial diversity (Trivedi et al., 2012). Further, as described later, changes in microbial communities during infection by pathogenic *Fusarium* species have been documented in several studies.

Similar microbial taxa were identified in association with healthy plants under different cropping systems and across variable geographic locations (Shen et al., 2019a; Xiong et al., 2017). One study focused on continuous cropping systems of banana (*Musa* sp.) where high levels of Fusarium wilt disease caused by *Fusarium oxysporum* f. sp. *cubense* race 4 (FOC 4) have emerged and resulted in reductions of banana yields in China (Shen et al., 2019a). Examinations of microbial communities associated with healthy and diseased bananas found that higher fungal richness was significantly correlated to higher disease incidence; however, potential biocontrol taxa, *Flavobacterium*, *Mortierella*, and Acidobacteria, were negatively correlated with disease incidence (Shen et al., 2019a). In Central America, healthy banana plants were associated with Gammaproteobacteria (Köberl et al., 2017). In China, a reduced level of *Funneliformis*, an AMF in the family Glomeraceae, was also associated with lower yields and higher diseases indexes, which reflects the importance of the AMF in resistance to banana wilt disease. Interestingly, *Mortierella* and Acidobacteria were also found to be important in soils suppressive against Fusarium wilt of vanilla caused by *Fusarium oxysporum* f. sp. *vanilla*. Suppressive soils also had increased levels of general Basidiomycota taxa and taxa within the Zygomycota genus *Mortierella*, which made up 37% of all the DNA sequence reads from suppressive soils (Xiong et al., 2017). Communities associated with Fusarium wilts of banana and vanilla were both associated with higher fungal abundance, richness, and phylogenetic diversity. Similarly, bacterial communities in Fusarium wilt-conducive soils also had higher values for all indices compared to suppressive soils (Xiong et al., 2017). In contrast, banana fields with low incidence of Fusarium wilt of banana in Costa Rica and Nicaragua were characterized with higher microbial diversity in the rhizosphere (Köberl et al., 2017). Further, these fields were also characterized with higher levels of Gammaproteobacteria, in addition to potentially plant-beneficial *Pseudomonas* and *Stenotrophomonas*. Fields with higher levels of disease severity were associated with more *Enterobacteriaceae* taxa (*Erwinia*, *Enterobacter*, and others) (Köberl et al., 2017). Interestingly, soils suppressive for Fusarium wilt of vanilla also had enrichments in some bacterial phyla, including Acidobacteria, Verrucomicrobia, Actinobacteria, and Firmicutes in disease-suppressive soils, while Proteobacteria and Bacteroidetes were more prevalent in the disease-conducive soils (Xiong et al., 2017). Previous studies have documented the importance of Actinobacteria and Firmicutes in the *Rhizoctonia*-suppressive soils (Mendes et al., 2011). Further, Rosenzweig et al. (2012) also found that Acidobacteria were found at higher levels in potato soils that were less conducive to common scab caused by *Streptomyces* spp. (Rosenzweig et al., 2012). Bacterial taxa from the phyla Verrucomicrobia, Actinobacteria, and Firmicutes are known to produce high levels of secondary metabolites and are likely play a role in disease suppression (Kim et al., 2011; Palaniyandi et al., 2013; Shen et al., 2019b).

In other *Fusarium* pathosystems, a different group of bacterial species have also been identified as important in disease suppression. Isolates of *Bacillus* and *Pseudomonas* collected from Lauraceae woody plant species were found to inhibit mycelial growth of *Fusarium solani* and *Fusarium kuroshium*; *Bacillus* sp. collected from the rhizosphere of *Aiouea effusa* produced diffusible compounds that reduced mycelia growth by 62.5% for *F. solani* and 73.6% for *Fusarium kuroshium* (Báez-Vallejo et al., 2020). This study also identified a suite of cloud-forest bacterial taxa, including representatives from *Bacillus*, *Pseudomonas*, *Curtobacterium*, *Microbacterium*, *Arthrobacter*, *Methylobacterium*, *Erwinia*, and *Hafnia*, that produce antifungal compounds that also reduced mycelial growth of *Fusarium* species (Báez-Vallejo et al., 2020).

Other studies have also identified additional taxa that may be important in disease-suppressive soils in other pathosystems. One study examining the microbial taxa within *Rhizoctonia*-suppressive soils identified that members of the Pseudomonadaceae protect plants via the production of putative chlorinated lipopeptides encoded by NRPS genes (Mendes et al., 2011). Burkholderiaceae, Xanthomonadales, and Actinobacteria were also found within suppressive soils and have been documented to have activities against soil-borne pathogenic fungi (Mendes et al., 2011). High levels of *Pseudomonas* were also found in grapevine asymptomatic to esca syndrome, a disease similar to replant diseases of fruit and nut trees (Saccá et al., 2019). The microbial community of barley challenged with *Fusarium graminearum* showed an enrichment of fluorescent pseudomonads and two antifungal/bacterial genes, *phID* and *hcnAB*, that encode proteins that synthesis 2,4 diacetylphloroglucinol and hydrogen cyanide, respectively (Dudenhöffer et al., 2016; Weller et al., 2002). It is hypothesized that plants can actively enrich disease suppression-associated bacterial taxa by exuding chemicals that attract these bacteria via chemotaxis (Dudenhöffer et al., 2016). *Bacillus megaterium* has also been shown to produce hydrogen cyanide and was demonstrated to be present in potato soils that suppress the bacterial pathogen, *Dickeya*, which causes black leg disease of potatoes (Mao et al., 2019).

Rhizosphere soils associated with nursery plantings of diseased and healthy black spruce (*Picea mariana*) were compared via cloned rRNA to determine differences in bacterial and fungal microbial communities, and the microbial communities associated with healthy and diseased seedlings varied considerably (Filion et al., 2004). Interestingly, some fungal operational taxonomic units (OTUs) occurred at high frequency, suggesting that the rhizosphere within nursery soils is dominated by a relatively few fungal species, including *Ossicaulis lignitalis*, *Thelephora* sp., and *Tricholoma myomyces*. A different pattern was observed within the bacterial communities, where more diversity was observed and no dominant OTUs were identified. Rhizosphere communities from healthy plants were dominated by Deltaproteobacteria (Proteobacteria), Eurothiomycetes (Ascomycota), and Pezizomycetes (Ascomycota). In contrast, rhizosphere soils associated with diseased plants were dominated by Nitrospirae (bacteria), Verrucomicrobia (bacteria), Dothideomycetes

(Ascomycota), mitosporic Ascomycota, and Heterobasidiomycetes (Basidiomycota). Overall, reasons for these patterns were difficult to discern due to high diversity within the soils; however, similar to many of the studies reviewed here, bacterial taxa related to *Pseudomonas*, *Bacillus*, and *Paenibacillus*, known biocontrol agents, were only found in association with healthy seedlings (Filion et al., 2004).

Lastly, actinomycetes have been found to be important in suppression of soil-borne pathogens. Across forest soils of scrubby sclerophyll forests, savannah woods, and wetter sclerophyll forests of Australia, reductions of actinomycetes in the presence of *Phytophthora cinnamomi* were correlated with decreasing levels of plant health (Weste and Vithanage, 1978). A survey resulted in the collection of 267 actinomycetes strains, and 15 strains were identified that exhibited antimicrobial properties against *Pythium ultimum* under both laboratory and greenhouse conditions (Crawford et al., 1993). In addition, some actinomycete isolates have displayed antagonistic activities against *Colletotrichum capsici* and *Fusarium oxysporum* (Ashokvardhan et al., 2014), and a *Streptomyces coelicolor* isolate from mangroves in India caused growth reductions of *Aspergillus niger*, *Aspergillus flavus*, *Aspergillus fumigatus*, and *Penicillium* sp. (Gayathri and Muralikrishnan, 2013).

7. Changes in fungal diversity as a response to root pathogens

Some studies also report a higher diversity and/or richness of fungal taxa in the presence of disease. One study examined microbial community differences in Norway spruce asymptomatic and symptomatic to Heterobasidion root disease, caused by an important fungal genus of root pathogens found in Europe and in North America (Kovalchuk et al., 2018). This study compared the microbial communities associated with multiple parts of trees, including the needles, wood, bark, and root systems. In the wood tissues near infection sites, fungal communities and mycelial growth of *Heterobasidion* were significantly different than asymptomatic trees, but differences between asymptomatic and symptomatic trees were not evident in microbial communities of the needles or roots. It is likely that *Heterobasidion* infections progress slowly and microbial changes are localized to the infection area instead of occurring systemically throughout the trees. Additional evidence suggests that *Heterobasidion* infections of Norway spruce in Finland may increase coinfection with other wood-degrading fungi (Kovalchuk et al., 2018). For example, *Inonotus*, a genus of known heartwood decayers, was observed in several of the symptomatic trees. In a study comparing microbial communities associated with *Armillaria solidipes*, a known root pathogen of conifers, and *Armillaria altimontana*, a species that was recently described as a potential antagonist against *Armillaria solidipes* (Warwell et al., 2019), greater fungal diversity was observed in soils associated with the pathogen, *Armillaria solidipes*, which is likely attributable to the increase in saprotrophic fungi (Lalande, 2019). Interestingly, in both studies, little change in the mycorrhizal community was observed, which concurs with findings of Gaitnieks et al. (2016) that infections of *Heterobasidion annosum* and *Heterobasidion parviporum* had limited effects on mycorrhizal colonization or ECM community structure. In addition, Zampieri et al. (2017) examined the effects of infection by *Heterobasidion irregulare*, a nonnative pathogen, and *Heterobasidion annosum*, a native pathogen, on the development of mycorrhizal associations of *Tuber borchii* with Italian stone pine (*Pinus pinea*), and found that infection by either species did not result in a reduction of the number or density of ECM colonizations. Further, another study examining the microbial communities associated with the medicinal perennial herbaceous plant, *Gastrodia elata*, cropping systems found that the presence of *Armillaria* spp., as a mycorrhizal fungus, increased organic matter and decaying wood materials; this symbiosis likely led to increased fungal diversity, but less influence on bacterial diversity was noted (Chen et al., 2019).

8. Changes in bacterial diversity as a response to root pathogens

Root pathogens (fungal or bacterial) can also drive changes in the bacterial microbial community, and unlike the fungal groups, many studies report a decrease in overall bacterial diversity in response to root pathogen. A study examining shifts in bacterial communities of wild rocket (*Diplotaxis tenuifolia*) in suppressive soils and *Fusarium*-infected plants found that overall bacterial diversity decreased in presence of Fusarium disease, with greater number of *Massilia*. After the addition of suppressive soils, a quantitative increase in beneficial bacteria, *Rhizobium*, *Bacillus*, *Paenibacillus*, and *Streptomycetes* spp. was observed after only 3 days (Klein et al., 2012). Stump removal as a management tool for stands infested with the forest root pathogen *Armillaria solidipes* (=North American *A. ostoyae*) significantly increased the biodiversity of bacterial communities, suggesting that bacterial diversity is decreased in the presence of *Armillaria* (Modi et al., 2020). When measuring root pathogen effects on microbial communities, most studies focus on root tissues; however, Ren et al. (2019) studied if *Heterobasidion* infections induced bacterial changes in Norway spruce needle, stem, bark, and roots tissues. Interestingly, those authors found that significant changes only occurred in the needle microbial communities. It was hypothesized that bacterial community changes were more apparent in needle tissues because these tissues also had lower species richness, diversity, and evenness compared to the bark, roots, and stem tissues. In that study, the root microbiome was the most diverse and had the largest number of unique taxa (not found in the other tissues), and it was suggested that slow growth of the pathogen may limit change in the bacterial community.

9. Microbiomes linked taxa to the pathobiome

Historically, plant disease was considered to be the result of the interaction between one host and one pathogen. The disease triangle, which highlights the connections among one host, one pathogen, and conducive environmental conditions to cause disease, is a prevailing paradigm in plant pathology that implies that the pathogens work alone when causing disease on a host. However, with the advent of microbiome technologies, more complex relationships among pathogens and their microbial associates have been described as part of the disease process. Amplicon metagenomics (or metabarcoding), for example, allow analyses of complex microbial communities associated with soil, organic debris, rhizosphere, root/stem/ foliar endophytes, and phyllosphere in association with forest health and ecological processes (Terhonen et al., 2019). Studies have found that several fungal or bacterial taxa are typically closely connected to the disease status of a host. In some cases, it remains unknown which additional taxa benefit from the diseased status of the host or if these taxa are part of a coinfection process. A complex mix of relationships among additional microbes and disease likely exist for many or most pathosystems. Some studies suggest that the representation of one host and one pathogen in plant pathology should not be a triangle, but rather a square or diamond would better describe the initiation of disease whereby microbial taxa associated with pathogens and their hosts are necessary for disease to occur (Fig. 15.3). Rovenich et al. (2014) suggest that pathogens can utilize effectors of coinhabitants on host tissue for disease expression. Effectors are proteins that plant pathogens secrete to interact with their hosts during invasion (Koeck et al., 2011). Current microbiome sequencing technologies allow the detection of polymicrobial interactions associated with plant diseases. Acute oak decline in Europe, first recognized in the United Kingdom in the 1980s, represents a significant threat to oaks, especially *Quercus robur* and *Quercus petraea*. Isolations from bleeding cankers identified three causal agents: *Brenneria goodwinii*, *Rahnella victoriana*, and an unnamed *Pseudomonas* sp. (Denman et al., 2012, 2014; Brady et al., 2014; Adeolu et al., 2016; Sapp et al., 2016). Bacterial communities associated with the oak decline disease compared to healthy trees also found that *Lonsdalea quercina* spp. *britannica*, *Gibbsiella quercinecans*, and *Rahnella victoriana* were present in high numbers in diseased samples, suggesting their potential roles in disease. All three species were shown through genomics and metatranscriptomic studies to have the genomic capabilities to cause disease and this capability has been demonstrated via Koch's postulates for *Brenneria goodwinii* and *Gibbsiella quercinecans* (Denman et al., 2018). As researchers continue to explore microbial communities associated with plant diseases, more disease complexes will likely be identified, whereby fungal and bacterial species/communities work in concert with each varying component providing specific factors required to cause disease in the hosts.

10. Impact of natural secondary metabolites on pathobiome composition

Secondary metabolites, also called natural products, are organic compounds of low molecular mass that are produced by bacteria (e.g., *Bacillus* spp., *Pseudomonas* spp., and *Streptomyces* spp.), fungi (e.g., *Penicillium* spp., *Aspergillus* spp., *Trichoderma* spp.), and plants of certain taxonomic groups. These metabolites often act as key factors that either

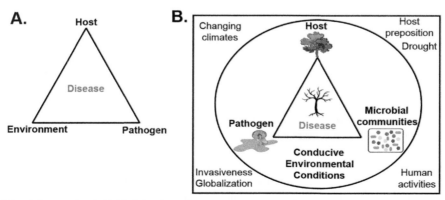

FIG. 15.3 (A) The traditional disease triangle highlighting that a disease can occur when a host and a pathogen on that host are present under conducive environmental conditions. (B) An adapted disease triangle that describes that microbial taxa associated with pathogens and their hosts, in the conducive environment, are necessary for disease initiation. Other factors within the square can also affect a host's susceptibility to disease.

enhance and suppress other organisms (e.g., bacteria, fungi, amoebae, plants, insects, and large animals), which can increase the survival of the organisms that produce them (Demain and Fang, 2000; Peterson et al., 2020). The production of secondary metabolites is frequently triggered by specific interactions between the organisms under specific circumstances (Peterson et al., 2020). Microorganism- and plant-produced secondary metabolites can influence disease development and suppression, as has been discussed in other parts of this chapter. It has become increasingly clear that these natural secondary metabolites can alter the structure, composition, and ecological function of the pathobiome, resulting in the increase or suppression of plant disease. A recent review paper emphasized that secondary metabolites (e.g., coumarins, benzoxazinoids, camalexin, and triterpenes) are responsible for shaping the composition and function of the plant microbiome. Diverse microbial responses (e.g., microbiome composition, nutrient mobilization, pathogen suppression, and hormonal signaling) can be elicited by variations in secondary metabolite abundance (Jacoby et al., 2021). Research is just beginning to understand the role of secondary metabolites in forest pathosystems, but previous studies indicate that some secondary metabolites could play a crucial role in biocontrol of forest pathogens.

11. Microbial changes in other anatomic regions of trees

Bacterial communities associated with bleeding canker caused by *Pseudomonas syringae* pv. *aesculi* on horse chestnut (*Aesculus hippocastanum*) were significantly different than communities from healthy trees, and overall bacterial diversity was lower in infected trees (Koskella et al., 2017). The authors further suggest that the greater diversity in the bacterial communities of healthy tree hosts may provide protective benefits against bleeding canker. In addition, bacterial communities of olive (*Olea europaea*) change in the presence of olive knot disease caused by *Pseudomonas savastanoi* pv. *savastaonoi* (Pss) (Mina et al., 2020). Comparisons of endophytic and epiphytic bacterial communities showed opposing trends. Endophytic communities changed in response to olive cultivar and the presence of Pss, whereas epiphyte communities remained largely unaffected. The growth behavior of Pss, which occurs in clusters aligned within biofilm-like layers, influenced a change in the endophytic bacterial communities (Mina et al., 2020).

12. Considerations

Soil is one of the most challenging ecosystems to perform metagenomic analyses because of the vast microbial diversity, rhizosphere influences, wide variation in soil properties, and other biotic/abiotic environmental factors that influence microbial diversity and function. This review highlights the extraordinary variation in the number and types of taxa that can even be found in association with the same pathosystems, but from different geographical regions. For this reason, the downstream effects of root pathogens to the overall communities or taxa can vary, depending on the individual ecosystem. Soil physical and chemical properties, such as soil type, pH, and organic matter, must also be considered when comparing results across datasets from different studies and sites. Thus far, few large-scale inferences have been identified from studies on the effects of root disease pathogens on microbial communities, especially within forest ecosystems. However, greater insights will develop as more researchers continue to focus on the complex interactions among microbial communities and root disease pathogens.

Recent developments on co-cultivation of phytopathogens with mixed bacteria-fungi may provide effective strategies for discovering antimicrobial agents with roles in reducing plant disease (Chagas et al., 2013; Vinale et al., 2017). Specific communication between the microorganisms through co-cultures may elicit or induce the expression of previously unexpressed microbial gene clusters associated with producing novel secondary metabolites with potential bioactivity of interest (Bertrand et al., 2014; Netzker et al., 2015). Advanced knowledge of microorganism co-culture, involving the cultivation of multiple microorganisms that are associated with the same host, could be applied to manage plant disease and increase crop and forest productivity (Costa et al., 2019). This biocontrol approach for suppressing pathogens is based on developing disease management methods to favor naturally occurring, biocontrol agents (i.e., microorganisms) and the associated microbes that elicit the expression of genes associated with biological control.

Innovative and integrative approaches are essential for assessing the complex interactions and functions of the microbial communities in forest disease processes in relation to influences of diverse of other environmental factors. Understanding the myriad of interactions among hosts, pathogens, microbial communities, and environments requires unprecedented integration of novel methodologies. Results from such integrated information will help develop novel approaches to manage forest disease and improve forest health by promoting conditions that suppress disease or enhance beneficial microbial and ecological processes.

List of trees and plants

Vanilla (*Vanilla planifolia*)
Banana (*Musa acuminata*)
Western white pine (*Pinus monticola*)
Sugar beet (*Beta vulgaris*)
Barley (*Hordeum vulgare*)
European chestnut (*Castanea sativa*)
Poplar (*Populus* spp.)
Alfalfa (*Medicago sativa*)
Common vetch (*Vicia sativa*)
Citrus
Lauraceae
Aiouea effuse
Potato (*Solanum tuberosum*)
Black spruce (*Picea mariana*)
Norway spruce (*Picea abies*)
Italian stone pine (*Pinus pinea*)
Sessile oak (*Quercus petraea*)
English oak (*Quercus robur*)
Horse chestnut (*Aesculus hippocastanum*)
Olive (*Olea europaea*)

List of microorganisms

Banana wilt pathogen (*Fusarium oxysporum* f. sp. *cubense*) race 4
Fusarium oxysporum f. sp. *vanilla*
Armillaria root disease (*Armillaria solidipes*)
Rhizoctonia solani
Burkholderiaceae
Xanthomonadales
Actinobacteria
Firmicute
Fusarium oxysporum f. sp. *cucumerinum*
Fusarium oxysporum f. sp. *fragariae*
Funneliformis mosseae
Gaeumannomycetes graminis
Laccaria laccata
Hebeloma crustuliniforme
Hebeloma sinapizans
Paxillus involutus
Hebeloma mesophaeum
Melampsora laricis-populina
Pisolithus tinctorius
Tricholoma pessundatum
Trichoderma asperellum
Armillaria mellea
Fusarium oxysporum
Rhizoctonia
Mortierella
Ceratobasidium
Gymnopus
Sclerotium rolfsii
Hypholoma fasciculare
Armillaria ostoyae

Huanglongbing (HLB) of citrus, caused by *Candidatus Liberibacter asiaticus*
Proteobacteria
Bacteroidetes
Streptomyces
Verrucomicrobia
Bacillus
Pseudomonas
Fusarium solani
Fusarium kuroshium
Curtobacterium
Microbacterium
Arthrobacter
Methylobacterium
Erwinia
Hafnia
Bacillus megaterium
Dickeya
Ossicaulis lignitalis
Thelephora sp.
Tricholoma myomyces
Deltaproteobacteria
Eurothiomycetes
Pezizomycetes
Nitrospirae
Paenibacillus
Phytophthora cinnamomi
Pythium ultimum
Colletotrichum capsica
Fusarium oxysporum
Aspergillus niger
Aspergillus flavus
Aspergillus fumigatus
Penicillium sp.
Heterobasidion
Armillaria altimontana
Alternaria alternata
Pencillium charlesii
Heterobasidion annosum
Heterobasidion parviporum
Heterobasidion irregulare
Tuber borchii
Gastrodia elata
Armillaria mellea
Brenneria goodwinii
Rahnella victoriana
Lonsdalea quercina spp. *britannica*
Gibbsiella quercinecans
Pseudomonas savastanoi pv. *savastaonoi*
Pseudomonas syringae pv. *aesculin*
Alternaria brassicicola

References

Abdel-Fattah, G.M., El-Haddad, S.A., Hafez, E.E., Rashad, Y.M., 2011. Induction of the defense responses in common bean plants by arbuscular mycorrhizal fungi. Microbiol. Res. 166, 268–281.

Adeolu, M., Alnajar, S., Naushad, S., Gupta, R., 2016. Genome based phylogeny and taxonomy of the 'Enterobacteriales': proposal for Enterobacterales ord. nov. divided into the families Enterobacteriaceae Erwiniaceae fam. nov., Pectobacteriaceae fam. nov., Yersiniaceae fam. nov., Hafniaceae fam. nov., Morganellaceae fam. nov., and Budviciaceaefam. nov. Int. J. Syst. Evol. Microbiol. 66, 5575–5599.

Aira, M., Gómez-Brandón, M., Lazcano, C., Baath, E., Domínguez, J., 2010. Plant genotype strongly modifies the structure and growth of maize rhizosphere microbial communities. Soil Biol. Biochem. 42, 2276–2281.

Antunes, P.M., Miller, J., Carvalho, L.M., Klironomos, J.N., Newman, J.A., 2008. Even after death the endophytic fungus of Schedonorus phoenix reduces the arbuscular mycorrhizas of other plants. Funct. Ecol. 22, 912–918.

Ashokvardhan, T., Rajithasri, A.B., Prathyusha, P., Satyaprasad, K., 2014. Actinomycetes from Capsicum annuum L. rhizosphere soil have the biocontrol potential against pathogen fungi. Int. J. Curr. Microbiol. App. Sci. 3, 894–903.

Báez-Vallejo, N., Camarena-Pozos, D.A., Monribot-Villanueva, M.L., Ramírez-Vázquez, M., Carrión-Villarnovo, G.L., Guerrero-Analco, J.A., Partida-Martínez, L.P., Reverchon, F., 2020. Forest tree associated bacteria for potential biological control of Fusarium solani and of Fusarium kuroshium, causal agent of Fusarium dieback. Microbiol. Res. 235, 126440.

Bahram, R., Hildebrand, F., Forslund, S.K., Andersen, J.L., Soudzilovskaia, N.A., Bodegom, P.M., Bengtsson-Palme, J., Anslan, S., Coelho, L.P., Harend, H., Heurta-Cepas, J., Medema, M.H., Maltz, M.R., Mundra, S., Olsson, P.A., Pent, M., Põlme, S., Sunagawa, S., Ryberg, M., Tedersoo, L., Bork, P., 2018. Structure and function of the global topsoil microbiome. Nature 560, 233–273.

Baker, K.F., Cook, R.J., 1974. Biological Control of Plant Pathogens. Freeman, San Francisco, CA.

Baldrian, P., 2017. Forest microbiome: diversity, complexity, and dynamics. FEMS Microbiol. Rev. 41, 109–130.

Bass, D., Stentiford, G.D., Wang, H.-C., Koskella, B., Tyler, C.R., 2019. The pathobiome in animal and plant diseases. Trends Ecol. Evol. 34, 996–1008.

Benizri, E., Piutti, S., Verger, S., Pagès, L., Vercambre, G., Poessel, J.L., et al., 2005. Replant diseases: bacterial community structure and diversity in peach rhizosphere as determined by metabolic and genetic fingerprinting. Soil Biol. Biochem. 37, 1738–1746.

Berg, G., Grube, M., Schloter, M., Smalla, K., 2014. Unraveling the plant microbiome: looking back and future perspectives. Front. Microbiol. 5, 148.

Bertrand, S., Bohnia, N., Schnee, S., Schumpp, O., Gindro, K., Wolfender, J.-L., 2014. Metabolite induction via microorganism co-culture: a potential way to enhance chemical diversity for drug discovery. Biotechnol. Adv. 32, 1180–1204.

Bitas, V., Kim, H.-S., Bennett, J.W., Kang, S., 2013. Sniffing on microbes: diverse roles of microbial volatile organic compounds in plant health. Mol. Plant-Microbe Interact. 26, 835–843.

Bitas, V., McCartney, N., Li, N., Demers, J., Kim, J.E., Kim, H.S., Brown, K.M., Kang, S., 2015. Fusarium oxysporum volatiles enhance plant growth via affecting auxin transport and signaling. Front. Microbiol. 6, 1248. https://doi.org/10.3389/fmicb.2015.01248.

Bonan, G.B., 2008. Forests and climate change: forcings, feedbacks, and the climate benefits of forests. Science 320, 1444–1449.

Brady, C., Hunter, G., Kirk, S., Arnold, D., Denman, S., 2014. Rahnella victoriana sp. nov., Rahnella bruchi sp. nov., Rahnella woolbedingensissp. nov., classification of Rahnella genomospecies 2 and 3 as Rahnella variigena sp. nov. and Rahnella inusitata sp. nov., respectively and emended description of the genus Rahnella. Syst. Appl. Microbiol. 37, 545–552.

Branzanti, M.B., Rocca, E., Pisi, A., 1999. Effect of ectomycorrhizal fungi on chestnut ink disease. Mycorrhiza 9, 103–109.

Buée, M., Reich, M., Murat, C., Morin, E., Nilsson, R.H., Uroz, S., Matrin, F., 2009. 454 Pyrosequencing analyses of forest soils reveal an unexpectedly high fungal diversity. New Phytol. 184, 449–456.

Cameron, D.D., Neal, A.L., van Wees, S.C.M., Ton, J., 2013. Mycorrhiza-induced resistance: more than the sum of its parts? Trends Plant Sci. 18, 539–545.

Cárdenas, L., Martínez, A., Sánchez, F., Quinto, C., 2008. Fast, transient and specific intracellular ROS changes in living root hair cells responding to Nod factors (NFs). Plant J. 56, 802–813.

Carrión, V.J., Perez-Jaramillo, J., Cordovez, V., Tracanna, V., de Hollander, M., Ruiz-Buck, D., Mendes, L.W., van Ijcken, W.F.J., Gomez-Exposito, R., Elsayed, S.S., Mohanraju, P., Arifah, A., van der Oost, J., Paulson, J.N., Mendes, R., van Wezel, G.P., Medema, M.H., Raaijmakers, J.M., 2019. Pathogen-induced activation of disease-suppressive functions in the endophytic root microbiome. Science 366, 606–612.

Cha, J.Y., Han, S., Hong, H.J., Cho, H., Kim, D., Kwon, Y., Kwon, S.K., Crusemann, M., Bok Lee, Y., Kim, J.F., Giaever, G., Nislow, C., Moore, B.S., Thomashow, L.S., Weller, D.M., Kwak, Y.-S., 2016. Microbial and biochemical basis of a Fusarium wilt-suppressive soil. ISME J. 10, 119–129.

Chagas, F.O., Dias, L.G., Pupo, M.T., 2013. A mixed culture of endophytic fungi increases production of antifungal polyketides. J. Chem. Ecol. 39, 1335–1342.

Chaparro, J.M., Badri, D.V., Bakker, M.G., Sugiyama, A., Manter, D.K., Vivanco, J.M., 2013. Root exudation of phytochemicals in Arabidopsis follows specific patterns that are developmentally programmed and correlate with soil microbial functions. PLoS One 8, e55731.

Chapelle, E., Mendes, R., Bakker, P.A.H.M., Raaijmakers, J.M., 2016. Fungal invasive of the rhizosphere microbiome. ISME J. 10, 265–268.

Chapman, B., Xiao, G., 2000. Inoculation of stumps with Hypholoma fasciculare as a possible means to control armillaria root disease. Can. J. Bot. 78 (1), 129–134. https://doi.org/10.1139/b99-170.

Chen, L., Wang, Y.-C., Qin, L.-Y., He, H.-Y., Yu, X.-L., Thang, M.-Z., Zhang, H.-B., 2019. Dynamics of fungal communities during Gastrodia elata growth. BMC Microbiol. 19, 158.

Chong, J., Xia, J., 2017. Computation approaches for integrative analysis of the metabolome and microbiome. Metabolites 7, 62.

Costa, R.G., Newton, C.M., Kroegerrecklenfort, E., Opelt, K., Berg, G., Smalla, K., 2007. Pseudomonas community structure and antagonistic potential in the rhizosphere: insights gained by combining phylogenetic and functional gene-based analyses. Environ. Microbiol. 9, 2260–2273.

Costa, J.H., Wassano, C.I., Angolini, C.F.F., Scherlach, K., Hertweck, C., Fill, T.P., 2019. Antifungal potential of secondary metabolites involved in the interaction between citrus pathogens. Sci. Rep. 9, 18647.

Crawford, C.L., Lynch, J.M., Whipps, J.M., Ousley, M.A., 1993. Isolation and characterization of actinomycete antagonists of a fungal root pathogen. Appl. Environ. Microbiol. 59, 3899–3905.

De Souza, E.M., Granada, C.E., Sperotto, R.A., 2016. Plant pathogens affecting the establishment of plant-symbiont interaction. Front. Plant Sci. 7, 15.

De Vries, F.T., Bardgett, R.D., 2012. Plant-microbial linkages and ecosystem nitrogen retention: lessions for sustainable agriculture. Front. Ecol. Environ. 10, 425–432.

Dechassa, N., Schenk, M.K., 2004. Exudation of organic anions by roots of cabbage, carrot, and potato as influenced by environmental factors and plant age. J. Plant Nutr. Soil Sci. 167, 623–629.

Demain, A.L., Fang, A., 2000. The natural functions of secondary metabolites. In: Fiechter, A. (Ed.), History of Modern Biotechnology I. Advances in Biochemical Engineering/Biotechnology. vol. 69. Springer, Berlin, Heidelberg.

Denman, S., Brady, C., Kirk, S., Cleenwerck, I., Venter, S., Coutinho, T., De Vos, P., 2012. *Brenneria goodwinii* sp. nov., associated with acute oak decline in the UK. Int. J. Syst. Evol. Microbiol. 62, 2451–2456.

Denman, S., Brown, N., Kirk, S., Jeger, M., Webber, J., 2014. A description of the symptoms of Acute Oak Decline in Britain and a comparative review on causes of similar disorders on oak in Europe. Forestry 87, 535–551.

Denman, S., Doonan, J., Ransom-Jones, E., Broberg, M., Plummer, S., et al., 2018. Microbiome and infectivity studies reveal complex polyspecies tree disease in Acute Oak Decline. ISME J. 12, 386–399.

Dinkelaker, B., Hengeler, C., Marschner, H., 1995. Distribution and function of proteoid roots and other root clusters. Bot. Acta 108, 183–200. https://doi.org/10.1111/j.1438-8677.1995.tb00850.x.

Dodds, P.N., Rathjen, J.P., 2010. Plant immunity: towards an integrated view of plant–pathogen interactions. Nat. Rev. Genet. 11, 539–548.

Dudenhöffer, J.-H., Scheu, S., Jousset, A., 2016. Systemic enrichment of antifungal traits in the rhizsosphere microbiome after pathogen attack. J. Ecol. 104, 1566–1575.

Ezquer, I., Li, J., Ovecka, M., Baroja-Fernandez, E., Munoz, F.J., Montero, M., de Cerio, J.D., Hidalgo, M., Sesma, M.T., Bahaji, A., Etxeberria, E., Pozueta-Romero, J., 2010. Microbial volatile emissions promote accumulation of exceptionally high levels of starch in leaves in monoand dicotyle-donous plants. Plant Cell Physiol. 51, 1674–1693.

Fierer, N., 2017. Embracing the unknown: disentangling the complexities of the soil microbiome. Nat. Rev. Microbiol. 15, 579–590.

Filion, M., Hamelin, R.C., Bernier, L., St-Arnaud, M., 2004. Molecular profiling of rhizosphere microbial communities associated with healthy and dis-eases black spruce (*Picea mariana*) seedling grown in a nursery. Appl. Environ. Microbiol. 70, 3541–3551.

Gaitnieks, T., Klavina, D., Muiznieks, I., Pennanen, T., Velmala, S., Vasaitis, R., Menkis, A., 2016. Impact of *Heterobasidion* root-rot on fine root mor-phology and associated fungi in *Picea abies* stands on peat soils. Mycorrhiza 26, 465–473.

Garbelotto, M., Lowell, N., Chen, I.Y., Osmundson, T.W., 2019. Evidence for inhibition of a fungal biocontrol agent by a plant microbiome. J. Plant Pathol. 101, 457–466.

Gayathri, P., Muralikrishnan, V., 2013. Isolation and characterization of endophytic actinomycetes from mangrove plant for antimicrobial activity. Int. J. Curr. Microbiol. App. Sci. 2, 78–89.

Grinstein, A., Elad, Y., Katan, J., Chet, I., 1979. Control of sclerotium rolshii by means of a herbicide and *Trichoderma harzianum*. Plant Dis. Rep. 63, 823–825.

Guerrero, R., Margulis, L., Berlanga, M., 2013. Symbiogenesis: the holobiont as a unit of evolution. Int. Microbiol. 16, 133–143.

Hahl, T., van Moorsel, S.J., Schmid, M.W., Zuppinger-Dingley, D., Schmid, B., Wagg, C., 2020. Plant responses to diversity-driven selection and associ-ated rhizosphere microbial communities. Funct. Ecol. 34, 707–722.

Hartmann, M., Niklaus, P.A., Zimmerman, S., Schmutz, S., Kremer, J., Abarenkov, K., Lüscher, P., Widmer, F., Brey, B., 2013. Resistance and resilience of the forest soil microbiome to logging-associated compaction. ISME J. 8, 226–244.

Hassani, M.A., Durán, P., Hacquard, S., 2018. Microbial interactions within the plant holobiont. Microbiome 6, 58.

Hayat, Q., Hayat, S., Irfan, M., Ahmad, A., 2010. Effect of exogenous salicylic acid under changing environments: a review. Environ. Exp. Bot. 68, 14–25.

Holtsmark, I., Eijsink, V.G., Brurberg, M.B., 2008. Bacteriocins from plant pathogenic bacteria. FEMS Microbiol. Lett. 280, 1–7.

Jacoby, R.P., Koprivova, A., Kopriva, S., 2021. Pinpointing secondary metabolites that shape the composition and function of the plant microbiome. J. Exp. Bot. 72, 57–69. https://doi.org/10.1093/jxb/eraa424.

Jansson, J.K., Hofmockel, K.S., 2018. The soil microbiome—from metagnomics to metaphenomics. Curr. Opin. Microbiol. 43, 162–168.

Jansson, J.K., Hofmockel, K.S., 2020. Soil microbiomes and climate change. Nat. Rev. Microbiol. 18, 35–46.

Jones, J.D., Dangl, J.L., 2006. The plant immune system. Nature 444, 323–329.

Khaosaad, T., Garcia-Garrido, J.M., Steinkellner, S., Vierheilig, H., 2007. Take-all disease is systemically reduced in root of mycorrhizal barley plants. Soil Biol. Biochem. 39, 727–734.

Kim, Y.C., Leveau, J., Gardener, B.B.M., Pierson, E.A., Pierson, L.S., Ryu, C.-M., 2011. The multifactorial basis for plant health promotion by plant-associated bacteria. Appl. Environ. Microbiol. 77, 1548–1555.

Klein, E., Ofek, M., Katan, J., Minz, D., Gamliel, A., 2012. Soil suppressiveness to Fusarium disease: shifts in root microbiome associated with reduction of pathogen root colonization. Phytopathology 103, 23–33.

Köberl, M., Dita, M., Martinuz, A., Staver, C., Berg, G., 2017. Member of Gammaproteobacteria as indicator species of healthy banana plants on Fusarium wilt-infested fields in Central America. Sci. Rep. 7, 45318.

Koeck, M., Hardham, A.R., Dobbs, P.N., 2011. The role of effectors of biotrophic and hemibiotrophic fungi in infection. Cell. Microbiol. 13, 1849–1857.

Kope, H.H., Fortin, J.A., 1989. Inhibition of phytopathogenic fungi in vitro by cell free culture media of ectomycorrhizal fungi. New Phytol. 113, 57–63.

Koskella, B., Meaden, S., Crowther, W.J., Leimu, R., Metcalf, J.E., 2017. A signature of tree health? Shifts in the microbiome and the ecological drivers of horse chestnut bleeding canker. New Phytol. 215, 737–746.

Kottb, M., Gigolashvili, T., Grobkinsky, D.K., Piechulla, B., 2015. Tirchoderma volatiles effecting Arabidopsis: from inhibition to protection against phytopathogenic fungi. Front. Microbiol. 6, 995.

Kovalchuk, A., Mukrimin, M., Zeng, Z., Raffaello, T., Liu, M., Kasanen, R., Sun, H., Asiegbu, F.O., 2018. Mycobiome analysis of asymptomatic and symptomatic Norway spruce trees naturally infected by the conifer pathogens *Heterobasidion* spp. Environ. Microbiol. Rep. 10, 532–541.

Kramer, R., Abraham, W.-R., 2012. Volatile sesquiterpenes from fungi: what are they good for? Phytochem. Rev. 11, 15–37.

Kusumoto, N., Zhao, T., Swedjemark, G., Ashitani, T., Takahashi, K., Borg-Karlson, A.K., 2014. Antifungal properties of terpenoids in *Picea abies* against *Heterobasidion parviporum*. For. Pathol. 44, 353–361.

Lalande, B., 2019. Abiotic and Biotic Factors Influencing Western United States Coniferous Forests. Colorado State University, Fort Collins, CO.

Lareen, A., Burton, F., Schäfer, P., 2016. Plant root-microbe communication in shaping root microbiomes. Plant Mol. Biol. 90, 575–587.

Lehmann, J., Hansel, C.M., Kaiser, C., Kleber, M., Maher, K., Manzoni, S., Nunan, N., Reichstein, M., Schimel, J.P., Torn, M.S., Wieder, W.R., Kögel-Knabner, I., 2020. Persistence of soil organic carbon caused by functional complexity. Nat. Geosci. 13, 529–534.

Li, H.-Y., Yang, G.-D., Shu, H.-R., Yang, Y.-T., Ye, B.-X., Nishida, I., Zheng, C.-C., 2006. Colonization by the arbuscular mycorrhizal fungus *Glomus versiforme* induces a defense response against the root-knot nematode *Meloidogyne incognita* in the grapevine (*Vitis amurensis* Rupr.), which includes transcriptional activation of the class III chitinase gene *VCH3*. Plant Cell Physiol. 47, 154–163.

Li, Z., Song, Z., Singh, B.P., Wang, H., 2019. The impact of crop residue biochars on silicon and nutrient cycles in croplands. Sci. Total Environ. 659, 673–680.

Lohar, D.P., Haridas, S., Gantt, J.S., VandenBosch, K.A., 2007. A transient decrease in reactive oxygen species in roots leads to root hair deformation in the legume–rhizobia symbiosis. New Phytol. 173, 39–49.

Ma, B., Zhoa, K., Lv, X., Su, W., Dai, Z., Gilbert, J.A., Brooks, P.C., Faust, K., Xu, J., 2018. Genetic correlation network prediction of forest soil microbial function organization. ISME J. 12, 2492–2505.

Mabood, F., Smith, D., 2007. The role of salicylates in Rhizobium-legume symbiosis and abiotic stresses in higher plants. In: Salicylic Acid, A Plant Hormone. Springer Publishers, Dordrechts, Netherlands.

Mao, L., Chen, Z., Xu, L., Zhang, H., Lin, Y., 2019. Rhizosphere microbiota compositional changes reflect potato blackleg disease. Appl. Soil Ecol. 140, 11–17.

Martínez-Abarca, F., Herrera-Cervera, J.A., Bueno, P., Sanjuan, J., Bisseling, T., Olivares, J., 1998. Involvement of salicylic acid in the establishment of the *Rhizobium meliloti*–Alfalfa symbiosis. Mol. Plant-Microbe Interact. 11, 153–155.

Mendes, R., Kruijt, M., de Bruijn, I., Dekkers, E., van der Voort, M., Schneider, J.H.M., Piceno, Y.M., Desantis, T.Z., Andersen, G.L., Bakker, P.A.H.M., Raaijmakers, J.M., 2011. Deciphering the rhizosphere microbiome for disease-suppressive bacteria. Science 27, 1097–1111.

Mina, D., Pereira, J.A., Lino-Neto, T., Baptista, P., 2020. Impact of plant genotype and plant habitat in shaping bacterial pathobiome: a comparative study in olive tree. Sci. Rep. 10, 3475.

Modi, D., Simard, S., Lavkulich, L., Hamelin, R.C., Grayston, S.J., 2020. Stump removal and tree species composition promote a bacterial microbiome that may be beneifical in the suppression of root disease. FEMS Microbiol. Ecol. https://doi.org/10.1093/femsec/fiaa213.

Moe, L.A., 2013. Amino acids in the rhizosphere: from plants to microbes. Am. J. Bot. 100, 1692–1705.

Montiel, J., Nava, N., Cárdenas, L., Sánchez-López, R., Arthikala, M.K., Santana, O., et al., 2012. A *Phaseolus vulgaris* NADPH oxidase gene is required for root infection by Rhizobia. Plant Cell Physiol. 53, 1751–1767.

Muñoz, N., Robert, G., Melchiorre, M., Racca, R., Lascano, R, 2012. Saline and osmotic stress differentially affects apoplastic and intracellular reactive oxygen species production, curling and death of root hair during *Glycine max* L.—*Bradyrhizobium japonicum* interaction. Environ. Exp. Bot. 78, 76–83.

Netzker, T., Fischer, J., Weber, J., Mattern, D.J., König, C.C., Valiante, V., Schroeckh, V., Brakhage, A.A., 2015. Microbial communication leading to the activation of silent fungal secondary metabolite gene clusters. Front. Microbiol. 6, 299.

Palaniyandi, S.A., Yang, S.H., Zhang, L., Suh, J.-W., 2013. Effects of *actinobacteria* on plant disease suppression and growth promotion. Appl. Microbiol. Biotechnol. 97, 9621–9636.

Palansooriya, K.N., Wong, J.T.F., Hashimoto, Y., Huang, L., Rinklebe, J., Chang, S.X., Bolan, N., Wang, H., Ok, Y.S., 2019. Response of microbial communities to biochar-amended soils: a critical review. Biochar 1, 3–22.

Panke-Buisee, K., Poole, A., Goodrich, J., Ley, R., Kao-Kniffin, J., 2015. Selection on soil microbiomes reveals reproducible impacts on plant function. ISME J. 9, 980–989.

Pelsi, L., Schymanski, E.L., Wilmes, P., 2018. Dark matter in host-microbiome metabolomics: tackling the unknowns—a review. Anal. Chim. Acta 1037, 13–27.

Peterson, L.-E., Kellermann, M.Y., Schupp, P.J., 2020. Secondary metabolites of marine microbes: from natural products chemistry to chemical ecology. In: Jungblut, S., et al. (Eds.), YOUMARES 9—The Oceans: Our Research, Our Future. Springer Open, Cham, Switzerland, pp. 159–180.

Pfabel, C., Eckhardt, K.U., Baum, C., Struck, C., Frey, P., Weih, M., 2012. Impact of ectomycorrhizal colonization and rust infection on the secondary metabolism of poplar (*Populus trichocarpa* × *deltoides*). Tree Physiol. 32, 1357–1364.

Pozo, M.J., Cordier, C., Dumas-Gaudot, E., Gianinazzi, S., Barea, J.M., Azcón-Aguilar, C., 2002. Localized versus systemic effect of arbuscular mycorrhizal fungi on defence responses to *Phytopathora* infection in tomato plants. J. Exp. Bot. 53, 525–534.

Prescott, C.E., Grayston, S.J., 2013. Tree species influence on microbial communities in litter and soil: current knowledge and research needs. For. Ecol. Manag. 309, 19–27.

Raaijmakers, J.M., Mazzola, M., 2016. Soil immune responses. Science 352, 1392–1393.

Raaijmakers, J.M., Paulitz, T.C., Steinberg, C., Alabouvette, C., Moënne-Loccoz, Y., 2009. The rhizosphere: a playground and battlefield for soilborne pathogens and beneficial microorganisms. Plant Soil 321, 341–361.

Ren, F., Kovalchuk, A., Mukrimin, M., Liu, M., Zeng, Z., Grimire, R.P., Kivimaenpaa, M., Holopainen, J.K., Sun, H., Asiegbu, F.O., 2019. Tissue microbiome of Norway spruce affected by *Heterobasidion*-induced wood decay. Microb. Ecol. 77, 640–655.

Rosenzweig, N., Tiedje, J.M., Quensen III, J.F., Meng, Q., Hao, J.F., 2012. Microbial communities associated with potato common scab-suppressive soil determined by pyrosequencing analyses. Plant Dis. 96, 718–725.

Rovenich, H., Boshoven, J.C., Thomma, B.P.H.J., 2014. Filamentous pathogen effector functions: of pathogen, hosts and microbiomes. Curr. Opin. Plant Biol. 20, 96–103.

Saccá, M.L., Manici, L.M., Caputo, F., Frisullo, S., 2019. Changes in the rhizosphere bacterial communities associated with tree decline: grapevine esca syndrome case study. Can. J. Microbiol. 65, 930–943.

Sakure, S., Bhosale, S., 2019. *Actinobacteria* for biotic stress management. In: Sayyed, R. (Ed.), Plant Growth Promoting Rhizobacteria for Sustainable Stress Management. Microorganisms for Sustainability. vol. 13. Springer, Singapore.

Sapp, M., Lewis, E., Moss, S., Barrett, B., Kirk, S., Elphinstone, J.G., Denman, S., 2016. Metabarcoding of bacteria associated with the acute oak decline syndrome in England. Forests 7, 95.

Shen, Z., Penton, C.R., Lv, N., Xue, C., Yuan, X., Ruan, Y., Li, R., Qirong, S., 2019a. Banana Fusarium wilt disease incidence is influenced by shifts of microbial communities under different monoculture spans. Microb. Ecol. 75, 739–750.

Shen, Z., Xue, C., Penton, C.R., Thomashow, L.S., Zhang, N., Wang, B., Ruan, Y., Li, R., Shen, Q., 2019b. Suppression of banana Panama disease induced by soil microbiome reconstruction through an integrated agricultural strategy. Soil Biol. Biochem. 128, 164–174.

Song, Y.Y., Cao, M., Xie, L.J., Liang, X.T., Zeng, R.S., Su, Y.J., Huang, J.H., Wang, R.L., Luo, S.M., 2011. Induction of DIMBOA accumulation and systemic defense responses as a mechanism of enhanced resistance of mycorrhizal corn (Zea mays L.) to sheath blight. Mycorrhiza 21, 721–731.

Terhonen, E., Blumenstein, K., Kovalchuk, A., Asiegbu, F.O., 2019. Forest tree microbiomes and associated fungal endophytes: functional roles and impact on forest health. Forests 10, 42.

Trivedi, P., He, Z., Nostrand, V., Joy, D., Albrigo, G., Zhou, J., Wang, N., 2012. Huanglongbing alters the structure and functional diversity of microbial communities associated with citrus rhizosphere. ISME J. 6, 363–383.

Trivedi, P., Delgao-Baquerizo, M., Trivedi, C., Hamonts, K., Anderson, I.C., Singh, B.K., 2017. Keystone microbial taxa regulate the invasion of a fungal pathogen in agro-ecosystems. Soil Biol. Biochem. 111, 10–14.

van der Heijden, M.G., Streitwolf-Engel, R., Riedl, R., Siegrist, S., Neudecker, A., Ineichen, K., Boller, T., Wiemken, A., Sanders, I.R., 2006. The mycorrhizal contribution to plant productivity, plant nutrition and soil structure in experimental grassland. New Phytol 172, 739–752.

Vandenkoornhuyse, P., Wuaiser, A., Duhamel, M., LeVan, A., Duresne, A., 2015. The importance of the microbiome of the plant holobiont. New Phytol. 206, 1196–1206.

Vannier, N., Agler, M., Hacquard, S., 2019. Microbiota-mediated disease resistance in plants. PLoS Pathog. 15, e1007740.

Veresoglou, S.D., Rillig, M.C., 2011. Suppression of fungal and nematode plant pathogens through arbuscular mycorrhizal fungi. Biol. Lett. 8, 214–217.

Vinale, F., Nicoletti, R., Borrelli, F., Mangoni, A., Parisi, O.A., Marra, R., Lombardi, N., Lacatena, F., Grauso, L., Finizio, S., Lorito, M., Woo, S.L., 2017. Co-culture of plant beneficial microbes as source of bioactive metabolites. Sci. Rep. 7, 14330.

Warwell, M.V., McDonald, G.I., Hanna, J.W., Kim, M.S., Lalande, B.M., Stewart, J.E., Hudak, A.T., Klopfenstein, N.B., 2019. *Armillaria altimontana* is associated with healthy western white pine (*Pinus monticola*): potential *in situ* biological control of Armillaria root disease pathogen, *A. solidipes*. Forests 10, 294.

Weller, D.M., Raaijmakers, J.M., Gardener, B.B., Thomasshow, L.S., 2002. Microbial populations responsible for specific soil suppressiveness to plant pathogens. Annu. Rev. Phytopathol. 40, 309–348.

Weste, G., Vithanage, K., 1978. Effect of Phytophthora cinnamomic on microbial populations associated with roots of forest flora. Aust. J. Bot. 26, 153–167.

Weston, L.A., Mathesius, U., 2013. Flavonoids: their structure, biosynthesis and role in the rhizosphere, including allelopathy. J. Chem. Ecol. 39, 283–297.

Xiong, W., Li, R., Ren, Y., Liu, C., Zhoa, Q., Wu, H., Jousset, A., Shen, Q., 2017. Distinct roles for soil fungal and bacterial communities associated with suppression of vanilla Fusarium wilt disease. Soil Biol. Biochem. 107, 198–207.

Yang, J., Kloepper, J.W., Ryu, C.-.M., 2009. Rhizosphere bacteria help plants tolerate abiotic stress. Trends Plant Sci. 14, 1–4.

Yuan, J., Chaparro, J.M., Manter, D.K., Zhang, R., Vivanco, J.M., Shen, Q., 2015. Root from distinct plant development stages are capable of rapidly selecting their own microbiome without the influence of environmental and soil edaphic factors. Soil Biol. Biochem. 89, 206–209.

Zampieri, E., Giordano, L., Lione, G., Wizzini, A., Sillo, F., Balestrini, R., Gontheir, P., 2017. A nonnative and a native fungal pathogen similarly stimulate ectomycorrhizal development but are perceived differently by a fungal symbiont. New Phytol. 213, 1836–1849.

Zeng, Y., Abdo, Z., Charkowski, A., Stewart, J.E., Frost, K., 2019. Responses of Bacterial and Fungal Community structure to different rates of 1-3-dichloropropene fumigation. Phytobiomes 3, 212–223.

Zhang, Y., Xu, J., Riera, N., Jin, T., Li, J., Wang, N., 2017. Huanglongbing impairs the rhizosphere-to-rhizoplant enrichment process of the citrus root-associated microbiome. Microbiome 5, 97.

Chapter 16

Microbiome of forest soil

Zhao-lei Qu and Hui Sun

Department of Forest Protection, College of Forestry, Nanjing Forestry University, Nanjing, China

Chapter Outline

1. Introduction

Peatlands are formed due to an imbalance between plant production and litter degradation (Yule, 2010; Rydin et al., 2013). On a global scale, they are the main representative of the carbon (C) reserve (Yu et al., 2011) and play an important role in the global carbon cycle (Finlayson and Milton, 2018). A "peatland" is a general term for peat soil and aboveground plant communities. It is caused by the generation of an anaerobic environment due to the accumulation of water on the ground, which slows down the degradation rate of litter and allows the accumulation and storage of organic matter (Boyd, 2004). Peatlands are divided into tropical, temperate, and boreal peatlands according to the climatic setting, the hydrological dynamics, and the aboveground vegetation community. Tropical peatlands are characterized by high temperature, low acidity, and high precipitation (Dohong et al., 2017; Rydin et al., 2013) (Table 16.1). They only account for 12% of global peatlands, but contribute 20% of the global carbon storage of peatlands (Joosten et al., 2009). Boreal peatlands can be found where the mean annual precipitation is between 500 and 3000 mm, where mean annual biotemperatures are between 3°C and 6°C (Clymo, 1984; Clymo et al., 1998). They store about 180–277 Gt ($1\,Gt = 10^{9}\,t$) of carbon, representing about one-third of the global terrestrial C pool (Gorham, 1991).

Microorganisms in peatlands play important roles in controlling the turnover of organic carbon and participating in nutrient cycles. They can also promote the nutrient absorption of plants and maintain the stability of the ecological environment (Andersen et al., 2013). Understanding microbial community dynamics in peatlands, therefore, is important to better understand biogeochemical cycles such as C cycling and greenhouse gas production. Many studies have focused on plant ecology in the peatland forest (Thormann and Bayley, 1997; Wilson et al., 2007; Granath et al., 2010). However, information on the microbiome in peatland forest soil on a large scale is still limited. Therefore, in this chapter we summarize the knowledge on the microbial community in terms of microbial diversity and structure as well as the environmental factors contributing to changes in the community in tropical and boreal peatland forests.

2. Microbiome in a tropical peatland forest

2.1 Bacterial diversity in a tropical peatland forest

Bacteria play a key role in the nutrient cycle of tropical peatlands. The dynamics of bacterial communities can predict the future trends of tropical peatland ecosystems and their functions. The bacterial diversity and composition in tropical peatlands are affected by soil conditions (Fierer and Jackson, 2006; Rousk et al., 2010; Zhalnina et al., 2015). The bacterial diversity gradually decreases along the soil depth within 0–50 cm and shows spatial changes along the vertical profile (Jackson et al., 2009). The higher bacterial diversity in the upper soil layer may be due to the high input of fresh litter in the upper soil layer (Ong et al., 2015), whereas the lower soil pH and low oxygen levels in the deeper layers are the two

TABLE 16.1 The differences and similarities between tropical and boreal peatlands.

Types of peatland	Temperature	Hydrogeological condition	Forming and maintaining	Major global locations	Common feature
Tropical peatlands	High temperature	Precipitation	Continuous large litter inputs from evergreen trees into seasonally water-saturated peat	Mainland East Asia, Southeast Asia, the Caribbean, Central America, South America, and Central and Southern Africa	They contain large amounts of organic matter
Boreal peatlands	Low temperature	Glaciers and ground water related	Waterlogged conditions and peat-forming plants	Boreal and subarctic regions in western Siberia in the Russian Federation, central Canada, northwest Europe, and Alaska in the United States	

TABLE 16.2 Bacterial alpha diversity and composition in tropical and boreal peatlands.

Types of peatlands	Bacterial alpha diversity	Bacterial composition
Tropical peatlands	The bacterial diversity presented a parabolic trend along the vertical depth of 0–90 cm. (1) The upper soil layer (0–45 cm) showed a decline. (2) The lower soil layer (45–90 cm) showed an increase.	(1) Proteobacteria and Acidobacteria are the two dominant phyla. (2) The land-use management could significantly affect bacterial composition such as *Syntrophobacter, Opitutus,* and *Methylocystis.*
Boreal peatlands	(1) The bacterial biomass in the top soil layer (0–40 cm) increased along increasing soil depth and reached a peak at a buffer layer (40–50 cm), and then decreased afterward in the deep layer (50–60 cm). (2) The bacterial diversity decreased along increasing soil depth (0–60 cm).	(1) Proteobacteria and Acidobacteria are the dominant phyla, of which Alpha-proteobacteria and Acidobacteria_Gp1 are the main classes. (2) *Burkholderia* play an important role in boreal peatlands. (3) The land-use management may cause changes in the abundance of *Mucilagenibacter.*

main factors causing lower bacterial diversity (Jackson et al., 2009). However, a study conducted in tropical peatlands in Malaysia showed that the bacterial diversity presented a parabolic trend along the vertical depth of 0–90 cm, in which the upper soil layer (0–45 cm) showed a decline in bacterial diversity and the lower layer (45–90 cm) showed an increase (Too et al., 2018). Variations in the plant root network and organic matter content in different soil depths can be a possible explanation. Similar results were also found in boreal peatlands, that is, the bacterial diversity declined along the depth of soil within a certain depth and then increased afterward (Tsitko et al., 2014; Zhou et al., 2017). Compared to soil depth, the tree species has less effect on bacterial diversity in a tropical peatland (Too et al., 2018). This may be because the different tree species mainly cause changes in litter input and the bacteria have limited ability to degrade complex substrates, which more significantly affect the diversity of fungi than bacteria (Romaní et al., 2006). A previous study reported that environmental conditions rather than the tree species had a direct impact on bacterial diversity (Millard and Singh, 2010). For example, the soil pH correlates with microbial diversity across many soil types (Fierer and Jackson, 2006; Shen et al., 2013; Liu et al., 2014). The pristine tropical peatland soils had lower bacterial diversity compared to the disturbed peatland and mineral soils (Liu et al., 2020). At the same time, the pristine peatland had a normally lower soil pH than disturbed peatland and mineral soil (Table 16.2).

2.2 Bacterial composition in a tropical peatland forest

Proteobacteria and Acidobacteria are the two dominant phyla in tropical peatland soil, and they can account for more than half the community (Mishra et al., 2014). The most abundant phylum can be either Proteobacteria (Kanokratana et al., 2010; de Gannes et al., 2015; Liu et al., 2020) or Acidobacteria (Etto et al., 2012; Mishra et al., 2014), depending on the peatland type. The members of Acidobacteria can be widely distributed in tropical peatlands with insufficient nutrients and complex environments due to their high ability to adapt to decompose complex substrates (Barns et al., 1999; Sait et al., 2006; Ward et al., 2009; Kanokratana et al., 2010). Similarly, Proteobacteria also have high metabolic diversity in tropical peatland soils (Kersters et al., 2006). Many members from these two phyla can produce a variety of cellulolytic and amylolytic enzymes to degrade fresh litter in the soil (Béguin and Aubert, 1994; Pandey et al., 2000). Other bacterial phyla in tropical peatland soil include Verrucomicrobia, Planctomycetes, Actinobacteria, Firmicutes, and Bacteroidetes (Kanokratana et al., 2010; de Gannes et al., 2015; Liu et al., 2020). Land-use management in a tropical peatland could significantly affect the bacterial composition. At a lower taxonomic level, the abundance of some genera or species differed in different peatland soils. For example, *Schlesneria* and *Acidiphilium* were the more abundant in pristine peatland soil, whereas *Methylocystis*, *Telmatospirillum*, *Syntrophobacter*, *Sorangium,* and *Opitutus* had higher abundance in disturbed peatland soil in Indonesia (Liu et al., 2020). *Schlesneria* and *Acidiphilium* have an extremely high tolerance to an acidic environment and low-temperature conditions (Harrison, 1981; Hamamura et al., 2005; Kulichevskaya et al., 2007). *Schlesneria* may have a growth-promoting effect on plants and is correlated to the amount of plant litter (Peruzzi et al., 2017). When a disturbance happens in a pristine peatland, such as drainage and selective logging, the plant vegetation and the amount of litter will change, resulting in a shift in abundance of certain bacterial groups. In disturbed peatland soil, the most abundant genera (e.g., *Opitutus*, *Syntrophobacter*, and *Methylocystis*) have been shown to participate in the carbon degradation of litter and play an important role in the methane metabolism and the release of CO_2 (Bok et al., 2002; Chin and Janssen, 2002; McDonald et al., 2008; Müller et al., 2010). Therefore, the above results confirm that the disturbance of pristine tropical peatland can lead to changes in the bacterial community structure and the element cycle in tropical peatland soil (Table 16.2).

2.3   Fungal diversity in a tropical peatland forest

Fungi play an indispensable role in tropical peatlands due to their ability to degrade soil organic matter by producing hydrolytic and oxidative enzymes (Thormann, 2006a). Fungi are not as numerous and widely distributed as bacteria in tropical peatland and have less diversity than bacteria (de Gannes et al., 2015). They are more tolerant to low pH than bacteria (Liu et al., 2014) and are not as severely affected by the environment (Sun et al., 2017). The fungal diversity and soil pH in tropical peatlands have no correlation to each other (de Gannes et al., 2015). The disturbance of pristine peatland (deforestation and drainage) did not change the fungal diversity while the fungal richness is higher in pristine than in disturbed peatland soil (Liu et al., 2020). Similar results have been reported in tropical peatlands in Amuntai, Indonesia, where the number of fungi decreased significantly in the drained secondary peatland due to the moisture deficit (Paul and Clark, 1989). The peatland plantations with fertilization in Malaysia harbored higher fungal species richness than pristine peatland forests (Hujslová et al., 2013). The fertilization provided favorable conditions for fungal growth, resulting in higher fungal richness (Kerekes et al., 2013; Krashevska et al., 2014) (Table 16.3).

TABLE 16.3 Fungal alpha diversity and composition in tropical and boreal peatlands.

Types of peatland	Fungal alpha diversity	Fungal composition
Tropical peatlands	(1) The fungal diversity and soil pH in tropical peatland have no correlation to each other. (2) The land-use management will not change the diversity of fungi but will cause differences in richness.	(1) Ascomycota and Basidiomycota are the two dominant phyla, and Aspergillus is a common genus in tropical peatland. (2) Certain fungi with special wood degradation capabilities such as Hypocreales and Polyporales are abundant in tropical peatlands. (3) The land-use management will reduce saprophytic fungi groups.
Boreal peatlands	(1) The fungal spore numbers, biomass, and hyphal length decrease with increasing soil depth. (2) A drop in the water level will increase the fungal biomass, and climate warming would decrease the fungal richness.	(1) Ascomycota is the most abundant fungal phylum in the boreal peatland forest, followed by Basidiomycota and rare taxa of Zygomycota and Chytridiomycota. (2) The abundance of Ascomycota decreased significantly under elevated temperature, and permafrost thaw will reduce the abundance of mycorrhizal fungi decreases and saprotrophic and pathogenic fungi increases. (3) Nutrient input in boreal peatlands can significantly change the abundance of certain dominant genera such as *Phialophora*, *Tetrachaetum*, and *Anguillospora*.

2.4 Fungal composition in a tropical peatland forest

As they are in boreal peatlands, Ascomycota and Basidiomycota are the two dominant phyla in tropical peatlands, and they occupy half the fungal community in abundance (Liu et al., 2020). The advantage of Ascomycota in a tropical peatland is that most of its members are fast-growing fungi with a high sporulation rate (Kusai et al., 2018). Rare fungal phyla include Zygomycota and Glomeromycota (Kusai et al., 2018; Liu et al., 2020). Studies have reported that members of *Aspergillus* are ubiquitous in the Sarawak tropical peatlands (Kusai et al., 2018), which are known as participants in the carbon and nitrogen cycles (Dagenais and Keller, 2009). *A. fumigatus* is dominant in Sarawak primary peatland forests and *A. niger* is dominant in logging peatland forests. *A. niger* has been reported to play a role in the slow degradation of wood (Hamed, 2013). *Penicillium chrysogenum* is common in both primary and logging peatland forests (Kusai et al., 2018). All three fungal species have been reported to be tolerant to acidic environments (Hujslová et al., 2010; Rinu and Pandey, 2010). Certain fungi with special wood degradation capabilities, such as Hypocreales and Polyporales, are abundant in tropical peatlands; they can also be observed in boreal peatlands, but with low abundance (Thormann and Rice, 2007; Ovaskainen et al., 2013; Sun et al., 2016). The composition of fungal communities in tropical peatlands was found to be affected by nutrient contents. The phosphorus content in Panama tropical peatlands can significantly affect the fungal community structure and function (Morrison et al., 2020). With the decrease of phosphorus concentration, there are clear changes in fungal functional groups gradually from saprotrophic (Mortierellaceae and Mycenaceae) dominance to saprotrophic and parasitic coexistence (Cordycipitaceae, Annulatascaceae, Teratosphaeriaceae, and Sympoventuriaceae), and then to endophytic and epiphytic dominance (Clavicipitaceae) (Cannon and Kirk, 2007). Compared to disturbed (drainage and selective logging) tropical peatland soils, natural tropical peatland soil harbors more *Gliocephalotrichum* with saprotrophic ability (Lombard et al., 2014; Liu et al., 2020), whereas litter decomposers such as *Gymnopilus* have lower abundance in disturbed tropical peatlands (Capelari and Zadrazil, 1997) (Table 16.3).

3. Microbiome in a boreal peatland forest

3.1 Bacterial diversity in a boreal peatland forest

Bacteria are the most abundant microorganisms in boreal peatland soil, as in most other ecological environments (Roesch et al., 2007). They are the first microorganisms to colonize substrates with easier access (Schmidt et al., 2007; de Boer and van der Wal, 2008). Bacteria play an important role as a major participant in the nitrogen cycle in the northern peatlands. Microbial communities in a peatland forest are restricted due to special ecological environments such as an anaerobic environment caused by sufficient water and an acidic environment. They can also be affected by the aboveground plant communities (Borga et al., 1994; Fisk et al., 2003; Thormann et al., 2004), peatland types (Sun et al., 2016), hydrological conditions (Jaatinen et al., 2007; Peltoniemi et al., 2009), oxygen stress, nutrient availability, or peatland pH (Hobbie and Gough, 2004). The bacterial biomass in boreal peatlands increased along increasing soil depth (0–1 m), but the bacterial diversity gradually decreased (Morales et al., 2006). Subsequent studies showed that the bacterial biomass in the top soil layer (0–40 cm) increased along increasing soil depth and reached a peak at a buffer layer (40–50 cm), and then decreased afterward in the deep layer (50–60 cm) (Golovchenko et al., 2007; Jaatinen et al., 2007). The bacterial diversity decreased along increasing soil depth (0–60 cm) as well (Golovchenko et al., 2007). These observations are contrary to the results obtained in tropical peatlands, where the bacterial diversity showed a parabolic trend along soil depth (Too et al., 2018). In boreal peatland forests, the methanotrophs in the soil layer of 40–50 cm have high abundance due to the coexistence of CH_4 and O_2 (Hanson and Hanson, 1996). The bacteria become more specialized, resulting in decreased diversity along increasing soil depth (Morales et al., 2006). Both dominant tree species and peatland types affect bacterial diversity, in which the tree species seem to have a stronger impact on the diversity of bacteria than the peatland types (Sun et al., 2014). Peatland forests dominated by the Scots pine had higher bacterial diversity and richness compared to forests dominated by the Norway spruce (Sun et al., 2014) (Table 16.2).

In boreal peatland forests, the soil bacterial richness and diversity are higher than in mineral soil forests. The boreal peatland forest soils are characterized by an imbalance of C/N (Högberg et al., 2007) and a nitrogen deficit (Janssens et al., 2010). Interestingly, the bacterial diversity in boreal peatland soil has no correlation with soil pH (Preston et al., 2012; Sun et al., 2014), which is inconsistent with the fact that bacterial diversity and soil pH normally have a positive correlation with each other. In addition, the soil bacterial biomass increased with drainage (water level drawdown) in the boreal peatland forest (Jaatinen et al., 2008) (Table 16.2).

3.2 Bacterial composition in a boreal peatland forest

For bacterial composition, Proteobacteria and Acidobacteria are the dominant phyla in peatland forest soil, accounting for about 70%–85% of the population (Sun et al., 2014). Of these, Alpha-proteobacteria and Acidobacteria_Gp1 are the main classes. The abundance of Proteobacteria has been reported to be positively correlated with carbon availability (McCaig et al., 1999; Fazi et al., 2005; Fierer et al., 2007) and Acidobacteria are known to grow in acidic environments (Philippot et al., 2010; Bardhan et al., 2012). Therefore, the high abundance of the two bacterial phyla could reflect the high carbon and acidic condition of boreal peatland forests. *Burkholderia* are known as nitrogen-fixing bacteria and are related to litter degradation (Briglia et al., 1994), which plays an important role in the N cycle in boreal peatland forests (Uroz et al., 2007; Leveau et al., 2009). It was the most abundant genus in primitive and drained peatland forests in Finland (Sun et al., 2014). *Mucilaginibacter* have the ability to decompose cellulose (Straková et al., 2012) with a higher abundance in primitive peatland than in drained peatland (Sun et al., 2014); this can be used as an indicator for peat accumulation in a boreal peatland forest. Moreover, Actinobacteria respond much less to water level drawdown in a boreal peatland forest (Jaatinen et al., 2008). The changes in the ecological niche caused by drainage, such as a more aerobic environment, seem to have a greater impact on methanogens and methanotrophs in the soil, which have higher abundance in the deeper soil profiles (Francez et al., 2000) (Table 16.2).

3.3 Fungal diversity in a boreal peatland forest

The boreal peatland has accumulated a large amount of litter due to its unique environmental conditions (low temperature and rich in forest resources), in which fungi play crucial roles in litter degradation (Latter et al., 1967; Williams and Crawford, 1983), especially in the initial stage of decomposition (Newell et al., 1995; Kuehn et al., 2000). The fungi in boreal peatlands showed vertical stratification in which the fungal spore numbers, biomass, and hyphal length decreased with increasing soil depth (Golovchenko et al., 2002). Many fungi have the ability to degrade simple molecules, but they also have limited ability to degrade complex polymers (Thormann, 2006a). Therefore, the forest litter type is one of the main factors influencing the fungal decomposition process. With the increase of soil depth, more complex polymers or substrates are accumulated in the deep soil profiles, where the conditions are not favorable for fungal growth and spore germination (Turetsky et al., 2000). Boreal peatlands are widely distributed from wet fens to dry peatland forests. Hydrological changes can significantly affect the litter decomposition process in boreal peatlands, even on a global basis (Jaatinen et al., 2008). Hollows (wet depressions) and hummocks (drier raised areas) are common microtopographical features of northern peatlands. The fungal diversity, richness, and community structure in hollows differ from that in hummocks (Asemaninejad et al., 2017b). The differences are caused by several variables, including water table levels, vegetation structure, and litter properties. The more saturated conditions of hollows favor anaerobic and flagellated fungi (Chytridiomycota) and fungal species that normally grow in moist environments (Kurtzman et al., 2011; Gruninger et al., 2014). The fungal biomass in drier areas in boreal peatlands increased, suggesting that water level drawdown can benefit fungal growth (Jaatinen et al., 2008). It is possible that the substrate in drier areas becomes more recalcitrant, which provides a competitive advantage to fungi (Laiho, 2006). In addition, brown rot wood-degrading fungi may use reactive oxygen species (ROS) to break down plant cell walls and provide selective polysaccharide extraction (Castaño et al., 2018). These have also been proposed in other boreal peatland surveys (Jaatinen et al., 2007). Simulation experiments on climate change in boreal peatland forests proved that climate warming would decrease fungal richness, whereas the abundance of saprotrophic fungi would increase significantly, although this may be at the expense of reducing other types of fungi (Asemaninejad et al., 2017a) (Table 16.3).

3.4 Fungal composition in a boreal peatland forest

For fungal composition, Ascomycota is the most abundant fungal phylum in boreal peatland forests, followed by Basidiomycota (Thormann, 2006b; Sun et al., 2016), and the rare taxa of Zygomycota and Chytridiomycota (Peltoniemi et al., 2012; Sun et al., 2016). The abundance of Ascomycota increased over time over 12 months and decreased significantly under elevated temperature in a simulated climate warming experiment (Asemaninejad et al., 2017a). The elevated temperature increased the biomass of graminoids and ericaceous shrubs and decreased the biomass of *Sphagnum* mosses, resulting in an increase of endophytes and mycorrhizal root-associated Ascomycota (Dieleman et al., 2015). Permafrost thaw is a common phenomenon in boreal peatlands that not only induces the large-scale succession of plant communities (Schuur et al., 2007; Wolken et al., 2011), but also affects the structure of belowground fungal communities (Taş et al., 2014;

Hultman et al., 2015). Permafrost thaw can change the underground fungal community from beneficial to plant communities to harmful, which means that the relative abundance of mycorrhizal fungi decreases and the saprotrophic and pathogenic fungi increase (Schütte et al., 2019). Water saturation and the anaerobic environment caused by permafrost thaw may not be conducive to the growth of mycorrhizal fungi (Schütte et al., 2019). Nutrient input is a common way to promote plant growth and can supplement the lack of certain mineral nutrients in peatlands (Saarsalmi et al., 2004; Huotari et al., 2015). The application of wood ash in peatland soil can significantly change the fungal community structure (Peltoniemi et al., 2016). In unfertilized peatlands, *Phialophora*, *Tetrachaetum*, and *Anguillospora* were the dominant genera, and the abundance of these genera was reduced by the application of wood ash (Peltoniemi et al., 2016). Similar results were observed in which ash application decreased the amount of mold in surface peat in wetlands (Huikari, 1953). *Deconica*, *Galerina*, *Hypholoma*, and *Stropharia* were more abundant in peatland soil with a lower amount of wood ash application (Peltoniemi et al., 2016). Most members of *Galerina* are saprotrophic fungi that grow on certain woods and plant debris (Hibbett and Donoghue, 2001). Mycorrhizal fungi (e.g., *Russulaceae*) were more dominant than wood-decomposing fungi in the early stage of wood ash application (Peltoniemi et al., 2016). The predominant mycorrhizal fungi in the soil can promote the absorption and utilization of nutrients by plants (Mahmood et al., 2003; Peltoniemi et al., 2016). The higher pH caused by fertilization may promote the growth of certain mycorrhizal fungi and increase their colonization of plants (Silfverberg and Huikari, 1985). Moreover, the availability of phosphorus and potassium can also be beneficial to mycorrhizal fungi (Moilanen et al., 2002). The abundance of the predominant ectomycorrhizae, including *Russulaceae*, was significantly reduced after a 12-year application of wood ash in a drained peatland, indicating the negative impact of the long-term application of wood ash on mycorrhizal fungi in boreal peatlands (Klavina et al., 2016) (Table 16.3).

A peatland is a unique system that plays an important role in global nutrient cycling. Most previous microbial studies in peatlands were limited to phospholipid fatty acid (PLFA) and cultivation methods. The results obtained from these methods are not completely representative, as most of the microorganisms in peatland soils are not culturable. The recent advances in sequencing technologies, such as amplicon metagenome and metatranscriptome, provide new possibilities to assess the microbial population and activities in peatland forests, which could enable a better understanding of the true situation of the microbiome in this unique ecosystem. For example, methanotrophs and methanogens play important roles in CO_2 emission in boreal peatlands. However, there is little information available on the regulatory pathway of these two groups. With the new sequencing technologies, the methanogens or genes involved in CH_4 emission could potentially be identified.

References

Andersen, R., Chapman, S.J., Artz, R.R.E., 2013. Microbial communities in natural and disturbed peatlands: a review. Soil Biol. Biochem. 57, 979–994.

Asemaninejad, A., Thorn, R.G., Lindo, Z., 2017a. Experimental climate change modifies degradative succession in boreal peatland fungal communities. Microb. Ecol. 73, 521–531.

Asemaninejad, A., Thorn, R.G., Lindo, Z., 2017b. Vertical distribution of fungi in hollows and hummocks of boreal peatlands. Fungal Ecol. 27, 59–68.

Bardhan, S., Jose, S., Jenkins, M.A., Webster, C.R., Udawatta, R.P., Stehn, S.E., 2012. Microbial community diversity and composition across a gradient of soil acidity in spruce–fir forests of the southern Appalachian Mountains. Appl. Soil Ecol. 61, 60–68.

Barns, S.M., Takala, S.L., Kuske, C.R., 1999. Wide distribution and diversity of members of the bacterial kingdom *Acidobacterium* in the environment. Appl. Environ. Microbiol. 65, 1731.

Béguin, P., Aubert, J.-P., 1994. The biological degradation of cellulose. FEMS Microbiol. Rev. 13, 25–58.

Bok, F.A.M.d., Luijten, M.L.G.C., Stams, A.J.M., 2002. Biochemical evidence for formate transfer in syntrophic propionate oxidizing cocultures of *Syntrophobacter fumaroxidans* and *Methanospirillum hungatei*. Appl. Environ. Microbiol. 68, 4247–4252.

Borga, P., Nilsson, M., Tunlid, A., 1994. Bacterial communities in peat in relation to botanical composition as revealed by phospholipid fatty acid analysis. Soil Biol. Biochem. 26, 841–848.

Boyd, W., 2004. Peatlands and environmental change. Geoarchaeol. Int. J. 19, 505–507.

Briglia, M., Ri, E., Van Elsas, D.J., De Vos, W.M., 1994. Phylogenetic evidence for transfer of pentachlorophenol-mineralizing *Rhodococcus chlorophenolicus* PCP-I(T) to the genus *Mycobacterium*. Int. J. Syst. Bacteriol., 494498.

Cannon, P., Kirk, P., 2007. Fungal Families of the World.

Capelari, M., Zadrazil, F., 1997. Lignin degradation and in vitro digestibility of wheat straw treated with Brazilian tropical species of white rot fungi. Folia Microbiol. 42, 481–487.

Castaño, J.D., Zhang, J., Anderson, C.E., Schilling, J.S., 2018. Oxidative damage control during decay of wood by Brown rot fungus using oxygen radicals. Appl. Environ. Microbiol. 84, e01937-01918.

Chin, K.-J., Janssen, P.H., 2002. Propionate formation by *Opitutus terrae* in pure culture and in mixed culture with a hydrogenotrophic methanogen and implications for carbon fluxes in anoxic rice paddy soil. Appl. Environ. Microbiol. 68, 2089–2092.

Clymo, R.S., 1984. The limits to peat bog growth. Philos. Trans. R. Soc. Lond. Ser. B Biol. Sci. 303, 605–654.

Clymo, R.S., Turunen, J., Tolonen, K., 1998. Carbon accumulation in peatland. Oikos 81, 368–388.

Dagenais, T.R.T., Keller, N.P., 2009. Pathogenesis of *Aspergillus fumigatus* in invasive aspergillosis. Clin. Microbiol. Rev. 22, 447–465.

de Boer, W., van der Wal, A., 2008. Interactions between saprotrophic basidiomycetes and bacteria. In: Boddy, L., Frankland, J.C., van West, P. (Eds.), British Mycological Society Symposia Series. Academic Press, pp. 143–153 (Chapter 8).

de Gannes, V., Eudoxie, G., Bekele, I., Hickey, W.J., 2015. Relations of microbiome characteristics to edaphic properties of tropical soils from Trinidad. Front. Microbiol. 6.

Dieleman, C.M., Branfireun, B.A., McLaughlin, J.W., Lindo, Z., 2015. Climate change drives a shift in peatland ecosystem plant community: implications for ecosystem function and stability. Glob. Chang. Biol. 21, 388–395.

Dohong, A., Aziz, AA., Dargusch, P., 2017. A review of the drivers of tropical peatland degradation in south-east asia. Land Use Policy 69, 349–360.

Etto, R.M., Cruz, L.M., Jesus, E.C., Galvão, C.W., Galvão, F., Souza, E.M., Pedrosa, F.O., Steffens, M.B.R., 2012. Prokaryotic communities of acidic peatlands from the southern Brazilian Atlantic Forest. Braz. J. Microbiol. 43, 661–674.

Fazi, S., Amalfitano, S., Pernthaler, J., Puddu, A., 2005. Bacterial communities associated with benthic organic matter in headwater stream microhabitats. Environ. Microbiol. 7, 1633–1640.

Fierer, N., Jackson, R.B., 2006. The diversity and biogeography of soil bacterial communities. Proc. Natl. Acad. Sci. U. S. A. 103, 626.

Fierer, N., Bradford, M.A., Jackson, R.B., 2007. Toward an ecological classification of soil bacteria. Ecology 88, 1354–1364.

Finlayson, C.M., Milton, G.R., 2018. Peatlands. In: Finlayson, C.M., Milton, G.R., Prentice, R.C., Davidson, N.C. (Eds.), The Wetland Book: II: Distribution, Description, and Conservation. Springer, Netherlands, Dordrecht, pp. 227–244.

Fisk, M.C., Ruether, K.F., Yavitt, J.B., 2003. Microbial activity and functional composition among northern peatland ecosystems. Soil Biol. Biochem. 35, 591–602.

Francez, A.-J., Gogo, S., Josselin, N., 2000. Distribution of potential CO2 and CH4 productions, denitrification and microbial biomass C and N in the profileof a restored peatland in Brittany (France). Eur. J. Soil Biol. 36, 161–168.

Golovchenko, A., Semenova, T., Poliakova, A., Inisheva, L., 2002. The structure of the micromycete complexes of oligotrophic peat deposits in the southern taiga subzone of West Siberia. Mikrobiologiia 71, 667–674.

Golovchenko, A.V., Tikhonova, E.Y., Zvyagintsev, D.G., 2007. Abundance, biomass, structure, and activity of the microbial complexes of minerotrophic and ombrotrophic peatlands. Microbiology 76, 630–637.

Gorham, E., 1991. Northern peatlands: role in the carbon cycle and probable responses to climatic warming. Ecol. Appl. 1, 182–195.

Granath, G., Strengbom, J., Rydin, H., 2010. Rapid ecosystem shifts in peatlands: linking plant physiology and succession. Ecology 91, 3047–3056.

Gruninger, R.J., Puniya, A.K., Callaghan, T.M., Edwards, J.E., Youssef, N., Dagar, S.S., Fliegerova, K., Griffith, G.W., Forster, R., Tsang, A., McAllister, T., Elshahed, M.S., 2014. Anaerobic fungi (phylum Neocallimastigomycota): advances in understanding their taxonomy, life cycle, ecology, role and biotechnological potential. FEMS Microbiol. Ecol. 90, 1–17.

Hamamura, N., Olson, S.H., Ward, D.M., Inskeep, W.P., 2005. Diversity and functional analysis of bacterial communities associated with natural hydrocarbon seeps in acidic soils at Rainbow Springs, Yellowstone National Park. Appl. Environ. Microbiol. 71, 5943.

Hamed, S.A.M., 2013. In-vitro studies on wood degradation in soil by soft-rot fungi: *Aspergillus niger* and *Penicillium chrysogenum*. Int. Biodeterior. Biodegradation 78, 98–102.

Hanson, R.S., Hanson, T.E., 1996. Methanotrophic bacteria. Microbiol. Rev. 60, 439–471.

Harrison, a.P., 1981. *Acidiphilium cryptum* gen. nov., sp. nov., heterotrophic bacterium from acidic mineral environments. Int. J. Syst. Evol. Microbiol. 31, 327–332.

Hibbett, D.S., Donoghue, M.J., 2001. Analysis of character correlations among wood decay mechanisms, mating systems, and substrate ranges in homobasidiomycetes. Syst. Biol. 50, 215–242.

Hobbie, S.E., Gough, L., 2004. Litter decomposition in moist acidic and non-acidic tundra with different glacial histories. Oecologia 140, 113–124.

Högberg, M., Chen, Y., Högberg, P., 2007. Gross nitrogen mineralisation and fungi-to-bacteria ratios are negatively correlated in boreal forests. Biol. Fertil. Soils 44, 363–366.

Huikari, O., 1953. Tutkimuksia ojituksen ja tuhka-lannoituksen vaikutuksesta eräiden soiden pieneliöstöön.

Hujslová, M., Kubátová, A., Chudíčková, M., Kolařík, M., 2010. Diversity of fungal communities in saline and acidic soils in the Soos National Natural Reserve, Czech Republic. Mycol. Prog. 9, 1–15.

Hujslová, M., Kubátová, A., Kostovčík, M., Kolařík, M., 2013. *Acidiella bohemica* gen. et sp. nov. and *Acidomyces* spp. (Teratosphaeriaceae), the indigenous inhabitants of extremely acidic soils in Europe. Fungal Divers. 58, 33–45.

Hultman, J., Waldrop, M.P., Mackelprang, R., David, M.M., McFarland, J., Blazewicz, S.J., Harden, J., Turetsky, M.R., McGuire, A.D., Shah, M.B., VerBerkmoes, N.C., Lee, L.H., Mavrommatis, K., Jansson, J.K., 2015. Multi-omics of permafrost, active layer and thermokarst bog soil microbiomes. Nature 521, 208–212.

Huotari, N., Tillman-Sutela, E., Moilanen, M., Laiho, R., 2015. Recycling of ash—for the good of the environment? For. Ecol. Manag. 348, 226–240.

Jaatinen, K., Fritze, H., Laine, J., Laiho, R., 2007. Effects of short- and long-term water-level drawdown on the populations and activity of aerobic decomposers in a boreal peatland. Glob. Chang. Biol. 13, 491–510.

Jaatinen, K., Laiho, R., Vuorenmaa, A., Del Castillo, U., Minkkinen, K., Pennanen, T., Penttilä, T., Fritze, H., 2008. Responses of aerobic microbial communities and soil respiration to water-level drawdown in a northern boreal fen. Environ. Microbiol. 10, 339–353.

Jackson, C.R., Liew, K.C., Yule, C.M., 2009. Structural and functional changes with depth in microbial communities in a tropical Malaysian peat swamp forest. Microb. Ecol. 57, 402–412.

Janssens, I., Dieleman, W., Luyssaert, S., Subke, J.-A., Reichstein, M., Ceulemans, R., Ciais, P., Dolman, H., Grace, J., Matteucci, G., Papale, D., Piao, S., Schulze, E., Tang, J., Law, B., 2010. Reduction of forest soil respiration in response to nitrogen deposition. Nat. Geosci. 3, 315–322.

Joosten, H., Ernst-Moritz-Arndt-Universität Greifswald, Wetlands International, 2009. The Global Peatland CO2 Picture: Peatland Status and Drainage Related Emissions in All Countries of the World. Univ. Greifswald.

Kanokratana, P., Uengwetwanit, T., Rattanachomsri, U., Bunterngsook, B., Nimchua, T., Tangphatsornruang, S., Plengvidhya, V., Champreda, V., Eurwilaichitr, L., 2010. Insights into the phylogeny and metabolic potential of a primary tropical peat swamp forest microbial community by metagenomic analysis. Microb. Ecol. 61, 518–528.

Kerekes, J., Kaspari, M., Stevenson, B., Nilsson, R.H., Hartmann, M., Amend, A., Bruns, T.D., 2013. Nutrient enrichment increased species richness of leaf litter fungal assemblages in a tropical forest. Mol. Ecol. 22, 2827–2838.

Kersters, K., De Vos, P., Gillis, M., Swings, J., Vandamme, P., Stackebrandt, E., 2006. Introduction to the proteobacteria. In: Dworkin, M., Falkow, S., Rosenberg, E., Schleifer, K.-H., Stackebrandt, E. (Eds.), The Prokaryotes: Volume 5: Proteobacteria: Alpha and Beta Subclasses. Springer New York, New York, NY, pp. 3–37.

Klavina, D., Pennanen, T., Gaitnieks, T., Velmala, S., Lazdins, A., Lazdina, D., Menkis, A., 2016. The ectomycorrhizal community of conifer stands on peat soils 12 years after fertilization with wood ash. Mycorrhiza 26, 153–160.

Krashevska, V., Sandmann, D., Maraun, M., Scheu, S., 2014. Moderate changes in nutrient input alter tropical microbial and protist communities and belowground linkages. ISME J. 8, 1126–1134.

Kuehn, K.A., Lemke, M.J., Suberkropp, K., Wetzel, R.G., 2000. Microbial biomass and production associated with decaying leaf litter of the emergent macrophyte *Juncus effusus*. Limnol. Oceanogr. 45, 862–870.

Kulichevskaya, I.S., Ivanova, A.O., Belova, S.E., Baulina, O.I., Bodelier, P.L.E., Rijpstra, W.I.C., Sinninghe Damsté, J.S., Zavarzin, G.A., Dedysh, S.N., 2007. *Schlesneria paludicola* gen. nov., sp. nov., the first acidophilic member of the order *Planctomycetales*, from *Sphagnum*-dominated boreal wetlands. Int. J. Syst. Evol. Microbiol. 57, 2680–2687.

Kurtzman, C.P., Fell, J.W., Boekhout, T., 2011. Definition, classification and nomenclature of the yeasts. In: Kurtzman, C.P., Fell, J.W., Boekhout, T. (Eds.), The Yeasts, fifth ed. Elsevier, London, pp. 3–5.

Kusai, N., Ayob, Z., Maidin, M.S.T., Safari, S., Ali, D.S.R., 2018. Characterization of fungi from different ecosystems of tropical peat in Sarawak, Malaysia. Rendiconti Lincei. Scienze Fisiche e Naturali 29.

Laiho, R., 2006. Decomposition in peatlands: reconciling seemingly contrasting results on the impacts of lowered water levels. Soil Biol. Biochem. 38, 2011–2024.

Latter, P.M., Cragg, J.B., Heal, O.W., 1967. Comparative studies on the microbiology of four moorland soils in the Northern Pennines. J. Ecol. 55, 445–464.

Leveau, J., Uroz, S., de Boer, W., 2009. The bacterial genus *Collimonas*: mycophagy, weathering and other adaptive solutions to life in oligotrophic soil environments. Environ. Microbiol. 12, 281–292.

Liu, J., Sui, Y., Yu, Z., Shi, Y., Chu, H., Jin, J., Liu, X., Wang, G., 2014. High throughput sequencing analysis of biogeographical distribution of bacterial communities in the black soils of Northeast China. Soil Biol. Biochem. 70, 113–122.

Liu, B., Talukder, M.J.H., Terhonen, E., Lampela, M., Vasander, H., Sun, H., Asiegbu, F., 2020. The microbial diversity and structure in peatland forest in Indonesia. Soil Use Manag. 36, 123–138.

Lombard, L., Serrato-Diaz, L.M., Cheewangkoon, R., French-Monar, R.D., Decock, C., Crous, P.W., 2014. Phylogeny and taxonomy of the genus *Gliocephalotrichum*. Persoonia 32, 127–140.

Mahmood, S., Finlay, R.D., Fransson, A.-M., Wallander, H., 2003. Effects of hardened wood ash on microbial activity, plant growth and nutrient uptake by ectomycorrhizal spruce seedlings. FEMS Microbiol. Ecol. 43, 121–131.

McCaig, A.E., Glover, L.A., Prosser, J.I., 1999. Molecular analysis of bacterial community structure and diversity in unimproved and improved upland grass pastures. Appl. Environ. Microbiol. 65, 1721–1730.

McDonald, I.R., Bodrossy, L., Chen, Y., Murrell, J.C., 2008. Molecular ecology techniques for the study of aerobic methanotrophs. Appl. Environ. Microbiol. 74, 1305.

Millard, P., Singh, B.K., 2010. Does grassland vegetation drive soil microbial diversity? Nutr. Cycl. Agroecosyst. 88, 147–158.

Mishra, S., Lee, W., Hooijer, A., Reuben, S., Sudiana, I.M., Idris, A., Swarup, S., 2014. Microbial and metabolic profiling reveal strong influence of water table and land-use patterns on classification of degraded tropical peatlands. Biogeosciences 11.

Moilanen, M., Silfverberg, K., Hokkanen, T.J., 2002. Effects of wood-ash on the tree growth, vegetation and substrate quality of a drained mire: a case study. For. Ecol. Manag. 171, 321–338.

Morales, S.E., Mouser, P.J., Ward, N., Hudman, S.P., Gotelli, N.J., Ross, D.S., Lewis, T.A., 2006. Comparison of bacterial communities in New England sphagnum bogs using terminal restriction fragment length polymorphism (T-RFLP). Microb. Ecol. 52, 34–44.

Morrison, E.S., Thomas, P., Ogram, A., Kahveci, T., Turner, B.L., Chanton, J.P., 2020. Characterization of bacterial and fungal communities reveals novel consortia in tropical oligotrophic peatlands. Microb. Ecol., 1–14.

Müller, N., Worm, P., Schink, B., Stams, A.J.M., Plugge, C.M., 2010. Syntrophic butyrate and propionate oxidation processes: from genomes to reaction mechanisms. Environ. Microbiol. Rep. 2, 489–499.

Newell, S.Y., Moran, M.A., Wicks, R., Hodson, R.E., 1995. Productivities of microbial decomposers during early stages of decomposition of leaves of a freshwater sedge. Freshw. Biol. 34, 135–148.

Ong, C.S.P., Juan, J.C., Yule, C.M., 2015. Litterfall production and chemistry of *Koompassia malaccensis* and *Shorea uliginosa* in a tropical peat swamp forest: plant nutrient regulation and climate relationships. Trees 29, 527–537.

Ovaskainen, O., Schigel, D., Ali-Kovero, H., Auvinen, P., Paulin, L., Nordén, B., Nordén, J., 2013. Combining high-throughput sequencing with fruit body surveys reveals contrasting life-history strategies in fungi. ISME J. 7, 1696–1709.

Pandey, A., Nigam, P., Soccol, C., Thomaz-Soccol, V., Singh, D., Mohan, R., 2000. Advances in microbial amylases. Biotechnol. Appl. Biochem. 31 (Pt 2), 135–152.

Paul, E.A., Clark, F.E., 1989. Soil microbiology and biochemistry in perspective. In: Paul, E.A., Clark, F.E. (Eds.), Soil Microbiology and Biochemistry. Academic Press, San Diego, pp. 1–10.

Peltoniemi, K., Fritze, H., Laiho, R., 2009. Response of fungal and actinobacterial communities to water-level drawdown in boreal peatland sites. Soil Biol. Biochem. 41, 1902–1914.

Peltoniemi, K., Straková, P., Fritze, H., Iráizoz, P.A., Pennanen, T., Laiho, R., 2012. How water-level drawdown modifies litter-decomposing fungal and actinobacterial communities in boreal peatlands. Soil Biol. Biochem. 51, 20–34.

Peltoniemi, K., Pyrhönen, M., Laiho, R., Moilanen, M., Fritze, H., 2016. Microbial communities after wood ash fertilization in a boreal drained peatland forest. Eur. J. Soil Biol. 76, 95–102.

Peruzzi, E., Franke-Whittle, I.H., Kelderer, M., Ciavatta, C., Insam, H., 2017. Microbial indication of soil health in apple orchards affected by replant disease. Appl. Soil Ecol. 119, 115–127.

Philippot, L., Andersson, S.G.E., Battin, T.J., Prosser, J.I., Schimel, J.P., Whitman, W.B., Hallin, S., 2010. The ecological coherence of high bacterial taxonomic ranks. Nat. Rev. Microbiol. 8, 523–529.

Preston, M.D., Smemo, K.A., McLaughlin, J.W., Basiliko, N., 2012. Peatland microbial communities and decomposition processes in the james bay lowlands, Canada. Front. Microbiol. 3, 70.

Rinu, K., Pandey, A., 2010. Temperature-dependent phosphate solubilization by cold- and pH-tolerant species of *Aspergillus* isolated from Himalayan soil. Mycoscience 51, 263–271.

Roesch, L.F.W., Fulthorpe, R.R., Riva, A., Casella, G., Hadwin, A.K.M., Kent, A.D., Daroub, S.H., Camargo, F.A.O., Farmerie, W.G., Triplett, E.W., 2007. Pyrosequencing enumerates and contrasts soil microbial diversity. ISME J. 1, 283–290.

Romaní, A.M., Fischer, H., Mille-Lindblom, C., Tranvik, L.J., 2006. Interactions of bacteria and fungi on decomposing litter: differential extracellular enzyme activities. Ecology 87, 2559–2569.

Rousk, J., Bååth, E., Brookes, P.C., Lauber, C.L., Lozupone, C., Caporaso, J.G., Knight, R., Fierer, N., 2010. Soil bacterial and fungal communities across a pH gradient in an arable soil. ISME J. 4, 1340–1351.

Rydin, H., Jeglum, J.K., Bennett, K.D., 2013. The Biology of Peatlands, 2e. OUP Oxford.

Saarsalmi, A., Mälkönen, E., Kukkola, M., 2004. Effect of wood ash fertilization on soil chemical properties and stand nutrient status and growth of some coniferous stands in Finland. Scand. J. For. Res. 19, 217–233.

Sait, M., Davis, K.E.R., Janssen, P.H., 2006. Effect of pH on isolation and distribution of members of subdivision 1 of the phylum Acidobacteria occurring in soil. Appl. Environ. Microbiol. 72, 1852–1857.

Schmidt, S., Costello, E., Nemergut, D., Cleveland, C., Reed, S., Weintraub, M., Meyer, A.F., Martin, A., 2007. Biogeochemical consequences of rapid microbial turnover and seasonal succession in soil. Ecology 88, 1379–1385.

Schütte, U.M.E., Henning, J.A., Ye, Y., Bowling, A., Ford, J.D., Genet, H., Waldrop, M., Turetsky, M.R., White, J.R., Bever, J.D., 2019. Effect of permafrost thaw on plant and soil fungal community in the boreal forest: does fungal community change mediate plant productivity response? J. Ecol. 107, 1737–1752.

Schuur, E.A.G., Crummer, K.G., Vogel, J.G., Mack, M.C., 2007. Plant species composition and productivity following permafrost thaw and thermokarst in Alaskan Tundra. Ecosystems 10, 280–292.

Shen, C., Xiong, J., Zhang, H., Feng, Y., Lin, X., Li, X., Liang, W., Chu, H., 2013. Soil pH drives the spatial distribution of bacterial communities along elevation on Changbai Mountain. Soil Biol. Biochem. 57, 204–211.

Silfverberg, K., Huikari, O., 1985. Tuhkalannoitus metsäojitetuilla turvemailla.

Straková, P., Penttilä, T., Laine, J., Laiho, R., 2012. Disentangling direct and indirect effects of water table drawdown on above- and belowground plant litter decomposition: consequences for accumulation of organic matter in boreal peatlands. Glob. Chang. Biol. 18, 322–335.

Sun, H., Terhonen, E., Koskinen, K., Paulin, L., Kasanen, R., Asiegbu, F.O., 2014. Bacterial diversity and community structure along different peat soils in boreal forest. Appl. Soil Ecol. 74, 37–45.

Sun, H., Terhonen, E., Kovalchuk, A., Tuovila, H., Chen, H., Oghenekaro, A., Heinonsalo, J., Kohler, A., Kasanen, R., Vasander, H., Asiegbu, F., 2016. Dominant tree species and soil type affect fungal community structure in a boreal peatland forest. Appl. Environ. Microbiol. 82, 03858-03815.

Sun, S., Li, S., Avera, B.N., Strahm, B.D., Badgley, B.D., 2017. Soil bacterial and fungal communities show distinct recovery patterns during forest ecosystem restoration. Appl. Environ. Microbiol. 83, e00966-00917.

Taş, N., Prestat, E., McFarland, J.W., Wickland, K.P., Knight, R., Berhe, A.A., Jorgenson, T., Waldrop, M.P., Jansson, J.K., 2014. Impact of fire on active layer and permafrost microbial communities and metagenomes in an upland Alaskan boreal forest. ISME J. 8, 1904–1919.

Thormann, M., 2006a. Diversity and Function of Fungi in Peatlands: A Carbon Cycling Perspective.

Thormann, M.N., 2006b. The role of fungi in boreal peatlands. In: Wieder, R.K., Vitt, D.H. (Eds.), Boreal Peatland Ecosystems. Springer Berlin Heidelberg, Berlin, Heidelberg, pp. 101–123.

Thormann, M.N., Bayley, S.E., 1997. Aboveground plant production and nutrient content of the vegetation in six peatlands in Alberta, Canada. Plant Ecol. 131, 1–16.

Thormann, M., Rice, A., 2007. Fungi from peatlands. Fungal Divers. 24, 241–299.

Thormann, M.N., Bayley, S., Currah, R., 2004. Microcosm tests of the effects of temperature and microbial species number on the decomposition of *Carex aquatilis* and *Sphagnum fuscum* litter from southern boreal peatlands. Can. J. Microbiol. 50 (10), 793–802.

Too, C.C., Keller, A., Sickel, W., Lee, S.M., Yule, C.M., 2018. Microbial community structure in a Malaysian tropical peat swamp Forest: the influence of tree species and depth. Front. Microbiol. 9.

Tsitko, I., Lusa, M., Lehto, J., Parviainen, L., Ikonen, A., Lahdenperä, A.-M., Bomberg, M., 2014. The variation of microbial communities in a depth profile of an acidic, nutrient-poor boreal bog in southwestern Finland. Open J. Ecol. 4, 832–859.

Turetsky, M.R., Wieder, R.K., Williams, C.J., Vitt, D.H., 2000. Organic matter accumulation, peat chemistry, and permafrost melting in peatlands of boreal Alberta. Écoscience 7, 379–392.

Uroz, S., Calvaruso, C., Turpault, M.P., Pierrat, J.C., Mustin, C., Frey-Klett, P., 2007. Effect of the mycorrhizosphere on the genotypic and metabolic diversity of the bacterial communities involved in mineral weathering in a forest soil. Appl. Environ. Microbiol. 73, 3019–3027.

Ward, N.L., Challacombe, J.F., Janssen, P.H., Henrissat, B., Coutinho, P.M., Wu, M., Xie, G., Haft, D.H., Sait, M., Badger, J., Barabote, R.D., Bradley, B., Brettin, T.S., Brinkac, L.M., Bruce, D., Creasy, T., Daugherty, S.C., Davidsen, T.M., DeBoy, R.T., Detter, J.C., Dodson, R.J., Durkin, A.S., Ganapathy, A., Gwinn-Giglio, M., Han, C.S., Khouri, H., Kiss, H., Kothari, S.P., Madupu, R., Nelson, K.E., Nelson, W.C., Paulsen, I., Penn, K., Ren, Q., Rosovitz, M.J., Selengut, J.D., Shrivastava, S., Sullivan, S.A., Tapia, R., Thompson, L.S., Watkins, K.L., Yang, Q., Yu, C., Zafar, N., Zhou, L., Kuske, C.R., 2009. Three genomes from the phylum *Acidobacteria* provide insight into the lifestyles of these microorganisms in soils. Appl. Environ. Microbiol. 75, 2046.

Williams, R.T., Crawford, R.L., 1983. Microbial diversity of Minnesota peatlands. Microb. Ecol. 9, 201–214.

Wilson, D., Alm, J., Riutta, T., Laine, J., Byrne, K.A., Farrell, E.P., Tuittila, E.-S., 2007. A high resolution green area index for modelling the seasonal dynamics of CO2 exchange in peatland vascular plant communities. Plant Ecol. 190, 37–51.

Wolken, J.M., Hollingsworth, T.N., Rupp, T.S., Chapin Iii, F.S., Trainor, S.F., Barrett, T.M., Sullivan, P.F., McGuire, A.D., Euskirchen, E.S., Hennon, P.E., Beever, E.A., Conn, J.S., Crone, L.K., D'Amore, D.V., Fresco, N., Hanley, T.A., Kielland, K., Kruse, J.J., Patterson, T., Schuur, E.A.G., Verbyla, D.L., Yarie, J., 2011. Evidence and implications of recent and projected climate change in Alaska's forest ecosystems. Ecosphere 2, art124.

Yu, Z., Beilman, D.W., Frolking, S., MacDonald, G.M., Roulet, N.T., Camill, P., Charman, D.J., 2011. Peatlands and their role in the global carbon cycle. EOS Trans. Am. Geophys. Union 92, 97–98.

Yule, C.M., 2010. Loss of biodiversity and ecosystem functioning in Indo-Malayan peat swamp forests. Biodivers. Conserv. 19, 393–409.

Zhalnina, K., Dias, R., de Quadros, P.D., Davis-Richardson, A., Camargo, F.A.O., Clark, I.M., McGrath, S.P., Hirsch, P.R., Triplett, E.W., 2015. Soil pH determines microbial diversity and composition in the park grass experiment. Microb. Ecol. 69, 395–406.

Zhou, X., Zhang, Z., Tian, L., Li, X., Tian, C., 2017. Microbial communities in peatlands along a chronosequence on the Sanjiang Plain, China. Sci. Rep. 7, 9567.

Section E

Archaea and viruses in forest ecosystem and microbiota of forest nurseries and tree pests

Chapter 17

Mycobiome of forest tree nurseries

Marja Poteri[a], Risto Kasanen[b], and Fred O. Asiegbu[b]

[a]Natural Resources Institute Finland (Luke), Helsinki, Finland, [b]Department of Forest Sciences, Faculty of Agriculture and Forestry, University of Helsinki, Helsinki, Finland

Chapter Outline

1. Introduction

Planting of tree seedlings is one of the most extensive silvicultural operations with the objective to safeguard forest regeneration, increase timber yield, and shorten forest rotation time. Seedlings may originate from genetically improved seeds, with potential for producing more biomass, better quality as timber and resistance or tolerance to abiotic and biotic stress factors (Boshier and Buggs, 2015; Jansson et al., 2017). Planting is a preferable option in forest areas without natural seed dissemination or in areas where rich ground vegetation prohibits the development of a new forest from seeds. Forest tree seedlings are also targeted for afforestation of abandoned field lands as well as in areas suffering from erosion. Planting of seedlings further contributes to global effort in the sequestration of atmospheric carbon dioxide through reforestation (Wheeler et al., 2016).

2. Bare root and container seedling production

Growing of bare root seedling stock in outdoor open fields is based on agricultural technologies and requires sanitization of soil from weeds, harmful microorganisms, and other pests. Soil physical and chemical properties also have to be monitored and adjusted (Aldhous and Mason, 1994). Seeds are sown into seedbeds where seedlings grow at least one growing season. Thereafter, the seedlings are lifted and transplanted to provide more space for the shoot and root development of the young plants. The transplants are grown for a further 1–3 years in the field before outplanting. It is also possible to transplant small container seedlings into field and grow these as bare root seedlings at a greater spacing. Lifting and transplanting of seedlings in forest nurseries are now mechanized as machines have been developed to facilitate the process (Hallman, 1984). Bare root nurseries are open due to large transplant fields and in many cases wind protection with shelterbelt hedges or trees is used to prevent wind damages (Aldhous and Mason, 1994). In bare root production, maintaining good soil hygiene is

FIG. 17.1 Container forest tree seedlings are sown and grown during the first weeks in a plastic house before their transfer to outdoor fields where the seedlings develop winter hardiness. *(Photo: Fred O. Asiegbu.)*

important and a number of chemical fumigation treatments have been used (Sutherland, 1984). The ozone-depleting methyl bromide has been banned since 1989 due to its toxicity according to Montreal protocol (https://ozone.unep.org/treaties/montreal-protocol). To control soil-borne pests, finding novel alternative growing technique deserves to be further explored (Anon., 2005; Haase, 2009). So far, crop rotation and soilless cultivation are the most used alternatives, but new affordable methods are still needed (López-Aranda et al., 2016).

During the last decades, nurseries in northern latitudes have shifted from bare root to container seedling production (Nilsson et al., 2010). With this technique, seedlings are grown in nurseries for 1–2 years resulting in seedlings with smaller target size compared to bare roots (Fig. 17.1). The filling and sowing of seedlings in containers are facilitated by automated machines. For bigger container seedlings, automated transplanting robots often used for horticultural purposes could be adapted in tree nurseries. Container seedlings can tolerate handling, transportation, and storage stresses better than traditional bare root seedlings (Grossnickle and El-Kassaby, 2016).

2.1 Techniques in container seedling production

2.1.1 Growth substrates

Traditionally, peat has been commonly used in container seedling production (Puustjärvi, 1969; Schmilewski, 2008). Less decomposed peat with coarse structure and a high capacity to retain water is preferred in forest nurseries (Heiskanen, 1993). Tree seedlings should be grown in substrates with sufficient aeration to prevent water logging that hinders root development and predispose seedlings to root pathogens. The chemical and fungistatic properties of peat as well as its rich microflora may act as antagonists to root and damping-off pathogens (Tahvonen, 1982; Wolffhechel, 1988). In recent decades, coconut coir has become a more common substrate (Barrett et al., 2016), alone or as mixtures with peat. To replace fossil peat, the use of *Sphagnum* moss biomass as a growth substrate has also been investigated (Kämäräinen et al., 2018). Growth substrates are not directly usable without improvements for aeration qualities as well as adjustments for pH and nutrient content (Landis, 1990). New additives like biochar (Dumroese et al., 2018) have been tested in addition to traditionally used sawdust, perlite, and vermiculate. It is also important to screen growth substrates for impurities like seeds of weed, harmful chemical compounds (salts, heavy metals, etc.), or plant disease agents. To reduce risk of biotic diseases, peat fumigation treatments have been used (Landis, 1989).

2.1.2 Manipulation of growth rhythm

Seedlings are usually grown in natural day light; however, in container nurseries with variable sowing times additional light treatments are sometimes needed. In the temperate northern tree provenances, in early sowings, the long dark nighttime can be interrupted with short light pulses to prevent germinated seedlings becoming dormant (Dormling et al., 1968; Simak, 1975). In late growing season, the growing period of seedlings can be extended with additional light (Landis, 1992). A third application of light manipulation is a short-day treatment to enhance the growth cessation and winter hardening of seedlings (Colombo et al., 2001; Leikola, 1970).

2.1.3 Winter storage

Winter is a risky period for seedlings because there are diseases specialized to take advantage of seedling dormancy due to weak or inactive plant defense responses, stress, harsh and shifting winter conditions. In the north temperate climate, snow gives shelter against low temperatures. However, there are diseases specialized to colonize seedling tissues under snow. In early spring under deep snow cover, the temperature rises above zero, increasing the activity of snow blight fungi (Vuorinen and Kurkela, 1993).

Roots of container seedlings are unable to tolerate for too long the lowest winter temperatures (Colombo et al., 2001; Lindström and Stattin, 1994). In the absence of natural snow cover, nurseries use snow cannons to make artificial snow on the seedlings stored outdoors over the winter period. The other option to prevent root damages at wintertime is to pack seedlings into a freezer where the temperature can be controlled. In the freezers, temperature is kept above $-4°C$ but below $0°C$ to protect seedling roots and restrict the growth of mold fungi. During storage, the seedlings are packed into airtight cardboard boxes, after storage and during melting, the boxes should be opened to prevent mold problems (Luoranen et al., 2019).

2.2 Disease pressure in nurseries

The nursery environment differs from natural growing conditions, to which germinating seeds and seedlings are adapted. In nurseries, the natural growing and dormant phases of tree seedlings can be interfered by adverse sowing times and by artificial light treatments. With container seedlings, organic growth substrates as well as the communities of mycorrhizas differ from natural forest soils. Compared to natural environment, abiotic or biotic factors pose higher risk in nurseries as many seedlings are packed in a small growing area (Burdon and Chilvera, 1982). Several fungal disease agents that are effective in mycelial spreading or able to discharge plenty of spores could reach available seedlings in a relatively short distance.

In natural environment, the interaction dynamics of the host, pathogen, and environment has been conceptualized as the disease triangle (Gäumann, 1950). A modified version of the disease triangle that takes into consideration the human factor as could occur in nurseries is modeled as disease tetrahedron (Fig. 17.2) (Zadoks and Schein, 1979). In forest nurseries, a holistic approach takes into account both the seedling and its pathogens with emphasis on management and growing procedures in order to guarantee seedling health (Landis, 2000).

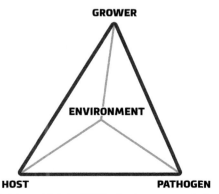

FIG. 17.2 A disease tetrahedron modified from Zadoks and Schein (1979) takes into account the human factor.

3. Mycobiome as disease agents of forest nursery

3.1 Seed pathogens and damping-off

Seed-transmitted diseases include species, which reduce germination or kill young emerged plants. Pitch canker (*Fusarium circinatum*) occurs mainly in pine forests (Pérez-Sierra et al., 2007) but may spread with contaminated seeds into nurseries (Bezos et al., 2017; Bracança et al., 2009). In nurseries, it causes damping-off, discoloration, and disintegration of the root cortex on pine seedlings (Vainio et al., 2019). The disease has mostly spread in subtropical climatic region but it has also recently spread to pine forests in Mediterranean area (Pérez-Sierra et al., 2007). As the fungus can cause extensive damage, "The European and Mediterranean Plant Protection Organization (EPPO)" has listed it as a quarantine pest (A2 list) (https://gd.eppo.int/taxon/gibbci/documents).

For nurseries, storage pathogens pose a great risk. Black rot (*Ciboria batschiana*) infects fallen acorns (Knudsen et al., 2004) and chestnuts (Migliorini et al., 2010) on the forest floor in the autumn. Infected acorns and chestnuts become mummified, which can result in important losses if infections are not detected before storage (Nef and Perrin, 1999).

In nurseries, the most important damping-off fungi belong to the genera of *Alternaria*, *Neonectria*, *Rhizoctonia*, *Pythium*, *Phytophthora*, and *Fusarium* (James et al., 1991; Lilja, 1994; Lilja et al., 1995; Mittal and Wang, 1993; Perrin and Sampangi, 1986; Sutherland et al., 2002; Sutherland and Davis, 1991). Some of these species can be seed transmitted but most of them are common saprobes, which are soil borne in nurseries. In preemergence damping-off, seed is decayed before the plant reaches the aboveground level. In postemergence damping-off, germination proceeds normally, but soon after small plants die due to a girdled and decayed root collar area (Fig. 17.3).

3.2 Root dieback

Root dieback symptoms are caused by slowly advancing root decay, which results in poor uptake of water and nutrients (Lilja, 1994). The emergence of the disease was attributed to the direct contact of containers to nursery soil surface (Venn et al., 1986). The stunting and yellowing of spruce seedlings are caused by oomycete species of *Pythium* and *Phytophthora* (Galaaen and Venn, 1979; Lilja, 1994; Lilja et al., 1992) or by *Rhizoctonia* spp. (Hietala, 1995, 1997; Hietala et al., 2001). Seedling roots stressed by water logging as well as by excessive exposure to fungicides are also predisposed to root infections and dieback caused by a common soil fungus genus *Neonectria* (*Cylindrocarpon*) (Beyer-Ericson et al., 1991; Unestam et al., 1989).

FIG. 17.3 Damping-off fungi decay the roots and root collar area of germinated seedlings. The fallen seedlings with green needles will be totally decayed as shown by the arrows. *(Photo: Marja Poteri.)*

3.3   Gray mold

The most common pathogenic mold fungus in the nurseries is gray mold, *Botrytis cinerea,* which infects young succulent seedlings at the germination stage and during shoot and needle growth (Hansen and Hamm, 1988; Mittal et al., 1987). Although seedlings are more tolerant to the fungus at the end of growing season (Petäistö et al., 2004), damages may also occur during winter storage (Petäistö, 2006; Venn, 1981). Several stress factors also predispose the seedlings to disease. These include high temperatures, drought episodes and low light intensity (Zhang and Sutton, 1994; Zhang et al., 1995), long lasting moisture and wet conditions (Peterson et al., 1988), cellular damages caused by frost, insects, and other physical damages (Sutherland and Davis, 1991). The conidial infections require free water on plant surface (Russell, 1990). The optimum temperature range for the germination of conidia is 10–20°C but gray mold may develop even at 2°C (Jarvis, 1980), which makes its control in nurseries challenging.

3.4   *Phytophthora* spp.

Several *Phytophthora* species are adapted to the nursery conditions where they infect seedling roots, root collar area, and young succulent tissues of stems and leaves (Hamm and Hansen, 1986). A reddish color in plant tissue is a typical *Phytophthora* symptom for some plant species usually at the early stages of the infection (Hamm and Hansen, 1982). On container-grown silver birch seedlings, necrotic stem lesions and top dying caused by *P. cactorum* (Fig. 17.4) have been a severe problem (Lilja et al., 1996). *Phytophthora* species are persistent in soil as oospores or chlamydospores, and they are adapted to spread in excess water as motile zoospores or transmitted as propagules in irrigation water (Ali-Shtayeh et al., 1991; Themann et al., 2002). With molecular identification methods, the earlier complex species have been distinguished to be new separate species (Jung and Burgess, 2009; Lilja and Poteri, 2013). Such species occurring in nurseries is *P. megasperma*, which has been divided to several subgroups. One group occurring on Douglas-fir seedlings is described as *P. sansomeana* (Hansen et al., 2009) whereas *P. citricola* complex includes the nursery pathogens *P. plurivora* and *P. multivora* (Jung and Burgess, 2009).

FIG. 17.4 *Phytophthora cactorum* is a serious pathogen of young succulent birch (*Betula* sp.) seedlings. Diseased seedlings have stem spots and withering of tops and leaves. *(Photo: Marja Poteri.)*

3.5 Scleroderris canker

Scleroderris canker is caused by *Gremmeniella abietina*, which is a severe pathogen in nurseries especially on pine seedlings (Gremmen, 1972). Among host species are mainly pines but spruce seedlings can also be infected (Børja et al., 2006; Petäistö, 2008). The fungus attacks seedling buds or current year shoots during growing season and in the autumn. Symptoms appear only during the following spring at the time of bud burst, when seedlings have dead resinous buds and the top needles turn brown at the base, die and fall off (Nef and Perrin, 1999). Small pine seedlings do not survive the infection (Fig. 17.5), but spruce seedlings with dried tops may recover and have a new growth from the buds in the lower shoot base (Petäistö, 2008). With diverse identification techniques, *G. abietina* has been divided into the geographical races of North America, Asia, and Europe (Dorworth and Krywienczyk, 1975; Hamelin et al., 1993; Lecours et al., 1994; Müller and Uotila, 1997). The disease persists in lower dying branches of mature pine trees and the sporulation is especially high after cool and rainy periods with low irradiation (Petäistö and Heinonen, 2003). Due to the slow development of the fungus (2–3 years), pycnidia are more common in container seedlings than later formed apothecia (Petäistö, 2008).

3.6 Sirococcus blight

Sirococcus conigenus infects both young germinating plants and succulent growing shoots of older seedlings. In nurseries, the fungus is pathogenic to both pine and spruce seedlings and its primary inoculum is from seeds originating from infested cones (Sutherland et al., 2002). Low temperature and irradiance connected with high moisture conditions predispose seedlings to the disease (Tian Fu and Uotila, 2002; Wall and Magasi, 1976). In containers, healthy seedlings quickly overgrow the infected ones. Later on, pycnidia are formed on those dead seedlings (Fig. 17.6), and the released conidia are able to spread the disease to older second-year seedlings (Sutherland, 1987; Sutherland et al., 1981). In second-year seedlings, cankers and bending of the shoot appear after midsummer (Lilja et al., 2005).

3.7 Sphaeropsis blight (Diplodia blight)

Diplodia pinea attacks mainly pine species but also several other conifers, e.g., species of *Abies*, *Picea*, and *Pseudotsuga menziesii* (Nef and Perrin, 1999; Stanosz et al., 1999). In nurseries, the pathogen can reduce seed germination and cause symptoms, which resemble pre and postemergency damping-off (Rees and Webber, 1988). Other disease symptoms are browning of current year needles, shoot blight, and shoot tips may also twist (Nef and Perrin, 1999; Stanosz et al., 1999). Various seedling stresses are known to predispose seedlings to the disease (Paoletti et al., 2001).

FIG. 17.5 Symptoms of *Gremmeniella abietina* infection appear at the time of bud burst. The apical bud of a one-year-old container Scots pine (*Pinus sylvestris*) seedling has died and the browning of needles starts from the needle base. *(Photo: Marja Poteri.)*

FIG. 17.6 Norway spruce (*Picea abies*) seedlings infected by *Sirococcus conigenus*. Later on, the conidia emerge on the dead seedlings and spread the disease to healthy ones. *(Photo: Marja Poteri.)*

3.8 Dothistroma needle blight

Dothistroma needle blight (red band needle cast) caused by *Mycosphaerella pini* (teleomorph *Dothistroma septosporum*) has been a serious pathogen of *Pinus radiata* seedlings in tropical nurseries (Ivory, 1994). The disease occurs in young native pine forests also in northern Europe (Hanso and Drenkhan, 2008; Müller et al., 2009). In UK, *Dothistroma* needle blight has been found in nurseries since 1950s but due to climate change the fast spreading of the disease necessitates the need to explore novel control strategies for nurseries (Piotrowska et al., 2017). *M. pini* can infect also larch, Douglas-fir, and spruce seedlings (Nef and Perrin, 1999).

3.9 Snow blight Phacidium infestans and brown felt blight (black snow mold) Herpotrichia juniperi

P. infestans attacks seedlings in autumn when ascospores are released from the dead needles infected in previous year. A prerequisite for the development of snow blight is the snow cover, under which the fungus spreads efficiently with hyphal colonization (Fig. 17.7). In spring after snowmelt, the symptoms appear in patches (Peace, 1962). The host species are mainly pine seedlings (Björkman, 1948; Kujala, 1950; Roll-Hansen, 1987) and spruce seedlings (Petäistö et al., 2013).

FIG. 17.7 The browning of Scots pine (*Pinus sylvestris*) needles caused by the snow blight fungus *Phacidium infestans* is visible immediately after snow melt. The coarse mycelium on the surface of diseased needles will disappear due to drying or being washed away by rain. *(Photo: Marja Poteri.)*

Brown felt blight *H. juniperi* attacks several conifer species (Schneider et al., 2009) and does not require snow cover but high humidity (Hanso and Tõrva, 1975; Von Bazzigher, 1976). The disease is more common in central and south Europe at high altitudes where it kills young seedlings and lower branches of conifers. It is equally a nursery pathogen causing significant losses for seedlings (Von Bazzigher, 1976). After the winter period, the diseased seedlings appear in patches, like in snow blight infections. The fungus is able to spread efficiently with its coarse mycelium in optimal moisture conditions, and growth at lower temperatures from $-3°C$ to $-5°C$ has been recorded (Von Bazzigher, 1976).

3.10 Needle diseases

Lophodermium seditiosum causes pine needle cast disease globally. It infects primarily current year pine needles (Diwani and Millar, 1987; Kowalski, 1993). The yellow spots symptoms of L. *seditiosum* infection are hardly visible during the autumn (Diwani and Millar, 1981, 1987). The actual disease symptoms appear during springtime, which coincides with the period of bud burst when the infected previous year needles turn brown and fall (Kurkela, 1979; Martinsson, 1975; Stenström and Ihrmark, 2005). The needle loss weakens the ability of the seedlings to tolerate and resist drought stress. The diseased seedlings hardly survive the planting shock (Lilja, 1986). Ascospores, which are released in autumn, can travel airborne long distances, which facilitates remote dissemination (Hanso, 1968; Kurkela, 1979). Larch needle cast caused by *Meria laricis* reduces seedling growth and increases mortality and can be a serious problem in nurseries (Nef and Perrin, 1999). The risk of the infection is higher in 2-year-old seedlings, which are infected from the previous year diseased needles (Lilja et al., 2010).

3.11 Powdery mildews and leaf spots

Powdery mildews are obligate parasites, which are highly specialized on their host plants (Glawe, 2008). In nurseries, oak powdery mildew (*Microsphaera alphitoides*) can be very devastating in the absence of control measures (Nef and Perrin, 1999). The pathogen covers leaf surfaces and shoots with its mycelium and as a result the diseased seedlings become forked and their resistance to cold is further weakened (Marçais and Desprez-Loustau, 2014). Seedlings can suffer several leaf spot diseases, which reduce growth and winter hardiness. Cherry leaf spot (*Phloeosporella padi*), which occurs on *Prunus* sp., overwinters in diseased fallen leaves and in spring infects the new leaves of seedlings (Nef and Perrin, 1999). During growing season, the infection proceeds from lower leaves to upper ones especially in high seedling densities which may cause severe defoliation and premature leaf-fall (Stanosz, 1992). Most common leaf spot species of birch seedlings are *Asteroma* sp., *Gloeosporium* sp., and *Marssonina betulae* (Lilja et al., 1997).

3.12 Rust diseases

Rust diseases have a minor role as nursery pathogens although in years when rust epidemics start early, also seedling damages may occur. Leaf rusts of poplars (*Melampsora* sp.), birch (*Melampsoridium betulinum*), and alder (*Melampsoridium hiratsukanum*) appear usually in late summer when low nighttime temperatures condensate water on leaf surfaces for uredospore germination (Hamelin et al., 1992; Krzan, 1980; Pinon et al., 2006). At this time of the growing season, seedling shoot growth has ceased and leaves are almost fully expanded. The heavy rust infection causes premature leaf fall, which prevents natural winter hardening of shoots and buds (Dooley, 1984), consequently, seedlings have a lower survival rate after planting (Lilja, 1973). Basidiospores of spruce needle rust, *Chrysomyxa ledi*, can infect young current year spruce needles if nurseries are located near the distribution range of the host species *Ledum palustre* (*Rhododendron tomentosum*) (Kaitera et al., 2017).

Pine twisting rust (*Melampsora pinitorqua*) occurs sporadically on Scots pine and maritime pine (*Pinus pinaster*) seedlings (Desprez-Loustau and Wagner, 1997; Kurkela and Lilja, 1984) but *P. contorta* seems to be resistant (Martinsson, 1985). Cherry spruce rust (*Thekopsora areolata*) may infect spruce seedlings (Hietala et al., 2008). Both rust species are spread by basidiospores in early summer when new succulent shoot tissues are susceptible to infection, which leads to twisting and canker formation of shoots (Fig. 17.8) (Desprez-Loustau and Wagner, 1997; Hietala et al., 2008). Rust infections are often accompanied with other fungal species, and in nurseries *Phomopsis* sp. (Hietala et al., 2008) and *B. cinerea* in young pine plantations (Domanski and Kowalski, 1988) have been reported in connection with rust infections.

FIG. 17.8 The cherry spruce rust fungus *Thekopsora areolata* has caused the twisting and dying of the top on a Norway spruce (*Picea abies*) seedling. *(Photo: Marja Poteri.)*

4. Mycobiota of forest nursery

4.1 Endophytic and saprotrophic mycobiota

Beneficial microbiome is the key to success of any living organisms (Carthey et al., 2020; Vandenkoornhuyse et al., 2015), but most of the roles of microbes are unknown in plant rhizosphere (Mendes et al., 2013) and also phyllosphere (Rodriguez et al., 2009). The microbiome in needle and shoot tissues of conifers apparently is modified by host genotype of buds (Elfstrand et al., 2020), physiological stage of needles (Rajala et al., 2014), and environment (Millberg et al., 2015).

The fungal life strategies are flexible, which has changed our views of the function of plant microbial partners. Increasing evidence indicates that the soil saprobes and mycorrhizae are not distinct groups but continuum (Selosse et al., 2018). The mycorrhizal symbiosis in basidiomycetes has evolved from saprotrophic ancestors multiple times (Hibbett et al., 2000). Several saprotrophs, such as three wood-decay fungi, *Phlebiopsis gigantea*, *Phlebia centrifuga*, and *Hypholoma fasciculare* were able to establish neutral interaction with root system of conifer seedling (Vasiliauskas et al., 2007). In addition, endophytic fungi may utilize saprotrophic life strategy in order to degradate the plant tissues such as needle litter (Boberg et al., 2011) or to pathogenic lifestyle in case the trees suffer abiotic stress (Stanosz et al., 1997). The dual lifestyle of pathogen and endophyte may exist in different tissues of plant, such as in needles and seeds (Ridout and Newcombe, 2018).

4.2 Potentials of mycorrhizas in nursery production

Mycorrhizal inoculation (Fig. 17.9) might have beneficial effects for survival and development of young trees such as in the case of *Laccaria laccata* in Sitka spruce (Thomas and Jackson, 1983) and even long-term growth performance of

FIG. 17.9 *Laccaria laccata* is an ectomycorrhizal fungus used in the inoculations of seedlings. White mycorrhizal hyphae on the surface of the root plug of a two-year-old Norway spruce (*Picea abies*) container seedling and an emerging sporocarp of L. *laccata* on top of the plug. *(Photo: Marja Poteri.)*

Douglas-fir provided by *Laccaria bicolor* (Selosse et al., 2000). However, the effects are not straightforward. In a study by Molina and Chamard (1983), inoculation with mycorrhiza did not affect the size of Douglas-fir seedlings but it significantly reduced growth of ponderosa pine under low fertility. Ectomycorrhizal fungus L. *laccata* performed well under high-fertility regimes. Yet the experiment resulted in out-plantable nursery stock with well-developed ectomycorrhizae. In a study of two-year-old seedlings of *Betula pendula* and *Acer pseudoplatanus*, mycorrhizae accounted for a small but significant proportion of the variation in growth of both hosts (Frankland and Harrison, 1985). Differences in the degree of infection between fresh state soils were highly significant. Infection by mycorrhizae was highly correlated with several soil factors (pH, organic matter, phosphorus, and iron), and with all the plant variables. The factors affecting mycorrhization are apparently complex. The results by Rajala et al. (2014) suggest that individual strategies of fine-root architecture of young seedlings may promote growth success later on.

One of the major obstacles of utilizing mycorrhizae is that some nursery species are frequently lost after the plant roots are exposed to natural environment and microbial flora. Villeneuve et al. (1991) identified mycorrhizal partners of seedlings and studied their colonization abilities. Ectomycorrhizal fungi of nursery (*Thelephora terrestris* and *Suillus* sp.) were not detected after out-planting. Inoculated and indigenous *Laccaria* species and *Cenococcum geophilum* were found on old roots and new roots, excluding natural mycobiome. *Laccaria* species colonized inoculated seedlings, whereas controls were dominated by the natural mycorrhizae. During the second year of growth, the root and shoot weight of inoculated seedlings increased. Mycorrhizae and the rate of root colonization had effect on growth of the seedlings. The decade-lasting persistency of *Laccaria bicolor* has been demonstrated in Selosse et al. (2000). Buschena et al. (1992) showed that roots colonized by L. *bicolor* remained colonized by that isolate for the period of 2.5-year test period. The competitiveness of the species was demonstrated when control and *Laccaria longipes*-treated seedlings on the same mineral soil site became colonized by indigenous isolates of L. *bicolor*.

The cultivation system also has effect on mycorrhizal communities (Menkis et al., 2005, Menkis and Vasaitis, 2011). Menkis et al. (2005) studied the effect of bare root cultivation, containerized plastic tray systems, and polyethylene rolls with direct internal transcribed spacer ribosomal DNA sequencing and isolation. In pine, the highest level of mycorrhizal colonization (48%) occurred in bare root systems, while in spruce, colonization was highest in polyethylene rolls (71%). In total, 93 fungal taxa, of which 27 were mycorrhizal were found. Characteristic mycorrhizas were *Phialophora finlandia*, *Amphinema byssoides*, *Rhizopogon rubescens*, *Suillus luteus*, and *Thelephora terrestris*. There was a moderate similarity in mycorrhizal communities between pine and spruce and among different cultivation systems.

Plant protection achieved by mycorrhizal inoculation would be desirable. However, the interactions are not yet fully understood. Velmala et al. (2018) tested if root colonization by ectomycorrhizal fungi (EMF) could alter the susceptibility of Norway spruce (*P. abies*) seedlings to root rot infection or necrotic foliar pathogens. Protection of seedlings was achieved with *Meliniomyces bicolor* that had showed strong in vitro antagonistic properties toward root rot-causing *Heterobasidion*. Interestingly, gray mold *Botrytis cinerea* infection was not affected but needle damage severity caused by shoot pathogen *Gremmeniella abietina* was high. In conclusion, the mycorrhizal inoculation failed to induce systemic resistance against a diverse and broad range of pathogens.

Majority of mycorrhizal studies deal with single host species. Rudawska et al. (2019) found highly diverse ECM fungal communities in bare root nurseries of admixture tree species. The roots of hornbeams and limes had most diverse ECM community (23 taxa), and birch had slightly less taxa (21). The most common fungi in the roots were *Tuber* (average relative abundance: 18.9%) and *Hebeloma* (average relative abundance: 18.4%) species. Some of the fungal species were observed for the first time in forest nursery (e.g., *Lactarius pubescens*, *Leccinum holopus*, *Pachyphloeus melanoxantha*, *Russula grata*, and *Pachyphloeus melanoxantha*, which is a rare species).

4.3 Saprotroph communities pose a risk in nurseries

In bulk forest soil, the dominant fungal group is Basidiomycota (Terhonen et al., 2019), which includes a number of saprotrophs. Garcia-Lemos et al. (2019) studied root-associated bacterial and fungal communities of small and tall three-year-old *Abies nordmanniana* bare root seedlings. Sequencing of 16S and 18S rRNA gene amplicon showed that Ascomycota was the most prevalent fungal phylum within plant tissues. The fungal biota included members of the *Agaricales*, *Hypocreales*, and *Pezizales*. Both bacterial and fungal communities indicate the significance of soil characteristics and climatic conditions for the composition of root-associated microbial communities. Major differences between communities from tall and small plants were the dominance of the pathogen *Fusarium* (*Hypocreales*) in the small plants, while *Agaricales*, that includes beneficial ectomycorrhizal fungi, dominated in the tall plants.

The results by Velmala et al. (2018) suggest that unsterile inoculum sources, such as the forest humus may predispose seedlings to needle pathogens. Menkis et al. (2016) also demonstrated the presence of latent pathogenic fungi in roots and the growth substrate, although their study showed that current management practices in forest nurseries produce healthy seedlings in general.

Menkis et al. (2016) studied the effect of environment on the mycobiome of decayed roots of conifer seedlings. The highest numbers of taxa were detected in forest nurseries (56 and 54). The afforested clear-cut areas had more species (36 and 40) than abandoned farmland (24 and 16). They also found that native soilborne fungi colonized roots of seedlings only 12 weeks after planting on clear-cuts and agricultural land. *Fusarium* spp. were the most common fungi in forest nurseries. *Nectria* spp. dominated the clear-cuts and abandoned agricultural land areas were characterized by *Penicillium* spp. and *Trichoderma* spp.

4.4 Endophytes in rhizosphere and phyllosphere may promote seedling health

4.4.1 Root endophytes and dark septate endophyte (DSE)

Ascomycota is the most prevalent fungal group within plant tissues (Terhonen et al., 2019). Commonly these fungi inhabit roots without forming specific morphological features of mycorrhizas or pathogens. The most studied taxa have darkly pigmented hyphal walls and have been referred to as dark septate endophytes (DSE) (Addy et al., 2005; Jumpponen, 2001). Alberton et al. (2010) have shown that nitrogen use and growth of Scots pine seedlings was enhanced under elevated CO_2 if dark septate root endophytic fungi (*Phialocephala fortinii*, *Cadophora finlandica*, *Chloridium paucisporum*, *Scytalidium vaccinii*, *Meliniomyces variabilis*, and *M. vraolstadiae*) were present in the seedlings. Under elevated CO_2, the biomass of seedlings inoculated with DSE fungi was on average 17% higher than in control seedlings. Simultaneously, belowground respiration doubled or tripled, and consequently carbon use efficiency by the DSE fungi significantly decreased. DSE fungi increase the efficiency of plant nutrient use and are therefore more beneficial to the plant under elevated CO_2.

DSEs might have an inhibitory effect against plant pathogens. Terhonen et al. (2016) found that the endophyte *Phialocephala sphaeroides* was able to inhibit phytopathogens (*Heterobasidion parviporum*, *Phytophthora pini*, *Botrytis cinerea*) and to protect Norway spruce seedlings against *H. parviporum* infection. Interestingly, *Cryptosporiopsis* sp. with strong inhibitory effect against *Heterobasidion* did not promote growth of Norway spruce seedlings. Berthelot et al. (2019) observed that DSE strains *Cadophora* sp., *Leptodontidium* sp., and *Phialophora mustea* decreased the growth of the root phytopathogens *Pythium intermedium*, *Phytophthora citricola*, and *Heterobasidion annosum*.

In natural ecosystem, plants can alter the nature of the interaction they have with fungi, therefore, in vitro detected antagonism may not necessarily always reflect the situation inside the plant. Similarly, the inhibitory effect of DSE fungus *Cadophora* sp. against soilborne pathogens of tomato under in vitro condition could not be observed during natural interaction with roots (Yakti et al., 2019).

4.4.2 Endophytes of phyllosphere

In addition to DSEs, endophytes in general may protect trees from herbivory and pathogenic invasion. Miller et al. (2002) reported that inoculations of toxigenic endophytes of seedlings from a breeding population of white spruce were successful across a range of genotypes. The needles colonized by a rugulosin-producing endophyte were found to contain rugulosin in concentrations that are effective in vitro at retarding the growth of spruce budworm larvae.

Arnold et al. (2003) showed that inoculation of endophyte-free leaves with endophytes isolated frequently from naturally infected, asymptomatic hosts significantly decreased both leaf necrosis and leaf mortality of *Theobroma cacao* seedlings when challenged with pathogen (*Phytophthora* sp.). Ganley et al. (2008) demonstrated that fungal endophytes from *Pinus monticola* facilitated seedling survival against *Cronartium ribicola*, the agent of white pine blister rust. Seedlings previously inoculated with fungal endophytes lived longer and showed decrease in disease severity.

Millberg et al. (2015) observed that needles with symptoms of disease hosted a more diverse mycobiota compared to healthy needles, presumably supporting more pathogenic or saprotrophic species. Similarly, Rajala et al. (2014) found that many endophytes of needles are facultative saprotrophs, because more than a third of the species occurred in both fresh and decomposing needles, together with fungal species of pathogenic potential.

The diversity and composition of European beech (*Fagus sylvatica*) leaf mycobiome correlated significantly with the origin of the trees (Unterseher et al., 2016). In forests, the mycobiome was more diverse at lower than at higher elevation in the German Alps, whereas fungal diversity was lowest in the tree nursery, indicating that local stand conditions are important for the structure of beech leaf mycobiome. Also the correlation of chlorophylls and flavonoids with habitat and fungal diversity highlights the connection of phyllosphere fungi and leaf physiology. This was also observed by Rajala et al. (2014), who found significant differences between endophyte communities of spruce clones with fast and slow growth rate. Interestingly, the endophyte community responded to temporal growth variations of trees.

5. Control approaches to minimize pathogenic infections in forest nursery

5.1 Alternatives for pesticides are needed

The internationally accepted principles of Integrated Pest Management (IPM) and the International Code of Conduct on Pesticide Management aim to enhance the development of alternatives for chemical pesticides (Anon., 2019). In forest nurseries, disease preventive methods can be combined with different growing and management strategies to reduce the dependence on pesticides.

5.2 Reduction of natural inoculum sources

Nursery seedlings have the same disease agents, which occur in natural forests or in adjacent older trees. If these trees are diseased, e.g., have *G. abietina* infections in canopy, such inoculum sources should if possible be eradicated. Conidia of *G. abietina* are released on short distances in water droplets (Petäistö and Heinonen, 2003) such that the seedlings next to diseased trees have a higher infection risk. Bigger trees can be used in nurseries as windbreak, but they may also create harmful moist-shadow microclimate for seedlings and predispose them to *Scleroderris* canker (Gremmen, 1972). The practice of packing seedlings into freezers over winter has reduced infection pressure of snow blight (*P. infestans*) in outdoors, but increased the risk during storage (Petäistö et al., 2013). Fresh logging slash of Scots pine increases the risk of snow blight infection (Hansson, 2006). Therefore the general advice to avoid harvesting trees during late autumn in areas where deep snow cover is expected should be applied close to nurseries. In addition, any harvesting residues and logging slash should be eradicated in the nursery. The same applies to the infected lower branches of naturally grown young pines along roadsides and next to the outdoor fields.

5.3 Seedling management and biostimulants

Attacks of mold fungi can be prevented by reducing the seedling damages caused by insects, frost, drought, and weeds. Excess free water on plant surfaces favors the germination of molds, whereas efficient aeration and irrigation of plants in the mornings minimize the time when there is free water on seedling surfaces (Mittal et al., 1987). To avoid mold problems, there are techniques available to brush off water droplets from the top of nursery seedlings after irrigation (James et al., 1995).

Besides the prevention of pathogen attacks, there are biocontrol products based on beneficial mycorrhizal species and other microorganisms, which can be applied in nurseries. These organisms together with versatile plant growth-promoting compounds form a group of biostimulants, which constitute a new era for plant disease control (du Jardin, 2015).

5.4 Hygiene

It is important to clean growing areas from dead plant material and debris, which form a growth substrate for mold fungi (*B. cinerea*). In addition, several other pathogenic fungi, like powdery mildews, certain leaf rusts, and needle casts continue their life cycle and sporulate in fallen leaves or needles. *Phytophthora* species can persist in plant debris in nurseries (Junker et al., 2016; Shishkoff, 2009). In the growing of larch seedlings, the infection risk of gray mold has been minimized by vacuum cleaning the fallen needles (Dumroese and Wenny, 1992).

Different heating treatments can be used to get rid of possible nursery pests in growth substrates. At the same time, with excessive temperatures there is a risk that beneficial organisms are lost and altered compounds harmful for seedlings are produced (Landis, 1989). Between sowings, hot water treatments are used to clean reusable containers from pathogen propagules, which may persist in soil or root particles (Iivonen et al., 1996; Kohman and Børja, 2002). Composted forest nursery waste is not usable as it may contain active pathogen propagules (Veijalainen et al., 2005).

For recycled irrigation water, it is recommended to test the presence of oomycete species such as *Pythium* and *Phytophthora* (Redekar et al., 2019). Forest nurseries use surface water that can be decontaminated (Landis, 1989), but so far very few forest nurseries have adopted water recycling technique. It is obvious that for environmental reasons the circulation of irrigation water will become more common among forest nurseries.

5.5 Chemical control of diseases

The chemical control of diseases is performed with the fungicides, which are approved to be used as plant protection products (PPP). National authorities regulate and control the use of commercial pesticide products. In EU and USA, the active ingredients in pesticides are evaluated and approved by EFSA (European Food Safety Association) and by EPA (Environmental Protection Agency), respectively.

The key issue in using fungicides is that the treatments are performed as needed. Each fungicide is used according to manufacturer's instructions where the approved target plants, the application concentration, and the information on application timing are indicated. As the grower has to make the decision about the fungicide use, it is crucial to have knowledge on the dissemination, infection biology, and disease cycles of the pathogen to be controlled. The predisposing factors to infections, such as seedling provenances, microclimate, current and previous year disease prevalence are critical for decision making. Unlike insects, which can be monitored at operational scale, there are hardly any direct methods for estimating the rate of possible infection pressure in forest nurseries (Petäistö and Heinonen, 2003).

The appearance of fungicide-resistant fungal strains has become a challenge in crop protection, which necessitates the development of new active ingredients (Hollomon, 2015). Information on the fungicide resistance is constantly collated by FRAC (Fungicide Resistance Action Committee, https://www.frac.info/home) for the use of all stakeholders. Besides the resistance aspects, the reductions in fungicide use also have impacts on human health and environment. This is a continuous task for the authorities and it is implemented as the withdrawals of products or as the revisions of the label texts are implemented. The plant protection operators in nurseries are expected to be familiar with the national up-to-date selection of registered fungicides and their appropriate use.

6. Impact of pesticides (fungicides) on mycobiota of forest nursery

Forest nursery is an excellent environment for phytopathogenic microbes. Occurrence of pests always causes economic losses for the nursery, which necessitates the use of pesticide for the plant protection. Pesticides by definition are chemical or biological agents that can be used to protect plants or plant products against pests or diseases. There are many types of pesticides, which include among others: fungicides, insecticides, antimicrobials, biocides, bactericides, herbicides, growth regulators, nematicides, etc. (https://ec.europa.eu/food/plant/pesticides_en). Pesticides including fungicides are increasingly used in the production of forest tree seedlings and in agriculture. Fungicide is a type of pesticide that kills or inhibits parasitic fungi. There are three kinds of fungicides, which can be grouped based on their mobility in plants (Mueller, 2006): systemic, contact, or translaminar. Systemic fungicides are absorbed by the plant and translocated within the plant through the xylem vessels (Beckerman, 2018). Contact fungicides on the other hand, are not absorbed into the plant tissue and are mainly used to control foliar diseases (Mueller, 2006). Translaminar fungicides are absorbed by the upper treated leaf surface and redistributed to the lower untreated leaf surface (Klittich and Ray, 2013); (https://www.growertalks.com/Article/?articleid=22487). Fungicides are also grouped based on their biochemical mode of action and a real-time list of their mode of action and resistance risk is available on FRAC internet pages (https://www.frac.info/docs/default-source/publications/frac-code-list/frac-code-list-2020-final.pdf?sfvrsn=8301499a_2). Inorganic fungicides such as sulfur or copper salts were the earlier chemicals used to control plant pathogens. The primary target of copper salts is to disrupt basic metabolic processes and many of them are phytotoxic (Deacon, 1997). Similarly, organic contact fungicides interrupt the fungal metabolic machinery. Systemic fungicides such as sterol biosynthesis inhibitors (SBIs) inhibit sterol synthesis in fungal cell membrane and as a result fungi are unable to synthesize ergosterol. Weete (1973) noted that SBIs are not toxic to mitosporic fungi that lack ergosterol. A wide and diverse range of fungicides are in commercial use. The FRAC-code list has at this moment almost 85 different chemical groups of fungicides. The majority of the fungicides affect some of the biosynthetic pathways in pathogens. There are eight compounds included in the group, which are known to enhance host plant defense reactions, five compounds are based on microbial activity and two on plant extracts. Fungicides are generally used to control a wide range of tree or nursery diseases. Many of the pathogens that cause diseases are already highlighted before and in previous subsections including a few notable ones: *Botrytis* blight, Cedar apple rusts (by *Gymnosporangium*), needle rust (by *Chrysomyxa, Coleosporium*), Pine-pine gall rust (*Endocronartium*), seedling blight (*Diplodia, Phomopsis*), seedling root rot (*Phytophthora*), etc. Fungicides containing propiconazole have been used since the beginning of 1980s (Tomlin, 2009) to prevent several fungal diseases. They have also been used to prevent winter time damages caused by snow blights (*P. infestans*) and Scleroderris canker (*G. abietina*) in forest nurseries. Propiconazole (1-[2,4-dichlorophenyl]-4-propyl-1,3-dioxolan-2ylmethyl]-1H-1,2,4-triazole (IUPAC, International Union of Pure and Applied Chemistry) is a broad-spectrum SBI systemic foliar fungicide having curative and protective action in plants (https://sitem.herts.ac.uk/aeru/ppdb/en/Reports/551.htm). In EU, the approval of propiconazole has not been renewed as it was withdrawn from the EU list of approved active ingredient in plant protection products at the end of January 2019 (http://www.fao.org/3/ca5344en/ca5344en.pdf; https://eur-lex.europa.eu/legal-content/en/ALL/?uri=CELEX:32018R1865). The major concern on the use of fungicides in a forest nursery is the potential effect of the long-term use on fungal species richness, microbial succession, fungal biodiversity, and on nontarget species (Klingberg, 2012; Terhonen et al., 2013; Vasiliauskas et al., 2004).

Majority of the studies on effects of pesticides including fungicides have mostly reported on the impact on communities of soil microbiota with limited information on the effects on the plant microbiome (Meena et al., 2020). Equally, literature reports on the adverse effects of fungicides on nontarget microbes have mostly focused on agronomic crops (Elmholt and Smedegaard-Petersen, 1988). Muturi et al. (2017) reported that, like in higher organisms, pesticides could have toxic effects on microbiome communities. Some of the indirect effects of pesticides are ecological alterations that could lead to shifts in communities of microbes with potential impact on their functions. Schaeffer et al. (2017) reported that treatment of almond (*Prunus dulcis*) flowers with two fungicides (metconazole and penthiopyrad) had negative impact on fungal richness and diversity, particularly on the relative abundance of OTU of nectar specialists such as the yeast *Metschnikowia reukaufii*. By contrast, fungicide treatment did not significantly affect bacteria community composition, diversity, or richness of almond flowers. Laforest-Lapointe et al. (2017) reported that anthropogenic pressures, such as heavy metal deposition, on trees led to reduction in the abundance of Alphaproteobacteria in urban environment. Özdemir and Erkiliç (2018) reported that application of fungicide (mancozeb, copper hydroxide) on citrus leaves led to 50-fold reduction in fungal population.

Manninen et al. (1998) reported that propiconazole not only reduced mycorrhiza infection of fine roots of Scots pine, but it also selectively killed ascomycete symbionts. Other additional side effects documented included reduced soil respiration with negative impact on soil microorganisms. Other authors have confirmed the potential harmful effects of propiconazole on the growth of ectomycorrhizal fungi (Laatikainen and Heinonen-Tanski, 2002). Similar negative impacts of propiconazole on nontarget fungi have previously been documented in wheat field where significant inhibitory effects were observed on the fungus *Cladosporium* sp. (Elmholt, 1991). Other authors have reported that application of propiconazole on the tropical tree (*Guarea guidonia*) reduced the number of endophytic fungal isolates compared to untreated plant (Gamboa et al., 2005). The use of the systemic fungicide hexaconazole on mango (*Mangifera indica*) was also demonstrated to decrease the frequency of foliar fungal endophytes (Mohandoss and Suryanarayanan, 2009).

Klingberg (2012) observed that needles of three-month-old Scots pine seedlings treated with the fungicide propiconazole had the highest frequency of endophytic fungal isolates compared to control untreated seedlings. The study also noted that in treated seedlings, frequencies of fungal endophytes increased from springtime to the autumn. Other authors have documented periodic variations in the abundance of endophytic fungi (Guo et al., 2008; Helander et al., 1994). Guo et al. (2008) did not document foliar endophytes in the needles of one-year-old seedlings, but rather observed that frequencies of the endophyte *Pinus tabulaeformis* were lower in May than in August. Several authors noted that age and physiological changes in the needles could be a determining factor in the frequency of endophytic isolates (Hata et al., 1998; Lehtijärvi and Barklund, 1999). Lehtijärvi and Barklund (1999) reported that in needles of mature Norway spruce, the frequency of isolation of the fungal endophyte (*Lophodermium piceae*) increased with cessation of tree growth in the autumn. According to Hata et al. (1998), the possible explanation for the higher frequencies could be attributed to health and physiological status of needles or to lower levels of antimicrobial compounds within the needles.

Interestingly, Klingberg (2012) reported that untreated control seedlings had more OTUs than propiconazole-treated seedlings. Some OTUs were found to be common to both treatments and included among others *Penicillium* sp. (<5%), *Sistotrema* sp. (<5%), *Lophodermium pinastri* (<5%), *Phoma herbarum* (<5%), *Phoma glomerata* (6.5%), and *Phoma* sp. (36.6%). Klingberg (2012) also reported that the frequency of epi-mycota was higher in control seedlings in September whereas in October fungicide-treated seedlings had higher frequency of epi-mycota. Among the epi-mycota, *Epicoccum nigrum* and *Sistotrema brinkmannii* were the most frequently isolated.

7. Concluding remarks

The nurseries in northern latitudes have shifted from bare root to container seedling production, which has resulted in improvements in hygiene and decrease in microbial diversity. Overwintering of seedlings in freezers has facilitated more efficient seedling logistics, but at the same time seedlings have become more challenged by storage pathogens. The diverse mycobiota of forest nurseries has direct or indirect effects on plant health. Although seedlings are commonly utilized as plant material for studies of plant–microbe interactions, actual nursery conditions have been seldom used. Comparative studies of mycobiomes of trees responding to diverse habitats and climate are still required. Increasing access to high-throughput sequencing (HTS) data offers the researchers chance to analyze mycobiome responses in detail. The nursery is very different environment in comparison to natural growth sites of seedlings. The high humidity created by dense growing and nutrition as well as growth substrates might harbor mycobiota with harmful effects on plants. However, the new possibilities to control environmental conditions (e.g., freezing and winter hardening) could help to facilitate efficient production of forest regeneration material but also maintenance of plant health.

References

Addy, H.D., Piercey, M.M., Currah, R.S., 2005. Microfungal endophytes in roots. Can. J. Bot. 83, 1–13.

Alberton, O., Kuyper, T.W., Summerbell, R.C., 2010. Dark septate root endophytic fungi increase growth of scots pine seedlings under elevated CO_2 through enhanced nitrogen use efficiency. Plant and Soil 328, 459–470.

Aldhous, J.R., Mason, W.L. (Eds.), 1994. Forest Nursery Practice. Forestry Commission Bulletin 111. HMSO for Forestry Commission, UK, p. 268.

Ali-Shtayeh, M.S., MacDonald, J.D., Kabashima, J., 1991. A method for using commercial ELISA test to detect zoospores of *Phytophthora* and *Pythium* species in irrigation water. Plant Dis. 75, 305–311.

Anon, 2005. Methyl Bromide-phase out and alternatives. PAN Europe pesticides action network Europe. Briefing no. 4, October 2005. https://www.pan-europe.info/old/Archive/publications/MethylBromide.htm.

Anon, 2019. Global Situation of Pesticide Management in Agriculture and Public Health. World Health Organization and Food and Agriculture Organization of the United Nations, Geneva. [2019]. ISBN 978-92-4-151688-4 (WHO) ISBN 978-92-5-131969-7 (FAO) http://www.fao.org/3/ca7032en/ca7032en.pdf.

Arnold, A.E., Mejia, L.C., Kyllo, D., Rojas, E.I., Maynard, Z., Robbins, N., Herre, E.A., 2003. Fungal endophytes limit pathogen damage in a tropical tree. Proc. Natl. Acad. Sci. U. S. A. 100, 15649–15654.

Barrett, G.E., Alexander, P.D., Robinson, J.S., Bragg, N.C., 2016. Achieving environmentally sustainable growing media for soilless plant cultivation systems—a review. Sci. Hortic. 212, 220–234. https://doi.org/10.1016/j.scienta.2016.09.030.

Beckerman, J., 2018. Fungicide Mobility for Nursery, Greenhouse, and Landscape Professionals. https://www.extension.purdue.edu/extmedia/bp/bp-70-w.pdf.

Berthelot, C., Leyval, C., Chalot, M., Blaudez, D., 2019. Interactions between dark septate endophytes, ectomycorrhizal fungi and root pathogens in vitro. FEMS Microbiol. Lett. 366. fnz158.

Beyer-Ericson, L., Damm, E., Unestam, T., 1991. An overview of root dieback and its causes in Swedish forest nurseries. Eur. J. For. Pathol. 21, 439–443.

Bezos, D., Martínez-Alvarez, P., Fernández, M., Diez, J.J., 2017. Epidemiology and management of pine pitch canker disease in Europe—a review. Balt. For. 23, 279–293.

Björkman, E., 1948. Studier över snöskyttesvampens (*Phacidium infestans*) biologi samt metoder for snöskyttets bekämpande. Summary: studies on the biology of the *Phacidium* blight (*Phacidium infestans* karst.) and its prevention. Meddelanden från Statens Skogsforskningsinstitut 37 (2), 1–136.

Boberg, J.B., Ihrmark, K., Lindahl, B.D., 2011. Decomposing capacity of fungi commonly detected in *Pinus sylvestris* needle litter. Fungal Ecol. 4, 110–114.

Børja, I., Solheim, H., Hietala, A.M., Fossdal, C.G., 2006. Etiology and real-time polymerase chain reaction-based detection of *Gremmeniella*- and *Phomopsis*-associated disease in Norway spruce seedlings. Phytopathology 96 (12), 1305–1314.

Boshier, D., Buggs, R.J.A., 2015. The potential for field studies and genomic technologies to enhance resistance and resilience of British tree populations to pests and pathogens. Forestry 88 (1), 27–40. https://doi.org/10.1093/forestry/cpu046.

Bracança, H., Diogo, E., Moníz, F., Amaro, P., 2009. First report of pitch canker on pines caused by *Fusarium circinatum* in Portugal. Plant Dis. 93, 1079.

Burdon, J.J., Chilvera, G.A., 1982. Host density as a factor in plant disease ecology. Annu. Rev. Phytopathol. 20, 143–166.

Buschena, C.A., Doudrick, R.L., Anderson, N.A., 1992. Persistence of *Laccaria* spp. as ectomycorrhizal symbionts of container-grown black spruce. Can. J. For. Res. 22, 1883–1887.

Carthey, A.J.R., Blumstein, D.T., Gallagher, R., Tetu, S.G., Gillings, M.R., 2020. Conserving the holobiont. Funct. Ecol. 34, 764–776.

Colombo, S.J., Menzies, M.I., O'Reilly, C., 2001. Influence of nursery cultural practices on cold hardiness of coniferous forest tree seedlings. In: Bigras, F.J., Colombo, S.J. (Eds.), Conifer Cold Hardiness. Kluwer Academic Publishers, pp. 223–252.

Deacon, J., 1997. Modern Mycology, third ed. Blackwell Publishing, UK, p. 303.

Desprez-Loustau, M.-L., Wagner, K., 1997. Influence of silvicultural practices on twisting rust infection and damage in maritime pine, as related to growth. For. Ecol. Manage. 98, 135–147.

Diwani, S.A., Millar, C.S., 1981. Biology of *Lophodermium seditiosum* in nurseries in N.E. Scotland. In: Millar, C.S. (Ed.), Current Research on Conifer Needle Diseases. Proceedings of IUFRO WP on Needle Diseases, Sarajevo, 1980. Aberdeen University Forest Department, Old Aberdeen, pp. 67–74.

Diwani, S.A., Millar, C.S., 1987. Pathogenicity of three *Lophodermium* species on *Pinus sylvestris* L. Eur. J. For. Pathol. 17, 53–58.

Domanski, S., Kowalski, T., 1988. Untypical dieback of the current season's shoots of *Pinus sylvestris* in Poland. Eur. J. For. Pathol. 18, 157–160.

Dooley, H.L., 1984. Temperature effects on germination of uredospores of *Melampsoridium betulinum* and on rust development. Plant Dis. 68 (8), 686–688.

Dormling, I., Gustavsson, Å., von Wettstein, D., 1968. The experimental control of the life cycle in *Picea abies* (L.) karst. 1. Some basic experiments on the vegetative cycle. Silva Genet. 17, 44–64.

Dorworth, C.E., Krywienczyk, J., 1975. Comparisons among isolates of *Gremmeniella abietina* by means of growth rate, conidia measurement and immunogenic reaction. Can. J. Bot. 53, 2506–2525.

du Jardin, P., 2015. Plant biostimulants: definition, concept, main categories and regulation. Sci. Hortic. 196, 3–14.

Dumroese, K.R., Wenny, D.L., 1992. Reducing *Botrytis* in container-grown Western larch by vacuuming dead needles. Tree Plant. Notes 43 (2), 30–32.

Dumroese, R.K., Pinto, J.R., Heiskanen, J., Tervahauta, A., McBurney, K.G., Page-Dumroese, D.S., Englund, K., 2018. Biochar can be a suitable replacement for *Sphagnum* peat in nursery production of *Pinus ponderosa* seedlings. Forests 9, 232. https://doi.org/10.3390/f9050232.

Elfstrand, M., Zhou, L., Baison, J., Olson, A., Lunden, K., Karlsson, B., Wu, H.X., Stenlid, J., Garcia-Gil, M.R., 2020. Genotypic variation in Norway spruce correlates to fungal communities in vegetative buds. Mol. Ecol. 29, 199–213.

Elmholt, S., 1991. Side effects of propiconazole (tilt 250 EC) on non-target soil fungi in a field trial compared with natural stress effects. Microb. Ecol. 22, 99–108.

Elmholt, S., Smedegaard-Petersen, V., 1988. Side-effects of field applications of 'propiconazol' and 'captafol' on the composition of non-target soil fungi in spring barley. J. Phytopathol. 123, 79–88.

Frankland, J.C., Harrison, A.F., 1985. Mycorrhizal infection of *Betula pendula* and *Acer pseudoplatanus*—relationships with seedling growth and soil factors. New Phytol. 101, 133–151.

Galaaen, R., Venn, K., 1979. *Pythium sylvaticum* Campbell & Hendrix and other fungi associated with root dieback of 2-0 seedlings of *Picea abies*. (L.) Karst. in Norway. Meddelelser fra Norsk Institutt for Skogforskning 34, 221–228.

Gamboa, M.A.G., Wen, S., Fetcher, N., Bayman, P., 2005. Effects of fungicides on endophytic fungi and photosynthesis in seedlings of a tropical tree, *Guarea Guidonia* (Meliaceae). Acta Biol. Colomb. 2, 41–48.

Ganley, R.J., Sniezko, R.A., Newcombe, G., 2008. Endophyte-mediated resistance against white pine blister rust in *Pinus monticola*. For. Ecol. Manage. 255, 2751–2760.

Garcia-Lemos, A.M., Grosskinsky, D.K., Stokholm, M.S., Lund, O.S., Nicolaisen, M.H., Roitsch, T.G., Veierskov, B., Nybroe, O., 2019. Root-associated microbial communities of *Abies nordmanniana*: insights into interactions of microbial communities with antioxidative enzymes and plant growth. Front. Microbiol. 10, 1937.

Gäumann, E., 1950. Principles of Plant Infection. Hafner Publ, NY.

Glawe, D.A., 2008. The powdery mildews: a review of the world's most familiar (yet poorly known) plant pathogens. Annu. Rev. Phytopathol. 46, 27–51.

Gremmen, J., 1972. *Scleroderris lagerbergii* Gr.: the pathogen and disease symptoms. Eur. J. For. Pathol. 2, 1–5.

Grossnickle, S., El-Kassaby, Y.A., 2016. Bareroot versus container stocktypes: a performance comparison. New For. 47, 1–51. https://doi.org/10.1007/s11056-015-9476-6.

Guo, L.D., Huang, G.R., Wang, Y., 2008. Seasonal and tissue age influences on endophytic fungi of *Pinus tabulaeformis* (Pinaceae) in the Dongling Mountains, Beijing. J. Integr. Plant Biol. 50 (8), 997–1003.

Haase, D.L., 2009. The latest on soil fumigation in bareroot forest nurseries. For. Nurs. Notes 29 (2), 22–25.

Hallman, R.G., 1984. Equipment for forest nurseries. In: Duryea, M.L., Landis, T.D., Perry, C.R. (Eds.), Forestry Nursery Manual: Production of Bareroot Seedlings. Forestry Sciences. vol. 11. Springer, Dordrecht, pp. 17–24.

Hamelin, R.C., Shain, L., Thielges, B.A., 1992. Influence of leaf wetness, temperature, and rain on poplar leaf rust epidemics. Can. J. For. Res. 22, 1249–1254.

Hamelin, R.C., Quellette, G.B., Bernier, L., 1993. Identification of *Gremmeniella abietina* races with random amplified polymorhic DNA markers. Appl. Environ. Microbiol. 59, 1752–1755.

Hamm, P.B., Hansen, E.M., 1982. Pathogenicity of *Phytophthora* species to Pacific northwest conifers. Eur. J. For. Pathol. 12, 167–174.

Hamm, P.B., Hansen, E.M., 1986. *Phytophthora* root rot in forest nurseries of the Pacific northwest. In: Landis, T.D. (Ed.), Proceedings: combined western forest nursery council and intermountain nursery association meeting, August 12–15, 1986, Tumwater, Washington, general technical report RM-137. US Department of Agriculture, Forest Service, Rocky Mountain Research Station, Fort Collins, Colorado, pp. 122–124.

Hansen, E.M., Hamm, P.B., 1988. Canker disease of Douglas-fir seedlings in Oregon and Washington bareroot nurseries. Can. J. For. Res. 18, 1053–1058.

Hansen, E.M., Wilcox, W.F., Reeser, P.W., Sutton, W., 2009. *Phytophthora rosacearum* and *P. sansomeana*, new species segregated from the *Phytophthora megasperma* 'complex'. Mycologia 101, 129–135.

Hanso, M., 1968. Microseente levimise fenoloogilisi vaatlusi männikutes. Eesti Pollumajanduse Akademia Teaduslike tööde Kogumik 50, 194–209 (in Estonian).

Hanso, M., Drenkhan, R., 2008. First observations of *Mycosphaerella pini* in Estonia. Plant Pathol. 57 (6), 1177. https://doi.org/10.1111/j.1365-3059.2008.01912.x.

Hanso, M., Tõrva, A., 1975. Okaspuu—nogihallitus Eestis [summary—black snow mould in Estonia]. Metsanduslikud Uurimused 12, 262–279 (in Estonian).

Hansson, P., 2006. Effects of small tree retention and logging slash on snowblight growth on scots pine regeneration. For. Ecol. Manage. 236, 368–374.

Hata, K., Futai, K., Tsuda, M., 1998. Seasonal and needle age-dependent changes of the endophytic mycobiota in *Pinus thunbergii* and *Pinus densiflora* needles. Can. J. Bot. 76 (2), 245–250.

Heiskanen, J., 1993. Variation in water retention characteristics of peat growth media used in tree nurseries. Silva Fenn. 27 (2), 77–97. https://doi.org/10.14214/sf.a15664.

Helander, M., Sieber, T., Petrini, O., Neuvonen, S., 1994. Endophytic fungi in scots pine needles: spatial variation and consequences of simulated acid rain. Can. J. Bot. 72 (8), 1108–1113.

Hibbett, D.S., Gilbert, L.B., Donoghue, M.J., 2000. Evolutionary instability of ectomycorrhizal symbioses in basidiomycetes. Nature 407, 506–508.

Hietala, A.M., 1995. Uni- and binucleate *Rhizoctonia* spp. co-existing on the roots of Norway spruce seedlings suffering from root dieback. Eur. J. For. Pathol. 25, 136–144.

Hietala, A.M., 1997. The mode of infection of a pathogenic uninucleate *Rhizoctonia* sp. in conifer seedling roots. Can. J. For. Res. 27, 471–480.

Hietala, A.M., Vahala, J., Hantula, J., 2001. Molecular evidence suggests that *Ceratobasidium bicorne* has an anamorph known as a conifer pathogen. Mycol. Res. 105, 555–562.

Hietala, A.M., Solheim, H., Fossdal, C.G., 2008. Real-time PCR-based monitoring of DNA pools in the tri-trophic interaction between Norway spruce, the rust *Thekopsora areolata*, and an opportunistic ascomycetous *Phomopsis* sp. Phytopathology 98, 51–58.

Hollomon, D.W., 2015. Fungicide resistance: facing the challenge. Plant Prot. Sci. 51, 170–176.

Iivonen, S., Lilja, A., Tervo, L., 1996. Juurilahoa aiheuttavan yksitumaisen *Rhizoctonia*-sienen torjunta kuumavesikäsittelyllä. Folia For. 1, 51–55 (in Finnish).

Ivory, M.H., 1994. Records of foliage pathogens of *Pinus* species in tropical countries. Plant Pathol. 43 (3), 511–518.

James, R.L., Dumroese, R.K., Wenny, D.L., 1991. *Fusarium* diseases of conifer seedlings. In: Sutherland, J.R., Glover, S.G. (Eds.), Proceedings of the First Meeting of IUFRO Working Party S2.07-09. Diseases and Insects in Forest Nurseries, August 23–30 1990, Victoria, BC. Information Report BC-X-331. Forestry Canada, Pacific and Yukon Region, Pacific Forestry Centre, Victoria, British Columbia, pp. 181–190.

James, R.L., Dumroese, R.K., Wenny, D.L., 1995. *Botrytis cinerea* carried by adult fungus gnats (Diptera: Sciaridae) in container nursery. Tree Plant. Notes 46 (2), 48–53.

Jansson, G., Hansen, J.K., Haapanen, M., Kvaalen, H., Steffenrem, A., 2017. The genetic and economic gains from forest tree breeding programmes in Scandinavia and Finland. Scand. J. For. Res. 32, 273–286.

Jarvis, W.R., 1980. Epidemiology. In: Coley-Smith, J.R., Verhoeff, K., Jarvis, W.R. (Eds.), The Biology of *Botrytis*. Academic Press Inc., London, pp. 219–250.

Jumpponen, A., 2001. Dark septate endophytes—are they mycorrhizal? Mycorrhiza 11, 207–211.

Jung, T., Burgess, T.I., 2009. Re-evaluation of *Phytophthora citricola* isolates from multiple woody hosts in Europe and North America reveals a new species, *Phytophthora plurivora* sp. nov. Persoonia 22, 95–110.

Junker, C., Goff, P., Wagner, S., Werres, S., 2016. Occurrence of *Phytophthora* species in commercial nursery production. Plant Health Prog. 17 (2), 64–67.

Kaitera, J., Kauppila, T., Hantula, J., 2017. New *Picea* hosts for *Chrysomyxa ledi* and *Thekopsora areolata*. For. Pathol. 47. https://doi.org/10.1111/efp.12365, e12365.

Kämäräinen, A., Simojoki, A., Linden, L., Jokinen, K., Silvan, N., 2018. Physical growing media characteristics of *Sphagnum* biomass dominated by *Sphagnum fuscum* (Schimp.) Klinggr. Mires Peat 21, 17. https://doi.org/10.19189/MaP.2017.OMB.278.

Klingberg, N.M., 2012. The Effect of Fungicide Treatment on the Non-target Foliar Mycobiota of *Pinus sylvestris*—Seedlings in Finnish Forest Nursery (M.Sc. Thesis). Department of Forest Sciences, University of Helsinki, Finland. https://helda.helsinki.fi/bitstream/handle/10138/34669/Ninni%20Klingberg%20Master's%20thesis.pdf?sequence=1.

Klittich, C.R.J., Ray, S.L., 2013. Effects of physical properties on the translaminar activity of fungicides. Pestic. Biochem. Physiol. 107, 351–359.

Knudsen, I.M.B., Thomsen, K.A., Jensen, B., Poulsen, K.M., 2004. Effects of hot water treatment, biocontrol agents, disinfectants and a fungicide on storability of English oak acorns and control of pathogen, *Ciboria batschiana*. For. Pathol. 34, 47–64.

Kohman, K., Børja, I., 2002. Hot-water treatment for sanitising forest nursery containers: effects of container microflora and seedling growth. Scand. J. For. Res. 17, 111–117.

Kowalski, T., 1993. Fungi in living symptomless needles of *Pinus sylvestris* with respect to some observed disease processes. J. Phytopathol. 139, 129–145.

Krzan, Z., 1980. Effect of climatic factors on the development of the disease caused by *Melampsora larici-populina*. Folia For. 422, 10–13.

Kujala, V., 1950. Über die Kleinpilze der Koniferen in Finnland. Commun. Inst. For. Fenn. 38 (4), 1–121 (in German).

Kurkela, T., 1979. *Lophodermium seditiosum* Minter et al. sienen esiintyminen männynkaristeen yhteydessä. Summary: Association of *Lophodermium seditiosum* Minter et al. with a needlecast epidemic on Scots pine. Folia For. 393, 1–11.

Kurkela, T., Lilja, S., 1984. Taimitarhan sienitauteja. Keskusmetsälautakunta Tapio, p. 15 (in Finnish).

Laatikainen, T., Heinonen-Tanski, H., 2002. Mycorrhizal growth in pure cultures in the presence of pesticides. Microbiol. Res. 157, 127–137.

Laforest-Lapointe, I., Messier, C., Kembel, S.W., 2017. Tree leaf bacterial community structure and diversity differ along a gradient of urban intensity. mSystems. https://doi.org/10.1128/mSystems.00087-17.

Landis, T.D., 1989. The disease and pest management. In: Landis, T.D., Tinus, R.W., McDonald, S.E., Barnett, J.P. (Eds.), The Container Tree Nursery Manual. The Biological Component: Nursery Pests and Mycorrhizae. Vol. 5. Agric. Handbk 674. U.S. Department of Agriculture, Forest Service, Washington, DC, pp. 1–99. 171 P.

Landis, T.D., 1990. Growing media. In: Landis, T.D., Tinus, R.W., McDonald, S.E., Barnett, J.P. (Eds.), The Container Tree Nursery Manual. Containers and Growing Media. Vol. 2. Agric. Handbk 674. U.S. Department of Agriculture, Forest Service, Washington, DC, pp. 41–85. 88.

Landis, T.D., 1992. Light. In: Landis, T.D., Tinus, R.W., McDonald, S.E., Barnett, J.P. (Eds.), The Container Tree Nursery Manual. Atmospheric Environment. vol. 3. USDA Forest Service, Washington DC, pp. 73–121. Agriculture Handbook 674. Vol. 3, 145.

Landis, T.D., 2000. Holistic nursery pest management: a new emphasis on seedling health. In: Lilja, A., Sutherland, J.R. (Eds.), Proceedings of the 4th Meeting of IUFRO Working Party 7.03.04. Research Papers 781. Finnish Forest Research Institute, pp. 5–15.

Lecours, N., Toti, L., Sieber, T.N., Petrini, O., 1994. Pectic enzyme pattern as a taxonomic tool for the characterization of *Gremmeniella* spp. isolates. Can. J. Bot. 72, 891–896.

Lehtijärvi, A., Barklund, P., 1999. Effects of irrigation, fertilization and drought on the occurrence of *Lophodermium piceae* in *Picea abies* needles. Scand. J. For. Res. 14 (2), 121–126.

Leikola, M., 1970. The effect of artificially shortened photoperiod on the apical and radial growth of Norway spruce seedlings. Ann. Bot. Fenn. 7, 193–202.

Lilja, S., 1973. Koivun ruoste ja sen torjuminen. Metsänviljelyn koelaitoksen tiedonantoja 9, 21–26 (in Finnish).

Lilja, S., 1986. Diseases and pest problems on *Pinus sylvestris* nurseries in Finland. Bull. EPP/EPPO Bull. 16, 561–564.

Lilja, A., 1994. The occurrence and pathogenicity of uni and binucleate *Rhizoctonia* and Pythiaceae fungi among conifer seedlings in Finnish forest nurseries. Eur. J. For. Pathol. 24, 181–192.

Lilja, A., Poteri, M., 2013. Seed, seedling and nursery diseases. In: Gonthier, P.G., Nicolotti, G. (Eds.), Infectious Forest Diseases. CABI Publications, Wallingford, UK, ISBN: 978-1780640402, pp. 567–590.

Lilja, A., Lilja, S., Poteri, M., Ziren, L., 1992. Conifer seedling root fungi and root dieback in Finnish forest nurseries. Scand. J. For. Res. 7, 547–556.

Lilja, A., Hallaksela, A.M., Heinonen, R., 1995. Fungi colonizing scots pine cone scales and seeds and their pathogenicity. Eur. J. For. Pathol. 25, 38–46.

Lilja, A., Rikala, R., Hietala, A., Heinonen, R., 1996. Fungi isolated from necrotic stem lesions of *Betula pendula* seedlings in forest nurseries and the pathogenicity of *Phytophthora cactorum*. Eur. J. For. Pathol. 26, 89–96.

Lilja, A., Lilja, S., Kurkela, T., Rikala, R., 1997. Nursery practices and management of fungal diseases in forest nurseries in Finland. Silva Fenn. 31 (1), 79–100.

Lilja, A., Poteri, M., Vuorinen, M., Kurkela, T., Hantula, J., 2005. Cultural and PCR-based identification of the two most common fungi from cankers on container grown Norway spruce seedlings. Can. J. For. Res. 35, 432–439.

Lilja, A., Poteri, M., Petäistö, R.-L., Rikala, R., Kurkela, T., Kasanen, R., 2010. Fungal diseases in forest nurseries in Finland. Silva Fenn. 44 (3), 525–545.

Lindström, A., Stattin, E., 1994. Root freezing tolerance and vitality of Norway spruce and scots pine seedlings; influence of storage duration, storage temperature, and prestorage root freezing. Can. J. For. Res. 24 (12), 2477–2484. https://doi.org/10.1139/x94-319.

López-Aranda, J.M., Domínguez, P., Miranda, L., de los Santos, B., Talavera, M., Daugovish, O., Soria, C., Chamorro, M., Medina, J.J., 2016. Fumigant use for strawberry production in Europe: the current landscape and solutions. Int. J. Fruit Sci. 16 (Suppl. 1), S1–S15. https://doi.org/10.1080/15538 362.2016.1199995.

Luoranen, J., Pikkarainen, L., Poteri, M., Peltola, H., Riikonen, J., 2019. Duration limits on field storage in closed cardboard boxes before planting of Norway spruce and scots pine container seedlings in different planting seasons. Forests 9–10. https://doi.org/10.3390/f10121126, 1126.

Manninen, A.-M., Laatikainen, T., Holopainen, T., 1998. Condition of scots pine roots and mycorrhiza after fungicide application and low-level ozone exposure in a 2-year field experiment. Trees 12, 347–355.

Marçais, B., Desprez-Loustau, M.-L., 2014. European oak powdery mildew: impact on trees, effects of environmental factors, and potential effects of climate change. Ann. For. Sci. 71, 633–642.

Martinsson, O., 1975. *Lophodermium pinastri* (needlecast)—an outline of the problem in Sweden. In: Stephan, B.R., Millar, C.S. (Eds.), *Lophodermium* an Kiefern. Mitteilungen der Bundesforschunganstalt für Forst- und Holzwirtschaft. vol. 108, pp. 131–135.

Martinsson, O., 1985. The influence of pine twist rust (*Melampsora pinitorqua*) on growth and development of scots pine (*Pinus sylvestris*). Eur. J. For. Pathol. 15, 103–110.

Meena, R.S., Kumar, S., Datta, R., Lal, R., Vijayakumar, V., Brtnicky, M., Sharma, M.P., Yadav, G.S., Jhariya, M.K., Jangir, C.K., Pathan, S.I., Dokulilova, T., Pecina, V., Marfo, T.D., 2020. Impact of agrochemicals on soil microbiota and management: a review. Land 9, 34. https://doi.org/10.3390/land9020034.

Mendes, R., Garbeva, P., Raaijmakers, J.M., 2013. The rhizosphere microbiome: significance of plant beneficial, plant pathogenic, and human pathogenic microorganisms. FEMS Microbiol. Rev. 37, 634–663.

Menkis, A., Vasaitis, R., 2011. Fungi in roots of nursery grown *Pinus sylvestris*: ectomycorrhizal colonisation, genetic diversity and spatial distribution. Microb. Ecol. 61, 52–63.

Menkis, A., Vasiliauskas, R., Taylor, A., Stenlid, J., Finlay, R., 2005. Fungal communities in mycorrhizal roots of conifer seedlings in forest nurseries under different cultivation systems, assessed by morphotyping, direct sequencing and mycelial isolation. Mycorrhiza 16, 33–41.

Menkis, A., Burokiene, D., Stenlid, J., Stenstrom, E., 2016. High-throughput sequencing shows high fungal diversity and community segregation in the rhizospheres of container-grown conifer seedlings. Forests 7, 44.

Migliorini, M., Funghini, L., Marinelli, C., Turchetti, T., Canuti, S., Zanoni, B., 2010. Study of water curing for the preservation of marrons (*Castanea sativa* Mill., Marrone fiorentino cv). Postharvest Biol. Technol. 56, 95–100.

Millberg, H., Boberg, J., Stenlid, J., 2015. Changes in fungal community of scots pine (*Pinus sylvestris*) needles along a latitudinal gradient in Sweden. Fungal Ecol. 17, 126–139.

Miller, J.D., Mackenzie, S., Foto, M., Adams, G.W., Findlay, J.A., 2002. Needles of white spruce inoculated with rugulosin-producing endophytes contain rugulosin reducing spruce budworm growth rate. Mycol. Res. 106, 471–479.

Mittal, R.K., Wang, B.S.P., 1993. Effects of some seed-borne fungi on *Picea glauca* and *Pinus strobus* seeds. Eur. J. For. Pathol. 23, 138–146.

Mittal, R.K., Singh, P., Wang, B.S.P., 1987. *Botrytis*: a hazard to reforestation. Eur. J. Plant Pathol. 17, 369–384.

Mohandoss, J., Suryanarayanan, T.S., 2009. Effect of fungicide treatment on foliar fungal endophyte diversity in mango. Sydowia 61 (1), 11–24.

Molina, R., Chamard, J., 1983. Use of the ectomycorrhizal fungus *Laccaria laccata* in forestry. 2. Effects of fertilizer forms and levels on ectomycorrhizal development and growth of container-grown Douglas-fir and Ponderosa pine seedlings. Can. J. For. Res. 13, 89–95.

Mueller, D.S., 2006. Fungicides: Terminology. Integrated Crop Management News, p. 1250. http://lib.dr.iastate.edu/cropnews/1250.

Müller, M., Uotila, A., 1997. The diversity of *Gremmeniella abietina* var. *abietina* FAST-profiles. Mycol. Res. 101, 169–175.

Müller, M.M., Hantula, J., Vuorinen, M., 2009. First observations of *Mycosphaerella pini* on scots pine in Finland. Plant Dis. 93 (3), 322.

Muturi, E.J., Donthu, R.K., Fields, C.J., Moise, I.K., Kim, C.-H., 2017. Effect of pesticides on microbial communities in container aquatic habitats. Sci. Rep. 7, 44565. https://doi.org/10.1038/srep44565.

Nef, L., Perrin, R. (Eds.), 1999. Damaging Agents in European Forest Nurseries. Practical Handbook. European Union Air 2-CT93-1694 Project, ISBN: 92-828-2803-4, p. 352.

Nilsson, U., Luoranen, J., Kolström, T., Örlander, G., Puttonen, P., 2010. Reforestation with planting in northern Europe. Scand. J. For. Res. 25 (4), 283–294. https://doi.org/10.1080/02827581.2010.498384.

Özdemir, S.K., Erkiliç, A., 2018. Effects of some fungicides and foliar fertilizers on epiphytic fungal and yeast population of citrus leaves. J. Nat. Appl. Sci. 22 (2), 429–434.

Paoletti, E., Danti, R., Strati, S., 2001. Pre- and post-inoculation water stress affects *Sphaeropsis sapinea* canker length in *Pinus halepensis* seedlings. For. Pathol. 31, 209–218.

Peace, T.R., 1962. Pathology of Trees and Shrubs. Clarendon Press, Oxford, pp. 299–302.

Pérez-Sierra, A., Landeras, E., León, M., Berbegal, M., García-Jiménez, J., Armengol, J., 2007. Characterization of *Fusarium circinatum* from *Pinus* spp. in northern Spain. Mycol. Res. 111, 832–839.

Perrin, R., Sampangi, R., 1986. La fonte des semis en pépinière forestière [Damping off in forest nurseries]. Eur. J. For. Pathol. 16, 309–321.

Petäistö, R.-L., 2006. *Botrytis cinerea* and Norway spruce seedlings in cold storage. Balt. For. 12, 24–33.

Petäistö, R.-L., 2008. Infection of Norway spruce container seedlings by *Gremmeniella abietina*. For. Pathol. 38, 1–15.

Petäistö, R.-L., Heinonen, J., 2003. Conidial dispersal of *Gremmeniella abietina*: climatic and microclimatic factors. For. Pathol. 33, 1–11.

Petäistö, R.-L., Heiskanen, J., Pulkkinen, A., 2004. Susceptibility of Norway spruce seedlings to gray mold in the greenhouse during the first growing season. Scand. J. For. Res. 19, 30–37.

Petäistö, R.-L., Lilja, A., Hantula, J., 2013. Artificial infection and development of snow mold fungus (*Phacidium infestans*) in container-grown Norway spruce seedlings. Balt. For. 19, 31–38.

Peterson, M.J., Sutherland, J.R., Tuller, S.E., 1988. Greenhouse environment and epidemiology of grey mould of container-grown Douglas-fir seedlings. Can. J. For. Res. 18, 974–980.

Pinon, J., Frey, P., Husson, C., 2006. Wettability of poplar leaves influences dew formation and infection by *Melampsora larici-populina*. Plant Dis. 90 (2), 177–184. https://doi.org/10.1094/PD-90-0177.

Piotrowska, M.J., Ennos, R.A., Riddell, C., Hoebe, P.N., 2017. Fungicide sensitivity of *Dothistroma septosporum* isolates in the UK. For. Pathol. 47, e12314. https://doi.org/10.1111/efp.12314.

Puustjärvi, V., 1969. Water-air relationships of peat in peat culture. Peat Plant Yearbook 4, 43–55.

Rajala, T., Velmala, S.M., Vesala, R., Smolander, A., Pennanen, T., 2014. The community of needle endophytes reflects the current physiological state of Norway spruce. Fungal Biol. 118, 309–315.

Redekar, N., Eberhart, J.L., Parke, J.L., 2019. Diversity of *Phytophthora*, *Pythium*, and *Phytopythium* species in recycled irrigation water in a container nursery. Phytobiomes J. 3, 31–45.

Rees, A.A., Webber, J.F., 1988. Pathogenicity of *Sphaeropsis sapinea* to seed, seedlings and saplings of some Central American pines. Trans. Br. Mycol. Soc. 91 (2), 273–277.

Ridout, M., Newcombe, G., 2018. *Sydowia polyspora* is both a foliar endophyte and a preemergent seed pathogen in *Pinus ponderosa*. Plant Dis. 102, 640–644.

Rodriguez, R.J., White Jr., J.F., Arnold, A.E., Redman, R.S., 2009. Fungal endophytes: diversity and functional roles. New Phytol. 182, 314–330.

Roll-Hansen, F., 1987. *Phacidium infestans* and *Ph. abies*. Hosts, especially *Abies* species in Norwegian nurseries. Eur. J. For. Pathol. 17, 311–315.

Rudawska, M., Kujawska, M., Leski, T., Janowski, D., Karlinski, L., Wilgan, R., 2019. Ectomycorrhizal community structure of the admixture tree species *Betula pendula*, *Carpinus betulus*, and *Tilia cordata* grown in bare-root forest nurseries. For. Ecol. Manage. 437, 113–125.

Russell, K., 1990. Gray mold. In: Hamm, P.B., Campbell, S.J., Hansen, E.M. (Eds.), Growing Healthy Seedlings. Identification and Management of Pests in Northwest Forest Nurseries. Forest Pest Management, U.S. Department of Agriculture, Forest Service, Pacific Northwest Region; and Forest Research Laboratory, College of Forestry, Oregon State University, pp. 10–13.

Schaeffer, R.N., Vannette, R.L., Brittain, C., Williams, N.M., Fukami, T., 2017. Non-target effects of fungicides on nectar-inhabiting fungi of almond flowers. Environ. Microbiol. Rep. 9 (2), 79–84. https://doi.org/10.1111/1758-2229.12501.

Schmilewski, G., 2008. The role of peat in assuring the quality of growing media. Mires Peat 3, 1–8.

Schneider, M., Grünig, C.R., Holdenrieder, O., Sieber, T.N., 2009. Cryptic speciation and community structure of *Herpotrichia juniperi*, the causal agent of brown felt blight of conifers. Mycol. Res. 113, 887–896.

Selosse, M.A., Bouchard, D., Martin, F., Le Tacon, F., 2000. Effect of *Laccaria bicolor* strains inoculated on Douglas-fir (*Pseudotsuga menziesii*) several years after nursery inoculation. Can. J. For. Res. 30, 360–371.

Selosse, M.-A., Schneider-Maunoury, L., Martos, F., 2018. Time to re-think fungal ecology? Fungal ecological niches are often prejudged. New Phytol. 217, 968–972.

Shishkoff, N., 2009. Propagule production by *Phytophthora ramorum* on lilac (*Syringa vulgaris*) leaf tissue left on the surface of potting mix in nursery pots. Plant Dis. 93, 475–480.

Simak, M., 1975. Kort nattbelysning av skogsplantor i plastväxthus Ger bättre odlingsmaterial. Summary: intermittent light treatments of forest plants in a plastic greenhouse produces better plant material. Sveriges Skogsvårdsförbunds Tidskrift 4, 373–381.

Stanosz, G.R., 1992. Effect of cherry leaf spot on nursery black cherry seedlings and potential benefits from control. Plant Dis. 76, 602–604.

Stanosz, G.R., Smith, D.R., Guthmiller, M.A., Stanosz, J.C., 1997. Persistence of *Sphaeropsis sapinea* on or in asymptomatic shoots of red and jack pines. Mycologia 89, 525–530.

Stanosz, G.R., Swart, W.J., Smith, D.R., 1999. RAPD marker and isozyme characterization of *Sphaeropsis sapinea* from diverse coniferous hosts and locations. Mycol. Res. 103, 1193–1202.

Stenström, E., Ihrmark, K., 2005. Identification of *Lophodermium seditiosum* and *L. pinastri* in Swedish forest nurseries using species-specific PCR primers from the ribosomal ITS region. For. Pathol. 35, 163–172.

Sutherland, J.R., 1984. Pest management in Northwest Bareroot nurseries. In: Duryea, M.L., Landis, T.D. (Eds.), Forest Nursery Manual: Production of Bareroot Seedlings. Martinus Nijhoff/Dr W. Junk Publishers, The Hague/Boston/Lancaster, for Forest Research Laboratory, Oregon State University, Corvallis, pp. 203–210.

Sutherland, J.R., 1987. *Sirococcus* blight. In: Sutherland, J.R., Miller, T., Rodolfo, S., Quinard, R.S. (Eds.), Cone and Seed Diseases of North American Conifers. Victoria NAFC Publ. 1, pp. 34–41.

Sutherland, J.R., Davis, C., 1991. Diseases and insects in forest nurseries in Canada. In: Sutherland, J.R., Glover, S.G. (Eds.), Proceedings of the First Meeting of IUFRO Working Party S2.07–09. Diseases and Insects in Forest Nurseries. Victoria, British Columbia, Canada. August 23–30, 1990. Forestry Canada, Pacific and Yokon Region, Pacific Forestry Centre. BC-X-331, pp. 25–32.

Sutherland, J.R., Lock, W., Farris, S.H., 1981. *Sirococcus* blight: a seed-borne disease of container-grown spruce seedlings in coastal British Columbia forest nurseries. Can. J. Bot. 59, 559–562.

Sutherland, J.R., Diekmann, M., Berjak, P., 2002. Forest tree seed health for germplasm conservation. In: IPGRI Technical Bulletin 6. Italy, Rome, ISBN: 92-9043-515-1, p. 85.

Tahvonen, R., 1982. Preliminary experiments into the use of *Streptomyces* spp. isolated from peat in the biological control of soil and seed-borne diseases in peat culture. J. Sci. Agric. Soc. Finl. 4, 357–369.

Terhonen, E., Sun, H., Buée, M., Kasanen, R., Paulin, L., Asiegbu, F.O., 2013. Effects of the use of biocontrol agent (*Phlebiopsis gigantea*) on fungal communities of *Picea abies* stumps. For. Ecol. Manage. 310, 428–433.

Terhonen, E., Sipari, N., Asiegbu, F.O., 2016. Inhibition of phytopathogens by fungal root endophytes of Norway spruce. Biol. Control 99, 53–63.

Terhonen, E., Blumenstein, K., Kovalchuk, A., Asiegbu, F.O., 2019. Forest tree microbiomes and associated fungal endophytes: functional roles and impact on forest health. Forests 10, 42.

Themann, K., Werres, S., Lüttmann, R., Diener, H.-A., 2002. Observations of *Phytophthora* spp. in water recirculation systems in commercial hardy ornamental nursery stock. Eur. J. Plant Pathol. 108, 337–343.

Thomas, G.W., Jackson, R.M., 1983. Growth-responses of Sitka spruce seedlings to mycorrhizal inoculation. New Phytol. 95, 223–229.

Tian Fu, W., Uotila, A., 2002. Observation of *Sirococcus conigenus* and it's pathogenicity. Research Papers 829, Finnish Forest Research Institute, pp. 30–34.

Tomlin, C.D.S. (Ed.), 2009. The Pesticide Manual. A World Compendium, fifteenth ed. BCPC Publications, Hampshire, ISBN: 978-1-901396-18-8.

Unestam, T., Beyer-Ericson, L., Strand, M., 1989. Involvement of *Cylindrocarpon destructans* in root death of *Pinus sylvestris* seedlings: pathogenic behaviour and predisposing factors. Scand. J. For. Res. 4, 521–536.

Unterseher, M., Siddique, A.B., Brachmann, A., Persoh, D., 2016. Diversity and composition of the leaf mycobiome of beech (*Fagus sylvatica*) are affected by local habitat conditions and leaf biochemistry. PLoS One 11, e0152878.

Vainio, E.J., Bezos, D., Bragança, H., Cleary, M., Fourie, G., Georgieva, M., Ghelardini, L., Hannunen, S., Ioos, R., Martín-García, J., Martínez-Álvarez, P., Mullett, M., Oszako, T., Papazova-Anakieva, I., Piškur, B., Romeralo, C., Sanz-Ros, A.V., Steenkamp, E.T., Tubby, K., Wingfield, M.J., Diez, J.J., 2019. Sampling and detection strategies for the pine pitch canker (PPC) disease pathogen *Fusarium circinatum* in Europe. Forests 10, 723. https://doi.org/10.3390/f10090723.

Vandenkoornhuyse, P., Quaiser, A., Duhamel, M., Le Van, A., Dufresne, A., 2015. The importance of the microbiome of the plant holobiont. New Phytol. 206, 1196–1206.

Vasiliauskas, R., Stenlid, J., Lygis, V., Thor, M., 2004. Impact of biological (rotstop) and chemical (urea) treatments on fungal community structure in freshly cut *Picea abies* stumps. Biol. Control 31, 405–413.

Vasiliauskas, R., Menkis, A., Finlay, R.D., Stenlid, J., 2007. Wood-decay fungi in fine living roots of conifer seedlings. New Phytol. 174, 441–446.

Veijalainen, A.-M., Lilja, A., Juntunen, M.-L., 2005. Survival of uninucleate *Rhizoctonia* species during composting of forest nursery waste. Scand. J. For. Res. 20, 206–212.

Velmala, S.M., Vuorinen, I., Uimari, A., Piri, T., Pennanen, T., 2018. Ectomycorrhizal fungi increase the vitality of Norway spruce seedlings under the pressure of *Heterobasidion* root rot *in vitro* but may increase susceptibility to foliar necrotrophs. Fungal Biol. 122, 101–109.

Venn, K., 1981. Winter vigour in *Picea abies* (L.) karst. VIII. Moldiness and injury to seedlings during overwinter cold storage. Meddelelser fra Norsk Institutt for Skogforskning 36, 1–28.

Venn, K., Sandvik, M., Langerud, B., 1986. Nursery routines, growth media and pathogens affect growth and root dieback in Norway spruce seedlings. Meddelelser fra Norsk Institutt for Skogforskning 39, 314–328.

Villeneuve, N., Letacon, F., Bouchard, D., 1991. Survival of inoculated *Laccaria bicolor* in competition with native ectomycorrhizal fungi and effects on the growth of outplanted Douglas-fir seedlings. Plant and Soil 135, 95–107.

Von Bazzigher, G., 1976. Der schwarze Schneeschimmel der Koniferen [*Herpotrichia juniperi* (Duby) Petrak und *Herpotrichia coulteri* (Peck) Bose]. Eur. J. For. Pathol. 6, 109–122.

Vuorinen, M., Kurkela, T., 1993. Concentration of CO_2 under snow cover and the winter activity of the snow blight fungus *Phacidium infestans*. Eur. J. For. Pathol. 23, 441–447.

Wall, R.E., Magasi, L.P., 1976. Environmental factors affecting *Sirococcus* shoot blight of black spruce. Can. J. For. Res. 6, 448–452.

Weete, J.D., 1973. Sterols of the fungi: distribution and biosynthesis. Phytochemistry 12, 1842–1864.

Wheeler, C.E., Omeja, P.A., Chapman, C.A., Glipin, M., Tumwesigye, C.T., Lewis, S.L., 2016. Carbon sequestration and biodiversity following 18 years of active tropical forest restoration. For. Ecol. Manage. 373, 44–55. https://doi.org/10.1016/j.foreco.2016.04.025.

Wolffhechel, H., 1988. The suppressiveness of sphagnum peat to *Pythium* spp. Acta Hortic. (221), 217–222.

Yakti, W., Kovacs, G.M., Franken, P., 2019. Differential interaction of the dark septate endophyte *Cadophora* sp. and fungal pathogens in vitro and in planta. FEMS Microbiol. Ecol. 95. fiz164.

Zadoks, J.C., Schein, R.D., 1979. Epidemiology and Plant Disease Management. Oxford University Press, New York. 427 p.

Zhang, P.G., Sutton, J.C., 1994. High temperature, darkness, and drought predispose black spruce seedlings to gray mold. Can. J. Bot. 72, 135–142.

Zhang, P.G., Sutton, J.C., Hopkin, A.A., 1995. Low light intensity predisposes black spruce seedlings to infection by *Botrytis cinerea*. Can. J. Plant Pathol. 17, 13–18.

Chapter 18

Microbiome of forest tree insects

Juliana A. Ugwu[a,b], Riikka Linnakoski[c], and Fred O. Asiegbu[b]

[a]*Forestry Research Institute of Nigeria, Ibadan, Nigeria,* [b]*Department of Forest Sciences, Faculty of Agriculture and Forestry, University of Helsinki, Helsinki, Finland,* [c]*Natural Resources Institute Finland (Luke), Helsinki, Finland*

Chapter Outline

1. Introduction

Insects are arthropods (Insecta; Arthropoda) that have three pairs of legs, three body segments (head, thorax, and abdomen), a pair of antennae, and pairs of wings (Fig. 18.1). Insects are evolutionary old, and they are estimated to have originated about 480 million years ago (Furniss and Carolin, 1977). Today, they are major global components of ecosystems. They are extremely rich in species diversity, constituting approximately 66% of all known animals or 82% of arthropods (Zhang, 2013), and form more than three-quarters of today's global biodiversity (Kim, 1993). Currently, approximately 1 million insect species are known to science (Zhang, 2013). Many insects remain unknown, as less than 20% of the species has been formally described thus far (Samways, 1993; Stork, 2018). Based on the estimates, there might be about 8 million insect species existing on Earth (Samways, 2005).

Insects are found in almost all ecosystems, but are underrepresented in marine ecosystems. They live in close association with other organisms, including microorganisms (bacteria, archaea, fungi, protists, viruses). Many of the associated microbiota have important impacts on insect ecology and evolution (Feldhaar, 2011; Gurung et al., 2019; Hammer et al., 2017; Kaufman et al., 2000). The majority of insect species that live on/in forest trees are key components in maintaining vital ecosystem functions such as facilitating pollination and as detritivores. Considering their importance to forestry, some insects can be at least occasionally damaging to tree health and forestry products such as timber (Kim, 1993). They can also be moved in global trade inside wooden packaging material and with living plants.

Forest Microbiology. https://doi.org/10.1016/B978-0-12-822542-4.00018-8

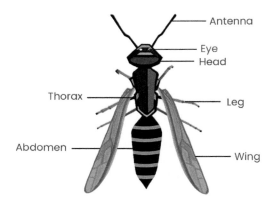

FIG. 18.1 Artistic illustration of insect body parts.

2. Insects as vital components of forest ecosystems

Insects are vital components of many ecosystems where they have a significant ability to alter their environment (Schowalter, 2016). They are major contributors to biodiversity, with ecological functions in nutrient cycling, maintaining community composition and structure. Insects exhibit wide disparity among species in their biology and are quite abundant in all terrestrial ecosystems (Gullan and Cranston, 2010). From a human point of view, insects are largely considered as pests or potential pests and their ecological value is commonly neglected. The net primary productivity of a forest or a tree stand can be increased due to the activity of phytophagous insects (Mattson and Addy, 1975). Insects additionally represent diverse trophic niches and play several different ecological roles, as outlined below.

3. Ecological roles of insects in forest ecosystems

The primary ecological role of insects in forest ecosystems is maintaining forest ecosystem balance. Insects represent the dominant type of animal biomass and life on Earth with diverse roles in their natural ecosystem such as herbivores, carnivores, and detritivores.

The ecological roles of insects in forest ecosystems include

a. ecosystem cycling and herbivory
b. pollination
c. predation
d. decomposition
e. seed dispersal
f. insects as a key link in the food chain in the forest

3.1 Ecosystem cycling and herbivory

Insect herbivores are important agents of ecosystem processes through the process of transforming plant biomass into frass and shading of excess water from wet leaves (through fall) (Metcalfe et al., 2014). Through their feeding activities, they alter the timing, quantity, and quality of plant detrital inputs and can hypothetically have large impacts on ecosystem cycling (Belovsky and Slade, 2000; Mattson and Addy, 1975). Grasshopper herbivory was found to increase plant abundance because of greater nitrogen (N) availability (Belovsky and Slade, 2000). The grasshopper herbivory activities significantly increased the amount of below ground to above ground phosphorus (P) and N fluxes of the entire ecosystems (Metcalfe et al., 2014). Insects have a large effect on ecosystem functions during outbreaks of species such as the gypsy moth, *Lymantria dispar* (Lepidoptera; Lymantriidae) or *Oporinia autumnata* (Lepidoptera; Geometridae). The proportion of leaf area removed by the extremely high population densities of herbivorous insect species can reach 100% during outbreaks, which has immediate and large effects on nutrient fluxes (Christenson et al., 2002; Kosola et al., 2001). According to Schowalter (2000), herbivorous insects have an indirect regulatory control on ecosystem processes. Insects affect nutrient cycling both directly and indirectly. Direct effects include a reduction of net primary productivity (NPP) by herbivores and the breakdown of litter by detritivores (Weisser and Siemann, 2008). The indirect effects include matter fluxes due to changes in plant species composition mediated by insect herbivory. Interactions of insects with other organisms also impact matter flux in ecosystems (Weisser and Siemann, 2008).

3.2   Pollination

Insects are the major agents of pollination in many forest ecosystems. In temperate regions, the majority of wild flowering plants and other plant species depend on insects for pollination (Jankielsohn, 2018). Insects pollinate 80% of all the plants worldwide. The most important insect pollinators are flies, beetles, butterflies, and bees (Schoonhoven et al., 2005). In tropical rain forests, the vast majority of plant species are pollinated by insects. The major insect pollinators in tropical rain forests are bees. Bees constitute the most important insect group in number and diversity of plant species they pollinate. In neotropical lowland rain forests, a large number of plant species in many families such as Burseraceae, Euphorbiaceae, Clusiaceae, Fabaceae, Flacourtiaceae, Lecythidaceae, Melastomataceae, Orchidaceae, and Sapotaceae are pollinated by bees. The bee pollination system is mainly predominant in canopy trees (Bawa, 1990). Other insect pollinators include beetles, flies, ants, moths, butterflies, and wasps. Butterflies and moths (Lepidopterans) are essential pollinators of flowering plants in forest ecosystems and nonforest ecosystems such as parks and yards (Ostiguy, 2011). Butterflies and moths have different niches; butterflies are active during the day while moths are active in the evening and at night (Ostiguy, 2011). Insects thus contribute greatly to plant diversity and affect animal biodiversity indirectly through pollination (Jankielsohn, 2018).

3.3   Predation/parasitism

As secondary or tertiary consumers (parasitic, predators), insects occupy the higher trophic levels, which enables them to control the population increase of phytophagous organisms or primary consumers. Insect predators and parasitoids play a role in the natural control of those insects that have the potential of becoming pests (e.g., herbivorous insects) (van Lenteren, 2012). Predatory insects are dragonflies, lacewings, bugs, tiger beetles, ladybird beetles, mantids, robber flies, ants, bees, and wasps. These insects are predators, either in the adult stage or larvae or at both stages. Some groups of insects are parasitoids. They parasitize the eggs, larvae, or adults of other insects. Examples of this include *Aphytis lingnanensis* (red scale parasite) that parasitizes scale insects; different species of parasitoid wasps such as *Aphidius ervi, Aphelinus asychis, Aphidius matricariae, Aphelinus varipes, Aphidius colemani*, and *Diaeretiella rapae* that parasitize cereal aphids; and *Trichogramma* parasitic wasps that attack Lepidopteran eggs.

3.4   Decomposition: Organic waste

Organic waste decomposition is a vital ecosystem process that is principally done by microorganisms and some insect species. Many dung beetle species have been reported to play an important role in the decomposition of manure. Dung beetles live in many habitats, including planted forests. Many dung beetles feed on decaying fruits, leaves, and mushrooms while also eating the dung of omnivores and herbivores. Furthermore, the larvae of beetles, termites, ants, and flies clean up dead plant matter by breaking it down through feeding by gut microbes. Animal tissues provide food to diverse groups of insect detritivores such as flies and beetles (Merritt and De Jong, 2015). Many of these insects contribute to soil health by improving the essential nutrients, micronutrients or total protein content in the soil (Macfadyen et al., 2015).

3.5   Seed dispersal

Seed dispersal is an important factor for biodiversity conservation. Insect species play a significant role in spreading the seeds and fruits of different plants. Dung beetles and ants are important insects that are involved in seed dispersal; they usually redistribute seeds from the dung of larger primary dispersers (Hardesty, 2011; Martinez-Mota et al., 2006). Ants disperse small seeds, particularly of plant species adapted for dispersal by ants (Leal et al., 2015). The process of seed dispersal by ants is known as "**myrmecochory**," and it occurs mostly in angiosperm, particularly in temperate forest herbs. About 11,000 angiosperm species (4.5%) from 334 genera and 77 families are dispersed by ants (Lengyel et al., 2009).

3.6   Insects as a key link in the food chain in the forest

Insects constitute a very important link in the food chain in forest ecosystems. Many insects are sources of food to other insects as well as other wild species of animals such as amphibians, reptiles, birds, and mammals. In many countries, insects are part of the human diet. Apart from honey from honeybees, which is a valued food item worldwide, many types of insects such as locusts, grasshoppers, termites, and lepidopteran larvae and pupae form part of the human diet for many rural communities in many countries. In some parts of Africa (e.g., Nigeria), the *Anaphe venata* larvae, a pest of a tree species—*Triplochiton scleroxylon*, are roasted and consumed by local tribes (Ashiru, 1988). In Uganda, the grasshopper

Homorocoryphus nitidulus, which periodically swarms in large numbers, is eaten either raw or cooked (Hill, 1997). In Asia (e.g., Indonesia), the pupae of the *Tectona grandis* (teak) defoliator *Hyblaea puera* are consumed. Grasshoppers that are roasted are often available at roadside food stalls in Thailand (Nair, 2007). The larvae of the palm weevil *Rhynchophorus* species such as *R. ferrugineus*, *R. phoenicis*, and *R. palmarum* are consumed in Asia, Africa, and Latin America, respectively (Huis, 2013). Currently, a recent trend is the commercial production of insects as sources of protein worldwide. The most consumed insect groups are beetles, caterpillars, the larvae of Lepidoptera (moths or butterflies), Hymenoptera (bees, wasps, and ants), and Orthoptera (grasshoppers, crickets, and locusts), Hemiptera (cicadas, leafhoppers, plant hoppers, and true bugs). Termites, dragonflies, flies, and other insects each comprise less than 3% of the groups that are consumed (Jongema, 2017). Presently, most edible insects are harvested in the wild from the forest. The domestication, rearing, and/ or farming of insects for direct human consumption began a few years ago (Huis, 2013). Insect rearing as a food source is a common practice in Southeast Asia as well as Central and Southern Africa (Durst and Hanboonsong, 2015; Gahukar, 2016; Kelemu et al., 2015).

Insects are also an important source of nutrition for some plants. A typical example is the pitcher plant (*Nepenthes* spp.), which lives in soil with poor nutrients. This plant uses trapped insects as a dietary supplement. The modified leaves of this plant are able to hold a liquid in a special cavity. Insects such as flies are attracted by the plant color or odor. The insects are unable to escape when they fall into the plant cavity. They are sunk in the liquid, which contains digestive enzymes secreted by the plant. The digested nutrients from the insects are then absorbed by the plant (Nair, 2007).

3.7 The detrimental roles of insects in the forest

Insects can influence forest tree health negatively in several ways. Some beetle species such as the European spruce bark beetle (*Ips typographus*), the major pest in coniferous forests in Europe, colonize mainly stressed and weakened trees when beetle populations are low. Another example is the sharp-dentated bark beetle (*Ips acuminatus*), severity in European Scots pine (*Pinus sylvestris*) stands has increased in Europe during the past decade (Columbari et al., 2013; Siitonen, 2014).

They can mass attack large numbers of healthy trees once their populations are high (Hlásny et al., 2019). Through their feeding activities, insects degrade wood and timber products and cause deformities by boring into tree stems during feeding. Insect attacks on trees predisposes the plants to secondary infestations through the creation of entry points for pathogens or by weakening tree defenses. Furthermore, insects can be vectors of diseases that affect forest trees, including some of the most devastating global forest health problems such as Dutch elm disease fungal pathogens vectored by elm bark beetles (Ploetz et al., 2013). Another contemporary example is the polyphagous shot hole borer (PSHB), an ambrosia beetle that transmits the *Fusarium euwallaceae* fungal disease to many tree species and genera (Paap et al., 2018).

4. Classification of forest insects

More than 1 million species of insects have so far been described and their taxonomical classification can be complex (Stork, 2018). It is, however, important to correctly identify insect species so they can be reliably studied. The presented classification is restricted to the insect order level (Table 18.1)

5. Microbiome of diverse forest insect orders

Insects can live in close association with diverse microorganisms (Fig. 18.2). Microbes such as viruses, bacteria, archaea, fungi, and protists may be permanently or transitorily associated with their host insects and such a relationship may be neutral, mutual, or harmful to the insect (Feldhaar, 2011; Hammer et al., 2017; Kaufman et al., 2000). Most endosymbiont microbes tend to depend on their insect hosts for the nutrients necessary for their growth and development. Such symbiotic associations contribute to weakening or strengthening defenses and consequent protection from the pathogens, parasites, and environmental stresses of their hosts (Mereghetti et al., 2017). Conversely, microbes might also be pathogenic, causing morbidity and reducing viability. Moreover, physiological costs can also be incurred through inhabiting endosymbionts (Krams et al., 2017).

Insect-associated bacteria may influence the elicitation of defensive reactions in the host plant (Frago et al., 2012; Sugio et al., 2015; Zhu et al., 2014). Many fungal species associated with insect pests comprise species of filamentous fungi belonging to the genera *Pandora*, *Aspergillus*, *Isaria*, *Beauveria*, *Cordyceps*, and *Metarhizium* (Shang et al., 2015). Entomophthoromycota is one of the most important groups of all entomopathogens that are mainly pathogens of insects. They frequently occur as epizootic, killing many insects in small coverings of forest or agricultural systems (Araújo and Hughes, 2016). A few genera in Basidiomycota are known to be entomopathogenic, including *Septobasidium*, *Uredinella*, and *Fibularhizoctonia*.

TABLE 18.1 Insect orders and their general characteristics.

Sub class	Order	Examples	General characteristics	References
Apterygota (wingless insects)	Microcoryphia (Archaeognatha)	Bristletails	Include many members that live in grassy or wooded habitats where they are most likely to be found in leaf litter, under bark, among stones, or near the upper tidal line in coastal areas. They are mostly active at night, feeding as herbivores or scavengers on algae, mosses, lichens, or decaying organic matter. Bristletails are common inhabitants of forest leaf litter. They are part of the community of decomposers that breaks down and recycles organic nutrients	Barnard (2011)
	Thysanura	Silverfish/firebrats	Members are fast-running insects that hide under stones or leaves during the day and emerge after dark to search for food. A few species are resistant to desiccation and well-adapted to survive in domestic environments such as basements and attics. Silverfish are scavengers or browsers; they survive on a wide range of food, but seem to prefer a diet of algae, lichens, or starchy vegetable matter	Barnard (2011)
Pterygota			Members of this insect group have wings. They undergo distinct stages of development before they become adults: complete metamorphosis and incomplete metamorphosis	Resh and Carde (2003)
Endopterygot (complete metamorphosis)	Coleoptera	Ambrosia beetles, beetles, ladybugs fireflies, stag beetles weevils, burying beetles rove beetles, click beetles, rose chafers, ground beetles, leaf beetles, sap beetles, pleasing fungus beetles, soldier beetles, bark beetles	Insects in this order are vital in decomposition, nutrient cycling, pollination, seed dispersal, and the biological control (predators) of animal pests. Many members (e.g., beetles) are indicators of soil properties, temperature, and humidity variation of the environments and forest disturbances. Beetles are found in nearly all climates and latitudes, except in extreme environments such as in Antarctica and at the highest altitudes	Lawrence and Britton (1991) Costa (2000) and Davis et al. (2001) Davis et al. (2001), Dunxiao et al. (1999), Stork and Eggleton (1992), and Gressitt (2013)
	Lepidoptera	Butterflies and moths: armyworms, corn earworm, cutworms, spruce budworm, western spruce budworm, western hemlock looper, Douglas-fir tussock moth. In the tropical forest, the lepidopteran insects causing severe defoliation include Teak defoliators, three spot yellow butterflies, common grass yellow butterfly attack *Acacia*, *Paraserianthes*, pine caterpillar. Some members of this order are shoot and cone borers causing severe damage to forest trees, such as mahogany moths	Lepidoptera is estimated to have 160,000 named species and thus is one of the two or three largest insect orders. Most of them are highly destructive agricultural pests globally. They are scaly winged insects and are well identified for their beauty; they have attractive wings of different colors. They hold their wings vertically when they at rest as they bear beautiful wings of various colors. Over time, their wings fade, their scales wilt, and their color vanishes. Lepidopterans are found in every environment and serve as a bioindicator for health and beauty. They are useful as well as harmful, and have prodigious esthetic and commercial values. Their larvae are very destructive to groups of plant-eating organisms, which is very economically important in agriculture, horticulture, and forestry. Many insects of this order are forest defoliators	Powell (2009) and Sree and Varma (2015) Perveen and Khan (2014) Perveen and Fazal (2013) Khan and Perveen (2015) Speight (2016)
	Hymenoptera	Bees, wasps, and ants	Hymenopterans is among the five largest diverse insect orders and likely the richest species of insect order, although this is not true at temperate latitudes. Hymenoptera typically have two pairs of wings, a large fore pair and a smaller hind pair. The most noticeable Hymenoptera are the fast-flying, often black and yellow ones such as bees and wasps. The members of this insect order are very important in pollination, predation, and seed dispersal processes in forest ecosystems	Quicke (2009)

Continued

TABLE 18.1 Insect orders and their general characteristics.—cont'd

Sub class	Order	Examples	General characteristics	References
	Diptera		Commonly called true flies or two-winged flies, they include the tsetse fly, mosquitoes, fruit flies, house flies, gall midges, blow flies, flesh flies, etc. The Dipterans are among the most diverse insect orders with about 150,000 species described. Flies are important detritivores in forest ecosystems and play an essential role in material circulation and energy flow. They also play various roles such as herbivory, pollination, and biological control (natural enemies) and are important food sources for birds, mammals, amphibians, fish, and reptiles. Dipteran members of the families Mycetophilidae, Sciaridae, and Cecidomyiidae are rich in species and predominant prevail in the decay of logs. The family Tephritidae, the true fruit flies, comprise more than 4000 species that are among the most economically important insect pests attacking soft fruits globally	Cranston and Penny (2009) and Hövemeyer and Schauermann (2003) Lee et al. (2015) Økland (1994) and Irmler et al. (1996) White and Elson-Harris (1992)
	Megaloptera	Alderflies, dobsonflies, and fish flies	They are a small order of neuropterans insects with 328 pronounced species. The larvae are aquatic predators and they are mostly found in temperate regions rather than tropical. Adult alderflies are usually found amid the vegetation lining of aquatic habitats. They are sometimes attracted to lights at night. The adult alderflies are sources of food to a variety of insectivorous animals such as fly-catching birds and spiders. Their larvae eat smaller invertebrates or organic detritus and are eaten by larger aquatic organisms, such as crayfish and fish Dobson flies live under stones or submerged vegetation and feed on a variety of small aquatic organisms. They are found in North and South America, Asia, Australia, and Africa	Norman (2009)
Exopterygota			This group comprises about 15 insect orders with about three major orders of forestry importance	Cardé (2009)
	Orthoptera	Grasshoppers, crickets, and locusts	The order comprises two suborders: Caelifera (grasshoppers, locusts, and mole crickets) and Ensifera (true crickets, katydids, and bush crickets). Orthopteran insects are widely distributed globally and can generally be recognized by their ability to produce sound and are known for their jumping skills. Orthopterans are usually described as herbivores, but many species are scavengers in addition to herbivores (feeding on plants). Major Families in the Order include; Gryllidae—true or field crickets, Acrididae—short-horned grasshoppers, Tetrigidae—grouse locusts or pygmy grasshoppers, Gryllotalpidae—mole crickets and Tettigoniidae—long-horned grasshoppers and katydids. Members of this order play vital roles in nutrient cycling in forest ecosystems	Debbie (2018) Parsons (2010)
	Isoptera	Termites	They have bead-like antennae and strong biting mouthparts with which they chew seeds, wood, or leaves. They are social insects that usually live in colonies and a colony consist of three castes: workers, soldiers, and swarmers. Workers and soldiers are wingless and never leave the colony. There are around 3000 species in seven families. Termites are among the most important of soil fauna in relation to their impact on soil structure and on decomposition processes. They are a key group of insects for the dynamics of tropical forests where they are primary decomposers and drivers of nutrient cycling. The lower termite species of the genera Zootermopsis, Heterotermes, Schedorhinotermes, Reticulitermes, Coptotermes, and members of the Mastotermitidae and Kalotermitidae feed primarily on wood of living trees. Members of the family Kalotermitidae have a gut flora of protozoans that enables them to digest cellulose	Ackerman et al. (2009), Bandeira and Vasconcellos (2004), Coleman et al. (2004), and Lee and Wood (1971)
	Hemiptera	Truebugs, which include shield bugs, plant bugs, bed bugs, pond skaters, cicadas, water bugs, aphids, and scale insects	Hemiptera is ranked as the fifth-largest order of insects, being by far the largest order of Hemimetabola. Hemiptera consist of approximately 100,000 species. Members include Fulgoromorpha, Cicadomorpha, Coleorrhyncha, Sternorrhyncha (formerly included in the order Homoptera), and Heteroptera. Some families of this order are Reduviidae (assassin bugs), Tingidae (sugarcane lace bug, avocado lace bug, lace bugs), Lygaeidae (seed bugs, big eyed bugs, milkweed bugs), Coreidae (leaf-footed bugs, squash bugs), Pentatomidae (stink bugs), and Aphids (Aphididae). Many insects in this group, especially aphid species, are important pests in agriculture and forestry, causing direct damage through feeding on plants and indirect damage as a vector of several plant pathogens	Adler and Foottit (2009) Beutel et al. (2014), Blackman and Eastop (2000), van Emden and Harrington (2007), and Liu et al. (2015)

FIG. 18.2 Illustration of diverse microbiota associated with forest insects.

Septobasidium and *Uredinella* attack scale insects while *Fibularhizoctonia* attacks the eggs of the termite genus *Reticulitermes* (Araújo and Hughes, 2016). *Uredinella* attacks single insects among the scale insects and *Septobasidium* attacks whole colonies of plant-feeding insects, with as many as 250 insects infected by one fungus (Couch, 1938).

Some fungal species within the group *Ascosphaera* are selective saprotrophs on honey, cocoons, larval feces, or nest materials such as leaves, mud, or bee wax (Wynns et al., 2012). Some species are fungal disease agents infecting and causing "chalk-brood" diseases in numerous species of solitary and social bees (Klinger et al., 2013). Hypocrella (*Archersonia* species) are known to infect whiteflies and scale insects in tropical forests (Chaverri et al., 2008). They are responsible for the majority of infections with about 92 species infecting scale insects [Coccidae, Lecaniidae, and whiteflies (Aleyrodidae)] (Araújo and Hughes, 2016; Table 18.2). Scale insects are usually infected by the podnecteria species, which covers their body surface with cotton layers that produce perithecia and spores with many septa that do not separate into spores (Kobayasi and Shimizu, 1977). The larvae of lepidopteran and coleopteran larvae have been found to be associated with Cordyceps and Ophiocordyceps (entomopathogenic fungi) (Araújo and Hughes, 2016). Also, several yeasts belonging to the genera *Cryptococcus*, *Saccharomyces*, *Metschnikowia*, *Pichia*, *Debaryomyces*, *Kluyveromyces*, *Hanseniaspora*, and *Candida* have equally been reported (Nguyen et al., 2007; Piper et al., 2017; Stefanini, 2018; Suh et al., 2008; Table 18.2). Some yeasts are known to benefit from their association with insects, as this facilitates their dispersal to diverse environments. Insect guts may also provide a conducive environment for yeasts to reside for prolonged periods.

Insects are also vectoring many viruses. Some viruses can be pathogenic to their insect hosts; thus, some viruses are potential biological control agents of insect pests (Lacey et al., 2001; Winstanley and Rovesti, 1993). Viruses in the taxa Rhabdoviridae, Baculoviridae, Bunyavirales, Parvoviridae, Togaviridae, Flaviviridae, and Ascoviridae have been commonly known to be associated with insects (Asgari and Johnson, 2010).

5.1 Microbiome of Hemiptera (bugs)

Members of Hemiptera are insects that ingest plant or animal fluids with their sucking mouthparts and are known to exhibit several associations with gut microorganisms (Kuechler et al., 2012; Table 18.2). Numerous heteropteran insects that feed on plant fluids have mid-guts with caeca that are filled with a large population of symbiotic bacteria (Baumann, 2005). The family Pentatomidae in this order that attacks forests or wild host plants include the stink bug—*Halyomorpha halys*, which attacks heaven, catalpa, yellowwood, paulownia, cherry, walnut, redbud, and grape trees (Bakken et al., 2015); and the forest bug (red-legged shieldbug, *Pentatoma rufipes*), which feeds on deciduous tree species such as alder (*Alnus* spp.), birch (*Betula* spp.), hornbeams (*Carpinus* spp.), hazel (*Corylus* spp.), beech (*Fagus* spp.), and oak (*Quercus* spp.) (Wachmann et al., 2008). Adelgids of the family Adelgidae (*Adelges nordmannianae/piceae*) are severe pests of conifers in the northern hemisphere.

TABLE 18.2 The microbiome of different insect orders and types of association they exhibit.

Microbiome (bacteria)	Insect order	Insect genera/species/stage	Association	References
Buchnera aphidicola	Hemiptera	Aphids	Primary endosymbiotic	Satar (2019)
Serratia symbiotica, Hamiltonella defense, Regiella insecticola, Rickettsia, Wolbachia, Spiroplasma, Arsenophonus	Hemiptera	Aphids	Secondary symbionts	Oliver et al. (2010) and Tsuchida et al. (2002)
Sitobion miscanth	Hemiptera	Aphids	L-type symbiont	Li et al. (2016)
Staphylococcus, Pseudomonas Acinetobacter, Pantoea	Hemiptera	Pea aphid, *Acyrthosiphon pisum*	–	Leroy et al. (2011) and Stavrinides et al. (2009, 2010)
Carsonella ruddii	Hemiptera	Psyllids	Primary endosymbiotic	Thao et al. (2000)
Candidatus Sulcia muelleri, Candidatus Baumannia cicadellinicola, Candidatus Zinderia insecticola, Candidatus Nasuia, Hodgkinia cicadicola	Hemiptera	Leafhoppers	Obligate endosymbionts	Koga et al. (2013)
Wolbachia, Cardinium	Hemiptera	Leaf hoppers	Facultative endosymbionts	Zhang et al. (2017)
Portiera aleyrodidarum	Hemiptera	All whitefly species	Primary symbiont	Zchori-Fein (2009)
Hamiltonella, Rickettsi, Wolbachia, Cardinium, Arsenophonus, Fritschea	Hemiptera	*Bemisia tabaci*	Primary symbiont	Zchori-Fein (2009)
Gammaproteobacteria	Hemiptera	Stinkbugs (Pentatomoidea)	Primary symbiont	Kikuchi et al. (2008)
Ishikawaella capsulatus (Proteobacterium)	Hemiptera	Plataspid bug (Plataspidae)	Primary symbiont	Fukatsu and Hosokawa (2002) and Hosokawa et al. (2006)
Burkholderia sp. (Proteobacterium)	Hemiptera	Alydid bug *Riptortus clavatus*	Primary symbiont	Kikuchi et al. (2007, 2012)
Burkholderia	Hemiptera	Bean bug *Riptortus pedestris*	Primary symbiont	(Kikuchi et al. 2005, 2011)
Rhodococcus rhodnii	Hemiptera	Kissing bug *Rhodnius prolixus*	Primary symbiont	Beard et al. (2002), Eichler and Schaub (2002), and Engel and Moran (2013)
Pseudomonas Bradyrhizobium	Lepidoptera	Eastern spruce budworm *Choristoneura fumiferana*	Primary symbiont	van Frankenhuyzen et al. (2010) and Landry et al. (2015)
Pantoea and Citrobacter (Proteobacteria)	Lepidoptera	*Spodoptera littoralis* (cotton leafworm)/early instar stage	Primary symbiont	Shao et al. (2014)
Enterococcus and Clostridium spp.	Lepidoptera	*Spodoptera littoralis* (cotton leafworm)/late instar stage	Primary symbiont	Shao et al. (2014) and Tang et al. (2012)
Enterococcus	Lepidoptera	Gypsy moth (*Lymantria dispar*)/larva midgut Cotton bollworm (*Helicoverpa armigera*)/larva Tobacco hornworm (*Manduca sexta*)/egg	Primary symbiont	Broderick et al. (2004), Brinkmann et al. (2008), and Priya et al. (2012)

Bacteria	Order	Host (insect)/stage	Symbiont type	Reference
Proteobacteria, Firmicutes, Cyanobacteria, Bacteroidetes, Actinobacteria, Nitrospirae	Lepidoptera	Diamond back moth, *Plutella xylostella*—different stages	Primary symbiont	Xia et al. (2017)
Acetobacteraceae, Moraxellaceae, Enterobacteriaceae, Enterococcaceae, Streptococcaceae	Lepidoptera	Red postman (*Heliconius erato*)/larval, pupal, and adult stages	Primary symbiont	Hammer et al. (2014)
Paenibacillus sp., Bacillus safensis, Pseudomonas sp., Bacillus pseudomycoides, Corynebacterium variabile, Enterococcus sp., Gordonia sp., Acinetobacter calcoaceticus, Arthrobacter sp., Micrococcus sp., Bacillus cereus	Lepidoptera	Agave red worm (*Comadia redtenbacheri*)/larvae	Primary symbiont	Hernández-Flores et al. (2015)
Klebsiella, Stenotrophomonas, Microbacterium, Bacillus, Enterococcus	Lepidoptera	Sugarcane stalks borer (*Diatraea saccharalis*)/larvae	Primary symbiont	Dantur et al. (2015)
Pseudomonas aeruginosa, Brevundimonas aurantiaca, Chryseobacterium formosense, Acinetobacter sp., Microbacterium thalassium, Bacillus megaterium, Serratia sp., Ochrobactrum sp., Variovorax paradoxus, Corynebacterium glutamicum, Paenibacillus sp., Alcaligenes faecalis, Microbacterium testaceum, Leucobacter sp., Leucobacter sp., Serratia marcescens	Lepidoptera	European corn borer (*Ostrinia nubilalis*)/larvae	Primary symbiont	Secil et al. (2012)
Bacillus, Enterococcus, Klebsiella, Brevibacillus spp.	Lepidoptera	African maize stem bore (*Busseola fusca*)/larvae	Primary symbiont	Snyman et al. (2016)
Staphylococcus aureus, Klebsiella cloacae, Streptococcus faecalis, Bacillus thuringiensis, Pseudomonas chlororaphis, Klebsiella granulomatis	Lepidoptera	*Bombyx mori* (silkworm)/larvae	Primary symbiont	Mohanta et al. (2014) and Tao et al. (2011)
Staphylococcus pasteuri, Bacillus pumilus, Achromobacter marplatensis, Paucisalibacillus globulus, Staphylococcus warneri, Bacillus sp., Bacillus subtilis	Lepidoptera	Muga silkworm (*Antheraea assamensis*)/larvae 1st to 5th instar stage	Primary symbiont	Haloi et al. (2016)
Paucisalibacillus aeruginosa, Ornithinibacillus bavariensis, Achromobacter xylosoxidans, Staphylococcus aureus, Bacillus thuringiensis	Lepidoptera	Muga silkworm (*Antheraea assamensis*)/silkworm from 3rd instar to 5th instar (diseased)	Primary symbiont	Haloi et al. (2016)
Propionibacterium acnes	Lepidoptera	Noctuid moth (*Helicoverpa armigera*)/larvae	Primary symbiont	Ranjith et al. (2016)
Klebsiella, Citrobacter freundii, Enterobacter spp., Pantoea spp., Pectobacterium spp., Providencias tuartii, Pseudomonas spp.	Diptera	Mediterranean fruit fly, *Ceratitis capitata*/adult gut	Primary symbiont	Behar et al. (2008a,b)
Actinobacteria, Bacteroidetes, Cyanobacteria, Firmicutes, Proteobacteria, Tenericutes, Planctomycetes	Diptera	Fruit flies (*Bactrocera carambolae* and *Bactrocera dorsalis*)/adult gut	Primary symbiont	Yong et al. (2017)
Fusobacteria	Diptera	*B. carambolae*	Primary symbiont	Yong et al. (2017)
Enterobacteriaceae, Acetobacteraceae, Streptococcaceae, Enterococcaceae	Diptera	*B. neohumeralis, B. carambolae, B. jarvisi*, and *C. capitata*	Primary symbiont	Morrow et al. (2015) Malacrinò et al. (2018), and Yong et al. (2017)
Enterobacteriaceae, Acetobacteraceae, Streptococcaceae, Enterococcaceae	Diptera	Queensland fruit fly (*Bactrocera tryoni*)	Primary symbiont	Malacrinò et al. (2018) and Yong et al. (2017)

Continued

TABLE 18.2 The microbiome of different insect orders and types of association they exhibit.—cont'd

Microbiome (bacteria)	Insect order	Insect genera/species/stage	Association	References
Fungi				
Podonectria species	Hemiptera	Scale insects	Pathogenic	Kobayasi and Shimizu (1977)
Hypocrella (Archersonia sp.)	Hemiptera	Scale insects	Pathogenic	Chaverri et al. (2008)
Septobasidium, Uredinella, Fibularhizoctonia	Hemiptera	Scale insects	Pathogenic	Araújo and Hughes (2016)
Hypocrella (Archersonia spp.)	Hemiptera	White flies	Pathogenic	Chaverri et al. (2008)
Sporodiniella umbellata	Hemiptera	Tree hoppers (membracids)	Pathogenic	Evans and Samson (1977)
Metharizium anisopliae	Hemiptera	Glasshouse whitefly (Trialeurodes vaporariorum)	Pathogenic	Jankevica (2004)
Massospora cicadina	Hemiptera	Periodical cicadas (Magicicada spp.) nymph and adult	Pathogenic	Cooley et al. (2018)
Neozygites fresenii	Hemiptera	Cotton Aphids (Aphis gossypii) nymph and adult	Pathogenic	Steinkraus et al. (1995)
Cordyceps, Ophiocordyceps	Lepidoptera	Ghostmoth (Thitarodes spp.)	Pathogenic	Araújo and Hughes (2016)
Beauveria sp.	Lepidoptera	Turnipmoth (Agrotissegetum)	Pathogenic	Jankevica (2004)
Beauveria sp., Zoophthoraradicans	Lepidoptera	Cabbagebutterfly (Pieris brassicae)	Pathogenic	Jankevica (2004)
Beauveria brongniartii	Lepidoptera	Lackeymoth (Malacosoma neustria)	Pathogenic	Jankevica (2004)
Beauveria bassiana	Lepidoptera	Millmoth (Ephestia kuehniella)/larvae	Pathogenic	Alali et al. (2019)
Furia virescens	Lepidoptera	True armyworm (Mythimna unipuncta)/larvae	Pathogenic	Steinkraus et al. (1993)
Entomophaga maimaiga	Lepidoptera	Gypsy moth (Lymantria dispar)/larvae	Pathogenic	Hajek and Soper (1991)
Ascomycota (Saccharomycetaceae)	Diptera	Drosophila species/adults		Hamby et al. (2012) and Morais et al. (1995)
Hanseniaspora uvarum	Diptera	Drosophila species/adults, Drosophila suzukii/adults		Chandler et al. (2012), Hamby et al. (2012), and Phaff et al. (1956)
Candida inconspicua, Alcaligenes faecalis, Aspergillus flavus, A. fumigattus, A. niger, Fusarium equiseti/oxysporum, Geotrichum candidum	Diptera	Drosophila melanogaster/adults	Obligate symbiont	Ramirez-Camejo et al. (2017)
Entomophthora muscae	Diptera	Cabbage fly (Delia brassicae)	Pathogenic	(Jankevica, 2004)
Viruses				
Potyvirus	Hemiptera	Aphids	Parasitic	Whitfield et al. (2015)
Begomovirus (Geminiviridae), Crinivirus, Closterovirus, Ipomovirus, Carlavirus (Betaflexiviridae)	Hemiptera	White fly	Parasitic	Jones (2003)
Baculoviridae	Hemiptera	Sawflies	Pathogenic	Williams et al. (2017)
Baculovirus-nucleopolyhedrovirus (MNPV)	Lepidoptera	Cabbage looper (Trichoplusia ni) larvae	Pathogenic	Grasela et al. (2008)
Nuclear polyhedral viruses (NPVS)	Lepidoptera	Corn earworm Heliothis armigera	Pathogenic	Dhandapani et al. (1993)
Nuclear polyhedral viruses (NPVS)	Lepidoptera	Fall armyworm Spodoptera frugiperda	Pathogenic	Fuxa et al. (1992)

The gut symbionts of the superfamily Pentatomoidea consist of Gammaproteobacteria. They are transmitted by host insects vertically through the symbiont association of posthatch transmission mechanisms, egg surface contamination, coprophagy, or symbiont capsule provisioning (Kikuchi et al., 2008). The Adelgids *Adelges nordmannianae/piceae* transmit gammaproteobacterial *Candidatus Steffania adelgidicola* and *Candidatus Ecksteinia adelgidicola* vertically from the mother to the offspring (Toenshoff et al., 2012). The firebug *Pyrrhocoris apterus* (Pyrrhocoridae) is a pest of the linden tree seed *Tilia cordata* and *Tilia platyphyllos*. It maternally transmits several transient bacteria in its mid-gut (Actinobacteria and Firmicutes) (Salem et al., 2013). The green stink bug *Nezara viridula* (Pentatomidae) is a highly polyphagous pest of mostly herbaceous, annual plant species (Cruciferae, Poaceae, Malvaceae, and Solanaceae). It also has a specific symbiont bacterium in the gut crypts and is acquired through the environment in each generation (Engel and Moran, 2013; Prado et al., 2006). Obligate symbionts transmitted by aphids typically provide essential amino acids that are scarce in the plant phloem on which they feed. Equally, a wide variety of facultative symbionts are known to be beneficial to aphid hosts (Douglas, 1998; Oliver et al., 2010; Simon et al., 2011). *Serratia symbiotica*, a facultative symbiont commonly associated with aphids, is extracellularly transmitted to future generations, potentially via contamination with honeydew (Henry et al., 2015; Pons et al., 2019). The pea aphid *Acyrthosiphon pisum* carries *Staphylococcus*, *Pseudomonas*, *Acinetobacter*, and *Pantoea*, which are transmitted via the environment during feeding on plant sap (Leroy et al., 2011; Stavrinides et al., 2009, 2010).

Plataspid bugs (Plataspidae) transmit *Ishikawaella capsulatus* (Proteobacterium) vertically through the maternal (egg capsule) route (Fukatsu and Hosokawa, 2002; Hosokawa et al., 2006). The *Burkholderia* species is transmitted by *Riptortus clavatus* (alydid bug) via the environment (Kikuchi et al., 2007, 2012). The bean bug *Riptortus pedestris* acquires a specific *Burkholderia* symbiont that forms thick clusters in the mid-gut caeca every generation from the environment (Kikuchi et al., 2005, 2011). The vector of trypanosome parasites, the kissing bug (*Rhodnius prolixus*), transmits *Rhodococcus rhodnii* (actinobacteria) that forms huge populations on the lumen of the anterior mid-gut via coprophagy (Beard et al., 2002; Eichler and Schaub, 2002; Engel and Moran, 2013).

5.2 Microbiome of Lepidoptera (moth and butterflies)

The gut microbiome of lepidopterans differs between and within species; this has resulted in a series of arguments on the functional significance of microbes in the guts of insect species in this order (Paniagua Voirol et al., 2018). The enormous inconsistency in the gut microbiome of lepidopterans might be influenced by diverse factors such as the environment, the insect diet, the insect developmental stage, and the gut physiology; these factors may act alone or in combination (Paniagua Voirol et al., 2018). The bacterial communities of lepidopteran insects of the same species vary significantly between their life stages (larvae and adults) due to their different diets (Staudacher et al., 2016; Xia et al., 2017). Almost all lepidopteran species feed on plant tissues at their early life stage (larva), though most species feed on flower nectars at the adult stage (Strong et al., 1984). Some bacteria may persevere throughout the complete life cycle of the insect (Hammer et al., 2014). The microbiome of lepidopteran insects is presented in Table 18.2.

The transmission mode of symbiotic bacteria in lepidopterans might be from one generation to the next, known as vertical transmission; straight via contact among species; or through feeding from the environment, known as horizontal transmission (Hammer et al., 2017; Hurst, 2017). The possible vertical transmission or horizontal transmission of some species can be determined by the presence of major bacteria cohorts in the lepidopteran guts, but acquisition from the environment is more possible in other species (Staudacher et al., 2016). In *Galleria mellonella*, translocation of the gut bacteria is via the oocytes (Freitak et al., 2014). The gypsymoth caterpillar (*Lymantria dispar*), a highly polyphagous pest of agricultural and forest crops (oak, larch, birch, linden, alder, etc.), acquires *Pseudomonas*, *Enterobacter*, *Pantoea*, *Staphylococcus*, *Serratia*, and *Bacillus* from the environment (plant leaves) during feeding (Broderick et al., 2004, 2009; Mason et al., 2011). Some gut bacteria have been isolated from the eggs of some lepidopteran insects (Mason and Raffa, 2014; Tang et al., 2012). But the metabolic activity of the bacteria (*Enterococcus*) during the egg stage is only confirmed in the tobacco hornworm (*Manduca sexta*) (Brinkmann et al., 2008). The metabolic activity of the bacteria at the egg stage denotes a great metabolic compliance that enhances bacterial survival on the egg and larvae gut (Brinkmann et al., 2008). The development of bacteria at the egg stage may enhance their chances to inhabit the newly hatched instar larvae as well as their environment such as the soil or host plant, which could eventually promote horizontal transmission (Paniagua Voirol et al., 2018). Lepidopteran insects also acquire intra- and extracellular symbionts through their host plants, which indicates the vast adaptability of the bacteria to various habitats (Chrostek et al., 2017; Flórez et al., 2017; Li et al., 2016).

5.3 Microbiomes of Coleoptera (beetles)

Some of the most devastating forest tree diseases that emerged during the last century resulted from the interactions between fungi and wood-boring beetles (Curculionidae; Coleoptera) in the subfamilies Scolytinae (bark beetles and ambrosia beetles) and Platypodinae (ambrosia beetles) (Six, 2003; Hulcr and Dunn, 2011; Ploetz et al., 2013). The bark beetles and ambrosia beetles differ in their feeding preferences. Bark beetles feed on the host tree phloem tissue (the vascular tissue responsible for the transport of sugars) while most ambrosia beetles bore deeper into the xylem (sapwood and heartwood) and rely on fungi as their sole source of nutrition in this otherwise nutrition-poor niche (Batra, 1966, 1967; Beaver, 1989; Farrell et al., 2001). Although bark beetles may have access to more readily available nutrients, many species also feed on symbiotic fungi as a supplement in their diets (Six and Paine, 1998; Ayres et al., 2000; Bleiker and Six, 2007).

Bark beetles include many aggressive (primary) tree pests that can cause significant economic losses to forestry and forest ecosystems. However, the majority of bark beetle species infest mainly already dead, dying, or stressed trees in their native environments and are thus usually harmless to healthy living trees. They are abundant and important components of forest ecosystems (Martikainen et al., 1999). A characteristic that has fascinated researchers is their widespread association with microorganisms, including fungi, bacteria, and metazoans (mites and nematodes). The first reports of associations between bark beetles and fungi and their roles in timber staining were recognized already in the 19th century (Schmidberger, 1836; Hartig, 1844, 1878). It was not until the early decades of the 20th century, with the expansion and mechanization of forestry and forest product industries as well as modern globalization, that the economic importance of bark beetles and the fungi they carry became evident and were recognized as serious risks to forest health.

Due to their economic and ecological importance as well as scientific curiosity to understand symbiotic interactions and their evolutionary histories, research on the bark beetle microbiome (Fig. 18.1) has been active during the past century. It has also strongly been focused on certain wood-inhabiting fungi. Bark beetles are found in association with diverse fungi, mainly ascomycetous species that are members of the fungal orders Hypocreales, Ophiostomatales, Microascales, and Saccharomycetales (Ploetz et al., 2013). The most investigated of these microorganisms are the so-called blue-stain fungi, commonly known also as "ophiostomatoid fungi" (the term used later in the text to refer to the assemblage of morphologically similar fungi adapted for arthropod dispersal) (Wingfield, 1993). Many ophiostomatoid fungi have pigmented hyphae that colonize freshly exposed sapwood and cause grey, black, or brown discoloration of wood, downgrading the value of timber and resulting in economic losses (Uzunovic and Byrne, 2013). The damage to wood is cosmetic in contrast to the structural damage of wood caused by rot fungi (Seifert, 1993). Research on ophiostomatoid fungi has also been fueled by the fact that some fungal species are aggressive tree pathogens when accidentally introduced into new environments, and examples such as Dutch elm disease have caused major losses to forestry and greatly impacted natural forest ecosystems worldwide (Ploetz et al., 2013).

Since the earliest studies, controversy has surrounded the taxonomic placement of ophiostomatoid fungi and the role they potentially play in bark beetle lives and the tree killing processes of primary bark beetles. The difficulties in providing correct species identification and confusion over the taxonomy of these fungi are because the species share similar, minute, and overlapping morphological characteristics; the simultaneous presence of various life stages; and the sharing of the same or similar ecological niches in beetle galleries. The typical morphological features of ophiostomatoid fungi are their spore-forming structures, which are considered adaptations for dispersal by arthropod vectors (Malloch and Blackwell, 1993). The spore-forming structures of both the asexual and sexual states of these fungi are typically long stalks or necks, which bear spores in their apices in slimy masses that provide a mechanism to reach and attach to the bodies of passing arthropod vectors for transport to new host trees. The ecological and morphological similarities of these fungi have evolved more than once during evolution and are thus examples of convergent evolution (Malloch and Blackwell, 1993). Phylogenetic analyses have shown that ophiostomaid fungi are members of Sordariomycetes (Ascomycota) that reside in two distinct orders, the Ophiostomatales and Microascales (Hibbett et al., 2007; De Beer et al., 2013; Ploetz et al., 2013). In the single family Ophiostomataceae, 10 genera are currently included (De Beer et al., 2013; Bateman et al., 2017; van der Linde et al., 2016). Microascales is comprised of five families, of which the Gondwanamycetaceae, Graphiaceae, and Ceratocystidaceae include ophiostomatoid fungi (De Beer et al., 2014, 2017; Mayers et al., 2015; Nel et al., 2017). In the Ophiostomatales, all the ophiostomatoid fungi reside in the Ophiostomataceae (De Beer and Wingfield, 2013). Considering their importance as forest tree pathogens, only the families Ceratocystidaceae and Ophiostomataceae include important tree pathogens, the majority of which reside in *Ceratocystis*, *Endoconidiophora* (Microascales, Ceracystidaceae), *Leptographium*, *Ophiostoma*, and *Raffaelea* (Ophiostomatales, Ophiostomataceae) (Jacobs and Wingfield, 2001; Harrington et al., 2010; Ploetz et al., 2013; Seifert et al., 2013; De Beer et al., 2014). The advances in molecular genetic tools, especially DNA sequence comparisons, have greatly enhanced the accurate and reliable identification of these fungi while accelerating the taxonomic work, delineation of species boundaries, and understanding of the true species diversity. The scientific research has been active, resulting in the discovery of numerous species novel to science and taxonomic revisions. In many cases, several changes in the nomenclature of these fungi have occurred. If you are not a specialist in the field, it is advised to check the synonyms and confirm the currently valid name, especially when dealing with quarantine or phytosanitary issues.

Unlike bark beetles, ambrosia beetles usually infest only dead or stressed trees and their fungal associates are tree pathogens only in rare cases, typically connected to their introduction into new environments (Ploetz et al., 2013). An example of a recent such event is the damage caused by the fungal symbiont *Fusarium euwallaceae* together with its invasive host, the polyphagous shot hole borer (PSHB) (Paap et al., 2018). Interactions between ambrosia and bark beetles and ophiostomatoid fungi are among the most intensively investigated insect-fungi relationships in forest ecosystems. A range of different types of associations exist, varying from mutualistic interactions that benefit both the fungus and the beetle to occasional relationships that likely do not have importance for the beetle, but benefit the fungus as a means of facilitated transport to a new host tree. There is increasing evidence that some fungal symbionts can facilitate beetle colonization success and amplify insect damage (Wadke et al., 2016; Zhao et al., 2018). In some cases, the fungi are obligate nutritional symbionts of the ambrosia beetles (Batra, 1966, 1967; Beaver, 1989; Farrell et al., 2001; Ploetz et al., 2013) while the others are saprotrophs inhabiting the beetle galleries on wood. The associations also vary depending on the family to which the fungi belong. The majority of ophiostomatoid fungi in Ophiostomataceae are typically more specific with certain beetle species compared to species of Ceratocystidaceae (Harrington et al., 2010; Jacobs and Wingfield, 2001; Kirisits, 2004; Paine et al., 1997). *Ceratocystis* and *Endoconidiophora* species also attract various other insect vectors than beetles by producing strong aromas to attract their vectors (Kile, 1993).

While most studies have focused particularly on fungi associated with beetles, rather little is known about other microbes involved in these interactions. Yeasts have been recognized as constant components in bark and ambrosia beetle galleries (Siemaszko, 1939; Davis, 2015), but they have probably been overlooked in the majority of the previous collections that are mainly based on the culturable fraction of fungal diversity. Particularly common seem to be ascomycetous yeasts, which are more dependent on vectors to move to new host trees compared to the basidiomycetous species (Kurtzman et al., 2011). Molds (e.g., *Penicillium*, *Aspergillus*, Mucoromycetes) are also abundant and likely overlooked (Davis, 2015; Silva et al., 2015; Kasson et al., 2016; Li et al., 2018; Hofstetter et al., 2015).

Other organisms are also involved in these interactions, and increasing evidence indicates that bark beetle-associated mites are important vectors of fungi present in beetle galleries (Moser et al., 1989; Chang et al., 2017; Vissa and Hofstetter, 2017). The mites can carry fungal spores in their bodies or specialized structures called sporothecae (Moser, 1985). Some mite species feed on fungi and can thus promote the growth of certain fungi in beetle galleries (Hofstetter and Moser, 2014). Nematodes are also common associates of beetles, found in beetle galleries and on the bodies of beetles. Some of them are beetle parasites and have also been studied as potential biological control agents of pest beetles (Grucmanová and Holuša, 2013). Nematodes consume other microorganisms, including fungi and bacteria (Yeates et al., 1993; Ledón-Rettig et al., 2008). As for other insects, bacterial symbionts are also common and diverse, including enterobacteria as the most prevalent ones (Dohet et al., 2016).

In summary, microbiomes associated with bark and ambrosia beetles are very diverse and complex. Only a few have been extensively studied, and generally, our knowledge of these associations is still limited and biased toward forest ecosystems in parts of Europe and North America. Although research has been active, it has remained in the discovery phase for a long time, at least partially because the vast majority of microbes associated with bark and ambrosia beetles remain uninvestigated. The development of high-throughput sequencing methods has provided tools to explore whole microbial communities, but only a few studies thus far have focused on those associated with bark and ambrosia beetles (Kostovcik et al., 2015).

5.4 Microbiome of Isopteran (termites)

Microbiomes of termites consist of a gut microbiome involved in cellulose degradation, nutritional symbionts, and other interactions with wood-inhabiting microbes that share the same habitat (Amburgey, 1979; Aanen and Eggleton, 2005; Bignell, 2010). Termites are capable of feeding on wood, which is a rare characteristic for a majority of organisms. Wood as a source of nutrition is complex in structure and thus difficult to exploit, and many organisms rely on more readily available carbohydrates. The ability to survive in this nutritionally poor niche requires an enzymatic capacity from the organism to break down wood cellulose. The rather unusual ability to feed on wood has allowed termites to become one of the most abundant insects, resulting in them having an important ecological role in the recycling of wood and other plant material in nature (Waller, 1988; Aanen et al., 2002).

However, termites are not capable of cellulose degradation themselves. The only organisms capable of direct cellulose digestion include bacteria, archaea, and fungi. All other cellulose-digesting species, including termites, are dependent on their symbiotic gut microbes that break down cellulose. The termite microbiome and its evolution have fascinated researchers for more than a century. The interest has been driven by the termite's exceptional capacity for lignocellulose degradation (Liu et al., 2011). The microbiome in the gut is highly complex when compared to any other animals studied; more than

1000 species of flagellate protists (in lower termites), bacteria, and archaea (only in termites) have been detected. The gut microbiota and its role have been reviewed in detail in several studies (Brune and Ohkuma, 2011; Hongoh, 2011; Ohkuma and Brune, 2011; Brune and Dietrich, 2015).

The gut microbiome, especially the cellulolytic flagellates unique to the guts of lower termites, have likely had a key role in the speciation of termites and the development of social behavior (Bourguignon et al., 2015; Nalepa, 2015). The majority of the gut flagellates are members of the parabasalids (phylum Parabasalia) (Ohkuma and Brune, 2011). Parabasalids have a key role in cellulose digestion in lower termite guts. Molecular studies have contributed greatly to understanding the diversity of termite gut flagellates, revealing a rich, species-specific diversity as well as the presence of currently undescribed species and lineages (Brune and Dietrich, 2015). Bacterial gut symbionts are restricted to a few dominant phyla such as Spirochaetes, Bacteroidetes, Firmicutes, Proteobacteria, Elusimicrobia, and Fibrobacteres (Brune and Dietrich, 2015). They differ in their host preference and abundance, as some are found in all termites while others such as Spirochaetes are common in the guts of wood-feeding termites (Breznak and Leadbetter, 2006) and bacteroidetes in fungus-cultivating termites (Hongoh, 2010; Makonde et al., 2013). Archaea are most typically present in the guts of higher termites (especially in the subfamily Termitinae), especially in the soil-feeding lineages.

Fungus farming (the ability to grow fungi for food) represents a fascinating example of obligatory symbiosis between insects and fungi. It independently evolved about 40–60 million years ago in three distinct insect lineages: termites (Isopteran), ants (Hymenopteran), and beetles (Coleopteran) (Mueller and Gerardo, 2002; Mueller et al., 2005). Termite fungi culture is the most studied of the nutritional symbioses (Mueller and Gerardo, 2002; Aanen et al., 2002). A nutritional symbiont for numerous termite species is a specific basidiomycete white rot fungus, *Termitomyces* (Batra and Batra, 1979; Wood and Thomas, 1989; Aanen et al., 2002). The termites are maintaining (gardening) the fungus on a special structure, a fungus comb, in their nests, providing predigested plant material substrate that the fungus helps to degrade. Fungus-farming termites are found to be ecologically dominant in savannas (Wood and Sands, 1978). Recent studies have provided strong evidence that termite agriculture originated in the African rain forest (Viana-Junior et al., 2018).

Fungus farming and gut symbionts are examples of symbioses, but termites also come in contact and share the same environment with numerous other wood-inhabiting microbes. Some of these associations, especially those that include decay fungi and ophiostomatoid fungi, are beneficial for termites by means of increasing their resource consumption, aggregation behavior, and survival (Viana-Junior et al., 2018). On the contrary, termites avoid wood if certain fungi already inhabit it (Amburgey, 1979; Viana-Junior et al., 2018).

5.5 Microbiome of Diptera

The microbiomes of Dipteran insects are highlighted in Table 18.2. The medfly (*Ceratitis capitata*) microbiome, like that of other insects, is reported to be shaped by phylogenetic, metabolic, and taxonomic diversities during formation and development (morphogenesis) (Aharon et al., 2013). The microbiome is a major mediator of fitness in tephritid flies (Andongma et al., 2015; Yong et al., 2017). For example, some medfly microbiomes are known to have probiotic effects on their host by enhancing the sexual performance of *Ceratitis capitata* males at emergence (Hamden et al., 2013). *Ceratitis capitata* is a highly invasive species infesting many forest trees such as *Chrysophyllum viridifolium* (fluted milkwood), *Chrysophyllum albidum* (African star apple), *Diospyros abyssinica* (giant diospyros), *Olea woodiana* (forest olive), *Podocarpus elongatus* (African yellow wood), etc. (CABI/EPPO, 2015).

Symbiotic bacteria play an important role in the mating preference of *Drosophila melanogaster* through altering the levels of sex pheromones (Sharon et al., 2010). Many studies have identified the structure and function of the gut microbiota of different dipteran insects. The fruit fly, *Drosophila melanogaster*, acquires *Lactobacillus* spp. (Acetobacteraceae) via the environment through feeding on decaying fruits (Broderick and Lemaitre, 2012).

6. Functional roles of insect symbionts

6.1 Functional roles of bacterial symbionts of insects

Bacterial symbionts of insects that are maternally transmitted include two types, obligate (primary) and facultative (secondary) endosymbionts (Philip and Nancy, 2013). Obligate endosymbiosis in the cytosol of a specialized host cell provides deficient nutrients for the hosts for growth and development (Baumann, 2005). The facultative or secondary symbionts may provide conditional adaptive advantages to their host by shielding the host against pathogens and natural enemies (Oliver et al., 2005; Guay et al., 2009). Additionally, they could improve insecticide resistance by supporting adaptation to the plant (Guidolin and Cônsoli, 2017), facilitating insect host metabolism and biosynthesis (Benoit et al., 2017) and ameliorating the harmful effects of heat (Montllor et al., 2002). Phytophagous hemipteran insects, which usually feed on nutritionally deficient xylem or phloem diets, have endosymbionts that provide essential amino acids and other nutrients (Redak et al., 2004).

In aphids, the symbiont *Buchnera aphidicola* also supplies additional nutrients that are lacking in their main plant sap diets (Hansen and Moran, 2011; Poliakov et al., 2011). *Hamiltonella defensa* (facultative symbionts) in aphids protects its host against pathogen or parasite invasion through a bacteriophage-encoded mechanism (Degnan and Moran, 2008; Moran et al., 2005). It also induces tolerance to high temperatures (Russell and Moran, 2006), thus enabling its own survival and transmission (Oliver et al., 2010). In some cases, symbionts may support their own spread without profiting their hosts, as in the case of *Wolbachia pipientis*, which can cause sex ratio biases or inconsistencies in reproduction that reduce host health or fitness (Werren et al., 2008). However, in the *Drosophila* species, *Wolbachia pipientis* symbionts confer beneficial services by protecting their host against some RNA virus infections (Hedges et al., 2008; Teixeira et al., 2008).

The bacterial symbionts of *Bactrocera oleae* (olive fruit fly) aid in the digestion of green olives, specifically in protein hydrolysis (Pavlidi et al., 2017). *Candidatus Erwinia dacicola* in the larvae of *Bactrocera oleae* offers vital amino acids and permits the larvae to grow in unripe olives containing oleuropein, which hinders the growth of other insects (Ben-Yosef et al., 2015) and thus enhances reproduction in *Bactrocera oleae* (Pasternak et al., 2014). *Enterobacter* spp. in the lepidopteran insect gut can provide reactive oxygen species (ROS) detoxifying enzymes such as superoxide dismutase or catalase (Xia et al., 2017). Similarly, *Enterococcus faecalis* in the gypsy moth acidifies its local environment to colonize alkaline niches. This perhaps shields the gut from pathogens that are active in alkaline environments such as *Bacillus thuringiensis* (Broderick et al., 2004). *Enterococcus mundtii*, in the gut of *Spodoptera littoralis*, yields an antimicrobial compound against Gram-positive *Listeria* but not against resident gut bacteria (Shao et al., 2017).

The community of nitrogen-fixing bacteria (e.g., Enterobacteriaceae) improves development and reproduction in *Ceratitis capitata* (Mediterranean fruit fly) (Behar et al. 2008a,b). Gut symbionts of bark beetles and weevils feeding on conifers are linked with the detoxification of organic compounds such as terpenoids that permit their host to survive the toxigenic contents in their plant diets (Adams et al., 2013; Wang et al., 2013; Berasategui et al., 2017). *Rhodococcus* in the gut of the gypsy moth helps to degrade monoterpenes at high alkalinity, which allows the insects to digest food rich in monoterpenes (Broderick et al., 2004; Powell and Raffa, 1999; van der Vlugt-Bergmans and van der Werf, 2001).

6.2 Functional roles of fungi associated with insects

As previously highlighted, many fungi associated with insects fulfill ecological roles as mutualistic endosymbionts that aid in nutrition or as pathogens (Suh et al., 2005). Bark beetles are known to carry wood-decaying fungi, further mediating nutrient and carbon cycling in a forest ecosystem (Jacobsen et al., 2017). Some fungi serve as sources of food for insects such as ants (Currie et al., 2003) or as a vector of many vertically transmitted parasites (Lucarotti and Klein, 1988). Several others are pathogenic with prominent effects on insect host species (entomopathogenic) (Araújo and Hughes, 2016; Evans and Samson, 1984) and a few are commensals (DeKesel, 1996). Yeast-like symbionts are involved in amino acid and fatty acid metabolisms within the digestive tracts, thereby providing vital nutritional benefits to the phloem-feeding plant hoppers (e.g., *Nilaparvata)* and many other insect species (Horgan and Ferrater, 2017; Gurung et al., 2019). The absence of yeast symbionts could lead to incomplete metamorphosis (Carvalho et al., 2010).

6.3 Functional roles of viruses associated with insects

Many plant viruses depend on insect vectors to transmit them to other plants, and these interactions can be specific with certain plant virus genera. For instance, Potyvirus is mostly transmitted by aphids (Whitfield et al., 2015). The whitefly *Bemisia tabaci* (Gennadius) can transmit more than 100 viral species belonging mostly to the genus *Begomovirus* (Geminiviridae). Other viruses that are transmitted by white flies include members of the genera Carlavirus (Betaflexiviridae), Crinivirus, Ipomovirus, and Closterovirus (Jones, 2003). Some viruses in the Baculoviridae family have been used as biocontrols against insect pests (Lacey et al., 2001; Williams et al., 2017; Winstanley and Rovesti, 1993). By contrast, some viruses have been reported to hinder pest management strategies by influencing resistance to commonly used insecticides (Yoshikawa et al., 2018). In some cases, the infection of pest larvae by viruses could prevent the parasitoids from parasitizing the larvae (Robertson et al., 2013). Some viruses provide protection to their insect hosts, for example wasps that harbor Polydnaviridae (viral symbiont) confer protection to them during reproduction. The viruses infect the ovaries and reproductive tract of the wasp, which then transmits the virus to its host during egg laying (Herniou et al., 2013; Strand and Burke, 2015).

7. Transmission route in insects

7.1 Fungal transmission route in insects

Some fungi are dimorphic in their life stages (e.g., spore, hyphal, or yeast-like forms), which they utilize during transmission to their different hosts, especially to insect hosts. Entomopathogenic fungi are basically transmitted horizontally through the environment. In *Metarhizium*, two modes of transmission to the insect host from the soil exist: the sleeper form and the creeper form (Angelone et al., 2018). In the sleeper form, the fungus produces large amounts of conidia that remain in the soil until they are picked up by a suitable host to initiate infection in the host. The creeper form produces fewer conidia with more mycelia, which allows hyphae to grow through the soil, thus increasing their chances of associating with a suitable host for infection onset (Angelone et al., 2018). Soil is a reservoir for the resting spores of insect pathogenic fungi of the order Entomophthorales (van Oers and Eilenberg, 2019). Most species in the fungal subphylum Entomophthoromycotina produce long-lived spores that reside in the environment to initiate infections in insects when susceptible host stages are available (Hajek et al., 2018). Most entomophthoralean species have two types of spores, conidia and resting spores (Hajek et al., 2018). In *Entomophaga maimaiga*, both conidia and resting spores are sometimes produced from one cadaver, but only one type of spore is usually produced from the cadaver of one host individual in most cases. For many fungal entomopathogens, the production of aerially dispersing spores is associated with the relative humidity of the environment. Fungal pathogens spread over various distances in numerous ways; for instance, in constrained measures, infected insects can be active and continue to move around until the point of death while spores can be spread by rain splash (Fernández-García and Fitt, 1993; Lloyd et al., 1982). The movement of soil-inhabiting fungal entomopathogens can be limited, but spores can be washed through the soil or in some cases fungal mycelia originating in insect cadavers can spread through the soil within the environment (Keller et al., 1989; Studdert and Kaya, 1990). In some cases, insect pathogenic fungal spores inhabiting soil that infect insects living aboveground can move away from the soil environment by attaching to the growing plant surfaces, making it easier for the insects that fly or hop around the plant surfaces to get infected (Iijima et al., 1993).

7.2 Bacterial symbionts and transmission routes in insects

Symbiotic bacteria in insects can either be intracellular or extracellular (Engel and Moran, 2013). The symbionts could be regarded as obligate or facultative. Obligate symbionts, also known as primary symbionts, play an important role in the nutrition of insects that feed on cellulose (Baumann, 2005). On the other hand, facultative symbionts (secondary) have a much wider range of effects varying from mutualism to manipulation of the reproductive capacity of their hosts (Moran et al., 2008). Bacterial symbionts of insects are transmitted by either intracellular transmission or extracellular transmission. The extracellular transmission route comprises contamination from the environment, acquisition from feces known as coprophagy, social acquisition, rubbing of the egg surface, and transmission through jelly secretions.

7.3 Viral transmission route in insects

There are two transmission routes of viruses, horizontal and vertical. Horizontal transmission occurs among individuals in the same generation via the environment while vertical transmission occurs from parent to offspring (Andreadis, 1987; Rothman and Myers, 2000). Viral pathogens of lepidopterans such as granulosis viruses (GVs) and nuclear polyhedrosis viruses (NPVs) are mainly transmitted between and within host generations via the release of occlusion bodies to the environment after the death of the infected host (Myers, 1993). Less-lethal viral pathogens are transmitted mainly in feces and through regurgitation. This is commonly found in small RNA viruses, Oryctes virus, entomopox viruses (EPVs) of some Lepidoptera, NPVs of sawflies, and cytoplasmic polyhedrosis viruses (CPVs) infecting bees (Myers and Rothman, 1995). Horizontal transmission normally takes place through the feeding of contaminated food (foliage or egg chorion on hatching), as found in the iridescent viruses (IVs) of *Aedes taeniarhynchus* and *Tipula oleracea* (Carter, 1973a, b; Linley and Nielson, 1968b) or through cannibalism, as seen in NPV of *Heliothis armigera* (Dhandapani et al., 1993).

Vertical transmission occurs by contamination of the egg surface, known as transovum, or within the ovary, which is called transovarian (Rothman and Myers, 2000). Transovum transmission is important among the CPVs of Lepidoptera. The larvae of *Heliothis virescens* can be infected when they feed on CPV-contaminated chorion during hatching (Sikorowski et al., 1973). Transovarian transmission could be vital for viruses that attack flies, such as the iridescent virus of the mosquito *Aedes taeniorhynchus* (Linley and Nielson, 1968a; Woodard and Chapman, 1968) and the sigma virus of fruit fly *Drosophila melanogaster*, which is spread through female and male sex cells (Fleuriet and Periquet, 1993).

8. Interactions of insect microbiomes with forest trees and their environment

Forests are one of the key global vegetation types, and their health is essential to wildlife as well as to humans (Myers, 1997; Wingfield et al., 2015). Recent studies have revealed alarming effects, especially due to invasive forest pests, pathogens, and their interactions (Fisher et al., 2012; Wingfield et al., 2016; Wong and Daniels, 2016; Burke et al., 2017; Pyšek et al., 2020). Environmental conditions associated with climate change are amplifying the impact of these biotic threats on forests globally. The majority of studies have focused on geographic distribution and the productivity of forest tree species of economic importance (e.g., Iverson and Prasad, 1998; Bakkenes et al., 2002; Hanewinkel et al., 2013) while much less information is available on the distribution of insects and the microorganisms they carry (Fisher et al., 2012; Thakur et al., 2019).

The challenge in studying microorganisms is their small size, overlapping features, and complex lifecycles. The difficulties in providing correct species identification and confusion in taxonomy surround many species, making accurate identification impossible in many cases. This together with the fact that baseline microbial diversity existing in native environment is not well investigated. This further complicates the determination on whether the microorganism is native or alien and in evaluating its possible risks to forest health (Fisher et al., 2012; Thakur et al., 2019; Pyšek et al., 2020).

Climate change can enable native insects and microorganisms to expand their ranges into areas where they could previously persist only at low densities or not at all. Examples include insects such as the mountain pine beetle (*Dendroctonus ponderosae*), which together with its microbial associates has destroyed large areas of pine forests in North America (Raffa et al., 2013). The mountain pine beetle and its microbiome represent a rare case where the effects of climate change to insect-microbe associations are understood on some level. Climate change affects diseases such as ash dieback and myrtle and white pine blister rust, in which altered host-pathogen relationships have facilitated their geographical and host plant range expansions (Helfer, 2013; Evans, 2016; Ghelardini et al., 2016; Goberville et al., 2016). Elevated temperatures and CO_2 levels may also directly influence the species and their reproduction or replication rates. Such effects have been seen with insect pests such as the European spruce bark beetle (*Ips typographus*) (Hoffmann, 2017) and with rust fungi due to their short generation times and sensitivity to temperature and CO_2 levels (Chakraborty and Datta, 2003; Helfer, 2013; Boddy et al., 2014; Váry et al., 2015).

The potential risks related to invasive insect pests and their associated microorganisms are typically unpredictable, and thus have been recognized as major threats to forest ecosystems (Liu et al., 2011; Ploetz et al., 2013; Wingfield et al., 2016). Their life cycles are dependent on fresh wood tissues, and they are either together or separately capable of moving across national boundaries such as via the trade of inappropriately treated timber, wood products, live plants, and soil (Aukema et al., 2010; Ferreira et al., 2011; Liebhold et al., 2012). The risks involved in these interactions are related also to their shared habitat with native forest insects. As a result of the introduction to new environments, novel insect-microorganism interactions are possible. The introduced invasive microorganism may come in contact with a native insect species, and an invasive insect pest may become a vector for native microorganisms (Haack, 2006; Wingfield et al., 2016).

9. Challenges and constraints in the study of the insect microbiome

Novel technological advances in the field of high-throughput sequencing technology as well as "omics" approaches have facilitated our understanding of the diverse microbiota associated with forest insects. However, the vast majority of studies on the insect microbiome are focused on bacteria and fungi, and the remaining groups of microbes (e.g., protists) remain largely underexplored. Protists are also important components of the insect microbiota. Also, the technical challenge using contemporary amplicon metabarcoding of fungal communities is that it largely relies on the internal transcribed spacer region of the ribosomal RNA, which is an unreliable marker for many beetle symbiotic fungi (Harrington et al., 2011; Nilsson et al., 2019). Their correct and accurate identification is the basis for communication and applied research, but this is challenged by the fact that many of them remain novel to science or their taxonomic placement is uncertain. Only recently have shared best practices for studying these interactions started to develop (Hulcr et al., 2020).

There is also a lack of basic or baseline global information on the insect microbiome. The limited information indirectly undermines or affects phytosanitary measures for the unintentional transport of wood and plant products across international boundaries. Currently, there are limited studies in the research field of the insect microbiome in developing countries in places such as Africa, Asia, and South America, partly due to the lack of facilities and resources. Microbiome research is an emerging research topic that requires advanced facilities and funding. Thus, few reports on the insect microbiome from developing countries are documented due to limited research in this field. This has constrained our understanding of the microbiomes of many forest insect species from Africa and other emerging economies of the world.

Another constraint is the limited expertise in the field. The lack of entomologists, mycologists and taxonomists to embark on microbiome research is also a contributory factor that has hindered research on insect microbiomes, especially in developing countries.

References

Aanen, D.K., Eggleton, P., 2005. Fungus-growing termites originated in African rain forest. Curr. Biol. 15 (9), 851–855. https://doi.org/10.1016/j.cub.2005.03.043.

Aanen, D.K., Eggleton, P., Rouland-Lefèvre, C., Guldberg-Frøslev, T., Rosendahl, S., Boomsma, J.J., 2002. The evolution of fungus-growing termites and their mutualistic fungal symbionts. PNAS 99 (23), 14887–14992. https://doi.org/10.1073/pnas.222313099.

Ackerman, I.L., Constantino, R., Gauch Jr., H.G., Lehmann, J., Riha, S.J., Fernandes, E., 2009. Termite (Insecta: Isoptera) species composition in a primary rain forest and agroforests in Central Amazonia. Biotropica 41 (2), 226–233. https://doi.org/10.1111/j.1744-7429.2008.00479.x.

Adams, A.S., Aylward, F.O., Adams, S.M., Erbilgin, N., Aukema, B.H., Currie, C.R., et al., 2013. Mountain pine beetles colonizing historical and naïve host trees are associated with a bacterial community highly enriched in genes contributing to terpene metabolism. Appl. Environ. Microbiol. 79, 3468–3475. https://doi.org/10.1128/AEM.00068-13.

Adler, P.H., Foottit, R.G., 2009. Introduction. In: Foottit, R.G., Adler, P.H. (Eds.), Insect Biodiversity. Science and Society. Willey-Blackwell, Oxford, pp. 1–6.

Aharon, Y., Pasternak, Z., Ben Yosef, M., Behar, A., Lauzon, C., Yuval, B., et al., 2013. Phylogenetic, metabolic, and taxonomic diversities dhape mediterranean fruit fly microbiotas during ontogeny. Appl. Environ. Microbiol. 79 (1), 303–313. https://doi.org/10.1128/AEM.02761-12.

Alali, S., Mereghetti, V., Faoro, F., Bocchi, S., Al Azmeh, F., Montagna, M., 2019. Thermotolerant isolates of Beauveria bassiana as potential control agent of insect pest in subtropical climates. PLoS One 14 (2), e0211457. https://doi.org/10.1371/journal.pone.0211457.

Amburgey, T.L., 1979. Review and checklist of the literature on interactions between wood-inhabiting fungi and subterranean termite: 1960-1978. Sociobiology 4 (2), 279–296.

Andongma, A.A., Wan, L., Dong, Y.C., Desneux, N., White, J.A., Niu, C.Y., 2015. Pyrosequencing reveals a shift in symbiotic bacteria populations across life stages of Bactrocera dorsalis. Sci. Rep. 5 (1), 1–6. https://doi.org/10.1038/srep09470.

Andreadis, T.G., 1987. Transmission. In: Fuxa, J.R., Tanada, Y. (Eds.), Epizootiology of Insect Diseases. Wiley, New York, pp. 159–176.

Angelone, S., Piña-Torres, I., Padilla-Guerrero, I., Bidochka, M., 2018. "Sleepers" and "Creepers": a theoretical study of colony polymorphisms in the fungus Metarhizium related to insect pathogenicity and plant rhizosphere colonization. Insects 9 (3), 104. https://doi.org/10.3390/insects9030104.

Araújo, J.P.M., Hughes, D.P., 2016. Diversity of entomopathogenic fungi. In: Lovett, B., Leger, R.J. (Eds.), Genetics and Molecular Biology of Entomopathogenic Fungi. Elsevier, pp. 1–39, https://doi.org/10.1016/bs.adgen.2016.01.001.

Asgari, S., Johnson, K.N. (Eds.), 2010. Insect Virology. Caister Academic Press. https://www.caister.com/insect-virology. (Accessed 11 June 2020).

Ashiru, M.O., 1988. The food value of the larvae of anaphevenata butler (Lepidoptera: Notodontidae). Ecol. Food Nutr. 22 (4), 313–320. https://doi.org/10.1080/03670244.1989.9991080.

Aukema, J.E., McCullough, D.G., Von Holle, B., Liebhold, A.M., Ritton, K., Frankel, S.J., et al., 2010. Historical accumulation of nonindigenous forest pests in the continental United States. Bioscience 60 (11), 886–897. https://doi.org/10.1525/bio.2010.60.11.5.

Ayres, M.P., Wilkens, R.T., Ruel, J.J., Lombardero, M.J., Vallery, E., 2000. Nitrogen budgets of phloem-feeding bark beetles with and without symbiotic fungi (Coleoptera: Scolytidae). Ecology 81 (8), 2198–2210. https://doi.org/10.1890/0012-9658(2000)081[2198:NBOPFB]2.0.CO;2.

Bakken, A.J., Schoof, S.C., Bickerton, M., Kamminga, K.L., Jenrette, J.C., Malone, S., et al., 2015. Occurrence of brown marmorated stink bug (Hemiptera: Pentatomidae) on wild hosts in nonmanaged woodlands and soybean fields in North Carolina and Virginia. Environ. Entomol. 44 (4), 1011–1021. https://doi.org/10.1093/ee/nvv092.

Bakkenes, M., Alkemade, J.R.M., Ihle, F., et al., 2002. Assessing effects of forecasted climate change on the diversity and distribution of European higher plants for 2050. Glob. Chang. Biol. 8 (4), 390–407. https://doi.org/10.1046/j.1354-1013.2001.00467.x.

Bandeira, A., Vasconcellos, A., 2004. Efeitos de distúrbiosflorestaissobre as populações de cupins (Isoptera) do brejo dos cavalos, Pernambuco. In: Porto, K., Cabral, J., Tabareli, M. (Eds.), Brejos de altitude em Pernambuco e Paraíba: História Natural, Ecologia e Conservação Brasília, D.F. Série Biodiversidade 9, Ministério do Meio Ambiente, Brasília, pp. 145–152.

Barnard, P.C., 2011. The Royal Entomological Society Book of British Insects. Wiley. 396 p https://www.wiley.com/en-68.

Bateman, C., Huang, Y.T., Simmons, D.R., Kasson, M.T., Stanley, E.L., Hulcr, J., 2017. Ambrosia beetle Premnobiuscavipennis (Scolytinae: Ipini) carries highly divergent ascomycotan ambrosia fungus, Afroraffaelea ambrosiae gen. nov. et sp. nov. (Ophiostomatales). Fungal Ecol. 25, 41–49. https://doi.org/10.1016/j.funeco.2016.10.008.

Batra, L.R., 1966. Ambrosia fungi: extent of specificity to ambrosia beetles. Science 153 (37329), 193–195. https://doi.org/10.1126/science.153.3732.193.

Batra, L.R., 1967. Ambrosia fungi: a taxonomic revision and nutritional studies of some species. Mycologia 59 (6), 976–1017. https://doi.org/10.2307/3757271.

Batra, L.R., Batra, S.W.T., 1979. In: Batra, L.R. (Ed.), Insect-Fungus Symbiosis. Allenheld and Osmum, Montclair, NJ, pp. 117–163.

Baumann, P., 2005. Biology bacteriocyte-associated endosymbionts of plant sap-sucking insects. Annu. Rev. Microbiol. 59, 155–189. https://doi.org/10.1146/annurev.micro.59.030804.121041.

Bawa, K.S., 1990. Plant-pollinator interactions in tropical rain forests. Annu. Rev. Ecol. Syst. 21, 399–422. https://www.jstor.org/stable/209703.

Beard, C.B., Cordon-Rosales, C., Durvasula, R.V., 2002. Bacterial symbionts of the Triatominae and their potential use in control of Chagas disease transmission. Annu. Rev. Entomol. 47, 123–141. https://doi.org/10.1146/annurev.ento.47.091201.145144.

Beaver, R.A., 1989. Insect–fungus relationships in bark and ambrosia beetles. In: Wilding, N., Collins, N.M., Hammond, P.M., Webber, J.F. (Eds.), Insect–Fungus Interactions. Academic Press, London, pp. 121–143.

Behar, A., Jurkevitch, E., Yuval, B., 2008a. Bringing back the fruit into fruit fly–bacteria interactions. Mol. Ecol. 17 (5), 1375–1386. https://doi.org/10.1111/j.1365-294X.2008.03674.x.

Behar, A., Yuval, B., Jurkevitch, E., 2008b. Gut bacterial communities in the mediterranean fruit fly (Ceratitis capitata) and their impact on host longevity. J. Insect Physiol. 54 (9), 1377–1383. https://doi.org/10.1016/j.jinsphys.07.011.

Belovsky, G.E., Slade, J.B., 2000. Insect herbivory accelerates nutrient cycling and increases plant production. Proc. Natl. Acad. Sci. 97, 14412–14417. https://doi.org/10.1073/pnas.250483797.

Benoit, J.B., Vigneron, A., Broderick, N.A., Wu, Y., Sun, J.S., Carlson, J.R., et al., 2017. Symbiont-induced odorant binding proteins mediate insect host hematopoiesis. eLife 6, e19535. https://doi.org/10.7554/eLife.19535.

Ben-Yosef, M., Pasternak, Z., Jurkevitch, E., Yuval, B., 2015. Symbiotic bacteria enable olive fly larvae to overcome host defences. R. Soc. Open Sci. 2, 150170. https://doi.org/10.1098/rsos.150170.

Berasategui, A., Salem, H., Paetz, C., Santoro, M., Gershenzon, J., Kaltenpoth, M., et al., 2017. Gut microbiota of the pine weevil degrades conifer diterpenes and increases insect fitness. Mol. Ecol. 26 (15), 4099–4110. https://doi.org/10.1111/mec.14186.

Beutel, R.G., Friedrich, F., Ge, S.Q., Yang, X.K., 2014. Insect Morphology and Phylogeny. Walter de Gruyter, Berlin.

Bignell, D.E., 2010. Morphology, physiology, biochemistry and functional design of the termite gut: an evolutionary wonderland. In: Bignell, D., Roisin, Y., Lo, N. (Eds.), Biology of Termites: A Modern Synthesis. Springer, Dordrecht, pp. 375–412.

Blackman, R.L., Eastop, V.F., 2000. Aphids on the World's Crops: An Identification and Information Guide, second ed. John Wiley & Sons, Chichester, England. 466 pp.

Bleiker, K.P., Six, D.L., 2007. Dietary benefits of fungal associates to an eruptive herbivore: potential implications of multiple associates on host population dynamics. Environ. Entomol. 36 (6), 1384–1396. https://doi.org/10.1603/0046-225x(2007)36[1384:dbofat]2.0.co;2.

Boddy, L., Buntgen, U., Egli, S., Gange, A.C., Heegaard, E., Kirk, P.M., et al., 2014. Climate variation effects on fungal fruiting. Fungal Ecol. 10, 20–33. https://doi.org/10.1016/j.funeco.2013.10.006.

Bourguignon, T., Lo, N., Cameron, S.L., Sobutník, J., Hayashih, Y., Shigenobu, S., et al., 2015. The evolutionary history of termite as inferred from 66 mitochondrial genomes. Mol. Biol. Evol. 32 (2), 406–421. https://doi.org/10.1093/molbev/msu308.

Breznak, J.A., Leadbetter, J.R., 2006. Termite gut spirochetes. In: Dworkin, M., Falkow, S., Rosenber, E., Schleifer, K.-H., Stackebrandt, E. (Eds.), The Prokaryotes, Vol. 7: Proteobacteria: Delta and Epsilon Subclasses. Deeply Rooting Bacteria, third ed. Springer, New York, pp. 318–329.

Brinkmann, N., Martens, R., Tebbe, C.C., 2008. Origin and diversity of metabolically active gut bacteria from laboratory-bred larvae of Manduca sexta (Sphingidae, Lepidoptera, Insecta). Appl. Environ. Microbiol. 74 (23), 7189–7196. https://doi.org/10.1128/AEM.01464-08.

Broderick, N.A., Lemaitre, B., 2012. Gut-associated microbes of Drosophila melanogaster. Gut Microbes 3 (4), 307–321. https://doi.org/10.4161/gmic.19896.

Broderick, N.A., Raffa, K.F., Goodman, R.M., Handelsman, J., 2004. Census of the bacterial community of the gypsy moth larval midgut by using culturing and culture-independent methods. Appl. Environ. Microbiol. 70 (1), 293–300. https://doi.org/10.1128/AEM.70.1.293-300.2004.

Broderick, N.A., Robinson, C., McMahon, M., Holt, J., Handelsman, J., Raffa, K., 2009. Contributions of gut bacteria to Bacillus thuringiensis-induced mortality vary across a range of Lepidoptera. BMC Biol. 7, 11. https://doi.org/10.1186/1741-7007-7-11.

Brune, A., Dietrich, C., 2015. The gut microbiota of termites: digesting the diversity in the light of ecology and evolution. Annu. Rev. Microbiol. 69, 145–166. https://doi.org/10.1146/annurev-micro-092412-155715.

Brune, A., Ohkuma, M., 2011. Role of termite gut microbiota in symbiotic digestion. In: Bignell, D.E., Roisin, Y., Lo, N. (Eds.), Biology of Termites: A Modern Synthesis. Springer, Dordrecht, Netherlands, pp. 439–475.

Burke, J.L., Bohlmann, J., Carroll, A.L., 2017. Consequences of distributional asymmetry in a warming environment: invasion of novel forests by the mountain pine beetle. Ecosphere 8 (4), e01778. https://doi.org/10.1002/ecs2.1778.

CABI/EPPO, 2015. Ceratitis capitata [Distribution map]. Distribution Maps of Plant Pests, No. December. CABI, Wallingford, UK. Map 1 (5th revision).

Cardé, R.T., 2009. Exopterygota. In: Encyclopedia of Insects. Elsevier Publishers, https://doi.org/10.1016/B978-0-12-374144-8.00102-8. 339 p.

Carter, J.B., 1973a. The mode of transmission of Tipula iridescent virus. I. Source of infection. Invertebr. Pathol. 21, 123–130.

Carter, J.B., 1973b. The mode of transmission of Tipula iridescent virus. II. Route of infection. Invertebr. Pathol. 21, 136–143.

Carvalho, M., Schwudke, D., Sampaio, J.L., Palm, W., Riezman, I., Dey, G., et al., 2010. Survival strategies of a sterol auxotroph. Development 137, 3675–3685. https://doi.org/10.1242/dev.044560.

Chakraborty, S., Datta, S., 2003. How will plant pathogens adapt to host plant resistance at elevated CO_2 under a changing climate? New Phytol. 159 (3), 733–742. https://doi.org/10.1046/j.1469-8137.2003.00842.x.

Chandler, J.A., Eisen, J.A., Kopp, A., 2012. Yeast communities of diverse Drosophila species: comparison of two symbiont groups in the same hosts. Appl. Environ. Microbiol. 78, 7327–7336.

Chang, R., Duong, T.A., Taerum, S.J., Wingfield, M.J., Zhou, X., de Beer, Z.W., 2017. Ophiostomatoid fungi associated with conifer-infesting beetles and their phoretic mites in Yunnan, China. MycoKeys 28, 19–64. https://doi.org/10.3897/mycokeys.28.21758.

Chaverri, P., Liu, M., Hodge, K.T., 2008. A monograph of the entomopathogenic genera Hypocrella, Moelleriella, and Samuelsia gen. nov. (Ascomycota, Hypocreales, Clavicipitaceae), and their aschersonia-like anamorphs in the Neotropics. Stud. Mycol. 60, 1–66. https://doi.org/10.3114/sim.2008.60.01.

Christenson, L., Lovett, G.M., Mitchell, M.J., Groffman, P.M., 2002. The fate of nitrogen in gypsy moth frass deposited to an oak forest floor. Oecologia 131, 444–452. https://doi.org/10.1007/s00442-002-0887-7.

Chrostek, E., Pelz-Stelinski, K., Hurst, G.D.D., Hughes, G.L., 2017. Horizontal transmission of intracellular insect symbionts via plants. Front. Microbiol. 8, 2237. https://doi.org/10.3389/fmicb.2017.02237.

Coleman, D.C., Crossley, D.A., Hendrix, P.F., 2004. Secondary production: activities of heterotrophic organisms—the soil fauna. In: Coleman, D.C., Crossley, D.A., Hendrix, P.F. (Eds.), Fundamentals of Soil Ecology, second ed. Academic Press, Burlington, pp. 79–185, https://doi.org/10.1016/B978-012179726-3/50005-8.

Columbari, F., Schroeder, M.L., Battisti, A., Faccoli, M., 2013. Spatio-temporal dynamics of an Ips acuminatus outbreak and implications for management. Agric. For. Entomol. 15 (1), 34–42. https://doi.org/10.1111/j.1461-9563.2012.00589.x.

Cooley, J.R., Marshall, D.C., Hill, K.B.R., 2018. A specialized fungal parasite (Massosporacicadina) hijacks the sexual signals of periodical cicadas (Hemiptera: Cicadidae: Magicicada). Sci. Rep. 8 (1), 1432. https://doi.org/10.1038/s41598-018-19813-0.

Costa, C., 2000. Estado de conocimiento de los Coleoptera Neotropicales. In: Martín-Piera, F., Morrone, J.J., Melia, A. (Eds.), Hacia un Proyecto CYTED para el Inventario y Estimación de la Diversidad Entomológicaen Iberoamérica. Sociedad Entomológica Aragonesa, Zaragoza. 326 p.

Couch, J.N., 1938. The Genus Septobasidium. The University of North Carolina Press, USA.

Cranston, P.S., Penny, J.G., 2009. Chapter 199—phylogeny of insects. In: Resh, V.H., Ring, T. (Eds.), Cardé Encyclopedia of Insects, second ed. Academic Press, San Diego, pp. 780–793, https://doi.org/10.1016/B978-0-12-374144-8.00208-3.

Currie, C.R., Wong, B., Stuart, A.E., Schultz, T.R., Rehner, S.A., Mueller, U.G., et al., 2003. Ancient tripartite coevolution in the attine ant-microbe symbiosis. Science 299 (5605), 386–388. https://doi.org/10.1126/science.1078155.

Dantur, K.I., Enrique, R., Welin, B., Castagnaro, A.P., 2015. Isolation of cellulolytic bacteria from the intestine of Diatraea saccharalis larvae and evaluation of their capacity to degrade sugarcane biomass. AMB Express 5 (1), 15. https://doi.org/10.1186/s13568-015-0101-z.

Davis, T.S., 2015. The ecology of yeasts in the bark beetle holobiont: a century of research revisited. Microb. Ecol. 69 (4), 723–732. https://doi.org/10.1007/s00248-014-0479-1.

Davis, A.J., Holloway, J.D., Huijbregts, H., Krikken, J., Kirk-Spriggs, A.H., Sutton, S.L., 2001. Dung beetles as indicators of change in the forests of northern Borneo. J. Appl. Ecol. 38 (3), 593–616. https://doi.org/10.1046/j.1365-2664.2001.00619.x.

De Beer, Z.W., Duong, T.A., Barnes, I., Wingfield, B.D., Wingfield, M.J., 2014. Redefining ceratocystis and allied genera. Stud. Mycol. 79, 187–219. https://doi.org/10.1016/j.simyco.2014.10.001.

De Beer, Z.W., Marincowitz, S., Duong, T.A., Wingfield, M.J., 2017. Bretziella, a new genus to accommodate the oak wilt fungus, Ceratocystis fagacearum. MycoKeys 27, 1–19. https://doi.org/10.3897/mycokeys.27.20657.

De Beer, Z.W., Seifert, K.A., Wingfield, M.J., 2013. The ophiostomatoid fungi: their dual position in the Sordariomycetes. In: Seifert, K.A., De Beer, Z.W., Wingfield, M.J. (Eds.), The Ophiostomatoid Fungi: Expanding Frontiers. CBS, Utrecht, The Netherlands, pp. 1–19.

De Beer, Z.W., Wingfield, M.J., 2013. Emerging lineages in the ophiostomatales. In: Seifert, K.A., De Beer, Z.W., Wingfield, M.J. (Eds.), The Ophiostomatoid Fungi: Expanding Frontiers. CBS, Utrecht, The Netherlands, pp. 21–46.

Debbie, H., 2018. Grasshoppers, Crickets, and Katydids, Order Orthoptera. Available at: https://www.thoughtco.com/grasshoppers-crickets-katydids-order-orthoptera-1968344. (Accessed 28 April 2020).

Degnan, P.H., Moran, N.A., 2008. Diverse phage-encoded toxins in a protective insect endosymbiont. Appl. Environ. Microbiol. 74, 6782–6791. https://doi.org/10.1128/AEM.01285-08.

DeKesel, A., 1996. Host specificity and habitat preference of Laboulbenia slackensis. Mycologia 88 (4), 565–573. https://doi.org/10.1080/00275514.1996.12026687.

Dhandapani, M., Jayaraj, S., Rabindra, R.J., 1993. Cannibalism on nuclear polyhidrosis virus infected larvae by *Heliothisarmigera* (Hubn.) and its effect on viral infection. Insect Sci. Appl. 14 (4), 427–430. https://doi.org/10.1017/S1742758400014089.

Dohet, L., Grégoire, J.-L., Berasategui, A., Kaltenpoth, M., Biedermann, P.H.W., 2016. Bacterial and fungal symbionts of parasitic Dendroctonus bark beetles. FEMS Microbiol. Ecol. 92 (9), fiw129. https://doi.org/10.1093/femsec/fiw129.

Douglas, A.E., 1998. Nutritional interactions in insect-microbial symbioses: aphids and their symbiotic bacteria Buchnera. Annu. Rev. Entomol. 43, 17–37. https://doi.org/10.1146/annurev.ento.43.1.17.

Dunxiao, H., Chunru, H., Yaling, X., Banwang, H., Liyuan, H., Paoletti, M.G., 1999. Relationship between soil arthropods and soil properties in a Suburb of Qianjiang City, Hubei, China. Crit. Rev. Plant Sci. 18 (3), 467–473. https://doi.org/10.1016/S0735-2689(99)00378-0.

Durst, P.B., Hanboonsong, Y., 2015. Small-scale production of edible insects for enhanced food security and rural livelihoods: experience from Thailand and Lao People's Democratic Republic. J. Insects Food Feed 1 (1), 25–31. https://doi.org/10.3920/JIFF2014.0019.

Eichler, S., Schaub, G.A., 2002. Development of symbionts in triatomine bugs and the effects of infections with trypanosomatids. Exp. Parasitol. 100 (1), 17–27. https://doi.org/10.1006/expr.2001.4653.

Engel, P., Moran, N.A., 2013. The gut microbiota of insects—diversity in structure and function. FEMS Microbiol. Rev. 37 (5), 699–735. https://doi.org/10.1111/1574-6976.12025.

Evans, A.M., 2016. The speed of invasion: rates of spread for thirteen exotic forest insects and diseases. Forests 7 (5), 99. https://doi.org/10.3390/f7050099.

Evans, H.C., Samson, R.A., 1977. Sporodiniella umbellata, an entomogenous fungus of the Mucorales from cocoa farms in Ecuador. Can. J. Bot. 55 (23), 2981–2984. https://doi.org/10.1139/b77-334.

Evans, H.C., Samson, R.A., 1984. Cordyceps species and their anamorphs pathogenic on ants (Formicidae) in tropical forest ecosystems II. The Camponotus (Formicinae) complex. Trans. Br. Mycol. Soc. 82, 127–150.

Farrell, B.D., Sequeira, A.S., O'Meara, B.C., Normark, B.B., Chung, J.H., Jordal, B.H., 2001. The evolution of agriculture in beetles (Curculionidae: Scolytinae and Platypodinae). Evolution 55 (10), 2011–2027. https://doi.org/10.1111/j.0014-3820.2001.tb01318.x.

Feldhaar, H., 2011. Bacterial symbionts as mediators of ecologically important traits of insect hosts. Ecol. Entomol. 36 (5), 533–543. https://doi.org/10.1111/j.1365-2311.2011.01318.x.

Fernández-García, E., Fitt, B.D.L., 1993. Dispersal of the entomopathogen Hirsutella cryptosclerotium by simulated rain. J. Invertebr. Pathol. 61 (1), 39–43. https://doi.org/10.1006/jipa.1993.1007.

Ferreira, M.A., Harrington, T.C., Alfenas, A.C., Mizubuti, E.S.G., 2011. Movement of genotypes of Ceratocystis fimbriata within and among Eucalyptus plantations in Brazil. Phytopathology 101 (8), 1005–1012. https://doi.org/10.1094/PHYTO-01-11-0015.

Fisher, M., Henk, D.A., Briggs, C.J., Brownstein, J.S., Madoff, L.C., Mccraw, S.L., et al., 2012. Emerging fungal threats to animal, plant and ecosystem health. Nature 484, 186–194. https://doi.org/10.1038/nature10947.

Fleuriet, A., Periquet, G., 1993. Evolution of the *Drosophila melanogaster-sigma* virus systemin natural populations from Languedoc (southern France). Arch. Virol. 129 (1–4), 131–143. https://doi.org/10.1007/BF01316890.

Flórez, L.V., Scherlach, K., Gaube, P., Ross, C., Sitte, E., Hermes, C., et al., 2017. Antibiotic-producing symbionts dynamically transition between plant pathogenicity and insect-defensive mutualism. Nat. Commun. 8, 15172. https://doi.org/10.1038/ncomms15172.

Frago, E., Dicke, M., Godfray, H.C.J., 2012. Insect symbionts as hidden players in insect-plant interactions. Trends Ecol. Evol. 27 (12), 705–711. https://doi.org/10.1016/j.tree.2012.08.013.

Freitak, D., Schmidtberg, H., Dickel, F., Lochnit, G., Vogel, H., Vilcinskas, A., 2014. The maternal transfer of bacteria can mediate trans-generational immune priming in insects. Virulence 5 (4), 547–554. https://doi.org/10.4161/viru.28367.

Fukatsu, T., Hosokawa, T., 2002. Capsule-transmitted gut symbiotic bacterium of the Japanese common plataspid stinkbug, Megacopta punctatissima. Appl. Environ. Microbiol. 68 (1), 389–396. https://doi.org/10.1128/AEM.68.1.389-396.2002.

Furniss, R.L., Carolin, V.M., 1977. Western Forest Insects. U.S.D.A. Forest Service Misc. Pub. No. 1339, Washington, DC, 654 p.

Fuxa, J.R., Weidner, E.H., Richter, A.R., 1992. Polyhedra without virions in a vertically transmitted nuclear polyhidrosis virus. J. Invertebr. Pathol. 60 (1), 53–58. https://doi.org/10.1016/0022-2011(92)90153-U.

Gahukar, R.T., 2016. Edible insects farming: efficiency and impact on family livelihood, food security, and environment compared with livestock. In: Dossey, A.T., Morales-Ramos, J.A., Guadalupe Rojas, M. (Eds.), Insects as Sustainable Food Ingredients: Production, Processing and Food Applications. Elsevier, London, UK, pp. 85–111.

Ghelardini, L., Pepori, A.L., Luchi, N., Capretti, P., Santini, A., 2016. Drivers of emerging fungal diseases of forest trees. For. Ecol. Manage. 381, 235–246. https://doi.org/10.1016/j.foreco.2016.09.032.

Goberville, E., Hautekèete, N.C., Kirby, R.R., Piquot, Y., Luczak, C., Beaugrand, G., 2016. Climate change and the ash dieback crisis. Sci. Rep. 6, 35303. https://doi.org/10.1038/srep35303.

Grasela, J.J., McIntosh, A.H., Shelby, K.S., Long, S., 2008. Isolation and characterization of a baculovirus associated with the insect parasitoid Wasp, *Cotesia marginiventris*, or its host, *Trichoplusia ni*. J. Insect Sci. 8, 1–19. https://doi.org/10.1673/031.008.4201.

Gressitt, J.L., 2013. Notes on arthropod populations in the Antarctic Peninsula-South Shetland Islands-South Orkney Islands Area. AGU Books Publishers, USA. https://doi.org/10.1029/AR010p0373.

Grucmanová, S., Holuša, J., 2013. Nematodes associated with bark beetles, with focus on the genus *Ips* (Coleoptera: Scolytinae) in Central Europe. Acta Zool. Bulg. 65, 547–554.

Guay, J.F., Boudreault, S., Michaud, D., Cloutier, C., 2009. Impact of environmental stress on aphid clonal resistance to parasitoids: role of Hamiltonelladefensa bacterial symbiosis in association with a new facultative symbiont of the pea aphid. J. Insect Physiol. 10, 919–926. https://doi.org/10.1016/j.jinsphys.2009.06.006.

Guidolin, A.S., Cônsoli, F.L., 2017. Symbiont diversity of aphis (Toxoptera) citricidus (Hemiptera: Aphididae) as influenced by host plants. Microb. Ecol. 73, 201–210. https://doi.org/10.1007/s00248-016-0892-8.

Gullan, P.J., Cranston, P.S., 2010. The Insects: An Outline of Entomology. Blackwell Publishing, Hoboken, NJ. 584 p.

Gurung, K., Wertheim, B., Salles, J.F., 2019. The microbiome of pest insects: it is not just bacteria. Entomol. Exp. Appl. 167 (3), 156–170. https://doi.org/10.1111/eea.12768.

Haack, R.A., 2006. Exotic bark- and wood-boring Coleoptera in the United States: recent establishments and interceptions. Can. J. For. Res. 36, 269–288. https://doi.org/10.1139/x05-249.

Hajek, A.E., Soper, R.S., 1991. Within-tree location of gypsy moth, Lymantria dispar, larvae killed by Entomophagamaimaiga (Zygomycetes: Entomophthorales). J. Invertebr. Pathol. 58 (4), 468–469.

Hajek, A., Steinkraus, D., Castrillo, L., 2018. Sleeping beauties: horizontal transmission via resting spores of species in the Entomophthoromycotina. Insects 9, 102. https://doi.org/10.3390/insects9030102.

Haloi, K., Kalita, M.K., Nath, R., Devi, D., 2016. Characterization and pathogenicity assessment of gut-associated microbes of Muga Silkworm Antheraea assamensis Helfer (Lepidoptera: Saturniidae). J. Invertebr. Pathol. 138, 73–85. https://doi.org/10.1016/j.jip.06.006.

Hamby, K.A., Hernández, A., Boundy-Mills, K., Zalom, F.G., 2012. Associations of yeasts with spotted-wing Drosophila (*Drosophila suzukii*; Diptera: Drosophilidae) in cherries and raspberries. Appl. Environ. Microbiol. 78 (14), 4869–4873.

Hamden, H., M'saad Guerfali, M., Fadhl, S., Saidi, M., Chevrier, C., 2013. Fitness improvement of mass-reared sterile males of Ceratitis capitate (Vienna 8 strain) (Diptera: Tephritidae) after gut enrichment with probiotics. J. Econ. Entomol. 106 (2), 641–647. https://doi.org/10.1603/EC12362.

Hammer, T.J., Janzen, D.H., Hallwachs, W., Jaffe, S.P., Fierer, N., 2017. Caterpillars lack a resident gut microbiome. Proc. Natl. Acad. Sci. USA 114, 9641–9646. https://doi.org/10.1073/pnas.1707186114.

Hammer, T.J., McMillan, W.O., Fierer, N., 2014. Metamorphosis of a butterfly-associated bacterial community. PLoS One 9, e86995. https://doi.org/10.1371/journal.pone.0086995.

Hanewinkel, M., Cullmann, D.A., Schelhass, M.J., Nabuurs, G.-J., Zimmermann, N.E., 2013. Climate change may cause severe loss in the economic value of European forest land. Nat. Clim. Chang. 3, 203–207. https://doi.org/10.1038/nclimate1687.

Hansen, A.K., Moran, N.A., 2011. Aphid genome expression reveals host-symbiont cooperation in the production of amino acids. Proc. Natl. Acad. Sci. USA 108 (7), 2849–2854. https://doi.org/10.1073/pnas.1013465108.

Hardesty, B.D., 2011. Effectiveness of seed dispersal by ants in a Neotropical tree. Integr. Zool. 6, 222–226. https://doi.org/10.1111/j.1749-4877.2011.00246.x.

Harrington, T.C., Aghayeva, D.N., Fraedrich, S.W., 2010. New combinations in Raffaelea, Ambrosiella, and Hyalorhinocladiella, and four new species from the redbay ambrosia beetle, Xyleborusglabratus. Mycotaxon 111, 337–361. https://doi.org/10.5248/111.337.

Harrington, T.C., Yun, H.Y., Lu, S.-S., Goto, H., Aghayeva, D.N., Fraedrich, S.W., 2011. Isolations from the redbay ambrosia beetle, Xyleborusglabratus, confirm that the laurel wilt pathogen, Raffaelealauricola, originated in Asia. Mycologia 103 (5), 1028–1036. https://doi.org/10.3852/10-417.

Hartig, T., 1844. Abrosia des Bostrichusdispar. Allg. Forst. Jagdztg. 13, 73.

Hartig, T., 1878. Die Zersetzungserscheinungen des Holzes der Nadelbäume und der Eiche in forstlicher, botanischer und chemischer Richtung. Springer, Berlin, Germany.

Hedges, L.M., Brownlie, J.C., O'Neill, S.L., Johnson, K.N., 2008. Wolbachia and virus protection in insects. Science 322 (5902), 702. https://doi.org/10.1126/science.1162418.

Helfer, S., 2013. Rust fungi and global change. New Phytol. 201 (3), 770–780. https://doi.org/10.1111/nph.12570.

Henry, L.M., Maiden, M.C.J., Ferrari, J., Godfray, H.C.J., 2015. Insect life history and the evolution of bacterial mutualism. Ecol. Lett. 18, 516–525. https://doi.org/10.1111/ele.12425.

Hernández-Flores, L., Llanderal-Cázares, C., Guzmán-Franco, A.W., Aranda-Ocampo, S., 2015. Bacteria present in Comadia redtenbacheri larvae (Lepidoptera: Cossidae). J. Med. Entomol. 52 (5), 1150–1158. https://doi.org/10.1093/jme/tjv099.

Herniou, E.A., Huguet, E., Thézé, J., Bézier, A., Periquet, G., Drezen, J.M., 2013. When parasitic wasps hijacked viruses: genomic and functional evolution of polydnaviruses. Philos. Trans. R. Soc. Lond. B Biol. Sci. 368 (1626). 20130051. 10.1098/rstb.2013.0051.

Hibbett, D.S., Binder, M., Bishoff, J.F., Blackwell, M., Cannon, B.F., Erikssson, O.E., et al., 2007. A higher-level phylogenetic classification of the fungi. Mycol. Res. 111 (5), 509–547. https://doi.org/10.1016/j.mycres.2007.03.004.

Hill, D.S., 1997. The Economic Importance of Insects. Chapman & Hall, United Kingdom. 399 p.

Hlásny, T., Krokene, P., Liebhold, A., Montagné-Huck, C., Müller, J., Raffa, K., et al., 2019. Living with bark beetles: impacts, outlook and management options. From Science to Policy 8, European Forest Institute, https://doi.org/10.36333/fs08.

Hoffmann, A.A., 2017. Rapid adaptation of invertebrate pests to climatic stress? Curr. Opin. Insect. Sci. 21, 7–13. https://doi.org/10.1016/j.cois.2017.04.009.

Hofstetter, R.W., Dinkins-Bookwalter, J., Davis, T.S., Klepzig, K.D., 2015. Symbiotic associations of bark beetles. In: Vega, F.E., Hofstetter, R.W. (Eds.), Bark Beetles. Academic Press, San Diego, pp. 209–245.

Hofstetter, R.W., Moser, J.C., 2014. The role of mites in insect-fungus associations. Annu. Rev. Entomol. 59, 537–557. https://doi.org/10.1146/annurev-ento-011613-162039.

Hongoh, Y., 2010. Diversity and genomes of uncultured microbial symbionts in the termite gut. Biosci. Biotechnol. Biochem. 74 (86), 1145–1151. https://doi.org/10.1271/bbb.100094.

Hongoh, Y., 2011. Toward the functional analysis of uncultivable, symbiotic microorganisms in the termite gut. Cell. Mol. Life Sci. 68 (8), 1311–1325. https://doi.org/10.1007/s00018-011-0648-z.

Horgan, F., Ferrater, J., 2017. Benefits and potential trade-offs associated with yeast-like symbionts during virulence adaptation in a phloem-feeding planthopper. Entomol. Exp. Appl. 163 (1), 112–125. https://doi.org/10.1111/eea.12556.

Hosokawa, T., Kikuchi, Y., Nikoh, N., Shimada, M., Fukatsu, T., 2006. Strict host-symbiont cospeciation and reductive genome evolution in insect gut bacteria. PLoS Biol. 4, e337. https://doi.org/10.1371/journal.pbio.0040337.

Hövemeyer, K., Schauermann, J., 2003. Succession of Diptera on dead beech wood: a 10-year study. Pedobiologia 47, 61–75. https://doi.org/10.1078/0031-4056-00170.

Huis, A.V., 2013. Edible Insects: Future Prospects for Food and Feed Security. Food and Agriculture Organization of the United Nations, Rome, pp. 1–154. Forestry Paper, 171.

Hulcr, J., Barnes, I., De Beer, Z.W., Duong, T.A., Gazis, R., Johnson, A.J., et al., 2020. Bark beetle mycobiome: collaboratively defined research priorities on a widespread insect-fungus symbiosis. Symbiosis 81, 101–113. https://doi.org/10.1007/s13199-020-00686-9.

Hulcr, J., Dunn, R.R., 2011. The sudden emergence of pathogenicity in insect–fungus symbioses threatens naive forest ecosystems. Proc. R. Soc. B Biol. Sci. 278, 2866–2873. https://doi.org/10.1098/rspb.2011.1130.

Hurst, G.D.D., 2017. Extended genomes: symbiosis and evolution. Interface Focus 7. https://doi.org/10.1098/rsfs.2017.0001. 20170001.

Iijima, R., Kurata, S., Natori, S., 1993. Purification, characterization, and eDNA cloning of an antifungal protein from the hemolymph of Sarcophaga peregrina (flesh fly) larvae. J. Biol. Chem. 268 (16), 12055–12061.

Irmler, U., Helier, K., Warning, J., 1996. Age and tree species as factors influencing the populations of insects living in dead wood (Coleoptera, Diptera: Sciaridae, Mycetophilidae). Pedobiologia 40, 134–148.

Iverson, L.R., Prasad, A.M., 1998. Predicting abundance of 80 tree species following climate change in the eastern United States. Ecol. Monogr. 68, 465–485.

Jacobs, K., Wingfield, M.J., 2001. Leptographium Species: Tree Pathogens, Insect Associates, and Agents of Blue-Stain. American Phytopathological Society (APS Press), St. Paul.

Jacobsen, R.M., Kauserud, H., Sverdrup-Thygeson, A., Bjorbækmo, M.M., Birkemoe, T., 2017. Wood-inhabiting insects can function as targeted vectors for decomposer fungi. Fungal Ecol. 29, 76–84.

Jankevica, L., 2004. Ecological association between entomopathogenic fungi and pest insects recorded in Latvia. Lativ. Entomol. 41, 60–65.

Jankielsohn, A., 2018. The importance of insects in agricultural ecosystems. Adv. Entomol. 6, 62–73. https://doi.org/10.4236/ae.2018.62006.

Jones, D.R., 2003. Plant viruses transmitted by whiteflies. Eur. J. Plant Pathol. 109, 195–219. https://doi.org/10.1023/A:1022846630513.

Jongema, Y., 2017. List of Edible Insects of the World, 2017. Available at: https://www.wur.nl/en/Research-Results/Chairgroups/Plant-Sciences/Laboratory-of-Entomology/Edible-insects/Worldwide-species-list.htm. (Accessed 9 December 2019).

Kasson, M.T., Wickert, K.L., Stauder, C.M., Macias, A.M., Berger, M.C., Simmons, D.R., et al., 2016. Mutualism with aggressive wood-degrading Flavodonambrosius (Polyporales) facilitates niche expansion and communal social structure in Ambrosiophilus ambrosia beetles. Fungal Ecol. 23, 86–96. https://doi.org/10.1016/j.funeco.2016.07.002.

Kaufman, M.G., Walker, E.D., Odelson, D.A., Klug, M.J., 2000. Microbial community ecology & insect nutrition. Am. Entomol. 46 (3), 173–185. https://doi.org/10.1093/ae/46.3.173.

Kelemu, S., Niassy, S., Torto, B., Fiaboe, K., Affognon, H., Tonnang, H., et al., 2015. African edible insects for food and feed: inventory, diversity, commonalities and contribution to food security. J. Insects Food Feed 1, 103–119. https://doi.org/10.1016/j.funeco.2016.07.002.

Keller, S., Zimmermann, G., Wilding, N., Collins, N.M., Hammond, P.M., Webber, J.F., 1989. Mycopathogens of soil insects. In: Insect-Fungus Interactions. Academic Press, United States, pp. 239–270.

Khan, H., Perveen, F., 2015. Distribution of butterflies (Family Nymphalidae) in Union Council Koaz Bahram Dheri, Khyber Pakhtunkhwa, Pakistan. Soc. Basic Sci. Res. Rev. 3 (1), 52–57.

Kikuchi, Y., Hayatsu, M., Hosokawa, T., Nagayama, A., Tago, K., Fukatsu, T., 2012. Symbiont-mediated insecticide resistance. Proc. Natl. Acad. Sci. USA 109 (22), 8618–8622. https://doi.org/10.1073/pnas.1200231109.

Kikuchi, Y., Hosokawa, T., Fukatsu, T., 2007. Insect-microbe mutualism without vertical transmission: a stinkbug acquires a beneficial gut symbiont from the environment every generation. Appl. Environ. Microbiol. 73, 4308–4316. https://doi.org/10.1128/AEM.00067-07.

Kikuchi, Y., Hosokawa, T., Fukatsu, T., 2008. Diversity of bacterial symbiosis in stinkbugs. In: Dijk, T.V. (Ed.), Microbial Ecology Research Trends. Nova Science Publishers, New York, NY, pp. 39–63.

Kikuchi, Y., Hosokawa, T., Fukatsu, T., 2011. An ancient but promiscuous host-symbiont association between Burkholderia gut symbionts and their heteropteran hosts. ISME J. 5, 446–460. https://doi.org/10.1038/ismej.2010.150.

Kikuchi, Y., Meng, X.Y., Fukatsu, T., 2005. Gut symbiotic bacteria of the genus Burkholderia in the broad-headed bugs *Riptortus clavatus* and *Leptocorisa chinensis* (Heteroptera: Alydidae). Appl. Environ. Microbiol. 71, 4035–4043.

Kile, G.A., 1993. Plant diseases caused by species of Ceratocystis sensustricto and Chalara. In: Wingfield, M.J., Seifert, K.A., Webber, J. (Eds.), Ceratocystis and Ophiostoma: Taxonomy, Ecology and Pathogenicity. APS Press, St. Paul, pp. 173–184.

Kim, K.C., 1993. Biodiversity, conservation and inventory: why insects matter. Biodivers. Conserv. 2, 191–214. https://doi.org/10.1007/BF00056668.

Kirisits, T., 2004. Fungal associates of European bark beetles with special emphasis on the ophiostomatoid fungi. In: Lieutier, F, Day, KR, Battisti, A, Gregoire, JC, Evans, HF (Eds.), Bark and Wood Boring Insects in Living Trees in Europe, A Synthesis. Kluwer Academic Publishers, Dordrecht, pp. 181–235.

Klinger, E.G., James, R.R., Youssef, N.N., Welker, D.L., 2013. A multi-gene phylogeny provides additional insight into the relationships between several Ascosphaera species. J. Invertebr. Pathol. 112 (1), 41–48. https://doi.org/10.1016/j.jip.2012.10.011.

Kobayasi, Y., Shimizu, D., 1977. Two new species of Podonectria (Clavicipitaceae). Bull. Natl. Mus. Nat. Sci., Ser. B, Bot. 017574592.

Koga, R., Bennett, G.M., Cryan, J.R., Moran, N.A., 2013. Evolutionary replacement of obligate symbionts in an ancient and diverse insect lineage. Environ. Microbiol. 15, 2073–2081. https://doi.org/10.1111/1462-2920. 12121.

Kosola, K.R., Dickmann, D.I., Paul, E.A., Parry, D., 2001. Repeated insect defoliation effects on growth, nitrogen acquisition, carbohydrates, and root demography of poplars. Oecologia 129, 65–74. https://doi.org/10.1007/s004420100694.

Kostovcik, M., Bateman, C.C., Kolarik, M., Stelinski, L.L., Jordal, B.H., Hulcr, J., 2015. The ambrosia symbiosis is specific in some species and promiscuous in others: evidence from community pyrosequencing. ISME J. 9, 126–138. https://doi.org/10.1038/ismej.2014.115.

Krams, I.A., Kecko, S., Jõers, P., Trakimas, G., Elferts, D., Krams, R., et al., 2017. Microbiome symbionts and diet diversity incur costs on the immune system of insect larvae. J. Exp. Biol. 220, 4204–4212. https://doi.org/10.1242/jeb.169227.

Kuechler, S.M., Renz, P., Dettner, K., Kehl, S., 2012. Diversity of symbiotic organs and bacterial endosymbionts of lygaeoid bugs of the families Blissidae and Lygaeidae (Hemiptera: Heteroptera: Lygaeoidea). Appl. Environ. Microbiol. 78, 2648–2659. https://doi.org/10.1128/AEM.07191-11.

Kurtzman, C.D., Fell, J.W., Boekhout, T. (Eds.), 2011. The Yeasts: A Taxonomic Study. Elsevier, USA.

Lacey, L.A., Frutos, R., Kaya, H.K., Vail, P., 2001. Insect pathogens as biological control agents: do they have a future? Biol. Control 21 (3), 230–248. https://doi.org/10.1006/bcon.2001.0938.

Landry, M., Comeau, A.M., Derome, N., Cusson, M., Levesque, R.C., 2015. Composition of the spruce budworm (*Choristoneura fumiferana*) midgut microbiota as affected by rearing conditions. PLoS One 10, e0144077. https://doi.org/10.1371/journal.pone.0144077.

Lawrence, J.F., Britton, E.B., 1991. Coleoptera. In: Naumann, I. (Ed.), The Insects of Australia. Cornell University Press, New York, pp. 543–683.

Leal, I., Leal, L., Andersen, A., 2015. The benefits of myrmecochory: a matter of stature. Biotropica 47 (3), 281–285. https://doi.org/10.1111/btp.12213.

Ledón-Rettig, C.C., Moczek, A.P., Ragsdale, E.J., 2008. Diplogastrellus nematodes are sexually transmitted mutualist that alter the bacterial and fungal communities of their beetle host. Proc. Natl Acad. Sci. USA 115 (42), 10696–10701. https://doi.org/10.1073/pnas.1809606115.

Lee, C.M., Kwon, T.-S., Ji, O.Y., Kim, S.-S., Park, G.-E., Lim, J.-H., 2015. Prediction of abundance of forest flies (Diptera) according to climate scenarios RCP 4.5 and RCP 8.5 in South Korea. J. Asia-Pacific Biodiv. 8, 349–370. https://doi.org/10.1016/j.japb.2015.10.009.

Lee, K.E., Wood, T.G., 1971. Termites and Soils. Academic Press, London.

Lengyel, S., Gove, A.D., Latimer, A.M., Majer, J.D., Dunn, R.R., 2009. Ants sow the seeds of global diversification in flowering plants. PLoS One 4 (5), e5480. https://doi.org/10.1371/journal.pone.0005480.

Leroy, P.D., Sabri, A., Heuskin, S., Thonart, P., Lognay, G., Verheggen, F.J., et al., 2011. Microorganisms from aphid honeydew attract and enhance the efficacy of natural enemies. Nat. Commun. 2, 348. https://doi.org/10.1038/ncomms1347.

Li, Y., Huang, Y.-T., Kasson, M.T., Macias, A.M., Skelton, J., Carlson, P.S., et al., 2018. Specific and promiscuous ophiostomatalean fungi associated with Platypodinae ambrosia beetles in the southeastern United States. Fungal Ecol. 35, 42–50. https://doi.org/10.1016/j.funeco.2018.06.006.

Li, T., Wu, X.-J., Jiang, Y.-L., Zhang, L., Duan, Y., Miao, J., et al., 2016. The genetic diversity of SMLS (*Sitobion miscanthi* L type symbiont) and its effect on the fitness, mitochondrial DNA diversity and *Buchneraaphidicola* dynamic of wheat aphid, Sitobion miscanthi (Hemiptera: Aphididae). Mol. Ecol. 25, 3142–3151. https://doi.org/10.1111/mec.13669.

Liebhold, A.M., Brockerhoff, E.G., Garrett, L.J., Parke, J.L., Britton, K.O., 2012. Live plant imports: the major pathway for forest insect and pathogen invasions of the US. Front. Ecol. Environ. 1 (3), 135–143. https://doi.org/10.1890/110198.

Linley, J.R., Nielson, H.T., 1968a. Transmission of a mosquito iridescent virus in Aedes taeniorhynchus I. Laboratory experiments. J. Invertebr. Pathol. 12, 7–26.

Linley, J.R., Nielson, H.T., 1968b. Transmission of a mosquito iridescent virus in Aedes taeniorhynchus II. Experiments related to transmission in nature. J. Invertebr. Pathol. 12, 17–24.

Liu, W., Gray, S., Huo, Y., Li, L., Wei, T., Wang, X., 2015. Proteomic analysis of interaction between a plant virus and its vector insect reveals new functions of Hemipteran cuticular protein. Mol. Cell. Proteomics 4 (8), 2229–2242. https://doi.org/10.1074/mcp.M114.046763.

Liu, N., Yan, X., Zhang, M., Xie, L., Wang, Q., Huang, Y., et al., 2011. Microbiome of fungus-growing termites: a new reservoir for lignocellulase genes. Appl. Environ. Microbiol. 77, 48–56. https://doi.org/10.1128/AEM.01521-10.

Lloyd, M., White, J.A., Stanton, N., 1982. Dispersal of fungus-infected cicadas to new habitat. Environ. Entomol. 11, 852–858.

Lucarotti, C.J., Klein, M.B., 1988. Pathology of *Coelomomyces stegomyiae* in adult *Aedes aegypti* ovaries. Can. J. Bot. 66, 877–884.

Macfadyen, S., Kramer, E.A., Parry, H.R., Schellhorn, N.A., 2015. Temporal change in vegetation productivity in grain production landscapes: linking landscape complexity with pest and natural enemy communities. Ecol. Entomol. 40, 56–69. https://doi.org/10.1111/een.12213.

Makonde, H.M., Boga, H.I., Osiemo, Z., Mwirichia, R., Mackenzie, L.M., et al., 2013. 16S-rRNA-based analysis of bacterial diversity in the gut of fungus-cultivating termites (Microtermes and Odontotermes species). Antonie Van Leeuwenhoek 104, 869–883. https://doi.org/10.1007/s10482-013-0001-7.

Malacrinò, A., Campolo, O., Medina, R.F., Palmeri, V., 2018. Instar-and host-associated differentiation of bacterial communities in the Mediterranean fruit fly Ceratitiscapitata. PLoS One 13, e0194131. https://doi.org/10.1371/journal.pone.0194131.

Malloch, D.W., Blackwell, M., 1993. Dispersal biology of the ophiostomatoid fungi. In: Wingfield, M.J., Seifert, K.A., Webber, J. (Eds.), Ceratocystis and Ophiostoma: Taxonomy, Ecology and Pathogenicity. APS Press, St. Paul, pp. 195–206.

Martikainen, P., Siitonen, J., Kaila, L., Punttila, P., Rauh, J., 1999. Bark beetles (Coleoptera, Scolytidae) and associated beetle species in mature managed and old-growth boreal forests in southern Finland. For. Ecol. Manag. 116, 233–245. https://doi.org/10.1016/S0378-1127(98)00462-9.

Martinez-Mota, R., Serio-Silva, J., Rico-Gray, V., 2006. The role of canopy ants in removing icus perforate seeds from howler monkey (Alouatta palliata mexicana) Feces at Los Tuxtlas, México. Biotropica 36, 429–432. https://doi.org/10.1111/j.1744-7429.2004.tb00338.x.

Mason, C.J., Raffa, K.F., 2014. Acquisition and structuring of midgut bacterial communities in gypsy moth (Lepidoptera: Erebidae) larvae. Environ. Entomol. 43, 595–604.

Mason, K.L., Stepien, T.A., Blum, J.E., Holt, J.F., Labbe, N.H., Rush, J.S., et al., 2011. From commensal to pathogen: translocation of Enterococcus faecalis from the midgut to the hemocoel of Manduca sexta. MBio2. https://doi.org/10.1128/mBio.00065-11. e00065-11.

Mattson, W.J., Addy, N.D., 1975. Phytophagous insects as regulators of forest primary production. Science 190, 515–522. https://doi.org/10.1126/science.190.4214.515.

Mayers, C.G., McNew, D.L., Harrington, T.C., Roeper, R.A., Fraedrich, S.W., Biedermann, P.H.W., et al., 2015. Three genera in the Ceratocystidaceae are the respective symbionts of three independent lineages of ambrosia beetles with large, complex mycangia. Fungal Biol. 119, 1075–1092. https://doi.org/10.1016/j.funbio.2015.08.002.

Mereghetti, V., Chouaia, B., Montagna, M., 2017. New insights into the microbiota of moth pests. Int. J. Mol. Sci. 18 (11), 2450. https://doi.org/10.3390/ijms18112450.

Merritt, R.W., De Jong, G.D., 2015. Arthropod communities in terrestrial environments. In: Benbow, M.E., Tomberlin, J.K., Tarone, A.M. (Eds.), Carrion Ecology, Evolution, and Their Applications. CRC Press, Boca Raton, FL, pp. 65–91.

Metcalfe, D.B., Asner, G.P., Martin, R.E., Silva Espejo, J.E., Huasco, W.H., Farfán Amézquita, F.F., et al., 2014. Herbivory makes major contributions to ecosystem carbon and nutrient cycling in tropical forests. Ecol. Lett. 17, 324–332. https://doi.org/10.1111/ele.12233.

Mohanta, M.K., Saha, A.K., Saleh, D.K.M.A., Islam, M.S., Mannan, K.S.B., Fakruddin, M., 2014. Characterization of Klebsiella granulomatis pathogenic to silkworm, Bombyx mori L. 3 Biotech 5 (4), 577–583. https://doi.org/10.1007/s13205-014-0255-4.

Montllor, C.B., Maxmen, A., Purcell, A.H., 2002. Facultative bacterial endosymbionts benefit pea aphids Acyrthosiphonpisum under heat stress. Ecol. Entomol. 27, 189–195. https://doi.org/10.1046/j.1365-2311.2002.00393.x.

Morais, P.B., Rosa, C.A., Hagler, A.N., Hagler, L.C.M., 1995. Yeast communities as descriptor of habitat use by the Drosophila fasciola subgroup (repleta group) in Atlantic rain forests. Oecologia 104, 45–51.

Moran, N.A., Degnan, P., Santos, S.R., Dunbar, H.E., Ochman, H., 2005. The players in a mutualistic symbiosis: insects, bacteria, viruses, and virulence genes. Proc. Natl Acad. Sci. USA 102, 16919–16926. https://doi.org/10.1073/pnas.0507029102.

Moran, N.A., McCutcheon, J.P., Nakabachi, A., 2008. Genomics and evolution of heritable bacterial symbionts. Annu. Rev. Gen. 42 (1), 165–190.

Morrow, J.L., Frommer, M., Shearman, D.C., Riegler, M., 2015. The microbiome of field-caught and laboratory-adapted Australian tephritid fruit fly species with different host plant use and specialisation. Microb. Ecol. 70 (2), 498–508. https://doi.org/10.1007/s00248-015-0571-1.

Moser, J.C., 1985. Use of sporothecae by phoretic *Tarsonemus mites* to transport ascospores of coniferous bluestain fungi. Trans. Br. Mycol. Soc. 84, 750–753.

Moser, J.C., Perry, T.J., Solheim, H., 1989. Ascospores hyperphoretic on mites associated with Ips typographus. Mycol. Res. 93, 513–517. https://doi.org/10.1016/S0953-7562(89)80045-0.

Mueller, U.G., Gerardo, N., 2002. Fungus-farming insects: multiple origins and diverse evolutionary histories. PNAS 99 (24), 15247–15249. https://doi.org/10.1073/pnas.242594799.

Mueller, U.G., Gerardo, N.M., Aanen, D.K., Six, D.L., Schultz, T.R., 2005. The evolution of agriculture in insects. Annu. Rev. Ecol. Evol. Syst. 36, 563–595. https://doi.org/10.1146/annurev.ecolsys.36.102003.152626.

Myers, J.H., 1993. Population outbreaks in forest Lepidoptera. Am. Sci. 81, 240–251.

Myers, N., 1997. The worlds forests and their ecosystem services. In: Daily, G. (Ed.), Nature's Services: Societal Dependence on Natural Ecosystems. Island Press, Washington, DC, pp. 215–235.

Myers, J.H., Rothman, L.E., 1995. Virulence and transmission of infectious diseases in humans and insects: evolutionary and demographic patterns. Trends Ecol. Evol. 10, 194–198.

Nair, K., 2007. Ecology of insects in the forest environment. In: Tropical Forest Insect Pests: Ecology, Impact, and Management. Cambridge University Press, Cambridge, pp. 57–77, https://doi.org/10.1017/CBO9780511542695.005.

Nalepa, C.A., 2015. Origin of termite eusociality: trophallaxis intergrates the social, nutritional, and microbial environments. Ecol. Entomol. 40 (4), 323–335. https://doi.org/10.1111/een.12197l.

Nel, W.J., Duong, T.A., Wingfield, B.D., Wingfield, M.J., De Beer, Z.W., 2017. A new genus and species for the globally important, multi-host root pathogen Thielaviopsis basicola. Plant Pathol. 67 (4), 871–882. https://doi.org/10.1111/ppa.12803.

Nguyen, N.H., Suh, S.O., Blackwell, M., 2007. Five novel Candida species in insect-associated yeast clades isolated from Neuroptera and other insects. Mycologia 99 (6), 842–858. https://doi.org/10.3852/mycologia.99.6.842.

Nilsson, R.H., Anslan, S., Bahram, M., Wurzbacher, C., Baldrian, P., Tedersoo, L., 2019. Mycobiome diversity: high-throughput sequencing and identification of fungi. Nat. Rev. Microbiol. 17, 95–109. https://doi.org/10.1038/s41579-018-0116-y.

Norman, H.A., 2009. Megaloptera: alderflies, fishflies, hellgrammites, dobsonflies. In: Resh, V.H., Cardé, R.T. (Eds.), Encyclopedia of Insects, second ed. Academic Press, San Diego, pp. 620–623, https://doi.org/10.1016/B978-0-12-374144-8.00173-9.

Ohkuma, M., Brune, A., 2011. Diversity, structure, and evolution of the termite gut microbial community. In: Bignell, D.E., Roisin, Y., Lo, N. (Eds.), Biology of Termites: A Modern Synthesis. Springer, Netherlands, pp. 413–438.

Økland, B., 1994. Mycetophilidae (Diptera), an insect group vulnerable to forestry practices? A comparison of clearcut, managed and seminatural spruce forests in southern Norway. Biodivers. Conserv. 3 (1), 68–85. https://doi.org/10.1007/bf00115334.

Oliver, K.M., Degnan, P.H., Burke, G.R., Moran, N.A., 2010. Facultative symbionts in aphids and the horizontal transfer of ecologically important traits. Annu. Rev. Entomol. 55, 247–266. https://doi.org/10.1146/annurev-ento-112408-085305.

Oliver, K.M., Moran, N.A., Hunter, M.S., 2005. Variation in resistance to parasitism in aphids is due to symbionts not host genotype. Proc. Natl Acad. Sci. USA 102, 12795–12800. https://doi.org/10.1073/pnas.0506131102.

Ostiguy, N., 2011. Pests and pollinators. Nat. Educ. Knowl. 3 (10), 3. Pakistan. J. Entomol. Zool. Stud. 2(1) 56–69.

Paap, T., de Beer, Z.W., Migliorini, D., Nel, W.J., Wingfield, M.J., 2018. The polyphagous shot hole borer (PSHB) and its fungal symbiont Fusarium euwallaceae: a new invasion in South Africa. Australas. Plant Pathol. 47, 231–237. https://doi.org/10.1007/s13313-018-0545-0.

Paine, T., Raffa, K., Harrington, T., 1997. Interactions among scolytid bark beetles, their associated fungi, and live host conifers. Annu. Rev. Entomol. 42, 179–206. https://doi.org/10.1146/annurev.ento.42.1.179.

Paniagua Voirol, L.R., Frago, E., Kaltenpoth, M., Hilker, M., Fatouros, N.E., 2018. Bacterial symbionts in Lepidoptera: their diversity, transmission, and impact on the host. Front. Microbiol. 9. https://doi.org/10.3389/fmicb.2018.00556.

Parsons, Y.M., 2010. Orthopteran behavioral genetics. In: Breed, M.D., Moore, J. (Eds.), Encyclopedia of Animal Behavior. Academic Press, Oxford, pp. 604–609, https://doi.org/10.1016/B978-0-08-045337-8.00271-0.

Pasternak, Z., Jurkevitch, E., Yuval, B., 2014. Symbiotic bacteria enable olive flies (B actroceraoleae) to exploit intractable sources of nitrogen. J. Evol. Biol. 27 (12), 2695–2705. https://doi.org/10.1111/jeb.12527.

Pavlidi, N., Gioti, A., Wybouw, N., Dermauw, W., Ben-Yosef, M., Yuval, B., et al., 2017. Transcriptomic responses of the olive fruit fly Bactroceraoleae and its symbiont Candidatus Erwinia dacicola to olive feeding. Sci. Rep. 7, 42633. https://doi.org/10.1038/srep42633.

Perveen, F., Fazal, F., 2013. Biology and distribution of butterfly fauna of Hazara University, Garden Campus, Mansehra, Pakistan. Open J. Anim. Sci. 3 (2), 28–36.

Perveen, F., Khan, A.S., 2014. Characteristics of butterfly (Lepidoptera) fauna from Kabal, Swat, Pakistan. J. Entomol. Zool. Stud. 2 (1), 56–59.

Phaff, H.J., Miller, M., Recca, J., Shifrine, M., Mrak, E., 1956. Yeasts found in the alimentary canal of Drosophila. Ecology 37, 533–538.

Philip, E., Nancy, A.M., 2013. The gut microbiota of insects—diversity in structure and function. FEMS Microb. Rev. 37 (5), 699–735. https://doi.org/10.1111/1574-6976.12025.

Piper, A.M., Farnier, K., Linder, T., Speight, R., Cunningham, J.P., 2017. Two gut-associated yeasts in a Tephritid fruit fly have contrasting effects on adult attraction and larval survival. J. Chem. Ecol. 43 (9), 891–901. https://doi.org/10.1007/s10886-017-0877-1.

Ploetz, R.C., Hulc, J., Wingfield, M.J., de Beer, Z.W., 2013. Destructive tree diseases associated with ambrosia and bark beetles: black swan events in tree pathology? Plant Dis. 97 (7), 856–872. https://doi.org/10.1094/PDIS-01-13-0056-FE.

Poliakov, A., Russell, C.W., Ponnala, L., Hoops, H.J., Sun, Q., Douglas, A.E., et al., 2011. Large-scale label-free quantitative proteomics of the pea aphid-Buchnera symbiosis. Mol. Cell. Proteomics 10 (6). https://doi.org/10.1074/mcp.M110.007039.

Pons, I., Renoz, F., Noël, C., Hance, T., 2019. New insights into the nature of symbiotic associations in aphids: infection process, biological effects and transmission mode of cultivable Serratia symbiotica bacteria. Appl. Environ. Microbiol. 85. e02445-18. 10.1128/AEM.02445-18.

Powell, J.A., 2009. Lepidoptera. In: Resh, V.H., Cardé, R.T. (Eds.), Encyclopedia of Insects, second (illustrated) ed. Academic Press, ISBN: 978-0-12-374144-8, pp. 557–587.

Powell, J.S., Raffa, K.F., 1999. Effects of selected Larix laricina terpenoids on Lymantria dispar (Lepidoptera: Lymantriidae) development and behavior. Environ. Entomol. 28, 148–154. https://doi.org/10.1093/ee/28.2.148.

Prado, S.S., Rubinoff, D., Almeida, R.P.P., 2006. Vertical transmission of a pentatomid caeca-associated symbiont. Ann. Entomol. Soc. Am. 99 (3), 577–585. https://doi.org/10.1603/0013-8746(2006)99[577:VTOAPC]2.0.CO;2.

Priya, N.G., Ojha, A., Kajla, M.K., Raj, A., Rajagopal, R., 2012. Host plant induced variation in gut bacteria of Helicoverpa armigera. PLoS One 7 (1). https://doi.org/10.1371/journal.pone.0030768.

Pyšek, P., Hulme, P.E., Simberloff, D., Bacher, S., Blackburn, T.M., Carlton, J.T., et al., 2020. Scientists' warning on invasive alien species. Biol. Rev. 95, 1511–1534. https://doi.org/10.1111/brv.12627.

Quicke, D.L.J., 2009. Hymenoptera: ants, bees, wasps. In: Resh, V.H., Cardé, R.T. (Eds.), Encyclopedia of Insects, second ed. Academic Press, San Diego, pp. 473–484, https://doi.org/10.1016/B978-0-12-374144-8.00136-3.

Raffa, K.F., Powell, E.N., Townsend, P.A., 2013. Temperature-driven range expansion of an irruptive insect heightened by weakly coevolved plant defenses. Proc. Natl. Acad. Sci. 110, 2193–2198. https://doi.org/10.1073/pnas.1216666110.

Ramírez-Camejo, L.A., Maldonado-Morales, G., Bayman, P., 2017. Differential microbial diversity in *Drosophila melanogaster*: are fruit flies potential vectors of opportunistic pathogens? Int. J. Microbiol. 2017, 8526385.

Ranjith, M.T., Chellappan, M., Harish, E.R., Girija, D., Nazeem, P.A., 2016. Bacterial communities associated with the gut of tomato fruit borer, Helicoverpa armigera (Hübner) (Lepidoptera: Noctuidae) based on illumina next-generation sequencing. J. Asia Pac. Entomol. 19 (2), 333–340. https://doi.org/10.1016/j.aspen.2016.03.007.

Redak, R.A., Purcell, A.H., Lopes, J.R.S., Blua, M.J., Mizell, R.F., Andersen, P.C., 2004. The biology of xylem fluid-feeding insect vectors of Xylella fastidiosa and their relation to disease epidemiology. Annu. Rev. Entomol. 49, 243–270. https://doi.org/10.1146/annurev.ento.49.061802.123403.

Resh, V.H., Carde, R.T., 2003. Encyclopedia of Insects. Academic Press, ISBN: 978-0-08-054605-6, p. 64.

Robertson, J., Tsubouchi, A., Tracey, W.D., 2013. Larval defense against attack from parasitoid wasps requires nociceptive neurons. PLoS One 8, e78704. https://doi.org/10.1371/journal.pone.0078704.

Rothman, L.D., Myers, J.H., 2000. Ecology of insect viruses. In: Hurst, C.J. (Ed.), Viral Ecology. Academic Press, New York, USA, pp. 385–412.

Russell, J.A., Moran, N.A., 2006. Costs and benefits of symbiont infection in aphids: variation among symbionts and across temperatures. Proc. R. Soc. Lond. B 273, 603–610. https://doi.org/10.1098/rspb.2005.3348.

Salem, H., Kreutzer, E., Sudakaran, S., Kaltenpoth, M., 2013. Actinobacteria as essential symbionts in firebugs and cotton stainers (Hemiptera, Pyrrhocoridae). Environ. Microbiol. 15 (7), 1956–1958. https://doi.org/10.1111/1462-2920.12001.

Samways, M.J., 1993. Insects in biodiversity conservation: some perspectives and directives. Biodivers. Conserv. 2, 258–282. https://doi.org/10.1007/BF00056672.

Samways, M.J., 2005. Insect Diversity Conservation. Cambridge University Press, Cambridge, UK.

Satar, G., 2019. Phylogenetics of Buchneraaphidicola based on 16S rRNA amplified from seven aphid species. Turk. J. Entomol. 43 (2), 227–237. https://doi.org/10.16970/entoted.527118.

Schmidberger, J., 1836. Naturgeschichte des Apfelborkenkäfers Apate dispar. In: Beiträgezur Obstbaumzucht und zur Naturgeschichte der den Obstbäumen shcädlichen Insekten. Cajetan Haslinger Linz, Germany, pp. 213–230.

Schoonhoven, L.M., van Loon, J.J.A., Dicke, M., 2005. Insect-Plant Biology. Oxford University Press, Oxford, UK. 400 p.

Schowalter, T.D., 2000. Insect Ecology: An Ecosystem Approach. Academic Press, San Diego.

Schowalter, T.D., 2016. Summary and synthesis. In: Schowalter, T.D. (Ed.), Insect Ecology, fourth ed. Academic Press, pp. 597–608 (Chapter 18).

Secil, E., Sevim, A., Demirbag, Z., Demir, I., 2012. Isolation, characterization and virulence of bacteria from Ostrinia nubilalis (Lepidoptera: Pyralidae). Biologia 67 (4), 767–776. https://doi.org/10.2478/s11756-012-0070-5.

Seifert, K.A., 1993. Sapstain of commercial lumber by species of Ophiostoma and Ceratocystis. In: Wingfield, M.J., Seifert, K.A., Webber, J.F. (Eds.), Ceratocystis and Ophiostoma: Taxonomy, Ecology, and Pathogenicity. American Phytopathological Society, St. Paul, Minnesota, USA, pp. 141–151.

Seifert, K.A., De Beer, Z.W., Wingfield, M.J., 2013. The Ophiostomatoid Fungi: Expanding Frontiers. CBS Biodiversity Series, vol. 12 CBS-KNAW Biodiversity Centre, Utrecht, The Netherlands.

Shang, Y., Feng, P., Wang, C., 2015. Fungi that infect insects: altering host behavior and beyond. PLoS Pathog. 11, e1005037. https://doi.org/10.1371/journal.ppat.1005037.

Shao, Y., Arias-Cordero, E., Guo, H., Bartram, S., Boland, W., 2014. In vivo Pyro-SIP assessing active gut microbiota of the cotton leafworm, Spodoptera littoralis. PLoS One 9, e85948. https://doi.org/10.1371/journal.pone.0085948.

Shao, Y., Chen, B., Sun, C., Ishida, K., Hertweck, C., Boland, W., 2017. Symbiont-derived antimicrobials contribute to the control of the lepidopteran gut microbiota. Cell Chem. Biol. 24, 66–75.

Sharon, G., Segal, D., Ringo, J.M., Hefetz, A., Zilber-Rosenberg, I., Rosenberg, E., 2010. Commensal bacteria play a role in mating preference of Drosophila melanogaster. Proc. Natl. Acad. Sci. 107, 20051–20056. https://doi.org/10.1073/pnas.1009906107.

Siemaszko, W., 1939. Zespoly grzbow towarzyszacych kornikom polskim. Planta Pol. 7, 1–54.

Siitonen, J., 2014. Ips acuminatus kills pines in southern Finland. Silva Fenn. 48 (4). https://doi.org/10.14214/sf.1145.

Sikorowski, P.P., Andrews, G.L., Broome, J.R., 1973. Trans-ovum transmission of a cytoplasmic polyhidrosis virus of *Heliothis virescens* (Lepidoptera: Noctuidae). J. Invertebr. Pathol. 21, 41–45.

Silva, X., Terhonen, E., Sun, H., Kasanen, R., Heliövaara, K., Jalkanen, R., Asiegbu, F.O., 2015. Comparative analyses of fungal biota carried by the pine shoot beetle (Tomicus piniperda L.) in northern and southern Finland. Scan. J. Forest Res. 30 (6), 497–506. https://doi.org/10.1080/02827581.2015.1031824.

Simon, J.-C., Boutin, S., Tsuchida, T., Koga, R., Le Gallic, J.-F., Frantz, A., et al., 2011. Facultative symbiont infections affect aphid reproduction. PLoS One 6, e21831. https://doi.org/10.1371/journal.pone.0021831.

Six, D.L., 2003. Bark beetle-fungus symbioses. In: Bourtzis, K., Miller, T. (Eds.), Insect Symbiosis. CRC Press, Boca Raton, FL, pp. 97–114.

Six, D.L., Paine, T.D., 1998. Effects of mycangial fungi and host tree species on progeny survival and emergence of Dendroctonusponderosae (Coleoptera: Scolytidae). Environ. Entomol. 27 (6), 1393–1401. https://doi.org/10.1093/ee/27.6.1393.

Snyman, M., Gupta, A.K., Bezuidenhout, C.C., Claassens, S., van den Berg, J., 2016. Gut microbiota of Busseola fusca (Lepidoptera: Noctuidae). World J. Microbiol. Biotechnol. 32 (7), 115. https://doi.org/10.1007/s11274-016-2066-8.

Speight, M.R., 2016. Insects and other animals in tropical forests. In: Pancel, L., Köhl, M. (Eds.), Tropical Forestry Handbook. Springer, Berlin, Heidelberg, pp. 2607–2657.

Sree, K.S., Varma, A., 2015. Biocontrol of Lepidopteran Pests: Use of Soil Microbes and Their Metabolites. Springer, Berlin, https://doi.org/10.1007/978-3-319-14499-3.

Staudacher, H., Kaltenpoth, M., Breeuwer, J.A., Menken, S.B., Heckel, D.G., Groot, A.T., 2016. Variability of bacterial communities in the moth Heliothis virescens indicates transient association with the host. PLoS One 11, e0154514. https://doi.org/10.1371/journal.pone.0154514.

Stavrinides, J., McCloskey, J.K., Ochman, H., 2009. Pea aphid as both host and vector for the phytopathogenic bacterium Pseudomonas syringae. Appl. Environ. Microbiol. 75, 2230–2235. https://doi.org/10.1128/AEM.02860-08.

Stavrinides, J., No, A., Ochman, H., 2010. A single genetic locus in the phytopathogen Pantoeastewartii enables gut colonization and pathogenicity in an insect host. Environ. Microbiol. 12 (1), 147–155. https://doi.org/10.1111/j.1462-2920.2009.02056.x.

Stefanini, I., 2018. Yeast-insect associations: it takes guts. Yeast 35 (4), 315–330. https://doi.org/10.1002/yea.3309.

Steinkraus, D.C., Hollingsworth, R.G., Slaymaker, P.H., 1995. Prevalence of Neozygitesfresenii (Entomophthorales: Neozygotaceae) on cotton aphids (Homoptera: Aphididae) in Arkansas cotton. Environ. Entomol. 24, 465–474.

Steinkraus, D.C., Mueller, A.J., Humber, R.A., 1993. Furiavirescens (Thaxter) Humber (Zygomycetes: Entomophthoraceae) infections in the army-worm, Pseudaletia unipuncta (Haworth) (Lepidoptera: Noctuidae) with notes on other natural enemies. J. Entomol. Sci. 28 (4), 376–386. https://doi.org/10.18474/0749-8004-28.4.376.

Stork, N.E., 2018. How many species of insectsand other terrestrialarthropods are there on earth? Annu. Rev. Entomol. 63, 31–45.

Stork, N.E., Eggleton, P., 1992. Invertebrates as determinants and indicators of soil quality. Am. J. Altern. Agric. 7 (1–2), 38–47.

Strand, M.R., Burke, G.R., 2015. Polydnaviruses: from discovery to current insights. Virology 479, 393–402. https://doi.org/10.1016/j.virol.2015.01.018.

Strong, D.R., Lawton, J.H., Southwood, S.R., 1984. Insects on Plants. Community Patterns and Mechanisms. Blackwell Scientific Publications, London.

Studdert, J.P., Kaya, H.K., 1990. Water potential, temperature, and soil type on the formation of Beauveria bassiana soil colonies. J. Invertebr. Pathol. 56, 380–386.

Sugio, A., Dubreuil, G., Giron, D., Simon, J.-C., 2015. Plant–insect interactions under bacterial influence: ecological implications and underlying mechanisms. J. Exp. Bot. 66, 467–478. https://doi.org/10.1093/jxb/eru435.

Suh, S.O., McHugh, J.V., Pollock, D.D., Blackwell, M., 2005. The beetle gut: a hyperdiverse source of novel yeasts. Mycol. Res. 109, 261–265. https://doi.org/10.1017/s0953756205002388.

Suh, S.O., Nguyen, N.H., Blackwell, M., 2008. Yeasts isolated from plant-associated beetles and other insects: seven novel Candida species near Candida albicans. FEMS Yeast Res. 8 (1), 88–102. https://doi.org/10.1111/j.1567-1364.2007.00320.x.

Tang, X., Freitak, D., Vogel, H., Ping, L., Shao, Y., Cordero, E.A., et al., 2012. Complexity and variability of gut commensal microbiota in polyphagous Lepidopteran larvae. PLoS One 7, e36978. https://doi.org/10.1371/journal.pone.0036978.

Tao, H.P., Shen, Z.Y., Zhu, F., Xu, X.F., Tang, X.D., Xu, L., 2011. Isolation and identification of a pathogen of silkworm Bombyx mori. Curr. Microbiol. 62 (3), 876–883. https://doi.org/10.1007/s00284-010-9796-x.

Teixeira, L., Ferreira, A., Ashburner, M., 2008. The bacterial symbiont Wolbachia induces resistance to RNA viral infections in Drosophila melanogaster. PLoS Biol. 6, e2. https://doi.org/10.1371/journal.pbio.1000002.

Thakur, M.P., van der Putten, W.H., Cobben, M.M.P., van Kleunen, M., Geisen, S., 2019. Microbial invasions in terrestrial ecosystems: from processes to impacts and implications. Nat. Rev. Microbiol. 17, 621–631. https://doi.org/10.1038/s41579-019-0236-z.

Thao, M.L., Moran, N.A., Abbot, P., Brennan, E.B., Burckhardt, D.H., Baumann, P., 2000. Cospeciation of psyllids and their primary prokaryotic endosymbionts. Appl. Environ. Microbiol. 66, 2898–2905. https://doi.org/10.1128/AEM.66.7.2898-2905.2000.

Toenshoff, E.R., Penz, T., Narzt, T., Collingro, A., Schmitz-Esser, S., Pfeiffer, S., et al., 2012. Bacteriocyte-associated gammaproteobacterial symbionts of the Adelges nordmannianae/piceae complex (Hemiptera: Adelgidae). ISME J. 6, 384–396. https://doi.org/10.1038/ismej.2011.102.

Tsuchida, T., Koga, R., Shibao, H., Matsumto, T., Fukatsu, T., 2002. Diversity and geographic distribution of secondary endosymbiotic bacteria in natural populations of the pea aphid, Acyrthosiphonpisum. Mol. Ecol. 11, 2123–2135. https://doi.org/10.1046/j.1365-294X.2002.01606.x.

Uzunovic, A., Byrne, T., De Beer, Z.W., 2013. Wood market issues relating to blue-stain caused by ophiostomatoid fungi in Canada. In: Seifert, K.A., Wingfield, M.J. (Eds.), The Ophiostomatoid Fungi: Expanding Frontiers. CBS, Utrecht, The Netherlands, pp. 1–19.

van der Linde, J.A., Six, D.L., De Beer, Z.W., Wingfield, M.J., Roux, J., 2016. Novel ophiostomatalean fungi from galleries of Cyrtogenius africus (Scolytinae) infesting dying Euphorbia ingens. Antonie Van Leeuwenhoek 109, 589–601. https://doi.org/10.1007/s10482-016-0661-1.

van der Vlugt-Bergmans, C.J.B., van der Werf, M.J., 2001. Genetic and biochemical characterization of a novel monoterpene εlLactone hydrolase from Rhodococcus erythropolis DCL14. Appl. Environ. Microbiol. 67, 733–741. https://doi.org/10.1128/AEM.67.2.733-741.2001.

van Emden, H.F., Harrington, R. (Eds.), 2007. Aphids as Crop Pests. CAB International, Oxford. 717 p.

van Frankenhuyzen, K., Liu, Y., Tonon, A., 2010. Interactions between Bacillus thuringiensis subsp. kurstaki HD-1 and midgut bacteria in larvae of gypsy moth and spruce budworm. J. Invertebr. Pathol. 103 (2), 124–131. https://doi.org/10.1016/j.jip.2009.12.008.

van Lenteren, J.C., 2012. Internet Book of Biological Control. International Organization for Biological Control, Zürich, Switzerland. http://tinyurl.com/zk3rdrr.

van Oers, M.M., Eilenberg, J., 2019. Mechanisms underlying the transmission of insect pathogens. Insects 10, 194. https://doi.org/10.3390/insects10070194.

Váry, Z., Mullins, E., McElwain, J.C., Doohan, F.M., 2015. The severity of wheat diseases increases when plants and pathogens are acclimatized to elevated carbon dioxide. Glob. Chang. Biol. 21 (7), 2661–2669. https://doi.org/10.1111/gcb.12899.

Viana-Junior, A.B., Côrtes, M.O., Cornelissen, T.G., de Siqueir, N., 2018. Interactions between wood-inhabiting fungi and termites: a meta-analytical review. Arthropod Plant Interact. 1, 229–235. https://doi.org/10.1007/s11829-017-9570-0.

Vissa, S., Hofstetter, R.W., 2017. The role of mites in bark and ambrosia beetle-fungal interactions. In: Shields, V. (Ed.), Insect Physiology and Ecology. InTech, Rijeka, pp. 135–156.

Wachmann, E., Melber, A., Deckert, J., 2008. Wanzen. In: Band 4. Pentatomomorpha II. Cydnidae, Thyreocoridae, Plataspidae, Acanthosomatidae, Scutelleridae, Pentatomidae. Die Tierwelt Deutschlands und der angrezenden Meeresteile. 81. Teil. Goecke & Evers, Keltern.

Wadke, N., Kandasamy, D., Vogel, H., Lah, L., Wingfield, B.D., Paetz, C., et al., 2016. The bark beetle-associated fungus, Endoconidiophora polonica, utilizes the phenolic defense compounds of its host as a carbon source. Plant Physiol. 171, 914–931. https://doi.org/10.1104/pp.15.01916.

Waller, D.A., 1988. Ecological similarities of fungus-growing ants (Attini) and termites. In: Trager, J.C. (Ed.), Advances in Myrmecology. E. J. Brill, Leiden, UK, pp. 337–345.

Wang, Y., Lim, L., DiGuistini, S., Robertson, G., Bohlmann, J., Breuil, C., 2013. A specialized ABC efflux transporter GcABC-G1 confers monoterpene resistance to Grosmanniaclavigera, a bark beetle-associated fungal pathogen of pine trees. New Phytol. 197, 886–898. https://doi.org/10.1111/nph.12063.

Weisser, W.W., Siemann, E., 2008. The various effects of insects on ecosystem functioning. In: Weisser, W.W., Siemann, E. (Eds.), Insects and Ecosystem Function. Springer, Berlin, Heidelberg, pp. 3–24, https://doi.org/10.1007/978-3-540-74004-9_1.

Werren, J.H., Baldo, L., Clark, M.E., 2008. Wolbachia: master manipulators of invertebrate biology. Nat. Rev. Microbiol. 6, 741–751. https://doi.org/10.1038/nrmicro1969.

White, I.M., Elson-Harris, M.M., 1992. Fruit Flies of Economic Significance: Their Identification and Bionomics. CAB International, Wallingford, Oxon, UK. 601 pp.

Whitfield, A.E., Falk, B.W., Rotenberg, D., 2015. Insect vector-mediated transmission of plant viruses. Virology 479 (11), 278–289. https://doi.org/10.3390/v8110303.

Williams, T., Virto, C., Murillo, R., Caballero, P., 2017. Covert infection of insects by baculoviruses. Front. Microbiol. 8, 1337. https://doi.org/10.3389/fmicb.2017.01337.

Wingfield, M.J., 1993. Problems in delineating the genus Ceratocystiopsis. In: Wingfield, M.J., Seifert, K.A., Webber, J.F. (Eds.), Ceratocystis and Ophiostoma: Taxonomy, Ecology and Pathogenicity. American Phytopathological Society Press, St Paul, pp. 20–24.

Wingfield, M.J., Brockerhoff, E.G., Wingfield, B.D., Slippers, B., 2015. Planted forest health: the need for a global strategy. Science 349 (6250), 832–836. https://doi.org/10.1126/science.aac6674.

Wingfield, M.J., Garnas, J.R., Hajek, A., Hurley, B.P., de Beer, Z.W., Taerum, S.J., 2016. Novel and co-evolved associations between insects and microorganisms as drivers of forest pestilience. Biol. Invasions 18, 1045–1056. https://doi.org/10.1007/s10530-016-1084-7.

Winstanley, D., Rovesti, L., 1993. Insect viruses as biocontrol agents. In: Jones, D.G. (Ed.), Exploitation of Microorganisms. Springer, Dordrecht, Netherlands, pp. 105–136.

Wong, C.M., Daniels, L.D., 2016. Novel forest decline triggered by multiple interactions among climate, an introduced pathogen and bark beetles. Glob. Chang. Biol. 23, 1926–1941. https://doi.org/10.1111/gcb.13554.

Wood, T.G., Sands, W.A., 1978. The role of termites in ecosystems. In: Brian, M.V. (Ed.), Production Ecology of Ants and Termites. Cambridge Univ. Press, Cambridge, pp. 245–292.

Wood, T.G., Thomas, R.J., 1989. The mutualistic association between Macrotermitinae and Termitomyces. In: Wilding, N., Collins, N.M., Hammond, P.M., Webber, J.F. (Eds.), Insect-Fungus Interaction. Academic Press, London, England, pp. 69–92.

Woodard, D.B., Chapman, H.C., 1968. Laboratory studies with the mosquito iridescent virus (MIV). J. Invertebr. Pathol. 11, 296–301.

Wynns, A.A., Jensen, A.B., Eilenberg, J., James, R., 2012. Ascosphaera subglobosa, a new spore cyst fungus from North America associated with the solitary bee Megachile rotundata. Mycologia 104 (1), 108–114. https://doi.org/10.3852/10-047.

Xia, X., Gurr, G.M., Vasseur, L., Zheng, D., Zhong, H., Qin, B., et al., 2017. Metagenomic sequencing of diamondback moth gut microbiome unveils key holobiont adaptations for herbivory. Front. Microbiol. 8, 663. https://doi.org/10.3389/fmicb.2017.00663.

Yeates, G.W., Bongers, T., De Goede, R.G.M., Freckman, D.W., Georgieva, S.S., 1993. Feeding habits in soil nematode families and genera-an outline for soil ecologists. J. Nematol. 25 (3), 315–331.

Yong, H.S., Song, S.L., Chua, K.O., Lim, P.E., 2017. Microbiota associated with Bactroceracarambolae and B. dorsalis (Insecta: Tephritidae) revealed by next-generation sequencing of 16S rRNA gene. Meta Gene 11, 189–196. https://doi.org/10.1016/j.mgene.2016.10.009.

Yoshikawa, K., Matsukawa, M., Tanaka, T., 2018. Viral infection induces different detoxification enzyme activities in insecticide-resistant and-susceptible brown planthopper Nilaparvata lugens strains. J. Pestic. Sci. 43 (1), 10–17. https://doi.org/10.1584/jpestics.D17-052.

Zchori-Fein, E., 2009. Diversity of prokaryotes associated with Bemisiatabaci (Gennadius) (Hemiptera: Aleyrodidae). Ann. Entomol. Soc. Am. 95, 711–718. https://doi.org/10.1603/0013-8746(2002)095[0711:DOPAWB]2.0.CO;2.

Zhang, Z.Q., 2013. Animal biodiversity: an update of classification and diversity in 2013. In: Zhang, Z.-Q. (Ed.), Animal Biodiversity: An Outline of Higher-Level Classification and Survey of Taxonomic Richness. (Addenda 2013) Zootaxa 3703(1), 5–11.

Zhang, P., Liu, Y., Liu, W., Massart, S., Wang, X., 2017. Simultaneous detection of wheat dwarf virus, northern cereal mosaic virus, barley yellow striate mosaic virus and rice black-streaked dwarf virus in wheat by multiplex RT-PCR. J. Virol. Methods 249, 170–174. https://doi.org/10.1016/j.jviromet.2017.09.010540-74004-9.

Zhao, T., Kandasamy, D., Krokene, P., Chen, J., Gershenzon, J., Hammerbacher, A., 2018. Fungal associated of the tree-killing bark beetle, Ips typographus, vary in virulence, ability to degrade conifer phenolics and influence bark beetle tunneling behaviour. Fungal Ecol. 38, 71–79. https://doi.org/10.1016/j.funeco.2018.06.003.

Zhu, F., Poelman, E.H., Dicke, M., 2014. Insect herbivore-associated organisms affect plant responses to herbivory. New Phytol. 204 (2), 315–321. https://doi.org/10.1111/nph.12886.

Further reading

Baker, W.L., 1972. Eastern Forest Insects. U.S.D.A. Forest Service Misc. Pub. No. 1175, Washington, DC.

Buchs, W., 2003. Biodiversity and agri-environmental indicators-general scopes and skills with special reference to the habitat level. Agric. Ecosyst. Environ. 98 (1–3), 35–78. https://doi.org/10.1016/S0167-8809(03)00070-7.

Chouaia, B., Rossi, P., Epis, S., Mosca, M., Ricci, I., Damiani, C., et al., 2012. Delayed larval development in Anopheles mosquitoes deprived of Asaia bacterial symbionts. BMC Microbiol. 12, S2. https://doi.org/10.1186/1471-2180-12-S1-S2.

Daly, H.V., Doyen, J.T., Purcell, A.H., 1998. Introduction to Insect Biology and Diversity. Oxford University Press, Oxford. 680 p.

De Beer, Z.W., Duong, T.A., Wingfield, M.J., 2016. The divorce of Sporothrix and Ophiostoma: solution to a problematic relationship. Stud. Mycol. 83, 165–191. https://doi.org/10.1016/j.simyco.2016.07.001.

Eggleton, P., Bignell, D.E., Sands, W.A., Mawdsley, N.A., Lawton, J.H., Wood, T.G., et al., 1996. The diversity, abundance and biomass of termites under differing levels of disturbance in the Mbalmayo Forest Reserve, southern Cameroon. Philos. Trans. R. Soc. Lond. B. Biol. Sci. 351 (1335), 51–68. https://doi.org/10.1098/rstb.1996.0004.

Fagundes, C.K., Di Mare, R.A., Wink, C., Manfio, D., 2011. Diversity of the families of Coleoptera captured with pitfall traps in five different environments in Santa Maria, RS, Brazil. Braz. J. Biol. 71 (2), 381–390. https://doi.org/10.1590/S1519-69842011000300007.

Halffter, G., Favila, M.E., 1993. The Scarabaeidae Insecta: Coleoptera an animal group for analyzing, inventorying and monitoring biodiversity in tropical rainforest and modified landscapes. Biol. Int. 27, 15–21.

Hlásny, T., Krokene, P., Liebhold, A., Montagné-Huck, C., Müller, J., Qin, H., et al., 2012. Polyphyly of gut symbionts in stinkbugs of the family Cydnidae. Appl. Environ. Microbiol. 78 (13), 4758. https://doi.org/10.1128/AEM.00867-12.

Horák, J., 2017. Insect ecology and veteran trees. J. Insect Conserv. 21, 1–5. https://doi.org/10.1007/s10841-017-9953-7.

Hunt, T., Bergsten, J., Levkanicova, Z., Papadopoulou, A., St. John, O., Wild, R., et al., 2007. A comprehensive phylogeny of beetles reveals the evolutionary origins of a superradiation. Science 318 (5858), 1913–1916. https://doi.org/10.1126/science.1146954.

James, D.Z.W., Paul, A.M., 2002. Impact of pests on forest health. In: Wear, D.N., Greis, J.G. (Eds.), Southern Forest Resource Assessment. U.S. Department of Agriculture, Forest Service, Southern Research Station, Asheville, NC. Gen. Tech. Rep. SRS-53. 635 p.

Judson, L.G., 2018. Coleopteran, Encyclopædia Britannica. https://www.britannica.com/animal/beetle. (Accessed 11 December 2019).

Lemaitre, B., Hoffmann, J., 2007. The host defense of Drosophila melanogaster. Annu. Rev. Immunol. 25, 697–743. https://doi.org/10.1146/annurev.immunol.25.022106.141615.

Li, S.-J., Ahmed, M.Z., Lv, N., Shi, P.-Q., Wang, X.-M., Huang, J.-L., Qiu, B.-L., 2017. Plantmediated horizontal transmission of Wolbachia between whiteflies. ISME J. 11, 1019–1028. https://doi.org/10.1038/ismej.2016.164.

Miller, K.E., Hopkings, K., Inward, D.J.G., Vogler, A.P., 2016. Metabarcoding of fungal communities associated with bark beetles. Ecol. Evol. 6, 1590–1600. https://doi.org/10.1002/ece3.1925.

Nouri, S., Matsumura, E.E., Kuo, Y.W., Falk, B.W., 2018. Insect-specific viruses: from discovery to potential translational applications. Curr. Opin. Virol. 33, 33–41. https://doi.org/10.1016/j.coviro.2018.07.006.

Pinto-Tomás, A.A., Sittenfeld, A., Uribe-Lorío, L., Chavarría, F., Mora, M., Janzen, D.H., et al., 2011. Comparison of midgut bacterial diversity in tropical caterpillars (Lepidoptera: Saturniidae) fed on different diets. Environ. Entomol. 40, 1111–1122. https://doi.org/10.1603/EN11083.

Ricci, I., Valzano, M., Ulissi, U., Epis, S., Cappelli, A., Favia, G., 2012. Symbiotic control of mosquito borne disease. Pathog. Glob. Health 106 (7), 380–385. https://doi.org/10.1179/2047773212Y.0000000051.

Stoops, J., Crauwels, S., Waud, M., Claes, J., Lievens, B., van Campenhout, L., 2016. Microbial community assessment of mealworm larvae (Tenebrio molitor) and grasshoppers (Locusta migratoria migratorioides) sold for human consumption. Food Microbiol. 53, 122–127. https://doi.org/10.1016/j.fm.2015.09.010.

Teh, B.-S., Apel, J., Shao, Y., Boland, W., 2016. Colonization of the intestinal tract of the polyphagous pest Spodoptera littoralis with the GFP-tagged indigenous gut bacterium Enterococcus mundtii. Front. Microbiol. 7. https://doi.org/10.3389/fmicb.2016.00928.

Zchori-Fein, E., Brown, J.K., 2002. Diversity of prokaryotes associated with Bemisiatabaci (Gennadius) (Hemiptera: Aleyrodidae). Ann. Entomol. Soc. Am. 95 (6), 711–718. https://doi.org/10.1603/0013-8746(2002)095[0711,DOPAWB]2.0.CO;2.

Zucchi, T.-D., Prado, S.S., Cônsoli, F.L., 2012. The gastric caeca of pentatomids as a house for acitomycetes. BMC Microbiol. 12, 101. https://doi.org/10.1186/1471-2180-12-101.

Chapter 19

Archaea as components of forest microbiome

Kim Yrjälä[a] and Eglantina Lopez-Echartea[b]

[a]Zhejiang A & F University, Hangzhou, China, [b]University of Chemistry and Technology, Prague, Czech Republic

Chapter Outline

1. What are Archaea

Archaea means old in Greek. That was the first opinion about them when they were found in 1977 as one type of microbe. Methanogens (methane-producing Archaea) and other types of microorganisms, including halobacteria (halophilic Archaea) and thermophiles, had already been discovered, but misclassified to the domain Bacteria (Woese and Fox, 1977). Carl Woese found that through preference for extreme environments, they were phylogenetically related. Three years later, this domain was divided into Crenarchaeota and Euryarchaeota. They were to have great importance in biology as knowledge about them was increasing thanks to the tedious and long-lasting work of Carl Woese. The work with Archaea as a separate type of organism started through molecular genetic studies; those studies are still continuing mainly due to the difficulty in culturing these organisms (Fig. 19.1).

A technical development with the sequencing of oligonucleotides and their cataloging caught Woese's eye, so he started using this technique for studies of 16S rRNA and the genes encoding it. Following him since the 1980s, microbial ecologists have been using that as a barcode for microbes. On the basis of this gene, Woese found that Archaeal 16S rRNA had some sequence signatures that were lacking in bacteria. Later, it was found that they were different from Eukaryotes as well. Based on the sequencing of thousands of 16S rRNA genes, Woese proposed the division of life on Earth into three domains: Archaea, Eukaryotes, and Bacteria (Woese et al., 1990). He brought genetics into the classification of organisms, but was not immediately appreciated by the conservative scientific field that favored phenotypic classification. It took some time for the microbiology field to appreciate this great discovery. It took until the beginning of this millennium for other fields (Hebert et al., 2003) to make use of this idea in distinguishing species isolates. By 2004, it was finally made clear that the ribosomal gene is a suitable marker for taxonomic studies to distinguish the huge diversity of microbes on Earth; the existence of three domains of life was to be accepted as fact (Woese, 2004). This event was going to revolutionize the field of molecular systematics, which continues to bring us new knowledge on the origins of life on Earth.

1.1 The three domains of life

The discovery of a new separation of life domains was underlined by other molecular and biochemical signatures combined with the ecological features of Archaea. Thus, the archaeal concept was established and the five kingdoms of life classification scheme (Whittaker, 1969) was replaced by the three domains of life classification (Woese et al., 1990). The genomic representation of microbial biodiversity, particularly of Archaea, has since expanded significantly. This is largely due to

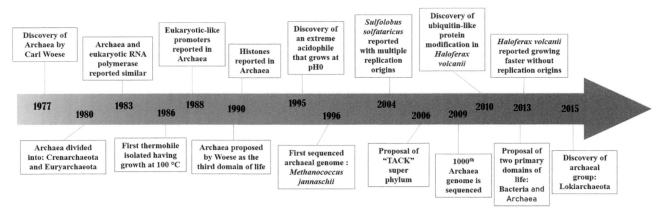

FIG. 19.1 Timeline of the milestones in Archaea research from discovery to 2015.

advances in environmental genome sequencing and to the sampling of microbial DNA directly from the environment without the need for culturing (Rinke et al., 2013; Sunagawa et al., 2015).

Text Box 1   Assembly of genomes and multiloci phylogenetic trees.

The metagenomic sequencing of DNA isolated directly from soil and differentiated computational methods have made it possible to assemble genomes in large numbers from the environment. Acquired sequences can be assembled with improved binning to form partial genomes (Parks et al., 2015; Nurk et al., 2017; Alneberg et al., 2014). These methodologies have spurred an array of new archaeal genomes, providing a basis for robust phylogenetic analyses. The 16S rRNA gene phylogenetics have been updated by multiloci phylogenetic trees using ribosomal proteins and conserved proteins such as tRNA synthetases to demarcate the phyla among superphyla (Castelle et al., 2015; Brown et al., 2015). The advantages with these proteins are that they have low lateral gene transfer, are somewhat congruent with ribosomal RNA phylogeny, and have a wide-ranging existence within the tree of life. As many as 27 phyla have been proposed using concatenated protein phylogeny. They have helped distinguish diversity, but also enabled us to resolve deep-branching topology, and have given new insight into the kinship of Archaea to Eukaryotes (Spang et al., 2015; Zaremba-Niedzwiedzka et al., 2017; Sorek et al., 2007). A wide-ranging set of 3599 uncultured and cultured archaeal genomes was utilized to update the archaeal tree of life (Baker et al., 2020) with robust phylogenomic analyses of a set of conserved ribosomal marker proteins.

The improvement of sequencing and bioinformatics methods has resulted in a new, amended understanding of the tree of life. Fig. 19.2 presents an updated view of the three-domain tree that includes new sequences from DPANN archaea.

2.   Archaea in boreal forests

Thanks to the groundbreaking work of Woese, Archaea in the 1990s were distinguished from Bacteria in the scientific literature (Woese et al., 1990). At that time, the general knowledge was that these small one-celled organisms were methane-producing, often halophilic Archaea of the Euryarchaeota phyla or thermophiles of the Crenarchaeota phyla. Methanogens are the only microorganisms producing methane, most typically in water-submerged soils such as tropical, subtropical, and boreal peatlands (Finn et al., 2020; Kotsyurbenko et al., 2019; Bae et al., 2018; Hedderich and Whitman, 2013). Boreal peatlands emitting methane have annual temperatures that range from freezing temperatures in the winter to 17°C in the summer, which is far from thermophilic conditions (Juottonen et al., 2005). It was then not such a big surprise that Archaea were also found in typical boreal forests, but these findings at the end of the 1990s (Jurgens and Saano, 1999; Jurgens et al., 1997) challenged the thought of Archaea being some kind of extremophile. These novel Crenarchaeota were found in a mixed forest of Norway spruce and Scots pine and the 16S rRNA FFS-uni and FFS-6 Crenarchaeota clusters were evolutionary different from the Amazon and Wisconsin Crenarchaeota that previously had been detected in soil. Later, more Crenarchaeota were detected in Finnish boreal forest humus (Yrjälä et al., 2004) that were sensitive to Cd-containing ash, a side product from wood burning. These Crenarchaeota have further been detected at the bottom of the peat layer in the boreal fen of Siikaneva, representing a boreal peatland that started developing after the last ice age with an approximate age between 8000 and 10,000 years (Putkinen et al., 2009).

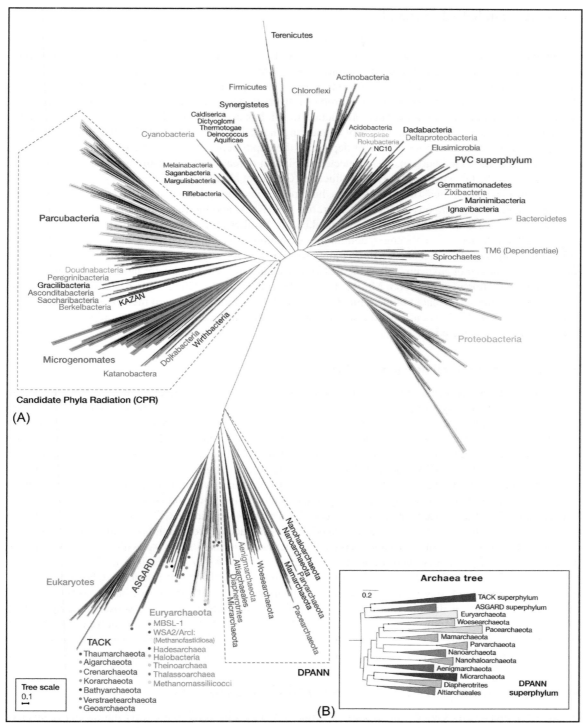

FIG. 19.2 Phylogenetic tree. (A) An updated tree from (Hug et al., 2016) with the addition of new DPANN archaeal sequences. (B) Reconstruction of a phylogenetic tree with only the archaeal sequences using bacteria as an outgroup in which the monophyly of the DPANN clade is retrieved (Castelle and Banfield, 2018).

2.1 Methanogens in boreal peatlands

In the search for functions to the boreal forest Crenarchaeota, the isolation of these microorganisms became a general goal. However, the isolation of pure cultures of Archaea from soil is very tedious, and all trials failed to come up with pure isolates that could have been studied for their basic functions in forest soils (Lehtovirta et al., 2009). The cultivation of an obligate acidophilic ammonia oxidizer from a nitrifying acid soil was finally successful in 2011 (Lehtovirta-Morley et al., 2011). The best and most reasonable way to study Archaeal diversity and function has been by using molecular ecology tools. In boreal peatlands, the potential methane production in soil has been measured and correlated to the occurrence of methane-producing Archaea sequences, in this way connecting archaeal diversity to function (Galand et al., 2002, 2003; Juottonen et al., 2008). Their dispersion with soil depth and the depth dependence of diversity was first shown in 2002, which also was the first time methanogens were detected in Finnish boreal peatlands using the genetic marker gene $mcrA$ (methyl coenzyme M reductase) needed for CH_4 production (Galand et al., 2002). The diversity of methanogens was connected to microsites in a boreal fen showing that flarks, hummocks, and lawns have distinct archaeal communities producing CH_4 (Galand et al., 2003). The phylogenetic marker, the 16S rRNA, and the functional marker $mcrA$ gene (Juottonen et al., 2006) were used to depict methanogens in a drained bog in Northern Finland. Ash fertilization had been used to improve tree growth that affected the methanogen community structures. A special type of methanogen was for the first time recognized and published in 2004 that clustered separately from other known types of methanogens such as those from rice fields. These peculiar methanogens were called the FEN-cluster Euryarchaeota and were later found in all studied boreal peatlands in Finland (Juottonen et al., 2005; Merilä et al., 2005; Galand et al., 2005b). The first cultured representative of this group of methanogens was isolated in the United States in 2006 and published with the name Methanoregula (Bräuer et al., 2011) as a species preferring acidic conditions similar to the prevailing conditions in boreal peatlands where the fen cluster methanogens were detected.

2.2 Crenarchaeota and methanogens in rhizospheres

The influence of soil temperature on the archaeal populations associated with different boreal forest tree species, Scots pines, silver birches, and Norway spruce seedlings was studied in forest humus microcosms at 7–11.5°C in the bulk soil, rhizosphere, and mycorrhizosphere (Bomberg et al., 2011). In this study, the occurrence of Archaea was studied by the nested PCR of the archaeal 16S rRNA with denaturing gradient gel electrophoresis (DGGE) profiling. Typically, methanogenic Euryarchaeota belonging to *Methanolobus* sp. and *Methanosaeta* sp. were detected on the roots and mycorrhiza. The most abundant archaeal 16S rRNA gene, however, was affiliated with group I.1c Crenarchaeota, typically found in boreal and alpine forest soils. There was a difference in archaeal occurrence within tree species. Scots pine roots and mycorrhizas harbored Crenarchaeota and only some Euryarchaeota, but silver birch roots and mycorrhizas mostly harbored Euryarchaeota. Regarding the Euryarchaeota, it was found that just 15% were affiliated with the *Halobacteria*. A minority of the sequences obtained from the mycorrhizas belonged to Euryarchaeota (order Halobacteriales). In general, the in situ conditions of oxic forest soils and litter are not conducive to methanogenesis. Still, in tropical forests CH_4 emissions have been reported from low O_2 sites that were high (22–101 nmolCH_4 m − 2 s − 1) and similar to CH_4 emissions of natural wetlands (Teh et al., 2005). Methanogenic Archaea possess some degree of oxygen tolerance. CH_4 turnover in a boreal fen ecosystem was studied after long-term water table drawdown where several types of methanogens were found in a dry pine-dominated mesotrophic peatland forest using PCR amplification of the mcrA gene and subsequent fingerprinting with T-RFLP. Methanosarcinaceae, Methanocellales, fen cluster (Methanomicrobiales), and Methanobacteriaceae were identified from sequenced TRfs (Yrjälä et al., 2011). A shift in the archaeal populations from dominating crenarchaeota toward a higher abundance of euryarchaeotal methanogenic taxa may take place in a longer-term perspective.

3. Evolving taxonomy of methanogens and Crenarchaeota

The taxonomy of methanogens started with groups of similar phylogenetic depth with fairly unrelated organisms (Boone et al., 1993; Whitman et al., 2006). The methanogens are genetically extremely diverse. Many of them appear to perform the same way, but their genetical pathways are different, showing varying types of methanogenesis. The more detailed physiologic studies have supported this interpretation, finding big differences in cellular structure, metabolic pathways, and regulation. Not so many methanogens have been cultured, representing only a minority of the diversity, so knowledge from cultured isolates is thus very limited (Hedderich and Whitman, 2013). All methanogens were thought to belong to two euryarchaeal clades, Class I and Class II methanogens, until quite recently (Bapteste et al., 2005). The majority of methanogens belonging to Class I/II grow by reducing CO_2 to form methane using H_2 as the electron donor (Whitman et al., 2006).

Methanogens of Methanosarcinales (Class II methanogens) can use additional substrates such as acetate and methylated compounds (Kendall and Boone, 2006). The Class 1 methanogens, Methanosphaera spp., solely reduce methanol with H_2 (Oren, 2014). All Class I/II methanogens have the H4MPT methyl-branch of the Wood-Ljungdahl pathway (m-WL), the N5-methyltetrahydromethanopterin-coenzyme M-methyltransferase complex (MtrABCDEFGH or MTR), and the methyl-coenzyme M-reductase complex (McrABG or MCR) (Borrel et al., 2016).

The knowledge regarding the diversity and metabolic versatility of methanogenic Archaea is rapidly growing with the accessibility of assembled genomes (MAGs) from environmental DNA and from some new isolates (Borrel et al., 2016; Adam et al., 2017; Spang et al., 2017). Additional archaeal lineages distantly related to Class I/II methanogens, comprising Methanomassiliicoccales (Borrel et al., 2013), Methanofastidiosales (Nobu et al., 2016), Methanonatronarchaeia (Sorokin et al., 2017) (Sorokin et al., 2017), and Verstraetearchaeota (Vanwonterghem et al., 2016) have been described (Table 19.1). A remarkable common characteristic of these methanogens is the absence of the methyltransferase MTR complex, and the presence of specific methyltransferases for the metabolism of methylated compounds (Borrel et al., 2019).

3.1 From Crenarchaea to Thaumarchaeota

There has been a growing pace of Archaea genome sequencing while at the same time, very few new isolates have been reported. This has led to an intriguing situation in Archaea phylogeny to decide placements for new uncultured lineages. The importance of this task is obvious in light of the fundamental origin of this domain of life and its relationship to eukaryotes (Brochier-Armanet et al., 2008). Large concatenated datasets of ribosomal (R) proteins have become an alternative to SSU rRNA (Matte-Tailliez et al., 2002). In particular, ribosomal protein analyses have helped to clarify the phylogenetic positions of "lonely" archaeal species such as *Nanoarchaeum equitans*. R proteins and other protein markers suggest that this species is not an early archaeal offshoot, but a fast-evolving Euryarchaeal lineage possibly related to Thermococcales (Brochier et al., 2005).

The examination of larger genome fragments of uncultured I.1b Crenarchaeota and the genome sequences of the I.1a Crenarchaeote *Cenarchaeum symbiosum* (Hallam et al., 2006) have led to the proposal of Group I Crenarchaeota as a novel Phylum of the archaeal domain. This new Phylum was given the name Thaumarchaeota (Brochier-Armanet et al., 2008). Thaumarchaeota (former Group I Crenarchaeota) include the ecologically versatile aerobic ammonia-oxidizing marine Group I.1a Archaea (Cenarchaeum, Nitrosoarchaeum, Nitrosopumilales, Nitrosopumilus), and in soil environments Group I.1b Archaea (Nitrososphaerales, Nitrososphaera) (Pester et al., 2011). Increasing genomic data from new thaumarchaeal lineages has provided substantial insights into the still largely unexplored metabolic versatility of Thaumarchaeota.

Thaumarchaeota have been reported from poplar plantations in the temperate climate zone in Northern China. Archaea were relatively rare in all soil layers of poplar plantations, but Thaumarchaeota were detected with decreased relative abundance with depth (Feng et al., 2019). Soil_Crenarchaeotic_Group_SCG was the dominant class of detected Archaea sequences. Based on genome studies, Thaumarchaeota have been firmly connected to ammonia oxidation and carbon sequestration activities (Könneke et al., 2014), but they also exhibit high denitrification potential under hypoxic conditions (Walker et al., 2010). In poplar plantations, the relative abundance of Euryarchaeota increased significantly with depth within the first four layers, which was reported to be due to their preference for anaerobic environments. The characterized Thaumarchaeota ammonia oxidizers of Groups 1.1a and 1.1b are mainly aerobic, but deeply rooted Thaumarchaeota may not always be aerobic and their metabolic potential is not known. Actually, many reports have given evidence for both aerobic and anaerobic metabolisms. This was studied more closely in controlled soil mesocosms with oxic and anoxic Scottish pine forest soil (Biggs-Weber et al., 2020). The aim was to study the effect of oxygen on the mesophilic Groups 1.1c and 1.3 Thaumarchaeota community structure (Fig. 19.3).

The authors concluded that mesophilic deeply rooted Thaumarchaeota in Scottish pine soil included both aerobes and anaerobes that have a well-developed ability to thrive in alternate optimal environmental conditions.

3.2 The TACK superphylum

The increasing amount of assembled new archaeal genomes promoted the description of a new phylum, the TACK, in 2011. This phylum revealed phylogenetic proximity to eukaryotes and shared signatures with eukaryotes (Guy and Ettema, 2011). The name is derived from the included Thaumarchaeota, the Aigarchaeota, the Crenarchaeota, and the Korarchaeota. Geoarchaeota was additionally proposed (Kozubal et al., 2013), but to represent a deep-branching lineage of the Crenarchaeota (Guy et al., 2014). Even a kingdom-level clade has been proposed so that TACK would be named Proteoarchaeota (Fig. 19.4).

TABLE 19.1 Newly named archaeal taxa from 2008 to 2017.

Original name	New/proposed name	Reference
New phyla		
Group I Crenarchaeota	Thaumarchaeota	Brochier-Armanet et al. (2008)
Novel archaeal group 1 (NAG1)	Geoarchaeota (basal Crenarchaeota)	Kozubal et al. (2013)
Miscellaneous Crenarchaeotal Group (MCG)	Bathyarchaeota	Meng et al. (2014)
Terrestrial Miscellaneous Crenarchaeotal Group (TMCG)	Verstraetearchaeota	Vanwonterghem et al. (2016)
DSAG and AAG-related	Heimdallarchaeota	Zaremba-Niedzwiedzka et al. (2017)
Hot Water Crenarchaeotal Group (HWCG I)	Aigarchaeota	Nunoura et al. (2011)
Deep Sea Archaeal Group (DSAG)	Lokiarchaeota	Spang et al. (2015)
Marine Benthic Group B (MBG-B)	Thorarchaeota	Seitz et al. (2016)
New classes		
South African Gold Mine Euryarchaeotic Group (SAGMEG)	Hadesarchaea	Baker et al. (2016)
Mediterranean Seafloor Brine Lake Group 1 (MSBL-1)	Persephonarchaea (proposed)	Mwirichia et al. (2016)
WSA2/ArcI	Ca. Methanofastidiosa	Nobu et al. (2016)
Marine Benthic Group D (MBG-D)	Izemarchaea (proposed)	Lloyd et al. (2013)
Marine Group II (MG-II)	Thalassoarchaea	Martin-Cuadrado et al. (2015)
Marine Group III (MG-III)	Pontarchaea (proposed)	Li et al. (2015)
Z7ME43	Theionarchaea	Lazar et al. (2016)
New orders		
Sippenauer Moor 1 (SM1 Euryarchaeon)	Altiarchaeales	Probst and Moissl-Eichinger (2015)
Rice Cluster I (RC-I)	Methanocellales	Sakai et al. (2008)
GoM-Arch87	Syntropharchaeales (proposed)	Laso-Pérez et al. (2016)
Anaerobic Methanotroph 1 (ANME-1)	Methanophagales (proposed)	Meyerdierks et al. (2010)
Rumen Cluster C (RCC)/Rice Cluster III (RC-III)	Methanomassiliicoccales	Iino et al. (2013)
New family		
Rice Cluster II (RC-II)	Methanoflorentaceae	Mondav et al. (2014)
New superclasses		
Methanopyrales, Methanobacteriales, and Methanococcales	Methanomada	Petitjean et al. (2015b)
MG-II, MG-III, DHVE2, RCC/RC-III, TMEG, and Thermoplasmata	Diaforarchaea	Petitjean et al. (2015b)
Methanogens class 2, Halobacteria, ANME-1, GoM-Arch87, Archaeoglobi	Methanotecta (proposed)	
Hadesarchaea and MSBL-1	Stygia (proposed)	
New superphyla		
Lokiarchaeota, Thorarchaeota, Heimdallarchaeota, Odinarchaeota	Asgard	Zaremba-Niedzwiedzka et al. (2017)
Diapherotrites, Parvarchaeota, Aenigmarchaeota, Nanohaloarchaeota, and Nanoarchaeota	DPANN	Rinke et al. (2013)
Thaumarchaeaota, Aigarchaeota, Crenarchaeota, Geoarchaeota and Korarchaeota	TACK/Proteoarchaeota	Guy and Ettema (2011) and Petitjean et al. (2015a)

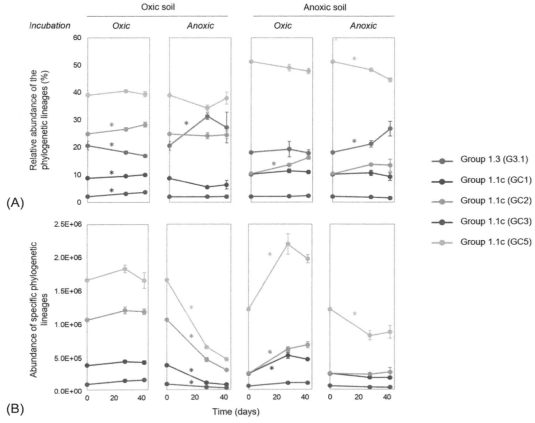

FIG. 19.3 Mesophilic thaumarchaeotal Group 1.1c and Group 1.3 community analysis in oxic and anoxic pine forest soils that were incubated either in oxic or anoxic conditions for 42 days. (A) The relative abundance of Group 1.1c and 1.3 analyzed by MiSeq sequencing using primers for Thaumarchaeota. (B) Absolute abundance of the five Group 1.1c phylogenetic clusters. It was calculated as the product of Group 1.1c 16S rRNA gene qPCR abundance and the relative abundance of each phylogenetic cluster received by MiSeq sequencing.

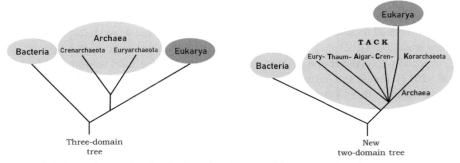

FIG. 19.4 Schematic universal phylogenetic trees showing the three-domain tree and the new two-domain tree.

4.   Archaea in alpine forests

The relative abundance of Archaea and Bacteria has been estimated in alpine forests. Four sites with different elevations revealed that the archaeal 16S rRNA gene copy number did not significantly ($P > .05$) vary between sites nor correlate with environmental and chemical properties (Siles and Margesin, 2016). Thaumarchaeota clearly dominated the Archaea in all samples, showing no significant difference in relative abundance between the four sites (ANOVA, $P > .05$). These Thaumarchaeota sequences belonged to the Thaumarchaeota class and the *Nitrososphaera* genus. Some Euryarchaeota and Crenarchaeota phyla sequences were additionally identified, with no closer affiliation given.

The gradient displayed no clear shifts in the archaeal 16S rRNA gene copy number and no significant correlation between the Archaea abundance and altitude. This was similar to the results described along another altitudinal gradient in Tibetan Plateau (Wang et al., 2015). The archaeal community only represented between 1.6×10^{-4} and 8.9×10^4 of the total 16S rRNA gene copy number, which was from 1/63 to 1/146 of the archaeal abundance reported in the Tibetan Plateau. Wang et al. (2015) reported that the archaeal community accounted for 1%–13% of the total prokaryotic abundance in the Tibetan altitude gradient. This very big difference of archaeal abundance could be due to the fact that in the Tibetan study, separate Bacteria-specific V1–V3 region and Archaea specific V3–V6 region primers were used. In the alpine study, the same primer pair was used for both Bacteria and Archaea (V4–V5 region primers). The archaeal community was dominated by Thaumarchaeota phylum. They were affiliated with Nitrososphaera (genus), which suggests the dominance of ammonium oxidizers in the archaeal community. The Thaumarchaeotal sequences were identified as *Nitrososphaera* (genus), which speaks for the dominance of ammonium oxidizers among the archaeal community at the studied sites. Some sequences belonged to Euryarchaeota and Crenarchaeota phyla, although a more detailed identification of these sequences was not possible.

The relative abundance of archaeal OTUs at the phylum level was variable among the two plant rhizospheres and the bulk soil in the Qinghai-Tibetan plateau (Zhang et al., 2020). The most dominant archaeal phyla across all samples were Thaumarchaeota, unclassified, and Euryarchaeota, accounting for 92.5%–98.0%, 1.4%–6.0%, and 0.6%–1.2% of the pyrosequencing reads, respectively. A significant enrichment of Thaumarchaeota in the rhizosphere microbiota of two plant species was reported (Dunnett test, $P < .05$). Conversely, the relative abundance of unclassified and Euryarchaeota decreased in the rhizosphere microbiota of two plant species but did not show significant differences compared with bulk soil.

5. Archaea in tropical forests

The archaeal domain is often compared to the Bacterial domain as basic entities to study in forest soils. In Amazonian rainforest bulk soil, it displayed a higher relative abundance at the phylum and family levels than in the rhizosphere (Fonseca et al., 2018). The specific abundant archaeal taxa were *Nitrosopumilus* of the Thaumarchaeota phylum. They are known as ammonia-oxidizing archaea with nitrite reductase activity. The species *Nitrosopumilus maritimus* was isolated as a mesophilic crenarchaeon that grows chemolithoautotrophically by aerobically oxidizing ammonia to nitrite, which was the first observation of nitrification in the Archaea (Könneke et al., 2005). Thaumarchaeota have been shown to play a significant role in geochemical cycles (Brochier-Armanet et al., 2008), and are known to have copper-dependent nitrite reductases (NirK) (Horell et al., 2017) Sulfur-metabolizing Desulphurococcaceae were detected in the Amazonian rainforest that may participate in diverse processes related to energy metabolism. The observed shift in archaeal abundance from the rhizosphere to bulk soil was explained in terms of soil depth. The low abundance compared to other domains of fungi and bacteria was further explained as a potential compartmentalized role in the microbiome. Significantly enriched Archaea in bulk soil were somehow related to acidic soils because the detected Archaea were the acidophilic Ferroplasmaceae and Picrophilaceae (Fonseca et al., 2018).

Land use change, especially the conversion of Amazon rainforest to agriculture, was studied for microbial community structures (Meyer et al., 2017). Forest and pasture sites were compared for their methane turnover. The forest site had been reported to have negative methane flux by methane consumption, even during the wet season (Fernandes et al., 2002). The cattle pasture had again been reported to have a positive methane flux with methane emission, pertaining also to the dry season (Fernandes et al., 2002). In the Amazon study, a response of methane-cycling microorganisms to land use change was discerned, where the response was dominated by methane-consuming microorganisms. A primary result was that the proportion of methanotrophs compared to methanogens was significantly higher in the forest. These responses included a reduction in the relative abundance of methanotrophs and a significant decrease in the abundance of genes encoding a particulate methane monooxygenase. A primary result was that the proportion of methanotrophs compared to methanogens was significantly higher in the forest (Meyer et al., 2017). The changes to methanotroph community suggested that the conversion of forest to pasture dealt with changes in methanotrophy. Methane-cycling microorganisms displayed sensitivity to land use change, and that sensitivity was suggested to result in changes of methane flux in land use change in the Amazon. The composition of methangogenic Archaea at the operational taxonomic unit (OTU) level did significantly differ between forest and pasture (Bray-Curtis R2 = 0.61, $P < .001$).

The average pairwise dissimilarity was higher in the forest soils, showing a more varied community in the forest, compared to the pasture, but then again they observed no significant differences in diversity or evenness for these sites. The proportion of known acetoclastic methanogens (methane produced from acetate and H_2) was significantly higher in the pasture than in the forest ($P < 0.01$), showing a slight increase in acetoclasts and a decrease (but not significant) in hydrogenotrophs (methane produced from CO_2 and H_2) (Galand et al., 2005a). Separately, the diversity patterns also varied across the forest and pasture. Both groups differed significantly in composition in pasture and forest (acetoclast: Euclidean R2=0.30, $P < .01$, hydrogenotroph: Bray R2=0.64, $P < .05$). The acetoclasts were reported to have significantly higher species richness in the pasture ($P < .05$). Looking closer at the orders of hydrogenotrophs, they had a differential response to land use change. The abundant Methanopyrales (hydrogenotrophic order) in the forest were not detected in the pasture while Methanocellales, being the most abundant hydrogenotroph in the pasture, were much lower in abundance in the forest. This result was explained as the variable life history strategy, which could be studied further. Functional genes were studied showing that the overall abundance of methanogenesis genes did not significantly differ between the forest and pasture. Still, the relative abundance of *mcrA* genes (encoding the common marker enzyme, methyl coenzyme M reductase) was significantly lower in the forest ($P < .01$). Genes encoding methanogenesis from methylated compounds (an alternative methane production pathway) were significantly less abundant in the forest ($P < .05$).

6. Archaea adapting to environments causing energy stress

Archaea as a domain of life is still today a bit of a mystery because we have so few real isolates of them to study. In the current study, the relation of Archaea to forests has been elaborated in an effort to elucidate their functions and roles. Five broad physiological groups of cultivated archaea are halophiles, thermophiles, acidophiles, metabolic nitrifiers, and methanogens. Bacterial occurrence is often compared to Archaeal occurrence in environmental gradients and changes. It has been put forward that functionally, Archaea differ from Bacteria in chronic energy stress situations related to maintenance energy (ME) and the biological energy quantum (BEQ) (Valentine, 2007). ME is the minimum energy flux from the catabolism needed to maintain cellular activity distinct from the energy required for growth or survival (Hoehler, 2004). The BEQ is the minimum catabolic energy yield that is required for maintenance, which involves a chemiosmotic potential important for anaerobes. The inability to attain ME leads to starvation, whereas the inability to attain BEQ leads to the decoupling of energy conservation from the catabolism (Valentine, 2007). In this current study, we have brought forward the latest developments about how life started on our planet and the idea that in the beginning there were Bacteria and Archaea, and that Eukaryotes developed from Archaea (Fig. 19.2). According to Valentine (2007), the evolutionary divergence of the Archaea and Bacteria ascended as a result of the selective pressure of energy stress. The archaeal branch is then an adaptation to extreme temperature and acidity. This capacity of the Archaea to thrive with chronic energy stress has defined the subsequent evolution of this group to occupy other niches. We will now look at some examples of life strategies of Archaea in forests in connection to their adaptation to environmental niches characterized by energy stress.

6.1 Ammonium oxidizers in forests

Ammonium oxidation was studied in the forests of southern China. Nitrification was mostly carried out by NH_3-oxidizing archaea, and less by heterotrophic nitrifiers and NH_3-oxidizing bacteria in the subtropical acidified forest (Isobe et al., 2018). The two studied acidic forests in southern China are close to Guanzhou, a city with more than 10 million inhabitants. They had high levels of N deposition and their pH was 3.7–4.2. The NH_3-oxidizing Archaea were suggested to be acidophilic with low AmoA diversity. In the study of N transformations, the AmoA abundance was found to be the most important factor in the acidified forests. The NH_4^+ production rate in acidified forests was estimated and shown not to differ between N-saturated and non-N-saturated forests (Fig. 19.5). At the same time, the nitrification rate was higher in N-saturated forests. The study showed that both the high abundance as well as the NH_3 oxidation activity of AmoA resulted in high nitrification rates, leading to NO_3^- leaching in the N-saturated forest. The authors (Isobe et al., 2018) expressed the need for a deeper understanding of forest N saturation from the perspective of microbial ecology.

This study connects well to the cultivated Thaumarchaeota, *Nitrosotalea devanaterra*, which was isolated from acidic agricultural soil by the research group at the University of Aberdeen (Lehtovirta-Morley et al., 2011). This isolated strain functioned best at pH 4–5 and could not actually function at a pH higher than 5.5, disparate from all previously isolated ammonia oxidizers. All functions, growth, ammonia oxidation, and autotrophy ensued during nitrification in soil at the low pH. The discovery of an ammonium oxidizer provided a surprising explanation for the high rates of nitrification in acidic soils and confirmed the vital role of Thaumarchaea in terrestrial nitrogen cycling. This organisms undergo the depicted low energy strategy within Archaea because they grow at extremely low ammonia concentrations (0.18 nM) at low pH (Lehtovirta-Morley et al., 2011).

Recent studies can be extrapolated to show that these AmoA archaea outcompete bacteria in those conditions where the energy supply is lowered, as in unfertilized forest soils (Leininger et al., 2006).

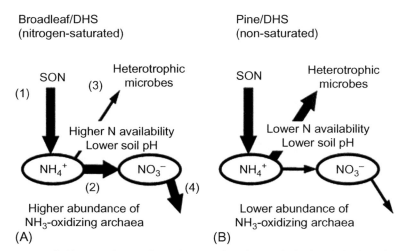

FIG. 19.5 N transformation processes in N-saturated (A) and non N-saturated (B) forest soils in the Dinghushan Biosphere Reserve, Southern China (Isobe et al., 2018).

6.2 Methane oxidizers in forests

The anaerobic methane oxidizer was a new metabolic group discerned through environmental, laboratory, and metagenomic analyses. They are closely related to methanogens, including at least three phylogenetic groups. The anaerobic methane oxidizer organisms may form syntrophic associations with sulfate- or nitrate-reducing bacteria when they grow by oxidizing methane (Orphan et al., 2002). Their growth mode yields only small amounts of energy and the growth is often extremely slow (Knittel and Boetius, 2009). These anaerobic microorganisms presumably experience chronic energy stress trying to achieve the BEQ and ME. This capacity seems to be exclusive to Archaea. The lipid-membrane structure of Archaea is thought to be the primary characteristic for adaptation to energy stress. According to Valentine (2007), these membrane structures can reduce energy loss at the cellular level, reducing their ME, which makes them differ from Bacteria. The evolution of their catabolic pathways has taken place by adaptation to continuing energy stress and their energy conservation mechanism has allowed Archaea to grow in energy-restricted environments. When comparing Archaea to Bacteria, their internal competition intensifies in moderate environmental conditions, where Archaea use a singular catabolism by which they outcompete bacteria, mostly in certain niches where energy availability is low. Methanogens, methane oxidizers, and possibly also nitrifiers have such singularity, where the catabolic emphasis is on a single well-defined pathway (Valentine, 2007).

6.2.1 Methane metabolism from an evolutionary perspective

Both methanogenic and methanotrophic Archaea belong to the phylum Euryarchaeota, where they share a genetically similar, interrelated pathway for methane metabolism. The key enzyme of this pathway, the methyl-coenzyme M reductase, catalyzes the last step in methanogenesis and the first step in methane oxidation. The discoveries of *mcr* and divergent *mcr*-like genes in new euryarchaeota as well as the archaeal phyla have challenged the views of the evolutionary origin of this metabolic capacity among the Euryarchaeota (Vanwonterghem et al., 2016). Recently divergent mcr-like genes have been reported to encode the oxidation of short-chain alkanes, which indicates that these gene structures have evolved to the metabolism of other substrates than methane (Laso-Pérez et al., 2016). Such methanotrophs belong to the orders of Methanosarcinales (ANME-2 and ANME-3) and Methanophagales (ANME-1). ANME evidently oxidizes methane to CO_2 through a reversed hydrogenotrophic methanogenesis pathway. The exception is the metF gene, which is substituted for the *mer* gene in some of these genomes (McGlynn, 2017; Timmers et al., 2017; Leu et al., 2020). Currently, several clades within the archaeal phylum Euryarchaeota have been shown to be capable of anaerobic methanotrophy, including ANME-1a-b, ANME-2a-c, 57 Methanoperedenaceae (formerly known as ANME-2d), and ANME-3 (Adam et al., 2017; Murrell, 2010).

7. Conclusions

The origin of Archaea is tied to the earliest history of our planet. The present knowledge even suggests that first there were Bacteria and Archaea and then at some point the Eukaryotes evolved within the proposed TACK superphylum from Archaea (Lokiarchaeota) (Figs. 19.2 and 19.4, Table 19.1). That event occurred before forests appeared on the Earth. The Archaea were first mentioned in 1776 with the experiments of Alessandro Volta using decaying vegetable residues in sediments, where he showed that the sediment gas was combustible. The identity of this gas was solved a century later. Carl Woese brought the archaebacteria to global attention in 1970. With new genetic methods, the three domains of life classification was published in 1990. Along with that, Crenarchaeota were some years later first found in nonthermophilic habitats such as boreal forests (Jurgens et al., 1997). With the same methodology (polymerase chain reaction and sequencing of archaeal genes), methanogens were observed in boreal peatlands using the *mcrA* marker gene (Galand et al., 2002) in connection to potential methane production at different depths of a boreal fen. The revelation of the unknown functions of forest Crenarchaeota had to wait some 10 years, when it was established that some of them are ammonium oxidizers (Leininger et al., 2006).

The Archaea abundance in forests is only a small percentage of the total prokaryotic abundance. In the current study, it was put forward that Archaea microorganisms are more oligotrophic in their lifestyle and occupy only certain niches in ecosystems where only small amounts of energy are available. They are mostly specialized to perform certain low-energy receiving processes in which they use a single, well-targeted pathway for growth and survival. Very few new isolates are reported to give new knowledge about the functions Archaea perform in forests, and most new knowledge comes from assembled genomes. It would be of great interest to find new smart ways of cultivating Archaea from the forest to learn much more about their specific functions in connection to tree rhizospheres and the mycorrhizosphere.

References

Adam, P.S., Borrel, G., Brochier-Armanet, C., Gribaldo, S., 2017. The growing tree of archaea: new perspectives on their diversity, evolution and ecology. ISME J. 11, 2407–2425.

Alneberg, J., Bjarnason, B.S., De Bruijn, I., Schirmer, M., Quick, J., Ijaz, U.Z., Lahti, L., Loman, N.J., Andersson, A.F., Quince, C., 2014. Binning metagenomic contigs by coverage and composition. Nat. Methods 11, 1144–1146.

Bae, H.-S., Morrison, E., Chanton, J.P., Ogram, A., 2018. Methanogens are major contributors to nitrogen fixation in soils of the Florida everglades. Appl. Environ. Microbiol. 84 (7), 1–16. https://doi.org/10.1128/AEM.02222-17.

Baker, B.J., Saw, J.H., Lind, A.E., Lazar, C.S., Hinrichs, K.-U., Teske, A.P., Ettema, T.J., 2016. Genomic inference of the metabolism of cosmopolitan subsurface Archaea, Hadesarchaea. Nat. Microbiol. 1, 1–9.

Baker, B.J., De Anda, V., Seitz, K.W., Dombrowski, N., Santoro, A.E., Lloyd, K.G., 2020. Diversity, ecology and evolution of archaea. Nat. Microbiol. 5, 887–900.

Bapteste, É., Brochier, C., Boucher, Y., 2005. Higher-level classification of the archaea: evolution of methanogenesis and methanogens. Archaea 1, 353–363.

Biggs-Weber, E., Aigle, A., Prosser, J.I., Gubry-Rangin, C., 2020. Oxygen preference of deeply-rooted mesophilic thaumarchaeota in fosrest soil. Soil Biol. Biochem. 148, 107848.

Bomberg, M., Münster, U., Pumpanen, J., Ilvesniemi, H., Heinonsalo, J., 2011. Archaeal communities in boreal forest tree rhizospheres respond to changing soil temperatures. Microb. Ecol. 62, 205–217.

Boone, D.R., Mathrani, I.M., Liu, Y., Menaia, J.A., Mah, R.A., Boone, J.E., 1993. Isolation and characterization of *Methanohalophilus Portucalensis* sp. nov. and DNA reassociation study of the genus *Methanohalophilus*. Int. J. Syst. Evol. Microbiol. 43, 430–437.

Borrel, G., O'toole, P.W., Harris, H.M., Peyret, P., Brugere, J.-F., Gribaldo, S., 2013. Phylogenomic data support a seventh order of methylotrophic methanogens and provide insights into the evolution of methanogenesis. Genome Biol. Evol. 5, 1769–1780.

Borrel, G., Adam, P.S., Gribaldo, S., 2016. Methanogenesis and the Wood–Ljungdahl pathway: an ancient, versatile, and fragile association. Genome Biol. Evol. 8, 1706–1711.

Borrel, G., Adam, P.S., Mckay, L.J., Chen, L.-X., Sierra-García, I.N., Sieber, C.M., Letourneur, Q., Ghozlane, A., Andersen, G.L., LI, W.-J., 2019. Wide diversity of methane and short-chain alkane metabolisms in uncultured archaea. Nat. Microbiol. 4, 603–613.

Bräuer, S.L., Cadillo-Quiroz, H., Ward, R.J., Yavitt, J.B., Zinder, S.H., 2011. *Methanoregula boonei* gen. nov., sp. nov., an acidiphilic methanogen isolated from an acidic peat bog. Int. J. Syst. Evol. Microbiol. 61, 45–52.

Brochier, C., Gribaldo, S., Zivanovic, Y., Confalonieri, F., Forterre, P., 2005. Nanoarchaea: representatives of a novel archaeal phylum or a fast-evolving euryarchaeal lineage related to thermococcales? Genome Biol. 6, R42.

Brochier-Armanet, C., Boussau, B., Gribaldo, S., Forterre, P., 2008. Mesophilic crenarchaeota: proposal for a third archaeal phylum, the Thaumarchaeota. Nat. Rev. Microbiol. 6, 245–252.

Brown, C.T., Hug, L.A., Thomas, B.C., Sharon, I., Castelle, C.J., Singh, A., Wilkins, M.J., Wrighton, K.C., Williams, K.H., Banfield, J.F., 2015. Unusual biology across a group comprising more than 15% of domain bacteria. Nature 523, 208–211.

Castelle, C.J., Banfield, J.F., 2018. Major new microbial groups expand diversity and alter our understanding of the tree of life. Cell 172, 1181–1197.

Castelle, C.J., Wrighton, K.C., Thomas, B.C., Hug, L.A., Brown, C.T., Wilkins, M.J., Frischkorn, K.R., Tringe, S.G., Singh, A., Markillie, L.M., 2015. Genomic expansion of domain archaea highlights roles for organisms from new phyla in anaerobic carbon cycling. Curr. Biol. 25, 690–701.

Feng, h., Guo, J., Wang, W., Song, X., Yu, S., 2019. Soil depth determines the composition and diversity of bacterial and archaeal communities in a poplar plantation. Forests 10, 550.

Fernandes, S.A.P., Bernoux, M., Cerri, C.C., Feigl, B.J., Piccolo, M.C., 2002. Seasonal variation of soil chemical properties and CO_2 and CH_4 fluxes in unfertilized and P-fertilized pastures in an Ultisol of the Brazilian Amazon. Geoderma 107, 227–241.

Finn, D.R., Ziv-El, M., Van Haren, J., Park, J.G., Del Aguila-Pasquel, J., Urquiza-Muñoz, J.D., Cadillo-Quiroz, H., 2020. Methanogens and methanotrophs show nutrient-dependent community assemblage patterns across tropical peatlands of the Pastaza-Marañón Basin, Peruvian Amazonia. Front. Microbiol. 11, 746.

Fonseca, J.P., Hoffmann, L., Cabral, B.C.A., Dias, V.H.G., Miranda, M.R., De Azevedo Martins, A.C., Boschiero, C., Bastos, W.R., Silva, R., 2018. Contrasting the microbiomes from forest rhizosphere and deeper bulk soil from an Amazon rainforest reserve. Gene 642, 389–397.

Galand, P.E., Saarnio, S., Fritze, H., Yrjälä, K., 2002. Depth related diversity of methanogen archaea in finnish oligotrophic fen. FEMS Microbiol. Ecol. 42, 441–449.

Galand, P.E., Fritze, H., Yrjälä, K., 2003. Microsite-dependent changes in methanogenic populations in a boreal oligotrophic fen. Environ. Microbiol. 5, 1133–1143.

Galand, P., Fritze, H., Conrad, R., Yrjälä, K., 2005a. Pathways for methanogenesis and diversity of methanogenic archaea in three boreal peatland ecosystems. Appl. Environ. Microbiol. 71, 2195–2198.

Galand, P., Juottonen, H., Fritze, H., Yrjälä, K., 2005b. Methanogen communities in a drained bog: effect of ash fertilization. Microb. Ecol. 49, 209–217.

Guy, L., Ettema, T.J., 2011. The archaeal 'TACK' superphylum and the origin of eukaryotes. Trends Microbiol. 19, 580–587.

Guy, L., Spang, A., Saw, J.H., Ettema, T.J., 2014. 'Geoarchaeote NAG1' is a deeply rooting lineage of the archaeal order Thermoproteales rather than a new phylum. ISME J. 8, 1353–1357.

Hallam, S.J., Konstantinidis, K.T., Putnam, N., Schleper, C., Watanabe, Y.-I., Sugahara, J., Preston, C., De La Torre, J., Richardson, P.M., Delong, E.F., 2006. Genomic analysis of the uncultivated marine crenarchaeote Cenarchaeum symbiosum. Proc. Natl. Acad. Sci. 103, 18296–18301.

Hebert, P.D., Cywinska, A., Ball, S.L., Dewaard, J.R., 2003. Biological identifications through DNA barcodes. Proc. R. Soc. London, Ser. B 270, 313–321.

Hedderich, R., Whitman, W., 2013. The Prokaryotes–Prokaryotic Physiology and Biochemistry.

Hoehler, T., 2004. Biological energy requirements as quantitative boundary conditions for life in the subsurface. Geobiology 2, 205–215.

Horell, S., Kekilli, D., Strange, R.W., Hough, M.A., 2017. Recent structural insights into the function of copper nitrite reductases. Metallomics 9, 1470–1482. https://doi.org/10.1039/c7mt00146k.

Hug, L.A., Baker, B.J., Anantharaman, K., Brown, C.T., Probst, A.J., Castelle, C.J., Butterfield, C.N., Hernsdorf, A.W., Amano, Y., Ise, K., Suzuki, Y., Dudek, N., Relman, D.A., Finstad, K.M., Amundson, R., Thomas, B.C., Banfield, J.F., 2016. A new view of the tree of life. Nat. Microbiol. 1, 16048.

Iino, T., Tamaki, H., Tamazawa, S., Ueno, Y., Ohkuma, M., Suzuki, K.-I., Igarashi, Y., Haruta, S., 2013. *Candidatus* Methanogranum caenicola: a novel methanogen from the anaerobic digested sludge, and proposal of *Methanomassiliicoccaceae* fam. nov. and *Methanomassiliicoccales* ord. nov., for a methanogenic lineage of the class *Thermoplasmata*. Microbes Environ. 28 (2), 244–250. https://doi.org/10.1264/jsme2.me12189. Me12189.

Isobe, K., Ikutani, J., Fang, Y., Yoh, M., Mo, J., Suwa, Y., Yoshida, M., Senoo, K., Otsuka, S., Koba, K., 2018. Highly abundant acidophilic ammonia-oxidizing archaea causes high rates of nitrification and nitrate leaching in nitrogen-saturated forest soils. Soil Biol. Biochem. 122, 220–227.

Juottonen, H., Galand, P.E., Tuittila, E.S., Laine, J., Fritze, H., Yrjälä, K., 2005. Methanogen communities and Bacteria along an ecohydrological gradient in a northern raised bog complex. Environ. Microbiol. 7, 1547–1557.

Juottonen, H., Galand, P.E., Yrjälä, K., 2006. Detection of methanogenic Archaea in peat: comparison of PCR primers targeting the *mcr*A gene. Res. Microbiol. 157, 914–921.

Juottonen, H., Tuittila, E.-S., Juutinen, S., Fritze, H., Yrjälä, K., 2008. Seasonality of rDNA-and rRNA-derived archaeal communities and methanogenic potential in a boreal mire. ISME J. 2, 1157–1168.

Jurgens, G., Saano, A., 1999. Diversity of soil Archaea in boreal Forest before, and after clear-cutting and prescribed burning. FEMS Microbiol. Ecol. 29, 205–213.

Jurgens, G., Lindström, K., Saano, A., 1997. Novel group within the kingdom Crenarchaeota from boreal forest soil. Appl. Environ. Microbiol. 63, 803–805.

Kendall, M.M., Boone, D.R., 2006. The order Methanosarcinales. Prokaryotes 3, 244–256.

Knittel, K., Boetius, A., 2009. Anaerobic oxidation of methane: progress with an unknown process. Annu. Rev. Microbiol. 63, 311–334.

Könneke, M., Bernhard, A.E., de la Torre, J.R., et al., 2005. Isolation of an autotrophic ammonia-oxidizing marine archaeon. Nature 437, 543–546. https://doi.org/10.1038/nature03911.

Könneke, M., Schubert, D.M., Brown, P.C., Hügler, M., Standfest, S., Schwander, T., Von Borzyskowski, L.S., Erb, T.J., Stahl, D.A., Berg, I.A., 2014. Ammonia-oxidizing archaea use the most energy-efficient aerobic pathway for CO_2 fixation. Proc. Natl. Acad. Sci. 111, 8239–8244.

Kotsyurbenko, O.R., Glagolev, M.V., Merkel, A.Y., Sabrekov, A.F., Terentieva, I.E., 2019. Methanogenesis in soils, wetlands, and peat. In: Stams, A.J.M., Sousa, D.Z. (Eds.), Biogenesis of Hydrocarbons. Springer International Publishing, Cham.

Kozubal, M.A., Romine, M., Dem Jennings, R., Jay, Z.J., Tringe, S.G., Rusch, D.B., Beam, J.P., Mccue, L.A., Inskeep, W.P., 2013. Geoarchaeota: a new candidate phylum in the archaea from high-temperature acidic iron mats in Yellowstone National Park. ISME J. 7, 622–634.

Laso-Pérez, R., Wegener, G., Knittel, K., Widdel, F., Harding, K.J., Krukenberg, V., Meier, D.V., Richter, M., Tegetmeyer, H.E., Riedel, D., 2016. Thermophilic archaea activate butane via alkyl-coenzyme M formation. Nature 539, 396–401.

Lazar, C.S., Baker, B.J., Seitz, K., Hyde, A.S., Dick, G.J., Hinrichs, K.U., Teske, A.P., 2016. Genomic evidence for distinct carbon substrate preferences and ecological niches of Bathyarchaeota in estuarine sediments. Environ. Microbiol. 18, 1200–1211.

Lehtovirta, L.E., Prosser, J.I., Nicol, G.W., 2009. Soil PH regulates the abundance and diversity of Group 1.1 C Crenarchaeota. FEMS Microbiol. Ecol. 70, 367–376.

Lehtovirta-Morley, L.E., Stoecker, K., Vilcinskas, A., Prosser, J.I., Nicol, G.W., 2011. Cultivation of an obligate acidophilic ammonia oxidizer from a nitrifying acid soil. Proc. Natl. Acad. Sci. 108, 15892–15897.

Leininger, S., Urich, T., Schloter, M., Schwark, L., Qi, J., Nicol, G.W., Prosser, J.I., Schuster, S., Schleper, C., 2006. Archaea predominate among ammonia-oxidizing prokaryotes in soils. Nature 442, 806–809.

Leu, A.O., Cai, C., Mcilroy, S.J., Southam, G., Orphan, V.J., Yuan, Z., Hu, S., Tyson, G.W., 2020. Anaerobic methane oxidation coupled to manganese reduction by members of the Methanoperedenaceae. ISME J. 14, 1030–1041.

Li, M., Baker, B.J., Anantharaman, K., Jain, S., Breier, J.A., Dick, G.J., 2015. Genomic and transcriptomic evidence for scavenging of diverse organic compounds by widespread deep-sea archaea. Nat. Commun. 6, 1–6.

Lloyd, K.G., Schreiber, L., Petersen, D.G., Kjeldsen, K.U., Lever, M.A., Steen, A.D., Stepanauskas, R., Richter, M., Kleindienst, S., Lenk, S., 2013. Predominant archaea in marine sediments degrade detrital proteins. Nature 496, 215–218.

Martin-Cuadrado, A.-B., Garcia-Heredia, I., Molto, A.G., Lopez-Ubeda, R., Kimes, N., López-García, P., Moreira, D., Rodriguez-Valera, F., 2015. A new class of marine Euryarchaeota group ii from the Mediterranean deep chlorophyll maximum. ISME J. 9, 1619–1634.

Matte-Tailliez, O., Brochier, C., Forterre, P., Philippe, H., 2002. Archaeal phylogeny based on ribosomal proteins. Mol. Biol. Evol. 19, 631–639.

McGlynn, S.E., 2017. Energy metabolism during anaerobic methane oxidation in ANME archaea. Microbes Environ. 32, Me16166.

Meng, J., Xu, J., Qin, D., He, Y., Xiao, X., Wang, F., 2014. Genetic and functional properties of uncultivated mcg archaea assessed by metagenome and gene expression analyses. ISME J. 8, 650–659.

Merilä, P., Galand, P., Fritze, H., Tuittila, E.-S., Kukko-Oja, K., Laine, J., Yrjälä, K., 2005. Development of methanogen communities during a primary succession of mire ecosystems. In: Program and Abstracts of the Joint International Symposia for Subsurface Microbiology (ISSM 2005) and Environmental Biogeochemistry (ISEB XVII). August 14–19, 2005, Jackson Hole, Wyoming, USA.

Meyer, K.M., Klein, A.M., Rodrigues, J.L., Nüsslein, K., Tringe, S.G., Mirza, B.S., Tiedje, J.M., Bohannan, B.J., 2017. Conversion of Amazon rainforest to agriculture alters community traits of methane-cycling organisms. Mol. Ecol. 26, 1547–1556.

Meyerdierks, A., Kube, M., Kostadinov, I., Teeling, H., Glöckner, F.O., Reinhardt, R., Amann, R., 2010. Metagenome and MRNA expression analyses of anaerobic methanotrophic archaea of the anme-1 group. Environ. Microbiol. 12, 422–439.

Mondav, R., Woodcroft, B.J., Kim, E.-H., Mccalley, C.K., Hodgkins, S.B., Crill, P.M., Chanton, J., Hurst, G.B., Verberkmoes, N.C., Saleska, S.R., 2014. Discovery of a novel methanogen prevalent in thawing permafrost. Nat. Commun. 5, 1–7.

Murrell, J.C., 2010. The aerobic methane oxidizing bacteria (Methanotrophs). In: Timmis, K.N. (Ed.), Handbook of Hydrocarbon and Lipid Microbiology. Springer Berlin Heidelberg, Berlin, Heidelberg.

Mwirichia, R., Alam, I., Rashid, M., Vinu, M., Ba-Alawi, W., Kamau, A.A., Ngugi, D.K., Göker, M., Klenk, H.-P., Bajic, V., 2016. Metabolic traits of an uncultured archaeal lineage-MSBL1-from brine pools of the Red Sea. Sci. Rep. 6, 1–14.

Nobu, M.K., Narihiro, T., Kuroda, K., Mei, R., Liu, W.-T., 2016. Chasing the elusive Euryarchaeota class WSA2: genomes reveal a uniquely fastidious methyl-reducing methanogen. ISME J. 10, 2478–2487.

Nunoura, T., Takaki, Y., Kakuta, J., Nishi, S., Sugahara, J., Kazama, H., Chee, G.-J., Hattori, M., Kanai, A., Atomi, H., 2011. Insights into the evolution of Archaea and eukaryotic protein modifier systems revealed by the genome of a novel archaeal group. Nucleic Acids Res. 39, 3204–3223.

Nurk, S., Meleshko, D., Korobeynikov, A., Pevzner, P.A., 2017. MetaSPAdes: a new versatile metagenomic assembler. Genome Res. 27, 824–834.

Oren, A., 2014. The family Methanococcaceae. In: Rosenberg, E., Delong, E.F., Lory, S., Stackebrandt, E., Thompson, F. (Eds.), The Prokaryotes: Other Major Lineages of Bacteria and The Archaea. Springer Berlin Heidelberg, Berlin, Heidelberg.

Orphan, V.J., House, C.H., Hinrichs, K.-U., Mckeegan, K.D., Delong, E.F., 2002. Multiple archaeal groups mediate methane oxidation in anoxic cold seep sediments. Proc. Natl. Acad. Sci. 99, 7663–7668.

Parks, D.H., Imelfort, M., Skennerton, C.T., Hugenholtz, P., Tyson, G.W., 2015. CheckM: assessing the quality of microbial genomes recovered from isolates, single cells, and metagenomes. Genome Res. 25, 1043–1055.

Pester, M., Schleper, C., Wagner, M., 2011. The Thaumarchaeota: an emerging view of their phylogeny and ecophysiology. Curr. Opin. Microbiol. 14, 300–306.

Petitjean, C., Deschamps, P., López-García, P., Moreira, D., 2015a. Rooting the domain archaea by phylogenomic analysis supports the foundation of the new kingdom Proteoarchaeota. Genome Biol. Evol. 7, 191–204.

Petitjean, C., Deschamps, P., López-García, P., Moreira, D., Brochier-Armanet, C., 2015b. Extending the conserved phylogenetic core of archaea disentangles the evolution of the third domain of life. Mol. Biol. Evol. 32, 1242–1254.

Probst, A.J., Moissl-Eichinger, C., 2015. "Altiarchaeales": uncultivated archaea from the subsurface. Life 5, 1381–1395.

Putkinen, A., Juottonen, H., Juutinen, S., Tuittila, E.-S., Fritze, H., Yrjälä, K., 2009. Archaeal rRNA diversity and methane production in deep boreal peat. FEMS Microbiol. Ecol. 70, 87–98.

Rinke, C., Schwientek, P., Sczyrba, A., Ivanova, N.N., Anderson, I.J., Cheng, J.-F., Darling, A., Malfatti, S., Swan, B.K., Gies, E.A., 2013. Insights into the phylogeny and coding potential of microbial dark matter. Nature 499, 431–437.

Sakai, S., Imachi, H., Hanada, S., Ohashi, A., Harada, H., Kamagata, Y., 2008. *Methanocella paludicola* gen. nov., sp. nov., a methane-producing archaeon, the first isolate of the lineage 'Rice Cluster I', and proposal of the new archaeal order *Methanocellales* ord. nov. Int. J. Syst. Evol. Microbiol. 58, 929–936.

Seitz, K.W., Lazar, C.S., Hinrichs, K.-U., Teske, A.P., Baker, B.J., 2016. Genomic reconstruction of a novel, deeply branched sediment archaeal phylum with pathways for acetogenesis and sulfur reduction. ISME J. 10, 1696–1705.

Siles, J.A., Margesin, R., 2016. Abundance and diversity of bacterial, archaeal, and fungal communities along an altitudinal gradient in alpine forest soils: what are the driving factors? Microb. Ecol. 72, 207–220.

Sorek, R., Zhu, Y., Creevey, C.J., Francino, M.P., Bork, P., Rubin, E.M., 2007. Genome-wide experimental determination of barriers to horizontal gene transfer. Science 318, 1449–1452.

Sorokin, D.Y., Makarova, K.S., Abbas, B., Ferrer, M., Golyshin, P.N., Galinski, E.A., Ciordia, S., Mena, M.C., Merkel, A.Y., Wolf, Y.I., 2017. Discovery of extremely halophilic, methyl-reducing euryarchaea provides insights into the evolutionary origin of methanogenesis. Nat. Microbiol. 2, 17081.

Spang, A., Saw, J.H., Jørgensen, S.L., Zaremba-Niedzwiedzka, K., Martijn, J., Lind, A.E., Van Eijk, R., Schleper, C., Guy, L., Ettema, T.J., 2015. Complex archaea that bridge the gap between prokaryotes and eukaryotes. Nature 521, 173–179.

Spang, A., Caceres, E.F., Ettema, T.J., 2017. Genomic exploration of the diversity, ecology, and evolution of the archaeal domain of life. Science 357 (6351), 1–10. https://doi.org/10.1126/science.aaf3883. PMID: 28798101.

Sunagawa, S., Coelho, L.P., Chaffron, S., Kultima, J.R., Labadie, K., Salazar, G., Djahanschiri, B., Zeller, G., Mende, D.R., Alberti, A., 2015. Structure and function of the global ocean microbiome. Science 348 (6237). 126359-1–126359-9. https://doi.org/10.1126/science.1261359.

Teh, Y.A., Silver, W.L., Conrad, M.E., 2005. Oxygen effects on methane production and oxidation in humid tropical forest soils. Glob. Chang. Biol. 11, 1283–1297.

Timmers, P.H.A., Welte, C.U., Koehorst, J.J., Plugge, C.M., Jetten, M.S.M., Stams, A.J.M., 2017. Reverse methanogenesis and respiration in methanotrophic Archaea. Archaea 2017, 1654237.

Valentine, D.L., 2007. Adaptations to energy stress dictate the ecology and evolution of the Archaea. Nat. Rev. Microbiol. 5, 316–323.

Vanwonterghem, I., Evans, P.N., Parks, D.H., Jensen, P.D., Woodcroft, B.J., Hugenholtz, P., Tyson, G.W., 2016. Methylotrophic methanogenesis discovered in the archaeal phylum Verstraetearchaeota. Nat. Microbiol. 1, 1–9.

Walker, C.B., De La Torre, J., Klotz, M., Urakawa, H., Pinel, N., Arp, D., Brochier-Armanet, C., Chain, P., Chan, P., Gollabgir, A., 2010. *Nitrosopumilus maritimus* genome reveals unique mechanisms for nitrification and autotrophy in globally distributed marine crenarchaea. Proc. Natl. Acad. Sci. 107, 8818–8823.

Wang, J.-T., Cao, P., Hu, H.-W., Li, J., Han, L.-L., Zhang, L.-M., Zheng, Y.-M., He, J.-Z., 2015. Altitudinal distribution patterns of soil bacterial and archaeal communities along mt. Shegyla on the Tibetan plateau. Microb. Ecol. 69, 135–145.

Whitman, W.B., Bowen, T.L., Boone, D.R., 2006. The methanogenic Bacteria. Prokaryotes 3, 165–207.

Whittaker, R.H., 1969. New concepts of kingdoms of organisms. Science 163, 150–160.

Woese, C.R., 2004. The archaeal concept and the world it lives in: a retrospective. Photosynth. Res. 80, 361–372.

Woese, C.R., Fox, G.E., 1977. Phylogenetic structure of prokaryotic domain: the primary kingdoms. Proc. Natl. Acad. Sci. U. S. A. 74, 5088–5090.

Woese, C.R., Kandler, O., Wheelis, M.L., 1990. Towards a natural system of organisms: proposal for the domains Archaea, Bacteria, and Eucarya. Proc. Natl. Acad. Sci. 87, 4576–4579.

Yrjälä, K., Katainen, R., Jurgens, G., Saarela, U., Saano, A., Romantschuk, M., Fritze, H., 2004. Wood ash fertilization alters the forest humus archaea community. Soil Biol. Biochem. 36, 199–201.

Yrjälä, K., Tuomivirta, T., Juottonen, H., Putkinen, A., Lappi, K., Tuittila, E.S., Penttilä, T., Minkkinen, K., Laine, J., Peltoniemi, K., 2011. CH4 production and oxidation processes in a boreal fen ecosystem after long-term water table drawdown. Glob. Chang. Biol. 17, 1311–1320.

Zaremba-Niedzwiedzka, K., Caceres, E.F., Saw, J.H., Bäckström, D., Juzokaite, L., Vancaester, E., Seitz, K.W., Anantharaman, K., Starnawski, P., Kjeldsen, K.U., 2017. Asgard archaea illuminate the origin of eukaryotic cellular complexity. Nature 541, 353–358.

Zhang, M., Chai, L., Huang, M., Jia, W., Guo, J., Huang, Y., 2020. Deciphering the archaeal communities in tree rhizosphere of the Qinghai-Tibetan plateau. BMC Microbiol. 20, 1–13.

Further reading

Bomberg, M., Timonen, S., 2009. Effect of tree species and mycorrhizal colonization on the archaeal population of boreal forest rhizospheres. Appl. Environ. Microbiol. 75, 308–315.

Haidl, I., Albers, S., Allers, T., 2016. Archaea and the meaning of life. In: Microbiology Today. Microbiology Society, London.

Hallam, S.J., Putnam, N., Preston, C.M., Detter, J.C., Rokhsar, D., Richardson, P.M., Delong, E.F., 2004. Reverse methanogenesis: testing the hypothesis with environmental genomics. Science 305, 1457–1462.

Ma, B., Dai, Z., Wang, H., Dsouza, M., Liu, X., He, Y., Wu, J., et al., 2017. Distinct biogeographic patterns for archaea, bacteria, and fungi along the vegetation gradient at the continental scale in Eastern China. msystems.asm.org 2 (1), 1–14. https://doi.org/10.1128/mSystems.00174-16.

Truu, M., Nõlvak, H., Ostonen, I., Oopkaup, K., Maddison, M., Ligi, T., Espenberg, M., Uri, V., Mander, U., Truu, J., 2020. Soil bacterial and archaeal communities and their potential to perform N-cycling processes in soils of boreal forests growing on well-drained peat. Front. Microbiol. 11 (591358), 1–22. https://doi.org/10.3389/fmicb.2020.591358.

Tzanakakis, V.A., Taylor, A.E., Bakken, L.R., Bottomley, P.J., Myrold, D.D., Dörsch, P., 2019. Relative activity of ammonia oxidizing archaea and bacteria determine nitrification-dependent N_2O emissions in Oregon forest soils. Soil Biol. Biochem. 139 (107612), 1–6. https://doi.org/10.1016/j.soilbio.2019.107612.

Chapter 20

Viruses as components of forest microbiome

Jarkko Hantula and Eeva J. Vainio

Natural Resources Institute Finland (Luke), Helsinki, Finland

Chapter Outline

1. Introduction

In this chapter, we will cover viruses of forest-dwelling organisms as traditionally categorized based on their hosts: plants, arthropods, fungi, bacteria, and oomycetes. However, knowledge on virus diversity is currently expanding quickly via high-throughput sequencing (HTS), which has also challenged our understanding on virus host ranges. The HTS approaches used for virus detection are usually based on the analysis of various RNA fractions (Table 20.1). This is because amplicon sequencing or DNA-based HTS is usually not feasible for discovering the entire viral diversity due to high level of sequence divergence and variability in genome composition among viruses. As demonstrated in a groundbreaking study by Shi et al. (2016), similar virus sequences are being discovered by RNA sequencing from plants, arthropods, fungi, and oomycetes, and it seems plausible that cross-kingdom virus transmission does happen at an evolutionary timescale. The close association with plants and insects or fungi and oomycetes feeding or infecting them, respectively, could provide such opportunities, and examples of potential cases are given as follows. Bacteria are phylogenetically very distant from the eukaryotic groups mentioned before, and their virus communities are also very different phylogenetically, biologically, and structurally. However, mitochondrial viruses infecting fungi (family *Mitoviridae*) show phylogenetic resemblance to bacterial leviviruses, which has led to the intriguing hypothesis that these viruses were brought along during endosymbiosis, and therefore reflect the bacterial origin of mitochondria (Wolf et al., 2018). Fig. 20.1 gives a glimpse of the diversity of virus particles and nucleic acids in different hosts based on virus taxa discussed in more detail as follows. Further information on virus classification is available in taxonomy reports regularly updated by the International Committee on Taxonomy of Viruses (https://talk.ictvonline.org/ictv-reports/ictv_online_report/).

Forest Microbiology. https://doi.org/10.1016/B978-0-12-822542-4.00008-5

TABLE 20.1 Various RNA fractions used for virus detection by high-throughput sequencing and their potential to detect different virus groups.

Nucleic acid fraction used for analysis	Total RNA	Total RNA depleted of ribosomal RNA	Polyadenylated RNA	Viral dsRNA	Small interfering RNA	Encapsidated nucleic acids
Method of extraction	Extraction of all cytosolic and organellar RNAs from the host	Removal of rRNA from the total RNA pool by hybridization with complementary oligonucleotides	Trapping of RNAs with terminal poly(A) tracts using oligo(dT) affinity matrices	Cellulose affinity chromatography (specific precipitation of dsRNA)	Extraction of RNAs of ca. 20–30 bp by size fractionation	Harvesting viral particles cwith ultracentrifugation or depletion of naked nucleic acids by nucleases
Target viruses	dsRNA and ssRNA viruses, transcripts of DNA viruses	dsRNA and ssRNA viruses, transcripts of DNA viruses	Viruses with RNA genomes with terminal poly(A) tracts and/or polyadenylated transcripts	dsRNA viruses and ssRNA viruses with replicative dsRNA intermediates	Viruses targeted by host RNA silencing in plants, fungi, and invertebrates	Viruses with protein capsids

The role of viruses in forest ecology and evolution is highly variable and depends strongly on the host-virus combination. Many viruses cause diseases on trees, but also harmless viruses are known. Some viruses are considered as powerful modulators of insect populations, and thus have a role in phasing out epidemics of pests. In contrast, infections of fungal viruses are mostly considered symptomless, although also virulent and even symbiotic virus-fungus relationships are known. Bacterial viruses are the most common creatures on the globe, and as such may affect the overall dynamics of microbial communities also in forests. Furthermore, some viruses are used as biocontrol agents against insects and fungi due to their detrimental effects on these organisms causing significant economic losses.

2. Plant viruses

2.1 Introduction to plant viruses

The first plant virus known to science was the Tobacco mosaic virus (TMV; *Virgaviridae*). In the 1890s, Ivanovski in Russia reported the existence of "filterable pathogens" that remained infectious after passing through filters that retain bacterial pathogens in diseased tobacco plants. In 1898, Beijerinck named this infectious substance causing tobacco mosaic disease a "living infectious fluid."

Epidemiologically, plant viruses represent two main types: viruses causing acute infections and persistent viruses (Roossinck, 2010). Plant viruses causing acute infections need a way of transmission between immobile host plants and to penetrate the plant cell wall (Fig. 20.2). This is usually accomplished by vector organisms, such as plant feeding arthropods (aphids, thrips, whiteflies, mealy bugs, plant hoppers, grasshoppers, scales, and beetles), root feeding nematodes, and sometimes fungal organisms (*Olpidium* sp.). Mechanical transmission during gardening or animal grazing can also help to transmit plant viruses, and some of them even survive passage through the alimentary tract of mammals. Virus transmission may also occur via soil water and root contacts, as well as vertically in pollen and seeds. Plant viruses causing acute infections are often associated with diseases and even mortality, and the viruses may move between plant cells using special movement proteins. In turn, persistent viruses do not move between cells and do not generally cause diseases. They are passed on during cell division and transmit via seeds into the next plant generation. An even more intimate association is formed by endogenized viruses. These viruses are integrated in the host genome, and may either remain "silent" or be activated during stress conditions.

2.2 Taxonomic diversity

Plant viruses have genomes composed of either RNA or DNA. Most of the plant virus families with members infecting forest trees have positive-sense RNA genomes (Fig. 20.1). These include members of *Betaflexiviridae*, *Bromoviridae*,

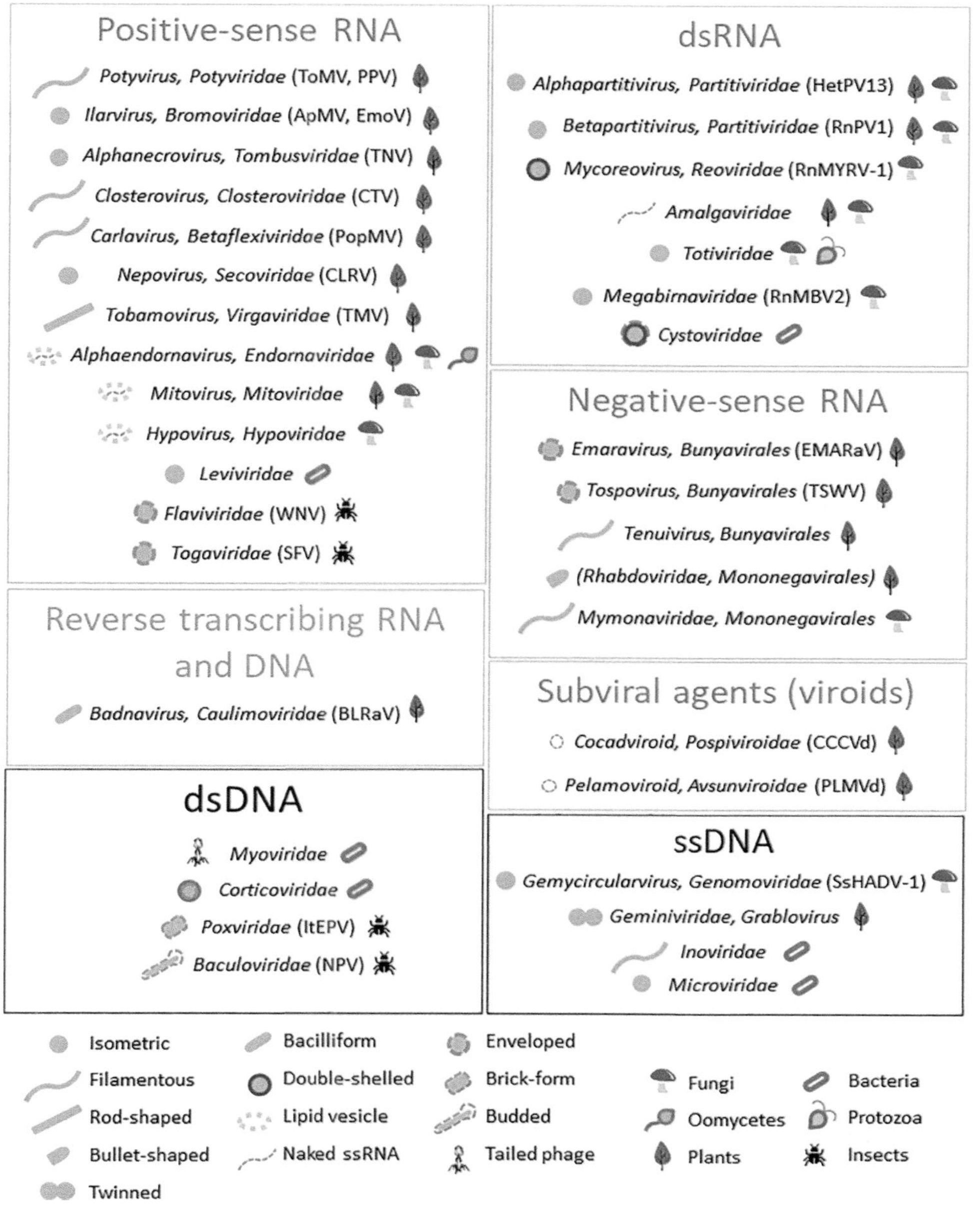

FIG. 20.1 Examples of virus morphologies and nucleic acid composition.

Closteroviridae, Potyviridae, Secoviridae, Tombusviridae, and *Virgaviridae*. The order *Bunyavirales* and family *Rhabdoviridae* accommodate plant viruses with negative-sense RNA genomes. Persistent plant viruses are poorly known although they are expected to be very common. Examples include members of family *Partitiviridae* with dsRNA genomes, and positive-sense RNA viruses of families *Endornaviridae* and *Mitoviridae*. Members of family *Geminiviridae* have circular ssDNA viruses, whereas the virus family *Caulimoviridae* accommodates reverse-transcribing DNA viruses that usually have an episomal replication cycle. However, certain caulimoviruses are endogenized into the host genome.

2.3 Host ranges

Plant viruses causing acute infections may be generalists with a wide host range, some of which can infect hundreds of plant species, or specialists that only infect very closely related plants. Several generalist viruses occur also in forest tree species. For example, Tomato mosaic virus (ToMV; *Potyviridae*) has been identified in *Picea* and *Salix*; Apple mosaic virus (ApMV; *Bromoviridae*) in *Aesculus*, *Betula*, and *Carpinus*; Tobacco necrosis virus (TNV; *Tombusviridae*) in *Betula*, *Pinus*, *Quercus*, *Salix*, and *Populus*; and TMV (*Virgaviridae*) in *Quercus*. Tomato spotted wilt virus (TSWV; *Tospovirus, Bunyavirales*) is

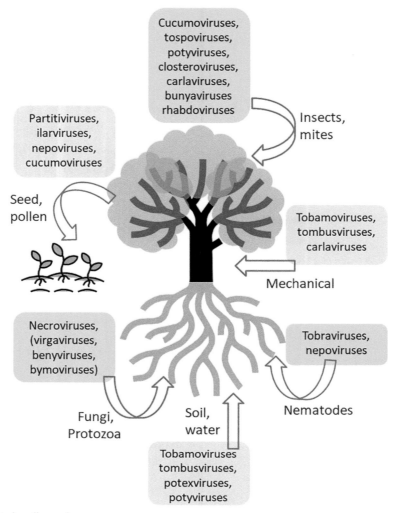

FIG. 20.2 Pathways of plant virus dispersal.

considered to have the broadest host range of all plant viruses and infects more than 1000 plant species spreading by thrips and has been identified in *Sorbus* and *Ulmus pumila*.

2.4 Economic importance of viruses in forest trees

Research on plant viruses has focused on economically important crop plants, and their diversity is less well known in natural ecosystems, including forests. Crop plant cultivations differ in many aspects from natural plant communities, most strikingly because the plants are typically cultivated in monoculture. This allows more efficient spread of specialist viruses and their vectors. On the other hand, many important woody crop plants, such as stone fruits, have numerous wild relatives in native forest ecosystems. Examples of notorious plant viruses with worldwide distribution causing significant economic losses in fruit trees include the Plum pox virus (PPV; *Potyviridae*), which infects also wild and ornamental *Prunus* species and other stone fruits, and the Citrus Tristeza virus (CTV; *Closteroviridae*) that led to the destruction of nearly 100 million citrus trees worldwide since the epidemic started in the 1930s in South America (Moreno et al., 2008).

Büttner and colleagues are among the few scientists investigating forest diseases caused by viruses and have included a thorough list of known viruses in a review article (Büttner et al., 2013). In natural forest habitats, broadleaved trees often exhibit visible symptoms during virus infection. These include growth decline, stem pitting, and various leaf symptoms (crinkling, discoloration, necrotic or chlorotic spots, and vein banding). Specialist viruses infecting only a few host species have been detected, for example in rowan (*Sorbus aucuparia*) that hosts the European mountain ash ringspot-associated emaravirus (EMARaV; *Bunyavirales*). This virus is highly common in both symptomatic and asymptomatic rowan trees, suggesting variability in host tolerance. In turn, Poplar mosaic virus (PopMV; *Betaflexiviridae*) has been shown to cause considerable growth decline in some poplar (*Populus* sp.) hybrids. The Elm mottle virus (EMoV; *Bromoviridae*) also has a relatively narrow host range in elms and a few woody ornamentals.

Conifers usually lack clear virus symptoms. Only a few viruses have been associated with the tree genera *Abies*, *Pinus*, and *Picea* that are widespread in the boreal forest zone and extensively used for timber production. Virus-like particles have been observed in all three host genera, but very few have been molecularly characterized. Some of these represent persistent virus types, for example members of *Partitiviridae* (*Pinus sylvestris* partitivirus) and *Amalgaviridae* (*Pinus patula* amalgavirus), whereas tenuiviruses with negative-sense RNA genomes have been detected in *Picea mariana*.

It should be noted that only a few tree species used for timber production have been studied for virus infections. This is particularly true in the case of many tropical and subtropical ecosystems where tree species diversity is considerably high. Taking into consideration that the estimated number of tree species in the world is 60,000 (Beech et al., 2017), the plant virus diversity in natural forest ecosystems can be considered mostly unknown. High-throughput RNA sequencing (RNA-Seq) has already revolutionized the detection and diagnosis of viruses in woody crop plants, such as grapevine (*Vitis vinifera*), rubber tree (*Hevea brasiliensis*), and apple (*Malus domestica*), and expanded the known host range of some virus taxa, for example identified geminiviruses in *Prunus* spp. In the future, this methodology can be used for natural forest trees and may be expected to reveal a high number of previously unknown viruses, including many persistent specialist viruses and asymptomatic endogenized viruses.

2.5 Case example: Birch leaf roll disease—New culprit revealed by HTS

Cherry leaf roll virus (CLRV; *Seoviridae*) has a wide host range among broadleaved trees and shrubs of genera *Aesculus*, *Betula*, *Malus*, *Prunus*, *Rubus*, *Sambucus*, *Ulmus*, *Sorbus*, and *Vaccinium*. It is spread by pollen and seed, as well as soil water (and possibly root connections). Symptoms corresponding to those caused by CLRV have been regularly observed in birch trees in Northern Europe, and mixed infections by different CLRV variants are detected even in single trees. However, recent HTS studies by Rumbou et al. (2019) have revealed that the virome of birch is actually more complex than thought before, and a badnavirus (*Caulimoviridae*) called Birch leaf roll-associated virus (BLRaV) may actually be the causative agent of the birch leaf roll. Moreover, also other previously unknown viruses were identified during the analysis, including putative members of genera *Carlavirus*, *Idaeovirus*, and *Capillovirus*. These observations highlight the effectiveness of HTS in revealing new viral diversity, which can even change our view on disease etiology.

2.6 Viroids: Subviral agents smaller than any known viruses

Viroids are small, only 246 to 401 nucleotides long, circular ssRNAs that replicate autonomously in higher plants but do not encode any proteins (Fig. 20.1). Thus far they have been detected in commercially cultivated fruit crop trees, but their natural host ranges remain mostly unknown. According to the 10th ICTV Report, species of the family *Avsunviroidae* infect avocado, apple, and peach (*Prunus persica*), and species of family *Pospiviroidae* infect apple, citrus, and pear (*Pyrus* sp.) as well as coconut (*Cocos nucifera*) and grapevine (*V. vinifera*). Despite being so simple nucleic acid entities, some viroids cause devastating plant diseases, as exemplified by the Coconut cadang-cadang viroid (CCCVd) that has killed millions of coconut palms in the Philippines. Peach latent mosaic viroid (PLMVd) is globally present in peach-growing areas, but varies greatly in symptom intensity, ranging from latent to symptomatic infections causing, for example leaf mosaic, fruit deformation, delayed foliation and flowering, and premature aging.

3. Entomopathogenic viruses

3.1 Low impact on insect populations, but still useful

Insects are the most common group of animals, but knowledge on their viruses is biased toward insect-borne viruses that cause serious diseases in plants, vertebrate animals, or humans. Examples of the latter include the West Nile fever virus (WNV), which belongs to the family *Flaviviridae* and is transmitted from birds to humans via mosquitoes feeding on both vertebrates. Another one, the Semliki forest virus (SFV) of the family *Togaviridae* even became a model of research on the viral life cycle and human diseases after its isolation in 1942 in Semliki forest, Uganda. SFV has also been used as a vector for genes encoding antigen components of vaccines and anticancer agents as well as a tool in gene therapy.

Although the WNV and SFV have played a major role in medicine and the development of basic research, in forest environments the most important insect viruses are probably those infecting forest pests, especially the ones capable of large-scale forest destruction. Interestingly, also the documented history of these viruses goes far back: there are historical records from the sixteenth century describing "wilting disease" in silkworm (*Bombyx mori*) larvae feeding on white mulberries (*Morus alba*), later shown to be caused by *Bombyx mori* nuclear polyhedrosis virus.

One of the most important forest pests, spruce bark beetle (*Ips typographus*), may cause vast epidemics after tree-felling storms or hot summers allowing the development of additional beetle generations in freshly damaged spruces (*Picea abies*). This beetle hosts Ips typographus entomopoxvirus (ItEPV), which is a double-stranded DNA virus of the family *Poxviridae* that produces large, enveloped virions, and replicates in the cytoplasm of the midgut epithelium cells. It is species specific and causes mortality by destroying the gut epithelium of the host and is thereafter transferred in galleries to wait for the next insect to be infected. The infection levels on spruce beetles are higher in unmanaged forests compared to managed ones, probably because the dead hosts are not removed during the sanitation cuttings. However, ItEPV has not been shown to play a significant role in stopping epidemics.

Neither has such an observation been made for most other forest pests, for example the roles of viruses attacking the nun moth (*Lymantria monacha*) or the autumnal moth (*Epirrita autumnata*) causing epidemics of birch trees (*Betula* sp.) in northern latitudes are considered negligible. However, the prolonged epidemics by the European pine sawfly (*Neodiprion sertifer*) have been associated with increasing numbers of virus-infected insects, which together with other parasites may have a role in ending epidemics on pines (*Pinus* sp.).

3.2 Nuclear polyhedrosis viruses as biological control agents

Despite the generally small effect of entomopathogenic viruses on forest insects in nature, there are cases of virus-based biocontrol (sometimes called virocontrol) applications against forest pests. Two of them are based on the nuclear polyhedrosis viruses (NPV), which are large dsDNA viruses belonging to the family *Baculoviridae* and infecting more than 600 invertebrate host species (Harrison and Hoover, 2012).

European pine sawfly populations are controlled by weather but also parasitic and predacious insects and disease-causing organisms. It was already in the 1950s when it was noticed that one of its viruses, the nuclear polyhedrosis virus, could be used as a biocontrol agent. In natural conditions these viruses may have infection rates of up to about 20%, but experiments with intentional spread of virus particles showed that more than 90% of larval colonies could be completely destroyed by deliberate control actions (Bird, 1953). A related virus has also been used in biological control of the gypsy moth (*Lymantria dispar*) on broadleaved trees. In practice, these viruses can be spread from airplanes over the infected forests during the epidemics.

The nuclear polyhedrosis viruses for biocontrol must be produced in host insect larvae fed by food contaminated with viruses, from which the developing virus particles are collected and partially purified. This material is then used by spreading it to insect larvae in the trees attacked by insects to be controlled. After being ingested the virus invades through the gut wall and reproduces in the internal tissues, causing disintegration of internal organs and death. Finally, the host ruptures exposing viral occlusion bodies to be ingested by the subsequent insects. Although highly efficient, the high cost of NPV production has restricted its use in controlling forest pests.

4. Fungal viruses

4.1 Fungi are hosting a diverse group of viruses

Fungi form a taxonomically and ecologically wide group of eukaryotes, which includes saprotrophs decomposing biological material, mycorrhizal and endophytic symbionts providing benefits to other organisms, as well as pathogens causing diseases in forest trees and other organisms. Economically most important forest fungi grow as mycelia composed of hyphae, but also single-celled yeasts are common.

The presence of viruses in fungi was observed in the year 1962, when Hollings showed that the La France disease of the cultivated mushroom, *Agaricus bisporus* was caused by viruses with a particle size of ca. 35 nm. Thereafter, the number of known fungal viruses, or mycoviruses, has increased rapidly, and today they are classified in 17 virus families, many of them observed also in fungal organisms associated with forest trees. However, HTS methods have only recently been utilized to search for new fungal viruses and based on those analyses the total diversity of mycoviruses exceeds considerably the current estimates (Marzano and Domier, 2016).

Many fungal viruses have similar particles as other viruses. However, some mycoviruses are free of protein capsids and are instead associated with intracellular membrane structures. The genomes of mycoviruses are relatively small and code only one or few genes. The genetic material of mycoviruses may be composed of various genetic materials, including single stranded RNA or DNA, and double stranded RNA. Classified fungal viruses with positive-sense RNA genomes include members of families *Alphaflexiviridae*, *Barnaviridae*, *Botourmiaviridae*, *Deltaflexiviridae*, *Endornaviridae*, *Gammaflexiviridae*, *Hypoviridae*, and *Mitoviridae* (Fig. 20.2). Viruses with dsRNA genomes include members of *Amalgaviridae*, *Chrysoviridae*, *Megabirnaviridae*, *Partitiviridae*, *Quadriviridae*, and *Reoviridae*. Members of family *Mymonaviridae* have negative-sense RNA genomes, whereas members of *Genomoviridae* are the only known mycoviruses with DNA genomes (specifically, circular ssDNA).

4.2 Intracellular lifestyle has consequences

Most mycoviruses differ from other viruses in their means to spread from one host mycelium to another. They do not form extracellular infective particles and are therefore not able to spread between separate fungal mycelia by themselves. The only exception for this is the *Sclerotinia sclerotiorum* hypovirulence-associated DNA virus 1 (SsHADV-1) of family *Genomoviridae*, which forms extracellular virus particles in its host, a plant pathogenic fungus *Sclerotinia sclerotiorum*.

In other mycoviruses viral transmission takes place via hyphal fusions between two mycelia. Such fusions are common and promiscuous among fungi and may be initiated even among strains belonging to different host species. This reaction is controlled by nuclear genes, which enable permanent fusion only if both strains possess the same vegetative compatibility alleles and thus belong to the same Vegetative Compatibility (VC) group. If the two fungal hyphae involved are not compatible, the cells participating in the fusion will soon get killed. This system allows almost free transmission of mycoviruses between mycelia belonging to the same Vegetative Compatibility (VC) groups but restricts efficiently transmission of mycoviruses between incompatible mycelia. This constraint is not complete as it takes some time before the vegetative incompatibility leads to cell deaths, and therefore viral transmissions may sometimes occur also between mycelia belonging to different VC groups, albeit at a lower frequency. There is even growing evidence of viral transmissions occurring between unrelated fungal species. For example, the same virus strains were detected in a conifer pathogen *Heterobasidion parviporum* and taxonomically distantly related saprotrophic or mycorrhizal fungi at the same forest site (Vainio et al., 2017).

The long distance spread of fungal viruses is possible only within mycelial pieces of their host fungi contaminating other organisms, or in asexual or sexual spores. However, the uptake of viruses into different spore types is variable. For example, no viruses have been observed in ascospore-derived cultures of the pine shoot pathogen *Gremmeniella abietina*, although they enter faithfully to the developing conidia, whereas mycoviruses enter at a low frequency into basidiospore-derived isolates of root-rotting fungi of the genus *Heterobasidion*, although only part of the conidia that develop from virus-infected mycelia include viruses (Botella and Hantula, 2018; Vainio and Hantula, 2016). The general trend is that viruses in the mycelium enter commonly into asexual spores, but only quite rarely into sexual basidio- or ascospores.

The lack of extracellular particles makes mycoviruses highly dependent on the welfare of their host mycelium. That may be the reason why (i) clear phenotypic effects are only rarely observed, and (ii) in addition to few harmful viruses, (iii) also symbiotic mycoviruses have been described. The host-virus relationship is very complex, and therefore single virus strains may have beneficial, cryptic, or detrimental effects on a single host isolate in different environmental and ecological conditions, and viruses may cause variable phenotypic effects on different host strains. This complexity is still increased in fungal strains infected by more than one virus: superinfections may affect viral transmission frequencies between mycelia as well as phenotypic effects of viruses in the hosts. Taken together, the close connection of mycoviruses and their hosts has resulted in a very complex coevolution that has led to a plethora of host-virus relationships often not in accordance with the common view of viruses as harmful parasites (Vainio and Hantula, 2018).

The role of fungal viruses in forest ecosystems is poorly understood. There are, however, few cases, where the importance of mycoviruses for the ecosystem and even fungal evolution has been shown and even utilized according to human interest.

4.3 Case example: Effects of viruses on Dutch elm disease

Elm trees (*Ulmus* sp.) in Europe and North America have been infected by an East Asian fungal pathogen *Ophiostoma ulmi* since years 1910 and 1927, respectively. In Europe this epidemic, named as Dutch elm disease, first caused serious damage, but leveled down by the 1940s. This decline of pathogenicity was later associated with detrimental mycovirus infections of the pathogen from an unknown source (Brasier and Buck, 2001).

In the 1940s a related fungus, *O. novo-ulmi*, appeared in Europe as a single clone (VC group), outcompeted the earlier pathogen populations, but at the same time got infected by mitoviruses, then referred to as d-factors, already present in *O. ulmi*. Furthermore, the two fungal species are partially interfertile and therefore introgression of genes occurred between them. The resulting hybrids were mostly of low fitness but despite that were able to further transmit VC group determining genes efficiently from *O. ulmi* to *O. novo-ulmi*. That resulted in an ultrafast evolution of VC groups in the latter species.

As an outcome, the spread of viruses in *O. novo-ulmi* was essentially reduced by incompatible anastomosis contacts between individual mycelia, and thus the virus load leveled down. That secured the pathogen health and ultimately led to destruction of almost all elm forests in Europe during the second wave of the Dutch elm disease.

The role of mycoviruses and gene introgression as key elements in the pathology of Dutch Elm Disease in Europe is supported by findings in North America and New Zealand. In North America the epidemic by *O. ulmi* never declined and later the virus load on *O. novo-ulmi* remained low. Introgression of VC genes, however, took place but in a considerably slower pace than in Europe. Dutch elm disease appeared in New Zealand only in the 1980s via an introduction of *O. novo-ulmi* and there the pathogen has never hosted viruses nor developed VC group diversity. Taken together, the findings in Europe, America, and New Zealand showed that mycovirus populations may have a crucial role on the development of fungal disease, and therefore also on the evolution of the pathogen and ultimately diversity of forest trees.

4.4 Case example: Control of a fungal disease by viruses

Chestnut blight is a devastating disease of Chestnut trees (Rigling and Prospero, 2018). The causative agent, an ascomycetous fungus *Cryphonectria parasitica*, was introduced from East Asia to United States several times around the turn of the 20th century. The disease practically wiped out American chestnut (*Castanea dentata*) trees over a period of 50 years.

The pathogen was introduced to Europe in the 1930s, and thereafter chestnut trees (*C. sativa*) seemed to follow the same track already seen in North America. However, the catastrophe in Europe was avoided thanks to observations of healing cankers in Italy made in the 1950s. The *C. parasitica* strains isolated from the healing cankers were later found to be morphologically different, had reduced virulence (hypovirulence) against the host trees, and when in contact to virulent isolates turned them to a hypovirulent phenotype. That was because of transmission of RNA viruses, which were the key elements responsible for the hypovirulent phenotype of the fungus, and therefore classified in family *Hypoviridae* and nicknamed as hypoviruses (Fig. 20.3).

The practical biocontrol agent, Cryphonectria hypovirus 1 acts by disrupting the gene expression of *C. parasitica* and altering its mitogen-activated protein kinase (MAPK) cascade. This cascade mediates the environmental information from the cell surface receptors to nucleus and regulates cell responses accordingly. In addition, hypoviruses suppress RNA silencing, which is a eukaryotic defense system against alien nucleic acids. As an outcome, hypoviruses survive in their hosts and at the same time reduce fungal virulence against the chestnut trees. Therefore in Europe, hypovirulent mycelia are today commonly inoculated to diseased trees in order to deliver the hypovirus to the canker-causing mycelium and ultimately to heal the tree.

The transmission of viruses between strains of the Chestnut blight fungus is efficiently restricted by VC groups, and therefore the hypoviruses to be used in the control of Chestnut blight must first be transmitted to an isolate of the same VC group as the canker-causing fungus in the tree or region. This makes practical control relatively laborious and expensive.

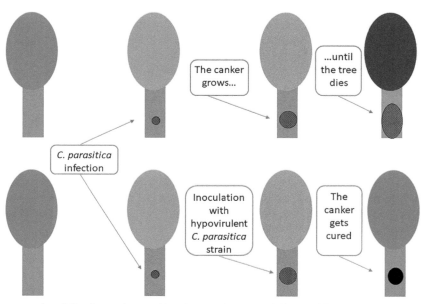

FIG. 20.3 Hypovirulence conversion of *Cryphonectria parasitica*, the causative agent of Chestnut blight, by a fungal virus.

However, the decades-long continuous activity of spreading hypoviruses intentionally has made them frequent in some areas, and the cankers are today healing commonly via natural viral infections without human assistance.

In contrast to Europe, the chestnut blight control in North America has been inefficient. That has been attributed to the high diversity of the host fungus in North America and possibly higher disease susceptibility of the North American chestnut compared to the European one. The problem has been addressed by cloning the viral genome to *Cryphonectria parasitica* and knocking down VC group determining genes in order to create a donor strain independent of VC group-based restriction. Unfortunately, these efforts have not resulted in a breakthrough to save the North American Chestnut.

4.5 Viruses of root rot pathogens

Tree root pathogens cause significant economic losses to both timber tree and fruit tree industries. Many of these fungi can persist for long periods of time (even decades) at the same site, forming large clonally spreading mycelia. During its long lifespan, the advancing fungal mycelium is exposed to many interacting microbes. Transmission of viruses between fungi is controlled by host cellular mechanisms regulating hyphal contact and anastomosis, i.e., antagonism, intersterility, vegetative incompatibility, and mating type incompatibility. Despite these mechanisms, viruses have been found to accumulate in individuals of tree root pathogens, suggesting that the systems are somewhat leaky, leading to events of horizontal virus transmission between incompatible individuals and even different species. This phenomenon suggests that the spread of root rot pathogens could be controlled by infecting the mycelial networks with viruses, provided that debilitation-associated viruses are found. In this regard, the most extensively studied fungi include the ascomycetous fruit tree pathogen *Rosellinia necatrix* (Suzuki, 2017), and basidiomycetous root rot fungi of genus *Heterobasidion* (Vainio and Hantula, 2016). Some promising results have been obtained, and viruses causing host debilitation have been identified in both species: Heterobasidion partitivirus 13 (HetPV13) reduces the growth and affects the gene expression of its *Heterobasidion* hosts, and Rosellinia necatrix reovirus (RnMYRV-1) mediates hypovirulence in *R. necatrix*.

However, the view unfolding is more complex than a simple virus-host interaction. Firstly, environmental conditions and other interacting microbes may affect the phenotypic outcome of a virus-host combination. Secondly, different strains of the same host species may respond very differently to the same virus infection. Thirdly, mixed virus infections may lead to different phenotypic effects compared to those caused by a single virus. As an example, a coinfection by Rosellinia necatrix partitivirus 1 (RnPV1) and an unclassified megabirnavirus (RnMBV2) cause hypovirulence in *R. necatrix*, whereas single infections by the viruses seem to cause no phenotypical alterations. The most extreme case of viral interactions in *R. necatrix* is the yado-kari virus. This capsidless positive-sense RNA virus is not able to replicate by its own, but when it hijacks the capsid protein of an unrelated dsRNA virus, the totivirus-like Yado-nushi virus, it is able to replicate as if it were a dsRNA virus. Furthermore, horizontal transmission of viruses in both *R. necatrix* and species of *Heterobasidion* is affected by the presence of preexisting viral infections in the recipient fungal mycelia. Taken together, the studies on root rot pathogens have shown that microbe-microbe interactions should be carefully considered when estimating the potential use of viruses as biocontrol agents.

4.6 Viruses of mycorrhizal (mutualistic) fungi

Mycorrhizal fungi involved in symbiotic relationships with forest trees represent mostly arbuscular and ectomycorrhizal (ECM) types. These fungi are essential for the functioning of forest ecosystems, and many of them also form edible fruiting bodies. Due to their high economic value, ascomycetous ECM-fungi of genus *Tuber* (true truffles) have been investigated for possible viral diseases. Several virus taxa have been identified, fortunately causing only asymptomatic infections. Viral particles or nucleic acids have also been identified in many other ECM and arbuscular mycorrhizal fungi, but the possible effects of these viruses on their host's phenotype or the symbiotic association remain virtually unknown (Sutela et al., 2019). However, it is intriguing to speculate whether viruses could have a regulatory effect on the intimate interaction between a tree host and its mycorrhizal partner, thereby affecting the functionality of the forest ecosystem. Such an interaction has been found in a totally different ecosystem, namely geothermal soils of the Yellowstone national park in the United States, where a strict three-way symbiosis allows a grass species, its endophytic fungus, and a mycovirus to survive in extreme conditions (Márquez et al., 2007).

5. Bacterial viruses

5.1 The most common biological entities

Interactions between bacteria and their viruses, called bacteriophages or phages, are among the most common ecological relationships in various ecosystems, including forests. These encounters have been successfully studied since their discovery was reported independently by Frederick W. Twort in 1915 and Felix d'Herelle in 1917.

Most of the scientific work on bacteriophages during the 20th century was conducted on artificial media in Petri plates or well-mixed liquid cultures. Such work has resulted in detailed understanding of lytic and lysogenic life cycles of phages, development of bacterial resistance, and constrained ability of viruses to follow the bacterial evolution in pure cultures. As a practical application, phage therapy against human bacterial infections has been practiced continuously in Eastern Europe since the 1920s and is today considered also in the western world a promising alternative to solve the challenge posed by the development of antibiotic resistance in human pathogens (Vandamme and Mortelmans, 2019).

Phages are the major component of global biodiversity. The total number of phage particles in aquatic environments has been estimated to be an astonishing 10^{31} particles belonging to at least 100 million species. Therefore they are undoubtedly also major components of forest ecosystems.

5.2 Phage taxonomy

The classical bacterial viruses such as members of the *Myoviridae* are complex particles with relatively large linear dsDNA genomes. There are, however, many other phage types with all forms of nucleic acid (dsDNA, ssDNA, dsRNA, and ssRNA) packed in icosahedral protein capsids (*Microviridae*), in virions including lipid envelopes inside (*Corticoviridae*) or outside (*Cystoviridae*) the protein capsids, or in filamentous particles (*Inoviridae*).

5.3 Surface-bound microbial communities

In forest ecosystems, both in soil and living trees or other plants, the bacteria and their phages are not mixed freely but are likely to occur predominantly within surface-bound communities. These communities, biofilms, are composed of a plethora of bacterial species in spatially constrained environments, where interactions with phages are largely limited to near neighbors. In such communities the phage-bacteria interactions are fundamentally different from well-mixed liquid, and e.g., promote protection of susceptible host cells from phage exposure ultimately weakening natural selection for phage resistance. Although there are no detailed studies of forest biofilms, phages attach to the mucus layers that are made up of glycoproteins that cover the surface of coral reefs and are there several times more abundant than in the surrounding water (Nguyen-Kim et al., 2015). Thus the frequency of phages in forest biofilms can be expected to be highly variable and affected by the properties of each surface inhabited.

Bacteria and phages in forest ecosystems are not pure cultures composed of one host and one virus. In contrast, broad-host-range phages are common, and in most environments a variety of narrow-range and broad-host-range phages coexists. Some of them are considered phenotypic and others genotypic broad-host-range phages. In the first case, individual phage particles can infect multiple hosts, whereas individual particles of genotypic broad-host-range phages are able to infect only one host. However, in the latter case, specific mutations may lead to rapid host switching between generations, resulting in broad-host-range quasispecies. The ecological parameters, such as host density and host diversity, probably determine the likelihood and richness of broad-host-range phages in a given environment.

At the same time, phages may control and modulate bacterial quality and quantity in nature and therefore have a role in recycling nutrients and determining carbon stocks and overall dynamics of microbial communities via lysis of bacterial cells. However, direct data from phages in forest environments is scarce.

5.4 Case example: Interactions of bacteria and phages on horse chestnut

Horse chestnut (*Aesculus hippocastanum*) is a broadleaved tree shown to host bacteria in its leaves. The interactions of these bacteria and their phages have been studied in detail by Britt Koskella and her colleagues at the University of Exeter, United Kingdom (Koskella and Meaden, 2013). Their work has shown that the susceptibility of bacteria for phages is changing over time: isolated strains are more resistant to phages collected 1 month earlier than bacteria, but highly susceptible to ones collected 1 month later. In other words, resistance develops on tree leaves, and therefore phages are changing the bacterial community over at least short timescales. Koskella's team has also shown that phages infect more often bacteria from their local tree than the bacterial hosts of the same species from other trees in the forest stand. These examples suggest that resistance to phages influences the successional dynamics of the bacterial community within a tree.

It is well known that only a small subset of bacteria is culturable, and therefore information from direct experiments on the interactions between bacteria and their phages is limited. However, at least four genera of culturable bacteria inhabit horse chestnut leaves, and phages in the same ecosystem are capable of infecting multiple bacterial species and genera, which makes the dynamics of bacteria and their phages considerably more complicated than in simple laboratory studies with a single host and a phage in a mixed chemostat.

The future of Horse chestnut is currently challenged by a bacterial disease caused by *Pseudomonas syringae* pv. *aesculi*, which is causing bleeding cankers on the trees and ultimately killing them. It was shown by deep sequencing followed by subsequent assembly of phage genomes that also this bacterium hosts phages. However, it remains to be seen whether they turn out to be useful in controlling the disease.

5.5   Case example: Biological control of *Xylella fastidiosa* by bacteriophages

Strains of *Xylella fastidiosa* are causing several agriculturally significant plant diseases, such as Pierce's Disease (PD) of grape, coffee leaf scorch, almond leaf scorch, citrus variegated chlorosis, and a serious olive tree (*Olea europaea*) disease called live quick decline syndrome. The appearance of the latter disease in Italy in 2013 has resulted in a devastating olive tree decline. Nothing is, however, known about the ecology of phages of *X. fastidiosa* in olive trees, but Das et al. (2015) have shown that phage cocktail composed of four virulent lytic phages reduced the number of these bacteria in grapevines and ceased the progress of the disease symptoms 1 week after the treatment.

During the following few years it was, however, observed that phage therapy is not able to completely eliminate *X. fastidiosa* from diseased plants. Therefore phage therapy developed against *X. fastidiosa* will not solve the problem of the decline of olive trees, which are expected to produce fruit for centuries. Despite this, the experience on *X. fastidiosa* has shown that bacteriophages are potential biocontrol agents against plant diseases, although further progress still needs to be done.

6.   Viruses of oomycetes: Examples from the genus *Phytophthora*

Oomycetes of the genus *Phytophthora* are notorious plant pathogens that infect forest trees, ornamental and fruit trees and seedlings, as well as herbaceous crop plants. Although they were for a long time considered to be members of kingdom Fungi, molecular phylogenetics has positioned them as members of kingdom Stramenopila, which is only distantly related to true fungi. Consequently, it could be assumed that viruses hosted by fungi and oomycetes would be very different. This is partly true, as highly divergent viruses that do not accommodate to any known virus family have been characterized in the infamous potato blight pathogen, *P. infestans* (Cai and Hillman, 2013). However, this pathogen hosts also viruses resembling plant and fungal viruses of genus *Mitovirus*.

In the case of forest diseases, one of the most devastating *Phytophthora* diseases is the sudden oak death caused by *P. ramorum*. This has motivated the search for viruses that could be used to control the disease. Thus far endornaviruses have been shown to occur commonly among *P. ramorum* isolates in Europe and also in the *Phytophthora* taxon "douglasfir" in the United States (Kozlakidis et al., 2010), but it remains to be investigated whether they affect the pathogenicity of the host. It should be noted that endornaviruses also occur in many plants, including avocado trees (*Persea americana*), again supporting the idea of occasional virus transmission between distantly related organisms at an evolutionary timescale. However, the plant endornaviruses are only distantly related to those found in isolates of *Phytophthora*.

7.   Complex host interactions shape the ecology of forest viruses

The viruses are described above according to their host phyla and kingdoms. The reality in forests is, however, considerably more complicated, and viruses may be able to move between distantly related organisms—in real time and in evolutionary timescale—as described in the following three examples.

Members of family *Partitiviridae* occur both in fungi and plants. In both cases, they cause persistent and mostly asymptomatic infections. Interestingly, the basidiomycete *Helicobasidium* purpureum and its host plant sugar beet (*Beta vulgaris*) are infected with highly similar partitiviruses (Szego et al., 2010). This example, and the occurrence of both fungal and plant-infecting partitiviruses in genera *Alphapartitivirus* and *Betapartitivirus*, suggests that partitiviruses have a capacity for occasional transmission between plant and fungal hosts. A potential route for virus transmission could be provided during intimate associations between plants and pathogenic, endophytic or symbiotic fungi (Roossinck, 2019).

Tobacco necrosis virus (TNV; *Tombusviridae*) is an example of a plant virus that is spread by a fungal vector of the genus *Olpidium*. It causes serious diseases on herbaceous plants like tulips (*Tulipa* sp.), beans (family Fabaceae), and cucumbers (*Cucumis sativus*). However, this generalist virus also dwells in the roots of many plant species without causing any visible symptoms or a systematic infection. In addition, it has also been identified in the roots of pine and spruce seedlings and based on inoculation studies it may cause debilitation in spruce seedlings.

Genomoviruses are examples of fungal viruses shown to be transmitted by insect vectors. The North American beetle causing the most extreme forest damage recorded thus far, the mountain pine beetle *Dendroctonus ponderosae*, was recently shown to host symbiotic fungi, which are infected by viruses of the family *Genomoviridae*. The effects of these virus infections are yet unknown, but another genomovirus has highly detrimental effects on a plant pathogen *Sclerotinia sclerotiorum*. This finding suggests that the impact of viruses may be more complex than can be deduced from the direct effects of those viruses in their host organisms. Recently, numerous fungal viruses have been identified by HTS in pooled arthropod samples, but it remains to be investigated to which extent arthropods are capable of acting as vectors for mycoviruses.

Overall, HTS has already revealed unprecedented viral diversity and expanded our knowledge on virus host ranges among insects, plants, and fungi. However, virus diagnosis is sometimes complicated by the complexity of microbial communities associated with eukaryotes, which may lead to difficulties in determining the true host of the viral sequences. Therefore a better understanding of viruses and their effects on host organisms still depends on detailed studies of viruses and isolated host organisms using experimental setups designed for hypothesis testing.

References

Beech, E., Rivers, M., Oldfield, S., Smith, P.P., 2017. GlobalTreeSearch: the first complete global database of tree species and country distributions. J. Sustain. For. 36, 454–489.

Bird, F.T., 1953. The use of virus disease in the biological control of European Pine Sawfly, *Neodiprion sertifer* (Geoffr.). Can. Entomol. 87, 124–127.

Botella, L., Hantula, J., 2018. Description, distribution, and relevance of viruses of the forest pathogen *Gremmeniella abietina*. Viruses 10, 614.

Brasier, C.M., Buck, K.W., 2001. Rapid evolutionary changes in a globally invading fungal pathogen (Dutch elm disease). Biol. Invasions 3, 223–233.

Büttner, C., von Bargen, S., Bandte, M., Mühlbach, H.P., 2013. Forests diseases caused by viruses. In: Gonthier, P., Nicolotti, G. (Eds.), Infectious Forest Diseases. APS Press, Boston, pp. 97–110.

Cai, G., Hillman, B., 2013. Phytophthora viruses. Adv. Virus Res. 86, 327–350.

Das, M., Bhowmick, T.S., Ahern, S.J., Young, R., Gonzalez, C.F., 2015. Control of Pierce's disease by phage. PLoS ONE, e0128902.

Harrison, R., Hoover, K., 2012. Baculoviruses and other occluded insect viruses. In: Vega, E., Kaya, H.K. (Eds.), Insect Pathology, second ed. Academic Press, Boston, pp. 73–131.

Koskella, B., Meaden, S., 2013. Understanding bacteriophage specificity in natural microbial communities. Viruses 5, 806–823.

Kozlakidis, Z., Brown, N.A., Jamal, A., Phoon, X., Coutts, R.H.A., 2010. Incidence of endornaviruses in *Phytophthora* taxon douglasfir and *Phytophthora ramorum*. Virus Genes 40, 130–134.

Márquez, L.M., Redman, R.S., Rodriguez, R.J., Roossinck, M.J., 2007. A virus in a fungus in a plant: three-way symbiosis required for thermal tolerance. Science 315, 513–515.

Marzano, S.Y.L., Domier, L.L., 2016. Novel mycoviruses discovered from metatranscriptomics survey of soybean phyllosphere phytobiomes. Virus Res. 213, 332–342.

Moreno, P., Ambrós, S., Albiach-Martí, M.R., Guerri, J., Peña, L., 2008. Citrus tristeza virus: a pathogen that changed the course of the citrus industry. Mol. Plant Pathol. 9, 251–268.

Nguyen-Kim, H., Bettarel, Y., Thierry Bouvier, T., Bouvier, C., Doan-Nhu, H., Nguyen-Ngoc, L., Nguyen-Thanh, T., Tran-Quang, H., Brune, J., 2015. Coral mucus is a hotspot for viral infections. Appl. Environ. Microbiol. 81, 5773–5783.

Rigling, D., Prospero, S., 2018. *Cryphonectria parasitica*, the causal agent of chestnut blight: invasion history, population biology and disease control. Mol. Plant Pathol. 19, 7–20.

Roossinck, M.J., 2010. Lifestyles of plant viruses. Philos. Trans. R. Soc. Lond. Ser. B Biol. Sci. 365, 1899–1905.

Roossinck, M.J., 2019. Evolutionary and ecological links between plant and fungal viruses. New Phytol. 221, 86–92.

Rumbou, A., Candresse, T., Marais, A., Svanella-Dumas, L., Landgraf, M., von Bargen, S., Büttner, C., 2019. Unravelling the virome in birch: RNA-Seq reveals a complex of known and novel viruses. BioRxiv. https://doi.org/10.1101/740092.

Shi, M., Lin, X.-D., Tian, J.-H., Chen, L.-J., Chen, X., Li, C.-X., Qin, X.-C., Li, J., Cao, J.-P., Eden, J.-S., Buchmann, J., Wang, W., Xu, J., Holmes, E.C., Zhang, Y.-Z., 2016. Redefining the invertebrate RNA virosphere. Nature 540, 539–543.

Sutela, S., Poimala, A., Vainio, E., 2019. Viruses of fungi and oomycetes in the soil environment. FEMS Microbiol. Ecol. 95, fiz119.

Suzuki, N., 2017. Frontiers in fungal virology. J. Gen. Plant Pathol. 83, 419–423.

Szego, A., Enünlü, N., Deshmukh, S.D., Veliceasa, D., Hunyadi-Gulyás, E., Kühne, T., Ilyés, P., Potyondi, L., Medzihradszky, K., Lukács, N., 2010. The genome of beet cryptic virus 1 shows high homology to certain cryptoviruses present in phylogenetically distant hosts. Virus Genes 40, 267–276.

Vainio, E.J., Hantula, J., 2016. Taxonomy, biogeography and importance of *Heterobasidion* viruses. Virus Res. 219, 2–10.

Vainio, E.J., Hantula, J., 2018. Fungal viruses. In: Hyman, P., Abedon, S.T. (Eds.), Viruses of Microorganisms. Caister Academic Press, U.K, pp. 193–209.

Vainio, E.J., Rajala, T., Pennanen, T., Hantula, J., 2017. Occurrence of similar mycoviruses in pathogenic, saprotrophic and mycorrhizal fungi inhabiting the same forest stand. FEMS Microbiol. Ecol. 93, fix003.

Vandamme, E.J., Mortelmans, K., 2019. A century of bacteriophage research and applications: impacts on biotechnology, health, ecology and the economy! J. Chem. Technol. Biotechnol. 94, 323–343.

Wolf, Y.I., Kazlauskas, D., Iranzo, J., Lucía-Sanz, A., Kuhn, J.H., Krupovic, M., Dolja, V.V., Koonin, E.V., 2018. Origins and evolution of the global RNA virome. MBio 9, e02329-18.

Further reading

Virus Taxonomy: The Classification and Nomenclature of Viruses, Online, 10th Report of the International Committee on Taxonomy of Viruses. https://talk.ictvonline.org/ictv-reports/ictv_online_report/.

Section F

Challenges and potentials

Chapter 21

Translational research on the endophytic microbiome of forest trees

Johanna Witzell[a], Carmen Romeralo[a], and Juan A. Martín[b]

[a]*Swedish University of Agricultural Sciences, Southern Swedish Forest Research Centre, Alnarp, Sweden,* [b]*School of Forest Engineering and Natural Resources, Technical University of Madrid (UPM), Madrid, Spain*

Chapter Outline

1. Introduction

Our knowledge about the endophytic microbiome, that is, fungi and bacteria that inhabit internal plant tissues (Hardoim et al., 2015), has increased greatly during the past decades. The advances in high-throughput DNA sequencing have acted as the main enabling technologies, allowing a taxonomic portrayal of microbial communities (Johnston et al., 2017; Nilsson et al., 2019). As a consequence, detailed studies describing the composition of the endophytic microbiome in forest trees are rapidly accumulating, and several of them associate microbiome traits with a specific disease resistance or susceptibility phenotype of the trees (Raghavendra and Newcombe, 2013; Martín et al., 2013; Busby et al., 2016a,b, 2017), or the overall vitality of the trees (Agostinelli et al., 2018). This mounting circumstantial evidence and the discovery of in vitro antagonistic and competitive interactions between individual microbes (Blumenstein et al., 2015; Romeralo et al., 2015b; Martínez-Arias et al., 2019) has encouraged interest in the potential of utilizing the endophytic microbiome in diagnostics, or as preventive or therapeutic measures in forest protection (Rabiey et al., 2019; Terhonen et al., 2019). However, while the potential impact of the endophytic microbiome on forest health seems indisputable, the realization of this potential to forest protection solutions necessitates an efficient translation of the knowledge from basic research into practical applications.

The contemporary concept of *translational research* derives from healthcare research that denotes a multidirectional flow of information and feedback between fundamental research, clinical research (focused on patients and populations), and policies and practices, with the ultimate goal of using scientific knowledge to achieve long-term health outcomes (Butler, 2008). The translational research process is generally characterized by a long lag time between gathering the evidence and the desired practical impacts (Rushmer et al., 2019). A considerable loss of knowledge and discoveries occurs along the way from basic research to applications, due to what is often referred to as the "knowledge/evidence-to-action gap" or the "science-to-society gap" (Kitson and Straus, 2010). The whole endeavor can be dissected into four subprocesses: *knowledge generation, interpretation, implementation*, and *impact*, and there is the potential to lose knowledge and discoveries between all the transitions (Aymerich et al., 2014; Fig. 21.1). The difficulties in translation have traditionally been explained by communication challenges between the involved actors: the knowledge producers or the experts, and end users who are seen as more passive receivers of the knowledge (Meinard and Quétier, 2013; Rushmer et al., 2019). Because multidisciplinary and cross-disciplinary approaches and collaborations are needed to facilitate knowledge translation (Rubio et al., 2010), communication can be challenging even within each of these groups. Recent analyses provide more nuanced explanations for the knowledge-to-action gaps, such as the lack of adequately specific hypotheses in basic research (Willis and Minot, 2020). It may also be oversimplified to think of the gaps as a void to be "bridged over" when a more helpful metaphor could be an evidence ecosystem with intersecting and overlapping boundaries (Rushmer et al., 2019).

Forest Microbiology. https://doi.org/10.1016/B978-0-12-822542-4.00015-2

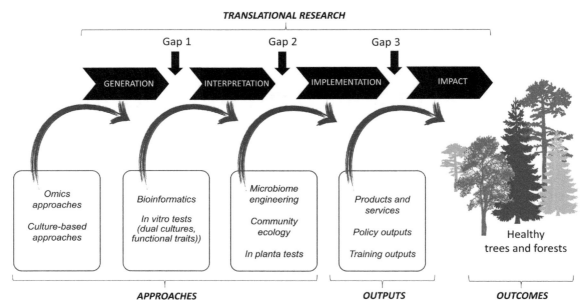

FIG. 21.1 The translational research process consists of four subprocesses: knowledge generation, interpretation, implementation, and impact. The gaps indicate the potential to lose knowledge and discoveries between all these transitions. The lower part of the illustration shows how different approaches and outputs in the forest tree microbiome research relate to these processes, when the desired impact is improved tree and forest health.

Similar to healthcare research, tree microbiome research struggles with the difficulty of translating basic research findings into impacts, that is, to practical improvements in forest protection and management for forest health. Knowledge and discoveries are lost in translation along the way from basic research to practical applications, and despite the numerous research papers implying a role for individual microbes or microbial communities in tree resistance, practical solutions based on microbiome utilization in forest protection are still rare. Earlier reviews provide overviews of the topic, focusing on management (Newcombe, 2011), ecological aspects (Witzell et al., 2014), and biocontrol potential (Witzell and Martín, 2018; Terhonen et al., 2019). In this review, we discuss some of the specific gaps in the knowledge translation process and illustrate them with authentic examples from tree microbiome studies. We also suggest measures to facilitate translational research on tree microbiomes to achieve impacts on forest health.

2. Translational research of forest tree microbiomes

In the following sections, we discuss each of the four subsequent but interlinked subprocesses of translational research and the gaps between as applied to research on forest tree microbiomes that aims at supporting and improving tree and forest health (Aymerich et al., 2014; Fig. 21.1). Our special attention is on endophytic microbes, but the discussion is relevant also for the whole microbiome.

2.1 Gap 1. From knowledge generation to interpretation

During the past decades, the rate of knowledge generation has accelerated (Foo et al., 2017). The diversity and structure of endophytes in forest trees has been actively explored using two basic techniques: cultivation, followed by morphological and/or molecular identification and more recently by high-throughput, next-generation sequencing (NGS). These approaches are not mutually exclusive; rather, the parallel use of traditional cultivation and high-throughput sequencing (HTS) techniques can help to obtain a complete qualitative and quantitative picture of endophyte community composition (Sun and Guo, 2012). The traditional methodology has a restricted capacity to the discovery of new endophytes due to the frequent occurrence of nonculturable species. However, cultures are needed for studies of the phenotypic and genotypic characteristics of individual species, such as when exploring their potential in biotechnological applications (Salazar-Cerezo et al., 2018; Mbareche, 2020) or antagonistic capacity against other microorganisms such as plant pathogens (Alabouvette et al., 2006; Stefani et al., 2015).

The recent advances in HTS technologies have enabled unprecedentedly detailed studies on microbial communities under specific environmental conditions and habitats (Beeck et al., 2015; Eevers et al., 2016). The massively parallel or

NGS platforms provide a means to overcome the crucial limitations of culture-based methods, enabling greater discovery of taxonomic richness and higher mutation resolution power, thus better reflecting the diversity and community structure closer to its natural state (Eevers et al., 2016; Romão et al., 2017). In contrast to qPCR, NGS offers a hypothesis-free method, without the need for prior knowledge of sequence information. Using a metabarcoding approach, it is possible to identify multiple species from a mixed sample based on the HTS of a specific DNA marker and overcome the constraints of conventional morphology-based species identification. The main disadvantage of the HTS approach is that potential species of interest cannot be cultured and explored for further study (Bullington and Larkin, 2015), and the results still provide only a point-in-time picture of the dynamic communities.

Metabarcoding faces some biases, especially regarding the bioinformatic pipelines used (Lindahl et al., 2013; Balint et al., 2014; Pauvert et al., 2019). While most NGS platforms have some data analysis functionality, the automated bioinformatics workflow is seldom adequate or directly applicable to specific cases (Ahmed, 2016). The limitations can provide an over- or underestimated fungal community, depending on the region of the genome and the primers used for amplification (Tedersoo et al., 2015), the platform used for sequencing (with Illumina platforms dominating the market), and the method to assemble reads (Nguyen et al., 2015). Furthermore, PCR errors in HTS are common and can produce chimeras, which are caused by the incomplete extension of DNA strands during amplification that make up a recombination between two sequences; this can cause biases in diversity results (Mbareche, 2020). Further steps in the bioinformatic pipeline can also prevent the accurate discovery of a fungal community, including the sequence clustering methods and the filters applied to the operational taxonomic unit (OTU) data (Bokulich et al., 2013; Pauvert et al., 2019).

The selection of bioinformatic procedure is dependent on the objective of the study. Some approaches detect fungal/bacterial strains with high accuracy, which can be useful for the identification of target species but it may also overestimate the community richness; other methods may more accurately retrieve the composition and richness and be more appropriate for community ecology studies (Pauvert et al., 2019). The major benefit of HTS technologies relies on the capacity to provide information about the main microbial colonizers in a large number of samples and on the changes in species/taxa abundance between those samples (Lindahl et al., 2013; Pauvert et al., 2019). Moreover, the characterization of core microbiomes specific to plant parts and organs (Lucaciu et al., 2019) may open new possibilities to implement generated knowledge in functional in planta settings.

2.2 Gap 2. From interpretation to implementation

In the context of tree microbiome research, implementation means making the research findings operational in forest trees. This involves microbiome engineering, such as manipulating the microbiome composition by microbiome transfer, synthetic microbiomes, or host-mediated artificial selection (Foo et al., 2017). One of the major challenges to implement biotechnological solutions based on microbiome engineering is the transition from in vitro antagonism tests or in planta controlled experiments to uncontrolled environments (e.g., field conditions). This transition is frequently assayed without a deeper knowledge of the composition, functions, and dynamics of the host microbiomes (Sessitsch et al., 2019). Yet, an in-depth understanding of the underlying factors could help to explain certain unstable results of microbiome manipulation under field conditions (Martín et al., 2015). The microbiome effects on host plants are context-dependent (Busby et al., 2016a), and failures in the establishment and/or functioning of inoculated microbes may occur depending on multiple factors, such as fast desiccation, incompatibility with the host genotype, or competition with the resident microbiome. Using native, locally adapted microbiomes to improve plant performance is expected to provide a more stable microbial establishment than using foreign microbes, as potential incompatible interactions with environmental and host factors are minimized (Frasz et al., 2014; Busby et al., 2016a). In practice, however, the manipulation of microbiomes through the inoculation of local microbes has limitations (Hacquard et al., 2015; Sessitsch et al., 2019). Compared to studies on the microbiome of annual agronomic or model crop plants (Busby et al., 2016a), the microbiomes of forest trees have been less studied (Fig. 21.2) and the detailed structure of the core microbiomes and the dynamics of microbial networks in forest trees remain largely unexplored (Witzell and Martín, 2018). In spite of this limitation, some practical applications based on beneficial symbionts have been successfully implemented in forest trees under field conditions (e.g., Quiring et al., 2019). A detailed analysis of both successful and unsuccessful works, and of available scientific knowledge from both herbaceous and woody plants, can help to disentangle key findings and gaps in microbiome engineering implementation.

While diverse strategies for microbiome engineering are thinkable in forest trees (Witzell and Martín, 2018), here we will focus on direct microbiome manipulation through artificial inoculation using selected microbes. The selection of beneficial microbes can originate from different exploratory studies, such as the phenotypic characterization of microbes (Blumenstein et al., 2015), metabarcoding studies coupled with microbial network analyses (Toju et al., 2018), or host-mediated selection according to adaptation to stress (Rodriguez and Durán, 2020). Otherwise, the microbiome can be

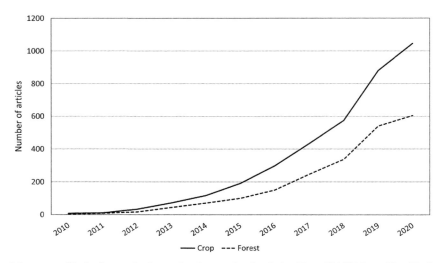

FIG. 21.2 Number of articles captured in the Scopus database using the search string "microbiome*" AND "crop*" and "microbiome*" AND "forest*" over the period 2000–20.

artificially engineered through an iterative selection process (e.g., Panke-Buisse et al., 2015; Swenson et al., 2000). Once a potential beneficial microbiome is selected, it is usually assayed in planta under controlled conditions (e.g., in vitro, greenhouse, or growth chamber studies) before implementation under field conditions. In planta experiments should follow specific inoculation protocols. Depending on the tree organ to be inoculated, the methods may include foliar spraying or misting (Arnold et al., 2003; Martínez-Arias et al., 2019), phloem exposure to microbes by artificial wounding (Romeralo et al., 2015a; Martínez-Álvarez et al., 2016), stem injection Martínez-Arias et al., 2021, soil inoculation with a suspension of selected microbes or soil slurry (Calderón et al., 2016), whole microbiome transplanting (Yergeau et al., 2015), or seed coating (Ownley et al., 2008; Parsa et al., 2018), among others. Inoculum dosage can be crucial to reach successful microbiome establishment and functioning (Howard et al., 2017) as well as the development of specific formulations that, for instance, prevent microbial desiccation (Sessitsch et al., 2019). Ideally, different protocols and dosages should be assayed in order to select the ones that provide the longer-term establishment of selected microbes in target tissues and maximize beneficial interactions with the host plant. However, in spite of the broad experience gathered on inoculation techniques, one of the major obstacles is that inoculated microbes can be rapidly outcompeted by resident microbes (Thomas and Sekhar, 2016). This adversity is particularly relevant in the case of single-strain inoculations (Tosi et al., 2020). Inoculation with mixtures or consortia of microbes that enables additive or synergistic effects between microorganisms may be a beneficial strategy, promoting the adaptation and survival of inoculated organisms (Ben Said and Or, 2017). Synergistic effects can be maximized when the whole microbiome inhabiting a certain plant organ is used as the inoculum source, and is transplanted into new hosts (Bell et al., 2016). Transplanting methods also have the advantage of including both culturable and unculturable microbes and a high diversity of functional assemblages. While inoculating a whole microbiome can be advantageous in some circumstances, its successful establishment in a plant harboring a well-established native microbiome (e.g., under field conditions) is not guaranteed. For instance, in the wheat rhizosphere, the inoculation of microbial communities had no consequences for bacterial composition and resulted in only negligible changes in fungal composition (Giard-Laliberté et al., 2019). These limitations are possibly exacerbated in forest stands, where higher microbial diversity and long-lasting microbe-microbe and plant-microbe interactions could lead to more stable, complex, and resilient microbiome assemblages.

A plausible approach to overcome the resilience of the resident microbiome to manipulation can be found in the priority effects. It is known that the order of species arrival influences the outcome of species interactions, and can drive community assembly, structure, and function (Drake, 1991). For instance, primary symbionts of winter wheat seed have shown strong, natural priority effects on microbial assemblages, and these effects are dependent on primary symbiont identity (Ridout et al., 2019). However, the artificial inoculation of primary symbionts in wheat seedlings did not result in similar priority effects, questioning the idea of directing microbiome assemblages by artificial methods. In the same line, bacterial seed endophytes seem to be highly conserved in some plant species and can provide the bulk of the species pool from which the seedling microbiome is recruited (Nelson, 2018, and refs. therein). Although more research is needed to disentangle the functional roles of primary symbionts and their manipulation possibilities, it seems clear that seed and seedling stages are critical windows for microbiome assembly (Donn et al., 2015). Microbial assemblages of trees are progressively established after seed germination, mainly through the horizontal spread of microbes from the surrounding environment.

Vertical transmission of seed-borne endophytes that occurs in some plant species (Barret et al., 2015) is assumed to be less important in forest trees, although it is not well studied. This process is largely influenced by environmental and nutritional factors (Berg and Koskella, 2018), along with microbe-microbe interactions and host genotype traits, such as the release of specific root exudates that modulate microbial recruitment (Walker et al., 2003). Once microbial assemblages are well established, there is little evidence for major changes in the microbiome (Lundberg et al., 2012), possibly due to the sessile nature of plants, which in combination with the rather stable soil-borne inoculum reduces the occurrence of major changes in communities (Hacquard et al., 2015). Interestingly, in an experiment on soil microbiome transplanting with a willow clone, Yergeau et al. (2015) found that different microbiome sources were initially divergent, but they converged on highly similar communities after 100 days, suggesting an active microbiome recruitment by the host plants. Thus, while the host effect on microbiome recruitment seems very relevant, more research should clarify the role of primary symbionts in such recruitment as well as the interaction between the host genotype and the identity of the primary symbionts. The fluctuations of tree microbiomes during the long lifespan of forest trees and after environmental disturbance (e.g., drought and flooding events; Martínez-Arias et al., 2020) should also be subjected to further research.

Although a few microbiome manipulation experiments have been conducted in forest trees under field conditions, some past experiences illustrate successful microbial inoculations during the first seedling stages, resulting in enhanced tree resilience to stress. For example, spruce seedlings inoculated with plant growth-promoting rhizobacteria (PGPR) and subsequently planted in the field showed a 32%–49% higher biomass than control plants 1 year after inoculation (Chanway, 1997, and references therein). The beneficial effect was highest in the poorest quality sites, suggesting that such inoculants could be prescribed for the reforestation of suboptimal terrain. In other studies, spruce seedlings were inoculated with a rugulosin-producing endophyte, reducing the growth and herbivory of the spruce budworm (Miller et al., 2002). The inoculated endophyte was shown to persist in plant tissues for at least 4 years under field conditions (Sumarah et al., 2008) while a subsequent study confirmed the presence of both the endophyte and rugulosin throughout the crown of a tree 11 years after inoculation (Frasz et al., 2014). In spite of such successful examples, we must be cautious about the positive results of inoculation published in the literature, as negative results are possibly underrepresented as well as the understudied risks of introducing microorganisms into an existing ecosystem (Tosi et al., 2020 and references therein). Possibilities for microbiome modification during later tree developmental stages are likely more restricted. For instance, Martínez-Arias et al. (2021) inoculated three fungal endophytes into the stem of 6-year-old elm trees under field conditions. While one of them reduced Dutch elm disease (DED) symptoms during the season, the abundance of the three endophytes in tree tissues markedly diminished 1 year after inoculation, suggesting the high resilience of the resident microbiome to manipulation. This result is supported by other works on elm trees reporting the limited spread and survival of inoculated biocontrol agents against DED (e.g., Bernier et al., 1996; Scheffer et al., 2008). Yet, emerging knowledge of core microbiomes, assembly dynamics, and host and environmental effects in such dynamics could open new possibilities for microbiome manipulation in forest trees at different ontogenetic stages. Research on direct microbiome manipulation should also be complemented with other studies on indirect methods for shaping tree microbiomes, such as agroforestry management practices (Wemheuer et al., 2020). In the future, both direct and indirect methods could be complementarily implemented to provide higher forest resilience.

2.3 Gap 3. From implementation to impacts

The final step for translational research is to surpass the gap between implementation and impact. According to Greenhalgh et al. (2016) "impact occurs when research generates benefits (health, economic, cultural) in addition to building the academic knowledge base." This definition articulates the anthropocentric and utilitarian perspective, but the impact may also be realized through the effects on other nonhuman organisms and the environment (Chandler, 2013). Even today, when the doubling time for human knowledge is calculated in hours (Chamberlain, 2020), these impacts take time to develop. A common time frame for the process from research findings to new and disruptive technologies is at least 10–15 years, and only a fraction of the research output will make a basis for radical innovation, commercialization tracks, and unique startups. Greenhalgh et al. (2016) emphasized that the linkage between research and impact is complex, and that it is usually easier to characterize and evaluate short-term proximate impacts. The benefits from, for example, the development of research infrastructure or key partnerships may become obvious only in the longer term, and are thus more difficult to capture through impact assessments that usually are based on surveys, interviews, or document analyses (Greenhalgh et al., 2016).

The academic impact of tree endophyte research is already tangible and demonstrated by the large number of publications from the research field. In addition to the accumulation of data and knowledge, the academic impact also materializes in relevant trainings for experts and infrastructures for the analyses of microbiomes. These are needed to meet the competence needs of the regulatory and risk assessment authorities (Kvakkestad et al., 2020). However, the research field is still relatively young, and for instance the concepts of the holobiont (Cregger et al., 2018; Simon et al., 2019) and the core microbiome (e.g., Noble et al., 2020) have only recently emerged as theoretical and experimental frameworks to study and understand the functional interactions between trees and their microbial communities. Thus, in order to expand and solidify the scientific foundation for microbiome-based forest protection solutions, more basic research focusing on the functional characters of microbes and their dynamic interactions with the biotic and abiotic environment is still needed. The integration of knowledge from different disciplines, such as community ecology, will be needed in this process.

Despite the growing academic impact, the step from in vivo experiments to practical, upscaled solutions implemented in a nursery or at a forest site has proven hard to take. For instance, as of now (October 2020), a search in the database Espacenet with the key words "microbiome" and "forest" or "trees" did not pick up transnational patents for the innovative use of the tree microbiome, whereas the transfer of knowledge regarding the soil microbiome seems somewhat better covered. Thus, the concrete and measurable environmental and economic impacts of tree microbiome research in forest settings are still impending. A fundamental underlying barrier to these impacts may be that the relevant policy frameworks have not embraced the potential of microbiomes in forest protection. Here, we understand policy as a statement of intent made by a regulatory body involving different relevant actors, describing the central problems, and outlining how they should be addressed (Evans and Cvitanovic, 2018). In policy formation, scientific evidence has an important role (Allio et al., 2006), although research output often tends to be used conceptually (for general information) or symbolically (to justify a chosen course of action) rather than instrumentally (feeding directly to decision-making) (Greenhalgh et al., 2016). Therefore, to increase the translational impact of tree microbiome research, more efforts should be given to widening and deepening the dialogue between policy makers and researchers. However, due to recent developments, the influence of the policy barrier may already be diminishing. For instance, the new plant health regulations of the European Union (EU) emphasize biological control as part of the integrated pest management strategies and a tool in reducing the use of environmentally harmful pesticides (EU, 2009, 2016). The official policy can translate into legally defined rights and restrictions that dictate the possibilities for use of microbiome-based solutions. In the near future, the increasing policy support may act as a strong incitement for the development of new and influential forest protection solutions based on the microbiome.

Legal and regulatory frameworks form another barrier to the translational process. In a study comparing the number of authorized microbial plant protection products in three Nordic countries, Kvakkestad et al. (2020) found that differences between the countries in microbial product availability were strongly dependent on the implementation of regulations. A fundamental problem is that the regulations concerning microbial plant protection products are often based on those developed for chemical pesticides, which has resulted in strong barriers for market entry, such as in the EU (Kvakkestad et al., 2020). However, this situation may be slowly changing. For example, the number of approved nonchemical active substances in the EU has increased since 2011, with microorganisms being the category of nonchemical alternatives that increased the most (Robin and Marchand, 2019). Still, the regulatory system for approval and registration of microbe-based plant protection solutions is often costly and tedious, which reduces the willingness of companies to start producing them (Kvakkestad et al., 2020).

The size and maturity of the markets for the microbiome-based solutions are other important determinants of the likelihood of impacts beyond academia. In order to result in a wider impact, the microbiome-based solutions (products, protocols, methods) need to be widely used by a critical mass of end users. However, microbiome-based solutions are often niche products, occupying narrow application areas. While a global market for microbe-based forest protection solutions may be increasing (Kvakkestad et al., 2020), the national and regional markets may be immature or simply too small to encourage companies to invest in them. The connection between the demand for and the supply of science-based solutions is an important precondition for a successful transfer of scientific knowledge across the science-impact gap (Sarewitz and Pielke, 2007). In the case of tree microbiome research, this connection may need considerable attention before any significant impacts can unfold. So far, the market pull for research output seems to have been rather ineffective, and more detailed demand analyses could be needed to identify the research needs and possibilities to increase the impact.

Market acceptance is one of the dimensions of social acceptance, which also includes sociopolitical and community acceptance (Wüstenhagen et al., 2007). The low transfer rate of research output into markets and society is also largely due to the fact that researchers are often not effective in pushing the findings to markets and practices, and therefore the majority of their innovations are suboptimally integrated and utilized in policy and practice. This is exemplified by the many findings showing the antagonistic potential of endophytic fungi against forest pathogens, but only a couple of them (*Phlebiopsis gigantea* and *Trichoderma* spp.) are actually being used as effective plant protection products in practical forestry. To analyze the barriers for the transfer of knowledge over the science-society interface, Böcher and Krott (2014) developed a theoretical framework that allows a structured analysis of how the research outputs are transferred across the science-society gap, and end up as practical solutions and science-based policy advice. This framework, called the RIU model, is an innovative approach that aims to meet the growing need for science-based practices and decision support, especially in fields such as food production, forestry, and climate change adaptation that are highly dependent on scientific knowledge. The model acknowledges the fact

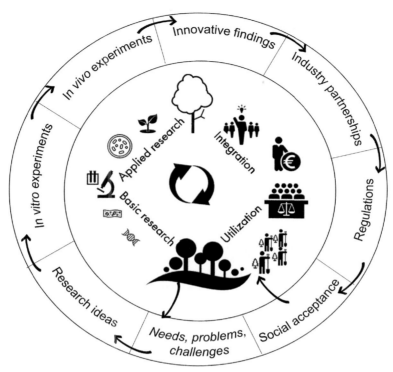

FIG. 21.3 A full cycle of translational research of tree microbiomes would proceed from identified problems and the formulation of research ideas to basic research, applied research, innovations (in a broad sense), and integration of the findings to relevant parts of society. Compatibility with regulations and social acceptance is a prerequisite for utilization.

that researchers themselves are often unable to push their findings to policy and practices. To overcome this barrier, the RIU model differentiates three groups of activities that are needed in the process of transferring scientific knowledge into policy and practice: research, integration, and utilization. The integration and the transfer of research to innovation is achieved by actively finding and forming alliances with external, interested actors (individuals or companies/organizations), who will then help the researchers by "pulling" their findings into policy and practices for the benefit of citizens and the environment (Fig. 21.3). Integration, such as the interface between science and utilization, is a relatively newly defined sphere that still lacks empirical evidence (Stevanov et al., 2013). Collaborative case studies exploring the barriers and possibilities in translational research on tree microbiomes could help to increase the flow across the academia-society interface.

3. Concluding remarks

In summary, the academic impact of the research on forest tree microbiomes is already strong and rapidly growing. While major translational impacts of this research on practical forest protection and production are still being awaited, it seems obvious that continued research on tree microbiomes will open interesting avenues for the development of diagnosis and environmentally friendly measures to promote and protect tree vitality and growth. In the coming decades, new insights on how to support beneficial tree microbiomes can be expected to arise from the deep sequencing and analysis of big data. Cross-disciplinary approaches that include the integration of knowledge from community ecology (e.g., priority effects) will be increasingly needed to understand the resilience and dynamics of tree microbiomes. The recent developments in policy such as in the EU create a positive momentum for translational research on tree microbiomes. Strategic partnerships with interested actors should be an integrated part of translational research initiatives, helping to bridge the science-society gap and ensuring that research on tree microbiomes will result in multiple benefits for humans and nature.

Acknowledgments

The work was supported by grants from the Swedish Research Council Formas (nr. 2016-00907) to J. Witzell, from the EU's Horizon 2020 research and innovation program MSCA (grant agreement no. 845419) to C. Romeralo, and from Ministerio de Ciencia e Innovación, Spain (PID2019-107256RB-I00) to J.A. Martín.

References

Agostinelli, M., Cleary, M., Martín, J.A., Albrectsen, B.R., Witzell, J., 2018. Pedunculate oaks (*Quercus robur* L.) differing in vitality as reservoirs for fungal biodiversity. Front. Microbiol. 9, 1758.

Ahmed, A., 2016. Analysis of metagenomics Next Generation Sequence data for fungal ITS barcoding: do you need advance bioinformatics experience? Front. Microbiol. 7, 1061.

Alabouvette, C., Olivain, C., Steinberg, C., 2006. Biological control of plant diseases: the European situation. Eur. J. Plant Pathol. 114, 329–341.

Allio, L., Ballantine, B., Meads, R., 2006. Enhancing the role of science in the decision-making of the European Union. Regul. Toxicol. Pharmacol. 44, 4–13.

Arnold, A.M., Mejía, L.C., Kyllo, D., Rojas, E.I., Maynard, Z., Robbins, N., et al., 2003. Fungal endophytes limit pathogen damage in a tropical tree. PNAS 100, 15649–15654.

Aymerich, M., Rodriguez-Jareño, M.C., Castells, X., Carrion, C., Zamora, A., Capellá, D., 2014. Translational research: a concept emerged from health sciences and exportable to education sciences. Ann. Transl. Med. Epidemiol. 1, 1005.

Balint, M., Schmidt, P.-A., Sharma, R., Thines, M., Schmitt, I., 2014. An illumina metabarcoding pipeline for fungi. Ecol. Evol. 4, 2642–2653.

Barret, M., Briand, M., Bonneau, S., Preveaux, A., Valiére, S., Bouchez, O., et al., 2015. Emergence shapes the structure of the seed microbiota. Appl. Environ. Microbiol. 81, 1257–1266.

Beeck, M.O.D., Ruytinx, J., Smits, M.M., Vangronsveld, J., Colpaert, J.V., Rineau, F., 2015. Below ground fungal communities in pioneer scots pine stands growing on heavy metal polluted and non-polluted soils. Soil Biol. Biochem. 86, 58–66.

Bell, T.H., Stefani, F.O.P., Abram, K., Champagne, J., Yergeau, E., Hijri, M., et al., 2016. A diverse soil microbiome degrades more crude oil than specialized bacterial assemblages obtained in culture. Appl. Environ. Microbiol. 82 (18), 5530–5541.

Ben Said, S., Or, D., 2017. Synthetic microbial ecology: engineering habitats for modular consortia. Front. Microbiol. 8, 1125.

Berg, M., Koskella, B., 2018. Nutrient- and dose-dependent microbiome-mediated protection against a plant pathogen. Curr. Biol. 28. 2487.e3–2492.e3.

Bernier, L., Yang, D., Ouellette, G.B., Dessureault, M., 1996. Assessment of *Phaeotheca dimorphospora* for biological control of the Dutch elm disease pathogens, *Ophiostoma ulmi* and *O. novo-ulmi*. Plant Pathol. 45, 609–617.

Blumenstein, K., Albrectsen, B.R., Martín, J.A., Hultberg, M., Sieber, T.N., Helander, M., Witzell, J., 2015. Nutritional niche overlap potentiates the use of endophytes in biocontrol of a tree disease. Biol. Control 60, 655–667.

Böcher, M., Krott, M., 2014. The RIU model as an analytical framework for scientific knowledge transfer: the case of the "decision support system forest and climate change". Biodivers. Conserv. 23, 3641–3656.

Bokulich, N.A., Subramanian, S., Faith, J.J., Gevers, D., Gordon, J.I., Knight, R., et al., 2013. Quality-filtering vastly improves diversity estimates from Illumina amplicon sequencing. Nat. Methods 10, 57–59.

Bullington, L.S., Larkin, B.G., 2015. Using direct amplification and next-generation sequencing technology to explore foliar endophyte communities in experimentally inoculated western white pines. Fungal Ecol. 17, 170–178.

Busby, P.E., Ridout, M., Newcombe, G., 2016a. Fungal endophytes: modifiers of plant disease. Plant Mol. Biol. 90, 645–655.

Busby, P.E., Peay, K.G., Newcombe, G., 2016b. Common foliar fungi of *Populus trichocarpa* modify *Melampsora* rust disease severity. New Phytol. 209, 1681–1692.

Busby, P.E., Soman, C., Wagner, M.R., Friesen, M.L., Kremer, J., Bennett, A., et al., 2017. Research priorities for harnessing plant microbiomes in sustainable agriculture. PLoS Biol. 15 (3), e2001793.

Butler, D., 2008. Translational research: crossing the valley of death. Nature 453, 840–842.

Calderón, K., Spor, A., Breuil, M.C., Bru, D., Bizouard, F., Violle, C., Barnard, R.L., Philippot, L., 2016. Effectiveness of ecological rescue for altered soil microbial communities and functions. ISME J. 11, 272–283.

Chamberlain, P., 2020. Knowledge is not everything. Design Health 4 (1), 1–3.

Chandler, C., 2013. What is the meaning of impact in relation to research and why does it matter? In: Denicolo, P. (Ed.), Achieving Impact in Research. Sage Publications, pp. 1–9. http://www.uk.sagepub.com/upm-data/58384_Denicolo__Achieving_Impact_in_Research.pdf. (Accessed 11 April 2020).

Chanway, C.P., 1997. Inoculation of tree roots with plant growth promoting soil bacteria: an emerging technology for reforestation. For. Sci. 43, 99–112.

Cregger, M.A., Veach, A.M., Yang, Z., Crouch, M.J., Vilgalys, R., Tuskan, C.W., Schadt, G.A., 2018. The *Populus* holobiont: dissecting the effects of plant niches and genotype on the microbiome. Microbiome 6, 31.

Donn, S., Kirkegaard, J.A., Perera, G., Richardson, A.E., Watt, M., 2015. Evolution of bacterial communities in the wheat crop rhizosphere. Environ. Microbiol. 17 (3), 610–621.

Drake, J.A., 1991. Community-assembly mechanics and the structure of an experimental species ensemble. Am. Nat. 137, 1–26.

Eevers, N., Beckers, B., Op de Beeck, M., White, J.C., Vangronsveld, J., Weyens, N., 2016. Comparison between cultivated and total bacterial communities associated with *Cucurbita pepo* using cultivation-dependent techniques and 454 pyrosequencing. Syst. Appl. Microbiol. 39, 58–66.

EU, 2009. Directive 2009/128/EC of the European Parliament and of the Council of 21 October 2009 establishing a framework for community action to achieve the sustainable use of pesticides. OJ L 309, 71–86. 24.11.2009.

EU, 2016. Regulation (EU) 2016/2031 on Protective Measures Against Plant Pests ("Plant Health Law"). Available at: https://ec.europa.eu/food/plant/plant_health_biosecurity/legislation/new_eu_rules_en. (Accessed 10 August 2020).

Evans, M.C., Cvitanovic, C., 2018. An introduction to achieving policy impact for early career researchers. Palgrave Commun. 4, 88.

Foo, J.L., Ling, H., Lee, Y.S., Chang, M.W., 2017. Microbiome engineering: current applications and its future. Biotechnol. J. 12, 1600099.

Frasz, S., Walker, A.K., Nsiama, T.K., Adams, G.W., Miller, J.D., 2014. Distribution of the foliar fungal endophyte *Phialocephala scopiformis* and its toxin in the crown of a mature white spruce tree as revealed by chemical and qPCR analyses. Can. J. For. Res. 44, 1138–1143.

Giard-Laliberté, C., Azarbad, H., Tremblay, J., Bainard, L., Yergeau, É., 2019. A water stress-adapted inoculum affects rhizosphere fungi, but not bacteria nor wheat. FEMS Microbiol. Ecol. 95 (7), fiz080.

Greenhalgh, T., Raftery, J., Hanney, S., Glover, M., 2016. Research impact: a narrative review. BMC Med. 14, 78.

Hacquard, S., Garrido-Oter, R., Gonzalez, A., Spaepen, S., Ackermann, G., Lebeis, S., et al., 2015. Microbiota and host nutrition across plant and animal kingdoms. Cell Host Microbe 17 (5), 603–616.

Hardoim, P.R., van Overbeek, L., Berg, G., Pirttilä, A., Compante, S., Campisano, A., et al., 2015. The hidden world within plants: ecological and evolutionary considerations for defining functioning of microbial endophytes. Microbiol. Mol. Biol. Rev. 79 (3), 293–320.

Howard, M.M., Bell, T.H., Kao-Kniffin, J., 2017. Soil microbiome transfer method affects microbiome composition, including dominant microorganisms, in a novel environment. FEMS Microbiol. Lett. 364, 1–8.

Johnston, P.R., Park, D., Smissen, R.D., 2017. Comparing diversity of fungi from living leaves using culturing and high-throughput environmental sequencing. Mycologia 109 (4), 643–654.

Kitson, A., Straus, S.E., 2010. The knowledge-to-action cycle: identifying the gaps. CMAJ 182 (2), E73–E77.

Kvakkestad, V., Sundbye, A., Gwynn, R., Klingen, I., 2020. Authorization of microbial plant protection products in the Scandinavian countries: a comparative analysis. Environ. Sci. Policy 106, 115–124.

Lindahl, B.D., Nilsson, R.H., Tedersoo, L., Abarenkov, K., Carlsen, T., Kjøller, R., et al., 2013. Fungal community analysis by high-throughput sequencing of amplified markers—a user's guide. New Phytol. 199, 288–299.

Lucaciu, R., Pelikan, C., Gerner, S.M., Zioutis, C., Köstlbacher, S., Marx, H., et al., 2019. A bioinformatics guide to plant microbiome analysis. Front. Plant Sci. 10, 1313.

Lundberg, D.S., Lebeis, S.L., Paredes, S.H., Yourstone, S., Gehring, J., Malfatti, S., et al., 2012. Defining the core *Arabidopsis thaliana* root microbiome. Nature 488, 86–90.

Martín, J., Witzell, J., Blumenstein, K., Rozpedowska, E., Helander, M., Sieber, T., Gil, L., 2013. Resistance to Dutch elm disease reduces xylem endophytic fungi presence in elms (*Ulmus* spp.). PLoS One 8 (2), e56987.

Martín, J.A., Macaya-Sanz, D., Witzell, J., Blumenstein, K., Gil, L., 2015. Strong in vitro antagonism by elm xylem endophytes is not accompanied by temporally stable in planta protection against a vascular pathogen under field conditions. Eur. J. Plant Pathol. 60, 655–667.

Martínez-Álvarez, P., Fernández-González, R.A., Sanz-Ros, A.V., Pando, V., Diez, J.J., 2016. Two fungal endophytes reduce the severity of pitch canker disease in *Pinus radiata* seedlings. Biol. Control 94, 1–10.

Martínez-Arias, C., Macaya-Sanz, D., Witzell, J., Martín, J.A., 2019. Enhancement of *Populus alba* tolerance to *Venturia tremulae* upon inoculation with endophytes showing in vitro biocontrol potential. Eur. J. Plant Pathol. 153, 1031–1042.

Martínez-Arias, C., Sobrino-Plata, J., Ormeño-Moncalvillo, S., Gil, L., Rodríguez-Calcerrada, J., Martín, J.A., 2021. Endophyte inoculation enhances *Ulmus minor* resistance to Dutch elm disease. Fungal Ecol. 50, 101024. https://doi.org/10.1016/j.funeco.2020.101024.

Martínez-Arias, C., Sobrino-Plata, J., Macaya-Sanz, D., Aguirre, N.M., Collada, C., Gil, L., et al., 2020. Changes in plant function and root mycobiome caused by flood and drought in a riparian tree. Tree Physiol. 40 (7), 886–903.

Mbareche, H., 2020. NGS in environmental mycology. A useful tool? In: Reference Module in Life Sciences. Elsevier, https://doi.org/10.1016/B978-0-12-809633-8.21045-5.

Meinard, Y., Quétier, F., 2013. Experiencing biodiversity as a bridge over the science–society communication gap. Conserv. Biol. 28 (3), 705–712.

Miller, J.D., MacKenzie, S., Foto, M., Adams, G.W., Findlay, J.A., 2002. Needles of white spruce inoculated with rugulosin-producing endophytes contain rugulosin reducing spruce budworm growth rate. Mycol. Res. 106 (4), 471–479.

Nelson, E.B., 2018. The seed microbiome: origins, interactions, and impacts. Plant Soil 422, 7–34.

Newcombe, G., 2011. Endophytes in forest management: four challenges. In: Pirttilä, A.M., Frank, A.C. (Eds.), Endophytes of Forest trees: Biology and Applications, Forestry Sciences 80. Springer, Berlin/Heidelberg/New York, pp. 251–262.

Nguyen, N.H., Smith, D., Peay, K., Kennedy, P., 2015. Parsing ecological signal from noise in next generation amplicon sequencing. New Phytol. 205, 1389–1393.

Nilsson, R.H., Anslan, S., Bahram, M., Wurzbacher, C., Baldrian, P., Tedersoo, L., 2019. Mycobiome diversity: high-throughput sequencing and identification of fungi. Nat. Rev. Microbiol. 17, 95–109.

Noble, A.S., Noe, S., Clearwater, M.J., Lee, C.K., 2020. A core phyllosphere microbiome exists across distant populations of a tree species indigenous to New Zealand. PLoS One 15 (8), e0237079.

Ownley, B.H., Griffin, M.R., Klingeman, W.E., Gwinn, K.D., Moulton, J.K., Pereira, R.M., 2008. *Beauveria bassiana*: endophytic colonization and plant disease control. J. Invertebr. Pathol. 98, 267–270.

Panke-Buisse, K., Poole, A.C., Goodrich, J.K., Ley, R.E., Kao-Kniffin, J., 2015. Selection on soil microbiomes reveals reproducible impacts on plant function. ISME J. 9, 980–989.

Parsa, S., Ortiz, V., Gómez-Jiménez, M.I., Kramer, M., Vega, F.E., 2018. Root environment is a key determinant of fungal entomo-pathogen endophytism following seed treatment in the com-mon bean, *Phaseolus vulgaris*. Biol. Control 116, 74–81.

Pauvert, C., Buée, M., Laval, V., Edel-Hermann, V., Fauchery, L., Gautier, A., et al., 2019. Bioinformatics matters: the accuracy of plant and soil fungal community data is highly dependent on the metabarcoding pipeline. Fungal Ecol. 41, 23–33.

Quiring, D., Flaherty, L., Adams, G., McCartney, A., Miller, D., Edwards, S., 2019. An endophytic fungus interacts with crown level and larval density to reduce the survival of eastern spruce budworm, *Choristoneura fumiferana* (Lepidoptera: Tortricidae), on white spruce (*Picea glauca*). Can. J. For. Res. 49, 221–227.

Rabiey, M., Hailey, L.E., Roy, S.R., Grenz, K., Al-Zadjali, M.A.S., Barrett, G.A., Jackson, R.W., 2019. Endophytes vs tree pathogens and pests: can they be used as biological control agents to improve tree health? Eur. J. Plant Pathol. 155, 711–729.

Raghavendra, A.K., Newcombe, G., 2013. The contribution of foliar endophytes to quantitative resistance to *Melampsora* rust. New Phytol. 197 (3), 909–918.

Ridout, M.E., Schroeder, K.L., Hunter, S.S., Styer, J., Newcombe, G., 2019. Priority effects of wheat seed endophytes on a rhizosphere symbiosis. Symbiosis 78, 19–31.

Robin, D.C., Marchand, P.A., 2019. Evolution of the biocontrol active substances in the framework of the European Pesticide Regulation (EC) no. 1107/2009. Pest Manag. Sci. 75, 950–958.

Rodriguez, R., Durán, P., 2020. Natural holobiome engineering by using native extreme microbiome to counteract the climate change effects. Front. Bioeng. Biotechnol. 8, 568.

Romão, D., Staley, C., Ferreira, F., Rodrigues, R., Sabino, R., Veríssimo, C., 2017. Next-generation sequencing and culture-based techniques offer complementary insights into fungi and prokaryotes in beach sands. Mar. Pollut. Bull. 119, 351–358.

Romeralo, C., Santamaría, O., Pando, V., Diez, J.J., 2015a. Fungal endophytes reduce necrosis length produced by *Gremmeniella abietina* in *Pinus halepensis* seedlings. Biol. Control 80, 30–39.

Romeralo, C., Witzell, J., Romeralo-Tapia, R., Botella, L., Diez, J., 2015b. Antagonistic activity of fungal endophyte filtrates against *Gremmeniella abietina* infections on Aleppo pine seedlings. Eur. J. Plant Pathol. 143, 691–704.

Rubio, D.M., Schoenbaum, E.E., Lee, L.S., Schteingart, D.E., Marantz, P.R., Anderson, K.E., et al., 2010. Defining translational research: implications for training. Acad. Med. 85 (3), 470–475.

Rushmer, R., Ward, V., Nguyen, T., Kuchenmüller, T., 2019. Knowledge translation: key concepts, terms and activities. In: Verschuuren, M., van Oers, H. (Eds.), Population Health Monitoring: Climbing the Information Pyramid. Springer International Publishing, Cham, pp. 127–150.

Salazar-Cerezo, S., Martinez-Montiel, N., Cruz-Lopez, M., Martinez-Contreras, R.D., 2018. Fungal diversity and community composition of culturable fungi in *Stanhopea trigrina* cast gibberellin producers. Front. Microbiol. 9, 612.

Sarewitz, D., Pielke, R.A., 2007. The neglected heart of science policy: reconciling supply of and demand for science. Environ. Sci. Policy 10, 5–16.

Scheffer, R.J., Voeten, J.G.W.F., Guries, R.P., 2008. Biological control of Dutch elm disease. Plant Dis. 92, 192–200.

Sessitsch, A., Pfaffenbichler, N., Mitter, B., 2019. Microbiome applications from lab to field: facing complexity. Trends Plant Sci. 24, 194–198.

Simon, J., Marchesi, J.R., Mougel, C., Selosse, M.A., 2019. Host-microbiota interactions: from holobiont theory to analysis. Microbiome 7, 5.

Stefani, F.O.P., Bell, T.H., Marchand, C., De La Providencia, I.E., El Yassimi, A., St-Arnaud, M., Hijri, M., 2015. Culture-dependent and -independent methods capture different microbial community fractions in hydrocarbon-contaminated soils. PLoS One 10 (6), e0128272.

Stevanov, M., Böcher, M., Krott, M., Krajter, S., Vuletic, D., Orlovic, S., 2013. The Research, Integration and Utilization (RIU) model as an analytical framework for the professionalization of departmental research organizations: case studies of publicly funded forest research institutes in Serbia and Croatia. Forest Policy Econ. 37, 20–28.

Sumarah, M.W., Adams, G.W., Berghout, J., Slack, G.J., Wilson, A.M., Miller, J.D., 2008. Spread and persistence of a rugulosin-producing endophyte in *Picea glauca* seedlings. Mycol. Res. 112 (6), 731–736.

Sun, X., Guo, L.-D., 2012. Endophytic fungal diversity: review of traditional and molecular techniques. Mycology 3, 65–76.

Swenson, W., Wilson, D.S., Elias, R., 2000. Artificial ecosystem selection. PNAS 97 (16), 9110–9114.

Tedersoo, L., Anslan, S., Bahram, M., Põlme, S., Riit, T., Liiv, I., et al., 2015. Shotgun metagenomes and multiple primer pair-barcode combinations of amplicons reveal biases in metabarcoding analyses of fungi. MycoKeys 10, 1–43.

Terhonen, E., Blumenstein, K., Kovalchuk, A., Asiegbu, F.O., 2019. Forest tree microbiomes and associated fungal endophytes: Functional roles and impact on forest health. Forests 10, 42.

Thomas, P., Sekhar, A.C., 2016. Effects due to rhizospheric soil application of an antagonistic bacterial endophyte on native bacterial community and its survival in soil: a case study with *Pseudomonas aeruginosa* from banana. Front. Microbiol. 7, 1–16.

Toju, H., Peay, K.G., Yamamichi, M., Narisawa, K., Hiruma, K., Naito, K., et al., 2018. Core microbiomes for sustainable agroecosystems. Nat. Plants 4 (5), 247–257.

Tosi, M., Mitter, E.K., Gaiero, J., Dunfield, K., 2020. It takes three to tango: the importance of microbes, host plant, and soil management to elucidate manipulation strategies for the plant microbiome. Can. J. Microbiol. 66, 413–433.

Walker, T.S., Bais, H.P., Grotewold, E., Vivanco, J.M., 2003. Root exudation and rhizosphere biology. Plant Physiol. 132 (1), 44–51.

Wemheuer, F., Berkelmann, D., Wemheuer, B., Daniel, R., Vidal, S., Daghela, H.B.B., 2020. Agroforestry management systems drive the composition, diversity, and function of fungal and bacterial endophyte communities in *Theobroma cacao* leaves. Microorganisms 8 (3), 405.

Willis, A.D., Minot, S.S., 2020. Strategies to facilitate translational advances from microbiome surveys. Trends Microbiol. 28 (5), 329–330.

Witzell, J., Martín, J.A., 2018. Endophytes and forest health. In: Pirttilä, A.M., Frank, A.C. (Eds.), Endophytes of Forest Trees, Forestry Sciences. vol. 86. Springer International Publishing AG, Springer Nature, pp. 261–282.

Witzell, J., Martín, J.A., Blumenstein, K., 2014. Ecological aspects of endophyte-based biocontrol of forest diseases. In: Verma, V.C., Gange, A.C. (Eds.), Advances in Endophytic Research. vol. 17. Springer India, pp. 321–333.

Wüstenhagen, R., Wolsink, M., Bürer, M.J., 2007. Social acceptance of renewable energy innovation: an introduction to the concept. Energy Policy 35 (5), 2683–2691.

Yergeau, E., Bell, T.H., Champagne, J., Maynard, C., Tardif, S., Tremblay, J., Greer, C.W., 2015. Transplanting soil microbiomes leads to lasting effects on willow growth, but not on the rhizosphere microbiome. Front. Microbiol. 6, 1436.

Chapter 22

Forest microbiome: Challenges and future perspectives

Fred O. Asiegbu

Department of Forest Sciences, Faculty of Agriculture and Forestry, University of Helsinki, Helsinki, Finland

The analysis of human, environmental, agricultural crop plants, and tree microbiome (Fig. 22.1) is a rapidly expanding scientific field. Many authors have highlighted that in addition to ecological and environmental factors, the balanced interactions between microbes and their plant hosts play a critical role in maintaining the host fitness (Guttman et al., 2014; Vandenkoornhuyse et al., 2015; Thompson et al., 2017). The disruption of this balanced dynamics and function could lead to a disease outcome. Recent studies have shown that perturbations of this balance may also lead to significant changes or reduction in the structure of microbial communities in asymptomatic and symptomatic host tree species (Kovalchuk et al., 2018). An astonishing diversity of microbiota in tissues of healthy and diseased Norway spruce trees has also been reported (Kovalchuk et al., 2018; Ren et al., 2018). The authors highlighted that differences in the tree microbiome composition and structure may exist depending on the tissue, age of the tree, and extent of pathogen infection. The available and accumulated evidence from many of the "omics" data suggests that plant microbiome has potential detrimental or beneficial effects on the environment and host fitness (Vandenkoornhuyse et al., 2015; Thompson et al., 2017). In laboratory conditions, demonstrated effects of plant-microbe interaction on gene regulation and metabolite profiles are equally evident. However, not much is known on the impact of biotic and abiotic factors on mechanistic aspects of microbial functions in natural environments (de Boer, 2017; Thompson et al., 2017). The major challenge in phytobiome or forest tree microbiome study is the potential practical applications or translational impact of the findings. One of the newly emerging approaches applied in agricultural plant systems is the use or reconstruction of synthetic microbial communities for validation of functional and ecological relevance of individual microbe in an environmental sample (Niu et al., 2017; Levy et al., 2018). This approach requires rigorous isolation and phenotypic analysis, which could provide insight on the interspecific microbe interaction combinations that might have positive effect on plant health and growth (Hartman et al., 2017). It could be particularly useful for identification of beneficial microbes as well as for investigation of the role of individual microbial isolates to the response of host plants to associated nutrient or biotic stresses. One of the successful examples of practical application of microbiome study is with respect to suppressive soils. Sergaki et al. (2018) in their review noted that the concept of "Suppressive soil" is an evidence that soil microbial communities could be deployed not only to protect plants against phytopathogens but also to enhance plant growth. Another strategy with future prospect is the potential to use breeding approach for selection of host tree genotypes with inherent genetic ability to interact with beneficial microorganisms.

It is imperative that for robust biological conclusions and potential translational impacts of microbiome study, accurate quantitative and qualitative measurements are essential (Poussin et al., 2018). Therefore great care is required for data acquisition and the success and meaningful conclusions of a microbiota analysis. Several authors emphasized that the success of microbiome study is dependent on a rigorous experimental study design supported by clear research objectives and scientific questions (Poussin et al., 2018; Goodrich et al., 2014; Kim et al., 2017; Laukens et al., 2016). The data analysis should be conducted with well thought-out computational strategies (Goodrich et al., 2014; Roumpeka et al., 2017; Oulas et al., 2015). Poussin et al. (2018) in their review article highlighted that under certain conditions, the makeup and characteristic of microbiome is influenced by several factors, including sampling time, health status, and relevant control groups. For forest microbiome studies, the structure of the microbial community could also be impacted by anthropogenic factors, longevity of trees and long timescale of their diseases, climatic variables as well as health status (Kovalchuk et al., 2018; Ren et al., 2018). Another overriding consideration in the design of microbiome study is the technique to be used for detection and quantification of microbes present in the environmental samples. The most common and widely used marker genes are internal transcribed space region (ITS) for fungi and 16S rRNA for archaea and bacteria. One of the major limitations of amplicon next generation sequencing (NGS) of 16S rRNA gene is the short sequences of the hypervariable

Forest Microbiology. https://doi.org/10.1016/B978-0-12-822542-4.00002-4

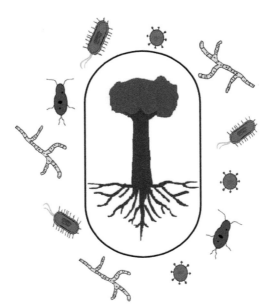

FIG. 22.1 Typical microbiota of a forest tree.

regions that considerably lower the sensitivity to identify bacterial taxa to species level compared to the use of full-length 16S rRNA gene. This problem could be mitigated using PacBio circular consensus sequencing (CCS) technology to sequence full-length bacterial 16S rRNA genes thereby facilitating high-fidelity species-level identification of microbiome data (Earl et al., 2018). Though ITS markers have been widely used as universal fungal DNA barcode sequence, there are some few noted limitations. Few primers used for ITS amplification (e.g., ITS2) could amplify more easily some fungal groups such as ascomycetes than basidiomycetes from environmental samples. Bellemain et al. (2010) reported this could be avoided by using ITS1 primers. By contrast, Tedersoo et al. (2015) reported less taxonomic bias with ITS2 than with ITS1 and this was attributed partly to lower length variation and more universal primer sites for the ITS2 subregion. In a follow-up study, the authors (Tedersoo et al., 2017) using the second-generation high throughput sequencing (HTS) noted that both 18S (SSU) and 28S (LSU) rRNA gene sequences (including ITS1 and ITS2 subregions) are better for taxonomic resolution. Using primers designed within the 5.8S (5.8s-Fun) and large subunit (LSU) ribosomal genes (ITS4-Fun), Taylor et al. (2016) were also able to successfully recover diverse operational taxonomic units (OTU) with wide coverage of the Fungal kingdom. Recently, Castaño et al. (2020) evaluated the efficiency of recovering 10 ITS fragments from artificially assembled fungal communities after sequencing with PacBio Sequel I, PacBio RS II, and Illumina MiSeq. Comparatively, they were able to recover more efficiently all fragments from the artificially assembled fungal community by sequencing with Pac Bio technologies than with Illumina MiSeq sequencing. Although majority of studies related to tree microbiome are focused on bacteria and fungi (Fig. 22.1), the remaining groups of microbes (e.g., protists) due to methodological difficulties remain largely underexplored (Glucksman et al., 2010). Recently, Sapp et al. (2018) highlighted the importance of protists as components of plant microbiota. Similarly, though protocols for virome analysis have been developed, viruses are still one of the understudied members of the microbial community (Fig. 22.1) (Thomas and Segata, 2019).

In addition to elucidating the microbiota community structures in diverse environments (Thompson et al., 2017), determining the functional role of each microbe is important, though a challenging task (Guttman et al., 2014). Novel approaches such as RNA-seq, metagenomics and computational modeling are therefore needed for better understanding of the ecological interactions and functional role of individual microbiome within forest biome. In the last few years, RNAseq (metatranscriptomics) has increasingly been used for taxonomic classification (Cox et al., 2017; Jiang et al., 2016; Messa et al., 2019). The RNAseq approach is however not without its own limitations or challenges as an aid to predicting transcriptionally active genes for application in taxonomic classification. One of the challenges is the predominance or contamination of host RNA (Bashiardes et al., 2016). Another limitation is the short half-life of mRNA which may constrain rapid detection of short-lived responses to environmental stimuli (Shirley et al., 2015). Apart from the use of RNAseq data for functional classification, one of the novel research approaches is the application of metagenome-wide association studies (MWAS) to establish a link between human microbiota and disease (Wang and Jia, 2016). Several analytical tools for metagenome datasets are well developed and effectively used for bacterial biota, partly due to the small size of their genome (Bulgarelli et al., 2015; Siegl et al., 2011; Woyke et al., 2006). Improved sequencing technology and computational methods will be needed to explore the diverse whole metagenome datasets of other microbial community particularly eukaryotes (Thomas and Segata, 2019). Other alternative and associated approaches for microbiome analysis include the use of data-driven network such as "species co-occurrence network" to study ecological interactions within microbial community. This could be complemented by use of "metabolism-based networks" to provide mechanistic understanding on the functional role of microbiome within an ecosystem (Bauer and Thiele, 2018). Predictive computational model (MelonnPan) for metabolome has recently been used for profiling ecological community of human microbiota (Mallick et al., 2019). The successful applications of both the network modeling and MWAS as well as metabolome predictive model in other organisms indicate that it can also be extended for tree microbiome study with potential practical relevance in forest management and tree health protection (Poudel et al., 2016; van der Heijden and Hartmann, 2016).

Furthermore, the data on microbiota of forest biome has mostly been generated from temperate and boreal forests; only very limited studies and data have been obtained from tropical flora and ecosystem. The habitats in the tropics host a remarkable diversity of microbial species. Tropical and temperate forests are different in many respects particularly in the number and diversity of the plant species. It is of vital importance to recognize that knowledge on the microbiome structure and composition from the tropics are equally immensely valuable. For many researchers from developing countries, sequencing costs are undoubtedly a major constraint in undertaking such microbiome studies. Additionally, inadequate technical manpower as well as lack of access to modern technologies are further limiting factors. The decreasing cost of NGS or HTS will hopefully facilitate in minimizing and overcoming the economic challenges together with collaboration and partnership with international organizations.

Generally, standardization, reproducibility, and quality controls are also critical components of microbiome study. For the human microbiome analysis, the international consortia on MicroBiome Quality Control (MBQC; https://www.mbqc.org/) set up the platform in order to tackle the issue of standardization and quality control in the evaluation of human microbiome. Many of the variables evaluated in MBQC include among other parameters: sample collection and handling, DNA extraction and 16S rDNA amplification, sequencing and bioinformatics analysis. A similar quality control platform is equally necessary for the profiling of microbiome of forest biomes. The Earth Microbiome Project (EMP, http://www.earthmicrobiome.org) and the Global Fungi Database (https://globalfungi.com/) would help to advance our understanding of microbiota of geographically diverse regions of the world. Equally, a global forest microbiome consortium would in future be necessary to provide better insight on the microbiota relevant for forest biomes from diverse habitats in the tropics and temperate countries.

References

Bashiardes, S., Zilberman-Schapira, G., Elinav, E., 2016. Use of metatranscriptomics in microbiome research. Bioinform. Biol. Insights 10, 19–25. https://doi.org/10.4137/BBI.S34610.

Bauer, E., Thiele, I., 2018. From network analysis to functional metabolic modeling of the human gut microbiota. mSystems 3, e00209–e00217. https://doi.org/10.1128/mSystems.00209-17.

Bellemain, E., Carlsen, T., Brochmann, C., Coissac, E., Taberlet, P., Kauserud, H., 2010. ITS as an environmental DNA barcode for fungi: an in silico approach reveals potential PCR biases. BMC Microbiol. 10, 189. https://doi.org/10.1186/1471-2180-10-189.

Bulgarelli, D., Garrido-Oter, R., Münch, P.C., Weiman, A., Dröge, J., Pan, Y., McHardy, A.C., Schulze-Lefert, P., 2015. Structure and function of the bacterial root microbiota in wild and domesticated barley. Cell Host Microbe 17 (3), 392–403.

Castaño, C., Berlin, A., Brandström Durling, M., Ihrmark, K., Lindahl, B.D., Stenlid, J., Clemmensen, K.E., Olson, Å., 2020. Optimized metabarcoding with Pacific biosciences enables semi-quantitative analysis of fungal communities. New Phytol. https://doi.org/10.1111/nph.16731.

Cox, J.C., Ballweg, R.A., Taft, D.H., Velayutham, P., Haslam, D.B., Porollo, A., 2017. A fast and robust protocol for metataxonomic analysis using RNAseq data. Microbiome 5, 7.

de Boer, W., 2017. Upscaling of fungal–bacterial interactions: from the lab to the field. Curr. Opin. Microbiol. 37, 35–41. https://doi.org/10.1016/j.mib.2017.03.007.

Earl, J.P., Adappa, N.D., Krol, J., Bhat, A.S., Balashov, S., Ehrlich, R.L., Palmer, J.N., Workman, A.D., Blasetti, M., Sen, B., Hammond, J., Cohen, N.A., Ehrlich, G.D., Mell, J.C., 2018. Species-level bacterial community profiling of the healthy sinonasal microbiome using Pacific biosciences sequencing of full-length 16S rRNA genes. Microbiome 6, 190.

Glucksman, E., Bell, T., Griffiths, R.I., Basset, D., 2010. Closely related protist strains have different grazing impacts on natural bacterial communities. Environ. Microbiol. 12, 3105–3113.

Goodrich, J.K., Di Rienzi, S.C., Poole, A.C., Koren, O., Walter, W.A., Caporaso, J.G., Knight, R., Ley, R.E., 2014. Conducting a microbiome study. Cell 158, 250–262.

Guttman, D.S., McHardy, A.C., Schulze-Lefert, P., 2014. Microbial genome-enabled insights into plant-microorganism interactions. Natl. Rev. 15, 797–813.

Hartman, K., van der Heijden, M.G.A., Roussely-Provent, V., Walser, J.C., Schlaeppi, K., 2017. Deciphering composition and function of the root microbiome of a legume plant. Microbiome 5, 2. https://doi.org/10.1186/s40168-0160220-z.

Jiang, Y., Xiong, X., Danska, J., Parkinson, J., 2016. Metatranscriptomic analysis of diverse microbial communities reveals core metabolic pathways and microbiome-specific functionality. Microbiome 4, 2.

Kim, D., Hofstaedter, C.E., Zhao, C., Mattei, L., Tanes, C., Clarke, E., Lauder, A., Sherill-Mix, S., Chehoud, C., Kelsen, J., Conrad, M., Collman, R.G., Baldassano, R., Bushman, F.D., Bittinger, K., 2017. Optimizing methods and dodging pitfalls in microbiome research. Microbiome 5, 52.

Kovalchuk, A., Mukrimin, M., Zeng, Z., Raffaello, T., Liu, M., Kasanen, R., Sun, H., Asiegbu, F.O., 2018. Mycobiome analysis of asymptomatic and symptomatic Norway spruce trees naturally infected by the conifer pathogen *Heterobasidion* sp. Environ. Microbiol. Rep. 10 (5), 532–541. https://doi.org/10.1111/1758-2229.12654.

Laukens, D., Brinkman, B.M., Raes, J., De Vos, M., Vandenabeele, P., 2016. Heterogeneity of the gut microbiome in mice: guidelines for optimizing experimental design. FEMS Microbiol. Rev. 40, 117–132.

Levy, A., Gonzalez, I.S., Mittelviefhaus, M., Clingenpeel, S., Paredes, S.H., Miao, J., Wang, K., Devescovi, G., Stillman, K., Monteiro, F., Alvarez, B.R., Lundberg, D.S., Lu, T.-Y., Lebers, S., Jin, Z., McDonlad, M., Klein, A.P., Feltcher, M.E., Rio, T.G., Grant, S.R., Doty, S.L., Ley, R.E., Zhao, B., Venturi, V., Pelletier, D., Vorholt, J.A., Tringer, S.G., Woyke, T., Dangl, J.L., Franzosa, M.H., McLver, E.A., et al., 2018. Genomic features of bacterial adaptation to plants. Nat. Genet. 50, 138–150. https://doi.org/10.1038/s41588-017-0012-9.

Mallick, H., Franzosa, E.A., McIver, L.J., et al., 2019. Predictive metabolomic profiling of microbial communities using amplicon or metagenomic sequences. Nat. Commun. 10, 3136. https://doi.org/10.1038/s41467-019-10927-1.

Messa, M., Slippers, B., Naidoo, S., Bezuidt, O., Kemler, M., 2019. Active fungal communities in asymptomatic *Eucalyptus grandis* stems differ between a susceptible and resistant clone. Microorganisms 7 (10), 375. https://doi.org/10.3390/microorganisms7100375.

Niu, B., Paulson, J.N., Zheng, X., Kolter, R., 2017. Simplified and representative bacterial community of maize roots. Proc. Natl. Acad. Sci. U. S. A. 114, E2450–E2459. https://doi.org/10.1073/pnas.1616148114.

Oulas, A., Pavloudi, C., Polymenakou, P., Pavlopoulos, G.A., Papanikolaou, N., Kotoulas, G., Arvanitidis, C., Iliopoulos, I., 2015. Metagenomics: tools and insights for analyzing next generation sequencing data derived from biodiversity studies. Bioinform. Biol. Insights 9, 75–88.

Poudel, R., Jumpponen, A., Schlatter, D.C., Paulitz, T.C., McSpaden Gardener, B.B., Kinkel, L.L., Garrett, K.A., 2016. Microbiome networks: a systems framework for identifying candidate microbial assemblages for disease management. Phytopathology 106, 1083–1096.

Poussin, C., Sierro, N., Boué, S., Battey, J., Scotti, E., Belcastro, V., Peitsch, M.C., Ivanov, N.V., Hoeng, J., 2018. Interrogating the microbiome: experimental and computational considerations in support of study reproducibility. Drug Discov. Today 23 (9), 1644–1657.

Ren, F., Kovalchuk, A., Mukrimin, M., Liu, M., Zeng, Z., Ghimire, R.P., Kivimäenpää, M., Holopainen, J.K., Sun, H., Asiegbu, F.O., 2018. Tissue microbiome of Norway spruce affected by *Heterobasidion*-induced Wood decay. Microb. Ecol. 77 (3), 640–650. https://doi.org/10.1007/s00248-018-1240-y.

Roumpeka, D.D., Wallace, R.J., Escalettes, F., Fotheringham, I., Watson, M., 2017. A review of bioinformatics tools for bio-prospecting from metagenomic sequence data. Front. Genet. 8, 23.

Sapp, M., Ploch, S., Fiore-Donno, A.M., Bonkowski, M., Roseet, L.E., 2018. Protists are an integral part of the *Arabidopsis thaliana* microbiome. Environ. Microbiol. 20, 30–43.

Sergaki, C., Lagunas, B., Lidbury, I., Gifford, M.L., Schäfer, P., 2018. Challenges and approaches in microbiome research: from fundamental to applied. Front. Plant Sci. 9, 1205. https://doi.org/10.3389/fpls.2018.01205.

Shirley, B., Alejandra, V.-L., Fernanda, C.-G., Rico, K., Canizales-Quinteros, S., Soberon, X., Pozo-Yauner, L.D., Ochoa-Leyva, A., 2015. Combining metagenomics, metatranscriptomics and viromics to explore novel microbial interactions: towards a systems-level understanding of human microbiomes. Comput. Struct. Biotechnol. J. 13, 390–401.

Siegl, A., Kamke, J., Hochmuth, T., Piel, J., Richter, M., Liang, C., Dandekar, T., Hentschel, U., 2011. Single-cell genomics reveals the lifestyle of Poribacteria, a candidate phylum symbiotically associated with marine sponges. ISME J 5, 61. https://doi.org/10.1038/ismej.2010.95.

Taylor, D.L., Walters, W.A., Lennon, N.J., Bochicchio, J., Krohn, A., Caporaso, J.G., Pennanen, T., 2016. Accurate estimation of fungal diversity and abundance through improved lineage-specific primers optimized for illumina amplicon sequencing. Appl. Environ. Microbiol. 82, 7217–7226. https://doi.org/10.1128/AEM.02576-16.

Tedersoo, L., Anslan, S., Bahram, M., Põlme, S., Riit, T., Liiv, I., Kõljalg, U., Kisand, V., Nilsson, H., Hildebrand, F., Bork, P., Abarenkov, K., 2015. Shotgun metagenomes and multiple primer pair-barcode combinations of amplicons reveal biases in metabarcoding analyses of fungi. MycoKeys 10, 1–43. https://doi.org/10.3897/mycokeys.10.4852.

Tedersoo, L., Bahram, M., Puusepp, R., Nilsson, R.H., James, T.Y., 2017. Novel soil-inhabiting clades fill gaps in the fungal tree of life. Microbiome 5, 42. https://doi.org/10.1186/s40168-017-0259-5.

Thomas, A.M., Segata, N., 2019. Multiple levels of the unknown in microbiome research. BMC Biol. 17, 48. https://doi.org/10.1186/s12915-019-0667-z.

Thompson, L.R., Sanders, J.G., McDonald, D., Amir, A., Ladau, J., Locey, K.J., Prill, R.J., Tripathi, A., Gibbons, S.M., Ackermann, G., Navas-Molina, J.A., Janssen, S., Kopylova, E., Vázquez-Baeza, Y., González, A., Morton, J.T., Mirarab, S., Xu, Z.Z., Jiang, L., Haroon, M.F., Kanbar, J., Zhu, Q., Song, S.J., Kosciolek, T., Bokulich, N.A., Lefler, J., Brislawn, C.J., Humphrey, G., Owens, S.M., Hampton-Marcell, J., Berg-Lyons, D., McKenzie, V., Fierer, N., Fuhrman, J.A., Clauset, A., Stevens, R.L., Shade, A., Pollard, K.S., Goodwin, K.D., Jansson, J.K., Gilbert, J.A., Knight, R., Agosto Rivera, J.L., Al-Moosawi, L., Alverdy, J., Amato, K.R., Andras, J., Angenent, L.T., Antonopoulos, D.A., Apprill, A., Armitage, D., Ballantine, K., Bárta, J., Baum, J.K., Berry, A., Bhatnagar, A., Bhatnagar, M., Biddle, J.F., Bittner, L., Boldgiv, B., Bottos, E., Boyer, D.M., Braun, J., Brazelton, W., Brearley, F.Q., Campbell, A.H., Caporaso, J.G., Cardona, C., Carroll, J.L., Cary, S.C., Casper, B.B., Charles, T.C., Chu, H., Claar, D.C., Clark, R.G., Clayton, J.B., Clemente, J.C., Cochran, A., Coleman, M.L., Collins, G., Colwell, R.R., Contreras, M., Crary, B.B., Creer, S., Cristol, D.A., Crump, B.C., Cui, D., Daly, S.E., Davalos, L., Dawson, R.D., Defazio, J., Delsuc, F., Dionisi, H.M., Dominguez-Bello, M.G., Dowell, R., Dubinsky, E.A., Dunn, P.O., Ercolini, D., Espinoza, R.E., Ezenwa, V., Fenner, N., Findlay, H.S., Fleming, I.D., Fogliano, V., Forsman, A., Freeman, C., Friedman, E.S., Galindo, G., Garcia, L., Garcia-Amado, M.A., Garshelis, D., Gasser, R.B., Gerdts, G., Gibson, M.K., Gifford, I., Gill, R.T., Giray, T., Gittel, A., Golyshin, P., Gong, D., Grossart, H.P., Guyton, K., Haig, S.J., Hale, V., Hall, R.S., Hallam, S.J., Handley, K.M., Hasan, N.A., Haydon, S.R.,

Hickman, J.E., Hidalgo, G., Hofmockel, K.S., Hooker, J., Hulth, S., Hultman, J., Hyde, E., Ibáñez-Álamo, J.D., Jastrow, J.D., Jex, A.R., Johnson, L.S., Johnston, E.R., Joseph, S., Jurburg, S.D., Jurelevicius, D., Karlsson, A., Karlsson, R., Kauppinen, S., Kellogg, C.T.E., Kennedy, S.J., Kerkhof, L.J., King, G.M., Kling, G.W., Koehler, A.V., Krezalek, M., Kueneman, J., Lamendella, R., Landon, E.M., Lanede Graaf, K., LaRoche, J., Larsen, P., Laverock, B., Lax, S., Lentino, M., Levin, I.I., Liancourt, P., Liang, W., Linz, A.M., Lipson, D.A., Liu, Y., Lladser, M.E., Lozada, M., Spirito, C.M., MacCormack, W.P., MacRae-Crerar, A., Magris, M., Martín-Platero, A.M., Martín-Vivaldi, M., Martínez, L.M., Martínez-Bueno, M., Marzinelli, E.M., Mason, O.U., Mayer, G.D., McDevitt-Irwin, J.M., McDonald, J.E., McGuire, K.L., McMahon, K.D., McMinds, R., Medina, M., Mendelson, J.R., Metcalf, J.L., Meyer, F., Michelangeli, F., Miller, K., Mills, D.A., Minich, J., Mocali, S., Moitinho-Silva, L., Moore, A., Morgan-Kiss, R.M., Munroe, P., Myrold, D., Neufeld, J.D., Ni, Y., Nicol, G.W., Nielsen, S., Nissimov, J.I., Niu, K., Nolan, M.J., Noyce, K., O'Brien, S.L., Okamoto, N., Orlando, L., Castellano, Y.O., Osuolale, O., Oswald, W., Parnell, J., Peralta-Sánchez, J.M., Petraitis, P., Pfister, C., Pilon-Smits, E., Piombino, P., Pointing, S.B., Pollock, F.J., Potter, C., Prithiviraj, B., Quince, C., Rani, A., Ranjan, R., Rao, S., Rees, A.P., Richardson, M., Riebesell, U., Robinson, C., Rockne, K.J., Rodriguezl, S.M., Rohwer, F., Roundstone, W., Safran, R.J., Sangwan, N., Sanz, V., Schrenk, M., Schrenzel, M.D., Scott, N.M., Seger, R.L., Seguinorlando, A., Seldin, L., Seyler, L.M., Shakhsheer, B., Sheets, G.M., Shen, C., Shi, Y., Shin, H., Shogan, B.D., Shutler, D., Siegel, J., Simmons, S., Sjöling, S., Smith, D.P., Soler, J.J., Sperling, M., Steinberg, P.D., Stephens, B., Stevens, M.A., Taghavi, S., Tai, V., Tait, K., Tan, C.L., Taş, N., Taylor, D.L., Thomas, T., Timling, I., Turner, B.L., Urich, T., Ursell, L.K., Van Der Lelie, D., Van Treuren, W., Van Zwieten, L., Vargas-Robles, D., Thurber, R.V., Vitaglione, P., Walker, D.A., Walters, W.A., Wang, S., Wang, T., Weaver, T., Webster, N.S., Wehrle, B., Weisenhorn, P., Weiss, S., Werner, J.J., West, K., Whitehead, A., Whitehead, S.R., Whittingham, L.A., Willerslev, E., Williams, A.E., Wood, S.A., Woodhams, D.C., Yang, Y., Zaneveld, J., Zarraonaindia, I., Zhang, Q., Zhao, H., 2017. A communal catalogue reveals Earth's multiscale microbial diversity. Nature 551, 457–463. https://doi.org/10.1038/nature24621.

van der Heijden, M.G., Hartmann, M., 2016. Networking in the plant microbiome. PLoS Biol. 14, e1002378.

Vandenkoornhuyse, P., Quaiser, A., Duhamel, M., Le Van, A., Dufresne, A., 2015. The importance of the microbiome of the plant holobiont. New Phytol. 206, 1196–1206.

Wang, J., Jia, H., 2016. Metagenome-wide association studies: fine-mining the microbiome. Nat. Rev. Microbiol. 14, 508–522.

Woyke, T., Teeling, H., Ivanova, N.N., Huntemann, M., Richter, M., Gloeckner, F.O., Boffelli, D., Anderson, I.J., Barry, K.W., Shapiro, H.J., 2006. Symbiosis insights through metagenomic analysis of a microbial consortium. Nature 443, 950. https://doi.org/10.1038/nature05192.

Index of Microorganisms

Note: Page numbers followed by *f* indicate figures, *t* indicate tables, and *b* indicate boxes.

Index of Plants and Trees

Note: Page numbers followed by *f* indicate figures and *t* indicate tables.

Index of Insects

Note: Page numbers followed by *t* indicate tables.

Index

Note: Page numbers followed by *f* indicate figures, *t* indicate tables, and *b* indicate boxes.